Biophysics

Searching for Principles

Biophysics

Searching for Principles

William Bialek

PRINCETON UNIVERSITY PRESS

Princeton and Oxford

Published by Princeton University Press, 41 William Street, Princeton, New Jersey 08540
In the United Kingdom: Princeton University Press, 6 Oxford Street, Woodstock,
Oxfordshire OX20 1TW

press.princeton.edu

ISBN 978-0-691-13891-6

Library of Congress Cataloging-in-Publication Data

Bialek, William S.
   Biophysics : searching for principles / William Bialek.
      p.   cm.
   Includes bibliographical references and index.
   ISBN 978-0-691-13891-6 (hardback : alk. paper)
   1. Biophysics.   I. Title.
   QH505.B455   2012
   571.4—dc23   2012014443

British Library Cataloging-in-Publication Data is available

This book has been composed in Birka with Myriad Pro display using ZzTEX
by Princeton Editorial Associates Inc., Scottsdale, Arizona.

Printed on acid-free paper. ∞

Printed in the United States of America

10 9 8 7 6 5 4 3 2

In memory of my parents, whose improbable journeys
made possible so many wonderful things.

Contents

Acknowledgments

A four-page paper in *Physical Review Letters* typically includes a full paragraph of acknowledgments. A naive scaling argument suggests that I may need upward of one hundred paragraphs in which to give thanks to the many people who have helped make this book possible. I don't propose to use quite that much space, but this book, and the ideas on which it is based, have occupied a substantial part of my life over many years, and I have accumulated many intellectual debts.

As I suspect will become clear rather soon upon entering the main text, the views of the field that I present here are very personal, and I don't want anyone else held responsible for my foibles. However, these views did not emerge in isolation. I am especially grateful to Rob de Ruyter van Steveninck, who introduced me to the wonders of close collaboration between theory and experiment. What began as a brief discussion about the possibility of measuring the precision of computation in a small corner of the fly's brain has become half a lifetime (so far!) of friendship and shared intellectual adventure.

My good fortune in finding wonderful experimental collaborators began with Rob but certainly did not end with him. A decade of conversations with Michael Berry, Allison Doupe, Steve Lisberger, and Leslie Osborne—sometimes reflected in joint papers and sometimes not—has influenced important parts of this book in ways I hope they will recognize. Special thanks to Michael and his group for the beautiful image that serves as the background of the chapter logos. After I moved to Princeton, David Tank, Eric Wieschaus, and I began a very different adventure, soon joined by Thomas Gregor. I have been amazed by how these interactions have so quickly reshaped my own thinking, leaving their mark on my view of the subject as a whole and hence on this text.

Theory itself is more fun in collaboration with others, even when not engaged with experimentalist friends. Different parts of the text trace their origins to joint work and discussions with Blaise Agüera y Arcas, Gurinder S. Atwal, Feraz Azhar, Naama Brenner, William J. Bruno, Curtis G. Callan, Jr., Andrea Cavagna, Denis Chigirev, Michael C. Crair, Michael DeWeese, Adrienne Fairhall, Irene Giardina, Robert F. Goldstein, Julian S. Joseph, Steven Kivelson, Roland Koberle, Dmitry Krotov, Leonid Kruglyak, Thierry Mora, Ilya Nemenman, José Nelson Onuchic, W. Geoffrey Owen, Stephanie Palmer, Fernando Pereira, Marc Potters, Kanaka Rajan, Fred Rieke, Daniel Ruderman, Elad Schneidman, Sima Setayeshgar, Tatyana Sharpee, Spyros Skourtis,

Noam Slonim, Nicholas Socci, Greg Stephens, Susanna Still, Steven Strong, Naftali Tishby, Gašper Tkačik, Aleksandra Walczak, David Warland, and Anthony Zee. I am hugely grateful to all of them. Many have become good friends, and I have continued to learn from them long after our original work together was done.

This book has its origins in a course I have taught at Princeton, and it is almost embarrassing to admit that I first taught the course a very long time ago, while I was still a member of the NEC Research Institute and a visiting lecturer at Princeton. Dawon Kahng and Joe Giordmaine were responsible for creating the enlightened environment at NEC, which lasted for a marvelous decade, while David Gross and Stew Smith made it possible for me to teach those early versions of the course at Princeton. The opportunity to interact with students while still enjoying the support of an industrial research laboratory dedicated to basic science was quite magical. During this period, frequent discussions with Albert Libchaber were also important, as he insisted that explorations at the interface of physics and biology be ambitious but still crisp and decisive—a demanding combination.

Although the wonders of life in industrial labs have largely disappeared, the pleasures of teaching at Princeton have continued and grown. I am especially grateful to my colleagues in the Physics Department for welcoming the intellectual challenges posed by the phenomena of life as being central to physics itself, rather than being "applications" of physics to another field. The result has been the coalescence of a very special community, and I hope that some of what I have learned from this community is recorded faithfully in this book. John Hopfield's role in making this happen—by setting an example for what could be done, by being an explicit and witty provocateur, and by being a quiet but persistent catalyst for change—cannot be overestimated; it is a pleasure to thank him. I don't think that even John imagined that there would eventually be a biophysics theory group at Princeton, but with Curt Callan and Ned Wingreen, we have managed to do it, and we have been joined by a steady stream of remarkable students, postdocs, and Fellows, all of whom have added enormously to our community. Curt deserves special thanks for his leadership and even more for the energy and enthusiasm he brings to seminars and discussions, engaging with the details but also reminding us that theoretical physics has lofty aspirations.

The community of physicists interested in the phenomena of life still is quite small, and it has been a pleasure to learn from my colleagues, even when we disagree— perhaps especially then. Beyond my circle of collaborators, I am grateful to Larry Abbott, Bob Austin, Steve Block, Henrik Flyvbjerg, Herbie Levine, Marcelo Magnasco, Markus Meister, Ken Miller, Phil Nelson, Rob Phillips, Boris Shraiman, Eric Siggia, Haim Sompolinsky, Chao Tang, Alessandro Treves, Chris Wiggins, and George Zweig, all of whom have taught me many things, some of which found their way into the text. A special thanks, as well, to my thesis adviser, Alan Bearden, whose own struggles to find an intellectual home at the interface of physics and biology never distracted him from the joys of science.

Even if I had the perfect idea for teaching a course, it would be meaningless without students. By now, hundreds of students have listened to the whole set of lectures and worked through the problems, providing feedback at every stage, as have several very able teaching assistants. At least as many students have heard pieces of the course in different venues, and every time I taught I learned something—at least, I hope, about how to say things more clearly. Less tangible, but even more important, the liveliness and engagement of the students have made teaching a pleasure.

At the same time that I have been teaching this course to advanced graduate students, I have also been involved in an experiment to renew our teaching of science to first-year undergraduates. The resulting stimulus to think carefully about what is fundamental, as opposed to conventional, has been incredibly valuable. Discussions with several of my partners in this enterprise have been especially thought provoking, and so it is a pleasure to thank David Botstein, Chase Broedersz, Curt Callan, Michael Desai, Jeremy England, Jay Groves, Michael Hecht, Matthias Kaschube, Leonid Kruglyak, Dan Marlow, Will Ryu, Eva-Maria Schötz, and Josh Shaevitz.

Everyone who has tried to write a book based on their teaching experience knows the enormous difference between a good set of lecture notes and the final product. I very much appreciate Arthur Wightman's suggestion, long ago, that this transition would be worth the effort. Ingrid Gnerlich, my editor at Princeton University Press, has consistently provided the right combination of encouragement and gentle reminders of looming (and passing) deadlines. The idea of actually finishing (!) started to crystallize during a wonderful sabbatical in Rome and has been greatly helped along by visiting professorships at the Rockefeller University and most recently at the Graduate Center of the City University of New York. Both in Rome and in New York, stimuli from colleagues and from the surrounding cities have proved delightfully synergistic. In particular, my final pass through the copyedited manuscript was done during a return visit to Rome, and I thank Andrea Cavagna and Irene Giardina for their warm hospitality, both scientific and personal.

Along the last segments of the trajectory from "almost done" to "done," several colleagues provided wonderful input on short notice, both in response to my pleas for help and in response to Ingrid's request for more official reports. Rob Phillips and Chris Wiggins brought objectivity and the proper amount of scathing humor, alerting me to a variety of problems. Pietro Cicuta and Philip Nelson, beyond their insights into technical matters, brought a deep concern for the eventual readership, both students and faculty. I hope I have been able to respond effectively to their concerns. Thomas Gregor, Justin Kinney, and Fred Rieke gave generously of their expertise, and Rob de Ruyter provided yet more of the insight, craftsmanship, and knowledge of scientific history that I have so much enjoyed in our long collaboration. My thanks to all of them. I have been able to attend to all this wise, last-minute advice because the process of converting the manuscript into a book has gone so smoothly, and for this I thank the staff at Princeton University Press as well as Peter Strupp and his capable colleagues at Princeton Editorial Associates. More generally, I have been able to concentrate on the book in part because of Lee Morgan's cheerful and effective management of many other things, and it is a pleasure to thank him here. Friends and collaborators also were remarkably patient with me during the last months of the project, even as those months seemed to grow longer.

It often is remarked that theory is a relatively inexpensive activity, so that we theorists are less dependent on raising money than are our experimentalist friends. But theory is a communal activity, and all members of the community need salaries. Because I have benefited so much from the stimulation provided by the scientists around me, I am especially grateful for the steady support my colleagues and I have received from the

National Science Foundation, and for the generosity of Princeton University in bringing all of us together. In particular, Denise Caldwell, Kenneth Whang, and especially Krastan Blagoev deserve our thanks for helping to ensure that this kind of science has a home at the National Science Foundation, even in difficult times. The Burroughs-Wellcome Fund, the W. M. Keck Foundation, and the Swartz Foundation also have been extremely generous, sometimes leaping in where the usual angels feared to tread.

Although the product of the scientific enterprise must have meaning beyond our individual feelings, the process of science is intensely personal. When we collaborate or even just learn from one another, we share not just our ideas about the next step in a small project, but our hopes and dreams for efforts that could occupy a substantial fraction of a lifetime. To make progress we admit to one another how little we understand, and how we struggle even to formulate the questions. Collaboration is, for want of a better word, an intimate activity. Colleagues become friends, friendships deepen, we come to care not just about ideas and results but about one another. It is, by any measure, a privileged life. If this text helps some readers to find their way to such enjoyment, I will have repaid a small fraction of my debt.

Family is a special category, here as always. Beyond the joyous foundations provided by thirty years of life together, Charlotte has provided advice about almost every major conceptual question in the organization and outlook of the book, as well as reviewing countless smaller things. She has shared my excitement about new ideas or results to be included and has reassured me when my worries about leaving things out threatened to become paralyzing. At the end, while correcting the proofs, we have literally read every word of the book together. For all this, and for so many other things, my love and gratitude. As the ideas in this book matured, Max and Fannie began to find their own intellectual voices, providing a new source of pleasure and, not infrequently, concrete advice. In this part of life as well I have been very privileged, and I am forever grateful.

EXPLORING THE PHENOMENA

Introduction

When a PhD student in physics picks up a textbook about elementary particles, or cosmology, or condensed matter, there is little doubt about what will be found inside the covers. There are questions, perhaps, about the level and style of presentation, or about the emphasis given to different subfields, but the overall topic is clear. The situation is very different for books or courses that attempt to bring the intellectual style of physics to bear on the phenomena of life. The problem is not just in how we teach, but also in how we do research. The community of physicists interested in biological problems is incredibly diverse, it spills over into more amorphously defined interdisciplinary communities, and individual physicists often are more connected to biologists working on the same system than they are to physicists asking the same conceptual question in other biological systems. None of this is necessarily good or bad, but it can be confusing for students.

1.1   About Our Subject

Ours is not a new subject, but over its long history, "biophysics" or "biological physics" has come to mean many different things to different communities.[1] At the same time, for many physicists today, biophysics remains new and perhaps a bit foreign. There is an excitement to working in a new field, and I hope to capture this excitement. Yet our excitement, and that of our students, sometimes is tempered by serious concerns, which can be summarized by naive questions: Where is the boundary between physics and biology? Is biophysics really physics or just the application of methods from physics to the problems of biology? My biologist friends tell me that "theoretical biology" is nonsense, so what would theoretical physicists be doing if they got interested in this field? In the interaction between physics and biology, what happens to chemistry? How much biology do I need to know to make progress? Why do physicists and biologists

---

1. The use of these two different words also is problematic. I think that, roughly speaking, "biophysics" can be used by people who think of themselves either as physicists or biologists, whereas "biological physics" is an attempt to carve out a subfield of physics, distinct from biology. The difficulty is that neither word really points to a set of questions that everyone can agree on. So, we need to dig in.

seem to be speaking such different languages? Can I be interested in biological problems and still be a physicist, or do I have to become a biologist? Although there has been much progress over the past decade, I still hear students (and colleagues) asking these questions, and so it seems worth a few pages to place the subject of this book into context.

To put things in perspective, we need to have in mind at least a cartoon sketch of the intellectual landscape that we are trying to explore. I am not a sociologist or historian of science, and cartoons can be, by definition, cartoonish, so there are myriad dangers here. Indeed, the dangers are sufficiently great that at least one colleague recommended scrapping this discussion completely. There is a school of thought that says, roughly, "I don't care if it's biology or physics, I just want to do interesting science," and according to this view one need not worry about the place of biophysics in relation to the larger, separate enterprises of physics and biology. This sounds good and hints at a nirvana in which disciplinary boundaries are erased and (by extension) universities have no departments. But this position is not interdisciplinary or multidisciplinary, it's antidisciplinary, and it assumes, implicitly, that the definition of "interesting" is objective, independent of our intellectual backgrounds. I find this very hard to believe.[2] One might hope that we could construct an objective definition of interesting science, but certainly practicing scientists don't automatically subscribe to such a thing. Indeed, faced with the same natural phenomena, my experience is that physicists and biologists (and mathematicians and chemists and engineers) will ask different questions. By the time we find answers, we ought to be able to convince one another that we have accomplished something. But at the stage where we are still formulating questions, I think that any search for unanimity about what is a "good" or "interesting" question would devolve into one group insisting that they alone have the right to determine what is relevant. I believe there really are physics problems motivated by the phenomena of life and that these problems can be different from those that engage our biologist friends—not better or worse, but different. So, with this in mind, let us draw that sketch, cartoonish though it may be.

Academic disciplines can define themselves either by their objects of study or by their style of inquiry. Physics is firmly in the second camp. Physicists make it their business to ask certain kinds of questions about Nature and to seek certain kinds of answers. The aspects of the world which capture the interest of the physics community can and do change, not least as new phenomena become accessible to the physicists' style of inquiry. Throughout these changes, "thinking like a physicist" is supposed to mean something, and it is this, above all else, that we try to convey to our students. We take ourselves (not without hubris) to be the intellectual heirs of Galileo, embracing his evocative claim that the book of Nature is written in the language of mathematics.

Biology translates from its Greek roots as the study of life. The style of inquiry may change, from studies of animal behavior and anatomy to genetics and molecular

---

2. Full disclosure: As an undergraduate at Berkeley, I took philosophy of science from Paul Feyerabend, who was a particularly eloquent (and amusing) critic of the idea that there is something objective that we could point to as a universally interesting scientific question; this was part of a larger critique of attempts to codify "the scientific method." I think his arguments were meant to make us look more carefully at our own assumptions about the nature of the scientific enterprise, and I think this worked for me. Some found him too much and construed his approach as antiscientific. I encourage you to explore for yourself in the references for this section at the end of the book.

structure, but the objects remain the same. It is especially important for physicists to appreciate the vastness of the enterprise that is labeled "biology" and the divisions within biology itself. A geneticist, for example, studying the dynamics of regulatory networks in a simple organism, such as yeast, has relatively little in common with a colleague who studies the dynamics of neural networks for the regulation of movement in higher organisms. Not only is biology defined by the objects of study, but also subfields of biology are similarly defined, so that networks of neurons and networks of genes are different subjects.

Differences in our view of the scientific enterprise translate rather directly into different educational structures. In physics, we (try to) teach principles and derive the predictions for particular examples. In biology, teaching proceeds (mostly) from example to example, system to system. Although physics has subfields, to a remarkable extent the physics community clings to the romantic notion that Physics is one subject. Not only is the book of Nature written in the language of mathematics, but also there is only one book, and we expect that if we really grasped its content, it could be summarized in very few pages. Where does biophysics fit into this view of the world?

There is something different about life, something that we recognize immediately as distinguishing the animate from the inanimate. But we no longer believe that there is a fundamental "life force" that animates a lump of inert stuff.[3] Similarly, there is no motive force that causes superfluid helium to crawl up the sides of a container and escape, or causes electrical current in a superconducting loop to flow forever; the phenomena of superfluidity and superconductivity emerge as startling consequences of well-known interactions among electrons and nuclei, interactions that usually have much more mundane consequences. As physicists studying the phenomena of life, we thus are not searching for a new force of Nature. Rather, we are trying to understand how the same forces that usually cause carbon-based materials to look like rocks or sludge can, under some conditions, cause some of this material to organize itself and walk (or swim or fly) out of the laboratory. What is special about the state of matter that we call life? How does it come to be this way? Different generations of physicists have approached these mysteries in different ways.

Looking Back

Some of the giants of classical physics—Helmholtz, Maxwell, and Rayleigh, to name a few—routinely crossed borders among disciplines that we now distinguish as physics, chemistry, biology, and even psychology. Some of their forays into the phenomena of life were driven by a desire to test the universality of physical laws, such as the conservation of energy. A very different motivation was that our own view of the world is determined by what we can see and hear, and more subtly by what we can reliably infer from the data that our eyes and ears collect. These physicists thus were drawn to the study of the senses; for them, there was no boundary between optics and vision, or between acoustics and hearing. Helmholtz in particular took a very broad view, seeing a path not just from acoustics to the mechanics of the inner ear and from the properties of light

---

3. This now obvious statement reflects centuries of hard work, and not a little real fighting. Einstein's explanation of Brownian motion, among other things, put to rest the idea that the spontaneous movements of microscopic particles reflected a life force.

to the optics of the eye, but all the way from the physical stimuli reaching our sense organs to the nature of our perceptions, to our ability to learn about the world, and even to what makes some sights or sounds more pleasing than others. Reading Helmholtz today I find myself struck by how much his insights still guide our thinking about vision and hearing, and how the naturalness of his cross-disciplinary discourse remains something that few modern scientists achieve, despite the current fanfare about the importance of multidisciplinary work. Most of all, I am struck by his soaring ambition that physics should not stop at the point where light hits our eyes or sound enters our ears, and that we should search for a physics that reaches all the way to our personal, conscious experience of the world in all its beauty.

The rise of modern physics motivated another wave of physicists to explore the phenomena of life. Fresh from the triumphs of quantum mechanics, they were emboldened to seek new challenges and brought new concepts. Bohr wondered aloud if the ideas of complementarity and indeterminacy would limit our ability to understand the microscopic events that provide the underpinnings of life. Delbrück was searching explicitly for new principles, hoping that a modern understanding of life would be as different from what came before as quantum mechanics was different from classical mechanics. Schrödinger, in his influential series of lectures titled *What Is Life?*, seized on the discovery that our precious genetic inheritance was stored in objects the size of single molecules, highlighting how surprising this is for a classical physicist, and contrasted the order and complexity of life with the ordering of crystals. Along the way he outlined a strikingly modern view of how nonequilibrium systems can generate structure out of disorder, continuously dissipating energy.

In one view of history, there is a direct path from Bohr, Delbrück, and Schrödinger to the emergence of molecular biology. Certainly, Delbrück did play a central role, not least because of his insistence that the community should focus (as the physics tradition teaches us) on the simplest examples of crucial biological phenomena, reproduction and the transmission of genetic information. The goal of molecular biology to reduce these phenomena to interactions among a countable set of molecules surely echoed the physicists' search for the fundamental constituents of matter, and perhaps the greatest success of molecular biology is the discovery that many of these basic molecules of life are universal, shared across organisms separated by hundreds of millions of years of evolutionary history. Where classical biology emphasized the complexity and diversity of life, the first generation of molecular biologists emphasized the simplicity and universality of life's basic mechanisms, and it is not hard to see this as an influence of the physicists who came into the field at its inception.

Another important idea at the start of molecular biology was that the structure of biological molecules matters. Although modern biology students, even in many high schools, are taught that "structure determines function," this was not always obvious. To imagine, in the years immediately after World War II, that all of classical biochemistry and genetics would be reconceptualized once we could see the actual structures of proteins and DNA was a revolutionary vision—one shared by only a handful of physicists and the most physical of chemists. Every physicist who visits the grand old Cavendish Laboratory in Cambridge should pause in the courtyard and realize that on that ground stood the Medical Research Council (MRC) hut, where Bragg nurtured a small group of young scientists who were trying to determine the structure of biological molecules through a combination of X-ray diffraction experiments and pure theory.

To make a long and glorious story short, they succeeded, perhaps even beyond Bragg's wildest dreams, and some of the most important papers of twentieth-century biology thus were written in a physics department.

Perhaps inspired by the successes of their intellectual ancestors, each subsequent generation of physicists offered a few converts. The idea, for example, that the flow of information through the nervous system might be reducible to the behavior of ion channels and receptors brought one group, armed with low-noise amplifiers, intuition about the interactions of charges with protein structure, and the theoretical tools to translate this intuition into testable quantitative predictions. The possibility of isolating a single complex of molecules that carried out the basic functions of photosynthesis brought another group, armed with the full battery of modern spectroscopic methods that had emerged in solid state physics. Understanding that the mechanical forces generated by a focused laser beam are on the same scale as the forces generated by individual biological molecules as they go about their business brought another generation of physicists to our subject. The sequencing of whole genomes, including our own, generated the sense that the phenomena of life could, at last, be explored comprehensively, and this inspired yet another group. Proximal stimuli were not always experimental: the idea that networks of neurons could be described in the language of statistical mechanics energized a whole theoretical community, and their results gradually fed back into the investigation of cells, circuits, and systems in the brain. These examples are far from complete, but they give a sense of the diversity of challenges that drew physicists toward problems that traditionally had been in the domain of biologists.

Through these many generations, some conventional views arose about the nature of science at the borders between physics and biology. First, there is a strong emphasis on technique. From X-ray diffraction to the manipulation of single molecules to functional imaging of the human brain, it certainly is true that physics has developed experimental techniques that allow much more direct exploration of questions raised by biologists. Second, there is a sense that in some larger classification system, biophysics is a biological science. Certainly when I was a student, and for many years afterward, physicists would speak (sometimes wistfully) of colleagues who were fascinated by the phenomena of life as having "become biologists." For their part, biologists would explain that physicists were successful in these explorations only to the extent that they appreciated what was "biologically important." Finally, biophysics has come to be organized along the lines of the traditional biological subfields. As a result, the dynamics of ion channels in single neurons and the collective behavior of large neural networks are separate subjects, and the generation of physicists exploring noise in the regulation of gene expression is disconnected from the previous generation that studied noise in ion channels.

Without taking anything away from what has been accomplished, I believe that much has been lost in the emergence of the conventional views about the nature of the interaction between physics and biology. By focusing on methods, we miss the fact that, faced with the same phenomena, physicists and biologists will ask different questions. In speaking of biological importance, we ignore the fact that physicists and biologists have different definitions of understanding. By organizing ourselves around structures that come from the history of biology, we lose contact with the dreams of our intellectual ancestors that the dramatic qualitative phenomena of life should be clues to deep theoretical insights, that there should be a physics of life and not just the physics of this or that particular system. It is, above all, these dreams that I want to rekindle in my students and in the readers of this book.

Looking Forward

At present, most questions about how things work in biological systems are viewed as questions that must be answered by experimental discovery. The situation in physics is very different, in that theory and experiment are more equal partners. In each area of physics we have a set of general theoretical principles, all interconnected, which define what is possible; the path to confidence in any of these principles is built on a series of beautiful quantitative experiments that have extended the envelope of what we can measure and know about the world. Beyond providing explanations for what has been seen, these principles provide a framework for exploring, sometimes playfully, what *ought* to be seen.[4] In many cases these predictions are sufficiently startling that to observe the predicted phenomena (a new particle, a new phase of matter, fluctuations in the radiation left over from the big bang, etc.) still constitutes a dramatic experimental discovery.

Can we imagine a physics of biological systems that reaches the level of predictive power that has become the standard in other areas of physics? Can we reconcile the physicists' desire for unifying theoretical principles with the obvious diversity of life's mechanisms? Could such theories engage meaningfully with the myriad experimental details of particular systems, yet still be derivable from succinct and abstract principles that transcend these details? For me, the answer to all these questions is an enthusiastic "yes." I hope that this book will convey both my enthusiasm and the reasons that lie behind it.

Although we aim at questions that have the generality and abstraction that are familiar from our experience in the rest of physics, our attention is attracted first not by

---

4. Reading an earlier draft, one friend found my use of "ought" here quite jarring, as if theoretical predictions were moral imperatives: children should mind their manners, and the Higgs boson should have a mass of 125 GeV. Surely physics does not make normative claims that the world should be one way or another, but rather it tries to describe the world as we find it. I was about to choose another word when I found myself reading David Foster Wallace's essay "Authority and American Usage." In what is nominally a book review of a dictionary (!), Wallace discusses many things, including the distinction between prescriptive and descriptive approaches to language. Descriptivists write dictionaries with the goal of cataloguing the language as it is used, without claiming that some patterns of usage are superior to others, only more common, whereas prescriptivists write dictionaries with the goal of defining what is correct, establishing norms for usage. According to Wallace, the descriptivists often claim their program to be the more scientific, being based on data about what native speakers actually do. One can hear echoes of this conflict in the interactions between the intellectual traditions of physics and biology. A more empirically minded biologist might insist that models be based on "everything that we know is there" in the real system. A more theoretically inclined physicist would feel comfortable writing down simplified models that leave out many details; indeed, much of the art of theorizing lies in separating what we suspect is essential from the ignorable details. Such theories thus make a normative claim, namely, that the world should behave in a way such that the things we leave out of our models aren't important. So, I think the use of "ought" in this context really gets at an essential aspect of what we do when we construct theories in the physics tradition: the theoretical physicist's relation to Nature is prescriptivist. I cannot resist noting that Wallace comes down on the side of prescriptivism, not least because a genuinely complete and unbiased descriptivism is impossible, much as it may be impossible to know everything about the components and interactions in a complex biological system. In the end, even descriptivists make normative claims about what constitutes a "good" data set, and so we might as well be honest about our prescriptions.

abstractions but by the concrete and dramatic phenomena of life, so we should start there. There are so many beautiful things about life, however, that it can be difficult to choose an appropriate starting point. Before explaining the choices I made in writing this book, I emphasize that there are many equally good choices. Indeed, if we choose almost any of life's phenomena—the development of an embryo, our appreciation of music, the ability of bacteria to live in diverse environments, the way ants find their way home in the hot desert—we can see glimpses of fundamental questions even in the seemingly most mundane events.

It is a remarkable thing that, pulling on the threads of one biological phenomenon, we can unravel so many general physics questions. In any one case, some problems will be presented in purer form than others, but in many ways everything is there. Thus, if we think hard about how crabs digest their food (to choose a particularly prosaic example), we will find ourselves worrying about how biological systems manage to find the right operating point in very large parameter spaces. This problem, as we will see in Chapter 5, arises in many different systems, across levels of organization from single protein molecules to short-term memory in the brain. Thus, in an odd way, everything is fair game. The challenge is not to find the most important or "fundamental" phenomenon but rather to see through any one of many interesting and beautiful phenomena to the deep physics problems that are hiding underneath the often formidable complexity of these systems.

The first problem, as noted above, is that there really is something different about being alive, and we'd like to know what this is—in the same way that we know what it is for a collection of atoms to be solid, for a collection of electrons to be superconducting, or for the vacuum to be confining (of quarks). This "what is life?" question harkens back to Schrödinger, and one might think that the molecular biology that arose in the decades after his manifesto would have answered his question, but this isn't clear. Looking around, we more or less immediately identify things that are alive, and the criteria that we use in making this discrimination between animate and inanimate matter surely have nothing to do with DNA or proteins. Even more strongly, we notice that things are alive long before we see them reproduce, so although self-reproduction might seem like a defining characteristic, it does not seem essential to our recognition of the living state. Being alive is a macroscopic state, whereas things like DNA and the machinery of self-reproduction are components of the microscopic mechanism by which this state is generated and maintained.[5] Although we have made much progress on identifying microscopic mechanisms, we have made rather less progress on identifying the "order parameters" that are characteristic of the macroscopic state.

Asking for the order parameters of the living state is a hard problem, and one that is not terribly well posed. One way to make progress is to realize that as we make more quantitative models of particular biological systems, these models belong to families: we can imagine a whole class of systems, with varying parameters, of which the one

---

5. More precisely, all the molecular components of life that we know about comprise *one way* of generating and maintaining the state that we recognize as being alive. We don't know if there are other ways, perhaps realized on other planets. This remark might once have seemed like science fiction, and perhaps it still is, but the discovery of planets orbiting distant stars has led many people to take these issues much more seriously. Designing a search for life on other planets gives us an opportunity to think more carefully about what it means to be alive.

we are studying is just one example. Presumably, most of these possible systems are not functional, living things. What then is special about the regions of parameter space that describe real biological systems? This is a more manageable question, and it can be asked at many different levels of biological organization. If there is a principle that differentiates the genuinely biological parts of parameter space from the rest, then we can elevate this principle to a theory from which the properties of the biological system could be calculated a priori, as we do in other areas of physics.

If real biological systems occupy only a small region in the parameter space of possible systems, we have to understand the dynamics by which parameters arrive at these special values. At one extreme, this is the problem of the origin of life. At the opposite extreme, we have the phenomena of physiological adaptation, whereby cells and systems adjust their behavior in relation to varying conditions or demands from the environment, sometimes in fractions of a second. In between we have learning and evolution. Adaptation, learning, and evolution represent very different mechanisms, on different but perhaps overlapping time scales, for accomplishing a common goal— tuning the parameters of a biological system to match the problems that organisms need to solve as they try to survive and reproduce. What is the character of these dynamics? Are the systems we see around us more or less "equilibrated" in these dynamics, or are today's organisms strongly constrained by the nature of the dynamics itself? Put another way, if evolution is an algorithm for finding better organisms, are the functional behaviors of modern biological systems significantly shaped by the algorithm itself, or can we say that the algorithm solves a well-defined problem, and what we see in life are the solutions to this problem?

To survive in the world, organisms do indeed have to solve a wide variety of problems. Many of these are really physics problems: converting energy from one form to another, sensing weak signals from the environment, controlling complex dynamical systems, transmitting information reliably from one place to another (or across generations), controlling the rates of thermally activated processes, predicting the trajectories of multidimensional signals, and so on. While it is obvious (now!) that everything that happens in living systems is constrained by the laws of physics, such physics problems in the life of the organism highlight these constraints and provide a special path for physics to inform our thinking about the phenomena of life.

Identifying all the physics problems that organisms need to solve is not so easy. On the one hand, thinking about how single-celled organisms, with sizes on the scale of one micron, manage to move through water, we quickly get to problems that have the look and feel of problems that we might find in Landau and Lifshitz. On the other hand, it was a truly remarkable discovery that all cells have built Maxwell demons, and that our description of a wide variety of biochemical processes can be unified by this observation (see Section 4.5). Efforts in this direction can be very rewarding, however, because they identify questions that connect functionally important behaviors—for which evolution might select—with basic physical principles. Physics shows us what is hard about these problems and where organisms face real challenges. Physics also places limits on what is possible, and this gives us an opportunity to put the performance of biological systems on an absolute scale. It makes precise our intuition that organisms are really good at solving some very difficult problems.

Let us conclude this discussion with a tentative summary. The business of life involves solving physics problems, and these problems provide us with a natural subject matter. In particular, these problems focus our attention on the concept of function, which is not part of the conventional physics vocabulary[6] but clearly is essential if we want to speak meaningfully about life. Of the possible mechanisms for solving these problems, most combinations of the available ingredients probably don't work, and specifying this functional ensemble provides a manageable approach to the larger question of what characterizes the living state. Adaptation, learning, and evolution allow organisms to find these special regions of parameter space, and the dynamics of these processes provide another natural set of problems.

1.2   About This Book

This book has its origins in a course that I have taught for several years at Princeton University. It is aimed at PhD students in physics, although a sizable number of brave undergraduates have also taken the course, as well as a handful of graduate students from biology, engineering, applied math, and other disciplines.[7] Bits and pieces have been tested in shorter courses, sometimes for quite different audiences, at the Marine Biological Laboratory, Les Houches, the Boulder Summer School on Condensed Matter Physics, "Sapienza" Università di Roma, the Rockefeller University, and the Graduate Center of the City University of New York.

Rationale

Many courses on biophysics address an audience drawn from multiple disciplines, trying both to introduce physics students to the intellectual challenges posed by the phenomena of life and to introduce biology students to the concepts and methods of physics. Doing this for graduate students is incredibly hard. In practice, making biophysics an interdisciplinary course means that the level of mathematics and physics that one draws on must be much lower than in other courses for physics graduate students. By itself this need not be a bad thing, and surely it makes the material more accessible. But if this is the only way that we teach the subject, then there is a danger that the lack of connection to deeper physics ideas will become a self-fulfilling prophecy. My plan in teaching was thus the opposite of the interdisciplinary course—to produce a biophysics course that serves as an advanced physics graduate course in the same way that my colleagues teach condensed matter physics, quantum field theory and particle physics, or astrophysics and cosmology.[8] When I was student, most physicists would have been skeptical about the feasibility of such a project, but things have changed. Physicists have explored the phenomena of life at many different levels, from single molecules to entire populations of organisms, and many have brought back insights

---

6. This is not quite fair. In thermodynamics we distinguish "useful work," which provides a notion of function, at least in the limited context of heat engines. But we need something more general if we want to capture the full range of problems that organisms have to solve.

7. With apologies to colleagues elsewhere in the world, I use the U.S. terms for students at different levels of their education.

8. By now this is more than a goal; it is something of a contractual obligation. Our graduate students are required to take several courses in different areas to demonstrate the breadth of their knowledge of physics, and biophysics is one of the courses on the list that can satisfy this requirement. For this plan to be meaningful, all these courses need to be taught on the same level.

that are exciting as physics, not just as applications of physics to other fields. Part of my point in teaching and in writing this book has been to celebrate this transformation.

In its earliest incarnations, the course consisted of a series of case studies—problems where physicists have tried to think about some particular biological system. The hope was that in each case study we might catch a glimpse of some deeper and more general ideas. As the course evolved, I tried to shift the balance from examples toward principles. The difficulty, of course, is that we don't know the principles—we just have candidates. At some point I decided that this was OK, and that trying to articulate the principles was important even if we get them wrong. I believe that, almost by definition, something we will recognize as a theoretical physics of biological systems will have to cut across the standard subfields of biology, organizing our understanding of very different systems as instantiations of the same underlying ideas.

Although we are searching for principles, we start by being fascinated with the *phenomena* of life. Thus, I start with one particular biological phenomenon that holds, I think, an obvious appeal for physicists: the ability of the visual system to count single photons. As we explore this phenomenon, we will meet some important facts about biological systems, see methods and concepts that have wide application, and identify and sharpen a series of questions that we can recognize as physics problems. The really beautiful measurements that people have made in this system also provide a compelling antidote to the physicists' prejudice that experiments on biological systems are necessarily messy; indeed, I think these measurements set a standard for quantitative experiments on biological systems that should be more widely appreciated and emulated.[9] Another crucial feature of the photon-counting problem is that it cuts across many levels of biological organization, from the quantum (really) dynamics of single molecules to the macroscopic dynamics of human cognition.

I think one of the most important aspects of our field is the process of digging around in the phenomena to find interesting physics problems. Obviously there is a matter of taste here, and even among physicists with similar educations different problems will leap out at us from the complexities of real biological systems. But teaching biophysics surely involves teaching this process of formulating problems, not by fiat but by going through real examples.[10] So, we will take an explicit pause to formulate problems and articulate candidate principles (Chapter 3). What emerges are three broad ideas: the importance of noise, the need for living systems to function without fine tuning of parameters, and the possibility that many different problems solved by living organisms are different aspects of one big problem about the representation of infor-

---

9. Perhaps surprisingly, many biologists share the expectation that their measurements will be noisy. Indeed, some biologists insist that physicists have to get used to this and that this is a fundamental difference between physics and biology. Certainly it is a difference between the sciences as they are practiced, but the claim that there is something essentially sloppy about life is deeper, and deserves more scrutiny. One not-so-hidden agenda in my course is to teach physics students that it is possible to uncover precise, quantitative facts about biological systems in the same way that we can uncover precise quantitative facts about nonbiological systems, and that this precision matters.

10. Perhaps teaching in any field should involve teaching the process of formulating problems, but in more mature areas of physics there are so many successful examples that we often feel justified in assuming that this process will become clear en passant. This may or may not be correct. One might also note that the success of theory changes this process, because one of the roles of theory is to define what would be interesting or surprising.

mation. Each of these ideas is something that many people have explored, and I hope to make clear that these ideas have generated real successes. The greatest successes, however, have been when these theoretical discussions are grounded in experiments on particular biological systems. As a result, the literature is fragmented along lines defined by the historical subfields of biology. The goal here is to present the discussion in the physics style, organized around principles from which we can derive predictions for particular examples.

My choice of candidate principles is personal, and I do not expect that everyone in the field will agree with me; indeed, by the time there is consensus, perhaps the field won't be as much fun. Even more than my choice of principles, my choice of examples is not meant to be canonical, but illustrative. In choosing these examples, I had three criteria. First, I had to understand what was going on, and of course this biases me toward cases that my friends and I have studied in the past. I apologize for this limitation and hope that I have been able to do justice at least to some fraction of the field. Second, I want to emphasize the tremendous breadth and depth of physics ideas that are relevant in thinking about the phenomena of life. Many students are given the impression, implicitly or explicitly, that to do biophysics one can get away with knowing less "real physics" than in other subfields, and I think this is a disastrous misconception. Third, if the whole program of finding principles is going to work, then it must be that a single principle really does illuminate the functioning of seemingly very different biological systems. Thus, I make a special effort to be sure that the set of examples for each principle cuts across the subfields of biology, in particular, across the great divide between molecular and cellular biology on the one hand and neurobiology on the other.[11] Ideally I would go even further, reaching toward the behavior of populations of organisms, the interactions among multiple species in natural ecologies, and so on. I can imagine how some of the same ideas and principles that I explore here connect to these larger scale phenomena, but I just don't know enough about these fields to do them justice, so I have to stop somewhere.

In trying to provide some perspective on our subject, in the previous section, I mentioned a number of now classic topics from across more than a century of interaction between physics and biology. I don't think it is right to teach by visiting these topics one after the other, for reasons which I hope are clear by now. However, it would be weird to take a whole course on biophysics and come out without having learned about these things. So I have tried to weave some of the classics into the conceptual framework of the course, perhaps sometimes in surprising places. There also are many beautiful things I have left out, and again I apologize to people who will find that I neglected matters close to their hearts. Sometimes the neglect reflects nothing more

---

11. I have also resisted the temptation to organize the book by physical scale, starting with molecules and ending (perhaps) with the brain. One can teach a great course in this way, but I think the principle is misleading. First, as I hope will be clear, there are actually physical principles about how biological systems function *as systems* that cut across these multiple scales. Second, ordering by scale encourages the illusion that we actually know how to build from our microscopic understanding of biological molecules "up" to the macroscopic behaviors of higher organisms. We certainly don't know how to do this, and it may even be misguided as a matter of principle—more is different, after all. Finally, starting with molecules is both historically and, if I may use the word, emotionally wrong. What intrigues us about life are the macroscopic behaviors of organisms, ourselves included, and careful quantitative investigations of these macroscopic phenomena set the stage for later molecular explorations.

than my ignorance, but I also felt strongly that everything I discuss should fit together into a larger picture and that it is almost disrespectful to give a laundry list of wonderful but undigested results. Thus, much was left unsaid.

User's Guide

As explained above, I am aiming at PhD students in physics, so I will assume that the reader has a strong physics background and is comfortable with the associated mathematical tools. Although many different areas of physics make an appearance, the most frequent references are to ideas from statistical mechanics. In practice, this is the area where at least U.S. students have the largest variance in their preparation. As a result, in places where my experience suggests that students will need help, I have not been shy to include (perhaps idiosyncratic) expositions of relevant physics topics that are not especially restricted to the biophysical context; this is, after all, a physics course. Some more technical asides are presented as an appendix, including both matters that students may have encountered in earlier courses and some opportunities for further exploration. Throughout the text, and especially in the appendix, I try very hard to avoid saying "it can be shown that"; the resulting text is longer but I hope more useful.[12]

While it is pretty clear how much physics I expect my readers to know, it might be less clear how much biology I am assuming. We can't get started without some grasp of the facts, but when we teach particle physics, we don't start by reading from the particle data book. Similarly, I won't start by reciting the "biological background." Good high school biology courses now are at the level of the introductory university courses from a generation ago. I think everybody getting a PhD knows that the instructions for building an organism are contained in DNA, even if they are unclear on exactly how these instructions are read out. So, rather than starting with a condensed pure biology course, I just plunge in, trying to wrestle with the biological phenomena that I find most exciting and stopping to explain things as needed.

I should warn the reader that I have never managed, despite a decade of good intentions, to cover all the material here in a one-semester course. Indeed, in writing the book I have been struck by those places where my lectures had "covered" a topic with a few remarks and some pointers to the literature, rather than a fully digested exposition. If this were a short book of lectures on limited topics, I might be forgiven for such casualness, but surely this book is now too long for such things. There still are pointers designed to encourage exploration, but these really should be pointing out to the unknown rather than pointing back toward incompletely explained background material.

If you can't cover everything, how should you use the book? Chapters 4, 5, and 6 present candidate principles, and each chapter discusses many examples. Some of these examples could be skipped, depending on your taste. Alternatively, you might think that the discussion of photon counting (Chapter 2) is a bit heavily biased toward the issues of noise, so you could skip Chapter 4 completely or at least use it very selectively (but don't miss kinetic proofreading, in Section 4.5). A completely different strategy would be to touch the material of every section in the book but not in too much depth,

---

12. Special thanks go to Phil Nelson, who noticed that an earlier draft rigorously avoided saying "it can be shown that," but suffered from a number of morally equivalent lapses. I hope I have caught them all!

compressing each section into a one-hour lecture—it is a quick pace, but possible, and still leaves time to pull in background from the appendix.

The most important practical comment about the structure of the book concerns the problems. I cannot overstate the importance of doing problems as a component of learning. One should go further, getting into the habit of calculating as one reads, checking that you understand all the steps of an argument and that things make sense when you plug in the numbers or make order-of-magnitude estimates.[13] For all these reasons, I have chosen (following Landau and Lifshitz) to embed the problems in the text, rather than relegating them to the ends of chapters. In some places the problems are small, really just reminding you to fill in some missing steps before going on to the next topic. At the opposite extreme, some problems are small research projects. Because progress in biophysics depends on intimate interaction between theory and experiment, some problems ask you to analyze real data, which can be found through http://press.princeton.edu/titles/9911.html.[14]

Finally, a few words about references. References to the original literature serve multiple functions, especially in textbooks. Most obviously, I should cite the papers that influenced my own thinking about the subject, acknowledging my intellectual debts. Because this text is based on a course for PhD students, citations also help launch the student into the current literature, marking the start of paths that can carry you well beyond digested, textbook discussions. In another direction, references point back to classic papers—papers worth reading decades after they were published, papers that can provide inspiration. Importantly, all these criteria are subjective, and it may be as important to explain the reasons for including a particular reference as to point to the reference itself. Thus, I have collected the references, with commentary, in an Annotated Bibliography at the end of the volume. Let me note that the reference list should *not* be viewed as a rough draft of the history of the subject nor as an attempt to establish objective priorities for some work over others.

---

13. In some sections I found it difficult to formulate manageable problems. I worry that this reflects poorly on my understanding. K.G. Wilson once remarked that if you really understand something, then you can do a series of calculations in which the accuracy of the prediction is in reasonable proportion to the effort of the calculation. A corollary seems to be that one can assign problems that allow students to test their own understanding.

14. In these problems, and also in problems that involve simulation rather than analytic calculations, I have chosen to phrase all the computations in MATLAB. This makes me slightly uneasy—although the language of mathematics is universal, programming languages really are not (at least not in the practical sense). Also, MATLAB is a commercial product, although there is a free alternative available at http://www.gnu.org/software/octave/. But numerical analysis of data certainly is essential for the field, and so there isn't really any way around making a choice from currently available options.

Photon Counting in Vision

The goal of this chapter is to introduce some of the beauty and richness to be found in the phenomena of life. As explained in Chapter 1, there are so many phenomena to choose from that we can easily become overwhelmed. The point here thus is not to enshrine particular phenomena as canonical examples but rather to provide a guide, encouraging your own explorations. We start with the remarkable ability of the visual system to detect single photons and then dig into the more microscopic phenomena that provide the basis for this ability. Some of these phenomena are examples of things that happen in many different biological contexts or that were discovered by methods of wider applicability, and I try to make these connections clear. There is a serious agenda here—we will look back at these phenomena with an eye toward formulating physics problems that will occupy us for most of the book—but we should start by letting ourselves enjoy what Nature has to offer.

2.1   A First Look

Imagine sitting quietly in a dark room, staring straight ahead. A light flashes. Do you see it? Surely if the flash is bright enough, the answer is "yes," but how dim can the flash be before we fail? Do we fail abruptly, so that there is a well-defined threshold—lights brighter than the threshold are always seen, lights dimmer than it are never seen—or is the transition from seeing to not seeing somehow more gradual? These questions are classical examples of "psychophysics," studies on the relationship between perception and the physical variables in the world around us, and they have a history reaching back at least into the nineteenth century.

In 1911, the physicist Lorentz was sitting in a lecture that included an estimate of the "minimum visible," the energy of the dimmest flash of light that could be consistently and reliably perceived by human observers. But by 1911 it was known that light was composed of photons, and if the light is of well-defined frequency (ω) or wavelength, then the energy E of the flash is equivalent to an easily calculable number of photons n, $n = E/\hbar\omega$. Doing this calculation, Lorentz found that just-visible flashes of light correspond to roughly 100 photons incident on the eyeball. Turning to his physiologist colleague Zwaadermaker, Lorentz asked whether much of the light incident on the cornea might get lost (scattered or absorbed) on its way through the gooey interior of the eyeball, or whether the experiments could be off by as much as a factor of ten. In other words, is it possible that the real limit to human vision is the counting of single photons?

Lorentz' suggestion is quite astonishing. If correct, it would mean that the boundaries of our perception are set by basic laws of physics, and that we reach the limits of what is possible. Further, if the visual system really is sensitive to individual light quanta, then some of the irreducible randomness of quantum events should be evident in our perceptions of the world around us.

If we watch a single molecule and ask whether it absorbs a photon, this is a random process, with a probability per unit time proportional to the light intensity. The emission of photons is similarly random, so that a typical light source does not deliver a fixed number of photons in a given interval of time. Thus, when we look at a flash of light, the number of photons that will be absorbed by the sensitive cells in our retina is a random number, not because biology is noisy but because of the physics governing the interaction of light and matter. One way of testing whether we can count single photons, then, is to see whether we can detect the signatures of this randomness in our perceptions. This line of reasoning came to fruition in experiments by Hecht, Shlaer, and Pirenne (in New York) and by van der Velden (in the Netherlands) in the early 1940s.

What we think of classically as the intensity of a beam of light is proportional to the *mean* number of photons per second that arrive at some region where they can be counted.[1] For most conventional light sources, however, the stream of photons is not regular, but completely random. Thus, in any very small window of time dt, there is a probability $r\,dt$ that a photon will be counted, where r is the mean counting rate or light intensity, and the events in different time windows are independent of one another. These are the defining characteristics of a Poisson process, which is the maximally random sequence of point events in time—if we think of the times at which photons are counted as being like the positions of particles, then the sequence of photon counts is like an ideal gas, with no correlations or "interactions" among the particles at different positions.[2]

---

1. More precisely, we can measure the mean number of photons per second per unit area.

2. I have used the verb "count" several times now, as if its meaning were obvious. At some practical level in the laboratory, it is obvious—if we have a photomultiplier, for example, then the arrival of each photon is registered (or not, as the device is not perfectly efficient) by a pulse of current, and we can measure and count these pulses with conventional electronics. Because the pulses are generated by photon absorption events, we say that we are counting photons. As we shall see, the receptor cells in the eye also generate pulses of current, and these are large enough that we—and, presumably, the brain—can detect and count them with high reliability, so we say that the system is counting photons, just as we do in the lab with the photomultiplier. But this colloquial discussion of counting photons smooths over some subtleties. In principle, when we open a photodetector to a beam of light, all that happens is that the quantum states of the detector become entangled with the quantum states of the light beam. The notion of counting is a classical idea, and so to really understand what we mean when we count photons, we have to think about issues in the quantum theory of measurement. In our retina as in photomultipliers, the key step is the absorption of the photon, where the interactions with (many) other degrees of freedom destroy the coherence between the two electronic states that are connected by the photon. We will see in Section 2.2 that taking care of these issues is not just a matter of being pedantic, because the molecular events that happen immediately after photon absorption are so fast that they actually compete with the loss of coherence. For now, however, we can rely on our usual intuitive understanding.

As explained in detail in Appendix A.1, if events occur as a Poisson process with rate r, then when we count the events over some time T, the mean number of counts will be $M = rT$, but the probability that we actually obtain a count of n is given by the Poisson distribution,

$$P(n|M) = e^{-M} \frac{M^n}{n!}. \tag{2.1}$$

In our case, the mean number of photons that will be counted at the retina is proportional to the classical intensity $\tilde{I}$ of the light flash, $M = \alpha \tilde{I} T$, where the constant α includes all the messy details of what happens to the light on its way through the eyeball. Thus, when we deliver the "same" flash of light again and again, the actual physical stimulus to the retina will fluctuate, and it is plausible that our perceptions will fluctuate as well.

In the simplest view, you would be willing to say "yes, I saw the flash" once you had counted K photons. Equation (2.1) gives us the probability of counting exactly n photons given the mean, and the mean is connected to the intensity of the flash. Separate experiments show that if we keep the duration of the flash shorter than ~ 100 ms, then we can trade the intensity $\tilde{I}$ against the duration T, and all that matters is the product, which we call (abusing terminology a bit) the "intensity of the flash" $I = \tilde{I} T$; then we have $M = \alpha I$. Thus, we predict that there is a probability of seeing a flash of intensity I given by

$$P_{\text{see}}(I) = \sum_{n=K}^{\infty} P(n|M = \alpha I) = e^{-\alpha I} \sum_{n=K}^{\infty} \frac{(\alpha I)^n}{n!}. \tag{2.2}$$

So, if we sit in a dark room and watch as dim lights are flashed, we expect that our perceptions will fluctuate—sometimes we see the flash and sometimes we don't—but there will be an orderly dependence of the *probability* of seeing on the intensity, given by Eq. (2.2). Importantly, if we plot P_{see} versus $\log I$, as in Fig. 2.1, then the *shape* of the curve depends crucially on the threshold photon count K, but changing the unknown constant α just translates the curve along the x-axis. So we have a chance to measure the threshold K by looking at the shape of the curve; more fundamentally we might say we are testing the hypothesis that the probabilistic nature of our perceptions is determined by the physics of photon counting.

Problem 1: Photon statistics, part one.   There are two reasons the arrival of photons might be described by a Poisson process. The first is a very general "law of small numbers" argument. Imagine a point process in which events occur at times $\{t_i\}$, with some correlations among the events. Assume that these correlations die away with time, so that events i and j are independent if $|t_i - t_j| \gg \tau_c$, where τ_c is the correlation time. Explain qualitatively why, if we select only a small fraction of events out of the original sequence at random, the resulting sparse sequence will be approximately Poisson. What is the condition for the Poisson approximation to be a good one? What does this have to do with why, for example, the light that reaches us from an incandescent light bulb comes in a Poisson stream of photons?

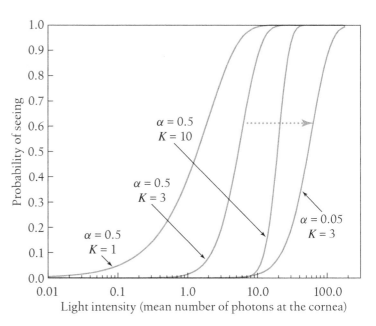

FIGURE 2.1

Probability of seeing calculated from Eq. (2.2), where the intensity I is measured as the mean number of photons incident on the cornea, so that α is dimensionless. Curves are shown for different values of the threshold photon count K and the scaling factor α. Note the distinct shapes for different K, but when we change α at fixed K, we just translate the curve along the log intensity axis, as shown by the blue dotted arrow.

Problem 2: How many sources of randomness?   The defining feature of a Poisson process is the independence of events at different times, and typical light sources generate a stream of photons whose arrival times approximate a Poisson process. But when we count these photons, we don't catch every one. Show that if the photon arrivals are a Poisson process with rate r, and we count a fraction f these, selected at random, then the times at which events are counted will also be a Poisson process, with rate fr. Why doesn't the random selection of events to be counted result in some "extra" variance beyond expectations for the Poisson process?

Hecht, Shlaer, and Pirenne did exactly the experiment we are analyzing. Subjects (the three co-authors) sat in a dark room and reported whether they did or did not see a dim flash of light. For each setting of the intensity, there were many trials, and responses were variable, but the subjects had to say "yes" or "no," with no "maybe." Thus, at each intensity it was possible to measure the probability that the subject would say "yes," and this is plotted in Fig. 2.2.

The first nontrivial result of these experiments is that human perception of dim light flashes really is probabilistic. No matter how hard we try, there is a range of light intensities in which our perceptions fluctuate from flash to flash of the same intensity, seeing one and missing another. Quantitatively, the plot of probability of seeing versus log (intensity) is fit very well by the predictions from the Poisson statistics of photon arrivals. In particular, Hecht, Shlaer, and Pirenne found a beautiful fit in the range from $K = 5$ to $K = 7$; subjects of different age had very different values for α (as must be true if light transmission through the eye gets worse with age) but similar values of K. In Fig. 2.2 I have shown all three observers' data fit to $K = 6$, along with error bars (absent in the original paper). Although one could do better by allowing each person to have a different value of K, it is not clear that this approach would be supported by the statistics. The different values of α, however, are quite important.

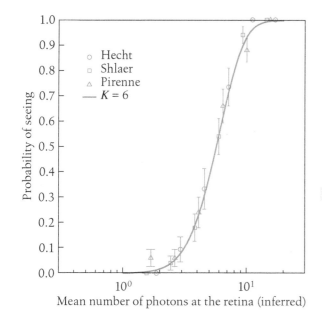

Mean number of photons at the retina (inferred)

FIGURE 2.2

Probability of seeing calculated from Eq. (2.2), with the threshold photon count $K = 6$, compared with experimental results from Hecht et al. (1942). For each observer we can find the value of α that provides the best fit and then plot all the data on a common scale as shown here. Error bars are computed on the assumption that each trial is independent, which probably generates errors bars that are slightly too small.

Details aside, the frequency-of-seeing experiment brings forward a beautiful idea: the probabilistic nature of our perceptions reflects the physics of random photon arrivals. An absolutely crucial point is that Hecht, Shlaer, and Pirenne chose stimulus conditions such that the five to seven photons needed for seeing were distributed across a broad area on the retina, an area that contains hundreds of photoreceptor cells. Thus, the probability of one receptor (rod) cell receiving more than one photon is very small. The experiments on human behavior therefore indicate that individual photoreceptor cells generate reliable responses to single photons. In fact, vision begins (as we discuss in more detail soon) with the absorption of light by the visual pigment rhodopsin, and so sensitivity to single photons means that each cell is capable of responding to a single molecular event. This is a wonderful example of using macroscopic experiments to draw conclusions about single cells and their microscopic mechanisms.

Problem 3: Simulating a Poisson process. Much of what we want to know about Poisson processes can be determined analytically (see Appendix A.1). Thus, if we do simulations we know what answer we should get (!). This provides us with an opportunity to exercise our skills, even if we don't get any new answers. In particular, doing a simulation is never enough; you have to analyze the results, just as you analyze the results of an experiment. Now is as good a time as any to get started. If you are comfortable doing everything in a programming language like C, that's great. However, high-level languages, such as MATLAB or Mathematica, have certain advantages. Here you should use MATLAB to simulate a Poisson process, and then analyze the results to be sure that you actually did what you expected to do.

(a) MATLAB has a command **rand** that generates random numbers with a uniform distribution from 0 to 1. Consider a time window of length T, and divide this window into many small bins of size dt. In each bin you can use **rand** to generate a number that you can compare with a threshold—if the random number is above threshold, you put an event in the bin, and you can adjust the threshold to set the average number of events in the window.

You might choose $T = 10^3$ sec and arrange that the average rate of the events is $\bar{r} \sim 10/s$: note that you should be able to relate the threshold to the mean rate $\bar{r}$ analytically. This procedure implements (in the limit $dt \to 0$) the definition of the Poisson process as independent point events.

(b) The next step is to check that the events you have made really do obey Poisson statistics. Start by counting events in windows of some size τ. What is the mean count? The variance? Do you have enough data to fill in the whole probability distribution $P_\tau(n)$ for counting n events in the window? How do all these things change as you change τ? What if you go back and make events with a different average rate? Do your numerical results agree with the theoretical expressions? In answering this question, you could try to generate sufficiently large data sets that the agreement between theory and experiment is almost perfect, but you could also make smaller data sets and ask whether the agreement is good within some estimated error bars; this will force you to think about how to put error bars on a probability distribution. You should also make a histogram (`hist` should help) of the times $\Delta t = t_{i+1} - t_i$ between successive events. The result should be an exponential, $P(\Delta t) \propto e^{-k\Delta t}$, and you should work to get it into a form where it is a properly normalized probability density. What is the connection between k and $\bar{r}$? Check this in your data.

(c) Instead of deciding about the presence or absence of an event in each bin, use the command `rand` to choose N random times in the big window T. Examine as before the statistics of counts in windows of size $\tau \ll T$. Do you still have an approximately Poisson process? Why? Do you see connections to the statistical mechanics of ideal gases and the equivalence of ensembles?

Problem 4: Photon statistics, part two.   The other reason we might find photon arrivals to be a Poisson process comes from a very specific quantum mechanical argument about coherent states.[3]

(a) Modes of the electromagnetic field (in free space, a cavity, or a laser) are described by harmonic oscillators. The Hamiltonian of a harmonic oscillator with frequency ω can be written as

$$\mathbf{H} = \hbar\omega(a^\dagger a + 1/2), \tag{2.3}$$

where $a^\dagger$ and a are the creation and annihilation operators that connect states with different numbers of quanta:

$$a^\dagger|n\rangle = \sqrt{n+1}|n+1\rangle, \tag{2.4}$$
$$a|n\rangle = \sqrt{n}|n-1\rangle. \tag{2.5}$$

There is a special family of states called coherent states, defined as eigenstates of the annihilation operator:

$$a|\alpha\rangle = \alpha|\alpha\rangle. \tag{2.6}$$

If we write the coherent state as a superposition of states with different numbers of quanta,

---

3. These ideas may or may not be familiar from your quantum mechanics courses. I have tried to write the problem so that it is self-contained, although obviously it is easier if you have seen these things before. See also the references for this section in the Annotated Bibliography.

$$|\alpha\rangle = \sum_{n=0}^{\infty} \psi_n |n\rangle, \tag{2.7}$$

then you can use the defining Eq. (2.6) to give a recursion relation for the ψ_n. Solve this, and show that the probability of counting n quanta in this state is given by the Poisson distribution, that is,

$$P_\alpha(n) \equiv \left| \langle n | \alpha \rangle \right|^2 = |\psi_n|^2 = e^{-M} \frac{M^n}{n!}, \tag{2.8}$$

where the mean number of quanta is $M = |\alpha|^2$.

(b) The specialness of the coherent states relates to their dynamics and their representation in position space. A quantum mechanical state $|\phi\rangle$ evolves in time as

$$i\hbar \frac{d|\phi\rangle}{dt} = \mathbf{H}|\phi\rangle. \tag{2.9}$$

Show that if the system starts in a coherent state $|\alpha(0)\rangle$ at time $t = 0$, it remains in a coherent state for all time. Find $\alpha(t)$.

(c) If we go back to the mechanical realization of the harmonic oscillator as a mass m hanging from a spring, the Hamiltonian is

$$\mathbf{H} = \frac{1}{2m} p^2 + \frac{m\omega^2}{2} q^2, \tag{2.10}$$

where p and q are the momentum and position, respectively, of the mass. What is the relationship between the creation and annihilation operators and the position and momentum operators $(\hat{q}, \hat{p})$? In position space, the ground state is a Gaussian wave function,

$$\langle q | 0 \rangle = \frac{1}{(2\pi\sigma^2)^{1/4}} \exp\left(-\frac{q^2}{4\sigma^2} \right), \tag{2.11}$$

where the variance of the zero point motion $\sigma^2 = \hbar/(4m\omega)$. The ground state is also a minimum uncertainty wave packet, so called because the variance of position and the variance of momentum have a product that is the minimum value allowed by the uncertainty principle; show that this is true. Consider the state $|\psi(q_0)\rangle$ obtained by displacing the ground state to a position q_0:

$$|\psi(q_0)\rangle = e^{iq_0\hat{p}/\hbar} |0\rangle. \tag{2.12}$$

Show that this is a minimum uncertainty wave packet and also a coherent state. Find the relationship between the coherent state parameter α and the displacement q_0.

(d) Put all these steps together to show that the coherent state is a minimum uncertainty wave packet with expected values of the position and momentum that follow the classical equations of motion.

The states of light inside a laser are, to a good approximation, coherent states. Thus, they are minimum uncertainty states and hence push against the limits of what is allowed by quantum mechanics, but at the same time they are very classical states, because they are described by the same parameters as a classical light field. For our purposes the important point is that these coherent states have exactly a Poisson distribution of photon numbers.

There is a very important point in the background of this discussion. By placing results from all three observers on the same plot and fitting with the same value of K, we are claiming that there is something reproducible, from individual to individual, about our perceptions. However, each observer has a different value for α, which means that there are individual differences, even in this simplest of tasks. Happily, what seems to be reproducible is something that feels like a fundamental property of the system, the number of photons we need to count to be sure that we saw something. But suppose that we just plot the probability of seeing versus the (raw) intensity of the light flash. If we average across individuals with different αs, we will obtain a result that does not correspond to the theory, and this failure might even lead us to believe that the visual system does not count single photons. This shows us that finding what is reproducible can be difficult, and averaging across an ensemble of individuals can be qualitatively misleading. Here we see these conclusions in the context of human behavior, but it seems likely that similar issues arise in the behavior of single cells. The difference is that techniques for monitoring the behavior of single cells (e.g., bacteria), as opposed to averages over populations of cells, have emerged much more recently. As an example, it still is almost impossible to monitor, in real time, the metabolism of single cells, whereas simultaneous measurements on many metabolic reactions, averaged over populations of cells, have become common. We still have much to learn from these older experiments.

Problem 5: Averaging over observers.   Go back to the original paper by Hecht et al. (1942)[4] and use their data to plot, versus the intensity of the light flash, the probability of seeing averaged over all three observers. Does this look anything like what you find for individual observers? Can you simulate this effect, say, in a larger population of subjects, by assuming that the factor α is drawn from a distribution? Explore this idea a bit, and see how badly misled you could be. This problem is deliberately open ended.

Before moving on, a few rather colloquial remarks about history are in order. It is worth noting that van der Velden's (1944) seminal paper was published in Dutch, a reminder of a time when anglophone cultural hegemony was not yet complete. Also (maybe more relevant for us), it was published in a physics journal. The physics community in the Netherlands during this period had a very active interest in problems of noise, and van der Velden's work was in this tradition. In contrast, Hecht was a distinguished contributor to understanding vision but had worked within a "photochemical" view, which he would soon abandon as inconsistent with the detectability of single photons and hence single molecules of activated rhodopsin. Parallel to this work, Rose and de Vries independently emphasized that noise due to the random arrival of photons at the retina also would limit the reliability of perception at intensities well above the point where things become barely visible. In particular, de Vries saw these issues as part of the larger problem of understanding the physical limits to biological function, and I think his perspective on the interaction of physics and biology was far ahead of its time.

---

4. As is true throughout the text, references are found in the Annotated Bibliography at the end of the book.

It took many years before anyone could measure directly the responses of photo-receptors to single photons. As with all cells in the nervous system, these cells respond by generating electrical signals, but the single-photon signals are especially small. They were first detected in the (invertebrate) horseshoe crab, and eventually by Baylor and co-workers in toads and then in monkeys, whose visual systems are very much like our own. The complication in the lower vertebrate systems is that the cells are coupled to-gether, so that the retina can do something like adjusting the size of pixels as a function of light intensity. Thus, the nice big current generated by one cell is spread as a small voltage in many cells, so the usual method of measuring the voltage across the mem-brane of one cell won't work; you have to suck the cell into a pipette and collect the current, as seen in Fig. 2.3.

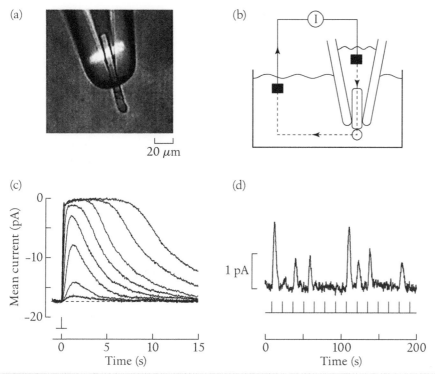

FIGURE 2.3

(a) A single rod photoreceptor cell from a toad, in a suction pipette. Viewing is with infrared light, and the bright bar is a stimulus of 500 nm light. (b) Equivalent circuit for recording the current across the cell membrane. Current flows into the rod cell outer segment, out through the inner segment, and the circuit is completed in the measurement apparatus; black boxes are electrodes that convert between ionic currents in the solution bathing the cell and electronic currents that can be measured and amplified in conventional ways. (c) Mean current in response to light flashes of varying intensity. Smallest response is to flashes that deliver a mean ~4 photons; successive flashes are brighter by factors of 4. (d) Current responses to repeated dim light flashes at times indicated by the tick marks. Note the apparently distinct classes of responses to zero, one, or two photons. Reprinted, with permission, from Rieke and Baylor (1998). Copyright © 1998 by the American Physical Society.

Problem 6: Gigaseals.   As we will see, the currents that are relevant in biological systems are on the order of picoamps. Although the response of rods to single photons is on the time scale of 1s, many processes in the nervous system occur on the millisecond time scale. Show that if we want to resolve picoamps in milliseconds, then the leakage resistance (e.g., between rod cell membrane and the pipette in Fig. 2.3b) must be $\sim 10^9$ ohm, to prevent the signal from being lost in Johnson noise.

In complete darkness, there is a standing current of ~ 20 pA flowing through the membrane of the rod cell's outer segment. You should keep in mind that currents in biological systems are carried not by electrons or holes, as in solids, but by ions moving through water; we will learn more about this soon. In the rod cell, the standing current is carried largely by sodium ions, although there are contributions from other ions as well. This is a hint that the channels in the membrane that allow the ions to pass are not especially selective for one ion over another. The current that flows across the membrane has to go somewhere, and the circuit is completed within the rod cell itself, so that what flows across the outer segment of the cell is compensated by flow across the inner segment. When the rod cell is exposed to light, the standing current is reduced, and with sufficiently bright flashes it is turned off all together.

As in any circuit, current flow generates changes in the voltage across the cell membrane. Near the bottom of the cell there are special channels that allow calcium ions to flow into the cell in response to these voltage changes, and calcium in turn triggers the fusion of vesicles with the cell membrane. These vesicles are filled with a small molecule, a neurotransmitter, which can then diffuse across a small cleft and bind to receptors on the surface of neighboring cells. These receptors then respond (in the simplest case) by opening channels in the membrane of this second cell, allowing currents to flow. In this way, currents and voltages in one cell are converted, via a chemical intermediary, into currents and voltages in the next cell, and so on through the nervous system. The place where two cells connect in this way is called a synapse, and in the retina the rod cells form synapses onto a class of cells called bipolar cells. More about this later, but for now you should keep in mind that the electrical signal in the rod cell is the first in a sequence of electrical signals that ultimately are transmitted along the cells in the optic nerve, connecting the eye to the brain and hence providing the input for visual perception.

Very dim flashes of light seem to produce a quantized reduction in the standing current, and the magnitude of these current pulses is roughly 1 pA, as seen in Fig. 2.3. When we look closely at the standing current, we see that it is fluctuating: there is a continuous background noise of ~ 0.1 pA, so the quantal events are easily detected. It takes a bit of work to convince yourself that these events really are the responses to single photons. Perhaps the most direct experiment is to measure the cross-section for generating the quantal events and compare this with the absorption cross-section of the rod, showing that a little more than 2/3 of the photons that are absorbed produce current pulses. In response to steady dim light, we can observe a continuous stream of pulses; the rate of the pulses is proportional to the light intensity, and the intervals between pulses are distributed exponentially, as expected if they represent the responses to single photons (see Appendix A.1).

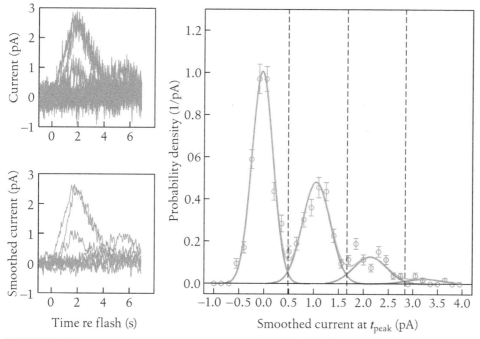

FIGURE 2.4

A closer look at the rod cell currents. At left, five instances in which the rod is exposed to a dim flash at $t = 0$. It looks as if two of these flashes delivered two or three photons (peak current $\sim 2-3$ pA), one delivered one photon (peak current ~ 1 pA), and two delivered zero. The top panel shows the raw current traces, and the bottom panel shows what happens when we smooth with a 100 ms window to remove some of the high-frequency noise. At right, the distribution of smoothed currents at the moment t_{peak} when the average current peaks; the data (circles) are accumulated from 350 flashes in one cell, and the error bars indicate standard errors of the mean due to this finite sample size. Blue line is the fit to Eq. (2.19), composed of contributions from $n = 0, n = 1, \cdots$ photon events, shown in orange. Dashed black lines divide the range of observed currents into the most likely assignments to different photon counts. These data are from unpublished experiments by F. M. Rieke at the University of Washington; many thanks to Fred for providing the data in raw form.

When you look at the currents flowing across the rod cell membrane, the statement that single-photon events are detectable above background noise seems pretty obvious, but it would be good to be careful about what we mean here. In Fig. 2.4 we take a closer look at the currents flowing in response to dim flashes of light. These data were recorded with a very high bandwidth, so you can see a lot of high-frequency noise. Nonetheless, for these five flashes, it is clear that twice the cell counted zero photons, once it counted one photon (for a peak current ~ 1 pA), and twice it counted two photons; this becomes even clearer if we smooth the data to get rid of some of the noise. Still, these results are anecdotal, and one would like to be more quantitative.

Problem 7: Are they really single-photon responses?   What is it about the experiments in Fig. 2.4 that provides the smoking gun for claiming that the current pulses are responses to single photons? Maybe it is something that is not shown. For example, if one pulse were generated by the coincidence of two photons, how would the distribution of peak currents shift with changing flash intensity? I have left this deliberately vague, hoping that you will explore. If you were doing these experiments, what would you do to seal the conclusion that you were seeing single-photon responses?

Even in the absence of light there are fluctuations in the current, and for simplicity let's assume that this background noise is Gaussian with some variance σ_0^2. The simplest way to decide whether we saw something is to look at the rod current at one moment in time, say, at $t = t_{peak} \sim 2$ s after the flash, when on average the current is at its peak. Then given that no photons were counted, this current i should be drawn out of the probability distribution

$$P(i|n = 0) = \frac{1}{\sqrt{2\pi\sigma_0^2}} \exp\left[-\frac{i^2}{2\sigma_0^2}\right]. \tag{2.13}$$

If one photon is counted, then there should be a mean current $\langle i \rangle = i_1$, but there is still some noise. Plausibly the noise has two pieces—the background noise still is present, with its variance σ_0^2, and in addition the single-photon response itself can fluctuate. We assume that these fluctuations are also Gaussian and independent of the background, so they just add σ_1^2 to the variance. Thus, we expect that, in response to one photon, the current will be drawn from the distribution

$$P(i|n = 1) = \frac{1}{\sqrt{2\pi(\sigma_0^2 + \sigma_1^2)}} \exp\left[-\frac{(i - i_1)^2}{2(\sigma_0^2 + \sigma_1^2)}\right]. \tag{2.14}$$

If each single photon event is independent of the others, then we can generalize this expression to get the distribution of currents expected in response to $n = 2$ photons,

$$P(i|n = 2) = \frac{1}{\sqrt{2\pi(\sigma_0^2 + 2\sigma_1^2)}} \exp\left[-\frac{(i - 2i_1)^2}{2(\sigma_0^2 + 2\sigma_1^2)}\right], \tag{2.15}$$

and more generally n photons,

$$P(i|n) = \frac{1}{\sqrt{2\pi(\sigma_0^2 + n\sigma_1^2)}} \exp\left[-\frac{(i - ni_1)^2}{2(\sigma_0^2 + n\sigma_1^2)}\right]. \tag{2.16}$$

In writing these equations, we are assuming that the response to each photon is being generated independently and so makes an independent, additive contribution not just to the mean response but also to the variance of the responses. Finally, because we know that the photon count n should be drawn out of the Poisson distribution, we can write the expected distribution of currents as

$$P(i) = \sum_{n=0}^{\infty} P(i|n)P(n) \tag{2.17}$$

$$= \sum_{n=0}^{\infty} P(i|n)e^{-\bar{n}}\frac{\bar{n}^n}{n!} \tag{2.18}$$

$$= \sum_{n=0}^{\infty} \frac{\bar{n}^n}{n!} \frac{e^{-\bar{n}}}{\sqrt{2\pi(\sigma_0^2 + n\sigma_1^2)}} \exp\left[-\frac{(i - ni_1)^2}{2(\sigma_0^2 + n\sigma_1^2)}\right], \tag{2.19}$$

where $\bar{n}$ is the mean photon count. In Fig. 2.4, we see that this equation really gives an excellent description of the distribution that we observe when sampling the currents in response to a large number of flashes.

Problem 8: Exploring the sampling problem. The data in Fig. 2.4 are not a perfect fit to our model. However, only 350 samples are used to estimate the shape of the underlying probability distribution. This is an example of a problem that you will meet many times in comparing theory and experiment; perhaps you have some experience from physics lab courses that is relevant here. We will return to these issues of sampling and fitting nearer the end of the book (Section 6.5), when we have some more powerful mathematical tools, but for now let me encourage you to play a bit. Use the model that leads to Eq. (2.19) to generate samples of the peak current, and then use these samples to estimate the probability distribution. For simplicity, assume that $i_1 = 1$, $\sigma_0 = 0.1$, $\sigma_1 = 0.2$, and $\bar{n} = 1$. Notice that because the current is continuous, you have to make bins along the current axis; smaller bins reveal more structure but also generate noisier results, because the number of counts in each bin is smaller. As you experiment with different bin sizes and different numbers of samples, try to develop some feeling for whether the agreement between theory and experiment in Fig. 2.4 really is convincing.

Seeing the distribution in Eq. (2.19), and especially seeing analytically how it is constructed, it is tempting to draw lines along the current axis in the troughs of the distribution and say that (for example) a current < 0.5 pA indicates zero photons. Is this the right way for us—or for the brain—to interpret these data? To be precise, suppose that we want to set a threshold for deciding between $n = 0$ and $n = 1$ photon. Where should we put this threshold to be sure that we get the right answer as often as possible?

Suppose we set the threshold at some current $i = \theta$. If there really were zero photons absorbed, then if by chance $i > \theta$, we will incorrectly say that there was one photon. This error has a probability

$$P(\text{say } n = 1 | n = 0) = \int_{\theta}^{\infty} di \, P(i|n=0). \tag{2.20}$$

In contrast, if there really were one photon, but by chance the current was less than the threshold, then we will say 0 when we should have said 1, and this error has a probability

$$P(\text{say } n = 0 | n = 1) = \int_{-\infty}^{\theta} di \, P(i|n=1). \tag{2.21}$$

There could be errors in which we confuse two photons for zero photons, but looking at Fig. 2.4, it seems that these higher order errors are negligible. So then the total probability of making a mistake in the $n = 0$ versus $n = 1$ decision is

$$P_{\text{error}}(\theta) = P(\text{say } n = 1 | n = 0) P(n = 0)$$
$$+ P(\text{say } n = 0 | n = 1) P(n = 1) \qquad (2.22)$$
$$= P(n = 0) \int_{\theta}^{\infty} di \; P(i | n = 0)$$
$$+ P(n = 1) \int_{-\infty}^{\theta} di \; P(i | n = 1). \qquad (2.23)$$

We can minimize the probability of error in the usual way by taking the derivative and setting the result to zero at the optimal setting of the threshold, $\theta = \theta^*$:

$$\frac{dP_{\text{error}}(\theta)}{d\theta} = P(n = 0) \frac{d}{d\theta} \int_{\theta}^{\infty} di \; P(i | n = 0)$$
$$+ P(n = 1) \frac{d}{d\theta} \int_{-\infty}^{\theta} di \; P(i | n = 1) \qquad (2.24)$$
$$= P(n = 0)(-1) P(i = \theta | n = 0)$$
$$+ P(n = 1) P(i = \theta | n = 1); \qquad (2.25)$$

$$\left. \frac{dP_{\text{error}}(\theta)}{d\theta} \right|_{\theta = \theta^*} = 0 \Rightarrow P(i = \theta^* | n = 0) P(n = 0) = P(i = \theta^* | n = 1) P(n = 1).$$
$$(2.26)$$

In particular, if one photon and zero photons are equally likely, then we should place the threshold θ at the value of the current i where $P(i | n = 0) = P(i | n = 1)$; that is, the observed current is equally likely to have been generated by either one or zero photons. When we observe $i > \theta$, it is more likely that what we observe was generated by a real photon ($n = 1$), and so we should say that this is what happened. Conversely, if we observe $i < \theta$, what we observe was more likely to have been generated by background noise alone ($n = 0$), and so this is our best guess. Choosing these "maximum likelihood" interpretations is the recipe for minimizing errors, and this is how we draw the boundaries shown by dashed lines in Fig. 2.4.

Problem 9: More careful discrimination.   You observe some variable x (e.g., the current flowing across the rod cell membrane) that is chosen either from the probability distribution $P(x|+)$ or from the distribution $P(x|-)$. Your task is to look at a particular x and decide whether it came from the $+$ or the $-$ distribution. Rather than just setting a threshold, as in the discussion above, suppose that when you see x you assign it to the $+$ distribution with a probability $p(x)$. You might think this is a good idea because, if you are not completely sure of the right answer, you can hedge your bets by a little bit of random guessing. Express the probability of a correct answer in terms of $p(x)$; this is a functional $P_{\text{correct}}[p(x)]$. Now solve the optimization problem for the function $p(x)$, maximizing P_{correct}. Show that the solution is deterministic [$p(x) = 1$ or $p(x) = 0$], so that if the goal is to be correct as often as possible, you shouldn't hesitate to make a crisp assignment even at values of x where you aren't sure. Hint: Usually, you would try to maximize P_{correct} by solving the variational equation $\delta P_{\text{correct}}/\delta p(x) = 0$. You should find that, in this case, this approach fails. What does this mean? Remember that $p(x)$ is a probability and hence can't take on arbitrary values.

If it is really dark outside, then the arrival of a photon is rather unlikely, and so we should have $P(n=1) \ll P(n=0)$. Intuitively, we should be able to use this knowledge to help us decide more effectively between the two alternatives and lower the error probability. Indeed, Eq. (2.26), which determines the optimal setting of the threshold, combines the likelihood that the current was generated by a photon versus background noise with the overall probability that we will see a photon (or not). How do we understand this combination?

What we are seeing in Eq. (2.26) is a first look at something much more general, something that will reappear many times, so let's try to give a more general formulation. Imagine that we have two variables $x \in X$ and $y \in Y$. There is some probability that we observe particular values of x and y together, and we will write this joint distribution as $P(x, y)$. In many cases, it makes sense to say that "x causes y," or perhaps "y is the response of the system to the input x." With this picture, it is natural to suppose that x is chosen out of some distribution $P_X(x)$, and then given this value of the input, the system generates a response that might be noisy and hence is chosen out of the conditional distribution $P(y|x)$. The probability of observing both x and y is then the probability of observing x multiplied by the probability of observing y given that we have already observed x:

$$P(x, y) = P(y|x)P_X(x). \qquad (2.27)$$

But we can take a different point of view. Imagine that we look first at the output y. We will find a value drawn out of the distribution $P_Y(y)$, and this distribution of outputs is determined by the other distributions:

$$P_Y(y) = \sum_x P(x, y) = \sum_x P(y|x)P_X(x). \qquad (2.28)$$

Now, having observed y, we surely know something about the value of x, because y is the response of the system to the input x. But this knowledge is imperfect, so we should describe the distribution of x values that are consistent with y, $P(x|y)$. The probability of observing both x and y is then the probability of observing y, multiplied by the probability of x given that we have already observed y:

$$P(x, y) = P(x|y)P_Y(y). \qquad (2.29)$$

We see that we have two different ways of decomposing the joint distribution, Eqs. (2.27) and (2.29). These have to be consistent with one another, and so we must have

$$P(x|y)P_Y(y) = P(y|x)P_X(x) \qquad (2.30)$$

$$\Rightarrow P(x|y) = \frac{P(y|x)P_X(x)}{P_Y(y)}. \qquad (2.31)$$

This seemingly innocent consequence of decomposing the joint probability in different ways is called Bayes' rule.

What does Eq. (2.31) have to do with our problem? Let's identify the input x with the number of photons n and the output y with the current i. Then Eq. (2.31) tells us that

$$P(n|i) = \frac{P(i|n)P(n)}{P(i)}.$$ (2.32)

We note that in Eq. (2.26) we have precisely the combinations $P(i|n)P(n)$ for $n = 0$ and $n = 1$. Now we see, from Bayes' rule, that the combination $P(i|n = 0)P(n = 0)$ is proportional to the probability that the observed current i was generated by counting $n = 0$ photons, and similarly the combination $P(i|n = 1)P(n = 1)$ is proportional to the probability that the observed current was generated by counting $n = 1$ photons. The optimal setting of the threshold, from Eq. (2.26), is when these two probabilities are equal. In other words, for each observable current i we should compute the probability $P(n|i)$, and then our best guess about the photon count n is the one that maximizes this probability. This guess is best in the sense that it minimizes the total probability of errors, and it is the best way combining the knowledge that we gain by observing the current with our prior knowledge about the probability of photons being present.

We will see this idea again, so let's try one more way of saying what it going on in Eqs. (2.26) and (2.31). With some grandeur, let's say that x is the state of the world, and y represents the data we can observe. We have a model of how different states of the world can generate the data, and because there is some noise, this model is described by a probability distribution $P(\text{data}|\text{state of the world})$. But the problem we usually face is to draw inferences about the world from our data, which means we really want to know the other conditional distribution, $P(\text{state of the world}|\text{data})$. Bayes' rule tells us how to construct this distribution:

$$P(\text{state of the world}|\text{data}) = \frac{P(\text{data}|\text{state of the world})P(\text{state of the world})}{P(\text{data})}.$$ (2.33)

If we want to minimize the probability of making an error in identifying the state of the world, we should maximize this posterior probability, which includes contributions both from the data and from our prior knowledge.[5] This technique is crucial in many examples of data analysis in the laboratory, and as we shall see, there is increasing evidence that the brain does computations that have this Bayesian structure.

Once we have found the decision rules that minimize the probability of error, we can ask about the error probability itself. As schematized in Fig. 2.5, we can calculate this by integrating the relevant probability distributions on the "wrong sides" of the threshold. For Fig. 2.4, this error probability is less than 3%. Thus, under these conditions, we can look at the current flowing across the rod cell membrane and decide whether we saw $n = 0, 1, 2 \ldots$ photons with a precision such that we are wrong only on a few flashes out of 100. In fact, we might even be able to do better if instead of looking at the current at one moment in time we look at the whole trajectory of current versus time, but to do this analysis, we need a few more mathematical tools. Even without a more sophisticated

---

5. $P(\text{state of the world})$ is a prior distribution, because it embodies knowledge that we have before collecting these data. Similarly, $P(\text{state of the world}|\text{data})$ is a posterior distribution, because it comes after our observations of the data.

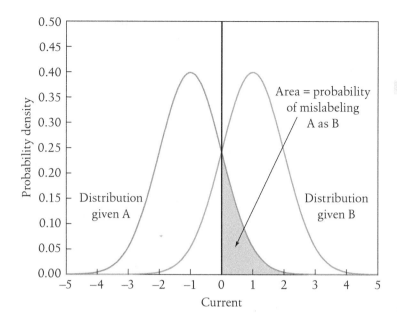

FIGURE 2.5

Schematic of discrimination in the presence of noise. We have two possible signals, A and B, and we measure something (e.g., the current flowing across a cell membrane). Given either A or B, the current fluctuates. As explained in the text, the overall probability of confusing A with B is minimized if we draw a threshold at the point where the probability distributions cross and identify all currents larger than this threshold as being B and all currents smaller than threshold as being A. Because the distributions overlap, it is not possible to avoid errors, and the area of the orange shaded region measures the probability that we will misidentify A as B.

analysis, it is clear that these cells really are acting as near-perfect photon counters, at least over some range of conditions.

Problem 10: Asymptotic error probabilities. Suppose that we measure y and are asked to distinguish between two alternatives A and B, as in Fig. 2.5. Assume that the distributions of y are Gaussians with different means but the same variance, so that

$$P(y|A) = \frac{1}{\sqrt{2\pi\sigma^2}} \exp\left[-\frac{(y - \bar{y}_A)^2}{2\sigma^2}\right], \tag{2.34}$$

$$P(y|B) = \frac{1}{\sqrt{2\pi\sigma^2}} \exp\left[-\frac{(y - \bar{y}_B)^2}{2\sigma^2}\right]. \tag{2.35}$$

Let's agree that $\bar{y}_B > \bar{y}_A$, and that you will guess B if you observe y to be above some threshold θ as in Eq. (2.20) et seq.

(a) Allowing that there may be different prior probabilities P_A and $P_B = 1 - P_A$ for the two alternatives, what is the optimal setting of the threshold for minimizing errors?

(b) Give an expression for the error probability when the threshold is set to its optimal value. Show that this minimum error P^*_{error} depends only on the signal-to-noise ratio $S \equiv (\bar{y}_A - \bar{y}_B)^2/\sigma^2$, not on the parameters $\bar{y}_A$, $\bar{y}_B$, and σ^2 separately.

(c) Find an approximate expression for $P^*_{\text{error}}(S)$ at large S.

A slight problem in our simple identification of the probability of seeing with the probability of counting K photons is that van der Velden found a threshold photon count of $K = 2$, which is completely inconsistent with the $K = 5$–7 found by Hecht, Shlaer, and Pirenne. Barlow explained this discrepancy by noting that even when counting single photons we may have to discriminate (as in photomultipliers) against a background of dark noise.

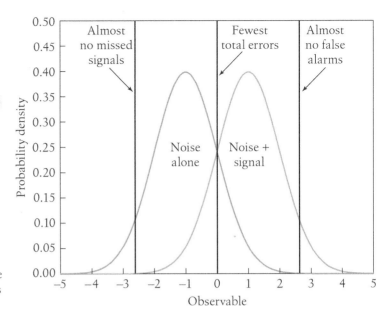

FIGURE 2.6

Trading of errors in the presence of noise. We observe some quantity that fluctuates even in the absence of a signal. When we add the signal these fluctuations continue, but the overall distribution of the observable is shifted. If we set a threshold, declaring the signal is present when the threshold is exceeded, then we can trade between the two kinds of errors. At low thresholds, we never miss a signal, but there will be many false alarms. At high thresholds, there are few false alarms, but we miss most of the signals, too. At some intermediate setting of the threshold, the total number of errors will be minimized.

Hecht, Shlaer, and Pirenne inserted blanks in their experiments to be sure that you almost never say "I saw it" when nothing is there, which means you have to set a high threshold to discriminate against any background noise. In contrast, van der Velden was willing to allow for some false positive responses, so his subjects could afford to set a lower threshold. Qualitatively, as shown in Fig. 2.6, this makes sense, but to be a quantitative explanation the noise has to be at the right level.

Before making quantitative comparisons, the qualitative picture of Fig. 2.6 tells us that the minimum number of photons needed to say "I saw it" should be reduced if we allow the observer the option of saying "I'm pretty sure I saw it," in effect taking control over the trade between misses and false alarms. Barlow showed that this worked, quantitatively.

One of the key ideas in the analysis of signals and noise is that of referring noise to the input. Imagine that we have a system to measure something (here, the intensity of light, but it could be anything), and it has a very small amount of noise somewhere along the path from input to output. In many systems we will also find, along the path from input to output, an amplifier that makes all signals larger. But the amplifier does not "know" which of its inputs are signal and which are noise, so everything is amplified. Thus, a small noise near the input can become a large noise near the output, but the size of this noise at the output does not, by itself, tell us how hard it will be to detect signals at the input. What we can do is to imagine that the whole system is noiseless and that any noise we see at the output really was injected at the input, and thus it followed exactly the same path as the signals we are trying to detect. Then we can ask how big this effective input noise needs to be to account for the output noise. Importantly, this effective noise is given in the same units as the input signal, and so we can compare signals and noise meaningfully. The bigger the effective noise, the larger the signal will need to be if we want to detect it with some criterion level of reliability.

In the case of counting photons, we can think of the effective input noise as being extra "dark" photons, also drawn from a Poisson distribution. Thus, if in the relevant window of time for detecting the light flash there are an average of 10 dark photons, for example, then because the variance of the Poisson distribution is equal to the mean, there will be fluctuations on the scale of $\sqrt{10}$ counts. To be very sure that we have seen something, we need an extra K real photons, with $K \gg \sqrt{10}$. Barlow's argument was that we could understand the need for $K \sim 6$ in the Hecht, Shlaer, and Pirenne experiments if indeed there were a noise source in the visual system that was equivalent to counting an extra ten photons over the window in time and area of the retina that was being stimulated. What could this noise be?

In the frequency-of-seeing experiments, as noted above, the flash of light illuminated roughly 500 receptor cells on the retina, and subsequent experiments showed that one could find essentially the same threshold number of photons when the flash covered many thousands of cells. Furthermore, experiments with different durations for the flash show that human observers are integrating over $\sim$0.1 s to make their decisions about whether they saw something. Thus, the dark noise in the system seems to be equivalent to $\sim$0.1 photon/receptor cell/s or less. To place this number in perspective, it is important to note that vision begins when the pigment molecule rhodopsin absorbs light and changes its structure to trigger some sequence of events in the receptor cell. In Sections 2.2 and 2.3, we will learn much more about the dynamics of rhodopsin and the cascade of events responsible for converting this molecular event into electrical signals that can be transmitted to the brain, but for now note that if rhodopsin can change its structure by absorbing a photon, there must also be some (small) probability that this same structural change or "isomerization" will happen as the result of a thermal fluctuation. If this does happen, then it will trigger a response that is identical to that triggered by a real photon. Further, such rare, thermally activated events really are Poisson processes (see Section 4.1), so that thermal activation of rhodopsin would contribute exactly a "dark light" of the sort we have been trying to estimate as a background noise in the visual system. But there are roughly one billion rhodopsin molecules per receptor cell, so that $\sim$0.1 s^{-1} cell^{-1} corresponds to a rate of once per $\sim$1000 years for the spontaneous isomerization of rhodopsin!

One of the key points here is that Barlow's explanation works only if people actually can adjust the threshold K in response to different situations. The realization that this is possible was part of the more general recognition that detecting a sensory signal does not involve a true threshold between, for example, seeing and not seeing. Instead, all sensory tasks involve a discrimination between signal and noise, and hence there are different strategies that provide different ways of trading off among the different kinds of errors. Notice that this picture matches what we know from the physics lab.

Problem 11: Simple analysis of dark noise.   Suppose that we observe events drawn out of a Poisson distribution, and we can count these events perfectly. Assume that the mean number of events has two contributions, $\bar{n} = \bar{n}_{\text{dark}} + \bar{n}_{\text{flash}}$, where $\bar{n}_{\text{flash}} = 0$ if there is no light flash and $\bar{n}_{\text{flash}} = N$ if there is a flash. As an observer, you have the right to set a criterion, so that you declare the flash to be present only if you count $n \geq K$ events. As you change K, you change the errors that you make—when K is small, you often say you saw something when nothing was there, but you hardly ever miss a real flash; at large K the situation is reversed. The conventional way of describing this is to plot the fraction of hits (probability that you correctly identify a real flash) against the probability of a false alarm (i.e., the probability that you say a flash is present when it isn't), with the criterion changing along the curve. Plot this "receiver operating characteristic" for the case $\bar{n}_{\text{dark}} = 10$ and $N = 10$. Hold $\bar{n}_{\text{dark}}$ fixed and change N to see how the curve changes. Explain which slice through this set of curves was measured by Hecht et al., and the relationship of this analysis to what we saw in Fig. 2.2.

There are classic experiments to show that people will adjust their thresholds automatically when we change the a priori probabilities of the signal being present, as expected for optimal performance. This can be done without any explicit instructions—you don't have to tell someone that you are changing the probabilities—and it works in all sensory modalities, not just vision. At least implicitly, then, people learn something about probabilities and adjust their criteria appropriately. Threshold adjustments also can be driven by changing the rewards for correct answers or the penalties for wrong answers. In this view, it is likely that Hecht et al. drove their observers to high thresholds by having a large effective penalty for false positive detections. Although it's not a huge literature, people have since manipulated these penalties and rewards in frequency-of-seeing experiments, with the expected results. Perhaps more dramatically, modern quantum optics techniques have been used to manipulate the statistics of photon arrivals at the retina, so that the trade-offs among the different kinds of errors are changed . . . again with the expected results.[6]

Not only did Baylor and co-workers detect the single-photon responses from toad photoreceptor cells, they also found that single receptor cells in the dark show spontaneous photon-like events roughly at the right rate to be the source of dark noise identified by Barlow. If you look closely, you can find one of these spontaneous events in the illustration of the rod cell responses to dim flashes in Fig. 2.3. Just to be clear, Barlow identified a maximum dark-noise level; anything higher and the observed reliable detection is impossible. The real rod cells have essentially this level of dark noise, so that the visual system is operating near the limits of reliability set by thermal noise in the input. It would be nice to have a more direct test of this idea.

In the lab we often lower the noise level of photodetectors by cooling them. This should work in vision too, because the rate of spontaneous photon-like events in the rod cell current is strongly temperature dependent, increasing by a factor of roughly four for every $10°$ increase in temperature.[7] Changing temperature is not so easy in humans, but it does work with cold-blooded animals like frogs and toads. To set the stage, it is worth noting that one species of toad in particular (*Bufo bufo*) manages to catch its prey under conditions so dark that human observers cannot see the toad, much less the prey. So, Aho et al. convinced toads to strike with their tongues at small wormlike objects illuminated by very dim lights, as shown in Fig. 2.7. Because one can actually

---

6. It is perhaps too much to go through all these results here, beautiful as they are. To explore, see the references for this section in the Annotated Bibliography.

7. The sign of the prediction is important. If we were looking for more reliable behaviors at higher temperatures, there could be many reasons for this, such as quicker responses of the muscles. Instead, the prediction is that we should see more reliable behavior as you cool down—all the way down to the temperature where behavior stops—and this is what is observed.

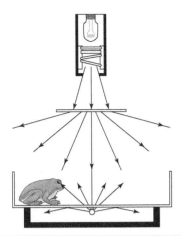
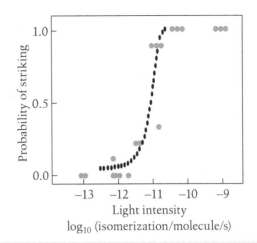

FIGURE 2.7

At left, a schematic of the apparatus used to measure the probability that toads will snap at dimly illuminated wormlike targets. Light from a shielded lamp falls on a diffuser screen, creating an artificial moon 35 cm above the toad. Below the platform where the toad stands there is a cross-section through which the targets are moved. At right, the probability of striking as a function of light intensity, calibrated as the probability per unit time that a rhodopsin molecule in the retina will absorb a photon and change its structure (isomerize). Redrawn from Aho et al. (1988).

make measurements on the retina itself, it is possible to calibrate light intensities as the rate at which rhodopsin molecules are absorbing photons and isomerizing. The toad's responses are almost deterministic once this rate is $r \sim 10^{-11}$ s^{-1} in experiments at 15° C, and responses are detectable at intensities a factor of three to five below this level. For comparison, the rate of thermal isomerizations at this temperature is $\sim 5 \times 10^{-12}$ s^{-1}.

If the dark noise consists of rhodopsin molecules spontaneously isomerizing at a rate r_d, then the mean number of dark events will be $n_d = r_d T N_r N_c$, where $T \sim 1$ s is the relevant integration time for the decision, $N_r \sim 3 \times 10^9$ is the number of rhodopsin molecules per cell in this retina, and $N_c \sim 4,500$ is the number of receptor cells that are illuminated by the image of the wormlike object. Similarly, the mean number of real events is $n = r T N_r N_c$, and reliable detection requires $n > \sqrt{n_d}$, or

$$r > \sqrt{\frac{r_d}{T N_r N_c}} \sim 6 \times 10^{-13} \text{ s}^{-1}. \tag{2.36}$$

Thus, if the toad knows exactly which part of the retina it should be looking at, then it should reach a signal-to-noise ratio of one at light intensities a factor of ten below the nominal dark-noise level. But there is no way to be sure where to look before the target appears, and the toad probably needs a rather higher signal-to-noise ratio before it is willing to strike. Thus, it is plausible that the threshold light intensities in this experiment should be comparable to the dark-noise level, as observed.

One can do an experiment very similar to the one with toads using human subjects (who say "yes" or "no," rather than sticking out their tongues), asking for a response to

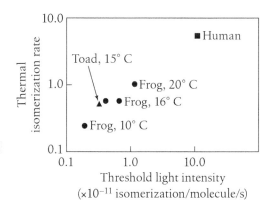

FIGURE 2.8

Comparison of spontaneous isomerization rates and threshold light intensities, from experiments on human, toads, and frogs at different temperatures. Redrawn from Aho et al. (1987, 1988).

small targets illuminated by steady, dim lights. In yet another similar experiment, frogs will spontaneously jump at a dimly illuminated patch of the ceiling, in an attempt to escape from an otherwise dark box. Combining all these experiments, with the frogs held at temperatures from 10° to 20° C, one can span a range of almost two orders of magnitude in the thermal isomerization rate of rhodopsin. It is not clear whether individual organisms hold their integration times T fixed as temperature is varied, or whether the experiments on different organisms correspond to asking for integration over a similar total number of rhodopsin molecules ($N_r N_c$). Nonetheless, it satisfying to see, in Fig. 2.8, that the threshold light intensity, where a response occurs 50% of the time, is varying systematically with the dark-noise level. It is certainly true that operating at lower temperatures allows the detection of dimmer lights, or equivalently, more reliable detection of the same light intensity, as expected if the dominant noise source were thermal in origin. These experiments support the hypothesis that visual processing in dim lights really is limited by input noise and not by any inefficiencies of the brain.

Problem 12: Getting a feel for the brain's problem.[8]   Let's go back to Problem 3, where you simulated a Poisson process.

(a) If you use the strategy of making small bins $\Delta\tau$ and testing a random number in each bin against a threshold, then it should be no problem to generalize this to the case where the threshold is different at different times, so you are simulating a Poisson process in which the rate is varying as a function of time. As an example, consider a 2 s interval in which the counting rate has some background (e.g., the dark noise in rods) value r_{dark} except in a 100 msec window where the rate is higher, say, $r = r_{dark} + r_{signal}$. Remember that for one rod cell, r_{dark} is ~0.02 s^{-1}, whereas humans can see flashes with $r_{signal} \sim 0.01$ s^{-1} if they can integrate over 1000 rods. Try to simulate events in this parameter range and actually look at examples, perhaps plotted with x marks to show you where the events occur in a single trial.

(b) Can you tell the difference between a trial where you have $r_{signal} = 0.01$ s^{-1} and one in which $r_{signal} = 0$? Does it matter whether you know when to expect the extra events? In

---

8. This problem has its origins in teaching at the Methods in Computational Neuroscience course at the Marine Biological Laboratory and was developed together with Rob de Ruyter van Steveninck.

effect these plots give a picture of the problem that the brain has to solve in the Hecht et al. experiment, or at least an approximate picture.

(c) Sitting in a dark room to repeat the Hecht et al. experiment would take a long time, but maybe you can go from your simulations here to design a psychophysical experiment simple enough that you can do it on one another. Can you measure the reliability of discrimination between the different patterns of x marks that correspond to the signal being present or absent? Do you see an effect of "knowing when to look"? Do people seem to get better with practice? Can you calculate the theoretical limit to how well one can do this task? Do people get anywhere near this limit? This is an open-ended problem.

Problem 13: A better analysis? Go back to the original paper by Aho et al. (1988) and see whether you can give a more compelling comparison between thresholds and spontaneous isomerization rates. From Eq. (2.36), we expect that the light intensity required for some criterion level of reliability scales as the square root of the dark-noise level, but it also depends on the total number of rhodospin molecules over which the subject must integrate. Can you estimate this quantity for the experiments on frogs and humans? Does this lead to an improved version of Fig. 2.8? Again, this is an open-ended problem.

The dominant role of spontaneous isomerization as a source of dark noise leads to a wonderfully counterintuitive result, namely, that the photoreceptor designed to maximize the signal-to-noise ratio for detection of dim lights will allow a significant number of photons to pass undetected. Consider a rod photoreceptor cell of length ℓ, with concentration C of rhodopsin; let the absorption cross-section of rhodopsin be σ. As a photon passes along the length of rod, the probability that it will be absorbed (and, presumably, counted) is $p = 1 - \exp(-C\sigma\ell)$, suggesting that we should make C or ℓ larger to capture more of the photons. But as we increase C or ℓ, we are increasing the number of rhodopsin molecules, $N_{rh} = CA\ell$, with A the area of the cell, so we also increase the rate of dark-noise events, which occurs at a rate r_{dark} per molecule.

If we integrate over a time τ, we will see a mean number of dark events (spontaneous isomerizations) $\bar{n}_{dark} = r_{dark}\tau N_{rh}$. The actual number will fluctuate, with a standard deviation $\delta n = \sqrt{\bar{n}_{dark}}$. If n_{flash} real photons are incident on the cell, the mean number counted will be $\bar{n}_{count} = n_{flash}p$. Putting these factors together, we can define a signal-to-noise ratio (SNR):

$$SNR \equiv \frac{\bar{n}_{count}}{\delta n} = n_{flash}\frac{[1 - \exp(-C\sigma\ell)]}{\sqrt{CA\ell r_{dark}\tau}}. \tag{2.37}$$

The absorption cross-section σ and the spontaneous isomerization rate r_{dark} are properties of the rhodopsin molecule, but as the rod cell assembles itself, it can adjust both its length ℓ and the concentration C of rhodopsin; in fact these factors enter together as the product $C\ell$. When $C\ell$ is larger, photons are captured more efficiently, which leads to an increase in the numerator, but there also are more rhodopsin molecules and hence more dark noise, which leads to an increase in the denominator. Viewed as a function of $C\ell$, the signal-to-noise ratio has a maximum at which these competing effects balance; working out the numbers, we find that the maximum is reached when $C\ell \sim 1.26/\sigma$. Note that all other parameters have dropped out. In particular, this means that the probability of an incident photon not being absorbed is

FIGURE 2.9

Results of experiments in which observers are asked to rate the intensity of dim flashes, including blanks, on a scale from 0 to 6. Main figure shows that the variance of the ratings at fixed intensity is equal to the mean, as expected if the ratings are Poisson distributed; the sizes of the bars in each cross show the error bars in the measurements along the two axes. Insets show that the full distribution is approximately Poisson (upper inset) and that the mean rating is linearly related to the flash intensity (lower inset), measured here as the mean number of photons delivered to the cornea. The mean rating is nonzero even for intensity zero (blanks), presumably because the subject is counting the spontaneous isomerizations of rhodopsin (dark noise). Redrawn from Sakitt (1972).

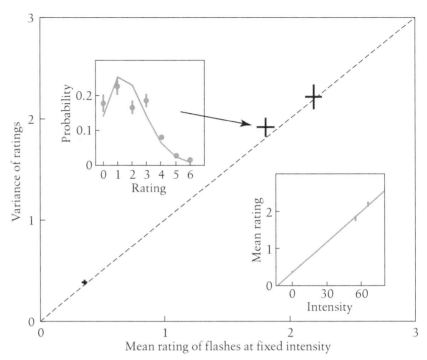

$1 - p = \exp(-C\sigma\ell) \sim e^{-1.26} \sim 0.28$. Thus, to maximize the signal-to-noise ratio for detecting dim flashes of light, nearly 30% of photons should pass through the rod without being absorbed!

Problem 14: Escape from the trade-off.   Derive for yourself the numerical factor $(C\ell)_{\mathrm{opt}} \sim 1.26/\sigma$. Can you see any way to design an eye that gets around this trade-off between more efficient counting and extra dark noise? Hint: Think about what you see when looking into a cat's eyes at night.

If this line of thought is correct, it should be possible to coax human subjects into giving responses that reflect the counting of individual photons rather than just the summation of multiple counts up to some threshold of confidence or reliability. Suppose we ask observers not to say "yes" or "no," but rather to rate the apparent intensity of the flash, say, on a scale from 0 to 6. Remarkably, as shown in Fig. 2.9, in response to very dim flashes interspersed with blanks, at least some observers will generate ratings that, given the intensity, are approximately Poisson distributed: the variance of the ratings is essentially equal to the mean, and even the full distribution of ratings over hundreds of trials is close to Poisson. Further, the mean rating is linearly related to the light intensity, with an offset that agrees with other measurements of the dark-noise level. Thus, the observer behaves exactly as if she can give a rating that is equal to the number of photons counted. This astonishing result would be almost too good to be true were it not that some observers deviate from this ideal behavior—they starting counting at two or three but otherwise follow all the same rules.

FIGURE 2.10

The fly's eye(s). This photograph, taken by H. L. Leertouwer at the Rijksuniversiteit Groningen, shows the hexagonal lattice of lenses in the compound eye. This is the blowfly *Calliphora vicina.*

Even though the phenomena of photon counting are very beautiful, one might worry that this represents just a very small aspect of vision. Does the visual system continue to count photons reliably even when it's not completely dark outside? To answer this question, let's look at vision in a rather different animal, as in Fig. 2.10. When you look down on the head of a fly, you see—almost to the exclusion of anything else— the large compound eyes. Each little hexagon that you see on the fly's head is a separate lens, and in large flies there are ∼5000 lenses in each eye, with approximately one receptor cell behind each lens, and ∼100 brain cells per lens devoted to the processing of visual information. The lens focuses light on the receptor, which is small enough to act as an optical waveguide. Each receptor sees only a small portion of the world, just as in our eyes; one difference between flies and us is that diffraction is much more significant for organisms with compound eyes—because the lenses are so small, flies have an angular resolution of ∼1°, whereas we do ∼100× better. It is worth emphasizing that the insect's compound eye is not so different than ours, as schematized in Fig. 2.11. In both cases the retina has a layer of photoreceptors that parse the world into pixels, but in our eyes all the photoreceptors "look" through the same large lens, whereas in the compound eye each cell has its own private, smaller lens.

The last paragraph is a little sloppy ("approximately one receptor cell"?), so let's try to be more precise. For flies there actually are eight receptors behind each lens. Two provide sensitivity to polarization and some color vision, which we ignore here. The other six receptors look out through the same lens in different directions, but as one moves to neighboring lenses one finds that there is one cell under each of six neighboring lenses that looks in the same direction. Thus, these six cells are equivalent to one cell with six times larger photon capture cross-section, and the signals from these

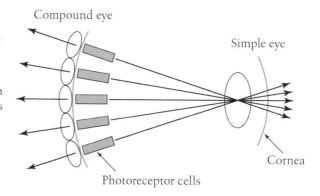

FIGURE 2.11

The similarity of simple and compound eyes. Both types of eyes have a single layer of photoreceptor cells in the retina. In the compound eye, each receptor cell has its own private lens through which it looks, and these lenses are arrayed in a lattice that mirrors that of the photoreceptors. In the simple eye, all photoreceptors look at the world through one large lens. What is missing from this image is that, in the simple eye, we can have a much higher density of receptors, because the point spread function of the larger lens is proportionally smaller.

cells are collected and summed in the first processing stage (the lamina); one can even see the expected sixfold improvement in signal-to-noise ratio.[9]

Because diffraction is such a serious limitation, one might expect that there would be fairly strong selection for eyes that make the most of the opportunities within these constraints. Indeed, there is a beautiful literature on optimization principles for the design of the compound eye; the topic even makes an appearance in Feynman's undergraduate physics lectures. Roughly speaking (Fig. 2.12), we can think of the fly's head as being a sphere of radius R and imagine that the lenses are pixels of linear dimension d on the surface.[10] Then the geometry determines an angular resolution (in radians) of $\delta\phi_{geo} \sim d/R$; resolution gets better if d becomes smaller. In contrast, diffraction through an aperture of size d creates a blur of angular width $\delta\phi_{diff} \sim \lambda/d$, where $\lambda \sim 500$ nm is the wavelength of the light we are trying to image; this limit improves as the aperture size d gets larger. Although one could try to give a more detailed theory, it seems clear that the optimum is reached when the two different limits are about equal, corresponding to an optimal pixel size

$$d_* \sim \sqrt{\lambda R}. \tag{2.38}$$

This is the calculation in the Feynman lectures, and Feynman notes that it gives the right answer to within 10% in the case of a honey bee.

A decade before Feynman's lectures, Barlow had derived the same formula and went into the drawers of the natural history museum at Cambridge University to find a variety of insects with varying head sizes, and he verified that the pixel size really does scale with the square root of the head radius, as shown in Fig. 2.13. I think this work

---

9. Although this is off our current topic, the structure of the fly's eye raises all sorts of interesting questions. The great anatomist Ramón y Cajal referred to the fly's visual system as a "neurocrystal," for its nearly perfect lattice structure. The pattern of six plus two receptor cells is repeated thousands of times over in a large fly, and the convergence of the six cells onto individual cells in the second layer of the retina is similarly repeated, apparently without errors. This is nontrivial, because the geometry of these connections actually is quite complicated, with the cells twisting around one another as they find their way to their proper targets. There is work being done on how these beautiful structures develop, but my guess is that we are just scratching the surface.

10. This is not the case of the spherical fly, but the case of the spherical head, which is a much better approximation.

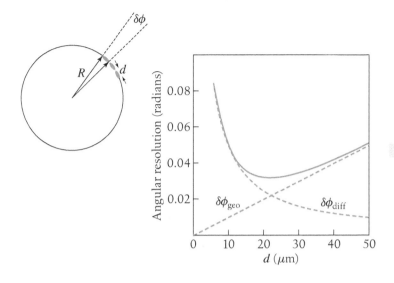

FIGURE 2.12

At left is a schematic of the compound eye, with lenses of width d on the surface of a spherical head with radius R. At right is the angular resolution of the eye as a function of lens size, showing the geometric ($\delta\phi_{geo} \sim d/R$) and diffraction ($\delta\phi_{diff} \sim \lambda/d$) contributions as dashed lines; the full resolution is shown by the solid line.

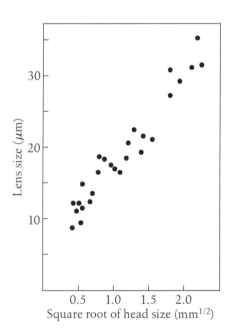

FIGURE 2.13

The size of lenses in compound eyes as a function of head size, across many species of insect. Redrawn from Barlow (1952).

should be more widely appreciated. It has several features we might like to emulate. First, it explicitly brings measurements on many species together in a quantitative way. Second, the fact that data from multiple species can be plotted on the same graph is not a phenomenological statement about, for example, scaling of one body part relative to another, but rather is based on a clearly stated physical principle. Finally, and most importantly for our later discussion, Barlow makes an important transition: rather than just asking whether a biological system approaches the physical limits to performance, he assumes that the physical limits are reached and uses this hypothesis to predict something else about the structure of the system. This is, to be sure, a simple example, but an early and interesting example nonetheless.[11]

Pushing toward diffraction-limited optics can't be the whole story, because at low light levels having lots of small pixels does not do much good—so few photons are captured in each pixel that there is a dramatic loss of intensity resolution. There must be some trade-off between spatial resolution and intensity resolution, and the precise form of this trade-off will depend on the statistical structure of the input images (if you are looking at clouds, it will be different than when looking at tree branches). The difficult question is how to quantify the relative worth of extra resolution in space versus intensity, and it has been suggested that the right way to do this is to count bits—design the eye not to maximize resolution but rather to maximize the information that can be captured about the input image. This approach was a semi-quantitative success, showing how insects that fly late at night or at very high speeds (leading to blurring by photoreceptors with finite time resolution) should have less than diffraction-limited spatial resolving power. I still think there are open questions here, however.

Coming back to the question of photon counting, we can record the voltage signals in the photoreceptor cells and detect single-photon responses, as in vertebrates. If we want to see what happens at higher counting rates, we have to be sure that the receptor cells are in a state where they don't "run down" too much because of the increased activity. In particular, the rhodopsin molecule itself has to be recycled after it absorbs a photon. In animals with backbones, this actually does not happen in the photoreceptor but in conjunction with other cells that form the pigment epithelium. In contrast, in invertebrates the resetting of the rhodopsin molecule occurs in the receptor cell and can even be driven by absorption of additional long-wavelength photons. Thus, if you want to do experiments at high photon flux on isolated vertebrate photoreceptors, there is a real problem of running out of functional rhodospin, but this problem doesn't happen in the fly's eye. Also, the geometry of the fly's eye makes it easier to do stable intracellular measurements without too much dissection.

To set the stage for experiments at higher counting rates, consider a simple model in which each photon arriving at time t_i produces a pulse $V_0(t - t_i)$, and these pulses just add to give the voltage

$$V(t) = V_{DC} + \sum_i V_0(t - t_i), \tag{2.39}$$

as shown schematically in Fig. 2.14; V_{DC} is the constant voltage that one observes across the cell membrane in the absence of light. In Appendix A.1, we can find the distribution

---

11. This example also raises an interesting question. In Fig. 2.13, each species of insect is represented by a single point. But not all members of the same species are the same size, as you must have noticed. Is the relationship between R and d that optimizes function preserved across the natural size variations among individuals? Does it matter whether the size differences are generated by environmental or genetic factors? These questions are about the reproducibility of spatial structures in development, a topic we will come back to (albeit in simpler forms) in Section 5.3. It would be good, though, if someone just measured the variations in eye dimensions across many individuals.

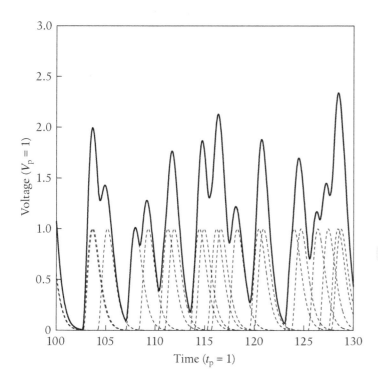

FIGURE 2.14

A Poisson stream of pulses sum to give the voltage across the cell membrane, as in Eq. (2.39). The units are chosen so that the elementary pulse peaks after one time unit, with an amplitude of one voltage unit. Dashed lines show the underlying responses to individual photons, and the solid line shows their summed effect.

of the arrival times $\{t_i\}$ using the hypothesis that the photons arrive as a Poisson process with a time-dependent rate $r(t)$. From Eq. (A.17) we have

$$P[\{t_i\}|r(t)] = \exp\left[-\int_0^T d\tau\, r(\tau)\right] \frac{1}{N!} r(t_1)r(t_2)\cdots r(t_N), \qquad (2.40)$$

where $r(t)$ is the rate of photon arrivals—the light intensity in appropriate units. To compute the average voltage response to a given time-dependent light intensity, we have to do a straightforward if tedious calculation:

$$\left\langle \sum_i V_0(t - t_i) \right\rangle = \sum_{N=0}^{\infty} \int_0^T d^N t_i\, P[\{t_i\}|r(t)] \sum_i V_0(t - t_i). \qquad (2.41)$$

This looks a terrible mess. Actually, it's not so bad, and one can proceed systematically to do all the integrals. The details are in Appendix A.1, along with all the other details about Poisson processes. The result, Eq. (A.69), is that the voltage responds linearly to the light intensity:

$$\langle V(t) \rangle = V_{DC} + \int_{-\infty}^{\infty} dt'\, V_0(t - t')r(t'). \qquad (2.42)$$

In particular, if we have some background photon counting rate $\bar{r}$ that undergoes fractional modulations $C(t)$, so that

$$r(t) = \bar{r}[1 + C(t)], \qquad (2.43)$$

then there is a linear response of the voltage to the "contrast" C,

$$\langle \Delta V(t) \rangle = \bar{r} \int_{-\infty}^{\infty} dt' V_0(t - t') C(t'). \tag{2.44}$$

Such integral relationships (convolutions) simplify when we use the Fourier transform. For a function of time $f(t)$, we define the Fourier transform with the conventions

$$\tilde{f}(\omega) = \int_{-\infty}^{\infty} dt \, e^{+i\omega t} f(t), \tag{2.45}$$

$$f(t) = \int_{-\infty}^{\infty} \frac{d\omega}{2\pi} e^{-i\omega t} \tilde{f}(\omega). \tag{2.46}$$

Then, for two functions of time $f(t)$ and $g(t)$, we have

$$\int_{-\infty}^{\infty} dt \, e^{+i\omega t} \left[\int_{-\infty}^{\infty} dt' \, f(t - t') g(t') \right] = \tilde{f}(\omega) \tilde{g}(\omega). \tag{2.47}$$

Armed with Eq. (2.47), we can write the response of the photoreceptor in the frequency domain,

$$\langle \Delta \tilde{V}(\omega) \rangle = \bar{r} \tilde{V}_0(\omega) \tilde{C}(\omega), \tag{2.48}$$

so that there is a transfer function, analogous to impedance relating current and voltage in an electrical circuit:

$$\tilde{T}(\omega) \equiv \frac{\langle \Delta \tilde{V}(\omega) \rangle}{\tilde{C}(\omega)} = \bar{r} \tilde{V}_0(\omega). \tag{2.49}$$

This transfer function is a complex number at every frequency, so it has an amplitude and a phase:

$$\tilde{T}(\omega) = |\tilde{T}(\omega)| e^{i\phi_T(\omega)}. \tag{2.50}$$

The units of $\tilde{T}$ are simply voltage per contrast, although contrast is formally dimensionless, because it measures a fractional change in intensity. The interpretation is that if we generate a time-varying contrast $C(t) = C \cos(\omega t)$, then the voltage will also vary at frequency ω:

$$\langle \Delta V(t) \rangle = |\tilde{T}(\omega)| C \cos[\omega t - \phi_T(\omega)]. \tag{2.51}$$

Problem 15: Convolutions.   Verify the "convolution theorem" in Eq. (2.47). If you need some reminders, see, for example, Lighthill (1958). Also, be sure that you can reproduce Eq. (2.51).

If every photon generates a voltage pulse $V_0(t)$, but the photons arrive at random, then the voltage must fluctuate. To characterize these fluctuations, let's use some of the general apparatus of correlation functions and power spectra. A review of these ideas is given in Appendix A.2.

We want to analyze the fluctuations $\delta V(t)$ of the voltage around its mean. By definition, the mean of this fluctuation is zero, $\langle \delta V(t) \rangle = 0$. There is a nonzero variance,

$\langle [\delta V(t)]^2 \rangle$, but to give a full description, we need to describe the covariance between fluctuations at different times, $\langle \delta V(t)\delta V(t') \rangle$. Importantly, we are interested in systems that have no internal clock, so this covariance or correlation can't depend separately on t and t', only on the difference. More formally, if we shift our clock by a time τ, this can't matter, so we must have

$$\langle \delta V(t)\delta V(t') \rangle = \langle \delta V(t+\tau)\delta V(t'+\tau) \rangle; \tag{2.52}$$

this is possible only if

$$\langle \delta V(t)\delta V(t') \rangle = C_V(t-t'), \tag{2.53}$$

where $C_V(t)$ is the correlation function of V. Thus, invariance under time translations restricts the form of the covariance. Another way of expressing time-translation invariance in the description of random functions is to say that any particular wiggle in plotting the function is equally likely to occur at any time. This is called stationarity, and we say that fluctuations that have this property are stationary fluctuations.

In Fourier space, the consequence of invariance under time translations can be stated more simply—if we compute the covariance between two frequency components, we find

$$\langle \delta \tilde{V}(\omega_1)\delta \tilde{V}(\omega_2) \rangle = 2\pi \delta(\omega_1+\omega_2)S_V(\omega_1), \tag{2.54}$$

where $S_V(\omega)$ is called the power spectrum (or power spectral density) of the voltage V. Remembering that $\delta \tilde{V}(\omega)$ is a complex number, it might be more natural to write this equation as

$$\langle \delta \tilde{V}(\omega_1)\delta \tilde{V}^*(\omega_2) \rangle = 2\pi \delta(\omega_1-\omega_2)S_V(\omega_1). \tag{2.55}$$

Time-translation invariance thus implies that fluctuations at different frequencies are independent.[12] This makes sense, because if (for example) fluctuations at 2 Hz and 3 Hz were correlated, we could form beats between these components and generate a clock that ticks every second. Finally, the Wiener-Khinchine theorem states that the power spectrum and the correlation function are a Fourier transform pair:

$$S_V(\omega) = \int d\tau\, e^{+i\omega\tau} C_V(\tau), \tag{2.56}$$

$$C_V(\tau) = \int \frac{d\omega}{2\pi} e^{-i\omega\tau} S_V(\omega). \tag{2.57}$$

The total variance in voltage can be written as

$$\langle [\delta V(t)]^2 \rangle \equiv C_V(0) = \int \frac{d\omega}{2\pi} S_V(\omega); \tag{2.58}$$

thus, we can think of each frequency component as having a variance $\sim S_V(\omega)$, and by summing these components, we obtain the total variance.

---

12. Caution: this is true only at second order; it is possible for different frequencies to be correlated when we evaluate products of three or more terms. See the next problem for an example.

Problem 16: More on stationarity.   Consider some fluctuating variable $x(t)$ that depends on time, with $\langle x(t) \rangle = 0$. Show that, because of time-translation invariance, higher order correlations among Fourier components are constrained:

$$\langle \tilde{x}(\omega_1)\tilde{x}^*(\omega_2)\tilde{x}^*(\omega_3) \rangle \propto 2\pi\delta(\omega_1 - \omega_2 - \omega_3), \tag{2.59}$$

$$\langle \tilde{x}(\omega_1)\tilde{x}(\omega_2)\tilde{x}^*(\omega_3)\tilde{x}^*(\omega_4) \rangle \propto 2\pi\delta(\omega_1 + \omega_2 - \omega_3 - \omega_4). \tag{2.60}$$

If you think of $\tilde{x}^*$ (or $\tilde{x}$) as being analogous to the operators for creation (or annihilation) of particles, explain how these relations are related to conservation of energy for scattering in quantum systems.

Problem 17: Brownian motion in a harmonic potential.   Consider a particle of mass m hanging from a spring of stiffness κ and surrounded by a fluid. The effect of the fluid is, on average, to generate a drag force, and in addition there is a Langevin force that describes the random collisions of the fluid molecules with the particle, resulting in Brownian motion. The equation of motion is

$$m\frac{d^2x(t)}{dt^2} + \gamma\frac{dx(t)}{dt} + \kappa x(t) = \eta(t), \tag{2.61}$$

where γ is the drag coefficient and $\eta(t)$ is the Langevin force. A standard result of statistical mechanics is that the correlation function of the Langevin force is

$$\langle \eta(t)\eta(t') \rangle = 2\gamma k_B T \delta(t - t'), \tag{2.62}$$

where T is the absolute temperature and $k_B = 1.36 \times 10^{-23}$ J/K is Boltzmann's constant.[13]

(a) Show that the power spectrum of the Langevin force is $S_\eta(\omega) = 2\gamma k_B T$, independent of frequency. Fluctuations with such a constant spectrum are called white noise.

(b) Fourier transform Eq. (2.61) and solve, showing how $\tilde{x}(\omega)$ is related to $\tilde{\eta}(\omega)$. Use this result to find an expression for the power spectrum of fluctuations in x, $S_x(\omega)$.

(c) Integrate the power spectrum $S_x(\omega)$ to find the total variance in x. Verify that your result agrees with the equipartition theorem:

$$\left\langle \frac{1}{2}\kappa x^2 \right\rangle = \frac{1}{2}k_B T. \tag{2.63}$$

Hint: The integral over ω can be done by closing a contour in the complex plane.

(d) Show that the power spectrum of the velocity, $S_v(\omega)$, is related to the power spectrum of position through

$$S_v(\omega) = \omega^2 S_x(\omega). \tag{2.64}$$

---

13. These delta function correlations (or white noise; see next part of the problem) are connected to the fact that the drag force $-\gamma(dx/dt)$ is related instantaneously to the velocity; the fluctuation-dissipation theorem tells us that the deterministic properties of the dissipative term dictate the form of the random noise. Thus, we are not making a "white-noise assumption," as this often is described. We are just being consistent with the hypothesized form of the drag. If we could look on sufficiently short time scales, we would see that the drag is not instantaneous and that the noise is not white.

Using this result, verify the other prediction of the equipartition theorem for this system:

$$\left\langle \frac{1}{2}mv^2 \right\rangle = \frac{1}{2}k_B T. \tag{2.65}$$

Now we have a language for describing the signals and noise in the receptor cell voltage by transforming to the frequency domain. What does this have to do with counting photons? The key point is that we can do a calculation similar to the derivation of Eq. (2.48) for $\langle \Delta V(t) \rangle$ to show that, even at zero contrast ($C = 0$), the voltage will undergo fluctuations—responding to the random arrival of photons—with power spectrum

$$N_V(\omega) = \bar{r}|\tilde{V}_0(\omega)|^2. \tag{2.66}$$

We call this N_V, because it is noise. The noise has a spectrum shaped by the pulses V_0, and the magnitude is determined by the photon counting rate; again see Appendix A.1 for details.

Both the transfer function and noise spectrum depend on the shape of the pulse $V_0(t)$. In particular, because this pulse has finite width in time, the transfer function becomes smaller at higher frequencies. Thus, if you watch a flickering light, the strength of the signal transmitted by your photoreceptor cells will decrease with increasing frequency. Roughly speaking, this filtering is why you can watch movies: the flicker associated with the discreteness of the frames is strongly attenuated, although this is far from the whole story—your brain does a lot of interpretation, leaning heavily on the expectation that things vary continuously in time.

The crucial point is that, for an ideal photon counter, although higher-frequency signals are attenuated, the signal-to-noise ratio actually does not depend on frequency. Thus, if we form the ratio

$$\frac{|\tilde{T}(\omega)|^2}{N_V(\omega)} = \frac{|\bar{r}\tilde{V}_0(\omega)|^2}{\bar{r}|\tilde{V}_0(\omega)|^2} = \bar{r}, \tag{2.67}$$

we just recover the photon counting rate, independent of details. Because this rate is proportional to the signal-to-noise ratio for detecting contrast modulations $\tilde{C}(\omega)$, real photodetectors will give less that this ideal value.

Problem 18: Frequency versus counting rate.   If we are counting photons at an average rate $\bar{r}$, you might think that it is easier to detect variations in light intensity at a frequency $\omega \ll \bar{r}$ than at higher frequencies, $\omega \gg \bar{r}$; after all, in the high-frequency case, the light changes from bright to dim and back even before (on average) a single photon has been counted. But Eq. (2.67) states that the signal-to-noise ratio for detecting contrast in an ideal photon counter is independent of frequency, counter to this intuition. Can you produce a simple simulation to verify the predictions of Eq. (2.67)? I've left this problem deliberately open ended. Hint: if you are looking for light intensity variations of the form $r(t) = \bar{r}[1 + C\cos(\omega t)]$, you should process the photon arrival times $\{t_i\}$ to form a signal $s = \sum_i \cos(\omega t_i)$.

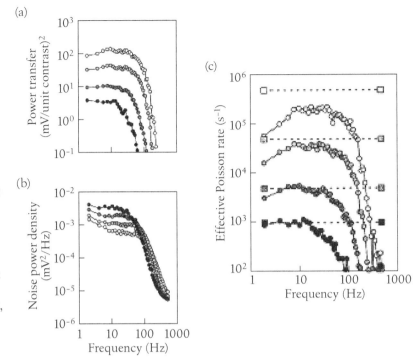

FIGURE 2.15

Signal and noise in fly photoreceptors, with experiments at four different mean light intensities. Redrawn from de Ruyter van Steveninck and Laughlin (1996b). (a) Transfer function $|\tilde{T}(\omega)|^2$ from contrast to voltage. (b) Power spectrum of voltage noise, $N_V(\omega)$. (c) The ratio $|\tilde{T}(\omega)|^2/N_V(\omega)$, which would equal the photon counting rate if the system were ideal; dashed lines show the actual counting rates.

So now we have a way of testing the photoreceptors: measure the transfer function $\tilde{T}(\omega)$ and the noise spectrum $N_V(\omega)$, form the ratio $|\tilde{T}(\omega)|^2/N_V(\omega)$, and compare this with the actual photon counting rate $\bar{r}$. This was done for the fly photoreceptors, with the results shown in Fig. 2.15. What we see is that, over some range of frequencies, the performance of the fly photoreceptors is close to the level expected for an ideal photon counter. It is interesting to see how this performance evolves as we change the mean light intensity, as shown in Fig. 2.16. The performance of the receptors tracks the physical optimum up to counting rates of $\bar{r} \sim 10^5$ photons/s. Because the integration time of the receptors is $\sim$10 ms, the cell can count, almost perfectly, up to $\sim$1000.

Problem 19: Calibrating the photon counting rates. Go back to the original experiments from which Fig. 2.15 is drawn and explain how it was possible to know the photon counting rates—that is, how it was possible to draw the dashed lines in Fig. 2.15c.

An important point about these results is that they would not work if the simple model were literally true. At low photon counting rates $\bar{r}$, the pulse V_0 has an amplitude of several millivolts, as you can work out from panel (a) in Fig. 2.15. If we count $\sim 10^3$ events, this should produce a signal of several volts, which is absolutely impossible in a real cell. What happens is that the system has an automatic gain control that reduces the size of the pulse V_0 as the light intensity is increased. Remarkably, this gain control or adaptation occurs while preserving (indeed, enabling) nearly ideal photon counting. Thus, as the lights go up, the response to each photon becomes smaller (and, if you look closely, faster) but no less reliable.

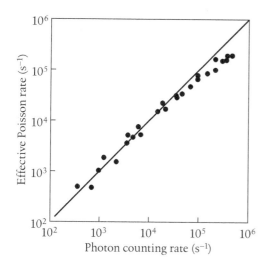

FIGURE 2.16

Performance of fly photoreceptors versus light intensity. Having measured the quantity $|\tilde{T}(\omega)|^2/N_V(\omega)$, as in Fig. 2.15, we plot the maximum value (typically at relatively low frequencies) versus the actual photon counting rate $\bar{r}$. We see that, over an enormous dynamic range, the signal-to-noise ratio tracks the value expected for an ideal photon counter.

Problem 20: Looking at the data.   Explain how the data in Fig. 2.15 provide evidence for the adaptation of the pulse V_0 with changes in the mean light intensity.

These observations on the ability of the visual system to count single photons—down to the limit set by thermal noise in rhodopsin and up to counting rates of $\sim 10^5 \, \mathrm{s}^{-1}$—raise questions at several different levels:

- At the level of single molecules, we will see that the performance of the visual system depends crucially on the dynamics of rhodopsin itself. In particular, the structural response of the molecule to photon absorption is astonishingly fast, whereas the dark-noise level means that the rate of spontaneous structural changes is extremely slow.

- At the level of single cells, there are challenges in understanding how a network of biochemical reactions converts the structural changes of single rhodopsin molecules into macroscopic electrical currents across the rod cell membrane.

- Beyond the receptor cell, we would like to understand how these signals are integrated without being lost in the inevitable background of noise. Our discussion has also assumed, implicitly, that the brain knows what it is "looking for" when we try to detect dim flashes.

In the next sections we look at each of these questions in turn.

2.2   Dynamics of Single Molecules

To a remarkable extent, our ability to see in the dark is limited by the properties of rhodopsin itself, essentially because everything else works so well. Rhodopsin consists of a medium-sized organic pigment, retinal, enveloped by a large protein, opsin (see Fig. 2.17). The primary photo-induced reaction is isomerization of the retinal, which ultimately couples to structural changes in the protein. The effort to understand the

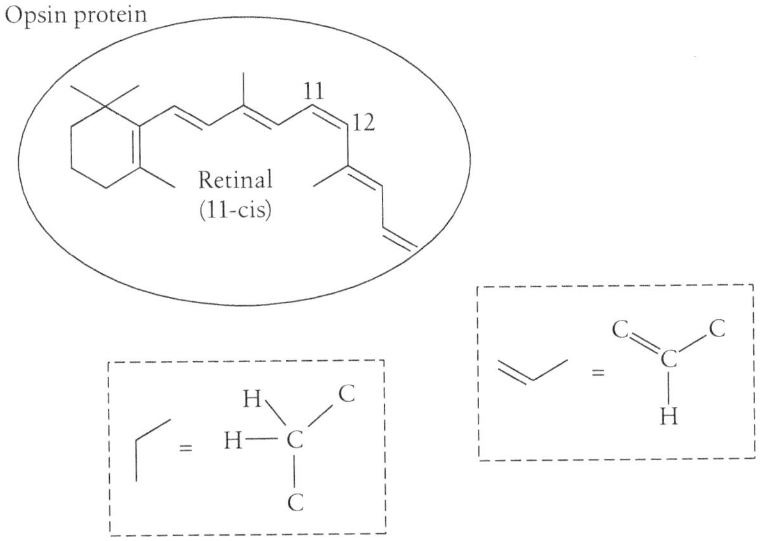

FIGURE 2.17

Schematic structure of rhodopsin, showing the organic pigment retinal nestled in a pocket formed by the surrounding opsin protein. This conformation of the retinal is called 11-cis, because there is a rotation around the bond between carbons numbered 11 and 12 (starting at the lower right in the ring). Insets illustrate the conventions in such chemical structures, with carbons at nodes of the skeleton. Hydrogen atoms are not shown, but you can add them in following the rule that each carbon must participate in four bonds.

dynamics of these processes goes back to Wald's isolation of retinal (a vitamin A derivative) in the 1930s, his discovery of the isomerization, and the identification of numerous states through which the molecule cycles. The field was given a big boost by the discovery that there are bacterial rhodopsins, some of which serve a sensory function, whereas others are energy-transducing molecules, using the energy of the absorbed photon to pump protons across the cell membrane. The resulting difference in electrochemical potential for protons is a universal intermediate in cellular energy conversion, not just in bacteria but in us as well; see, for example, the discussion of the flagellar motor in Section 4.2.

By now we know much more than Wald did about the structure of the rhodopsin molecule and the rod cell more generally, as schematized in Fig. 2.18. The outer segment of the rod cell is packed with "disks," which are closed but relatively flat spaces, bounded by a membrane not unlike the cell membrane itself. The rhodopsin molecules sit in the disk membrane and are rather densely packed but free to diffuse in two dimensions. Such proteins as opsin are polymers of amino acids, as discussed in more detail in Section 5.1, but unlike most polymers proteins can fold into compact, well-defined structures. In the case of opsin, the folded structure allows the polymer to snake back and forth across the membrane in seven segments, and along each segment the backbone of the polymer traces a helix, shown by the ribbon diagram in Fig. 2.18. The retinal is linked to the protein by a covalent bond, and sits nearer the inside of the disk than the outside. Because events in the rhodopsin molecule eventually have to be communicated to the rest of the cell—and, in particular, the cell membrane, across which the currents are flowing—events that start with a change in electronic state of the retinal must propagate through almost the entire span of the protein across the disk membrane.

Although there are many remarkable features of the rhodopsin molecule, we would like to understand those particular features that contribute to the reliability of photon counting. First among these is the very low spontaneous isomerization rate, roughly once per thousand years. As we have seen, these photon-like events provide the dominant noise source that limits our ability to see in the dark, so there is a clear advantage to having the lowest possible rate. When we look at the molecules themselves, puri-

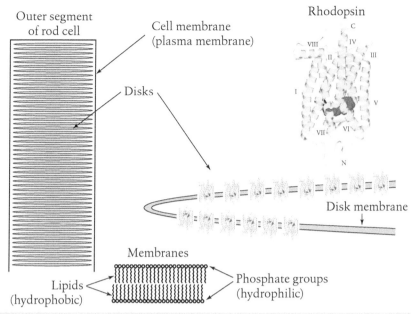

FIGURE 2.18

Schematic structure of the rod cell and rhodopsin. At left, the outer segment of the rod cell consists of a plasma membrane surrounding a dense stack of disks. Each disk, as shown at right, is a closed membrane, and all the membranes are made from bilayers of phospholipids that organize themselves to sequester their oily tails while allowing the charged head groups to access the surrounding water; there is lots of beautiful physics to be found in this self-organization, which can form a remarkable variety of structures. Rhodopsin is a protein that sits in the disk membrane, its seven helical segments snaking back and forth through the membrane. Rhodopsin structure from Stenkamp et al. (2002), with thanks to R. Stenkamp.

fied from the retina, we can "see" the isomerization reaction, because the initial 11-cis state and the final all-trans states (see Fig. 2.19) have different absorption spectra. For rhodopsin itself, the spontaneous isomerization rate is too slow to observe in a bulk experiment. If we isolate the pigment retinal, however, we find that it has a spontaneous isomerization rate of $\sim 1/\text{yr}$, so that a bottle of 11-cis retinal is quite stable, but the decay to all-trans is observable.

How can we understand that rhodopsin has a spontaneous isomerization rate $1000\times$ less than that of retinal? For that matter, how do we think about anything that rhodopsin does, including absorbing light? An intuitive picture is shown in Fig. 2.20. The fundamental tool we have for describing molecules is the Born-Oppenheimer approximation, which is based on the fact that (colloquially speaking), because nuclei are much heavier than electrons, they tend to move more slowly.[14] Thus, we can solve

---

14. I assume that most readers know something about the Born-Oppenheimer approximation, as it is a pretty classical subject. It is also one of the first adiabatic approximations in quantum mechanics. It took many years to realize that some very interesting things can happen in the adiabatic limit, notably, the appearance of nontrivial phase factors in the adiabatic evolution of wave functions. Some of these "complications" (to use a word from one of original papers) were actually discovered in the context of the Born-Oppenheimer approximation itself, but now we know that this circle of ideas is much larger, extending out to quantum optics and quite exotic field theories.

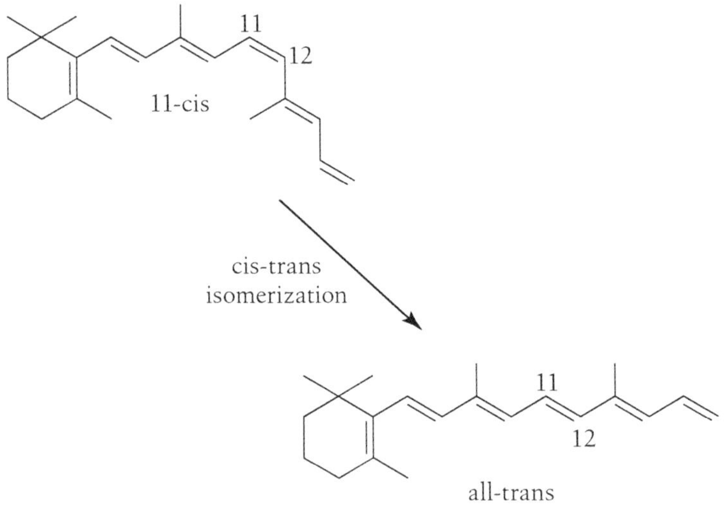

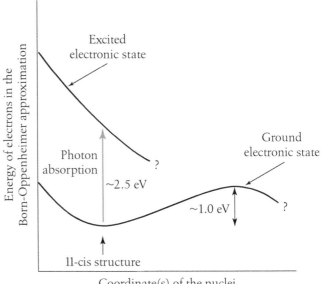

FIGURE 2.19

Isomerization of retinal, the primary event at the start of vision. The π bonds among the carbons favor planar structures, but there are still alternative conformations. The 11-cis conformation is the ground state of rhodopsin, and after photon absorption the molecule converts to the all-trans configuration. These different structures have different absorption spectra, as well as other, more subtle, differences. Thus, we can monitor the progress of the transition 11-cis → all-trans essentially by watching the molecule change color, albeit only slightly.

FIGURE 2.20

Energy surfaces for the rhodopsin molecule. We sketch the electronic ground state and the first excited state, in the Born-Oppenhemier approximation, as a function of the positions of the nuclei. The ground state has an energy minimum when the retinal is in the 11-cis structure. Photon absorption is maximal at a wavelength of $\lambda = 500$ nm or an energy of 2.5 eV, and the temperature dependence of spontaneous isomerization suggests that there is a barrier of ~1 eV for escape from the equilibrium ground-state structure. Question marks indicate parts of the diagram that we can't yet draw. Note that "coordinate(s) of the nuclei" is deliberately vague.

the quantum mechanics of the electrons at fixed positions of the nuclei, giving us a set of energy levels, and then follow these energy levels as the nuclei move. The electronic energy levels act as potential energy surfaces for the motion of the nuclei, which could then be approximately classical or fully quantum mechanical, depending on the details of the system.[15]

---

15. Because the electrons (mostly) follow the nuclei, I use "nuclei" and "atoms" interchangeably in what follows.

The spontaneous isomerization is thermally activated and has a large "activation energy" as estimated from the temperature dependence of the dark noise.[16] Thus, the dark isomerization rate behaves as $r = Ae^{-E_{\text{act}}/k_B T}$, and experimentally $E_{\text{act}} = 0.96 \pm 0.07$ eV, suggesting that there is a barrier of ~1 eV for escape from the ground-state structure.[17] The absorption of light by this molecule peaks at a wavelength of $\lambda = 500$ nm, corresponding to an energy of ~2.5 eV, and we will see shortly that this measures the "vertical" distance to the first excited state when the nuclear coordinates are fixed at the ground-state equilibrium structure. It is unlikely that the excited state has the same equilibrium structure as the ground state, so there should be a force on the molecule (slope of the energy surface) in the excited state. Although these few observations are far from enough to tell us how things work, it helps to sketch them in the schematic of Fig. 2.20, keeping in mind that this is just a sketch.

It seems reasonable that placing the retinal molecule into the pocket formed by the protein opsin would raise the activation energy for spontaneous isomerization, essentially because parts of the protein need to be pushed out of the way for the retinal to rotate and isomerize. In other words, for isolated retinal, the "coordinates of the nuclei" in our sketch just involve the retinal itself, whereas once bound to opsin there are contributions from the position of atoms in the protein. Although this sounds plausible, it is probably wrong: rhodopsin and retinal have very different thermal isomerization rates, but if we plot these rates versus temperature, we see that the activation energies E_{act} are the same within experimental error, and the big difference comes from the prefactor A. In contrast, if we look at photoreceptor cells that are used for daytime vision—the cones, which also provide us with sensitivity to colors, as discussed below—the dark noise level is higher (presumably, single-photon counting is unnecessary in bright light). This does arise from the activation energy, with molecules that absorb lower energy photons having lower activation energies, as one might have expected. Understanding prefactors is much harder than understanding activation energies,[18] and I think we don't have a really compelling theoretical picture that explains the difference between retinal and rhodopsin, although it seems plausible that we could understand the variations across the naturally occurring rhodopsins.

---

16. I am assuming here that the ideas of activation energy and Arrhenius behavior of chemical reaction rates are familiar. For more, see Section 4.1.

17. If you look in the original references in the Annotated Bibliography for this section, you will most often find such energies quoted in kilocalories per mole. I am assuming that physicists will find it easier to think about electron volts in relation to electronic energies, but if you are going to think more about these subjects, you will have to get used to converting among the different units used by different communities.

18. There can also be entropic factors in what is nominally an activation energy. More precisely, as discussed in Section 4.1, there are limits in which the motion of the molecule is on an effective potential surface determined by the free energy as a function of the relevant reaction coordinate. If we compute $k = Ae^{-F_{\text{act}}/k_B T}$, where the activation free energy is $F_{\text{act}} = E_{\text{act}} - k_B T S_{\text{act}}$, we see that making S_{act} increasingly negative slows the reaction rate without necessarily introducing an Arrhenius temperature dependence. Could this be the difference between retinal and rhodopsin? Perhaps. But the condition for thinking about free energies of activation is that all the degrees of freedom that contribute to the entropy are relaxing quickly relative to the motion of the reaction coordinate, and this does not seem right if the entropy is in rearrangements of the protein. So, although entropic factors may be important, I think it would be glib to say that they account for the slowing of the spontaneous isomerization without further discussion.

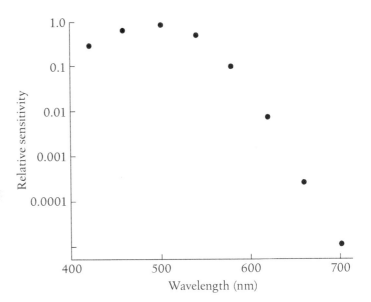

FIGURE 2.21

Sensitivity of the rod photoreceptor as a function of wavelength. This is measured, as explained in the text, by adjusting the intensity of light to give a criterion output, so that very low sensitivity corresponds to shining a bright light, rather than measuring a small output. Redrawn from Baylor et al. (1979a).

Before going any further we have to address the most obvious fact about rhodopsin —its very broad absorption spectrum. A physics education emphasizes the quantum mechanics of atoms, connecting the energy levels of electrons with the absorption and emission of light starting with Balmer and Bohr. But atoms have very narrow absorption spectra, so that we refer to "spectral lines." Rhodopsin, like many large molecules, has an absorption band that ranges over wavelengths from 400 to 600 nm and even beyond. There is a nice of way of measuring the absorption spectrum over a large dynamic range, and this is to use the rod cell itself as a sensor. Instead of asking how much light is absorbed, we can try assuming[19] that all absorbed photons have a constant probability of generating a pulse of current at the rod's output, and so we can adjust the light intensity at each wavelength to produce the same current. If the absorption is stronger, we need less light and conversely, more light if the absorption is weaker; results are shown in Fig. 2.21. It is beautiful that in this way one can follow the long wavelength tail of the spectrum down to cross-sections that are $\sim 10^{-5}$ of the peak. We also see that the width of the spectrum, say, at half maximum, is roughly 20% of the peak photon energy, which is enormous in contrast with atomic absorption lines.

As an aside, the fact that one can follow the sensitivity of the photoreceptor cell deep into the long wavelength tail opens the possibility of asking a very different question about the function of these cells (and all cells). In addition to the rod cells, there are three other types of photoreceptor cell in the retina: the three cones that provide the basis for our color vision at higher light intensities. The pigments in these cells are much like rhodopsin, a retinal molecule bound to a protein, but the proteins are slightly different in each type of cell, and these differences are responsible for the differences in absorption spectra. Proteins are polymers of amino acids, and there are 20 amino acids that Nature uses; each protein has a definite sequence, as discussed in more detail in Section 5.1. The sequence of amino acids for every protein is encoded by the sequence of

---

19. This assumption can also be checked. It is true, but I think there have not been very careful measurements in the long wavelength tail, where something interesting might happen.

bases in DNA, and every cell in our bodies has the same DNA and hence the instructions for making all possible proteins. In particular, all photoreceptor cells have the ability to make all visual pigments. But the different classes of receptors—rods and the three kinds of cones—make different proteins. This concept is much more general: what distinguishes cells with different functions in a complex, multicellular organism is that the different cells "read out" or "express" different proteins. If a single cone could not reliably turn on the expression of one rhodopsin gene, and turn off all others, then the retina would not be able to generate a mix of spectral sensitivities, and we would not see colors. But how off is "off"?

In a macaque monkey (not so different from us in these matters), "red" cones have their peak sensitivity at wavelength $\sim$570 nm, but at this wavelength the "blue" cones have sensitivities that are $\sim 10^5 \times$ reduced relative to their own peak.[20] Peak absorption cross-sections are comparable, so the relative concentration of red pigments in the blue cones must be less than 10^{-5}. That is, the cell makes at least $10^5 \times$ as much of the correct protein as it does of the incorrect proteins, which I have always thought is pretty impressive.[21]

Problem 21: A detour into DNA mechanics. DNA is a double-stranded linear polymer, with the famous double helical structure (see Appendix A.3). The distance from one base pair to the next along the helix is 0.34 nm, and three base pairs code for one amino acid. A typical protein has $\sim$200 amino acids, and even the simplest of single-celled organisms has enough DNA to code for almost 500 different proteins. Thus, a minimal genome is $\sim$3 $\times$ 10^5 base pairs, and in fact it needs to be even longer to allow space for segments of DNA that have a regulatory rather than a coding function. In round numbers, let's say that the minimum is $\sim$5 $\times$ 10^5 base pairs (bp).

(a) How does the length of the DNA compare with the $\sim$1 μm size of these small cells?

(b) Left to itself, a polymer like DNA will not adopt a straight structure but will take a path that looks more like a random walk. In the simplest cases, the long molecule acts like a series of connected segments, with free joints between the segments, so that the structure is exactly a random walk in three dimensions with steps of a size equal to the segment length. The picture of perfectly rigid segments and perfectly free joints is approximate but useful. For DNA the effective segment length, or persistence length, is $\ell_p \sim$ 50 nm, or $\sim$150 base pairs. Using this simple model, about how big is the "ball" that would be formed by a 5 $\times$ 10^5 bp length of DNA? What about the DNA of *E. coli*, which is roughly ten times longer?

(c) Hopefully you now see that the flexibility of DNA is not enough, by itself, to make it fit spontaneously into the container provided by the bacterial cell. But the configurations of random walks are random (!), so there is some probability that the polymer will be found in a structure that takes up much less space. Make a rough estimate of the probability that a random walk of N steps will stay inside a volume $\sim R^3$, where $R \ll \sqrt{N}\ell_p$.

(d) If we want to increase the probability of the polymer staying inside the volume R^3, we must decrease the number of available states, lowering the entropy. The only way to do this

---

20. Referring to cones as red, green, and blue is colloquial and should not be taken too seriously. A more precise scheme is to use "long," "medium," and "short" for the wavelengths of peak sensitivity.

21. Many thanks to Denis Baylor for reminding me of this argument. Because there are $\sim 10^9$ rhodopsins in one cell, errors of even one part in 10^5 would mean that there are thousands of "wrong" molecules floating around. I wonder whether this is true, or whether the errors are even smaller.

FIGURE 2.22

Schematic of the electronic states in a large molecule, highlighting their coupling to motion of the nuclei. The sketch show two states, with photon absorption (in blue) driving transitions between them. If we think in semiclassical terms, as explained in the text, then these transitions are "too fast" for the atoms to move and hence are vertical on such plots (the Franck-Condon approximation). Because the atomic coordinates fluctuate, as indicated by the Boltzmann distribution, the energy of the photon required to drive the transition also fluctuates, which broadens the absorption spectrum.

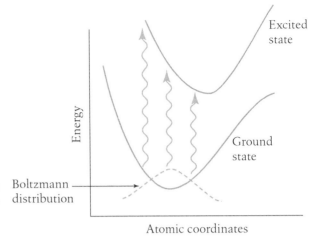

stably is to provide interactions between the DNA and other molecules that are energetically favorable when the polymer is in a more compact configuration. Make a rough estimate of this energy. Is it proportional to the length of the polymer?

Returning to the absorption spectrum itself, we can try to understand why it is so broad by taking the Born-Oppenheimer picture of Fig. 2.20 seriously. In particular, we need to add to this picture the fluctuating coordinates of the atoms, drawn from the Boltzmann distribution as in Fig. 2.22. In the ground state, we know that there is some arrangement of the atoms that minimizes the energy and that in the neighborhood of this minimum the potential surface must look roughly like that of a system of Hookean springs. Once we lift the electrons into the first excited state, there is again some configuration of the atoms that minimizes the energy (unless absorbing one photon is enough to break the molecule apart!), but this equilibrium configuration will be different than in the ground state. Hence in Fig. 2.22, the energy surfaces for the ground and excited states are shown displaced.

It is important to realize that sketches such as that in Fig. 2.22 are approximations in many senses. Most importantly, this sketch involves only one coordinate. You may be familiar with a similar idea in the context of chemical reactions, where out of all the atoms that move during the reaction, we focus on one "reaction coordinate" that forms a path from the reactants to products (for more about this, see Section 4.1). One view is that this approach is just a convenience—we can't draw in many dimensions, so we just draw one and interpret the figure cautiously. Another view is that the dynamics *are* effectively one dimensional, either because there is a separation of time scales, or because we can change coordinates to isolate, for example, a single coordinate that couples to the difference in energy between the ground and excited electronic states. The cost of this reduction in dimensionality might be a more complex dynamics along this one dimension, for example with a "viscosity" that is strongly frequency dependent, which again means that we need to be cautious in interpreting the picture that we draw.

In the limit that the atoms are infinitely heavy, they do not move appreciably during the time required for an electronic transition. However, the positions of the atoms still have to come out of the Boltzmann distribution, because the molecule is in equilibrium

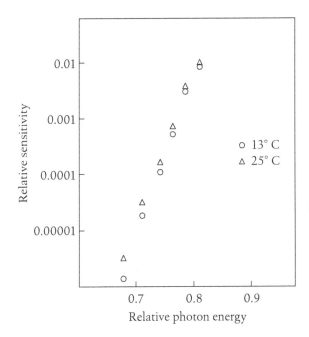

FIGURE 2.23

Temperature dependence of rod cell sensitivity in the long wavelength (low energy) limit. These experiments are done on red rods from the salamander, and error bars are roughly four times smaller than the symbols, including both measurement errors and variations from cell to cell. Notice that the difference in absolute temperature probed here is just $\Delta T/T \sim 0.04$. Redrawn from Luo et al. (2011).

with its environment at temperature T. In this limit, we can think of transitions between electronic states as occurring without atomic motion, corresponding to vertical lines on the schematic in Fig. 2.22. If the photon happens to arrive when the atomic configuration is a bit to the left of the equilibrium point, then as drawn the photon energy needs to be larger to drive the transition. If the configuration is a bit to the right, then the photon energy is smaller. In this way, the Boltzmann distribution of atomic positions is translated into a broadening of the absorption line. In particular, the transition can occur with a photon that has very little energy if we happen to catch a molecule in the rightward tail of the Boltzmann distribution: the electronic transition can be made up partly from the energy of the photon and partly from energy that is borrowed from the thermal bath. As a result, the absorption spectrum should have a tail at long wavelengths, and this tail will be strongly temperature dependent. This phenomenon is observed in rhodopsin and other large molecules, as shown in Fig. 2.23. Because our perception of color depends on the relative absorption of light by rhodopsins with different spectra, there must be wavelengths such that the apparent color of the light will depend on temperature.[22]

Concretely, if we imagine that the potential surfaces are perfect Hookean springs, but with displaced equilibrium positions, then we can relate the width of the spectrum directly to the magnitude of this displacement. In the ground state we have the potential

$$V_{\mathrm{g}}(q) = \frac{1}{2}\kappa q^2, \tag{2.68}$$

and in the excited state we have

$$V_{\mathrm{e}}(q) = \epsilon + \frac{1}{2}\kappa(q - \Delta)^2, \tag{2.69}$$

---

22. Various stories have been told about searching for this perceptual effect (in a hot tub).

where ϵ is the minimum energy difference between the two electronic states, and Δ is the shift in the equilibrium position, as indicated in Fig. 2.24. With q fixed, the condition for absorbing a photon is that the energy $\hbar\Omega$ match the difference in electronic energies,

$$\hbar\Omega = V_e(q) - V_g(q) = \epsilon + \frac{1}{2}\kappa\Delta^2 - \kappa\Delta q. \tag{2.70}$$

The probability distribution of q when molecules are in the ground state is given by

$$P(q) = \frac{1}{Z}\exp\left[-\frac{V_g(q)}{k_BT}\right] = \frac{1}{\sqrt{2\pi k_BT/\kappa}}\exp\left[-\frac{\kappa q^2}{2k_BT}\right], \tag{2.71}$$

so we expect the cross-section for absorbing a photon of frequency Ω to have the form

$$\sigma(\Omega) \propto \int dq\, P(q)\delta\left[\hbar\Omega - \left(\epsilon + \frac{1}{2}\kappa\Delta^2 - \kappa\Delta q\right)\right] \tag{2.72}$$

$$\propto \int dq\,\exp\left[-\frac{\kappa q^2}{2k_BT}\right]\delta\left[\hbar\Omega - \left(\epsilon + \frac{1}{2}\kappa\Delta^2 - \kappa\Delta q\right)\right] \tag{2.73}$$

$$\propto \exp\left[-\frac{(\hbar\Omega - \hbar\Omega_{\text{peak}})^2}{4\lambda k_BT}\right], \tag{2.74}$$

where the peak of the absorption is at

$$\hbar\Omega_{\text{peak}} = \epsilon + \lambda, \tag{2.75}$$

and

$$\lambda = \frac{1}{2}\kappa\Delta^2 \tag{2.76}$$

is the energy required to distort the molecule into the equilibrium configuration of the excited state if we stay in the ground state.

The energy λ is known, in different contexts, as the reorganization energy or the Stokes shift. If the molecule stays in the excited state for a long time, the distribution of coordinates will re-equilibrate to the Boltzmann distribution appropriate to $V_e(q)$, so that the most likely coordinate becomes $q = \Delta$. At this coordinate, if the molecule returns to the ground state by emitting a photon—fluorescence—the energy of this photon will be $\hbar\Omega_{\text{fluor}} = \epsilon - \lambda$. Thus, the peak fluorescence is at lower energies, or is red shifted from the absorption peak by an amount 2λ, as can be read off from Fig. 2.24. This connects the width of the absorption band to the red shift that occurs in fluorescence, and for many molecules this prediction is correct, quantitatively, giving us confidence in the basic picture.

In the case of rhodopsin, the peak absorption is at a wavelength of 500 nm or an energy of $\hbar\Omega_{\text{peak}} = 2.5$ eV. The width of the spectrum is described roughly by a Gaussian with a standard deviation of $\sim 10\%$ of the peak energy, so that $2\lambda k_BT \sim (0.25\,\text{eV})^2$, or $\lambda \sim 1.25$ eV. Surely we can't take this seriously, because this reorganization energy is enormous and would distort the molecule well beyond the point where we could describe the potential surfaces by Hookean springs. If we took this result literally, the peak fluorescence would be at zero energy! The correct conclusion is that there is a tremendously strong coupling between the excitation of the electrons and motion of

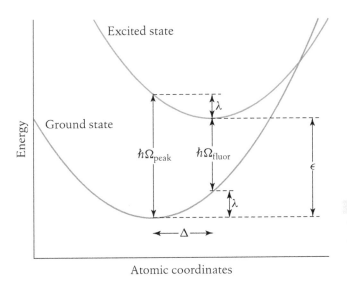

FIGURE 2.24

The potential surfaces of Fig. 2.22, redrawn in the special case where they are parabolic. Then, as in Eqs. (2.68)–(2.76), there are just a few key parameters that determine the shape of the absorption spectrum and also the fluorescence emission.

the atoms, and presumably this is related to the fact that photon absorption leads to very rapid structural changes.

Problem 22: Temperature-dependent spectra.   Equation (2.74) predicts that absorption spectra are temperature dependent.

(a) Figure 2.23 suggests that absorption spectra have an exponential dependence on photon energy at long wavelengths. Show that this dependence is consistent with the predictions of Eq. (2.74). Is it plausible that these data are at sufficiently long wavelengths that the asymptotic behavior should be seen?

(b) A small change in temperature should produce a proportionally small change in the slope of the log (absorption) versus energy plot. Show that this change in slope can be directly converted into an estimate of the reorganization energy λ. Are the data in Fig. 2.23 consistent with the rough estimates of λ from the width of the spectrum, as given above?

(c) Generalize the derivation of Eq. (2.74) to the case where the electronic transition is coupled to many coordinates, all still treated classically. Does this change anything about the predicted temperature dependence of the absorption spectra?

The strong coupling between excitation of the electrons and motions of the atoms means that a molecule that makes a transition to the excited state (at fixed atomic coordinates, in our approximation) will experience a large force. This force will push the molecular structure away from what was the equilibrium structure in the ground state. Because the function of this system requires the retinal to isomerize, rotating around the eleventh bond along the conjugated chain, it is tempting to think that this push is in the direction of the isomerization. Then a large force will lead to rapid motion and perhaps to rapid isomerization.

If we look again at isolated retinal and excite the system with a very short pulse of light, we can follow the resulting changes in absorption spectrum. What we see is the rise of an absorption spectrum that is consistent with the all-trans state, at a rate

$\sim 10^9 \, \text{s}^{-1}$. Although this is fast compared to the reactions that we can see by eye, it is actually slow enough to be comparable to the rate at which the molecule will re-emit the photon. Unless there are symmetries forbidding the transition, spontaneous emission of visible photons from electronic excited states typically occurs on the nanosecond time scale. Isomerization of retinal itself thus is not fast enough to prevent fluorescence and truly capture the energy of the photon with high probability.

Now fluorescence is a disaster for visual pigment—not only is the photon not counted where it was absorbed, it might also get counted somewhere else, blurring the image. In fact rhodopsin does not fluoresce: the quantum yield or branching ratio for fluorescence is $\sim 10^{-5}$. If we imagine the molecule sitting in the excited state, transitioning to the ground state via fluorescence at a rate $\sim 10^9 \, \text{s}^{-1}$, then to have a branching ratio of 10^{-5}, the competing process must have a rate of $\sim 10^{14} \, \text{s}^{-1}$. Thus, the rhodopsin molecule must leave the excited state by some process on a time scale of $\sim 10 \, \text{fs}$ (*femtoseconds*), which is extraordinarily fast. Indeed, for many years, every time people built faster pulsed lasers, they went back to rhodopsin to look at the initial events, culminating in the direct demonstration of femtosecond isomerization, making this one of the fastest molecular events ever observed.

The 11-cis and all-trans configurations of retinal have different absorption spectra, which is why we can observe the events following photon absorption as an evolution of the spectrum. The basic design of such experiments is to excite the molecules with a brief pulse of light, elevating them into the excited state, and then probe with another brief pulse after some delay. In the simplest version, one repeats the experiment many times with different choices of the delay and the energy or wavelength of the probe pulse. An example of the results from such an experiment are shown in Fig. 2.25. The first thing to notice is that the absorption at a wavelength of 550 nm, characteristic of the all-trans structure, rises very quickly after the pulse that excites the system, certainly within tens of femtoseconds. In fact this experiment reveals all sorts of interesting structure, to which we return below.

The combination of faster photon-induced isomerization and slower thermal isomerization means that the protein opsin acts as an electronic state selective catalyst: ground-state reactions are inhibited, and excited-state reactions accelerated, each by orders of magnitude. If these state-dependent changes in reaction rate did not occur—that is, if the properties of rhodopsin were those of retinal—then we simply would not be able to see in the dark of night.

Problem 23: What would vision be like if . . . ?   Imagine that the spontaneous isomerization rate and quantum yield for photo-isomerization in rhodopsin were equal to those in retinal. Estimate, quantitatively, what this would mean for our ability to see at night.

Before proceeding, it would be nice to do a calculation that reproduces the intuition of Figs. 2.22 and 2.24. This is done in Appendix A.4. The results of the calculation show, in more detail, how the coupling of electronic states to the vibrational motion of the molecule can shape the absorption spectrum. If there is just one lightly damped vibrational mode, then the single sharp absorption line that we expect from atomic physics becomes a sequence of lines, corresponding to changing electronic state and exciting one, two, three, or more vibrational quanta. If there are many modes, and these

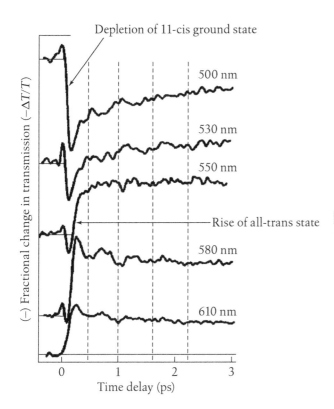

FIGURE 2.25

Femtosecond dynamics of rhodopsin. At time $t = 0$, the molecules are excited by a 35 fs pulse of 500 nm light. Transmission of light at several different wavelengths is monitored as a function of time, with ~10 fs resolution. We see the immediate depletion of the absorption associated with the initial 11-cis state, while an absorption band centered at ~550 nm rises within 200 fs. This absorption is characteristic of the all-trans state. In addition to the rapid rise of the all-trans state, there are noticeable wiggles in the absorption that persist for several picoseconds. We will see that these are systematic oscillations. Redrawn from Wang et al. (1994).

modes are damped by interaction with other degrees of freedom, these "vibronic" lines merge into a smooth spectrum that we can calculate in a semiclassical approximation.

The coupling of electronic transitions to vibrational motion generates the phenomenon of Raman scattering—a photon is inelastically scattered, making a virtual transition to the electronically excited state and dropping back down to the ground state, leaving behind a vibrational quantum, as in Fig. 2.26. The energy shifts of the scattered photons allow us to read off directly the frequencies of the relevant vibrational modes. With a bit more sophistication, we can connect the strength of the different lines to the coupling constants (e.g., the displacements Δ_i along each mode, generalizing the discussion above) that characterize the interactions between electronic and vibrational degrees of freedom. If everything works, it should be possible to reconstruct the absorption spectrum from these estimates of frequencies and couplings. This whole program has been carried out for rhodospin. To get everything right, however, one has to include motions that are effectively unstable in the excited state, presumably corresponding to the torsional motions that lead to cis-trans isomerization.

If we try to synthesize all these ideas into a single schematic, we might get something like Fig. 2.27. If we take this picture seriously, then after exciting the molecule with a pulse of light, we should see the disappearance of the absorption band associated with the 11-cis structure; the gradual appearance of the absorption from the all-trans state; and with a little luck, stimulated emission while the excited state is occupied. All these phenomena are seen. Looking closely (e.g., at Fig. 2.25), however, one sees that

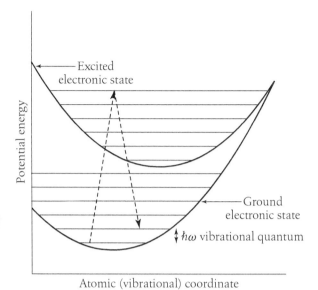

FIGURE 2.26

Resonant Raman scattering. A photon resonant with the combination of electronic and vibrational energy levels can be virtually absorbed and re-emitted, leaving behind a vibrational quantum.

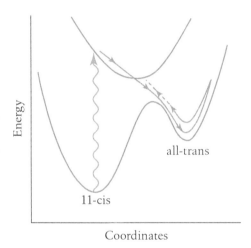

FIGURE 2.27

Schematic model of the energy surfaces in rhodopsin, revisited. The ground state has minima at both the 11-cis and the all-trans structures. A single excited state sits above this surface. At some intermediate structure, the surfaces approach each other. At this point, the Born-Oppenheimer approximation breaks down, and there will be some mixing between the two states. A molecule lifted into the excited state by absorbing a photon slides down the upper surface and can pass nonadiabatically into the potential well, whose minimum is at all-trans.

spectra are wiggling in time. This might have been measurement error, but it isn't—there is a systematic oscillation of the absorption across all wavelengths, dominated by a mode with frequency $\sim 60 \, \mathrm{cm}^{-1}$. Thus, rather than sliding irreversibly down the potential surfaces toward a local energy minimum, the atomic structure of the molecule oscillates. More remarkably, we can see a wavelength dependence of the phase of these oscillations that is a signature of quantum mechanical coherence in this vibrational motion (see Fig. 2.28).

Our usual picture of molecules and their transitions comes from chemical kinetics: there are reaction rates, which represent the probability per unit time for the molecule to make transitions among states that are distinguishable by some large-scale re-arrangement; these transitions are cleanly separated from the time scales for molecules to come to equilibrium in each state. The initial isomerization event in rhodopsin

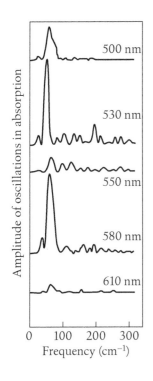

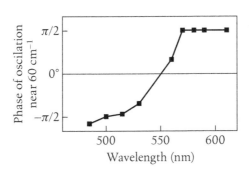

FIGURE 2.28

Oscillations in transient absorption (from Fig. 2.25) in the window from 200–3000 fs after the excitation pulse. After subtracting the smooth time dependence, the remaining "wiggles" are Fourier transformed, with the amplitude plotted at left. The frequency scale is the conventional one in vibrational spectroscopy: $1 \, cm^{-1} \equiv 3 \times 10^{10} \, s^{-1}$. Notice the peak at $\sim 60 \, cm^{-1}$ across the entire range of wavelengths. At right, the phase associated with this peak is shown as a function of wavelength. Redrawn from Wang et al. (1994).

is so fast that this approximation certainly breaks down. More profoundly, the time scale of the isomerization is so fast that it competes with the processes that destroy quantum mechanical coherence among the relevant electronic and vibrational states. The whole notion of an irreversible transition from one state to another necessitates the loss of coherence between these states (recall Schrödinger's cat), and so in this sense the isomerization is proceeding as rapidly as possible.

For this picture of transitions to work, the dynamics of rhodopsin must violate the Born-Oppenheimer approximation. Figure 2.27 indicates that we start in the electronic ground state, the photon lifts the molecule into an excited state, but then by the end of the process we have described, the system is back in the ground state without emitting a photon. Thus, motion of the atoms has actually caused a transition between electronic states, which is exactly what the Born-Oppenheimer approximation neglects. But the consistency of the Born-Oppenhemier approximation requires that the spacing between electronic energy levels at any value of the atomic coordinates be larger than the vibrational quantum energies, and evidently somewhere between the 11-cis and the all-trans structures this condition is violated and the different potential surfaces must approach each other quite closely.

At this point what we would like to do is a simplified calculation that generates the schematic in Fig. 2.27 and explains how the dynamics on these surfaces can be so fast. As far as I know, there is no clear answer to this challenge, although there are many detailed simulations, in the quantum chemical style, that probably capture elements of the truth. The central ingredient is the special nature of the π bonds along the retinal. In the ground state, electron hopping between neighboring p_z orbitals lowers the energy of the system, and this effect is maximized in planar structures where the orbitals are all in the same orientation. But this lowering of the energy depends on the character of the electron wave functions—in the simplest case of bonding between two atoms, the symmetric state (the bonding orbital) has lower energy in proportion to the hopping matrix element, whereas the antisymmetric state (antibonding orbital) has higher energy, again in proportion to the matrix element. Thus, if we excite the electrons, it is plausible that the energy of the excited state could be reduced by structural changes that reduce the hopping between neighboring carbons, which happens if the molecule rotates to become nonplanar. In this way we can understand why there is a force for rotation in the excited state, and why there is another local minimum in the ground state at the 11-cis structure.

Problem 24: Energy levels in conjugated molecules. The simplest model for a conjugated molecule is that the electrons that form the π orbitals can sit on each carbon atom with some energy that we can set to zero, and they can hop from one atom to its neighbors. Note that there is one relevant electron per carbon atom. If we write the Hamiltonian for the electrons as a matrix, then for a ring of six carbons (benzene), we have

$$\mathbf{H}_6 = \begin{pmatrix} 0 & -t & 0 & 0 & 0 & -t \\ -t & 0 & -t & 0 & 0 & 0 \\ 0 & -t & 0 & -t & 0 & 0 \\ 0 & 0 & -t & 0 & -t & 0 \\ 0 & 0 & 0 & -t & 0 & -t \\ -t & 0 & 0 & 0 & -t & 0 \end{pmatrix}, \tag{2.77}$$

where the "hopping matrix element" $-t$ is negative, because the electrons can lower their energy by being shared among neighboring atoms—the essence of chemical bonding! Models like this are called tight binding models in the condensed matter physics literature and Hückel models in the chemical literature. Notice that they leave out any direct interactions among the electrons. This problem is about solving Schrödinger's equation, $\mathbf{H}\psi = E\psi$, to find the energy eigenstates and the corresponding energy levels. Notice that for the case of benzene, if we write the wave function ψ in terms of its six components (one for each carbon atom), then Schrödinger's equation becomes

$$-t(\psi_6 + \psi_2) = E\psi_1, \tag{2.78}$$
$$-t(\psi_1 + \psi_3) = E\psi_2, \tag{2.79}$$
$$-t(\psi_2 + \psi_4) = E\psi_3, \tag{2.80}$$
$$-t(\psi_3 + \psi_5) = E\psi_4, \tag{2.81}$$
$$-t(\psi_4 + \psi_6) = E\psi_5, \tag{2.82}$$
$$-t(\psi_5 + \psi_1) = E\psi_6. \tag{2.83}$$

(a) Considering first the case of benzene, show that solutions to the Schrödinger equation are of the form $\psi_n \propto \exp(ikn)$. What are the allowed values of the "momentum" k? Generalize to an arbitrary N-membered ring.

(b) What are the energies corresponding to the states labeled by k? Because of the Pauli principle, the ground state of the molecule is constructed by putting the electrons two-by-two (spin up and spin down) into the lowest energy states; thus, the ground state of benzene has two electrons in each of the lowest three states. What is the ground state energy of benzene?

What about for an arbitrary N-membered ring (with N even)? Can you explain why benzene is especially stable?

(c) Suppose that the bonds between carbon atoms stretch and compress a bit, so that they become alternating single and double bonds rather than all being equivalent. To first order, if the bond stretches by an amount u, then the hopping matrix element should go down (the electron has farther to hop), so we write $t \to t - \alpha u$; conversely, if the bond compresses, so that u is negative, the hopping matrix element gets larger. If we have alternating long and short (single and double) bonds, then the Hamiltonian for a six-membered ring would be

$$
\mathbf{H}_6(u) = \begin{pmatrix}
0 & -t + \alpha u & 0 & 0 & 0 & -t - \alpha u \\
-t + \alpha u & 0 & -t - \alpha u & 0 & 0 & 0 \\
0 & -t - \alpha u & 0 & -t + \alpha u & 0 & 0 \\
0 & 0 & -t + \alpha u & 0 & -t - \alpha u & 0 \\
0 & 0 & 0 & -t - \alpha u & 0 & -t + \alpha u \\
-t - \alpha u & 0 & 0 & 0 & -t + \alpha u & 0
\end{pmatrix}. \quad (2.84)
$$

Find the ground-state energy of the electrons as a function of u, and generalize to the case of N-membered rings. Does the "dimerization" of the system ($u \neq 0$) raise or lower the energy of the electrons? Note that if your analytic skills (or patience) give out, this is a relatively simple numerical problem; feel free to use the computer, but be careful to explain what units you are using when you plot your results.

(d) To have bonds alternately stretched and compressed by an amount u, we need an energy $\frac{1}{2}\kappa u^2$ in each bond, where κ is the stiffness contributed by all the other electrons that are not tracked explicitly. Consider parameter values $t = 2.5$ eV, $\alpha = 4.1$ eV/Å, and $\kappa = 21$ eV/Å^2. Should benzene have alternating single and double bonds ($u \neq 0$), or should all bonds be equivalent ($u = 0$)?

(e) Peierls' theorem about one-dimensional electron systems predicts that, for N-carbon rings with N large, the minimum total energy will be at some nonzero u_*. Verify that this is true in this case, and estimate u_*. How large does N have to be for this prediction to hold? What do you expect for retinal?

Suppose that we succeed and have a semi-quantitative theory of the excited-state dynamics of rhodopsin, enough to understand why the quantum yield of fluorescence is so low, and what role is played by quantum coherence. We would then have to check that the barrier between the 11-cis and the all-trans structures in Fig. 2.27 comes out to have the right height to explain the activation energy for spontaneous isomerization. But then how do we account for the anomalously low prefactor in this rate, which is where, as discussed above, the protein acts to suppress dark noise? If there is something special about the situation in the environment of the protein that enables the ultrafast, coherent dynamics in the excited state, why does this special environment generate almost the same barrier as for isolated retinal?

It is clear that the ingredients for understanding the dynamics of rhodopsin—and hence for understanding why we can see into the darkening night—involve quantum mechanical ideas more related to condensed matter physics than to conventional biochemistry, a remarkably long distance from the psychology experiments on human subjects that we started with. Although Lorentz could imagine that people count single quanta, surely he couldn't have imagined that the first steps of this process are coherent.

Even though we have the ingredients, it is clear that we don't have them put together in quite the right way yet.

If rhodopsin were the only example of this "almost coherent chemistry" that would be good enough, but in fact the other large class of photon-induced events in biological systems—photosynthesis—also proceed so rapidly as to compete with loss of coherence, and the crucial events again seem to happen (pardon the partisanship) while everything is still in the domain of physics and not of conventional chemistry. Again there are beautiful experiments that present a number of theoretical challenges.[23] Why biology pushes to these extremes is a good question. How it manages to do all this with big floppy molecules in water at roughly room temperature also is a great question.

2.3   Biochemical Amplification

We have known for a long time that light is absorbed by rhodopsin and that light absorption leads to an electrical response detectable as a modulation in the current flowing across the photoreceptor cell membrane. Crucially, as seen already in Fig. 2.18, the rhodopsin molecules are in the disk membranes, which are physically separate from the cell membrane (also called the plasma membrane) across which the currents are flowing. Thus, there is a qualitative problem of how the signal gets from one place to the other in the cell. This is a sufficiently hard problem that one might worry about the evidence for the separation between the disks and the cell membranes, but eventually electron microscopy settled this question: in rod cells the disk membrane is closed and physically disconnected from the cell membrane. So there must be something that links the initial events in the disk to the final events in the plasma membrane.

A natural hypothesis is that the disk and plasma membranes are coupled by a molecule that diffuses in the interior of the rod cell. There is not much free space, so even if this molecule is small, the effective rates of diffusion will be slow. How can we identify this molecule? Suppose we start with a retina, grind it up, and try to identify the various small molecules that are floating around. In this process we will find all sorts of things, some specific to rod cells, some not. But we could do the same experiment twice, once in the dark and once and after exposing the retina to a bright light; hopefully, a much smaller number of molecules change their concentrations between these two conditions. The set of molecules whose concentration is sensitive to light provides us with a set of candidates for the messenger that carries signals between the disk and plasma membranes.

Once we have candidate messenger molecules, how do we test them? In the same way that we can make a tight seal around the entire rod cell with a glass pipette, we can make a seal between a smaller pipette and the surface of the cell membrane. In favorable cases, we can then rip away a piece of the membrane while maintaining the seal. Then all the current across that detached patch of membrane will flow up the pipette, where we can measure it, but what was the intracellular surface of the membrane is now exposed. We thus can insert this sample into solutions containing

---

23. As usual, a guide is found in the references in the Annotated Bibliography for this section; see also Section 4.1.

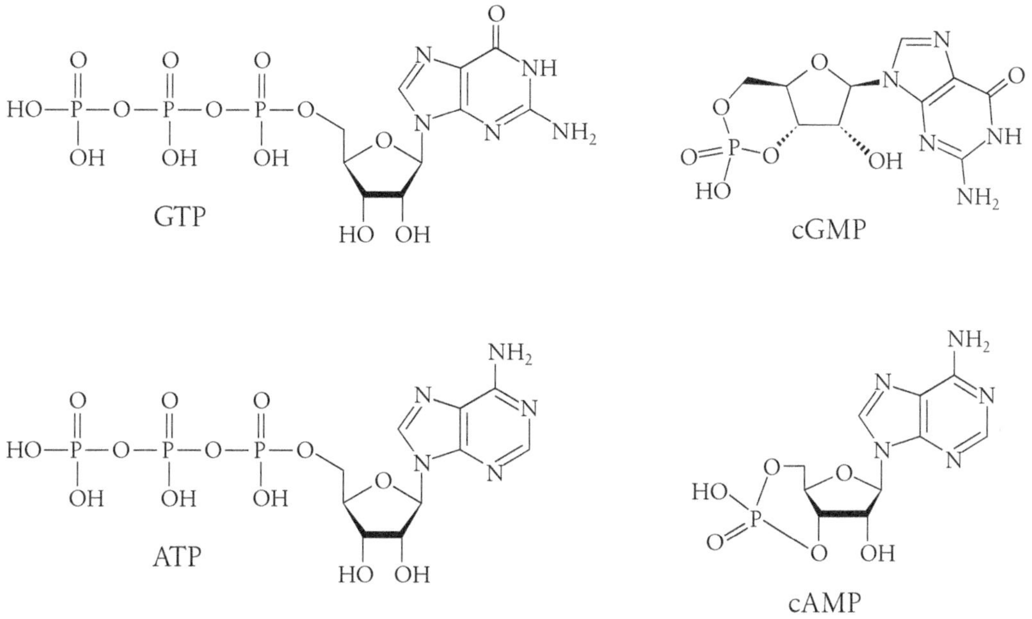

FIGURE 2.29

Messenger molecules and their precursors. Cyclic guanosine monophosphate (cGMP, top right) is made from guanosine triphosphate (GTP, top left). Cyclic adenosine monophosphate (cAMP, bottom right) is made from adenosine triphosphate (ATP, bottom left).

different candidate messenger molecules and see whether any of them causes a change in the current. Through experiments like these, we know that the internal messenger is the molecule cyclic guanosine monophosphate, or cGMP (Fig. 2.29).

Cyclic GMP is part of a family of cyclic nucleotides that are synthesized in cells by starting with nucleotide triphosphates (in this case, guanosine triphosphate, GTP). Adenosine triphosphate (ATP) is the energy currency of cells, a compound that is quite stable but releases a huge amount of energy when the bonds between the phosphates are broken; this reaction powers our muscles, for example. ATP and GTP are also the precursors of the "A" and "G" in DNA sequences (and there are CTP and TTP molecules as well, to round out the four bases). There are many processes where cells need to propagate signals from one place to another, and diffusion of cyclic nucleotides is a common way for them to do so.

As with almost all chemical reactions in cells, the synthesis of cGMP is catalyzed by the action of a specific protein molecule, and such protein catalysts are called enzymes. The concentration of cGMP reflects a balance between the synthesis reaction and a reaction in which the molecule is degraded, and this is also catalyzed by a specific enzyme. The synthesis enzyme in this case is guanylate cyclase (GC), and the degradation enzyme is a phosphodiesterase (PDE). With this background, there are several questions we can ask about how this signaling fits into the problem of photon counting:

1. How do the enzymes GC and PDE work? What features of these enzymes are critical for reliable photon counting?

2. Presumably, for photon absorption by rhodopsin to change the concentration of cGMP, there must be coupling from rhodopsin to one or both of the two enzymes. How does this work?

3. Why does changing the concentration of cGMP in the cell modulate the current flowing across the membrane?

4. Where, in this system, is the gain that converts the single molecular event—one rhodopsin absorbs a photon—into a macroscopic change in current?

There is no particular order in which we have to address these issues. Let's start by trying to understand current flow through the plasma membrane, and how this is modulated by changes in the cGMP concentration.

Biological systems contain no metallic or semiconductor components. Signals can still be carried by electrical currents and voltages, but now currents consist of ions, such as potassium or sodium, flowing through water or through specialized conducting pores. These pores, or channels, are also protein molecules, embedded in the cell membrane. They can thus respond to the electric field or voltage across the membrane as well as to the binding of small molecules. The coupled dynamics of channels and voltage turns each cell into a potentially complex nonlinear dynamical system.

Imagine a spherical molecule or ion of radius a; a typical value for this radius is 0.3 nm. From the Stokes formula we know that if this ion moves through the water at velocity v, it will experience a drag force $F = -\gamma v$, with the drag coefficient $\gamma = 6\pi \eta a$, where η is the viscosity; for water $\eta = 0.01$ poise, the cgs unit poise $= \text{gm}/(\text{cm} \cdot \text{s})$. The inverse of the drag coefficient is called the mobility, $\mu = 1/\gamma$, and the diffusion constant of a particle is related to the mobility and the absolute temperature by the Einstein relation or fluctuation-dissipation theorem, $D = k_B T \mu$, where k_B is Boltzmann's constant and T the absolute temperature. Because life operates in a narrow range of absolute temperatures, it is useful to remember that at room temperature ($25°$ C), $k_B T \sim 4 \times 10^{-21}$ J $\sim 1/40$ eV. Combining all these numbers we have, roughly,

$$D = k_B T \mu = k_B T \cdot \frac{1}{\gamma} = \frac{k_B T}{6\pi \eta a} \sim 10^{-9} \, \text{m}^2/\text{s} = 1 \, \mu\text{m}^2/\text{ms}. \qquad (2.85)$$

Ions and small molecules diffuse freely through water, but cells are surrounded by a membrane that functions as a barrier to diffusion. In particular, these membranes are composed of lipids (see Fig. 2.18), which are nonpolar and therefore cannot screen the charge of an ion that tries to pass through the membrane. The water does screen the charge, so pulling an ion out of the water and pushing it through the membrane would require surmounting a large electrostatic energy barrier. This barrier means that the membrane provides an enormous resistance to current flow between the inside and outside of the cell. If this were the whole story, there would be no electrical signaling in biology. In fact, cells construct specific pores or channels through which ions can pass, and by regulating the state of these channels, the cell can control the flow of electric current across the membrane.

Ion channels are proteins, composed of several thousand atoms in very complex arrangements. Let's try, however, to ask a simple question: If we open a pore in the cell membrane, how quickly can ions pass through? More precisely, since the ions carry current and will move in response to a voltage difference across the membrane, how large is the current in response to a given voltage (the electrical conductance) of one open channel?

Imagine that one ion channel serves as a hole in the membrane, and pretend that ion flow through this hole is essentially the same as through water. The electrical current that flows through the channel is

$$J = q_{\text{ion}} \cdot [\text{ionic flux}] \cdot [\text{channel area}], \tag{2.86}$$

where q_{ion} is the charge on one ion, and flux measures the rate at which particles cross a unit area, so that

$$\text{ionic flux} = \frac{\text{ions}}{\text{cm}^2\text{s}} = \frac{\text{ions}}{\text{cm}^3} \cdot \frac{\text{cm}}{\text{s}} \tag{2.87}$$

$$= [\text{ionic concentration}] \cdot [\text{velocity of one ion}]$$

$$= cv. \tag{2.88}$$

Major current carriers, such as sodium and potassium, are present at concentrations of $c \sim 100$ mM, or $c \sim 6 \times 10^{19}$ ions/cm^3.

The next problem is to compute the typical velocity of one ion. We are interested in a current, so this is not the velocity of random Brownian motion but rather the average of that component of the velocity directed along the electric field. In a viscous medium, the average velocity is related to the applied force through the mobility, or the inverse of the drag coefficient, as above. The force on an ion is in turn equal to the electric field times the ionic charge, and the electric field is (roughly) the voltage difference V across the membrane divided by the thickness ℓ of the membrane:

$$v = \mu F = \mu q_{\text{ion}} E \sim \mu q_{\text{ion}} \frac{V}{\ell} = \frac{D}{k_B T} q_{\text{ion}} \frac{V}{\ell}. \tag{2.89}$$

Putting the various factors together, we find the current

$$J = q_{\text{ion}} \cdot [\text{ionic flux}] \cdot [\text{channel area}]$$

$$= q_{\text{ion}} \cdot [cv] \cdot [\pi d^2/4] \tag{2.90}$$

$$= q_{\text{ion}} \cdot \left[c \cdot \frac{D}{\ell} \cdot \frac{q_{\text{ion}} V}{k_B T} \right] \cdot \frac{\pi d^2}{4} \tag{2.91}$$

$$= \frac{\pi}{4} q_{\text{ion}} \cdot \frac{c d^2 D}{\ell} \cdot \frac{q_{\text{ion}} V}{k_B T}, \tag{2.92}$$

where the channel has a diameter d. If we assume that the ion carries one electronic charge, as does sodium, potassium, or chloride, then $q_{\text{ion}} = 1.6 \times 10^{-19}$ C and $q_{\text{ion}} V/(k_B T) = V/(25 \text{ mV})$. Typical values for the channel diameter should be comparable to the diameter of a single ion, $d \sim 0.3$ nm, and the thickness of the membrane is $\ell \sim 5$ nm. Thus, we have

$$
\begin{aligned}
J &= \frac{\pi}{4} q_{\text{ion}} \cdot \frac{cd^2 D}{\ell} \cdot \frac{q_{\text{ion}} V}{k_B T} \\
&= \frac{\pi}{4} (1.6 \times 10^{-19} \text{ C}) \\
&\qquad \times \frac{(6 \times 10^{19} \text{ cm}^{-3})(3 \times 10^{-8} \text{ cm})^2 (10^{-5} \text{ cm}^2/\text{s})}{50 \times 10^{-8} \text{ cm}} \cdot \frac{V}{25 \text{ mV}}
\end{aligned}
\tag{2.93}
$$

$$
\sim 2 \times 10^{-14} \cdot \frac{V}{\text{mV}} \quad \text{C/s} \sim 2 \times 10^{-11} \frac{V}{\text{Volts}} \text{ A}, \tag{2.94}
$$

or

$$
J = gV, \tag{2.95}
$$
$$
g \sim 2 \times 10^{-11} \text{ A/V} = 20 \text{ pS}. \tag{2.96}
$$

So our order of magnitude argument leads us to predict that the conductance of an open channel is roughly 20 pS.[24] With a voltage difference across the membrane of ~ 50 mV, we thus expect that opening a single channel will cause ~ 1 pA of current to flow. Although incredibly oversimplified, this answer is basically right, as verified in experiments where one actually measures the currents flowing through single channel molecules.

Our first major problem is to reconcile this calculation (and the real properties of single channels, which are about the same), with what we see in response to single photons. The total change in current that results from a single photon arrival is also ~ 1 pA. But if this were just the effect of (closing) one channel, we would see "square edges" in the current trace as the single channels open or close. Also, if one photon changes the structure of one rhodopsin molecule, and this has an effect on one channel, we don't really have any gain. Maybe we don't need any gain—after all, the energy of the incoming photon is large, $\sim 100 k_B T$—but then why is there a complicated bit of biochemistry required to connect the rhodopsin and the channels? What's going on?

The answer turns out to be that the membrane contains many channels, and these channels flicker very rapidly between their open and closed states, as schematized in Fig. 2.30. Thus, on the relatively slow time scale of the rod response, one sees essentially a graded current proportional to the probability of a channel being open. So the population of channels in the rod cell membrane produces a current that depends continuously on the concentration of cGMP. Alternatively, the noise variance that is associated with the random binary variable open/closed has been spread over a very broad bandwidth, so that in the frequency range of interest (the single-photon response is on a time scale of ~ 1 s) the noise is much reduced. This idea is made precise in the following problem, which you can think of as an introduction to the analysis of noise in chemical systems where molecules fluctuate among multiple states.

---

24. Siemens are the units of conductance, which are inverse to units of resistance, ohms. In the old days, this inverse of resistance had the rather cute unit mho (℧, pronounced "moe," like the Stooge).

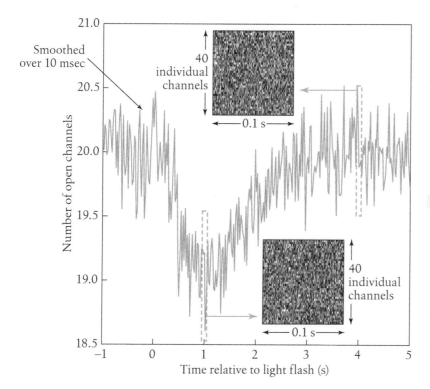

FIGURE 2.30

Forty flickering channels. A small simulation of 40 channels that open and close at random, with transition rates on the millisecond time scale. The response to a flash of light is a change in the probability of being in the open state, which lasts for ~2 s. Insets show the states of all individual channels, black for closed and white for open; the main figure shows the number of open channels averaged over a 10 ms window.

Problem 25: Flickering channels. Imagine a channel that has two states, open and closed. There is a rate k_{open} at which the molecule makes transitions from the closed to the open state, and conversely there is a rate k_{close} at which the open channels transition to the closed state. If we write the number of open channels as n_{open}, and similarly for the number of closed channels, then the deterministic kinetic equations are

$$\frac{dn_{open}}{dt} = k_{open}n_{closed} - k_{close}n_{open}, \tag{2.97}$$

$$\frac{dn_{close}}{dt} = k_{close}n_{open} - k_{open}n_{closed}, \tag{2.98}$$

or, because $n_{open} + n_{closed} = N$, the total number of channels,

$$\frac{dn_{open}}{dt} = k_{open}(N - n_{open}) - k_{close}n_{open} \tag{2.99}$$

$$= -(k_{open} + k_{close})n_{open} + k_{open}N. \tag{2.100}$$

For a single channel molecule, these kinetic equations should be interpreted as saying that an open channel has a probability $k_{close}dt$ of making a transition to the closed state in a small time dt, and conversely a closed channel has a probability $k_{open}dt$ of making a transition to the open state. We give a fuller account of noise in chemical systems in Chapter 4, but for now you should explore this simplest of examples.

(a) If we have a finite number of channels, then really the number of channels that make the transition from the closed to the open state in a small window dt is a random number. What is the mean number of these closed → open transitions? What is the mean number of open → closed transitions? Use your results to show that macroscopic kinetic equations, such as Eqs. (2.97) and (2.98), should be understood as equations for the mean numbers of open and closed channels:

$$\frac{d\langle n_{\text{open}}\rangle}{dt} = k_{\text{open}}\langle n_{\text{closed}}\rangle - k_{\text{close}}\langle n_{\text{open}}\rangle, \tag{2.101}$$

$$\frac{d\langle n_{\text{closed}}\rangle}{dt} = k_{\text{close}}\langle n_{\text{open}}\rangle - k_{\text{open}}\langle n_{\text{closed}}\rangle. \tag{2.102}$$

(b) Assuming that all channels make their transitions independently, what is the variance in the number of closed → open transitions in the small window dt? What is it in the number of open → closed transitions? Are these fluctuations in the number of transitions independent of one another?

(c) Show that your results in part (b) can be summarized by saying that the change in the number of open channels during the time dt obeys

$$n_{\text{open}}(t + dt) - n_{\text{open}}(t) = dt[k_{\text{open}}n_{\text{closed}} - k_{\text{close}}n_{\text{open}}] + \eta(t), \tag{2.103}$$

where $\eta(t)$ is a random number that has zero mean and a variance

$$\langle \eta^2(t)\rangle = dt[k_{\text{open}}n_{\text{closed}} + k_{\text{close}}n_{\text{open}}]. \tag{2.104}$$

Explain why the values of $\eta(t)$ and $\eta(t')$ are independent if $t \neq t'$.

(d) This discussion should remind you of the description of Brownian motion using a Langevin equation, in which the deterministic dynamics are supplemented by a random force that describes molecular collisions. In this spirit, show that, in the limit $dt \to 0$, you can rewrite your results in part (c) to give a Langevin equation for the number of open channels,

$$\frac{dn_{\text{open}}}{dt} = -(k_{\text{open}} + k_{\text{close}})n_{\text{open}} + k_{\text{open}}N + \zeta(t), \tag{2.105}$$

where

$$\langle \zeta(t)\zeta(t')\rangle = \delta(t - t')[k_{\text{open}}n_{\text{closed}} + k_{\text{close}}n_{\text{open}}]. \tag{2.106}$$

In particular, if the noise is small, show that $n_{\text{open}} = \langle n_{\text{open}}\rangle + \delta n_{\text{open}}$, where

$$\frac{d\delta n_{\text{open}}}{dt} = -(k_{\text{open}} + k_{\text{close}})\delta n_{\text{open}} + \zeta_s(t), \tag{2.107}$$

$$\langle \zeta_s(t)\zeta_s(t')\rangle = 2k_{\text{open}}\langle n_{\text{closed}}\rangle\delta(t - t'). \tag{2.108}$$

(e) Solve Eq. (2.107) to show that

$$\langle \delta n_{\text{open}}^2\rangle = Np_{\text{open}}(1 - p_{\text{open}}), \tag{2.109}$$

$$\langle \delta n_{\text{open}}(t)\delta n_{\text{open}}(t')\rangle = \langle \delta n_{\text{open}}^2\rangle \exp\left[-\frac{|t - t'|}{\tau_{\text{c}}}\right], \tag{2.110}$$

where the probability of a channel being open is $p_{\text{open}} = k_{\text{open}}/(k_{\text{open}} + k_{\text{close}})$, and the correlation time $\tau_{\text{c}} = 1/(k_{\text{open}} + k_{\text{close}})$. Explain how the result for the variance $\langle \delta n_{\text{open}}^2\rangle$ could be derived more directly.

(f) Give a critical discussion of the approximations involved in writing down these Langevin equations. In particular, in the case of Brownian motion of a particle subject to ordinary viscous drag, the Langevin force has a Gaussian distribution. Is that true here?

Problem 26: Averaging out the noise.   Consider a random variable, such as n_{open} in the previous problem, for which the noise has exponentially decaying correlations, as in Eq. (2.110). Imagine that we average over a window of duration τ_{avg} to form a new variable:

$$z(t) = \frac{1}{\tau_{\text{avg}}} \int_0^{\tau_{\text{avg}}} d\tau \, \delta n_{\text{open}}(t - \tau). \qquad (2.111)$$

Show that, for $\tau_{\text{avg}} \gg \tau_c$, the variance of z is smaller than the variance of δn_{open} by a factor of τ_{avg}/τ_c. Give some intuition for why this is true (e.g., how many statistically independent samples of n_{open} will you see during the averaging time?). What happens if your averaging time is shorter?

I think this example is fascinating, because evolution has selected for very fast channels to be present in a cell that signals very slowly. Our genome (as well as those of many other animals) codes for hundreds of different types of channels—thousands once one includes the possibility of alternative splicing. These different channels differ, among other things, in their kinetics. In the fly retina, for example, the dynamics of visual inputs looking straight ahead are very different from those looking to the side. In fact the receptor cells that look in these different directions have different kinds of channels—the faster channels respond to the more rapidly varying signals. In the vertebrate rod, signals are very slow, but the channels are fast, which makes sense only if the goal is to suppress noise.

Having understood a bit about the channels, let's take one step back and see how these channels respond to cGMP. Experimentally, with the rod outer segment sucked into the pipette for measuring current, one can break off the bottom of the cell and make contact with its interior, so that concentrations of small molecules inside the cell will equilibrate with concentrations in the surrounding solution. Because the cell makes cGMP from GTP, if we remove GTP from the solution, then there is no source other than the one that we provide, and now we can map current versus concentration. The results of such an experiment are shown in Fig. 2.31. The current I depends on the cGMP concentration G as

$$I = I_{\text{max}} \frac{G^n}{G^n + G_{1/2}^n}, \qquad (2.112)$$

with $n \approx 3$; I_{max} denotes the maximal current, and $G_{1/2}$ is the cGMP concentration at which the current is half maximal. Hill functions of the form shown in Eq. (2.112) often are interpreted to mean that n molecules bind together and trigger the output measured. This idea of cooperativity among multiple binding events is a common theme in many biological systems, and so it is worth taking a small detour to build our intuition.

To understand cooperative interactions in the binding of multiple ligands,[25] it is useful to start at the beginning, with the binding of a single ligand, especially as

---

25. "Ligand" is jargon for the molecule that binds to the receptor. In many cases the distinction is natural (e.g., at the surface of the cell, where the ligand is outside the cell and the receptor is embedded in the membrane). Here the channel is the receptor, and cGMP is the ligand.

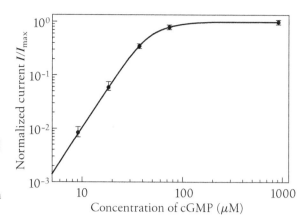

FIGURE 2.31

Current through the rod cell membrane as a function of cyclic guanosine monophosphate (cGMP) concentration. The fit is to Eq. (2.112), with $n = 2.9 \pm 0.1$ and $G_{1/2} = 45 \pm 4 \, \mu$M. Redrawn from Rieke and Baylor (1996).

many physics students do not have much experience with problems categorized as "chemistry." Suppose that we have a receptor molecule R to which some smaller ligand molecule L can bind. For simplicity let there be just the two states, R with its binding site empty, and RL with the binding site filled by an L molecule, and let us assume that every binding event is independent, so the different receptor molecules do not interact. To study the dynamics of this system, we keep track of the number of receptors in state R and the number in state RL; these numbers, n_R and n_{RL}, respectively, must add up to give the total number of receptors, N. We assume that N is large enough that we can write continuous equations for the kinetics of binding and unbinding. Alternatively, as we shall see very soon, we can think in terms of probabilities for a single receptor to be in different states.

The rate at which empty sites get filled ($R \rightarrow RL$) must be proportional to the number of empty sites and to the concentration c of the ligand. The rate at which filled sites become empty should just be proportional to the number of filled sites. Thus, we have

$$\frac{dn_{RL}}{dt} = k_+ c n_R - k_- n_{RL}, \tag{2.113}$$

where k_+ is the rate constant for binding and k_- is the rate constant for unbinding; note that these constants have different units. Because $n_R + n_{RL} = N$, this equation becomes

$$\frac{dn_{RL}}{dt} = k_+ cN - (k_- + k_+ c)n_{RL}. \tag{2.114}$$

The equilibrium state is reached when

$$n_{RL} = N \frac{k_+ c}{k_- + k_+ c}. \tag{2.115}$$

The fraction n_{RL}/N can also be interpreted microscopically as the probability that one receptor will be the state RL,

$$P_{RL} = \frac{k_+ c}{k_- + k_+ c} = \frac{c}{K + c}, \tag{2.116}$$

where the equilibrium constant (or dissociation constant) $K = k_-/k_+$.

From statistical mechanics, if we have a molecule that can be in two states, we should calculate the probability of it being in these states by knowing the energy of each state and using the Boltzmann distribution. Importantly, what we mean by "state," especially when discussing large molecules, often is a large group of microscopic configurations. Thus, saying that there are two states R and RL really means that we can partition the phase space of the system into two regions, and these regions are what we label as R and RL. Then what matters is not the energy of each state but the free energy. The free energy of the state R has one component from the receptor molecule itself, F_R, plus a component from the ligand molecules in solution. In the transition $R \to RL$, the free energy of the receptor changes to F_{RL}, and the free energy of the solution changes, because one molecule of the ligand is removed. The change in free energy when we add one molecule to the solution defines the chemical potential $\mu(c)$. Thus, up to an arbitrary zero of energy, we can consider the free energy of the two states to be F_R and $F_{RL} - \mu(c)$. Then the probability of being in the state RL is given by the Boltzmann distribution,

$$P_{RL} = \frac{1}{Z} \exp\left(-\frac{F_{RL} - \mu(c)}{k_B T}\right), \tag{2.117}$$

where the partition function Z is given by the sum of the Boltzmann factors over both available states:

$$Z = \exp\left(-\frac{F_R}{k_B T}\right) + \exp\left(-\frac{F_{RL} - \mu(c)}{k_B T}\right). \tag{2.118}$$

Putting the terms together, we have

$$P_{RL} = \frac{\exp\left[-(F_{RL} - \mu(c))/(k_B T)\right]}{\exp\left[-F_R/(k_B T)\right] + \exp\left[-(F_{RL} - \mu(c))/(k_B T)\right]} \tag{2.119}$$

$$= \frac{\exp\left[\mu(c)/k_B T\right]}{\exp\left[-(F_R - F_{RL})/(k_B T)\right] + \exp\left[\mu(c)/(k_B T)\right]}. \tag{2.120}$$

The only place where the ligand concentration appears is in the chemical potential $\mu(c)$. For this result to be consistent with the result from analysis of the kinetics in Eq. (2.116), we must have $e^{\mu(c)/k_B T} \propto c$, and you may recall that when concentrations are low—as in ideal gases and in ideal solutions—it is a standard result that

$$\mu(c) = k_B T \ln(c/c_0), \tag{2.121}$$

where c_0 is some reference concentration. Then we can also identify the equilibrium constant as

$$K = c_0 \exp\left(-\frac{F_{\text{bind}}}{k_B T}\right), \tag{2.122}$$

where $F_{\text{bind}} = F_R - F_{RL}$ is the change in free energy when the ligand binds to the receptor.

Now suppose we have a receptor to which two ligands can bind; there are four states, which we can think of as 00, 10, 01, and 11. If each binding event is identical and independent, then the free energies of these states are

$$F_{00} = F_R, \tag{2.123}$$

$$F_{01} = F_{10} = F_R - F_{\text{bind}} - \mu(c), \tag{2.124}$$

$$F_{11} = F_R - 2F_{\text{bind}} - 2\mu(c). \tag{2.125}$$

If we calculate, for example, the probability that both binding sites are occupied (i.e., that the molecule is in the state 11), we have

$$P_{11} = \frac{1}{Z} e^{-F_{11}/k_B T} \tag{2.126}$$

$$= \frac{\exp\left[-\frac{F_R - 2F_{\text{bind}} - 2\mu(c)}{k_B T}\right]}{\exp\left[-\frac{F_R}{k_B T}\right] + 2\exp\left[-\frac{F_R - F_{\text{bind}} - \mu(c)}{k_B T}\right] + \exp\left[-\frac{F_R - 2F_{\text{bind}} - 2\mu(c)}{k_B T}\right]} \tag{2.127}$$

$$= \frac{(c/K)^2}{1 + 2(c/K) + (c/K)^2} = \left(\frac{c}{c + K}\right)^2. \tag{2.128}$$

Thus, the probability of both sites being occupied is just the square of the probability that a single binding site will be occupied, as in Eq. (2.116). This makes sense, because we assumed that binding to each site is an independent event.

Problem 27: Counting bound molecules.   Rather than counting the fraction of molecules in the doubly bound state, count the number of ligands bound. Show that it is just $2 \times c/(c + K)$, and explain why.

In many cases we see experimentally that binding of multiple ligands to a protein molecule are not independent events. Suppose that we again have two binding sites, but the doubly bound state is stabilized (for as yet unspecified reasons) by an extra energy Δ. Then if we calculate the fraction of binding sites occupied, we have

$$f = \frac{1}{2}\left[P_{01} + P_{10} + 2P_{11}\right] \tag{2.129}$$

$$= \frac{1}{2} \frac{2\exp\left[-\frac{F_R - F_{\text{bind}} - \mu(c)}{k_B T}\right] + 2\exp\left[-\frac{F_R - 2F_{\text{bind}} - 2\mu(c) - \Delta}{k_B T}\right]}{\exp\left[-\frac{F_R}{k_B T}\right] + 2\exp\left[-\frac{F_R - F_{\text{bind}} - \mu(c)}{k_B T}\right] + \exp\left[-\frac{F_R - 2F_{\text{bind}} - 2\mu(c) - \Delta}{k_B T}\right]} \tag{2.130}$$

$$= \frac{c/K + J(c/K)^2}{1 + 2(c/K) + J(c/K)^2}, \tag{2.131}$$

where $J = \exp(\Delta/k_B T)$. Results are shown in Fig. 2.32. As the interaction energy increases, the binding sites can be occupied at lower concentration, but more importantly the steepness of the "switch" from empty to full sites is more abrupt. This abruptness is the signature of cooperativity.

What happens if the interactions are very strong, so that Δ becomes large (compared to $k_B T$) and hence the dimensionless J becomes even larger? If we define $\tilde{K} = K/\sqrt{J}$,

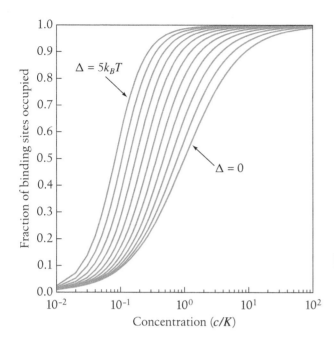

FIGURE 2.32

Cooperative binding, with two binding sites that interact. Curves show the predicted fraction of binding sites versus concentration for different values of the interaction energy Δ, as calculated from Eq. (2.131).

then we can rewrite the probability that a binding site is occupied from Eq. (2.131) as

$$f = \frac{J^{-1/2}(c/\tilde{K}) + (c/\tilde{K})^2}{1 + 2J^{-1/2}(c/\tilde{K}) + (c/\tilde{K})^2}. \tag{2.132}$$

As J becomes large, we see that the probability starts to approach $f = 1$ as $c/\tilde{K}$ grows; we can get within 1% of $f = 1$ by confining our attention to $0 < c/\tilde{K} < 10$. But so long as we have finite values for $c/\tilde{K}$, the terms $\propto J^{-1/2}$ become negligible as J becomes large. Thus, when the interactions are strong, the expression for the probability of binding becomes

$$\lim_{\Delta/k_B T \to \infty} f = \frac{(c/\tilde{K})^2}{1 + (c/\tilde{K})^2}, \tag{2.133}$$

which is a Hill function with $n = 2$.

The classic example of these ideas is the oxygen-binding protein hemoglobin in our blood. We now know that hemoglobin has four protein subunits, each of which has an iron atom that can bind one oxygen molecule. As Hill recognized in the early part of the twentieth century, the fraction of sites with bound oxygen behaves more nearly as if all four molecules had to bind together, so that

$$f = \frac{c^n}{c^n + K^n}, \tag{2.134}$$

with $n = 4$. As shown in Fig. 2.33, the binding is now sigmoidal, or more nearly switchlike at larger n. Because the natural quantity in statistical mechanics is the chemical potential and not the concentration, things look simpler on a logarithmic concentration axis. Few real systems are described exactly by the Hill model, but it is a

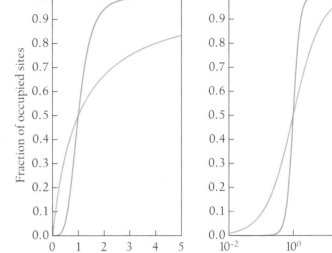

FIGURE 2.33

Cooperative binding in the Hill model. Blue lines show the predicted fraction of binding sites versus concentration when binding to each site is independent. Orange lines show the case of cooperative binding to four sites, as described by the Hill model in Eq. (2.134) with $n = 4$. At left, a linear concentration scale; at right, a logarithmic scale.

good approximation. We should also appreciate the power of Hill's intuition in seeing the connection of the sigmoidal binding curves to the number of protein subunits even before much was known about these molecules.

The Hill model suggests that there is some direct interaction between binding events that causes all the ligands to bind (or not to bind) simultaneously, which we can think of as a limiting case of the model above, with $\Delta \to \infty$. In some cases, including hemoglobin, there is little evidence for such a direct interaction. An alternative is to imagine that the whole system can be in two states. In the case of hemoglobin these states came to be called relaxed (R) and tense (T), but in other systems there are other natural choices. For the cGMP-sensitive channel in the rod cell membrane, the two natural states are channel open and channel closed, as shown schematically in Fig. 2.34.

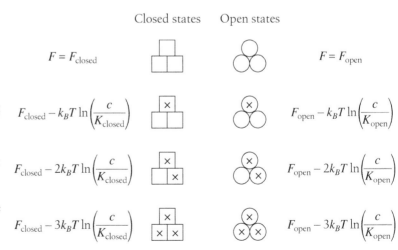

FIGURE 2.34

A model for binding of cyclic guanosine monophosphate (cGMP) to the channels in rod cells. Cooperativity arises not from direct interactions among the cGMP molecules but rather because the binding of each molecule contributes to stabilizing a different structure of the channel protein. In this case the two structures are just the open and closed states.

Let us assume that the channel can bind one, two, or three molecules of cGMP. If all the binding sites are empty, the free energies of the two states are F_{open} and F_{closed}. Given that the channel is closed, the binding of a single cGMP molecule lowers the energy by an amount F_{closed}^{bind}, but in addition this action takes one molecule out of the solution, and hence the free energy of the system also goes down by the chemical potential μ. So the total free energy of the state with the channel closed and one molecule bound is

$$F_{closed}(1) = F_{closed} - F_{closed}^{bind} - \mu \tag{2.135}$$

$$= F_{closed} - F_{closed}^{bind} - k_B T \ln(c/c_0) \tag{2.136}$$

$$= F_{closed} - k_B T \ln\left(\frac{c}{K_{closed}}\right), \tag{2.137}$$

and similarly for the open state,

$$F_{open}(1) = F_{open} - k_B T \ln\left(\frac{c}{K_{open}}\right). \tag{2.138}$$

The important point is that the binding energies to the open and closed states are different. Thus, using detailed balance, as the cGMP molecules bind, they will shift the equilibrium between open and closed. The two-state model was proposed, in a slightly different context, by Monod, Wyman, and Changeux (MWC). They made the simplifying assumption that the only source of cooperativity among the binding events was this shifting of equilibria, so that if the target protein is in one state, each binding event remains independent, and then the free energies work out as in Fig. 2.34.

Problem 28: Cooperativity in the MWC model. Show that the model in Fig. 2.34 is equivalent to the statement that the free-energy difference between open and closed states has a term proportional to the number of cGMP molecules bound. What is this proportionality constant in terms of the other parameters? Can you explain the connection between these two points of view on the model?

It is a useful exercise to work out the statistical mechanics of the MWC model more generally. The partition function has two classes of terms, coming from the two states of the protein. In each state, we have to sum over the occupied and unoccupied states of each binding site, but this is relatively easy, because the sites are independent. In the notation of Fig. 2.34, we find

$$Z = Z_{open} + Z_{closed}, \tag{2.139}$$

$$Z_{open} = \exp\left(-\frac{F_{open}}{k_B T}\right)\left(1 + \frac{c}{K_{open}}\right)^n, \tag{2.140}$$

$$Z_{closed} = \exp\left(-\frac{F_{closed}}{k_B T}\right)\left(1 + \frac{c}{K_{closed}}\right)^n, \tag{2.141}$$

where in the case of the cGMP-gated channels, $n = 3$. The probability of being in the open state is then

$$P_{\text{open}} = \frac{Z_{\text{open}}}{Z_{\text{open}} + Z_{\text{closed}}} \tag{2.142}$$

$$= \frac{\left(1 + c/K_{\text{open}}\right)^n}{\left(1 + c/K_{\text{open}}\right)^n + L\left(1 + c/K_{\text{closed}}\right)^n}, \tag{2.143}$$

where $k_B T \ln L = F_{\text{open}} - F_{\text{closed}}$ is the free-energy difference between open and closed states in absence of ligand binding. In the limit that binding is much stronger to the open state, $K_{\text{open}} \ll K_{\text{closed}}$, this equation simplifies to

$$P_{\text{open}} = \frac{\left(1 + c/K_{\text{open}}\right)^n}{L + \left(1 + c/K_{\text{open}}\right)^n} \tag{2.144}$$

$$= \frac{1}{1 + \exp\left[\theta - n \ln(1 + c/K_{\text{open}})\right]}, \tag{2.145}$$

where $\theta = \ln L$. This result is similar to the Hill model, but it is a little different in detail. Distinguishing the models from the equilibrium data alone is difficult, but clearly the MWC model predicts that binding has an extra kinetic step in which the protein makes the transition between its two states; if we are lucky we can catch the system after the first ligand molecules have bound but before this change in protein structure. Indeed, such experiments were critical to understanding the mechanism of cooperativity in hemoglobin.[26] For channels, all we can see is the open $\rightarrow$ closed transition, but if we analyze the kinetics at varying ligand concentrations, we can see that ligand binding and switching are separate events, as suggested by the MWC model.

Problem 29: Details of the MWC model.   Fill in the steps to Eqs. (2.139)–(2.141). Then compare the Hill model with MWC. Show that for $c \gg K_{\text{open}}$, Eq. (2.145) reduces to Eq. (2.134). What about at $c \ll K_{\text{open}}$? The MWC model, even in the limit $K_{\text{open}} \ll K_{\text{closed}}$, has one more

---

26. The idea that binding of oxygen is coupled to changes in protein structure has its origins in a remarkable story. Max Perutz, who would eventually solve the structure of hemoglobin by X-ray diffraction, was looking at very small crystals of the protein under a microscope. This was decades before the structure could be solved, and indeed not long after Bernal and Hodgkin had realized that to obtain clean diffraction patterns from protein crystals, one had to leave them bathed in the solution from which they crystallized. As Perutz looked at the tiny crystals, some air leaked in underneath the cover slip, and as this happened the crystals shattered, starting with the ones nearest the place where the air entered and moving across his field of view like a wave. Perutz realized what had happened: oxygen was diffusing into the crystals and binding to the protein, and the binding produced a change in structure large enough to break the (weak) bonds that held neighboring proteins in their crystalline arrangement. Thus, long before he could see the structures, he knew that binding of a tiny oxygen molecule to a large protein could change the structure of the protein substantially. After decades of work, he was able to see this by analysis of the X-ray diffraction data.

parameter than the Hill model; what does this freedom mean for the class of functions that the MWC model can realize?

So, now we understand how currents flow across the rod cell membrane and how they can be modulated by changes in the intracellular concentration of cGMP. The remaining problem is to connect the isomerization of rhodopsin with these concentration changes. To start, let's think about how enzymes work, in particular, the enzymes that synthesize and degrade the cGMP molecules.

Let's denote the enzyme molecule by E. This enzyme has a substrate S, which is the molecule that starts the reaction; for example, in the synthesis of cGMP, the substrate is GTP. The goal of the reaction is to turn the substrate into the product molecule P. There may be more than one substrate, and more than one product—in making cGMP from GTP an extra two phosphates must be floating around—but for now we ignore this. For the enzyme to do anything to the substrate, the two have to bind to each other, so the minimal scheme for the reaction is

$$E + S \rightleftharpoons E \cdot S \rightarrow E + P, \tag{2.146}$$

where $E \cdot S$ is the enzyme-substrate complex, and for simplicity we assume that once this complex is formed, products can be made and released in a single irreversible step. There are three reactions here, each with an independent parameter describing the rate:

$$E + S \xrightarrow{k_+} E \cdot S, \tag{2.147}$$

$$E \cdot S \xrightarrow{k_-} E + S, \tag{2.148}$$

$$E \cdot S \xrightarrow{V_{\max}} E + P. \tag{2.149}$$

The notation for the rate constants is a matter of convention: binding of substrate to the enzyme pushes the reaction forward and occurs with rate k_+, the reverse reaction occurs with k_-, and we will see that the rate $V_{\max}$ is the fastest rate at which product molecules can be generated per enzyme molecule.

To see the consequences of these reactions, we should write out the differential equations that describe the changes in the number of molecules in the different states.[27] Let's define N_E as the number of free enzyme (E) molecules, N_{ES} as the number of enzyme substrate complexes, $[S]$ as the concentration of the free substrate, and N_P as the number of product molecules. Then Eq. (2.149) tells us that

$$\frac{dN_P}{dt} = V_{\max}N_{ES}, \tag{2.150}$$

and Eqs. (2.147)–(2.149) combine to yield

$$\frac{dN_{ES}}{dt} = k_+ N_E[S] - k_- N_{ES} - V_{\max}N_{ES}. \tag{2.151}$$

27. Again, I am going slowly here, because of variability in how much chemistry the typical physicist remembers. Forgive me if this exposition is a bit pedantic.

There is a conservation law,

$$N_E + N_{ES} = N_E^{\text{total}}, \tag{2.152}$$

because the enzyme molecules are not created or destroyed by these reactions, so we can eliminate N_E to give

$$\frac{dN_{ES}}{dt} = k_+ N_E^{\text{total}}[S] - \left(k_- + V_{\text{max}} + k_+[S]\right) N_{ES}. \tag{2.153}$$

If the concentration of enzyme molecules is small, then even if they are very active, it is hard for their activity to change the concentration of substrate quickly, so it is plausible that the dynamics of N_{ES} in Eq. (2.153) will reach a steady state before $[S]$ changes significantly. Then we find

$$N_{ES} = N_E^{\text{total}} \frac{k_+[S]}{k_- + V_{\text{max}} + k_+[S]}, \tag{2.154}$$

and hence the overall rate of generating products (or depleting substrate) becomes

$$\frac{dN_P}{dt} = V_{\text{max}} N_E^{\text{total}} \frac{k_+[S]}{k_- + V_{\text{max}} + k_+[S]} = V_{\text{max}} N_E^{\text{total}} \frac{[S]}{[S] + K_M}, \tag{2.155}$$

where K_M is the Michaelis constant,

$$K_M = \frac{k_- + V_{\text{max}}}{k_+}, \tag{2.156}$$

and this whole scheme is referred to as Michaelis-Menten kinetics. Notice that if the catalytic step is slow, so that $V_{\text{max}} \ll k_-$, then K_M is just the equilibrium constant for the binding of substrate to the enzyme.

The behavior of an enzyme according to Michaelis-Menten kinetics is shown in Fig. 2.35. There is a proportional approach to the maximum rate as the substrate concentration is increased, so that when $[S] \sim 10 K_M$ the rate is still $\sim$10% below its maximum. At very low substrate concentrations, the rate is proportional to $[S]$, and the coefficient is

$$\lim_{[S] \ll K_M} \frac{dN_P}{dt} = N_E^{\text{total}} \frac{V_{\text{max}}}{K_M}[S] \leq N_E^{\text{total}} k_+[S]. \tag{2.157}$$

This behavior makes sense, because at low substrate concentrations the reaction is "waiting" for the enzyme and substrate to find each other, which happens with the rate $k_+[S]$ per enzyme molecule. Because molecules find each other through diffusion, this rate also can't be faster than it would be in a simplified model in which the reaction happens when the random walks taken by the two molecules intersect. As explained in more detail in Section 4.2, if molecules present targets of linear size a and have diffusion constants D, then the maximum $k_+ \sim Da$, which is also what one could guess from dimensional analysis.

Now let's use these ideas to think about the cGMP concentration in the rod cell. There are two enzymes, one of which synthesizes cGMP using GTP as a substrate S, and one of which degrades cGMP, so the cGMP molecule itself is the substrate. Let G stand for the concentration of cGMP, as before, and let the reaction be occurring in an

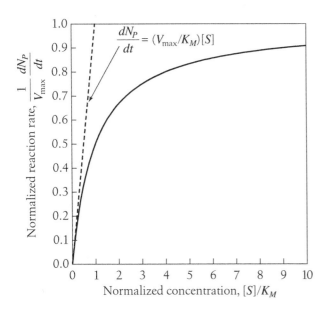

$$\frac{dN_P}{dt} = (V_{max}/K_M)[S]$$

FIGURE 2.35

Michaelis-Menten kinetics, following Eq. (2.155). The normalized reaction rate for a single enzyme molecule is plotted versus the normalized concentration of the substrate. The dashed line shows the asymptotic behavior at small substrate concentrations.

effective volume Ω, so we can convert between numbers of molecules and concentrations. Then, if both the synthesis and the degradation enzymes obey Michaelis-Menten kinetics, we have

$$\Omega \frac{dG}{dt} = V_s N_s^{\text{total}} \frac{[S]}{[S] + K_s} - V_d N_d^{\text{total}} \frac{G}{G + K_d}, \tag{2.158}$$

where subscript s refers to the synthesis enzyme GC and d refers to the degradation enzyme PDE. In steady state, the synthesis and degradation rates balance. Importantly, having a constant concentration of cGMP in the dark does not mean that nothing is happening. Rather, cGMP molecules are continuously being synthesized and degraded, and each synthesis reaction requires one molecule of GTP, which releases $\sim 1/2$ eV of energy. To keep this reaction working thus requires considerable energy expenditure. But the total rate of energy usage depends on parameters, and the cell could maintain the same steady state concentration by letting V_s and V_d both become smaller in proportion to each other. Why doesn't this happen?

Suppose that some signal in the system causes a change in G. How quickly can the system respond? We can answer this question by linearizing around the steady state, $G = G_{ss} + \delta G$, with δG small, so we obtain

$$\frac{d(\delta G)}{dt} = -V_d N_d^{\text{total}} \frac{G_{ss}}{G_{ss} + K_d} \cdot \frac{1}{\Omega G_{ss}} \left(1 - \frac{G_{ss}}{G_{ss} + K_d}\right) \delta G. \tag{2.159}$$

The combination ΩG_{ss} is the total number of molecules of cGMP in steady state. The combination $V_d N_d^{\text{total}} G_{ss}/(G_{ss} + K_d)$ gives the total number of molecules per second being degraded. The ratio of these two quantities defines an effective lifetime for each cGMP molecule in the cell, τ_{eff}. Thus, there is a time constant τ for the G to relax to its steady state, or equivalently to respond to external perturbations,

$$\frac{d(\delta G)}{dt} = -\frac{1}{\tau} \delta G, \tag{2.160}$$

and this response time is bounded by the effective lifetime, $\tau > \tau_{\text{eff}}$. Thus, such a system can respond only as quickly as the typical molecule is degraded, and this degradation must be balanced by synthesis: the rate of going around the cycle of synthesis and degradation sets the maximum rate at which the system can respond.

To get the quickest response given the rate of cycling, and hence of energy dissipation, the system should operate in the regime where $G \ll K_d$. In contrast, to get the most out of a limited number of synthesis enzymes, we should have $[S] \gg K_s$. But we should be more careful about counting the total number of enzymes. In the same way that channels can be in open and closed states, and hemoglobin can be in R and T states, enzymes can be in active and inactive states. The switch could be triggered by the (non-covalent) binding of another molecule to the enzyme or by the covalent modification of the enzyme (e.g., the attachment of a phosphate group drawn from ATP). So if we count active enzymes and work in the limits that seem to enhance efficiency, we have

$$\Omega \frac{dG}{dt} = V_s N_s^{\text{active}} - \frac{V_d}{K_d} N_d^{\text{active}} G. \tag{2.161}$$

Now the steady state is simple, $G_{\text{ss}} = K_d V_s N_s / (V_d N_d)$, and the relaxation time $\tau = K_d \Omega / (V_d N_d^{\text{active}})$.

The activation of rhodopsin by photon absorption reduces the current flowing across the rod cell membrane, which means that it reduces the concentration of cGMP. In principle, activated rhodopsin could act by reducing the number of active synthesis enzymes, but in fact it operates by increasing the number of active degradation enzymes. These enzymes, the PDEs, are associated with the disks, whereas the GC enzymes diffuse throughout the cell. A change in the number of active PDEs produces a change in the number of cGMP molecules:

$$\Delta N_G \equiv \Omega \Delta G_{\text{ss}} = -\frac{\Omega G_{\text{ss}}}{N_d^{\text{active}}} \Delta N_d^{\text{active}}. \tag{2.162}$$

The steady state gain $g_0 \equiv |\Delta N_G / \Delta N_d^{\text{active}}|$ depends on the concentration of active PDEs in the dark: the lower this concentration, the more molecules of cGMP are degraded per molecule of PDE that is activated. But the time constant of the response also depends on the concentration of PDEs in the same way. As a result, there is a quantity characterizing the performance of the system that is independent of how many PDE molecules the cell makes: the product of the gain and the bandwidth $1/\tau$ of the response, that is, the product of gain and "speed,"

$$g_0 \tau^{-1} = \frac{V_d}{K_d} G_{\text{ss}}. \tag{2.163}$$

This gain-bandwidth product is often used in characterizing manmade amplifiers. Here we see that it is determined by V_d / K_d, which is a property of the PDE enzyme. But it is not a property that can be infinitely adjusted; as noted above, this combination of parameters is limited by the rate at which cGMP molecules can find the enzyme via diffusion. With diffusion constants for free cyclic nucleotides in the range of $D \sim 4 \times 10^{-6} \text{ cm}^2/\text{s}$, and target sizes on the enzyme in the range of $a \sim 1 \text{ nm}$, cGMP concentrations of $G_{\text{ss}} \sim 1-10 \, \mu\text{M}$ suggest that the physical limit to the gain-bandwidth

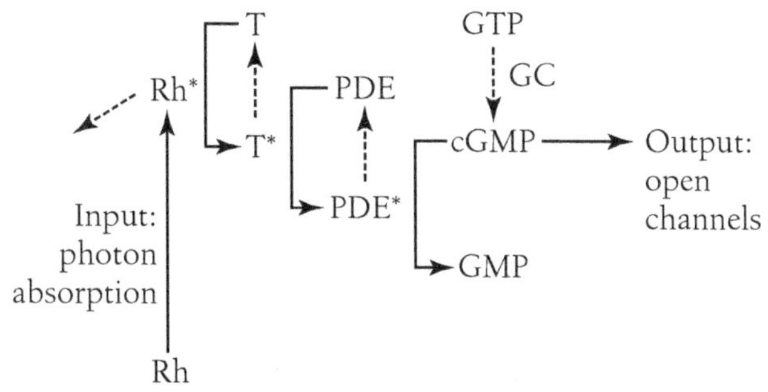

FIGURE 2.36

The cascade leading from photon absorption to ionic current flow in rod photoreceptors. Solid lines indicate forward steps that generate gain; dashed lines are the backward steps that shut off the process, and asterisks deonte activated states of the molecules. Rh stands for rhodopsin. T is the transducin molecule, a member of the broad class of G-proteins that couple receptors to enzymes. PDE is the enzyme phosphodiesterase, named for the particular bond that it cuts when it degrades cyclic guanosine monophosphate (cGMP) into GMP. GC is the guanylate cyclase that synthesizes cGMP from guanosine triphosphate (GTP).

product is $(g_0\tau^{-1})_{\max} \sim DaG_{ss} \sim 240–2400 \text{ s}^{-1}$. The rod cell operates on a time scale of $\tau \sim 1$ s. Thus, if the enzymes can operate near their diffusion-limited rates, then the transformation from PDE molecules to cGMP molecules can reach a gain on the order of 10^3.

In fact the path from rhodopsin to PDE involves an intermediate, and hence another opportunity for gain, as shown in Fig. 2.36. When the rhodopsin molecule changes its structure in response to photon absorption, it acts as a catalyst to change the structure of another protein, transducin (T). The activated T in turn activates the PDE, which breaks down cGMP. Finally, as we have seen, cGMP binds to channels in the cell membrane and opens the channels, allowing current to flow; breaking down the cGMP thus decreases the number of open channels and decreases the current.

It is useful to compare the rod cell to laboratory photon-counting devices. In a photomultiplier, for example, photon absorption results in the ejection of a primary photoelectron, and then the large electric field accelerates this electron, so that when it hits the next metal plate, it ejects many electrons, and the process repeats until at the output the number of electrons is sufficiently large that it constitutes a macroscopic current. Thus, the photomultiplier really is an electron multiplier. In the same way, the photoreceptor acts as a molecule multiplier, so that for one excited rhodopsin molecule there are many cGMP molecules degraded at the output of the enzymatic cascade.

This general scheme is ubiquitous in biological systems. Rhodopsin is a member of a family of proteins that share common structural features (seven alpha helices that span the membrane in which the protein is embedded) and act as receptors, usually activated by the binding of small molecules (e.g., hormones or odorants) rather than light. Proteins in this family interact with those from another family, the G proteins, of which T is an example, and the result of such interactions typically is the activation of yet another enzyme, often one that synthesizes or degrades a cyclic nucleotide. Cyclic nucleotides in turn are common intracellular messengers, not just opening ion channels but also activating or inhibiting a variety of enzymes. This universality of components means that understanding the mechanisms of photon counting in rod cells is not just a curiosity for physicists but an opportunity for us to provide a model for understanding an enormous range of biological processes.

Evidently generating gain in an enzymatic cascade is not a problem: we saw that even one step could generate a gain of 10^3. What, then, is the problem? It is not obvious that the mechanisms of gain won't also generate extra noise, which might be a problem. Indeed, there is evidence that the continuous background noise in rod cells results from the spontaneous activation of PDE. If we have more than one stage of gain, it should be possible to provide maximal insulation against noise by properly distributing the gain among the different stages.

Perhaps the most surprising aspect of the single-photon response in rods, however, is its reproducibility. If we look at the responses to dim light flashes and isolate those responses that correspond to a single photon (you have already done a problem to assess how easy or hard this is), we find that the amplitude of the response fluctuates by only $\sim$15–20%; see, for example, Fig. 2.37. To understand why this is surprising, we have to think about chemistry at the level of single molecules, specifically, the chemical reactions catalyzed by the single activated molecule of rhodopsin.

When we write that there is a rate k for a chemical reaction, what we mean is that for one molecule there is a probability per unit time k that the reaction will occur, as explored in Problem 25. Thus, when one molecule of rhodopsin is activated at time $t = 0$, if we imagine that deactivation is a simple chemical reaction, then the probability that the molecule is still active at time t obeys the usual kinetic equation

$$\frac{dp(t)}{dt} = -kp(t), \tag{2.164}$$

so that $p(t) = \exp(-kt)$. The probability density $P(t)$ that the molecule is active for exactly a time t is the probability that the molecule is still active at t times the probability per unit time of deactivation:

$$P(t) = kp(t) = k \exp(-kt). \tag{2.165}$$

This argument may seem pedantic, but it is important to be clear—and we will see that far from being obvious, there must be something wrong with this simple picture.

Given the probability density $P(t)$, we can calculate the mean and variance of the time spent in the active state:

$$\langle t \rangle \equiv \int_0^\infty dt\, P(t)\, t \tag{2.166}$$

$$= k \int_0^\infty dt\, \exp(-kt)t = 1/k; \tag{2.167}$$

$$\langle (\delta t)^2 \rangle \equiv \int_0^\infty dt\, P(t)\, t^2 - \langle t \rangle^2 \tag{2.168}$$

$$= k \int_0^\infty dt\, \exp(-kt)t^2 - 1/k^2 \tag{2.169}$$

$$= 2/k^2 - 1/k^2 = 1/k^2. \tag{2.170}$$

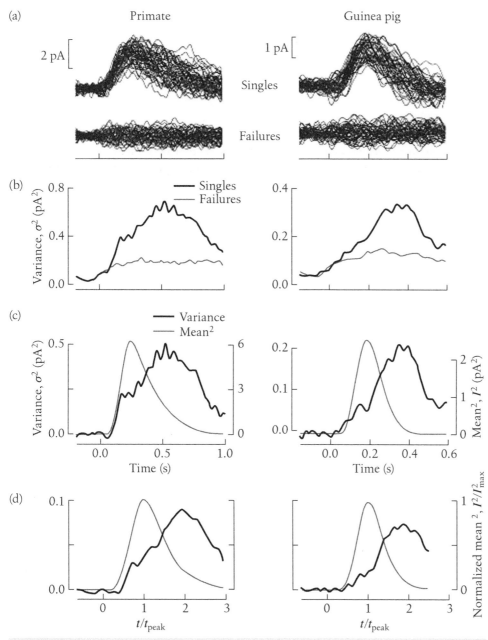

FIGURE 2.37

Reproducibility of the single-photon response. Reprinted from Field and Rieke (2002), with permission from Elsevier. (a) Examples of single-photon responses and failures from single mammalian rods. (b) Variances of the responses in panel (a). (c) Variance and square of mean response to one photon; variance in the response is defined as the difference in variance between responses and failures. (d) Mean of results as in panel (c) from eight primate rods and nine guinea pig rods; scales are normalized for each cell by the peak mean response and the time to peak. At the peak response, the relative variance is ~0.025, so the root-mean-square fluctuations are ~0.15.

Thus, we find that

$$\delta t_{\mathrm{rms}} \equiv \sqrt{\langle (\delta t)^2 \rangle} = 1/k = \langle t \rangle, \tag{2.171}$$

so that the root-mean-square fluctuations in the lifetime are equal to the mean.

How does this argument relate to the reproducibility of the single-photon response? The active rhodopsin molecule acts as a catalyst, activating T molecules. If this catalysis proceeds at some constant rate (presumably set by the time required for rhodopsin and T to find each other through diffusion in the disk membrane), then the number of activated T molecules is proportional to the time that rhodopsin spends in the active state—and hence we would expect that the number of active T molecules has root-mean-square fluctuations equal to the mean number. If the subsequent events in the enzymatic cascade again have outputs proportional to their input number of molecules, this variability will not be reduced, and the final output (the change in cGMP concentration) will again have relative fluctuations of order one, much larger than the observed 15–20%. If we measure the variance rather than the standard deviation, which seems more natural, we can't even claim to have an order of magnitude understanding of the reproducibility. To give an idea of the different possible solutions that people have considered, here I focus on simple versions of these ideas that we can explore analytically. At the end, we will look at the state of the relevant experiments.

One possibility is that although the lifetime of activated rhodopsin might fluctuate, the number of molecules at the output of the cascade fluctuates less because of saturation. For example, if each rhodopsin has access only to a limited pool of T molecules, most rhodopsins might remain active long enough to hit all the molecules in the pool. Let the total number of T molecules in the pool be N_{pool}, and let the number of activated Ts be n_T. When the rhodopsin is active, it catalyzes the conversion of inactive T molecules (of which there are $N_{\mathrm{pool}} - n_T$) into the active form at a rate r, so that (neglecting the discreteness of the molecules)

$$\frac{dn_T}{dt} = r(N_{\mathrm{pool}} - n_T). \tag{2.172}$$

If the rhodopsin molecule is active for a time t, then the number of activated Ts will be

$$n_T(t) = N_{\mathrm{pool}}[1 - \exp(-rt)]. \tag{2.173}$$

For small t the variations in t are converted into proportionately large variations in n_T, but for large t the saturation essentially cuts off this variation.

To be more precise, we can find the distribution of n_T by using the identity

$$P(n_T)dn_T = P(t)dt, \tag{2.174}$$

which applies when we have two variables that are related by a deterministic, invertible transformation. From Eq. (2.173) we have

$$t = -\frac{1}{r}\ln(1 - n_T/N_{\mathrm{pool}}), \tag{2.175}$$

and so, going through the steps explicitly:

$$P(n_{\mathrm{T}}) = P(t) \left| \frac{dn_{\mathrm{T}}}{dt} \right|^{-1} = \frac{k}{r N_{\mathrm{pool}}} \left(1 - \frac{n_{\mathrm{T}}}{N_{\mathrm{pool}}}\right)^{k/r-1}. \tag{2.176}$$

When the activation rate r is small, n_{T} always stays much less that N_{pool}, and the power law can be approximated as an exponential. When r is large, however, the probability distribution grows a power-law singularity at N_{pool}; for r finite this singularity is integrable, but as $r \to \infty$ it approaches a log divergence, which means that essentially all the weight will concentrated at N_{pool}. In particular, the relative variance of n_{T} vanishes as r becomes large, as promised.

This discussion has assumed that the limited number of target molecules is set, perhaps by some fixed structural domain. Depending on details, it is possible for such a limit to arise dynamically as a competition between diffusion and chemical reactions. In invertebrate photoreceptors (e.g., the flies we met in Fig. 2.10), there is actually a positive feedback loop in the amplifier that serves to ensure that each structural domain (which are more obvious in the fly receptor cells than in vertebrate receptor cells) fires a saturated, stereotyped pulse in response to each photon.

The next class of models are those that use feedback: if the output of the cascade is variable because the rhodopsin molecule does not "know" when to deactivate, why not count the molecules at the output and shut the rhodopsin molecule off when we reach some fixed count? When rhodopsin is active, it catalyzes the formation of some molecule (which might not actually be the T molecule itself) at rate r. Let the number of these output molecules be x, so that $x = rt$. Let's have the rate of deactivation of rhodopsin depend on x, so that instead of Eq. (2.164) we have

$$\frac{dp(t)}{dt} = k[x(t)]p(t). \tag{2.177}$$

For example, if deactivation is triggered by the cooperative binding of m x molecules (as in the discussion of cGMP-gated channels), we expect that

$$k[x] = k_{\mathrm{max}} \frac{x^m}{x_0^m + x^m}. \tag{2.178}$$

We can solve Eq. (2.177) and then recover the probability density for rhodospin lifetime as before:

$$p(t) = \exp\left(-\int_0^t d\tau \, k[x(\tau)]\right), \tag{2.179}$$

$$P(t) = k[x(t)] \exp\left(-\int_0^t d\tau \, k[x(\tau)]\right) \tag{2.180}$$

$$= k_{\mathrm{max}} \frac{x^m(t)}{x_0^m + x^m(t)} \exp\left(-k_{\mathrm{max}} \int_0^t d\tau \, \frac{x^m(t)}{x_0^m + x^m(t)}\right) \tag{2.181}$$

$$\approx k_{\mathrm{max}} \left(\frac{t}{t_0}\right)^m \exp\left[-\frac{k_{\mathrm{max}} t_0}{m+1} \left(\frac{t}{t_0}\right)^{m+1}\right], \tag{2.182}$$

where in the last step we identify $t_0 = x_0/r$ and assume that $t \ll t_0$.

To get a better feel for the probability distribution in Eq. (2.182), it is useful to rewrite it as

$$P(t) \approx k_{max} \exp\left[-G(t)\right], \tag{2.183}$$

$$G(t) = -m \ln\left(\frac{t}{t_0}\right) + \frac{k_{max}t_0}{m+1}\left(\frac{t}{t_0}\right)^{m+1}. \tag{2.184}$$

We can find the most likely value of the lifetime, $\bar{t}$, by minimizing G, setting the derivative to zero:

$$G'(t) = -\frac{m}{t} + k_{max}t_0 \cdot \frac{1}{t}\left(\frac{t}{t_0}\right)^{m+1}, \tag{2.185}$$

$$G'(t = \bar{t}) = 0 \Rightarrow k_{max}t_0 \cdot \frac{1}{\bar{t}}\left(\frac{\bar{t}}{t_0}\right)^{m+1} = \frac{m}{\bar{t}}, \tag{2.186}$$

$$\frac{\bar{t}}{t_0} = \left(\frac{m}{k_{max}t_0}\right)^{1/m}. \tag{2.187}$$

In particular, we see that for sufficiently large k_{max}, we have $\bar{t} \ll t_0$, consistent with the approximation above. What we really want to know is how sharp the distribution is in the neighborhood of $\bar{t}$, so we try a series expansion of $G(t)$:

$$P(t) \approx k_{max} \exp\left[-G(\bar{t}) - \frac{1}{2}G''(\bar{t})(t - \bar{t})^2 - \cdots\right], \tag{2.188}$$

$$G''(t) = \frac{m}{t^2} + (k_{max}t_0)m \cdot \frac{1}{t^2}\left(\frac{t}{t_0}\right)^{m+1} \approx \frac{m}{\bar{t}^2}, \tag{2.189}$$

where again in the last step we assume $\bar{t} \ll t_0$. Thus, the distribution of lifetimes, at least near its peak, takes the form

$$P(t) \approx P(\bar{t}) \exp\left[-\frac{m}{2\bar{t}^2}(t - \bar{t})^2 - \cdots\right], \tag{2.190}$$

which is a Gaussian with variance

$$\langle(\delta t)^2\rangle = \frac{1}{m} \cdot \bar{t}^2. \tag{2.191}$$

Thus, the relative variance is $1/m$ instead of 1 as in the original exponential distribution.

A concrete realization of the feedback idea can be built around the following facts: (1) the current flowing into the rod includes calcium ions, (2) the resulting changes in calcium concentration can regulate protein kinases (proteins that catalyze the attachment of phosphate groups to other proteins), and (3) rhodopsin shut-off is associated with phosphorylation at multiple sites. Calcium activation of kinases typically is cooperative, so $m \sim 4$ in the model above is plausible. Notice that in the saturation model the distribution of lifetimes remains broad, and the response to these variations is truncated; in the feedback model the distribution of lifetimes itself is sharpened.

A third possible model involves multiple steps in rhodopsin deactivation. Imagine that rhodopsin starts in one state and makes a transition to state 2, then from state 2 to state 3, and so on for K states, and then it is the transition from state K to $K + 1$ that actually corresponds to deactivation. Thus, there are K active states, and if the time spent in state i is t_i then the total time spent in activated states is

$$t = \sum_{i=1}^{K} t_i. \tag{2.192}$$

The mean value of t is the sum of the means of each t_i, and if the transitions are independent (again, independence is implied when you write the chemical kinetics with the arrows and rate constants), then the variance of t will also be the sum of the variances of the individual t_i:

$$\langle t \rangle = \sum_{i=1}^{K} \langle t_i \rangle, \tag{2.193}$$

$$\langle (\delta t)^2 \rangle = \sum_{i=1}^{K} \langle (\delta t_i)^2 \rangle. \tag{2.194}$$

Recall from above that for each single step, $\langle (\delta t_i)^2 \rangle = \langle t_i \rangle^2$. If the multiple steps occur at approximately equal rates, we can write

$$\langle t \rangle = \sum_{i=1}^{K} \langle t_i \rangle \approx K \langle t_1 \rangle, \tag{2.195}$$

$$\langle (\delta t)^2 \rangle = \sum_{i=1}^{K} \langle (\delta t_i)^2 \rangle = \sum_{i=1}^{K} \langle t_i \rangle^2 \approx K \langle t_1 \rangle^2, \tag{2.196}$$

$$\frac{\langle (\delta t)^2 \rangle}{\langle t \rangle^2} \approx \frac{K \langle t_1 \rangle^2}{(K \langle t_1 \rangle)^2} = \frac{1}{K}. \tag{2.197}$$

Thus, the relative variance declines as the inverse of the number of steps, and the relative standard deviation declines as the inverse of the square root of the number of steps. This is an example of how averaging K independent events causes a $1/\sqrt{K}$ reduction in the noise level.

A multistep scenario works only if the steps are irreversible. If there are significant reverse rates, then progress through the multiple states becomes more like a random walk, with an accompanying increase in variance. Thus, each of the (many) steps involved in rhodopsin shut-off must involve dissipation of a few $k_B T$ of energy to drive the whole process forward.

Problem 30: Getting the most out of multiple steps. Consider the possibility that rhodopsin leaves its active state through a two-step process. To fix the notation, let's say that the first step occurs with a rate k_1 and the second occurs with rate k_2:

$$\text{Rh}^* \xrightarrow{k_1} \text{Rh}^{**} \xrightarrow{k_2} \text{inactive.} \tag{2.198}$$

Assume that we are looking at one molecule, and at time $t = 0$ this molecule is in state Rh^*.

(a) Write out and solve the differential equations for the time-dependent probability of the molecule being in each of the three states.

(b) Use your results in part (a) to calculate the probability distribution for the time at which the molecule *enters* the inactive state. This is the distribution of lifetimes for the two active states. Compute the mean and variance of this lifetime as a function of the parameters k_1 and k_2.

(c) Is there a simple, intuitive argument that allows you to write down the mean and variance of the lifetime without solving any differential equations? Can you generalize this to a scheme in which inactivation involves N steps rather than two?

(d) Given some desired mean lifetime, is there a way of adjusting the parameters k_1 and k_2 (or, more generally, $k_1, k_2, \cdots, k_N$) to minimize the variance?

(e) Suppose that there is a back reaction $\text{Rh}^{**} \xrightarrow{k_{-1}} \text{Rh}^*$. Discuss what this reaction does to the distribution of lifetimes. In particular, what happens if the rate k_{-1} is very fast? Note that "discuss" is deliberately ambiguous; you could try to solve the relevant differential equations, or intuit the answer, or even do a small simulation.

To suppress the variance by a large factor requires many steps in the shutoff process. If we plot the free energy of the rhodopsin molecule as a function of its atomic coordinates, then the path from initial to final state will pass over many hills and valleys. Each valley must be a few $k_B T$ lower than the last, and the hills must be many $k_B T$ high to keep the rates in the right range. Thus, the energy surface is quite rugged. Now when we take one solid and slide it over another, the energy surface is rough on the scale of atoms, because in certain positions the atoms on each surface "fit" into the interatomic spaces on the other surface; then as we move by an angstrom or so, we encounter a very high barrier. If we step back and blur our vision a little bit, all this detailed roughness just becomes friction between the two surfaces. Formally, if we think about Brownian motion on a rough energy landscape and average over details on short length and time scales, the mobility or friction coefficient is renormalized, and then the system behaves on long time scales as if it were moving with this higher friction on a smooth surface.

So if the deactivation of rhodopsin is like motion on a rough energy surface, maybe we can think about the renormalized picture of motion on a smooth surface with high drag or low mobility. Suppose that the active and inactive states are separated by a distance ℓ along some direction in the space of molecular structures. Assume also that motion in this direction occurs with an effective mobility μ. If there is an energy drop ΔE between the active and deactivated states, then the velocity of motion is $v \sim \mu \Delta E / \ell$, and the mean time to make the deactivation transition is

$$\langle t \rangle \sim \frac{\ell}{v} \sim \frac{\ell^2}{\mu \Delta E}. \tag{2.199}$$

In contrast, diffusion over this time causes a spread in positions

$$\langle (\delta \ell)^2 \rangle \sim 2D \langle t \rangle = 2\mu k_B T \langle t \rangle, \tag{2.200}$$

where we make use of the Einstein relation $D = \mu k_B T$. Now (roughly speaking) because the molecule is moving in configuration space with typical velocity v, this spread in positions is equivalent to a variance in the time required to complete the transition to the deactivated state:

$$\langle (\delta t)^2 \rangle \sim \frac{\langle (\delta \ell)^2 \rangle}{v^2} \sim \frac{2\mu k_B T}{(\mu \Delta E / \ell)^2} \cdot \frac{\ell^2}{\mu \Delta E}. \tag{2.201}$$

If we express this variance as a fraction of the mean, we find

$$\frac{\langle (\delta t)^2 \rangle}{\langle t \rangle^2} \sim \frac{2\mu k_B T}{(\mu \Delta E / \ell)^2} \cdot \frac{\ell^2}{\mu \Delta E} \cdot \left(\frac{\mu \Delta E}{\ell^2} \right)^2 \sim \frac{2 k_B T}{\Delta E}. \tag{2.202}$$

Thus, when we look at the variability of the lifetime in this model, the effective mobility μ and the magnitude ℓ of the structural change in the molecule drop out, and the reproducibility is just determined by the amount of energy that is dissipated in the deactivation transition. Indeed, compared with the argument about multiple steps, our result here is the same as expected if the number of irreversible steps were $K \sim \Delta E / (2k_B T)$, consistent with the idea that each step must dissipate more than $k_B T$ to be effectively irreversible. To achieve a relative variance of 1/25 or 1/40 requires dropping $\sim 0.6-1.0$ eV (recall that $k_B T$ is 1/40 eV at room temperature), which is feasible, because the absorbed photon is ~ 2.5 eV.

Problem 31: Is there a theorem here?   The above argument hints at something more general. Imagine that we have a molecule in some state, and we ask how long it takes to arrive at some other state. Assuming that the molecular dynamics is that of overdamped motion plus diffusion on some energy surface, can you show that the fractional variance in the time required for the motion is limited by the free-energy difference between the two states?

How do we go about testing these different ideas? If saturation is important, we could try either by chemical manipulations or by genetic engineering to prolong the lifetime of rhodospin and see whether in fact the amplitude of the single-photon response is buffered against these changes. If feedback is important, we could make a list of candidate feedback molecules and try to manipulate the concentrations of these molecules. Finally, if there are multiple steps, we could try to identify the molecular events associated with each step and perturb these events again either with chemical or genetic methods. All these ideas are good and have been pursued by several groups.

An interesting hint about the possibility of multiple steps in the rhodopsin shut-off is the presence of multiple phosphorylation sites on the opsin proteins. In mice, there are six phosphorylation sites, and one can genetically engineer organisms in which some or all of these sites are removed. At a qualitative level it is quite striking that even knocking out one of these sites produces a noticeable increase in the variability of the single-photon responses, along with a slight prolongation of the mean response (Figs. 2.38b,c). When all but one or two sites are removed, the responses last a *very* long time, and start to look like on/off switches with a highly variable time in the "on" state (Fig. 2.38d,e). When there are no phosphorylation sites (Fig. 2.38f), rhodopsin can still turn off, presumably as a result of binding another molecule (arrestin). But now the time to shut-off is broadly distributed, as one might expect if there were a single step controlling the transition.

Remarkably, if we examine the responses quantitatively, the variance of the single-photon response seems to be inversely proportional the number of these sites, exactly as

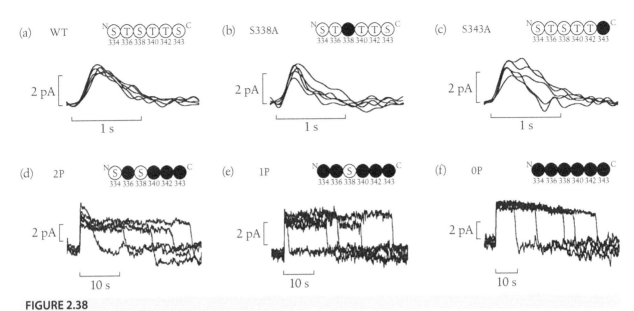

FIGURE 2.38

Variability in the single-photon response for genetically engineered rhodopsins. From Doan et al. (2006). Reprinted with permission from AAAS. (a) Wild-type responses from mouse rods. The schematic shows the six phosphorylation sites, which are serine or threonine residues. The remaining panels show responses when the number of phosphorylation sites has been reduced by mutating these residues to alanine, leaving five sites (b and c), two sites (d), one site (e), or none (f).

in the model where deactivation involved multiple steps, now identified with the multiple phosphorylations (Fig. 2.39). This result really is beautiful. One of the interesting things here is that, absent the discussion of precision and reproducibility, the multiple phosphorylation steps might just look like complexity for its own sake, the kind of thing that biologists point to when they want to tease physicists about our propensity to ignore details. In this case, however, the complexity seems to be the solution to a very specific physics problem.

So, do we understand how rod cells turn single molecular events into macroscopic responses? Certainly the idea of an enzymatic cascade explains how cells can achieve gain, and this mechanism seems quite general. Indeed, such cascades are common in many cellular processes where amplification is needed. In one system, at least, we even seem to understand why the gain is reproducible from event to event. One could object that the example chosen is one in which the degree of reproducibility is not as startling as in some other cases, and certainly the fact that removing phosphorylation sites increases variability does not mean that mechanisms other than phosphorylation do not contribute to maintaining reproducibility. More subtly, having multiple steps in the rhodopsin shut-off only reduces the variability if certain conditions are met, and we don't know how these are achieved.[28] Thus, it seems fair to say once again that we have the ingredients of understanding but perhaps not a complete picture. These ingredients,

---

28. In fact we don't have any independent evidence that these conditions are achieved, and it is difficult to see how to test this experimentally.

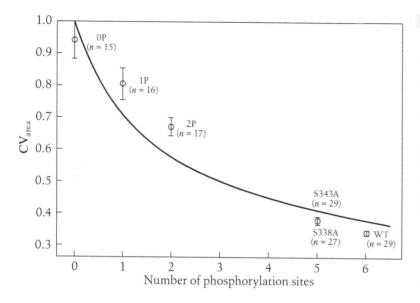

FIGURE 2.39

Coefficient of variation in the integral (area) of the single-photon response. The current traces in response to single photons are integrated, and the standard deviation of these integrals is normalized by the mean to define CV_{area}. As explained in the text, experiments are done on rod cells expressing genetically engineered rhodopsins that have varying numbers of phosphorylation sites, from the normal (WT) 6 down to zero (0P); the specific configuration of sites is shown in Fig. 2.38. Measurements are done on multiple cells, as indicated (e.g., $n = 29$ WT cells), and error bars are standard deviations of the mean. The solid line is $CV = 1/\sqrt{N_p + 1}$, where N_p is the number of phosphorylation sites. Redrawn from Doan et al. (2006).

however, have implications far beyond the rod cell because of the commonality of the molecular components in so many systems.

2.4 The First Synapse and Beyond

This is a good moment to remember a key feature of the Hecht, Shlaer, and Pirenne experiment (described in Section 2.1). In that experiment, observers saw flashes of light that delivered just a handful of photons spread over an area that includes many hundreds of photoreceptor cells. One consequence is that a single receptor cell has a very low probability of counting more than one photon, and this is how we know that these cells must respond to single photons. But it must also be possible for the retina to add up the responses of these many cells, so that the observer can reach a decision about detection. Importantly, there is no way to know in advance which cells will get hit by photons, so if we (sliding ourselves into the observer's brain) want to integrate the multiple photon counts, we have to integrate over all receptors in the area covered by the flash. This integration might be the simplest computation imaginable for a nervous system—just adding up a set of elementary signals, all given the same weight. In many retinas, a large part of the integration is achieved in the first step of processing, as many rod cells converge and form synapses onto a single bipolar cell, as shown schematically in Fig. 2.40.

If each cell generates an output n_i that counts the number of photons that have arrived, then the total photon count is $n_{total} = \sum_i n_i$. The problem is that the cells do not generate integers corresponding to the number of photons counted; they generate currents that have continuous variations. In particular, we have seen that the mean current in response to a single photon has a peak of $I_1 \sim 1\,pA$, but this signal rests on continuous background noise with an amplitude $\delta I_{rms} \sim 0.1\,pA$. Thus, in a single cell, the response to one photon stands well above the background, but if we try to sum the signals from many cells, we have a problem, as illustrated in Fig. 2.41.

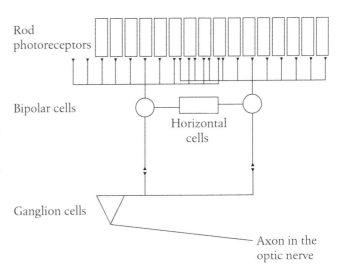

FIGURE 2.40

A schematic of the circuitry in the retina. Signals from the photoreceptors pass to bipolar cells and then to the ganglion cells, whose axons bundle together to form the optic nerve. There is lateral communication through the horizontal cells, as well as through amarcrine cells (not shown). This schematic is drawn to emphasize the role of bipolar cells in integrating signals from multiple rods; much else is also happening!

FIGURE 2.41

Simulation of the peak currents generated by $N = 500$ rod cells in response to a dim flash of light. At left, five of the cells actually detect a photon, each resulting in a current $I_1 \sim 1$ pA; at right is the response to a blank. All cells have an additive background noise, chosen from a Gaussian distribution with zero mean and standard deviation $\delta I_{rms} \sim 0.1$ pA. Although the single-photon responses stand clearly above the background noise, if we simply add up the signals generated by all the cells, then at left we find a total current $I_{total} = 1.85$ pA, whereas at right $I_{total} = 3.23$ pA—the summed background noise completely overwhelms the signal.

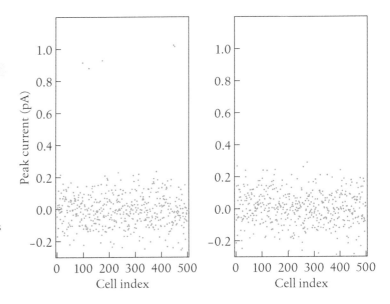

To make the problem precise, let's use x_i to denote the peak current generated by cell i:

$$x_i = I_1 n_i + \eta_i, \tag{2.203}$$

where n_i is the number of photons counted in cell i, and η_i is the background current noise; from what we have seen in the data, each η_i is chosen independently from a Gaussian distribution with a standard deviation δI_{rms}. Summing the signals generated by all the cells, we obtain

$$x_{\text{total}} \equiv \sum_{i=1}^{N_{\text{cells}}} x_i = I_1 \sum_{i=1}^{N_{\text{cells}}} n_i + \sum_{i=1}^{N_{\text{cells}}} \eta_i \qquad (2.204)$$

$$= I_1 n_{\text{total}} + \eta_{\text{eff}}, \qquad (2.205)$$

where the effective noise η_{eff} is the sum of N_{cells} independent samples of the η_i and hence has a standard deviation

$$\eta_{\text{eff}}^{\text{rms}} \equiv \sqrt{\langle \eta_{\text{eff}}^2 \rangle} = \sqrt{N_{\text{cells}}} \delta I_{\text{rms}}. \qquad (2.206)$$

The problem is that with $\delta I_{\text{rms}} \sim 0.1$ pA and $N_{\text{cells}} = 500$, we have $\eta_{\text{eff}}^{\text{rms}} \sim 2.24$ pA, which means that there is a sizable chance of confusing three or even five photons with a blank. In some species the number of cells over which the system integrates is even larger, and the problem becomes even more serious. Indeed, in primates like us, a single ganglion cell (one stage after the bipolar cells; see Fig. 2.40) receives input from ~ 4000 rods, while on a dark night we can see when just 1 in 1000 rods captures a photon. Simply put, summing the signals from many cells buries the clear single-photon response under the noise generated by those cells that did not see anything. This can't be the right way to do things!

Because the single-photon signals are clearly detectable in individual rod cells, we could solve the problem by making a decision for each cell—is there a photon present or not?—and then adding up the tokens that represent the outcome of this decision. Roughly speaking, this strategem means passing each rod's signal through some fairly strong nonlinearity, perhaps so strong that it has as an output only a 1 or a 0, and then pooling these nonlinearly transformed signals. In contrast, a fairly standard schematic of what neurons are doing throughout the brain is adding up their inputs and then passing this sum through a nonlinearity (Fig. 2.42). So perhaps the problems of noise in photon counting are leading us to predict that this very first step of neural computation in the retina has to be different from the standard schematic. Let's try to do a calculation that makes this idea precise.

Formally, the problem faced by the system is as follows. We start with the set of currents generated by all the rod cells, $\{x_i\}$, but we really aren't interested in the currents themselves. Ideally we want to know about what is happening in the outside world, but a first step would be to estimate the total number of photons that arrived, n_{total}. What

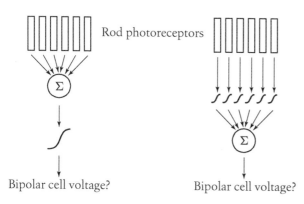

Rod photoreceptors

Bipolar cell voltage?   Bipolar cell voltage?

FIGURE 2.42

Schematic of summation and nonlinearity in the initial processing of rod cell signals. At left, a conventional model in which many rods feed into one bipolar cell; the bipolar cell sums its inputs and passes the results through a saturating nonlinearity. At right, an alternative model, suggested by the problems of noise, in which nonlinearities precede summation.

is the best estimate we can make? To answer this, we need to define what we mean by "best."

One simple idea, which is widely used, is that we want to make estimates that are as close as possible to the right answer, where closeness is measured by the mean-square error. That is, we want to map the data $\{x_i\}$ into an estimate of n_{total} through some function $n_{\text{est}}\left(\{x_i\}\right)$ such that

$$\mathcal{E} \equiv \left\langle \left[n_{\text{total}} - n_{\text{est}}\left(\{x_i\}\right)\right]^2 \right\rangle \tag{2.207}$$

is as small as possible. To find the optimal choice of the function $n_{\text{est}}\left(\{x_i\}\right)$ seems like a hard problem—maybe we have to choose some parameterization of this function, and then vary the parameters? In fact, we can solve this problem once and for all, which is part of the reason that this definition of "best" is popular.

When computing the average error, we are averaging over the joint distribution of the data $\{x_i\}$ and the actual photon count n_{total}:

$$\mathcal{E} \equiv \left\langle \left[n - n_{\text{est}}\left(\{x_i\}\right)\right]^2 \right\rangle$$
$$= \int \left[\prod_{i=1}^{N_{\text{cells}}} dx_i\right] \sum_n P(n, \{x_i\}) \left[n - n_{\text{est}}\left(\{x_i\}\right)\right]^2, \tag{2.208}$$

where, to simplify the notation, we drop the subscript "total." Now to minimize the error, we take the variation with respect to the function $n_{\text{est}}\left(\{x_i\}\right)$ and set the result equal to zero:

$$\frac{\delta \mathcal{E}}{\delta n_{\text{est}}\left(\{x_i\}\right)} = -\sum_n P(n, \{x_i\}) 2 \left[n - n_{\text{est}}\left(\{x_i\}\right)\right], \tag{2.209}$$

so setting this expression to zero gives (going through the steps carefully):

$$\sum_n P(n, \{x_i\}) n_{\text{est}}\left(\{x_i\}\right) = \sum_n P(n, \{x_i\}) n, \tag{2.210}$$

$$n_{\text{est}}\left(\{x_i\}\right) \sum_n P(n, \{x_i\}) = \sum_n P(n, \{x_i\}) n, \tag{2.211}$$

$$n_{\text{est}}\left(\{x_i\}\right) P(\{x_i\}) = \sum_n P(n, \{x_i\}) n, \tag{2.212}$$

$$n_{\text{est}}\left(\{x_i\}\right) = \sum_n \frac{P(n, \{x_i\})}{P(\{x_i\})} n, \tag{2.213}$$

and, finally,

$$n_{\text{est}}\left(\{x_i\}\right) = \sum_n P(n|\{x_i\}) n. \tag{2.214}$$

Thus, the optimal estimator is the mean value in the conditional distribution $P(n|\{x_i\})$. Because we did not use any special properties of the distributions, this result must be true in general, as long as "best" means to minimize mean-square error. We will use this result many times.

Notice that the relevant conditional distribution is the distribution of photon counts given the rod cell currents. From a mechanistic point of view, we understand the opposite problem; that is, given the photon counts, we know how the currents are being generated. More precisely, we know that, given the number of photons in each cell, the currents will be drawn out of a probability distribution, as implied by Eq. (2.203). To make this implication explicit, we have

$$P(\{x_i\}|\{n_i\}) \propto \exp\left[-\frac{1}{2}\sum_{i=1}^{N_{cells}}\left(\frac{x_i - I_1 n_i}{\delta I_{rms}}\right)^2\right]. \tag{2.215}$$

Again, this model tells us how the photons generate currents. But the problem of the organism is to use the currents to draw inferences about the photons. We expect that because the signals are noisy, this inference will be probabilistic, so really we would like to know $P(\{n_i\}|\{x_i\})$. As explained in Section 2.1 (see the discussion leading to Eq. (2.33)), the link between these distributions is given by Bayes' rule.

Problem 32: Just checking. Be sure that you understand the connection between Eq. (2.215) and Eq. (2.203). In particular, what assumptions are crucial in making the connection?

Applied to our current problem, Bayes' rule shows how to construct the probability distribution of photon counts given the rod currents:

$$P(\{n_i\}|\{x_i\}) = \frac{P(\{x_i\}|\{n_i\})P_{counts}(\{n_i\})}{P(\{x_i\})}. \tag{2.216}$$

Let's start with the case of just one rod cell, so we can drop the indices:

$$P(n|x) = \frac{P(x|n)P_{count}(n)}{P(x)}. \tag{2.217}$$

To keep things really simple, consider the case where the lights are very dim, so either there are zero photons or there is one photon. Then all we need is

$$P(1|x) = \frac{P(x|1)P_{count}(1)}{P(x)}, \tag{2.218}$$

and similarly for $P(0|x)$. In the denominator we have $P(x)$, which is the probability that we will see the current x without any conditions on what is going on in the world. We find this value by summing over all possibilities:

$$P(x) = \sum_n P(x|n)P_{count}(n) \tag{2.219}$$

$$= P(x|1)P_{count}(1) + P(x|0)P_{count}(0). \tag{2.220}$$

Putting the terms together, we have

$$P(1|x) = \frac{P(x|1)P_{count}(1)}{P(x|1)P_{count}(1) + P(x|0)P_{count}(0)}. \tag{2.221}$$

Now we can substitute for $P(x|n)$ from Eq. (2.215):

$$P(x|n) = \frac{1}{\sqrt{2\pi(\delta I_{\text{rms}})^2}} \exp\left[-\frac{(x - I_1 n)^2}{2(\delta I_{\text{rms}})^2}\right]. \tag{2.222}$$

Going through the steps, we have

$$P(1|x) = \frac{P(x|1)P_{\text{count}}(1)}{P(x|1)P_{\text{count}}(1) + P(x|0)P_{\text{count}}(0)} \tag{2.223}$$

$$= \frac{1}{1 + P(x|0)P_{\text{count}}(0)/[P(x|1)P_{\text{count}}(1)]} \tag{2.224}$$

$$= \frac{1}{1 + [P_{\text{count}}(0)/P_{\text{count}}(1)]\exp\left[-\frac{(x)^2}{2(\delta I_{\text{rms}})^2} + \frac{(x-I_1)^2}{2(\delta I_{\text{rms}})^2}\right]} \tag{2.225}$$

$$= \frac{1}{1 + \exp(\theta - \beta x)}, \tag{2.226}$$

where

$$\theta = \ln\left[\frac{P_{\text{count}}(0)}{P_{\text{count}}(1)}\right] + \frac{I_1^2}{2(\delta I_{\text{rms}})^2}, \tag{2.227}$$

$$\beta = \frac{I_1}{(\delta I_{\text{rms}})^2}. \tag{2.228}$$

Equation (2.226) has a familiar form—it is as if the two possibilities (zero and one photon) were two states of a physical system, and their probabilities were determined by a Boltzmann distribution. The energy difference between the two states shifts in proportion to the data x, and the temperature is related to the noise level.

The term θ in Eq. (2.227) has a contribution from the signal-to-noise ratio in a single cell, $(I_1/\delta I_{\text{rms}})^2$, so we expect that θ is large. Equation (2.226) then tells us that, if we observe a very small current x, the probability that there really was a photon present is small, $\sim e^{-\theta}$. As the observed current becomes larger, the probability that a photon was present goes up, and, gradually, as x becomes large, we become certain ($P(1|x) \to 1$). To build the best estimator of n from this one cell, our general result tells us that we should compute the conditional mean:

$$n_{\text{est}}(x) = \sum_n P(n|x)n \tag{2.229}$$

$$= P(0|x) \cdot (0) + P(1|x) \cdot (1) \tag{2.230}$$

$$= P(1|x). \tag{2.231}$$

Thus, the Boltzmann-like result Eq. (2.226) for the probability of a photon being counted is, in fact, the best estimator of the photon count in this limit where photons are rare. Further, in this limit one can show that the optimal estimator for the total photon count, which after all is the sum of the individual n_i, is just the sum of the individual estimators.

Problem 33: Summing after the nonlinearity.   Show that the optimal estimator for the total number of photons is the sum of estimators for the photon counts in individual rods, provided that the lights are very dim and hence photons are rare. The phrasing here is deliberately vague—you should explore the formulation of the problem and see exactly what approximations are needed to make things come out right.

    The end result of our calculations is that the optimal estimator of photon counts really is in the form shown at the right in Fig. 2.42: nonlinearities serve to separate signal from noise in each rod cell, and these cleaned signals are summed. How does this prediction compare with experiment? Careful measurements in the mouse retina, as shown in Fig. 2.43, show that the bipolar cells respond nonlinearly even to very dim flashes of light, in the range where the rods see single photons and respond linearly (two photons producing twice the response to one photon). The form of the nonlinearity is what we expect from the theory, a roughly sigmoidal function that suppresses noise and passes signals only above an amplitude threshold. Importantly, this nonlinearity is observed in one class of bipolar cells but not others, and this is the class that, on other grounds, one would expect is most relevant for processing of rod outputs at low light levels.

    Looking in more detail at the experiments, we can see discrete single-photon responses in the bipolar cells, and checking the values of N_{cells}, I_1, and δI_{rms} in this retina shows that, indeed, this would not be possible if rod signals were summed directly (as at right in Fig. 2.42). This strongly suggests that the observed nonlinearity in the synapse is playing a role in noise suppression, perhaps as predicted in the optimal estimators derived above. To test this, Field and Rieke considered models in which the nonlinearity has the form observed experimentally, but with an adjustable midpoint, analogous to the threshold θ in Eq. (2.227). They computed the signal-to-noise ratio at the bipolar cell output for the detection of brief flashes delivering a mean count of $\sim 10^{-4}$ photons/rod cell within the natural integration time set by the duration of the rod's response; this is, approximately, the point at which we can barely see something on a moonless night. Changing the threshold by a factor of two changes the signal-to-noise ratio by factors of several hundred. The measured value of the threshold is within 8% of the predicted optimal setting, certainly close enough to make us think that we are on the right track.

    An important point is that, as in all cases where we are trying to discriminate against a background of noise, optimal performance is not perfect performance; the best we can do is to trade among different kinds of errors (see, in particular, Fig. 2.6). In the effort to insulate the estimate of the summed photon count from the continuous background noise in the rods, there is no alternative but to set the threshold θ to a value for which, with some probability, real photons will be missed. Thus, the retina can solve the problem of spatial integration in the presence of noise, but only at the cost of lowering the overall quantum efficiency of the system. In fact this is (in some sense) good news. In the discussion of the Hecht, Shlaer, and Pirenne experiment (Section 2.1), we emphasized that it is difficult to estimate the factor that connects the mean number of photons at the cornea to the mean number of photons that will be counted in the retina. To be fair, however, if one infers this factor from the Hecht et al. data (or from the later data by Sakitt or other similar experiments; see Section 2.1), the answer seems

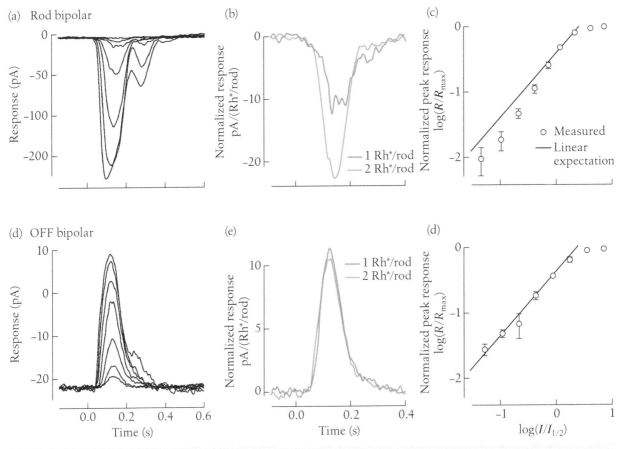

FIGURE 2.43

Nonlinear responses of bipolar cells in the mouse retina. Redrawn from Field and Rieke (2002). The top row shows bipolar cells involved in processing rod signals; the bottom row shows "off" bipolars, which do not participate in photon counting. (a,d) The currents required to hold the bipolar cell voltage fixed as they respond to 10 ms flashes at time $t = 0$; this "voltage clamp" is designed to avoid nonlinearities associated with the cell itself (from voltage-gated ion channels; see Section 5.2) and isolate the nonlinearities in the synapse. The dimmest flash is calibrated to isomerize an average of 0.25 rhodopsins per rod cell, and each successive flash was twice as bright. (b,e) The responses to two dim flashes, normalized to give the current per isomerized rhodopsin (Rh). Note the dramatic nonlinearity in the case of the rod bipolar cells. (c,f) The peaks of the current responses, normalized to the maximum that can be driven by the brightest flashes, versus flash intensity in units of the intensity that gives the half-maximal response $I_{1/2}$.

a bit small, perhaps smaller than is plausible. Indeed, this problem is large enough to have made some people wonder whether the whole picture of photon counting in vision is consistent. It now seems clear that some photons apparently lost between the cornea and the retina in fact are lost in the retina itself, not because the receptor cells don't detect them, but because they are missed in the retina's effort to discriminate against continuous background noise while integrating over large areas. This in turn suggests that the precise magnitude of the continuous background noise is much more important for perceptual behavior than one might have thought by looking at the rod cells in isolation.

The discussion thus far has emphasized separating signals from noise by their amplitudes. We also can see, by looking closely at the traces of current versus time, that signal and noise have different frequency content. Thus, we might also improve the signal-to-noise ratio by filtering. It is useful to think about a more general problem, in which we observe a time dependent signal $y(t)$ that is driven by some underlying variable $x(t)$; let's assume that the response of y to x is linear but noisy:

$$y(t) = \int_{-\infty}^{\infty} d\tau\, g(\tau) x(t - \tau) + \eta(t), \qquad (2.232)$$

where $g(\tau)$ describes the response function and $\eta(t)$ is the noise. What we would like to do is to use our observations on $y(t)$ to estimate $x(t)$.

Problem 34: Harmonic oscillator revisited. Just to be sure you understand what is going on in Eq. (2.232), think again about the Brownian motion of a damped harmonic oscillator, as in Problem 17, but now with an external force $F(t)$:

$$m\frac{d^2x(t)}{dt^2} + \gamma\frac{dx(t)}{dt} + \kappa x(t) = F(t) + \delta F(t). \qquad (2.233)$$

Show that

$$x(t) = \int_{-\infty}^{\infty} d\tau\, g(\tau) F(t - \tau) + \eta(t). \qquad (2.234)$$

Derive an explicit expression for the Fourier transform of $g(\tau)$, and find $g(\tau)$ itself in the limit of either small or large damping γ.

Because y is linearly related to x, we might guess that we can make estimates using some sort of linear operation. As we have seen already in the case of the rod currents, this approach might not be right, but let's try it anyway—we will need somewhat more powerful mathematical tools to sort out, in general, when linear versus nonlinear computations are the most useful. We don't have any reason to prefer one moment of time over another, so we should do something that is both linear and invariant under time translations, which means that the estimate must be of the form

$$x_{\text{est}}(t) = \int_{-\infty}^{\infty} dt'\, f(t - t') y(t'), \qquad (2.235)$$

where $f(t)$ is the filter that we hope will separate signal and noise. Following the spirit of the discussion above, the estimate should be as close as possible to the right answer in the sense of mean-square error. Thus, our task is to find the filter $f(t)$ that minimizes

$$\mathcal{E} = \left\langle \left[x(t) - \int_{-\infty}^{\infty} dt'\, f(t - t') y(t') \right]^2 \right\rangle. \qquad (2.236)$$

In taking the expectation value of the mean-square error, we average over possible realizations of the noise and the variations in the input signal $x(t)$. In practice this averaging can also be thought of as including an average over time.[29] Thus, we can also write

$$\mathcal{E} = \left\langle \int_{-\infty}^{\infty} dt \left[x(t) - \int_{-\infty}^{\infty} dt'\, f(t - t')y(t') \right]^2 \right\rangle. \tag{2.237}$$

This expression is useful, because we can use it to pass to the Fourier domain. As discussed in Appendix A.2, for any function $z(t)$ we have

$$\int_{-\infty}^{\infty} dt\, z^2(t) = \int_{-\infty}^{\infty} \frac{d\omega}{2\pi} |\tilde{z}(\omega)|^2, \tag{2.238}$$

and, for two functions, the Fourier transform of their convolution is the product of their transforms, as in Eq. (2.47),

$$\int_{-\infty}^{\infty} dt\, e^{+i\omega t} \int_{-\infty}^{\infty} dt'\, f(t - t')y(t') = \tilde{f}(\omega)\tilde{y}(\omega). \tag{2.239}$$

Putting things together, we can rewrite the mean-square error as

$$\mathcal{E} = \left\langle \int_{-\infty}^{\infty} \frac{d\omega}{2\pi} \left| \tilde{x}(\omega) - \tilde{f}(\omega)\tilde{y}(\omega) \right|^2 \right\rangle. \tag{2.240}$$

Now each frequency component of our filter $\tilde{f}(\omega)$ appears independently of all the others, so minimizing $\mathcal{E}$ is straightforward. The result is that

$$\tilde{f}(\omega) = \frac{\langle \tilde{y}^*(\omega)\tilde{x}(\omega) \rangle}{\langle |\tilde{y}(\omega)|^2 \rangle}. \tag{2.241}$$

Problem 35: Details of the optimal filter.   Fill in the steps leading to Eq. (2.241). Be careful about the fact that $f(t)$ is real, and so the transform $\tilde{f}(\omega)$ is not arbitrary.

To finish the calculation, we go back to Eq. (2.232), which in the frequency domain can be written as

$$\tilde{y}(\omega) = \tilde{g}(\omega)\tilde{x}(\omega) + \tilde{\eta}(\omega). \tag{2.242}$$

Thus, we have

$$\langle \tilde{y}^*(\omega)\tilde{x}(\omega) \rangle = \tilde{g}^*(\omega)\langle |\tilde{x}(\omega)|^2 \rangle, \tag{2.243}$$

$$\langle |\tilde{y}(\omega)|^2 \rangle = |\tilde{g}(\omega)|^2 \langle |\tilde{x}(\omega)|^2 \rangle + \langle |\tilde{\eta}(\omega)|^2 \rangle. \tag{2.244}$$

If all these variables have zero mean, then such quantities as $\langle |\tilde{x}(\omega)|^2 \rangle$ are the variances of Fourier components, which we know (see Appendix A.2) are proportional to power spectra. Finally, then, we can substitute into the expression for the optimal filter to find

$$\tilde{f}(\omega) = \frac{\tilde{g}^*(\omega)S_x(\omega)}{|\tilde{g}(\omega)|^2 S_x(\omega) + S_\eta(\omega)}, \tag{2.245}$$

---

29. More formally, if all relevant random variations are ergodic, then averaging over the distributions and averaging over time will be the same.

where, as before, S_x and S_η are the power spectra of x and η, respectively.

In the case where noise is small, we can let $S_\eta \to 0$, and we find

$$\tilde{f}(\omega) \to \frac{1}{\tilde{g}(\omega)}. \tag{2.246}$$

Thus, when noise can be neglected, the best way to estimate the underlying signal is just to invert the response function of our sensor, which makes sense. Notice that because $\tilde{g}$ generally serves to smooth the time dependence of $y(t)$ relative to that of $x(t)$, the filter $\tilde{f}(\omega) \sim 1/\tilde{g}(\omega)$ undoes this smoothing. This is important, because it reminds us of a very general point: smoothing in and of itself does not set a limit to time resolution; it is only the combination of smoothing with noise that obscures rapid variations in the signal.

Guided by the limit of high signal-to-noise ratio, we can rewrite the optimal filter as

$$\tilde{f}(\omega) = \frac{1}{\tilde{g}(\omega)} \cdot \frac{|\tilde{g}(\omega)|^2 S_x(\omega)}{|\tilde{g}(\omega)|^2 S_x(\omega) + S_\eta(\omega)} \tag{2.247}$$

$$= \frac{1}{\tilde{g}(\omega)} \cdot \frac{SNR(\omega)}{1 + SNR(\omega)}, \tag{2.248}$$

where we identify the signal-to-noise ratio at each frequency:

$$SNR(\omega) = \frac{|\tilde{g}(\omega)|^2 S_x(\omega)}{S_\eta(\omega)}. \tag{2.249}$$

Clearly, as the signal-to-noise ratio declines, so does the optimal filter—in the limit, if $SNR(\omega) = 0$, everything we find at frequency ω must be noise, and so it should zeroed out to minimize its corrupting effects on our estimates.

In the case of the retina, x is the light intensity, and the y values are the currents generated by the rod cells. When it is very dark, the signal-to-noise ratio is low, so that

$$\tilde{f}(\omega) \to \frac{\tilde{g}^*(\omega)}{S_\eta(\omega)} \cdot S_x(\omega). \tag{2.250}$$

The filter in this case has two pieces, one of which depends only on the properties of the rod cell,

$$\tilde{f}_1(\omega) = \frac{\tilde{g}^*(\omega)}{S_\eta(\omega)}, \tag{2.251}$$

and another piece that depends on the power spectrum of the time-dependent light intensity, $S_x(\omega)$. With a bit more formalism we can show that this first filter, $\tilde{f}_1(\omega)$, has a universal meaning, so that if instead of estimating the light intensity itself, we try to estimate something else (e.g., the velocity of motion of an object across the visual field), then the first step in the estimation process is still to apply this filter. So, it is a natural hypothesis that this filter will be implemented near the first stages of visual processing, in the transfer of signals from the rods to the bipolar cells.

Problem 36: Filtering the real rod currents.   The raw data that were used to generate Fig. 2.4 are available through http://press.princeton.edu/titles/9911.html, in `rodcurrents.mat`. The data consist of 395 samples of the rod current in response to dim flashes of light. The data are sampled in 10 ms bins, and the flash is delivered in the 100th bin. If the ideas above about filtering are sensible, we should be able to do a better job of discriminating between zero, one, and two photons by using the right filter. Notice that filtering of a response that is locked to a particular moment in time is equivalent to taking a weighted linear combination of the currents at different times relative to the flash. Thus, you can think of the current in response to one flash as a vector, and filtering amounts to taking the dot product of this vector with some template. As a first step, you should reproduce the results of Fig. 2.4, which are based just on averaging points in the neighborhood of the peak. Under some conditions, the best template would just be the average single-photon response. How well does this work? What conditions would make this work best? Can you do better? These data are from experiments by F. M. Rieke and collaborators at the University of Washington; my thanks to Fred for making them available.

The idea that the rod/bipolar synapse implements an optimal filter is interesting not least because this leads us to a prediction for the dynamics of this synapse, Eq. (2.251), which is written entirely in terms of the signal and noise characteristics of the rod cell itself. All these properties are measurable, so there are no free parameters in this prediction.[30] To get some feeling for how these predictions work, remember that the noise in the rod cell has two components—the spontaneous isomerizations of rhodopsin, which have the same frequency content as the real signal, and the continuous background noise, which extends to higher frequency. If we have only the spontaneous isomerizations, then $S_\eta \sim |\tilde{g}|^2$, and we are again in the situation where the best estimate is obtained by "unsmoothing" the response, essentially recovering sharp pulses at the precise moments when photons are absorbed. This unsmoothing, or high-pass filtering, is cut off by the presence of the continuous background noise, and the different effects combine to make $\tilde{f}_1$ a band-pass filter. By the time the theory was worked out, it was already known that something like band-pass filtering was happening at this synapse; among other things, this filtering speeds up the otherwise rather slow response of the rod. Figure 2.44 shows a more detailed comparison of theory and experiment.

Problem 37: Optimal filters, more rigorously.   Several things were left out of the optimal filter analysis above; let's try to put them back in.

(a) Assume that there is a signal $s(t)$, and we observe, in the simplest case, a noisy version of this signal, $y(t) = s(t) + \eta(t)$. Let the power spectrum of $s(t)$ be given by $S(\omega)$ and the power spectrum of the noise $\eta(t)$ be given by $N(\omega)$. Further, assume that both signal and noise have Gaussian statistics. Show that the distribution of signals given our observations is

$$P[s(t)|y(t)] = \frac{1}{Z} \exp\left[-\frac{1}{2} \int \frac{d\omega}{2\pi} \frac{|\tilde{s}(\omega) - \tilde{y}(\omega)|^2}{N(\omega)} - \frac{1}{2} \int \frac{d\omega}{2\pi} \frac{|\tilde{s}(\omega)|^2}{S(\omega)} \right]. \quad (2.252)$$

---

30. We should be a bit careful here. The filter, as written, is not causal. Thus, to make a real prediction, we need to shift the filter so that it does not have any support at negative times. To make a well-defined prediction, we adopt the minimal delay that makes this work. One could perhaps do better, studying the optimal filtering problem with explicitly causal filters and considering the trade-off between errors and acceptable delays. See Problem 37.

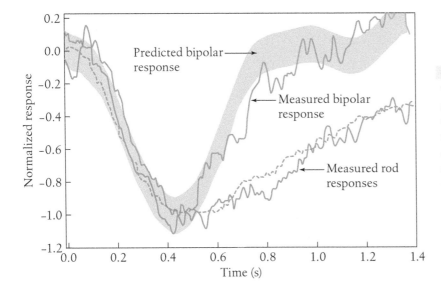

FIGURE 2.44

Voltage responses of rod and bipolar cells in the salamander retina compared with theory. The theory is that the transmission from rod currents to bipolar cell voltage implements the optimal filter as in Eq. (2.251). Measured responses are averages over many presentations of a flash at $t = 0$ that results in an average of five photons being counted. The predicted filter is computed from measured signal and noise properties of the rod cell, with no adjustable parameters. Redrawn from Rieke et al. (1991).

(b) Show that the most likely function $\tilde{s}(\omega)$ given the data on y is also the best estimate in the least-squares sense and is given by

$$\tilde{s}^{(\text{nc})}_{\text{est}}(\omega) = \frac{S(\omega)}{S(\omega) + N(\omega)} \tilde{y}(\omega);\qquad (2.253)$$

the superscript (nc) reminds us that this estimate does not respect causality. Show that this result is consistent with Eq. (2.245). Notice that you did not assume the optimal estimator was linear, so you have shown that it is (!). Which of the assumptions here are essential in obtaining this result?

(c) The noncausal estimator Eq. (2.253) is constructed by assuming that we have access to the entire function $y(t)$, with $-\infty < t < \infty$, as we try to estimate, for example, $s(t = 0)$. If we want the estimator to be something that we can build, then we must impose causality: the estimate of $s(t)$ can be based only on the history $y_- \equiv y(t' < t)$. In other words, we don't really know $y_+ \equiv y(t' > t)$, so we should average over this part of the trajectory. But the average should be computed in the distribution $P[y_+ | y_-]$. To construct this distribution, start by showing that

$$P[y_+, y_-] \equiv P[y(t)] = \frac{1}{Z_0} \exp\left[-\frac{1}{2}\int \frac{d\omega}{2\pi}\frac{|\tilde{y}(\omega)|^2}{S(\omega) + N(\omega)}\right].\qquad (2.254)$$

(d) When discussing causality, it is useful to think about the frequency ω as a complex variable. With this in mind, explain why we can write

$$\frac{1}{S(\omega) + N(\omega)} = |\tilde{\psi}(\omega)|^2,\qquad (2.255)$$

where $\tilde{\psi}(\omega)$ has no poles in the upper half of the complex ω plane. Verify that, with this decomposition,

$$\psi(t) = \int \frac{d\omega}{2\pi}e^{-i\omega t}\tilde{\psi}(\omega)\qquad (2.256)$$

is causal, that is, $\psi(t < 0) = 0$. Consider the case where the signal has a correlation time τ_c, so that $S(\omega) = 2\sigma^2\tau_c/[1 + (\omega\tau_c)^2]$, and the noise is white, $N(\omega) = N_0$. Construct $\tilde{\psi}(\omega)$ explicitly in this case.

(e) Putting Eq. (2.254) together with Eq. (2.255), we can write

$$P[y_+, y_-] = \frac{1}{Z_0} \exp\left[-\frac{1}{2}\int \frac{d\omega}{2\pi}\left|\tilde{y}(\omega)\tilde{\psi}(\omega)\right|^2\right]. \tag{2.257}$$

Show that

$$P[y_+, y_-] = \frac{1}{Z_0} \exp\left[-\frac{1}{2}\int_{-\infty}^{0} dt\left|\int \frac{d\omega}{2\pi}e^{-i\omega t}\tilde{y}_-(\omega)\tilde{\psi}(\omega)\right|^2\right.$$

$$\left.-\frac{1}{2}\int_{0}^{\infty} dt\left|\int \frac{d\omega}{2\pi}e^{-i\omega t}(\tilde{y}_-(\omega) + \tilde{y}_+(\omega))\tilde{\psi}(\omega)\right|^2\right], \tag{2.258}$$

and that

$$P[y_+|y_-] \propto \exp\left[-\frac{1}{2}\int_{0}^{\infty} dt\left|\int \frac{d\omega}{2\pi}e^{-i\omega t}(\tilde{y}_-(\omega) + \tilde{y}_+(\omega))\tilde{\psi}(\omega)\right|^2\right]. \tag{2.259}$$

Explain why averaging over the distribution $P[y_+|y_-]$ is equivalent to imposing the equation of motion

$$\int \frac{d\omega}{2\pi}e^{-i\omega t}(\tilde{y}_-(\omega) + \tilde{y}_+(\omega))\tilde{\psi}(\omega) = 0 \tag{2.260}$$

at times $t > 0$.

(f) Write the noncausal estimate Eq. (2.253) in the time domain as

$$s_{\text{est}}^{(\text{nc})}(t) = \int \frac{d\omega}{2\pi}e^{-i\omega t}S(\omega)\tilde{\psi}^*(\omega)\tilde{\psi}(\omega)\tilde{y}(\omega). \tag{2.261}$$

But the combination $\tilde{\psi}(\omega)\tilde{y}(\omega)$ is the Fourier transform of $z(t)$, which is the convolution of $\psi(t)$ with $y(t)$. Show that Eq. (2.260) implies that the average of $z(t)$ in the distribution $P[y_+|y_-]$ vanishes for $t > 0$, and hence averaging over y_+ is equivalent to replacing

$$\tilde{\psi}(\omega)\tilde{y}(\omega) \rightarrow \int_{-\infty}^{0} d\tau e^{i\omega\tau}\int \frac{d\omega'}{2\pi}\tilde{\psi}(\omega')\tilde{y}(\omega')e^{-i\omega'\tau} \tag{2.262}$$

in Eq. (2.261). Put all the pieces together to show that there is a causal estimate of $s(t)$ that can be written as

$$s_{\text{est}}(t) = \int \frac{d\omega}{2\pi}e^{-i\omega t}\tilde{k}(\omega)\tilde{y}(\omega), \tag{2.263}$$

where

$$\tilde{k}(\omega) = \tilde{\psi}(\omega)\int_{0}^{\infty} d\tau e^{i\omega\tau}\int \frac{d\omega'}{2\pi}e^{-i\omega'\tau}S(\omega')\tilde{\psi}^*(\omega'). \tag{2.264}$$

Verify that this filter is causal.

It is worth noting that we have given two very different analyses. In one, signals and noise are separated by linear filtering. In the other, the same separation is achieved by a static nonlinearity, applied in practice to a linearly filtered signal. Presumably there is some more general nonlinear dynamic transformation that really does the best job. The proper mix probably depends on the detailed spectral structure of the signals and noise, and on the relative amplitudes of the signals and noise, which might be why the different effects are clearest in retinas from very different species. Indeed, there is yet another approach emphasizing that the dynamic range of neural outputs is limited, which constrains how much information the second-order neuron can provide about visual inputs. Filters and nonlinearities can be chosen to optimize this information transmission across a wide range of background light intensities, rather than focusing only on the detectability of the dimmest lights. This approach has received the most attention in invertebrate retinas, such as the fly that we met near the end of Section 2.1. It would be nice to see all these ideas put together correctly. This is an open problem, surely with room for some surprises.

So far we have followed the single-photon signal from the single rhodopsin molecule to the biochemical network that amplifies this molecular event into a macroscopic current, and then traced the processing of this electrical signal as it crosses the first synapse. To claim to have said anything about *vision*, we at least have to follow the signal out of the retina and on its way to the brain along the axons of the ganglion cells schematized in Fig. 2.40.

In the photoreceptors and bipolar cells, we have seen discrete, pulselike electrical signals that reflect the discrete nature of photon arrivals. In the ganglion cells, as with most cells throughout the central nervous system, there are also discrete pulses of voltage across the cell membrane, but these are generated by the cell itself, no matter what the form of the inputs to the system. These pulses, called action potentials or "spikes," result from amplification mechanisms that allow them to propagate over long distances, for example, from our fingertips to our spinal cord or along the optic nerve from our eyes to locations deep inside our heads. We will learn more about the mechanisms of the action potential in Section 5.2, but for now all we need is the phenomenology: the signals that ganglion cells send to the brain—and hence everything we know about the visual world—consist of sequences of discrete identical spikes.

The nerve cells in our body are not hanging in empty space but are bathed in fluids. These fluids have high concentrations of various ions, and thus conduct electrical currents. As a result, the currents that flow in and out of cells as part of the action potential spread away from the cells and can be detected by an electrode placed near the cell. In the classical version of this experiment, one uses a single small wire placed as close to the cell as possible; as currents flow, the local voltage in the neighborhood of the wire changes by tens or hundreds of microvolts, and electrochemical reactions at the wire surface lead to a current flowing in the wire itself, which then is amplified. In a modern version of the experiment, as shown in Fig. 2.45, one can dissect the retina out of the animal (here a salamander) and place it on an array of electrodes, recording simultaneously from many ganglion cells. Related technologies allow the simultaneous recording from many cells in the intact animal.

The classic experiments on single-photon responses in retinal ganglion cells were done before it was possible to measure the responses of single rods. The spikes from

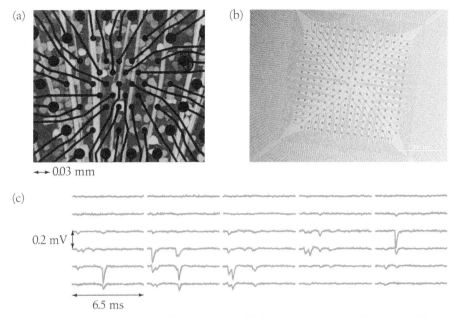

FIGURE 2.45

Recording action potentials from retinal ganglion cells. (a) Salamander retina on an array of electrodes. The electrodes, and the leads that carry signals away from the electrodes, are black features on a transparent slide; the ganglion cells of the retina have been filled with a green dye. Round green objects are cell bodies, and long lines are bundles of axons that eventually converge to form the optic nerve. Note that the number of electrodes is comparable to the number of cells. From Segev et al. (2004). (b) The next generation of electrode arrays. (c) Voltage traces from a selection of these electrodes during an experiment on the salamander retina. Blue traces are the actual voltages, and orange traces are a reconstruction of the voltages as a superposition of stereotyped waveforms—action potentials from individual neurons—learned from a different part of the data. Panels (b) and (c) from Amodei (2011).

single ganglion cells are relatively easy to record, and one can try to do something like the Hecht, Shlaer, and Pirenne experiment, but instead of "seeing" (as in Fig. 2.2), you just ask whether you can detect the spikes. There were hints in the data that a single absorbed photon generated more than one spike, so some care is required. As shown in Fig. 2.46, there are neurons that seem to count by threes—if you wait for three spikes, the probability of seeing is what you expect for setting a threshold of $K = 1$ photon; if you wait for six spikes it is as if $K = 2$, and so on. This simple linear relation between photons and spikes also makes it easy to estimate the rate of spontaneous photonlike events in the dark. Note that photons arrive as a Poisson process, but if each photon generates multiple spikes, then the spikes are not a Poisson process. This idea of Poisson events driving a second point process to generate non-Poisson variability has received renewed attention in the context of gene expression, where a single messenger RNA molecule (perhaps generated from a Poisson process) can be translated to yield multiple protein molecules; see Section 4.3 for more about noise in gene expression.

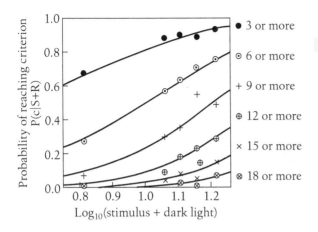

FIGURE 2.46

A frequency-of-seeing experiment with spikes. Recording from a single retinal ganglion cell, you can say you "saw" a flash when you detect 3, 6, 9, or more spikes within a small window of time (here, 200 ms). The probability of reaching this criterion is plotted versus the logarithm of the flash intensity, as in the original Hecht, Shlaer, and Pirenne experiments (Fig. 2.2), but here the intensity is adjusted to include a background rate of photonlike events ("dark light"). Curves are from Eq. (2.2), with the indicated values of the threshold K. Notice that three spikes correspond to one photon. Redrawn from Barlow et al. (1971).

Problem 38: Poisson-driven bursts. A characteristic feature of events drawn out of a Poisson process is that if we count the number of events, the variance of this number is equal to the mean. Suppose that each photon triggers exactly b spikes. What is the ratio of variance to mean (sometimes called the Fano factor) for spike counts in response to light flashes of fixed intensity? Suppose that the burst of spikes itself is Poisson distributed, with mean b. Now what happens to the Fano factor?

Before tracing the connections between individual spikes and photons, it was possible to do a different experiment: just counting spikes in response to flashes of different intensities and asking what the smallest value of the difference ΔI is such that intensities I and $I + \Delta I$ can be distinguished reliably. For sufficiently small I, the just noticeably different ΔI is constant. For large I, one finds $\Delta I \propto I$, so the just noticeable fractional change in intensity is constant; this behavior is common to many perceptual modalities and is called Weber's law. At intermediate intensities one can see $\Delta I \propto \sqrt{I}$. This last result, predicted by Rose and de Vries (see Section 2.1), is what you expect if detecting a change in intensity just requires discriminating against the Poisson fluctuations in the arrival of photons. At high intensities, many photons are being counted, and probably the system just can't keep up; then fluctuations in the gain of the response dominate, which can result in Weber's law. At the lowest intensities, the photons delivered by the flash are few in comparison with the thermal isomerizations of rhodopsin, and this constant noise source sets the threshold for discrimination. Happily, the rate of spontaneous isomerizations estimated from these sorts of experiments agrees with other estimates, including the (much later) direct measurements on rod cells discussed in Section 2.1. This work on discrimination with neurons also is important because it represents one of the first efforts to connect the perceptual abilities of whole organisms with the response of individual neurons.

If retinal ganglion cells generate three spikes for every photon, lights would not need to be very bright before the cells generate thousands of spikes per second, which is impossible—the spikes themselves are ∼1 ms in duration, and all neurons have a refractory period that defines a minimum time (like a hard core repulsion) between successive action potentials. The answer is something we have seen already in the voltage responses of fly photoreceptors (Fig. 2.15): as the background light intensity increases, the retina adapts and turns down the gain, in this case generating fewer spikes per photon. Of course this takes some time, so if we suddenly expose the retina to a bright light, there is very rapid spiking, which then adapts away to a much slower rate. If we imagine that our perceptions are driven fairly directly by the spikes, then our impression of the brightness of the light should similarly fade away. This is true not just for light (as you experience when you walk outside on a bright sunny day); almost all constant sensory inputs are adapted away—for instance, you don't feel the pressure generated by your shoes a few minutes after you put them on. But there are more subtle issues as well, involving the possibility that the coding strategy used by the retina adapts to the whole distribution of inputs rather than just to the mean. This behavior has been observed, and many subsequent experiments have been aimed at understanding the molecular and cellular mechanisms of these effects. The possibility that adaptation serves to optimize the efficiency of coding continuous signals into discrete spikes is something we return to in Chapter 6.

Problem 39: Adaptation in a cascade.   The simplest model of a biochemical cascade is that when activated, each component acts as a catalyst for activating the next component, and activated components decay. If there is a large supply of inactive components, the dynamics are linear:

$$\frac{dA^*}{dt} = -\frac{1}{\tau}A^* + rRh^*(t), \tag{2.265}$$

$$\frac{dB^*}{dt} = -\frac{1}{\tau}B^* + rA^*, \tag{2.266}$$

$$\frac{dC^*}{dt} = -\frac{1}{\tau}C^* + rB^*, \tag{2.267}$$

and so on; here the catalytic rates r and the decay times τ are the same for all steps. To avoid introducing any new parameters, we can assume that, on average, in response to a flash the population of activated rhodopsins $Rh(t) = Rh^*(0)e^{-t/\tau}$.

(a) Solve for the output $C^*(t)$ in response to a flash. The model here is a cascade with three steps. Show that you can generalize to the case of n steps.

(b) Define the gain g as the time integral of the response to a flash, in this case

$$g = \int_0^\infty dt\, C^*(t). \tag{2.268}$$

Evaluate the gain, also in n-step cascades.

(c) As background lights get brighter, responses of photoreceptors become faster, and the gain (e.g., current or voltage per photon) goes down. Show that this behavior is natural in such a cascade model.

(d) Explain at least two different ways to estimate the number of stages in the cascade from experiments on the response to flashes in varying backgrounds.

This is just a start. Interestingly, these ideas were worked out well before the molecular components of the enzymatic cascade were identified, and one can argue that such quantitative analyses provided the intellectual groundwork for the elucidation of the molecular details. See the papers by Baylor and Hodgkin and Baylor et al. in the references for this section in the Annotated Bibliography.

The problem of photon counting—or any simple detection task—also hides a deeper question: how does the brain "know" what it needs to do in any given task? Even in our simple example of setting a threshold to maximize the probability of a correct answer, the optimal observer must at least implicitly acquire knowledge of the relevant probability distributions. Along these lines, there is more to the frog-cooling experiment (Fig. 2.8) than a test of photon counting and dark noise. The retina has adaptive mechanisms that allow the response to speed up at higher levels of background light, in effect integrating for shorter times when the signal-to-noise ratio is certain to be high. The flip side of this mechanism is that the retinal response slows down dramatically in the dark. In moderate darkness (dusk or bright moonlight) the slowing of the retinal response is reflected directly in a slowing of the animal's behavior. It is as if the toad experiences an illusion, because images of its target are delayed, and it strikes at the delayed image. It is worth emphasizing that we see a closely related illusion.

Imagine watching a pendulum swinging while wearing glasses that have a neutral density filter over one eye, so the mean light intensity in the two eyes is different. The dimmer light results in a slower retina, so the signals from the two eyes are not synchronous. Recall that differences in the images between the right and left eyes are cues to the depth of an object. As we try to interpret these signals in terms of motion, we find that even if the pendulum is swinging in a plane parallel to the line between our eyes, what we see is motion in three dimensions. The magnitude of the apparent depth of oscillation is related to the neutral density and hence to the slowing of signals in the darkened retina. This is called the Pulfrich effect.

If the pattern of delay versus light intensity continued down to the light levels in the darkest night, it would be a disaster, because the delay would mean that the toad inevitably strikes behind the target. In fact, the toad does not strike at all in the first few trials of the experiment in dim light, and then it strikes well within the target. It is hard to escape the conclusion that the animal is learning about the typical velocity of the target and then using this knowledge to extrapolate and thereby correct for retinal delays.[31] Thus, performance in the limit where we count photons involves not only efficient processing of these small signals but also learning as much as possible about the world, so that these small signals become interpretable.

2.5   Coda

In the next chapter, we digest more fully what we have learned from the exploration of photon counting in vision. That discussion aims at identifying conceptual problems, so perhaps it is useful here to review a few basic facts.

All cells in a complex organism have the same DNA; cells differ because they "choose" to read out the instructions for making different proteins, in different amounts. Much of the business of cells is done by proteins, and many key proteins

---

31. As far as I know there are no further experiments that probe this learning more directly (e.g., by having the target move at variable velocities).

come in families, so that by choosing different members of the family, a cell can build very similar systems with almost continuously adjustable parameters. The example of rhodopsin shows that these choices can have a dynamic range of $\sim 10^5$ between making one protein and not making another.

Proteins interact with one another to form networks. We have seen interactions that are direct (rhodopsin activates T) and those that are mediated by small molecules (cGMP opens channels). Ion channels are special, because they can respond also to the voltage across the cell membrane, and currents flow through open channels, so there is an effective interaction among proteins mediated by the voltage itself. Interacting networks of proteins can carry out a variety of signal-processing tasks, although we have emphasized amplification. It is presumably some combination of proteins in the relevant cells and synapses that accomplish the nonlinearities and filtering in the transmission from rods to bipolar cells, and there are networks of neurons doing yet more complex computations.

As a practical matter, the nervous system is special, because the outputs of cells are currents and voltages, and we have great methods for measuring these in the lab. It is much more difficult to reach into the cell and monitor biochemical events. The rod cell is an interesting case, because we can perturb the biochemistry and monitor the consequences in the electrical output. Such experiments have been critical in sorting out the mechanisms of amplification.

Finally, remember that this chapter began with the exploration of human behavior. As our methods for uncovering molecular mechanisms explode, we should not forget the power of quantitative, behavioral—or, more generally, macroscopic, phenomenological—experiments to help us phrase questions and set the agenda for more mechanistic studies.

Lessons, Problems, Principles

What have we learned from our exploration of photon counting? In this brief essayistic chapter I draw some lessons, which, I hope, will prove to be more general. I gently nudge the discussion toward some candidate principles that emerge from the example of photon counting. These principles will occupy us for the remainder of the text. I also point you toward references for topics that are left out.

The first important point is to make explicit something that was taken for granted in everything I have said until now: photon counting in vision provides an example of a real biological phenomenon that is susceptible to the sorts of reproducible, quantitative experiments that we are used to in the rest of physics. This is not obvious, and runs counter to some fairly widespread prejudices.

Although things can get complicated,[1] it does seem that, with care, we can speak precisely about the properties of cells in the retina, not just on average over many cells but cell by cell, in enough detail that even the characteristics of the noise in the cellular response are reproducible from cell to cell, organism to organism. It is important that precision and reproducibility are not guaranteed—removing cells from their natural milieu can be traumatic, and every trauma is different. If you dig into the original papers, you will see glimpses of the many things that experimentalists need to get right to achieve the level of precision that we have emphasized in our discussion—the things one needs to do to turn the exploration of living systems into a physics experiment.

The nervous system, starting with the photoreceptor cells, is special in part because it uses electrical signaling. One might worry that what we learn about precision or reproducibility in photoreceptors thus will not generalize to other cellular processes, which are governed by changes in the concentration of different molecules rather than by the flow of electrical currents. But we know that in the rod cell the current flowing across the membrane really is determined by the internal concentration of cGMP, which in turn reflects the balance of phosphodiesterase and cyclase activities. Thus, the electrical signal is a direct readout from a more conventional biochemical network—indeed, an example of a family of biochemical networks that are involved

---

1. We have not explored, for example, the fact that the retina has many kinds of ganglion cells, and even bipolar cells come in multiple flavors.

in many different cellular processes. In the rod cell, the total current that flows in the dark is ~ 20 pA, and the response to a single photon is ~ 1 pA, corresponding to a 5% modulation. The continuous background noise is ~ 0.1 pA, less than 1% of the total, and we have seen that this scale is important as we try to understand the mechanisms of filtering and summation in the retina. This is prima facie evidence that ~ 1% changes, or less, in the activity of a biochemical network can be biologically significant.

Crucially, in the rod cell, effects on the 1% level are not quantitative decoration on basic processes that can be seen with measurements that are sensitive only to large changes. Our ability to see in the dark, and to see as well as we do, is intimately bound up with these tiny modulations of the cell's biochemical activity. There is no meaningful sense in which we could claim to understand vision in this regime without making precision measurements.

I don't want to hold up 1% as some sort of magic number setting a scale for biological relevance. One can imagine good reasons for the scale to be different in other systems. But if one's introduction to the phenomena of life comes by exploring a system in which the available measurement techniques are insensitive to such precise behavior, one can easily be led to think that precision is best left to the atomic physicists,[2] and that biology is a fundamentally messy business. By starting with the example of photon counting, we are alerted to how precise life's mechanisms can be and to the challenges that these systems pose to our experimental techniques. There are myriad ways in which a system can seem sloppier or less precise than it is, and we can't find the underlying precision unless we look for it.

The second point is something that has been emphasized already but bears repeating: the performance of these biological systems—something that results from mechanisms of incredible complexity—really is determined by the physics of the problem that the system has been selected to solve. If you plan on going out in the dark of night, there are obvious benefits to detecting dimmer sources of light, to making more reliable discriminations among subtly different intensities and, ultimately, to distinguishing more efficiently among different spatiotemporal patterns. You can't do better than to count every photon, and the reliability of photon counting by the system as a whole can't be better than the limits set by noise in the detector elements. That real visual systems reach these limits is extraordinary.

The third point concerns the nature of the explanations that we are looking for. We have discussed the currents generated in response to single photons, the filter characteristics and nonlinearities of synapses, and the spiking outputs of ganglion cells, and in each case we can ask why these properties of the system are as we observe them to be. Importantly, we can ask analogous questions about a wide variety of systems, from individual molecules to the regulation of gene expression in single cells to the dynamics of neural networks in our brains. But what are we doing when we look for an explanation of the data?

---

2. Corresponding to this remark about certain kinds of experiments is the quip about certain styles of theorizing: "elegance should be left to tailors and cobblers." This is variously (and, I must say, somewhat puzzlingly) attributed to Boltzmann and to Einstein. In the interests of full disclosure, I should note that my grandfather was a tailor.

When we ask "why" in relation to a biological system, we can imagine (at least) two very different kinds of answers.[3] First, we could plunge into the microscopic mechanisms. As we have seen (but not emphasized) in looking at the dynamics of biochemical amplification in the rod cell, what we observe as functional behavior of the system as a whole depends on a large number of parameters: the rates of various chemical reactions, the concentrations of various proteins, the density of ion channels in the membrane, the binding energies of cGMP to the channel, and so on. These obviously are not fundamental constants. On the contrary, almost all these parameters are under the organism's control.

Our genome encodes hundreds of different ion channels, and the parameters of the rod cell are set by its choice to read out the instructions for making some channels rather than others. Further, the cell can make more or less of these proteins, again adjusting the parameters of the system. A similar line of argument applies to other components (and many other systems), because many key molecules are organized into families such that the individual members have slightly different properties, and cells choose which member of the family will be expressed. More subtly, many of these molecules can be modified (e.g., by covalent attachment of phosphate groups, as with the shut-off of rhodopsin), and these modifications provide another pathway for adjusting parameters. Thus, saying that (for example) the response properties of the rod cell or the dynamics of transmission across a synapse are determined by the parameters of a biochemical network is very different from saying that the absorption spectrum of hydrogen is determined by the charge and mass of the electron. We would have to go into some alternative universe to change the properties of the electron, but most of the parameters of the biochemical network are under the control of the cell, these parameters can and do change in response to other signals, and in many cases we find that closely related organisms make similar but measurably different choices for these parameters. In the brain we can even find neighboring cells that are broadly of the same type but have made slightly different choices about almost all the relevant parameters.

---

3. My friend and collaborator Rob de Ruyter van Steveninck has an excellent way of talking about closely related issues. He once began a lecture by contrasting two different questions: Why is the sky blue? Why are trees green? The answer to the first question is a standard part of a good, high-level course on electromagnetism: when light scatters from a small particle—and molecules in the atmosphere are much smaller than the wavelength of light—the scattering is stronger at shorter wavelengths. This phenomenon is called Rayleigh scattering. Thus, red light (long wavelengths) moves along a more nearly straight path than does blue light (short wavelengths). The light that we see, which has been scattered, therefore has been enriched in the blue part of the spectrum, and this effect is stronger if we look farther away from the sun. So, the sky is blue because of the way in which light scatters from molecules. We can answer the question about the color of trees in much the same way that we answered the question about the color of the sky: leaves contain the molecule chlorophyll, which is quite a large molecule compared with the oxygen and nitrogen in the air, and this molecule actually absorbs visible light; the absorption is strongest for red and blue light, so what is reflected back to us is the (intermediate wavelengths) green light. Unlike the color of the sky, however, the color of trees could have a different explanation. Imagine trees of other colors—blue, red, perhaps even striped. Microscopically, this could happen because their leaves contain molecules other than chlorophyll, or even molecules related to chlorophyll but with slightly different structures. In fact we know that the biochemical processes that lead to the synthesis of such molecules as chlorophyll are complicated, and processes like these really do lead to the synthesis of other molecules with different absorption spectra. But trees of different colors will compete for resources, and some will grow faster than others. The forces of natural selection plausibly will cause one color of tree to win out over the others. In this sense, we can say that trees are green because green trees are more successful, or more fit, in their environment.

An explanation of functional behavior in microscopic terms, then, may be correct but somehow unsatisfying. Further, there may be more microscopic parameters than phenomenological parameters, and this may be critical in allowing the system to achieve nearly identical functional behaviors via several different mechanisms. But all of this casts doubt on the idea that we are "explaining" the functional behavior in molecular terms.

A second, very different kind of explanation is suggested by our discussion of the first synapse in vision, between the rod and bipolar cells. In that discussion (Section 2.4), we promoted the evidence of near-optimal performance in the problem of photon counting into a principle from which the functional properties of the system—nonlinearities and filter characteristics—could be derived. In this view, the system is the way it is because evolution has selected the best solution to a problem that is essential in the life of the organism. This principle does not tell us how the optimum is reached, but it can predict the observable behavior of the system. Evidently there are many objections to this approach, but it is familiar, because many different ideas in theoretical physics can be formulated as variational principles, from least action in classical mechanics to the minimization of free energy in equilibrium thermodynamics, among others.

Organizing our thinking about biological systems around optimization principles tends to evoke philosophical discussions, in the pejorative sense that scientists use this term. I would like to avoid discussions of this flavor. If we are going to suggest that "biological systems maximize X" is a principle, then rather than having everyone express their opinion about whether this is a good idea, we should discipline ourselves and insist on criteria that allow such claims to be meaningful and predictive. First, we have to understand why X can't be arbitrarily large—we need to have a theory that defines the physical limits to performance. Second, we should actually be able to measure X, and compare its value with this theoretical maximum. Finally, maximizing X should generate some definite predictions about the structure and dynamics of the system, predictions that can be tested in independent quantitative experiments. Perhaps the most important lesson from the example of photon counting is that we can carry through this program and maintain contact with real data. The challenge is to choose principles (candidate Xs) that are more generally applicable than the very specific idea that the retina maximizes the reliability of seeing on a dark night.

Before discussing the candidate Xs that will occupy us for the remainder of the text, we should be clear that the whole strategy of looking for variational principles is by no means guaranteed to succeed. That is, our attempts at theorizing might be wrong, not because we chose the wrong thing to maximize, but, more deeply, because the state of matter that we call alive is not characterized by being the extremum of *anything*. For inanimate systems, we can find nonequilibrium steady states that convert a flux of energy into various forms of structure (e.g., in Rayleigh-Bénard convection, where a layer of fluid heated from below generates a periodic spatial pattern of flows), and many people expressed the hope that these states could be characterized as minima of some potential function, generalizing the idea that equilibrium states are minima of the free energy. As far as I know, this hope has not been realized, except in special cases.

The alternative to variational principles is to look directly at dynamics. Perhaps the most fundamental example of this approach in relation to biological systems is the effort to describe the dynamics of evolution. The opposite of assuming that biological systems have found an extremum is to assume that, in their current environment, organisms have access to an essentially unlimited supply of beneficial mutations. In this limit, we expect a population of organisms to climb slowly up the "fitness gradient," accumulating beneficial mutations, in much the same way that conventional physical systems flow in response to forces. The natural question concerns the mobility: given that the typical mutation confers an advantage ϵ, and that such mutations have a probability p per generation, at what rate does a population of N organisms increase its fitness? Even in its simplest version, this problem is surprisingly subtle and depends in an essential way on fluctuations in the finite population. For this reason and others, evolutionary dynamics has attracted considerable attention from physicists over the past decade.

As noted in Chapter 1, evolution is one of several processes, along with learning and adaptation, by which biological systems adjust their parameters to be more effective, or more fit, in their environment. Learning and adaptation also have molecular mechanisms, and these mechanisms generate a nontrivial dynamics for these processes. Indeed, it is a very old observation that adaptation in many sensory systems is not characterized by a single time scale, but rather by distributed relaxation in which, after a switch from one environment to another, the system relaxes to its new steady state with a power-law time dependence. There are hints that there may be a similarly broad distribution of time scales in the elementary processes of learning at single synapses in the brain. As for evolution itself, we know that effective mutation rates are reduced by proofreading mechanisms (see Section 4.5), and that beyond mutations that alter single bases along the DNA, there are also recombination events, gene duplications, and gene transfers (at least among microorganisms) that change the pace of evolution. Interestingly, all the molecular components responsible for these different mechanisms are themselves encoded in the genome, meaning that the dynamics of evolution—and adaptation and learning—are themselves subject to evolutionary change.

There is a wonderful book to be written that would emphasize a dynamical view of biological systems. One would like to start, perhaps, with dynamical models for the origin of life itself, and proceed to the dynamics of evolutionary change in self-reproducing populations. We would like to understand how the dynamics generated by relatively small numbers of molecules can provide the building blocks for larger, functional networks, and the nature of the mapping between these functions and the underlying molecular parameters. We would see how dynamics emerges on many time scales and how organisms respond immediately to stimuli but also adapt and learn on longer time scales, tuning their responses in relation to their history. Central to all this would be a discussion—surely still incomplete—of how collective, macroscopic variables that embody biological function emerge from the many microscopic degrees of freedom. I would love to read this book.

I take the point of view that life is special, and that this specialness is expressible by statements of the flavor that "biological systems maximize X," for some well-chosen X. To first approximation, I ignore the question of how these systems find their way through the space of possible X values, and what guides them to the maximum. But, as noted above, I *do* discipline myself to speak primarily about systems where we have direct evidence for the maximization of X. In some sense, an appeal to variational principles is an attempt to take a path that goes around the wonderful dynamical questions I have outlined above. But if the real systems are operating at extremal values of some quantities that we can identify, then we can have a theory of these extremal states that is independent of the dynamics, much as equilibrium statistical mechanics is separate from the nonequilibrium variety. Even when the real systems are not quite at the maximal values of X, the idea of maximizing X might serve as a useful idealization. What, then, is X?

The first possibility is very much connected to what we have learned from photon counting. Real biological systems have been pushed by evolution to the point where crucial signals are not much larger than the minimum noise levels required by basic physical principles. A candidate principle, then, is that these systems have maximized their reliability in the presence of noise, or minimized the effective noise level until only the physically required noise sources remain significant. We will explore this possibility not just in the context of our sensory systems and the neural processing of sensory information, but also at the molecular and cellular level where signals are carried by the concentrations of specific molecules, or by the identity of individual molecules, as in DNA. In all these cases there is a clear physical limit, so that effective noise levels cannot be pushed to zero without expending infinite resources, and there is a program of quantitative experiments that allows us to compare the performance of real systems to these physical limits. Generalizing the results from photon counting, we will see that many systems—from bacteria to bats and from developing fruit fly embryos to human brains—operate at or near the relevant physical limits, motivating the elevation of minimizing noise or maximizing reliability to a more general principle. From this principle several surprising things can be derived, and many of these predictions are confirmed in independent experiments.

The second possibility is of a very different flavor. If we take seriously the attempt to build quantitative models for biological systems, then as noted above these models inevitably have many parameters. To some extent these parameter can be tuned by the organism, or over evolutionary time, to achieve functional behavior, but an implicit appeal to fine tuning is very dangerous. Concretely, there are many cases in which we can manipulate the parameters of complex biological systems or observe their natural variations—from changing the sequence of amino acids in a single protein to changing the numbers of molecules of particular proteins or even the strength of synapses in a network of neurons—and see that functional behavior is robust across these variations. At the same time, there is some obvious sense in which the behaviors characteristic of life are not generic behaviors of dynamical systems, and hence some tuning is required. Because these ideas have arisen in so many different contexts, it has been difficult to articulate a concise principle that captures everyone's intuition, but perhaps it is something like the claim that biological systems have been selected to maximize their tolerance to parameter variation.

Finally, there is the idea that what is important in biology is the flow of information. Organisms pass information from one generation to the next, but they also extract information from their environment to guide their actions and gather information about their own internal states as they do so. Although "information" as an informal concept has widespread currency in discussions of biological systems, there is less agreement about whether the formal mathematical versions of this concept are useful. Shannon taught us that there really is only one way to make the concept of information precise, and this connects deeply with ideas from statistical mechanics. Importantly, there is a physical cost to transmitting information, related to (but not quite the same as) the resource costs associated with reducing noise levels, and we will see that more biologically grounded principles of maximizing performance at fixed resource costs often have information-theoretic analogs. These ideas motivate the principle that biological systems maximize information transmission or the efficiency with which information is represented. Again there are clear physical limits, and experiments show that these limits are approached in many real systems, across a huge range of biological phenomena. Optimization principles based on information-theoretic quantities allow us to address problems from biochemical signaling in single cells up to learning and prediction in the brain in a single, more or less unified, mathematical framework.

This is not the place to argue that any one of these candidate principles is correct. All we need is the promise that the exploration of these principles will be productive, giving us a platform from which to discover, appreciate, and think more clearly about a wide range of beautiful phenomena. We will have a chance, at the end of the text, to look back at what we have learned, and forward to where we might be going (Chapter 7).

CANDIDATE PRINCIPLES

Noise Is Not Negligible

The great poetic images of classical physics are those of determinism and clockwork. In a clock, not only the output but also the internal mechanisms are models of precision. Strikingly, life seems very different. Interactions between molecules involve energies of just a few times the thermal energy. Biological motors, including the molecular components of our muscles, move in elementary steps that are on the nanometer scale, driven forward by energies that are larger than the thermal energies of Brownian motion but not much larger. Crucial signals inside cells often are carried by just a handful of molecules, and these molecules inevitably arrive randomly at their targets. Human perception can be limited by noise in the detector elements of our sensory systems, and individual elements in the brain, such as the synapses that pass signals from one neuron to the next, are surprisingly noisy. How do the obviously reliable functions of life emerge from under this cloud of noise? Are there principles at work that select, out of all possible mechanisms, the ones that maximize reliability and precision in the presence of noise?

In this chapter we take a tour of various problems involving noise in biological systems. This topic has always held a special fascination for me, and I firmly believe that there is something deep to be found in the exploration of these issues. We will see the problems of noise in systems ranging from the behavior of individual molecules to our subjective, conscious experience of the world. To address these questions, we will need a fair bit of mathematical apparatus, rooted in the ideas of statistical physics. I hope that, armed with this apparatus, you will develop a deeper view of many beautiful phenomena and a deeper appreciation for the problems that organisms have to solve.

4.1 Fluctuations and Chemical Reactions

To survive, living organisms must control the rates of many chemical reactions. Fundamentally, all reactions happen because of fluctuations. Put more strongly, chemical reactions are a nonperturbative consequence of molecular fluctuations. The rates of chemical reactions usually obey the Arrhenius law $k \propto e^{-E_{\mathrm{act}}/k_B T}$, where E_{act} is the activation energy. But the thermal energy $k_B T$ measures the mean-square amplitude of fluctuations, for example in the velocities of atoms. Thus, chemical reaction rates are

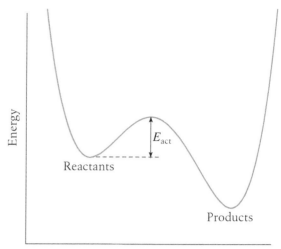

FIGURE 4.1

The simplest model of a chemical reaction. Along some molecular coordinate x, the potential energy $V(x)$ has two minima separated by a barrier. The height of the barrier is the activation energy E_{act}, which we expect will determine the rate of the reaction through the Arrhenius law $k \propto e^{-E_{act}/k_B T}$.

$\sim e^{-1/g}$, where g is the strength of the fluctuations. If we start by imagining a world in which there are no fluctuations, we can add them in piece by piece, but there is no way to get a chemical reaction rate as a perturbative series in g. So, our first order of business is to see how the Arrhenius law emerges, as an asymptotic result, for some real dynamical model, a problem first solved convincingly by Kramers in 1940. The details of this calculation are in Appendix A.5, but here it is important to understand how the problem is set up and how the result emerges. Once we have this more solid understanding, we will be ready to look at what might be special regarding the control of chemical reaction rates in biological systems.

Let us consider the simplest case, shown in Fig. 4.1. Here the molecules of interest are described by a single coordinate x, often called the reaction coordinate, and the potential energy $V(x)$ as a function of this coordinate has two wells that we can identify as reactant and product structures. Let's assume that motions along this coordinate are overdamped, so inertia is negligible.[1] Because the molecule is surrounded by an environment at temperature T, we really want to describe Brownian motion in this potential. So, the equation of motion is

$$\gamma \frac{dx}{dt} = -\frac{dV(x)}{dx} + \zeta(t), \tag{4.1}$$

where γ is the friction or drag coefficient, and the random or Langevin force $\zeta(t)$ reflects the random influences of all other degrees of freedom in the system. To ensure that the system eventually comes to equilibrium at temperature T, we must have

$$\langle \zeta(t)\zeta(t') \rangle = 2\gamma k_B T \delta(t - t'). \tag{4.2}$$

---

1. This really is just a simplifying assumption. We can also do everything in the case where inertia is significant, and none of the main results will be different. More precisely, we are going to go far enough to show that the Arrhenius law $k = Ae^{-E_{act}/k_B T}$ is true, and that the activation energy E_{act} corresponds to our intuition. The neglect of inertia would only change the prefactor A, which is in any case much more difficult to calculate.

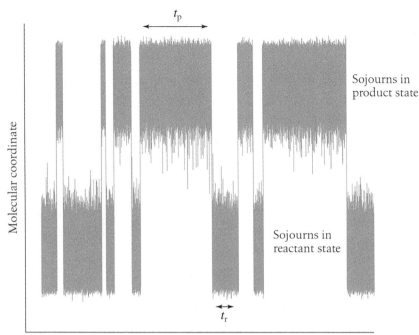

FIGURE 4.2

Example of the trajectories we expect to see in solving the Langevin equation, Eq. (4.1). Long sojourns in the reactant or product state are interrupted by rapid jumps from one potential well to the other. If we look at the times t_r spent in the reactant state, these should come from a probability distribution $P_r(t_r) = k_+ e^{-k_+ t_r}$, where k_+ is the rate of the chemical reaction from reactants to products. Similarly we should have $P_p(t_p) = k_- e^{-k_- t_p}$, where k_- is the rate of the reverse reaction.

The challenge is to see whether we can extract from these dynamics some approximate result that corresponds to our intuition about chemical reactions and in particular gives us the exponential dependence of the rate on the temperature.

When we solve Eq. (4.1), we find the coordinate as a function of time. What features of this trajectory correspond to the reaction rate k? If there really are only two states in the sense of chemical kinetics, then trajectories should look like those in Fig. 4.2. Specifically, we should see that trajectories spend most of their time in one potential well or the other, punctuated by rapid jumps between the wells. More precisely, there should be a clear separation of time scales between the dynamics in each well and the typical time between jumps. Further, if we look at the times spent in each well, between jumps, these times should be drawn from an exponential distribution $P(t) = k e^{-kt}$, and then k is the rate constant for the chemical reaction leading out of that well into the other state.

Problem 40: What's the alternative? You should think a bit about what was just described. Suppose, for example, that you don't know the potential, and I just give you samples of the trajectory $x(t)$. What would it mean if the trajectories paused at some intermediate point between reactants and products? How would you interpret nonexponential distributions of the time spent in each well?

Problem 41: Numerical experiments on activation over a barrier. Perhaps before launching into the long calculation that follows, you should get a feeling for the problem by doing a small simulation. Consider a particle at position x moving in a potential $V(x) = V_0[1 - (x/x_0)^2]^2$. This is a double well, with minima at $x = \pm x_0$ and a barrier of height V_0

between these minima. Consider the overdamped limit of Brownian motion in this potential, as in Eq. (4.1):

$$\gamma \frac{dx(t)}{dt} = \frac{4V_0}{x_0} \left(\frac{x}{x_0}\right) \left[1 - \left(\frac{x}{x_0}\right)^2\right] + \zeta(t). \tag{4.3}$$

We want to simulate these dynamics. The simplest approach is the naive one, in which we use discrete time steps separated by Δt and we approximate

$$\frac{dx(t)}{dt} \rightarrow \frac{x(n+1) - x(n)}{\Delta t}. \tag{4.4}$$

(a) To use this discretization we have to deal with the Langevin force. One (moderately) systematic approach is to integrate the Langevin equation over a small window of time Δt:

$$\gamma \int_t^{t+\Delta t} dt\, \frac{dx(t)}{dt} = -\int_t^{t+\Delta t} dt\, \frac{\partial V(x)}{\partial x} + \int_t^{t+\Delta t} dt\, \zeta(t), \tag{4.5}$$

$$\gamma \left[x(t+\Delta t) - x(t)\right] \approx -\Delta t \frac{\partial V(x)}{\partial x}\bigg|_{x=x(t)} + z(t), \tag{4.6}$$

where

$$z(t) = \int_t^{t+\Delta t} dt\, \zeta(t). \tag{4.7}$$

Using the correlation function of the Langevin force from Eq. (4.2), compute the variance of $z(t)$. Show also that the values of z at different times—separated at least by one discrete step Δt—are uncorrelated.

(b) Combine your results in part (a) with the equations above to show that this simple discretization is equivalent to

$$y(n+1) = y(n) + \alpha E^\dagger \cdot y(n) \cdot [1 - y^2(n)] + \sqrt{\frac{\alpha}{2}} \xi(n), \tag{4.8}$$

where $y = x/x_0$, the parameter $\alpha = 4k_B T \Delta t/(\gamma x_0^2)$ should be small, $E^\dagger = V_0/(k_B T)$ is the normalized activation energy for escape over the barrier, and $\xi(n)$ is a Gaussian random number with zero mean, unit variance, and no correlations among different time steps n.

(c) Implement Eq. (4.8), for example, in MATLAB. Note that MATLAB has a command `randn` that generates Gaussian random numbers.[2] You might start with a small value of $E^\dagger$ and experiment to see how small you need to make α before the results start to make sense. What do you check to see whether α is small enough?

(d) Explore what happens as you change the value of $E^\dagger$. For each value of $E^\dagger$, check that your simulation runs long enough so that the distribution of x actually is given by the Boltzmann distribution $P(x) \propto \exp[-V(x)/k_B T]$. As $E^\dagger$ increases, can you see that there are isolated discrete events corresponding to the chemical reaction in which the system jumps from one well to the other? Use your simulation to estimate the rate of these jumps, and plot the rate as a function of the activation energy $E^\dagger$. Can you verify the Arrhenius law?

---

2. More precisely, MATLAB *claims* that `randn` generates Gaussian random numbers that are independent. Maybe you should check this?

Problem 42: Effective potentials. We are discussing, for simplicity, a one-dimensional problem. Suppose that there are really many dimensions, not just x but also $y_1, y_2, \cdots, y_N \equiv \{y_j\}$. Then we have, again in the overdamped limit,

$$\gamma \frac{dx}{dt} = -\frac{\partial V(x; \{y_j\})}{\partial x} + \zeta(t), \tag{4.9}$$

$$\gamma_i \frac{dy_i}{dt} = -\frac{\partial V(x; \{y_j\})}{\partial y_i} + \xi_i(t), \tag{4.10}$$

where now the fluctuation-dissipation theorem tells us that the Langevin forces obey

$$\langle \xi_i(t)\xi_j(t')\rangle = 2k_B T \gamma_i \delta_{ij}\delta(t - t'). \tag{4.11}$$

Imagine now that x moves much more slowly than all the $\{y_i\}$.

(a) Verify that, from Eq. (4.10), the stationary distribution of $\{y_i\}$ at fixed x is the Boltzmann distribution:

$$P(\{y_j\}|x) = \frac{1}{Z(x)} \exp\left[-\frac{V(x; \{y_j\})}{k_B T}\right]. \tag{4.12}$$

(b) If x is slow compared with the $\{y_j\}$, it is plausible that we should average the dynamics of x in Eq. (4.9) over the stationary distribution $P(\{y_j\}|x)$. Show that this generates an equation in which x moves in an effective potential,

$$\gamma \frac{dx}{dt} = -\frac{\partial V_{\text{eff}}(x)}{\partial x} + \zeta(t), \tag{4.13}$$

and this effective potential is the free energy, $V_{\text{eff}}(x) = -k_B T \ln Z(x)$.

(c) Equations (4.9) and (4.10) still are not completely general, because we have taken the mobility tensor to be diagonal, so that forces on coordinate y_i lead to velocities only along this direction. Does the more general case, where γ_{ij} is an arbitrary symmetric matrix, present any new difficulties for the problem posed here?

In the point of view we are taking here, the probabilistic character of chemical reactions is fundamental: what we mean by a rate constant k *is* the probability per unit time for one molecule to make the transition over the barrier. If we have a macroscopic number of molecules, and we count the fraction that has not yet crossed, this fraction will decay as e^{-kt}, with the same k, so this is also the familiar rate constant from chemical kinetics. This connection between macroscopic kinetics and the probabilistic behavior of single molecules was explored in Problem 25, and this is a good time to review.

If you take the path-integral view of quantum mechanics, then when Planck's constant is small, so that tunneling is rare, there is a semiclassical approximation to the path integral that reproduces the WKB (Wentzel-Kramers-Brillouin) approximation to Schrödinger's equation. In this approximation the path integral is dominated by specific trajectories in the same way that a one-dimensional integral can be dominated by the point of stationary phase; these dominant trajectories have come to be called instantons. Instantons are precisely the jumps from one well to another, analogous to what we have drawn for the classical case in Fig. 4.2. In Appendix A.5 you can find the details of how to calculate these trajectories and the demonstration that such trajectories occur with probability $\propto \exp(-E_{\text{act}}/k_B T)$ less often than trajectories that simply fluctuate within one potential well or the other. This is the essence of the Arrhenius law.

Is there anything special about how these ideas play out in the case of biological molecules? If we try to draw Fig. 4.1 for a reaction involving a large biological molecule, such as an enzyme, we associate the reaction coordinate, (i.e., the molecular coordinate along which we see the double-well potential) with the motions that are involved in the chemical events themselves. Thus, if we are looking at the transfer of a hydrogen atom—breaking one bond and forming another—we might think that the relevant molecular coordinate is given by the position of the hydrogen atom itself.

Biological molecules—such as the proteins that act as enzymes, catalyzing specific chemical reactions of importance in the cell—are large and hence flexible. Certainly they can change reaction rates by holding the reactants in place. But because of their flexibility, there is also the possibility that, as they flex, the effective barrier for the reaction changes. In this case, the dominant path for the reaction might be for the protein to fluctuate into a favorable configuration, and then for the more local coordinates (e.g., the position of the hydrogen atom) to make their jump. In this way, the observed activation energy comes to have two components, the usual one that we measure along the reaction coordinate (which presumably is reduced by waiting for the protein to arrange itself properly) and then the energy of distorting the protein itself.

To be a little more formal, imagine that for every configuration Q of the protein, there is a different activation energy for the reaction, $E_{\text{act}}(Q)$. Then if the fluctuations in Q are fast, we expect to see an average rate constant

$$k = A \int dQ \, P(Q) \exp\left[-\frac{E_{\text{act}}(Q)}{k_B T} \right], \tag{4.14}$$

where $P(Q)$ is the distribution of protein structures. If we fix Q at its equilibrium position, we could find that $E_{\text{act}}(Q = Q_{\text{eq}})$ is large, which might make us think that the reaction will be slow. But by sampling nonequilibrium configurations, the protein can speed up the reaction.

It is important that reaction rates are exponentially suppressed, because then these rates are controlled not just by fluctuations but by the tails of the distribution of fluctuations in molecular structure. With reasonable guesses about the form of $P(Q)$ and $E_{\text{act}}(Q)$, the average rate constant in Eq. (4.14) thus will be dominated by the protein structures that are in the tail of $P(Q)$; in effect the reaction "waits" for fluctuation into the right structure. Unfortunately this means that experimental methods showing us the structure of proteins—methods that necessarily sample the center of the distribution $P(Q)$—will not show us the structures most relevant for the reaction itself.

Is there any evidence for such coupling of protein structural fluctuations to the modulations of chemical reaction rates? I think the strongest evidence is from the mid-1970s, in a beautiful series of experiments by Austin and colleagues. The idea is very simple. Suppose that we really do have the activation energy varying with the configuration of the protein. If we could stop the motion of the protein, then each molecule would be stuck with a different activation energy and hence a different reaction rate. Then, instead of seeing an average rate, each molecule reacts at its own rate, and if we count the total number of molecules that have not yet reacted, we should see

$$N(t) = \int dQ\, P(Q) \exp\left[-Ae^{-E_{\text{act}}(Q)/k_B T}t\right],\tag{4.15}$$

which definitely is not an exponential decay. In fact, if the fluctuations in Q generate very large variations in the activation energy, then this expression is very far from being an exponential decay.

Problem 43: Power-law decays. Suppose that the effect of the fluctuations in Q is to generate a distribution of activation energies

$$P(E) = \frac{1}{n!E_0}(E/E_0)^n e^{-E/E_0}.\tag{4.16}$$

Then, if the structures are frozen, we will have

$$N(t) = \int_0^\infty \frac{dE}{n!E_0}(E/E_0)^n e^{-E/E_0} \exp\left[-Ae^{-E/k_B T}t\right].\tag{4.17}$$

(a) Show that, at large t, there is a saddle-point approximation to this integral, and that this predicts a decay $N(t) \sim t^{-\alpha}$. What determines the power α? Are there corrections to this formula?

(b) Calculate the average rate constant, as in Eq. (4.14):

$$\bar{k} = A\int_0^\infty \frac{dE}{n!E_0}(E/E_0)^n e^{-E/E_0}\exp\left[-\frac{E}{k_B T}\right].\tag{4.18}$$

Does this mean rate obey the Arrhenius law? How large are the deviations? Is there a limit in which the Arrhenius law is recovered?

So, we have the dramatic prediction that if we could freeze the motion of the protein, we'd see something very far from the usual exponential decays. To test this prediction, we need the right model system. In particular, if we are literally going to freeze things, then molecules can't diffuse relative to one another, and most of what we usually think of as chemistry will stop. We need an example of a reaction that happens among molecules that are already together and ready to react. If things are frozen, then the usual trick of suddenly mixing the reactants together to start the reaction also isn't going to work.

In many organisms, including us, oxygen is essential for a wide variety of processes. We take in oxygen by breathing and need to distribute it to all our tissues. The way we do this is to have specific proteins to which oxygen binds, and then the proteins are transported, starting in the blood. The major such oxygen-transport protein in our blood is hemoglobin, and this molecule has four protein subunits, each of which can bind a single oxygen molecule. As noted in Section 2.3, hemoglobin provides the classic example of cooperativity in binding. In our muscles we find a simpler protein, with just one subunit, myoglobin. Myoglobin, hemoglobin, and the cytochromes that we discuss below all are members of the heme protein family, which are defined by the fact that they bind a rather large planar organic molecule, heme, with an iron atom at its center, as shown in Fig. 4.3. Myoglobin and hemoglobin were also the first proteins for which

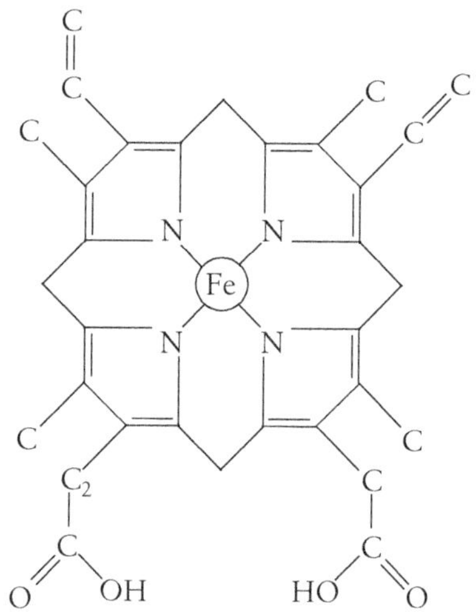

FIGURE 4.3

The heme group at the center of myoglobin, hemoglobin, and other heme proteins. Recall the convention (Fig. 2.17) that carbon atoms are at unmarked nodes of the skeleton, and the hydrogen atoms that complete the four bonds needed for each carbon are not shown. The iron atom at the center is also coordinated from below by a nitrogen from the protein, and oxygen or carbon monoxide can bind to the iron from above the plane. The large conjugated structure of the heme group endows the molecule with a strong absorption cross-section in the visible and ultraviolet range of the spectrum. Because the electronic states of the heme mix with the d orbitals of the iron, the absorption spectrum shifts upon oxygen binding.

the structures were solved using X-ray crystallography. These structures showed that the iron is held in the plane by nitrogens from the heme and from below by a nitrogen from the protein. Oxygen can bind to the iron from above the plane.

The iron atom, and hence the oxygen binding site, is buried deep inside the protein, as shown in Fig. 4.4. This is interesting in part because it tells us that the full process of binding and unbinding must involve some motion or "breathing" of the protein structure. Further, once oxygen binds, if we freeze the protein, it will be trapped, unable to escape. The conjugated electronic structure of the heme generates a strong optical absorption band, and because the electronic states of the heme mix with the orbitals of the iron, the absorption shifts when oxygen binds to the iron. An additional corollary of this mixing is that when a photon is absorbed by myoglobin with oxygen bound, there is some probability that the energy of the absorbed photon will be channeled into breaking the bond between the iron and the oxygen. Thus, if we let oxygen bind to myoglobin and then freeze the solution, we can knock the oxygen off the iron atom with a flash of light, and then we can watch the oxygen rebind after rattling around in the pocket formed by the protein.

In principle, motion of the oxygen molecule from the pocket to the iron atom need not be coupled to motions of the protein. But if this coupling does occur, we expect, from the discussion above, that the kinetics of the rebinding after a light flash will deviate strongly from an exponential decay. We can follow the kinetics by looking at the absorption spectrum, as shown in Fig. 4.5 for both oxygen and carbon monoxide binding to myoglobin. We see that once the solution is truly frozen solid (below ∼160 K in the glycerol-water mixtures used for these experiments), the fraction of molecules that have not reacted decays more nearly as a power law than an exponential. This behavior suggests that a very broad distribution of rate constants has been frozen in, and almost certainly this is because structural fluctuations in the protein couple to the rate constant.

Because proteins are large, the dynamics of the "floppiest" modes are shared across the entire molecule. Thus, the amplitude of fluctuations in the active site where chemical bonds are made and broken can be modulated by events occurring very far from

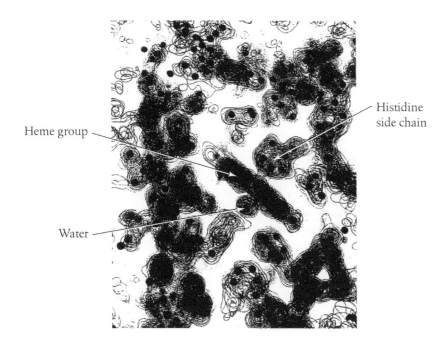

FIGURE 4.4

A slice through the electron density map of the myoglogin molecule, as inferred from X-ray diffraction data. This map is made from data at 1.4 Å resolution. In the center we see the heme group edge on. The histidine side chain from the protein coordinates the iron atom from below the plane of the heme, and in the crystals used in these experiments a water molecule binds to the iron atom in the position that would be taken by oxygen. Note that there is not much empty space in the structure, so that the protein actually has to "breathe" for oxygen to have access to the iron or to escape once it is bound. From Kendrew (1964).

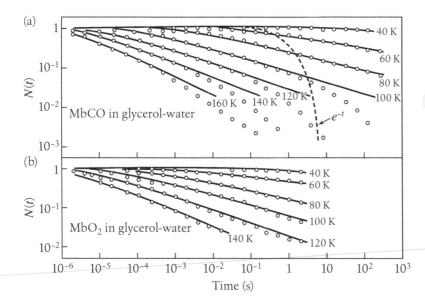

FIGURE 4.5

Rebinding of oxygen and carbon monoxide to myoglobin at low temperatures, following a flash of light to break the bond. Circles are data points, obtained by monitoring the absorption spectrum. Note that this is a logarithmic plot on both axes, so that an enormous range of times is shown. Lines are fits to the phenomenological power-law decay $N(t) = 1/[1 + (t/t_0)^n]$. The dashed line shows, for contrast, an exponential decay, $N(t) = e^{-kt}$, with $k = 1\,\text{s}^{-1}$. Redrawn from Austin et al. (1975).

the site, even if these events serve only to make the protein stiffer (or less stiff). As a result, there are many pathways for the rest of the cell to interact with a single protein molecule—binding small molecules, contacting other proteins, being phosphorylated or methylated, and so forth—and thereby control the pace of chemical events.

So far our discussion of chemical reactions has treated motion along the reaction coordinate as being completely classical. Is it possible that quantum effects could be relevant? Notice in Fig. 4.5 that as the temperature is lowered, the kinetics remain consistently nonexponential, but the typical time scale (e.g., the time required for the reaction to reach 90% completion) increases. If we keep lowering the temperature, eventually this slowing stops, and we see temperature-independent kinetics. Almost certainly this behavior arises because the reaction proceeds by quantum mechanical tunneling through the effective barrier rather than by thermal activation over the barrier. The observation of quantum mechanical effects in a biological system always triggers excitement, although this is tempered somewhat by the fact that, in this case, to see tunneling the temperatures have to be very low indeed (below 10 K). In fact, well before the work on myoglobin, there had been observations of temperature-independent kinetics in the photon-triggered electron-transfer reactions in photosynthesis. Although our immediate experience of photosynthesis involves plants, many of the key experiments on the dynamics of electron transfer were done in photosynthetic bacteria.

The basic business of photosynthesis is carried out by the reaction center, a huge complex of proteins that holds a collection of medium-sized organic molecules—chlorophylls, pheophytins (chlorophylls without the magnesium), and quinones, as shown in Fig. 4.6. Two of the chlorophylls are held in a special pair, and the electronic states of these two molecules are strongly mixed. If one purifies the reaction center away from all the accessory structures, the photochemistry is triggered when P absorbs a photon.

(a)   (b)

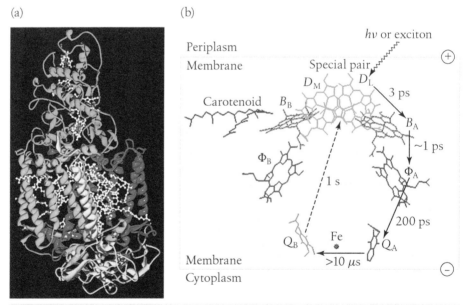

FIGURE 4.6

The photosynthetic reaction center from the bacterium *R. viridis*. (a) The complex of proteins and cofactors. Proteins are drawn as ribbons that trace the backbone of the polypeptide chain (for more details, see Section 5.1), with the cytochrome in green at top. Cofactors are shown as ball-and-stick structures. Image by T. McAnaney, Stanford University, based on coordinates in the Protein Data Bank. (b) The cofactors shown in isolation. In vivo this whole structure sits in a membrane, so that the electron transfers induced by photon absorption move charges across the membrane. Redrawn from http://metallo.scripps.edu/promise/PRCPB.html.

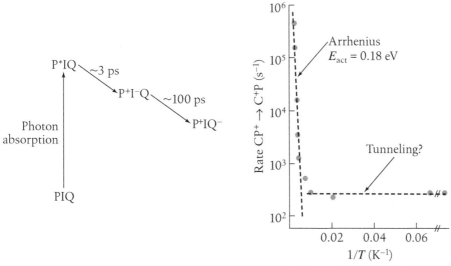

FIGURE 4.7

Schematic of the electron-transfer reactions in the reaction center of photosynthetic bacteria. The special pair P absorbs a photon, and transfers an electron from the excited state to an intermediate acceptor I (pheophytin), which then passes the electron to a quinone molecule Q; there is a second quinone, not shown. The hole on P is filled in by electron transfer from another protein, cytochrome c, (C) and the kinetics of the reaction $CP^+ \rightarrow C^+P$ provided the first evidence for quantum tunneling in a biological system, as shown. Redrawn from DeVault and Chance (1966).

From the excited state of the special pair P, an electron hops to states localized on the pheophytin I and then the quinone Q, as shown in Fig. 4.7. Because P and Q are held (by the protein scaffold) on opposite sides of a membrane, the net effect is to transfer charge across the membrane, capturing the energy of the absorbed photon. Quinones exist in multiple protonation states, so that the electron transfer can couple to proton transfer, and in this way the reaction center serves to drive protons across the membrane. The difference in electrochemical potential for protons provides a universal energy source that is used by other transmembrane proteins, for example to synthesize ATP, which all cells use to power other processes (including movement, as described in Section 4.2). In more complex organisms, including green plants, there are two kinds of reaction centers, one of which couples photon-driven electron transfer to the splitting of water to make most of the oxygen in our atmosphere.

To complete the cycle and reset the reaction center for the arrival of the next photon, the hole on P needs to be filled in, which happens by electron transfer from another protein, cytochrome c, which can also diffuse away from the membrane and interact with the rest of the cell's chemistry. It is this reaction that provided the first evidence for tunneling in a biological system. If the cytochrome is absent, as in purified reaction centers, one can observe the recombination reaction $P^+Q^- \rightarrow PQ$, which also has an anomalous temperature dependence, as discussed below. To connect with the discussion of myoglobin, this recombination reaction also exhibits nonexponential kinetics under some conditions, suggesting that it is possible to freeze some fluctuations in structure that normally provide rapid modulations of the reaction rate.

The key to experiments on the kinetics of photosynthetic electron transfer is that all molecules involved change their absorption spectra significantly when they gain or lose an electron. Not coincidentally, these spectra overlap with the spectrum of solar radiation and are concentrated in a range of wavelengths surrounding the visible—from the near infrared to the near ultraviolet. We can trigger the reactions with a pulsed laser tuned to the absorption band of P, and we can then monitor different spectral features that track the different components. This effort started in the 1950s and 1960s with time resolutions in the microsecond range, and evolved—with successive revolutions in the techniques for generating short laser pulses—down to picoseconds and femtoseconds. This development parallels the exploration of the visual pigments described in Section 2.2.

One key point about the photosynthetic reaction center is that all electron-transfer processes work even when the system is frozen, which tells us that there is no need for the different components to diffuse to find one another—all donor and acceptor sites are held in place by the protein scaffolding. This allows for investigation of the electron-transfer reactions over a wide range of temperatures, as was done to dramatic effect by DeVault and Chance in the mid-1960s, with the result shown in Fig. 4.7. Near room temperature, the electron transfer from cytochrome c back to the special pair exhibits a normal Arrhenius temperature dependence with an activation energy $E_{\rm act} \sim 0.18\,{\rm eV}$. Importantly, the temperature dependence is continuous as the system is cooled through the solvent's freezing point. But somewhere around $T \sim 100\,{\rm K}$, the temperature dependence stops, and the reaction rate remains the same down to liquid helium temperatures ($T \sim 4\,{\rm K}$). This behavior strongly suggests that the reaction proceeds by tunneling at low temperatures.

Problem 44: A wrong model.   If a reaction proceeds by activation over a barrier of height E, the rate is $k \propto \exp(-E/k_B T)$. If it proceeds by tunneling through the barrier, we expect $k \propto \exp(-2\sqrt{2mE}\ell/\hbar)$, where ℓ is the width of the barrier, and m is the effective mass of the tunneling particle. Be sure you understand where this comes from. For the DeVault-Chance reaction, there is a direct measurement of the activation energy $E \sim 0.18\,{\rm eV}$. If you imagine that it is the electron that has to go over or through this barrier, what value of ℓ is needed to explain the crossover from Arrhenius behavior to temperature independence near $T \sim 100\,{\rm K}$? Does this result make any sense?

After roughly a decade of confusion (including discussions of the model in Problem 44), a clearer understanding of tunneling in electron transfer emerged in the mid- to late 1970s.[3] The basic idea is schematized in Fig. 4.8. We have an electron donor D and an acceptor A; the reaction is DA → D$^+$A$^-$. The states DA and D$^+$A$^-$ are different *electronic* states of the system. From the Born-Oppenheimer approximation, we know that when a molecule shifts to a new electronic state, the nuclei move on a new potential surface. We usually describe these nuclear or atomic motions as molecular vibrations, so we refer to the relevant coordinates as vibrational coordinates. The simplest scheme,

3. The relevant physics here is essentially the same as in the discussion of absorption spectra in large molecules. See Section 2.2 and Appendix A.4.

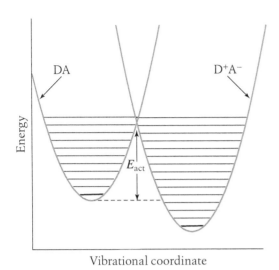

Energy

DA

D^+A^-

E_{act}

Vibrational coordinate

FIGURE 4.8

Electron transfer is coupled to vibrational motion. There are two electronic states, corresponding to localization of the electron on the donor D or the acceptor A, and these states mix only weakly. The strong coupling is to vibrational motion, because transfer of the electron is associated with a change in the equilibrium structure of the molecules. The activation energy of the transfer is E_{act}.

as in Fig. 4.8, is one in which the vibrations are approximately harmonic. Then when the electronic states change, we can imagine changes in the structure of the normal modes, changes in the frequencies of these modes, and shifts in the equilibrium positions along the modes. Barring symmetries, the last effect should be the dominant one, and certainly it is the simplest.

In the state DA, an electron is localized on the donor. In the state D^+A^-, this electron is localized on the acceptor. If the donor and acceptor sites are far apart, as is often the case in large biological molecules, then the wave functions of the electrons in these localized states will overlap only deep in their tails; thus, any matrix element that connects these two states must be very small. But if we want to have a transition between two states that are connected by only a small matrix element, then by Fermi's golden rule, we need these states to be of the same energy. As shown in Fig. 4.8, this happens only at special points, where the two potential energy surfaces for vibrational motion intersect. The rate of the reaction should then be proportional to the probability of finding the system at this crossing point. The key point, then, is that at high temperatures this probability is controlled by the thermal fluctuations in the vibrational coordinates, whereas at low temperatures the system can still reach the crossing point, but now the fluctuations are dominated by quantum zero-point motion. If the activation energy—the energy required to distort the molecule from its equilibrium structure in state DA to the crossing point—is large compared to the relevant vibrational quanta, then a zero-point fluctuation that carries the system to the crossing point necessarily involves sampling the tails of the ground-state wave function. Thus, the system moves into a region that would be forbidden to a classical particle, even granting that it has the zero-point energy to work with. Thus, at low temperatures, the reaction is controlled by tunneling of the vibrational degrees of freedom, whereas at high temperatures these degrees of freedom move classically over the barrier.

To make this argument a bit more precise, let's write the Hamiltonian corresponding to Fig. 4.8. We have two electronic states, which we can take as the up and down states of a spin one-half. There is an energy difference ϵ between these states and a weak matrix element Δ that mixes these states.[4] The vibrational coordinate is Q, and this coordinate moves in a potential that depends on the electronic state. Thus, we have

$$\mathbf{H} = \frac{\epsilon}{2}\sigma_z + \Delta\sigma_x + \frac{1}{2}\dot{Q}^2 + \frac{1+\sigma_z}{2}V_\uparrow(Q) + \frac{1-\sigma_z}{2}V_\downarrow(Q), \tag{4.19}$$

where σ_x and σ_z are the usual Pauli matrices for a spin one-half system, and $V_\uparrow(Q)$ $(V_\downarrow)$ is the potential surface in the upper (lower) state; ϵ is a Q-independent energy gap between the states. If we think semiclassically, then the vibrational coordinates hardly move at all during the electronic transition, and so from the golden rule, we should have the reaction rate

$$k \sim \frac{1}{\hbar}\Delta^2\left\langle\delta(E_\uparrow - E_\downarrow)\right\rangle = \frac{1}{\hbar}\Delta^2\left\langle\delta\left[\epsilon + V_\uparrow(Q) - V_\downarrow(Q)\right]\right\rangle, \tag{4.20}$$

where we have to average over the fluctuations of Q in the initial state DA. In the simplest case, where the potential surfaces are harmonic with stiffness κ, differing only by a shift Q_0 in their equilibrium positions, we have

$$V_\uparrow(Q) = \frac{\kappa}{2}Q^2, \tag{4.21}$$

$$V_\downarrow(Q) = \frac{\kappa}{2}(Q - Q_0)^2, \tag{4.22}$$

and hence $V_\uparrow(Q) - V_\downarrow(Q) = \kappa(Q_0 Q - Q_0^2/2)$, so that

$$k \sim \frac{1}{\hbar}\Delta^2\left\langle\delta\left(\epsilon - \frac{\kappa}{2}Q_0^2 + \kappa Q_0 Q\right)\right\rangle \tag{4.23}$$

$$= \frac{\Delta^2}{\hbar\kappa Q_0}P\left(Q = \frac{Q_0}{2} - \frac{\epsilon}{\kappa Q_0}\right). \tag{4.24}$$

If the particle moves in a harmonic potential with frequency ω, then in thermal equilibrium the distribution of Q is Gaussian. The variance is $\langle(\delta Q)^2\rangle = k_B T_{\text{eff}}/\kappa$, where

$$k_B T_{\text{eff}} = \hbar\omega\left[\frac{1}{2} + \frac{1}{e^{\hbar\omega/k_B T} - 1}\right]; \tag{4.25}$$

notice that as $T \to 0$, $k_B T_{\text{eff}}$ approaches the zero-point energy $\hbar\omega/2$, whereas for $k_B T \gg \hbar\omega$, we have $T_{\text{eff}} \approx T$. Putting all the terms together, we find

$$k \sim \frac{\Delta^2}{\hbar\sqrt{4\pi\lambda k_B T_{\text{eff}}}}\exp\left[-\frac{(\epsilon - \lambda)^2}{4\lambda k_B T_{\text{eff}}}\right], \tag{4.26}$$

where $\lambda = \kappa Q_0^2/2$ is the reorganization energy that would be required to distort the molecule from its equilibrium configuration in DA to the equilibrium configuration appropriate to D^+A^- if the electron did not actually transfer. Note the similarity to our results for the absorption spectra of large molecules in Eq. (2.74).

---

4. The use of Δ here is conventional. I hope it does not cause confusion with the use of Δ to describe atomic displacements in Section 2.2.

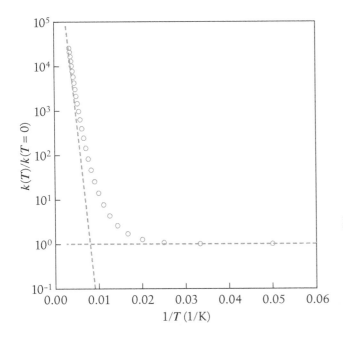

FIGURE 4.9

Temperature dependence of the electron transfer rate from Eq. (4.26). Parameters are chosen, as described in the text, to match the behavior of the DeVault-Chance reaction in Fig. 4.7. Circles are values of the rate computed at 20 K intervals, and dashed lines indicate the asymptotic behavior at high (activated) and low (tunneling) temperatures.

Figure 4.9 shows the predicted dependence of the electron-transfer rate on temperature in a parameter regime chosen to match the DeVault-Chance reaction. To have the transition between Arrhenius and tunneling behavior at the right temperature, we need a vibrational frequency[5] $\omega/2\pi \sim 200$ cm^{-1}. If we look at the Raman spectra of cytochrome c or related molecules, there is a vibrational mode near this frequency that corresponds to motions of the iron atom perpendicular to the plane of the heme group. This makes sense, because when we add or subtract an electron from the molecule, this charge is shared between the iron and the heme, and on average the iron is displaced relative to the heme when the molecule changes its oxidation state. The energy difference between reactants and products can be measured directly by separate electrochemical experiments ($\epsilon = 0.45$ eV), and then to get the activation energy right, we must have $\lambda \sim 0.14$ eV. If the relevant vibrational mode really is (mostly) the motion of the iron relative to the rest of the protein, then we know the mass associated with the mode and hence the stiffness $\kappa = m\omega^2$, so we can determine $Q_0 \sim 0.2$ Å. This value is consistent with the displacements found when comparing the oxidized and reduced structures of cytochrome c. So, this account of vibrational motion as controlling the temperature dependence of the reaction rate seems to make sense in light of everything else we know about these molecules, although admittedly it is a rough comparison.

---

5. Molecular vibrations contribute to the absorption of radiation in the infrared, and it is conventional to measure frequency in wavenumbers or inverse centimeters. To convert to the more usual hertz, just multiply by the speed of light, 3×10^{10} cm/s. Note that this convention is about units and is not a reference to the inverse wavelength in the medium used for the experiment, so there is no correction for the index of refraction. Once you start reading about molecular spectroscopy and chemical reactions (replete with calories and moles), you will have to get some practice at changing units!

Problem 45: Getting numbers out.   Convince yourself that the numbers in the preceding paragraph make sense. In particular, extract the estimate $Q_0 \sim 0.2$ Å for the motion of the iron atom relative to the protein.

There are many loose ends here. First, we have given a description in terms of one vibrational mode, but we have found an expression for the reaction rate that shows no sign of resonances when the energy difference ϵ between reactants and products is an integer multiple of the vibrational quantum $\hbar\omega$. Presumably the solution to the problem is the same as in our discussion of the absorption spectra of rhodopsin (Section 2.2): individual modes are damped, so that resonances are broadened, and there are many modes, so the broadened resonances overlap and smear into a continuum.

The second problem concerns the significance of all this machinery for biological function. It is very impressive to see quantum tunneling in a biological molecule, but our excitement should be tempered by the fact that we see this only at temperatures below 100 K, far out of the range where life actually happens. Measurements on the (much faster) initial steps of electron transfer, however, show that approximately temperature-independent reaction rates persist up to room temperature. Indeed if we look closely at the rates of $P^*I \to P^+I^-$ and $I^-Q \to IQ^-$ (see Fig. 4.7), we see a slight inverse temperature dependence, with the rate slowing by a factor of two or three as we increase the temperature from 4 to 300 K. If we isolate the reaction center and remove the second quinone, then we observe a slow recombination reaction $P^+Q^- \to PQ$, which also has a very weak, and perhaps even slightly inverse, temperature dependence up to room temperature. In fact, the theory as we have sketched it provides a possible explanation for this behavior: if we tune the energy difference between reactants and products so as to maximize the reaction rate, we have $\epsilon = \lambda$, and the exponential dependence of the reaction rate on T_{eff} disappears; all we have left is $k \propto 1/\sqrt{T_{\text{eff}}}$, which indeed is a weak inverse temperature dependence. This sort of fine tuning might make sense for the initial steps $P^*I \to P^+I^-$ and $I^-Q \to IQ^-$—perhaps evolution has selected for molecular parameters that maximize the electron-transfer rates—but it seems a stretch to assume that the very slow (and biologically irrelevant) reaction $P^+Q^- \to PQ$ has been "optimized" in this way. What's going on?

The structure of the reaction center is such that one can take out the quinone molecules and replace them with analogs that have different electron affinities, thus manipulating the value of ϵ. Perhaps surprisingly, increases in ϵ have very little effect on the rate constant for the recombination reaction $P^+Q^- \to PQ$, as shown in Fig. 4.10, or on the forward reaction $I^-Q \to IQ^-$; for all the values of ϵ probed, one sees an approximately temperature independent rate. This argues strongly against tuning of $\epsilon = \lambda$ as an explanation for the observed "activationless" behavior.

Suppose that instead of one vibrational mode, we have two—one at a low frequency ω (which we can treat by the semiclassical argument given above) and one at a high-frequency Ω that really needs a proper quantum mechanical description. The initial state of the high frequency mode is the ground state (because $k_B T \ll \hbar\Omega$), but in the final state we can excite one or more vibrational quanta, and the overall reaction rate will be a sum over terms corresponding to each of these possible final states. From the point of view of the low-frequency mode, if the system transitions into a state with n high-frequency quanta, the matrix element is normalized ($\Delta \to \Delta_n$) and the energy

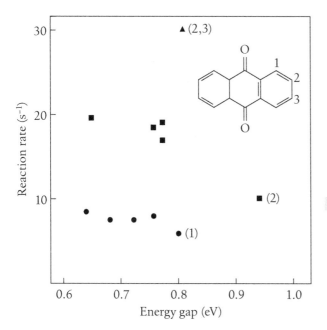

FIGURE 4.10

The rate of the recombination reaction $P^+Q^- \rightarrow PQ$ as a function of the energy gap ϵ between initial and final electronic states. The energy gap is changed by chemical modification of the quinone molecule, adding different groups (methyl, ethyl, etc.) to the sites 1 (circles), 2 (squares, triangle), and 3 (triangle), as indicated on the molecular structure. Redrawn from Gunner et al. (1986).

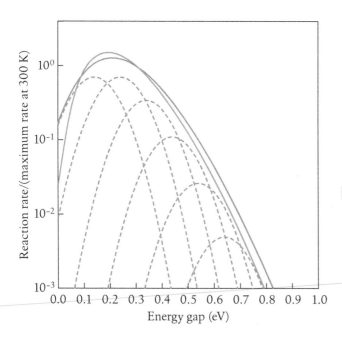

FIGURE 4.11

Electron transfer coupled to high-frequency vibrations, from Eq. (4.27) and Eq. (4.36). Dashed lines show contributions to the rate constant at $T = 300$ K from processes that leave behind $n = 0, 1, 2, \ldots$ quanta in the high-frequency mode. The total rate $k(T = 300$ K) is shown in the solid orange line, and $k(T = 30$ K) is in blue. The high-frequency mode has $\hbar\Omega = 0.1$ eV and $S = 1$ (see Eq. (4.36) for the definition of S).

gap is reduced ($\epsilon \rightarrow \epsilon - n\hbar\Omega$). Thus, the rate constant becomes

$$k = \sum_{n=0}^{\infty} \frac{\Delta_n^2}{\hbar\sqrt{4\pi\lambda k_B T_{\text{eff}}}} \exp\left[-\frac{(\epsilon - n\hbar\Omega - \lambda)^2}{4\lambda k_B T_{\text{eff}}}\right], \qquad (4.27)$$

where now λ refers only to the reorganization energy of the low-frequency mode. Results are shown in Fig. 4.11.

Problem 46: Renormalized matrix elements.   To complete the calculation in Eq. (4.27), we need to understand how the matrix elements are renormalized by coupling to the high-frequency modes. Each mode should be described by a harmonic oscillator, and for simplicity let's assume that the frequency of the oscillation does not change as the molecule transitions from one state to the other. There will be, however, a shift in the equilibrium position of the relevant coordinate; denote this shift by q_0. Then, for the reactants state we have a Hamiltonian

$$H_r = \frac{1}{2m}\hat{p}^2 + \frac{m\Omega^2}{2}\hat{q}^2, \tag{4.28}$$

where $\hat{p}$ and $\hat{q}$ are the momentum and position operators, respectively. For the products state we have

$$H_p = \frac{1}{2m}\hat{p}^2 + \frac{m\Omega^2}{2}(\hat{q} - q_0)^2. \tag{4.29}$$

Because the frequency Ω is large—in particular, $\hbar\Omega \gg k_B T$—we can assume that the system starts in the ground state of the reactant Hamiltonian, $|0_r\rangle$. The probability of "leaving behind" n quanta in the transition from reactants to products is then related to the overlap between the states $|0_r\rangle$ and $|n_p\rangle$,

$$P_n = |\langle n_p|0_r\rangle|^2, \tag{4.30}$$

and hence the renormalization we are looking for is $\Delta^2 \to \Delta^2 P_n$.

(a) Show that the wave function of the state $|n_p\rangle$ is related to the wave function of the state $|n_r\rangle$ by

$$\psi_n^{(\text{prod})}(q) \equiv \langle q|n_p\rangle = \psi_n^{(\text{reac})}(q + q_0). \tag{4.31}$$

That is, $\psi_n^{(\text{prod})}(q)$ is a spatially translated version of $\psi_n^{(\text{reac})}$.

(b) Use the explicit representation of the momentum operator as $\hat{p} = (-i\hbar)\partial/\partial x$ to show that translations are accomplished by the operator $e^{+iq_0\hat{p}/\hbar}$, that is,

$$e^{+iq_0\hat{p}/\hbar}\psi(q) = \psi(q + q_0) \tag{4.32}$$

for any wave function $\psi(q)$. Use this result to show that

$$P_n = |\langle n_p|0_r\rangle|^2 = |\langle n_r|e^{-iq_0\hat{p}/\hbar}|0_r\rangle|^2. \tag{4.33}$$

(c) Go back to Problem 4 for the definition of creation and annihilation operators, $a^\dagger$ and a. There you should have shown that

$$\hat{p} = i\sqrt{\frac{m\hbar\Omega}{2}}(a^\dagger - a). \tag{4.34}$$

Also, if $\hat{A}$ and $\hat{B}$ are two operators such that their commutator is a c-number, then

$$e^{\hat{A}+\hat{B}} = e^{-\frac{1}{2}[\hat{A},\hat{B}]}e^{\hat{A}}e^{\hat{B}}. \tag{4.35}$$

Use these ingredients to show that

$$P_n = e^{-S}\frac{S^n}{n!}, \tag{4.36}$$

and give a formula for S. What is the physical significance of this quantity?

We see that the possibility of exciting different numbers of vibrational quanta greatly broadens the dependence of the rate constant on the energy gap ϵ and provides a huge widening of the region over which we see very little (or even inverted) temperature dependence. This mechanism seems a more plausible and robust explanation of the observed activationless kinetics in the photosynthetic reaction center. Importantly, it relies in an essential way on the quantum behavior of the high-frequency vibrational motions that are coupled to the electron transfer, even at room temperature. There is no shortage of such high-frequency modes in the quinones, chlorophylls, and pheophytins; what is interesting is the way in which the interplay of these quantum modes with the lower frequency classical modes (including, presumably, modes of the protein scaffolding itself) shapes the observed functional behavior.

A third issue is that, although we are talking about electron-transfer reactions, we have said relatively little about the electrons themselves—there are two states, localized on the donor and acceptor sites, and there is a matrix element that connects these states, but that seems to be all. In fact we can say a bit more. Our use of perturbation theory obviously depends on the matrix element not being too large. If we go back to our simple model of the DeVault-Chance reaction (see Fig. 4.9) and try to fit the absolute rate constants as well as the temperature dependence, we find $\Delta \sim 10^{-4}$ eV. Certainly this value is small compared with the other energies in the problem (λ, $\hbar\omega$, $k_B T$, ϵ), which indicates that our use of perturbation theory is consistent. It requires substantial effort to calculate these matrix elements from the molecular structures, which seems daunting. One can make progress because one need not find the full electronic structure of these large molecules, just trace the decay of wave functions for an extra electron moving above the filled states in all the chemical bonds. There are interesting questions about the extent to which pathways can be defined for the electron transfer, as one might expect classically, or whether interference among multiple paths through the molecule is significant; these and other issues can be traced through the references for this section in the Annotated Bibliography.

The very first step of photoinduced electron transfer, $P^*I \rightarrow P^+I^-$, occurs on the picosecond time scale. This is fast enough that, as with the photoisomerization of rhodopsin, we need to reexamine the usual assumptions about the separation of time scales in chemical reactions. Our understanding of the Arrhenius law, for example, is based on the assumption that there is a large exponential suppression of the probability of trajectories that cross the barrier, so that reaction rates will be slow compared with the time scales of vibrational motion and relaxation in the potential wells. More simply, it takes time for molecular motions to equilibrate with their environment, and if reactions are too fast, the notion that the molecule is at a well-defined temperature will break down, especially if we are considering reactions triggered by the absorption of a photon. We also have to worry about the time scales for loss of quantum coherence.

Consider a simple version of our problem, as schematized in Fig. 4.12. There are two states of the molecules, one of which we can think of as P^*I and the other as P^+I^-. These states are connected by a matrix element Δ, and the final state decays at a rate Γ. In the limit that the decay is fast, we should think of the single final state as being broadened in energy with a width $\sim\Gamma$ ($\hbar = 1$), and so there is a density of states $\rho \sim 1/\Gamma$; thus, from Fermi's golden rule, the rate of the reaction is $k \sim \Delta^2/\Gamma$. In contrast, if the decay is slow, we will see coherent oscillations between the two states at a frequency Δ, and nothing irreversible will happen until the decay, so that any reasonable definition of the rate constant for the transition will give $k_{\text{eff}} \sim \Gamma$. Thus, if the matrix element is fixed,

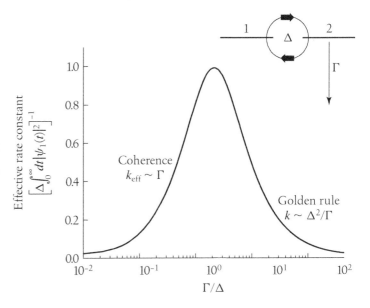

FIGURE 4.12

Competition between coherent mixing and dissipation. The decay rate is Γ, and Δ is the matrix element. For large dissipation rates, the transition $1 \to 2$ is described by a rate constant that we can calculate in perturbation theory, $k \sim \Delta^2/\Gamma$. For small dissipation rates, coherent mixing is evident, and irreversibility of the transition occurs only on the time scale of the dissipation itself. Optimal rates occur in the crossover region, $\Gamma/\Delta \sim 1$.

the fastest reaction occurs when $\Gamma \sim \Delta$, and in this regime we have a rate $k \sim \Delta$ as opposed to the golden rule of $k \propto \Delta^2$. Then reactions that are optimized for speed will sit at the threshold of coherence, where mixing and decay are happening on the same time scale. If we could perturb the system to slow the decays just a bit, we would then reveal the incipient coherence, and (as explained in the references for this section in the Annotated Bibliography) this seems to be what happens in the reaction $P^*I \to P^+I^-$. Similar ideas are emerging regarding the processes by which energy is collected in the "antenna pigments" and transferred to the reaction center.

Problem 47: Optimizing rates.   The model shown in Fig. 4.12 is described by the equations

$$i\frac{d\psi_1(t)}{dt} = -i\Gamma\psi_1(t) + \Delta\psi_2(t), \qquad (4.37)$$

$$i\frac{d\psi_2(t)}{dt} = \Delta\psi_1(t) - i\Gamma\psi_2(t), \qquad (4.38)$$

where again $\hbar = 1$.

(a) Solve this model for the time dependence of the probability to be in state 1, $|\psi_1(t)|^2$. Show that, for $\Gamma/\Delta \gg 1$, this probability decays exponentially with the rate you would have expected from the golden rule.

(b) To study the system outside the golden rule regime, define an effective lifetime of state 1 to be

$$\tau = \int_0^\infty dt \, |\psi_1(t)|^2, \qquad (4.39)$$

and the corresponding effective rate constant $k_{\text{eff}} = 1/\tau$. Use this to reproduce the results in the figure, in particular the asymptotic behavior for $\Gamma/\Delta \ll 1$.

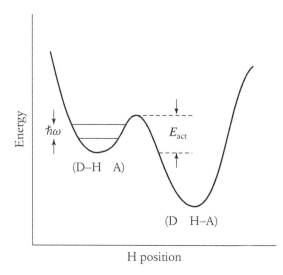

FIGURE 4.13

Transfer of a hydrogen atom H from donor D to acceptor A. The reaction coordinate is the position of the H atom, but we expect that quantization effects are significant.

All other things being equal, quantum effects are stronger for lighter particles. As we have seen, electrons essentially always tunnel—there are almost no chemical or biochemical reactions involving thermal activation of an electron over a barrier. Since the early days of quantum mechanics, people have wondered whether chemical reactions involving the next lightest particle—a proton or hydrogen atom—might also involve tunneling in a significant way. To be concrete, consider the situation in Fig. 4.13, where the reaction coordinate is the position of the H atom itself, moving from donor to acceptor atom. But, while still attached to the donor atom (e.g., a carbon) we can observe vibrations of the donor-hydrogen bond, and for C-H we know that the frequencies of these vibrations can be as high as $\nu \sim 2500$–3000 cm^{-1}. The vibrational quanta thus are $h\nu \sim 1/4$–$1/3$ eV. In fact the activation energies of many chemical reactions are not that much larger than this, perhaps 0.5–1.0 eV. Thus, as indicated in the figure, climbing up to the top of the barrier between reactants and products involves adding just two or three vibrational quanta. So, the reaction can't really be completely classical, if the reaction coordinate really is the stretching of the bond itself.

If we make the crude approximation that the barrier is rectangular with height E, then the rate of going over the barrier should be $k \propto e^{-E/k_B T}$, as before, while the rate of tunneling through the barrier is $k \propto e^{-2\sqrt{2mE}\ell/\hbar}$, where ℓ is the width of the barrier and m is the mass of tunneling particle. Although we could worry about the prefactors, the exponentials are probably the dominant effects, and so we might guess that tunneling is more important that classical thermal activation only if

$$e^{-E/k_B T} < e^{-2\sqrt{2mE}\ell/\hbar}, \tag{4.40}$$

or

$$T < T_0 \sim \frac{\hbar}{k_B}\sqrt{\frac{E}{8m}}\ell. \tag{4.41}$$

If the width of the barrier is $\ell \sim 1$ Å, and its height is $E \sim 50$ kJ/mole, then with m the mass of the proton we find $T_0 \sim 190$ K, well below room temperature. Thus, although it might be difficult to see the transfer of a proton as being completely classical, it is also true that the transfer reaction is unlikely to be dominated by tunneling at room temperature if the barrier is static.

In the interior of a protein, we can imagine that the donor and acceptor are held by different parts of the large molecule, and as the protein flexes and breathes, these sites will move. Effectively, the width of the barrier will fluctuate. On average, this increases the probability of tunneling through the barrier. If the fluctuations in ℓ are Gaussian, the tunneling probability becomes

$$e^{-2\sqrt{2mE}\ell/\hbar} \rightarrow \left\langle e^{-2\sqrt{2mE}\ell/\hbar} \right\rangle = \exp\left[-2\sqrt{2mE}\,\bar{\ell}/\hbar + 4mE\langle(\delta\ell)^2\rangle/\hbar^2\right], \quad (4.42)$$

where $\bar{\ell}$ is the average width of the barrier, and $\langle(\delta\ell)^2\rangle$ is the variance of this width. With the parameters as before, the enhancement of the tunneling probability involves the term

$$4mE\langle(\delta\ell)^2\rangle/\hbar^2 \sim 6\left\langle\left(\frac{\delta\ell}{0.1\,\text{Å}}\right)^2\right\rangle. \quad (4.43)$$

As described in Appendix A.3, measurements of Debye-Waller factors in X-ray diffraction from protein crystals provide estimates of the fluctuations in structure, and these structural fluctuations are easily several tenths of an angstrom. Thus, this term, which appears in the exponential, can be huge. It completely shifts the balance between tunneling and classical (thermal) activation, so that in the presence of fluctuations it becomes plausible that tunneling is dominant at room temperature.

Notice that the role of protein vibrational motions here is different than in our discussion of electron transfer. In electron transfer, there is a small matrix element that couples the two relevant states, and protein motions serve to bring these two states into degeneracy with each other. This effect presumably could happen in the case of proton transfer as well, but we have focused on the coupling of fluctuations to the tunneling matrix element. This coupling is especially interesting, because it generates exponential terms in the reaction rate that have a dependence on mass ($\ln k \propto m$) that is very different from the naive tunneling exponent ($\ln k \propto -\sqrt{m}$) or the zero-point corrections to the activation energy ($\ln k \propto 1/\sqrt{m}$; see Problem 48). Because this mass-dependent term also depends on the variance of structural fluctuations, it is also temperature dependent. Indeed, it was the discovery of anomalous temperature-dependent isotope effects in enzymatic proton-transfer reactions that prompted renewed discussion of these dynamical effects on tunneling.

Problem 48: Isotope effects.   Chemical reaction rates change when we substitute one isotope for another. There is a semiclassical theory of these isotope effects, which says that the reaction proceeds by conventional thermal activation, but the activation energy is reduced by the zero-point energy of vibrations along the reaction coordinate: $k \propto \exp\left[-(E_{\text{act}} - \hbar\omega/2)/k_B T\right]$.

(a) Vibrational frequencies are proportional to $1/\sqrt{m}$, with m the (effective) mass of the particle(s) moving along the mode with frequency ω. In the simple picture where all motion along the reaction coordinate is dominated by the motion of the proton, derive a relationship between the ratios of rate constants for hydrogen, deuterium, and tritium transfer.

(b) If the reaction coordinate involves motion of atoms other than transferred hydrogen, what happens to the predicted magnitude of the isotope effects? What about the relationship you derived in part (a)?

(c) Suppose that the proton tunnels, but the barrier fluctuates as some coordinate Q of the protein moves. Then by analogy with Eq. (4.14), the rate constant will be

$$k = A \int dQ \exp\left[-\frac{V(Q)}{k_B T} - 2\sqrt{m}\frac{S(Q)}{\hbar}\right], \qquad (4.44)$$

where $V(Q)$ is the effective potential for protein motion, and in the WKB approximation $S(Q)$ is the action for the tunneling of a proton ($m = 1$) through the barrier when the protein is in the structure defined by Q. If we are very careful with the WKB calculation, we can show that $A \propto 1/\sqrt{m}$. In the simplest case, motions of the protein are harmonic, and the barrier is square with a width determined by the protein coordinate, so that $V(Q) = \frac{1}{2}\kappa Q^2$ and $S = S_0(1 - \alpha Q)$, and you can compute k exactly. What is the predicted relationship among the rate constants for hydrogen, deuterium, and tritium transfer? Is there a parameter regime in which this relationship is not so far from the semiclassical results in part (a)? If you can measure rate constants accurately, over a modest range of temperatures, is there a way to test this model more precisely?

I hope that you take a few lessons away from this discussion. First, chemical reactions are the result of fluctuations at the molecular level. We can describe the nature of these fluctuations in some detail, because rare events, such as escape over a high barrier, are dominated by specific trajectories. In large biological molecules, the flexibility of the molecule provides another way for fluctuations to be important: the variations in protein structure, for example, couple to changes in the barrier for the relevant chemical rearrangements or bring weakly coupled electronic states into degeneracy. Finally, these fluctuations in protein structure can completely revise our view of whether the reaction itself proceeds by classical over-the-barrier motion or by quantum tunneling. These theoretical observations, and the experiments to which they connect, suggest that Nature exploits not just the structure of biological molecules but also the fluctuations in these structures to control the rates of chemical reactions.

4.2   Motility and Chemotaxis in Bacteria

If we look at a drop of pond water under a magnifying glass and then under a microscope, we see all sorts of living things moving around. Some are really quite large, complex organisms, such as the larvae of various insects, composed of tens of thousands of cells. We might also find clusters of hundreds or thousands of cells that are essentially all identical—colonies of unicellular organisms. Still smaller are unicellular organisms swimming around on their own, and these come in two varieties. If we look very closely, we will find that some of these single cells look like our cells, with their DNA packaged into a nucleus; they swim by waving or beating structures called cilia or flagella, which are essentially the same as the structures that we have in cells that help move or filter fluids as they pass through our gut and airways. The very smallest cells we can see are bacteria, many of which are on the order of one micron in size, and they don't have a nucleus. Cells with nuclei are called eukaryotic, and cells without them are called prokaryotic.

Although the motions in a drop of pond water may look a bit random, in fact even bacteria are navigating. They are endowed with sensory systems that allow them to move in response to a variety of signals from the environment—light, heat, mechanical forces, and the concentrations of various chemicals. A classical observation (from the nineteenth century) is that some bacteria, swimming in water on a microscope slide, under a cover slip, will collect at the center of cover slip, while others will collect at the edges. Those with more refined tastes will form a tight band that traces the outlines of the square cover slip. Oxygen diffuses into the water through the edges of the cover slip, and by collecting along a square, the bacteria have migrated to a place of constant (not maximal or minimal) oxygen concentration. It is plausible that this happens because they can sense the oxygen concentration and "know" the most comfortable value of this concentration, much as we might move to be the most comfortable distance from a fireplace in an otherwise unheated room.

That bacteria collect at nontrivial concentrations of different molecules really does not demonstrate that they sense the concentration. They might instead sense some internal consequences of the external variables, such as the accumulation of metabolic intermediates. In the 1960s Adler found mutants of *Escherichia coli* that cannot metabolize certain sugars or amino acids but will nevertheless migrate toward the sources of these molecules; also there are mutants that metabolize but can't migrate. This is convincing evidence that metabolism and sensing are separate systems, and thus begins the fruitful exploration of the sensory mechanisms of bacteria and the connection of these sensory mechanisms to motor output. This phenomenon is called chemotaxis.

I'll skip lots of the truly classical stuff and proceed with the modern biophysical approach, which begins around 1970. To a large extent this modern approach rests on the work of Howard Berg and collaborators. The first key step taken by Berg and Brown was to observe the behavior of individual bacteria. *Escherichia coli* are $\sim 1\,\mu$m in size and can be seen relatively easily under the light microscope, but because the bacteria swim at ~ 20 body lengths per second, they easily leave the field of view or the plane of focus. The solution is to build a tracking microscope.

Observations in the tracking microscope showed that the trajectories of individual bacteria consist of relatively straight segments interrupted by short intervals of erratic "motion in place" (Fig. 4.14). These have come to be called runs and tumbles, respectively. Tumbles last ~ 0.1 s, but the erratic motion during this brief time is sufficient to cause successive runs to be in almost random relative directions. Thus, the bacterium runs in one direction, then tumbles and chooses a new direction at random, and so on. Runs themselves are distributed in length, as if the termination of a run were itself a random process.

Closer examination of the runs shows how it is possible for this seemingly random motion to generate progress up the gradient of attractive chemicals. When the bacterium runs up the gradient, the mean duration of the runs becomes longer, biasing the otherwise random walk. Interestingly, these early experiments suggested that when bacteria swim down the gradient (of an attractant, or up the gradient of a repellent), there is relatively little change in the mean run length. Berg described this as a form of optimism: if things are getting better, keep going, but if things are getting worse, don't worry.

Because runs get longer when bacteria swim along a positive gradient, it is natural to ask whether the cell is responding to the spatial gradient itself or to the change in concentration with time along the path. As we will see, the spatial gradients to which the cell can respond are very small, and searching for a systematic difference (for example)

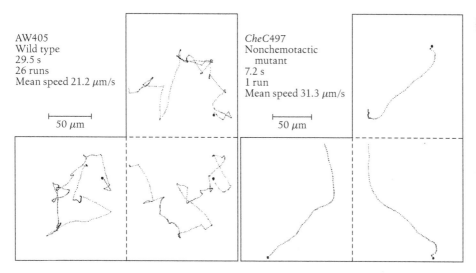

AW405
Wild type
29.5 s
26 runs
Mean speed 21.2 μm/s

50 μm

CheC497
Nonchemotactic
  mutant
7.2 s
1 run
Mean speed 31.3 μm/s

50 μm

FIGURE 4.14

Two paths of *Escherichia coli* as seen in the original tracking microscope experiments. The three panels in each case are projections of the path onto the three orthogonal planes (imagine folding the paper into a cube along the dashed lines). At left, a wild type bacterium, showing the characteristic runs and tumbles. At right, a nonchemotactic mutant that never manages to tumble. Reprinted by permission from Macmillan Publishers, Ltd.: Brown and Berg (1972).

between the front and back of the bacterium is unlikely to be effective just on physical grounds, independent of biological mechanisms. To test for a time-domain mechanism, one can expose the bacteria to concentrations that are spatially uniform but varying in time; if the sign of the change corresponds to swimming up a positive gradient, runs should be prolonged. The first such experiment used very large, sudden changes in concentration and found that cells that experience large positive signals could became trapped in extremely long runs. A more sophisticated experiment used enzymes to synthesize attractants from inert precursors, exposing the cells to gradual changes more typical of those encountered while swimming. Purely time-domain stimuli were sufficient to generate modulations of run length that agree quantitatively with those observed for bacteria experiencing spatial gradients.

Problem 49: Chemotaxis in one dimension. To make the intuition of the previous paragraphs more rigorous, consider a simplified problem of chemotaxis in one dimension. There are then two populations of bacteria, the $+$ population that moves to the right and the $-$ population that moves to the left, each at speed v. Let the probability of finding a $+$ [$-$] bacterium at position x be $P_+(x,t)$ [$P_-(x,t)$]. Assume that the rate of tumbling depends on the time derivative of the concentration along the bacterial trajectory as some function $r(\dot{c})$, where for the $\pm$ bacteria, we have $\dot{c} = \pm v\, dc/dx$, and that cells emerge from a tumble going randomly left or right.

(a) Show that the dynamics of the two probabilities obey

$$\frac{\partial P_+(x,t)}{\partial t} + v\frac{\partial P_+(x,t)}{\partial x}$$
$$= -r\left(v\frac{dc}{dx}\right)P_+(x,t) + \frac{1}{2}\left[r\left(v\frac{dc}{dx}\right)P_+(x,t) + r\left(-v\frac{dc}{dx}\right)P_-(x,t)\right], \quad (4.45)$$

$$\frac{\partial P_-(x,t)}{\partial t} - v\frac{\partial P_-(x,t)}{\partial x}$$
$$= -r\left(-v\frac{dc}{dx}\right)P_-(x,t) + \frac{1}{2}\left[r\left(v\frac{dc}{dx}\right)P_+(x,t) + r\left(-v\frac{dc}{dx}\right)P_-(x,t)\right]. \quad (4.46)$$

Explain the meaning of each of the terms in terms of what happens as cells enter into and emerge from tumbles. Note that in this approximation, tumbles themselves are instantaneous, which is pretty reasonable (0.1 s versus the $\sim$1–10 s for typical runs).

(b) To see whether the bacteria really migrate toward high concentrations, look for the steady state of these equations. If we simplify and assume that the rate of tumbling is modulated linearly by the time derivative of the concentration,

$$r(\dot{c}) \approx r(0) + \frac{\partial r}{\partial \dot{c}} \dot{c} + \cdots, \tag{4.47}$$

show that

$$P(x) = \frac{1}{Z} \exp\left[-\frac{\partial r}{\partial \dot{c}} c(x)\right]. \tag{4.48}$$

Thus, in these approximations, chemotaxis leads to a Boltzmann distribution of bacteria, in which the concentration acts as a potential. If the molecules are attractive, then $\partial r/\partial \dot{c} < 0$ and hence maxima of concentration are minima of the potential, conversely for repellents. The stronger the modulation of the tumbling rate is (as long as we stay in our linear approximation), the lower will be the effective temperature and the tighter the concentration of bacteria around the local maxima of concentration.

Problem 50: Nonlinearities.   In the simplified one-dimensional world of Problem 49, can you make progress without the approximation that $r(\dot{c})$ is linear? More specifically, what is the form of the stationary distribution $P(x)$ that solves Eq. (4.45) and Eq. (4.46) for nonlinear $r(\dot{c})$? Can you show that there still is an effective potential with minima located at places where the concentration is maximal?

Problem 51: A little more about the effectiveness of chemotaxis.

(a) In the one-dimensional model, what happens if the tumbling rate is modulated not just by the time derivative but also by the absolute concentration, so that the bacterium confuses "currently good" for "getting better"?

(b) Can you generalize this discussion to three dimensions? Instead of having just two groups + and −, one now needs a continuous distribution $P(\mathbf{\Omega}, x, t)$, where $\mathbf{\Omega}$ denotes the direction of swimming. Derive an equation for the dynamics of $P(\mathbf{\Omega}, x, t)$ in the same approximations used above, and see whether the Boltzmann-like solution obtains in this more realistic case.

All this description so far is about the phenomenology of swimming. But how does it actually work? The basic problem is that bacteria are too small to take advantage of inertia. When we swim, we can push off the wall of the pool and glide for some distance, even without moving our arms or legs; this gliding distance is on the order of one or two meters, roughly the length of our bodies. In contrast, if a bacterium stops running its motors, it will glide for a distance comparable not to its body length

($\sim 1\,\mu$m) but to the diameter of an atom. To see this, think about a small particle moving through a fluid, subject only to drag forces (the motors are off). If the velocities are small, we know the drag will be proportional to the velocity, so Newton's equation is just

$$m\frac{dv}{dt} = -\gamma v. \tag{4.49}$$

For a spherical object of radius r, the Stokes formula tells us that $\gamma = 6\pi\eta r$, where η is the viscosity of the fluid, and we also know that $m = 4\pi\rho r^3/3$, where ρ is the density of the object. The result is that

$$v(t) = v(0)\exp(-t/\tau), \tag{4.50}$$

where

$$\tau = \frac{m}{\gamma} = \frac{2\rho r^2}{9\eta}. \tag{4.51}$$

If we assume that the density of bacteria is roughly that of water, then it is useful to recall that η/ρ has units of a diffusion constant, and for water $\eta/\rho = 0.01$ cm^2/s. With $r \sim 1\,\mu$m $= 10^{-4}$ cm, this gives $\tau \sim 5 \times 10^{-7}$ s. If the initial velocity is $v(0) \sim 20\,\mu$m/s, the net displacement during this coasting is $\Delta x = v(0)\tau \sim 10^{-11}$ m; recall that a hydrogen atom has a diameter of ~ 1 Å $= 10^{-10}$ m.

The conclusion from such simple estimates is that bacteria can't coast. More generally, mechanics on the scale of bacteria is such that inertia is negligible, as if Aristotle (rather than Galileo and Newton) were right. This phenomenon really is about the nature of fluid flow on this scale. For an incompressible fluid (which is a good approximation here—surely the bacteria don't generate sound waves as they swim), the Navier-Stokes equations are

$$\rho\left[\frac{\partial \mathbf{v}}{\partial t} + \mathbf{v}\cdot\nabla\mathbf{v}\right] = -\nabla p + \eta\nabla^2\mathbf{v}, \tag{4.52}$$

where $\mathbf{v}$ is the local velocity of the fluid, p is the pressure, and as usual ρ is the density and η the viscosity. The pressure is not really an independent variable, but it needs to be there so we can enforce the condition of incompressibility,

$$\nabla\cdot\mathbf{v} = 0. \tag{4.53}$$

These equations need to be supplemented by boundary conditions, in particular, that the fluid moves with the same velocity as any object at the points where it touches that object. Thus, the velocity should be zero at a stationary wall and should be equal to the velocity of a swimmer at the swimmer's surface.

Problem 52: Understanding Navier-Stokes. This is not a fluid mechanics course, but you should be sure you understand what Eq. (4.52) is saying. In particular, it is nothing but Newton's $F = ma$. Explain.

Dimensional analysis is an enormously powerful tool in fluid mechanics. We are free to choose new units for length (ℓ) and time (t_0), and hence for velocity ($v_0 = \ell/t_0$), as well as for pressure p_0, and this gives us

$$\rho \left[\frac{v_0}{t_0} \frac{\partial \tilde{\mathbf{v}}}{\partial \tilde{t}} + \frac{v_0^2}{\ell} \tilde{\mathbf{v}} \cdot \tilde{\nabla} \tilde{\mathbf{v}} \right] = -\frac{p_0}{\ell} \tilde{\nabla} \tilde{p} + \eta \frac{v_0}{\ell^2} \tilde{\nabla}^2 \tilde{\mathbf{v}}, \tag{4.54}$$

$$\frac{\rho \ell v_0}{\eta} \left[\frac{\partial \tilde{\mathbf{v}}}{\partial \tilde{t}} + \tilde{\mathbf{v}} \cdot \tilde{\nabla} \tilde{\mathbf{v}} \right] = -\frac{p_0 \ell}{\eta v_0} \tilde{\nabla} \tilde{p} + \tilde{\nabla}^2 \tilde{\mathbf{v}}, \tag{4.55}$$

where $\tilde{t} = t/t_0$, $\tilde{\mathbf{v}} = \mathbf{v}/v_0$, and $\tilde{p} = p/p_0$. Now we can set $p_0 \ell / \eta v_0 = 1$, which gets rid of all the units, except we are left with a dimensionless combination

$$\mathrm{Re} \equiv \frac{\rho \ell v_0}{\eta}, \tag{4.56}$$

which is called the Reynolds number.[6] Notice that if we choose the unit of length to be the size of the objects that we are interested in, and v_0 to be the speed at which they are moving, then even the boundary conditions don't have any units, nor do they introduce any dimensionless factors that are far from unity. The conclusion is that all fluid mechanics problems with the same geometry (shapes) are the same if they have the same Reynolds number. In this sense, being smaller (reducing ℓ) is the same as living at increased viscosity.[7]

To make a long story short, we live at high Reynolds number, and bacteria live at low Reynolds number (Fig. 4.15), even though we are surrounded by the same fluid when we swim. To simulate the effect of being as small as bacteria on the human scale, we would have to swim through a fluid whose viscosity is roughly that of concrete just before it sets (!). Turbulence is a high–Reynolds number phenomenon, as is the more mundane gliding through the pool after we push off the wall. At low Reynolds number, life is very different. Inertia is absent, and so forces must balance at every instant of time. To say this more startlingly, if $\mathrm{Re} \to 0$, then time does not actually appear in the equations. Thus, as you swim, the distance that you move depends on the sequence of motions that you go through but not on the dynamics with which you execute them.

---

6. I admit that I was at first puzzled by the convention that this is "Reynolds number," not "Reynolds' number" (and certainly not "Reynold's number," although one sees this from time to time). The number is named after Osborne Reynolds (1842–1912), who emphasized its importance, although it had been introduced much earlier by Stokes. If it belongs to Reynolds, you might think it should be "Reynolds' number," but we also refer to "Bessel functions" and not "Bessel's functions." In a discussion that fascinated many of us in my student days, Jackson took the opportunity of a new edition of his *Classical Electrodynamics* to explain that, following this convention, we should talk about "Green functions" and not "Green's functions" (this "boggles some minds," he noted). The world of fluid mechanics abounds with such things—Prandtl number, Schmidt number, Nusselt number, Péclet number—all associated with proper names but not with the possessive construction.

7. It is worth reflecting on the level of universality that we have here. We could imagine starting with a molecular description of fluids, then figuring out that, on the relevant length and time scales, all we need to know are the density and viscosity. Now we see that even these quantities are tied up with our choice of units. If we want to know what happens in natural units (i.e., scaling to the size and speed of the objects of interest), then all that matters is a single dimensionless combination, Re.

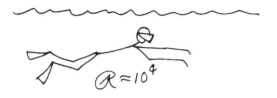

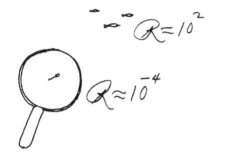

To use Purcell's evocative example, at high Reynolds number a scallop can propel itself by snapping shut, expelling a jet of water, and then opening slowly.[8] The jet will propel the scallop forward, and the drag of reopening can be made small by moving slowly. At low Reynolds number this strategy does not work, and the forward displacement generated by snapping shut will be exactly compensated by the drag on reopening. To have net movement from a cycle, the sequence of shapes that the swimmer goes through in the cycle must break time-reversal invariance, not just the trajectory.

So, how do bacteria evade the "scallop theorem"? If you watch them swimming, you can see that they have long filaments sticking out, and these seem to be waving. I emphasize that "see" is tough here. These filaments are very small, ~ 20 nm in diameter, much thinner than the wavelength of light. To see them, the easiest thing to do is to use dark-field microscopy, in which the sample is illuminated from the side and what you see is the light scattered by $\sim 90°$. These apparently waving appendages are called flagella, by analogy with the motile structures that project from eukaryotic cells, including some of the cells in our own bodies. The difference is that eukaryotic flagella are much thicker than bacterial flagella. If you slice through the tail of a sperm (a prime example of a eukaryotic flagellum) and take an electron micrograph, you find an enormously complex structure; if you analyze the system biochemically, you find it is made from many different proteins. Importantly, some of these proteins act as enzymes and eat ATP, which we know is a source of energy, for example, in our muscles. In contrast, the bacterial flagellum is small, with a relatively simple structure, and the biochemistry suggests that it is little more than a very long polymer made from one kind of protein; this protein is not an enzyme. How can this simple structure, with no ATPase activity, generate motions?

In experiments aimed at better ways to see the flagella, one can attach "flags" to them using viruses that would stick to the flagella by means of antibodies. Once in a

---

8. At least a hypothetical scallop. What real scallops do is less clear to me.

while, a virus with antibodies on both ends would stick to two flagella from different bacteria. When this happened, the bacterial cells could be seen rotating, which was a huge surprise. Eventually people figured out how to break off the flagella and stick the bacteria to a glass slide by the remaining stump, and then the bacterium rotates. One can also look at mutant bacteria that make the motor and "hook" at the base of the flagellum, projecting a short distance out of the cell, but no flagellum; these hook mutants are very convenient for sticking onto glass. Rotation can look like a wave if the flagellum is shaped like a corkscrew, and it is. Rotating a corkscrew obviously violates time-reversal invariance. If you have several corkscrews and you rotate them with the correct handedness, they can fit together into a bundle. If you rotate the other way, the corkscrews clash, and any bundle will be blown apart by this clashing. So, with many flagella projecting from their surface, we can imagine that by switching the direction of rotation, the bacterium switches between a bundle that can smoothly propel the cell forward, and many independently moving flagella that would cause the cell to tumble in place: runs and tumbles correspond to counterclockwise and clockwise flagellar rotation. If you find mutants that never tumble, and stick them down by their stumps, then they all rotate one way; similarly, mutants that tumble too often rotate the other way.

There is much more to say about the rotary engine sitting at the base of the flagella. It is powered not by ATP but by a difference in chemical potential for hydrogen ions between the inside and outside of the cell. This is an energy source that all cells use, albeit in different ways, because it allows chemical events at very different spatial locations to be coupled. Thus, as described in the preceding section, photosynthetic organisms use the energy of the absorbed photons to move electrons across a membrane and then compensate the charges by moving protons. The resulting difference in chemical potential can be used by other membrane-spanning enzymes to make ATP, which then need not be anywhere near the molecules that absorb the photon.[9] In fact, these enzymes that synthesize ATP also rotate as they let protons move down the gradient in their chemical potential, and the same enzymes are responsible for ATP synthesis in all cells. So, proton-driven rotary motors are at the heart of energy conversion in all organisms.

Problem 53: Switching in tethered bacteria.   As noted above, one way of studying bacterial motility and chemotaxis is to tether a bacterium by the stump of one flagellum, observing the rotation of the whole cell rather than the rotation of the flagellum. The file `omega.txt` contains a very long time series of the angular velocity from such an experiment done by Will Ryu, now at the University of Toronto.[10] The samples are taken sixty times per second, and the units of velocity are not quite arbitrary but are not really important either; you should be able to load this data set into MATLAB (`load omega.txt`).

(a) You should see that the velocity switches between positive and negative values, but these values are fairly constant. This observation is consistent with swimming by switching between runs and tumbles, with little or no modulation of the swimming speed. What is

9. You can imagine how confusing these phenomena were before people figured it out! It looked like a mysterious action at a distance.
10. Data can be found through http://press.princeton.edu/titles/9911.html.

the distribution of times spent during each segment of positive or negative (clockwise or counterclockwise) velocity?

(b) It usually is said that switching is a Poisson process, so that the distribution of intervals between switches should be exponential (see Appendix A.1). Are your results in part (a) consistent with this prediction?

(c) Look carefully at the velocity versus time in the data set. Are the data statistically stationary (time-translation invariant)? If you focus on segments of the data that are more clearly stationary, does it change your conclusions in part (b)?

(d) Sometimes the angular velocity makes a partial switch—a brief excursion away from the typical positive or negative value but not quite a full switch to the opposite direction of rotation. Qualitatively, what is happening in these cases? What would be the simplest model to describe the velocity versus time during such an event? Can you give a quantitative analysis of the data, fitting to your model? This line of inquiry is a bit open ended.

The flagellar motor is an engine, then, which converts the free-energy drop of protons crossing the membrane into mechanical work. How can we think about this energy conversion? The problem is really much more general. We have outlined a dynamical picture of a chemical reaction as Brownian motion over a barrier, from one well to another on an effective free-energy surface. To do work, we need to couple motions along the reaction coordinate to another mechanical coordinate, such as the rotation of the motor or flagellum. If we call the reaction coordinate Q and the mechanical coordinate x, it is tempting to write the generalization of Eq. (4.1),

$$\gamma_1 \frac{dQ}{dt} = -\frac{\partial V_{\text{eff}}(Q, x)}{\partial Q} + \zeta_1(t), \tag{4.57}$$

$$\gamma_2 \frac{dx}{dt} = -\frac{\partial V_{\text{eff}}(Q, x)}{\partial x} + \zeta_2(t), \tag{4.58}$$

where the Langevin forces are independent of one another, and each obeys

$$\langle \zeta_i(t)\zeta_i(t')\rangle = 2\gamma_i k_B T \delta(t - t'). \tag{4.59}$$

The problem is that these equations describe a system in equilibrium, and so there will be no net motion along either coordinate after the decay of initial transients. How do we describe a system held away from equilibrium, for example, by having different chemical potentials for protons on either side of the membrane?

In the case of the flagellar motor, the reaction coordinate must stand, schematically, for the motion of the proton through the membrane and for various internal degrees of freedom in the motor protein itself. If we imagine the proton moves relatively quickly, then the protein coordinates move on an effective free-energy surface obtained by averaging over the proton's motion (see Problem 42). But the protein, in effect, acts as a catalyst, so in the "chemical reaction" that moves one proton across the membrane, the coordinate Q should return to its initial position. Thus, it is natural to think of the reaction coordinate as periodic. But going once around the cycle of Q involves transporting a proton across the membrane, so if we unwrap the periodic coordinate, the potential must be tilted with a drop of ΔG per period, where ΔG is the free-energy change on moving one proton across the membrane. The problem of converting chemical energy into mechanical work is then the problem of why tilting the potential

$V_{\text{eff}}(Q, x)$ along the Q direction causes the x coordinate to move. But we are used to this: mechanical forces can cause charges to move (the piezoelectric effect), voltage differences can cause heat flows and vice versa (the various thermoelectric effects), and so on. So, in outline at least, we understand how chemo-mechanical energy conversion is possible. There is much more to say here, as can be found in the Annotated Bibliography for this section at the end of the volume.

There is also more to say about mechanics at low Reynolds number. Swimming involves changing shape, which provides the boundary conditions on the Navier-Stokes equations. A cycle of changing boundary conditions should lead to a net displacement. There is some subtlety here, because the space of shapes is not so easy to parameterize. If we think, for example, about a closed surface, shape is defined by three-dimensional position as a function of the two coordinates on the surface (e.g., latitude and longitude), but there is an arbitrariness to how we choose these coordinates; of course any physical quantity, such as the amount by which the swimmer moves forward, must be invariant to this choice. Looking more closely, the freedom to choose coordinates means that the natural formulation of the problem includes a gauge symmetry. Reluctantly, let's leave all this and go back to the problem of chemotaxis.

We are interested in the question of how sensitively the bacterium can respond to small concentration gradients. We suspect that, because individual molecular motions are random, there must be a limit, analogous to the shot noise in counting photons. In a classic paper, Berg and Purcell provided a clear intuitive picture of the noise in measuring chemical concentrations. Their argument, schematized in Fig. 4.16, was that if we have a sensor with linear dimensions a, then effectively the sensor samples a volume a^3. In this volume we expect to count an average of $N \sim ca^3$ molecules when the concentration is c. Each such measurement, however, is associated with a noise $\delta N_1 \sim \sqrt{N}$. Because the count of molecules is proportional to our estimate of the concentration, the fractional error will be the same, so from one observation we obtain a precision

$$\left.\frac{\delta c}{c}\right|_1 = \frac{\delta N_1}{N} = \frac{1}{\sqrt{N}} = \frac{1}{\sqrt{ca^3}}. \tag{4.60}$$

We can make more accurate measurements by averaging over time, although this is a bit tricky—we won't get a better estimate of the concentration by counting the same molecules over and over again. Thus, if we are willing to average over a time τ_{avg}, we can make K independent measurements, where $K \sim \tau_{\text{avg}}/\tau_c$, and the correlation time τ_c is the time we have to wait to ensure an independent sample of molecules.

How do we obtain independent samples? If we look in a small volume, the molecules exchange with the surroundings through diffusion. Thus, the time required to get an independent collection of molecules is the time required for molecules to diffuse in and out of the volume, $\tau_c \sim a^2/D$. Putting everything together, we have

$$\frac{\delta c}{c} = \frac{1}{\sqrt{K}} \cdot \left.\frac{\delta c}{c}\right|_1 \tag{4.61}$$

$$= \sqrt{\frac{\tau_c}{\tau_{\text{avg}}}} \cdot \frac{1}{\sqrt{ca^3}} = \sqrt{\frac{a^2}{D\tau_{\text{avg}}}} \cdot \frac{1}{\sqrt{ca^3}} \tag{4.62}$$

$$= \frac{1}{\sqrt{Dac\tau_{\text{avg}}}}. \tag{4.63}$$

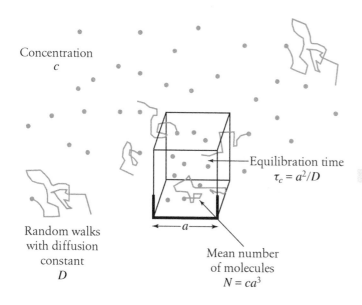

Concentration
c

Equilibration time
$\tau_c = a^2/D$

Random walks
with diffusion
constant
D

Mean number
of molecules
$N = ca^3$

FIGURE 4.16

Schematic of concentration measurements. A receptor of linear dimension a samples a volume a^3 and hence sees a mean number of molecules $N = ca^3$, where c is the concentration. These molecules random walk in and out of the sensitive volume with a diffusion constant D, corresponding to an equilibration or correlation time $\tau_c = a^2/D$.

This is a lovely result. It says that the limit to the accuracy of measurements depends on the absolute concentration (more molecules → more accuracy); on the size the detector (bigger detectors → more accuracy); on the time over which we are willing to average (more time → more accuracy); and finally on the diffusion constant of the molecules we are sensing, because faster diffusion lets us see more independent samples in the same amount of time. All these parameters combine simply, essentially in the only way allowed by dimensional analysis.

One way of understanding this result on the limits to precision is to think about the rate at which molecules find their target. For molecules at concentration c moving with diffusion constant D, the rate (number of molecules per second) that arrive at a target of size a should be proportional both to c and to D, and then by dimensional analysis, we need one factor of length, so the rate is $\sim Dac$ molecules per second. This result is used most often to talk about the diffusion-limited rate constant for a chemical reaction. If we have

$$A + B \xrightarrow{k_+} C, \tag{4.64}$$

then the second-order rate constant k_+ can never be bigger than $\sim Da$, where D is the diffusion constant of the molecules and a is their size (more precisely, the size of the region where they have to hit to react). But if the rate of molecular arrivals is $\sim Dac$, in a time τ_{avg} we will count $\sim Dac\tau_{avg}$ molecules, and if these molecules are arriving at random, then there will be the usual square root fluctuations, which leads us to Eq. (4.63). In this view, the Berg-Purcell limit is nothing but shot noise in molecular arrivals and thus is completely analogous to shot noise in photon arrivals. Photons propagate and molecules diffuse, but under most conditions, they both arrive at random; hence there is shot noise in counting.

Problem 54: Diffusion-limited rates, more carefully.   One can try a more careful calculation of the rate at which molecules find their target by diffusion. Imagine a sphere of radius a

such that all molecules that hit its surface are immediately absorbed. Outside the sphere, the concentration profile must obey the diffusion equation, and the absorption means that on the spherical surface, the concentration will be zero. Far from the sphere, the concentration should be equal to c. Thus, we have

$$\frac{\partial c(\mathbf{x}, t)}{\partial t} = D\nabla^2 c(\mathbf{x}, t); \tag{4.65}$$

$$c(|\mathbf{x}| = a, t) = 0, \tag{4.66}$$

$$c(\mathbf{x} \to \infty, t) = c. \tag{4.67}$$

The number of molecules arriving per second at the surface of the sphere is given by an integral of the diffusive flux over the surface,

$$\text{rate} = \int d^2 s \, \hat{\mathbf{n}} \cdot \left[-D\nabla c(\mathbf{x}, t) \right] \bigg|_{|\mathbf{x}| = a}, \tag{4.68}$$

where $d^2 s$ is an element of the surface area on the sphere, and $\hat{\mathbf{n}}$ is the unit vector normal to the sphere.

(a) Solve Eq. (4.65), with the boundary conditions in Eq. (4.66) and Eq. (4.67), in steady state. Note that as a first step you should transform to spherical coordinates; in three dimensions the Laplacian can be written as

$$\nabla^2 = \frac{1}{r^2} \frac{\partial}{\partial r} \left(r^2 \frac{\partial}{\partial r} \right) + \frac{1}{r^2} \left[\frac{1}{\sin^2 \theta} \frac{\partial^2}{\partial \phi^2} + \frac{1}{\sin \theta} \frac{\partial}{\partial \theta} \left(\sin \theta \frac{\partial}{\partial \theta} \right) \right], \tag{4.69}$$

where as usual r is the radius and θ and ϕ are the polar and azimuthal angles, respectively.

(b) Use your steady state solution to evaluate the rate at which molecules arrive at the sphere, using Eq. (4.68). Also, explain why simple dimensional analysis of these equations yields rate $\sim Dac$.

(c) What happens if you try to give a dimensional analysis argument for the rate in one or two dimensions? If there are problems, can you explain how these problems either go away or are made more precise by trying to solve the diffusion equation with appropriate boundary conditions? Hint: the two-dimensional case is a bit delicate; focus first on one dimension.

Bacteria, such as *E. coli*, have been observed to perform chemotaxis in environments where ambient concentrations of attractants (e.g., sugars or amino acids) are as low as ~ 1 nM, which is $\sim 10^{-9} \times (6 \times 10^{23})/10^3 = 6 \times 10^{11}$ molecules/cm^3. These small molecules diffuse through aqueous solution with $D \sim 10^{-5}$ cm^2/s, and the most generous assumption would be that the relevant size of the detector is the size of the whole bacterium, $a \sim 1\,\mu$m. Putting these factors together, we have $Dac \sim 600$ s^{-1}. Thus, if the bacterium integrates for $\tau_{\text{avg}} \sim 1.5$ s, the smallest concentration changes it can detect are $\delta c/c \sim 1/30$. If the cells were to detect this difference in concentrations across the $\sim 1\,\mu$m length of their bodies, then the concentration would be varying significantly on the scale of 30 μm, which is very short indeed. In real experiments (and, presumably, in the natural environment) the length scales of concentration gradients are one to two orders of magnitude longer. Thus, it is impossible—without integrating for minutes or hours—for bacteria to perform as they do by measuring a spatial gradient. The only possibility is to measure the concentration variation in time, along the trajectory that the bacterium takes through the gradient. Because the cells move at

$v = 10\text{--}20\,\mu\text{m/s}$, on times scales of $\tau_{\text{avg}} \sim 1.5\,\text{s}$, the signal is enhanced by a factor of 10–30, which brings it above the background of noise, allowing for reliable detection.

Although the comparisons are a bit rough,[11] we can draw several conclusions. First, real bacteria perform chemotaxis in response to small signals with a reliability close to the limits set by the physics of diffusion. Second, this is possible only if the cell measures the derivative of concentration versus time as it moves, not spatial gradients across its body. Finally, to reach a reasonable signal-to-noise ratio requires that the cell average over time for more than 1 s.

Why don't the bacteria integrate for longer and reduce the noise further? If you look closely at the trajectories of the bacteria, you can see that the longer runs curve a bit. In fact, the bacteria are sufficiently small that their own rotational Brownian motion disorients them on a time scale of 10–15 s. So, if you integrate for longer than this, you are no longer integrating something related to the gradient in a particular direction, or even the current direction of motion. Thus, there is a physical limit setting the longest useful integration time.

Berg and Purcell also argued that there is a minimum useful integration time. Recall that molecules moving by diffusion traverse a distance $x_{\text{diff}} \sim \sqrt{Dt}$ in a time t; in contrast, swimming at velocity v moves the bacterium by a distance $x_{\text{swim}} \sim vt$. For short times, diffusion, with its square root dependence on time, goes farther than ballistic swimming motion. Thus, on short time scales, the molecules that the bacterium sees along its path are the same molecules, and hence it really is not combining statistically independent measurements. So, there is a minimum useful integration time (assuming you want to improve the signal-to-noise ratio by integrating) of $\tau \sim D/v^2$, which works out to be $\sim 1\,\text{s}$.[12]

So, the strategy of *E. coli* for measuring gradients is incredibly constrained by physics. To reach the observed performance, it has to count nearly every molecule that arrives at its surface. Even with this nearly ideal behavior, it can work only by making comparsions across time, not space, and estimates of time derivatives have to be averaged for a few seconds, not more and not less. This set of predictions about chemotactic strategy is almost parameter free, even if not precisely quantitative.

What do real bacteria do? We have already seen that they make temporal comparisons. Does the detailed form of these comparisons agree with the Berg-Purcell predictions? Although one could probably do better with modern experimental techniques, the best test was done in the early 1980s. In these experiments, bacteria were tethered to a glass slide and exposed to changing concentrations of attractants or repellents. A long series of such observations is then combined to measure the probability that the flagellar motor is rotating counterclockwise (corresponding to running) as function of time relative to the changing concentration. A summary of these experiments is shown in Fig. 4.17. We see that the probability of running is modulated by the time derivative of the concentration, averaged over a window of a few seconds, exactly as predicted by the Berg-Purcell argument.

---

11. I think there is an opportunity for a better experiment here. One could imagine analyzing the moments of transition from run to tumble (and back) in the same way that we analyze the action potentials from sensory neurons (see Section 4.4), measuring the reliability of discrimination between small differences in concentration or reconstructing the concentration versus time along the trajectory of a freely swimming bacterium.

12. For a more rigorous version of this argument, see Problem 192 in Appendix A.6.

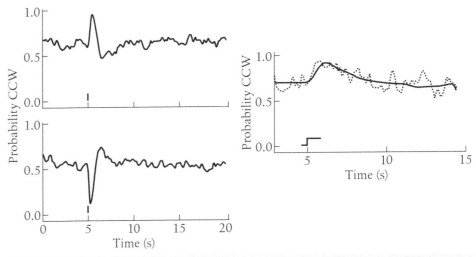

FIGURE 4.17

Impulse responses in bacterial chemotaxis. At left, changes in the probability of counter-clockwise (CCW) rotation of the motor, corresponding to running, as a function of time in response to a pulse of attractant (top) or repellent (bottom). The form of the response is equivalent to integrating the time derivative of the input over a window of several seconds. At right, the response to a step of attractant again has the form expected if we integrate the derivative over a short window. The real data are compared with a prediction based on integrating the response to impulses shown at left, and the agreement is good, as if the system were linear. Redrawn from Block et al. (1982).

Being sensitive to a derivative means that the response to a step comes back almost exactly to the baseline before the step, as seen at right in Fig. 4.17, so that the constant signal is ignored at times long after it was turned on. This gradual forgetting of a constant signal is common in biological systems, and such phenomena are called adaptation. All our sensory systems exhibit adaptation, the most familar being the experience of stepping into a dark movie theater or out into the bright sunlight; at first we are acutely aware of the large difference in overall light intensity, but after a while everything looks normal, and we are insensitive to the absolute photon flux. The case of bacteria is interesting, because it seems that the adaptation is nearly perfect.

Experiments of the sort pictured in Fig. 4.17 also make it possible to estimate the absolute sensitivity of the system in perhaps more compelling units. We now know how many receptors there are on the cell's surface, and so we can convert changes in concentration into changes in the number of occupied receptors. Indeed, one extra occupied receptor leads to a significant change in the probability of running versus tumbling. So, as expected, the bacterium is responding to individual molecular events.

This analysis seems a great success: much of bacterial behavior is understandable, semi-quantitatively, as a response to the physical constraints posed by life at low Reynolds number and the noise in molecular counting. One can go further and say that bacterial behavior is near optimal in relation to this noise. But several questions are left open.

First, can we turn the ideas about maximum and minimum useful integration times into a theory of optimal filtering that would predict, quantitatively, the form of the impulse responses in Fig. 4.17? We should be able to do it, but I don't think anyone has quite managed to get it right, convincingly. One might also wonder whether it even makes sense to formulate this problem for individual bacteria, as opposed to looking at competition or cooperation in a population; this is related to the question of what, precisely, is being optimized by the behavior. It seems likely that any theory of optimal strategies will predict that this optimum is context dependent. Here we should note that quantitative characterization of chemotactic behavior has not been pursued under a very wide range of stimulus conditions, so we may be missing the data needed to test such theories when they emerge.

The second question is about the mechanisms that make possible the extreme sensitivity of chemotaxis. Much progress has been made, although again, some issues are open. In many ways the problem is analogous to that in the rod photoreceptor cell. In rods, the structural changes caused by photon absorption in a single molecule of rhodopsin lead, eventually, to changes in the current flowing across the cell membrane. The link between these events is provided by a cascade of biochemical reactions that serve to amplify the initial single-molecule signal. In chemotaxis, something very similar is happening.

The amplification of individual molecular events by enzymes seems to be a very general mechanism for cells to sense small signals. One thing that is special about bacterial chemotaxis is that we actually can identify all components of the amplification cascade, and there are not too many of them. Part of the reason this can be done is that bacteria can live and reproduce without being able to do chemotaxis, as long as they get enough to eat—in fact they can live without swimming if they don't need to go looking for food. All the information that is needed to make the proteins that function in the signaling cascade is coded in the organism's DNA, so if this DNA gets damaged or modified—making a mutant organism—the function of the cascade can be modified or even lost completely. If we accelerate the process of mutation—by irradiating the cells, by adding chemicals that increase the rate of errors in copying the DNA from generation to generation, or by exploiting naturally occuring mechanisms to insert new sequence elements at random into the genome—we can screen the resulting mutants to look for cells that live but can't do chemotaxis. Such screens are how we know, as explained at the outset, that there really is a separate system for chemotaxis, rather than just some feedback from metabolism. In more detail, we can make a catalog of the mutants and try to show that the mutations impeding chemotaxis are not spread at random over the whole length of the DNA but are concentrated in a handful of genes that code for particular proteins. This is a big effort, even for a relatively simple system, such as chemotaxis, but it had led to convincingly complete lists of the molecular components.

Figure 4.18 shows a modern version of the search for genes involved in motility and chemotaxis. Conceptually the strategy is simple. One generates a very large number of mutant bacteria, in effect knocking out different genes. The cells are placed in a dense, highly viscous environment, so that at best they swim slowly. If this medium is filled with nutrients, then as the bacteria consume these and grow, they deplete their neighborhood and thus make chemical gradients. The bacteria that can do chemotaxis swim out toward greener pastures, and those who can't are stuck in the center. These less-able bacteria can be selected, and the process repeated several times. At the end of the process we have a collection of bacteria that are defective for chemotaxis. Now we can look at their genomes—imagine sequencing all the DNA and seeing which genes are missing or damaged. Such a global sequencing strategy is expensive and inefficient (although less so every year), but conceptually it is the right idea. The results of such

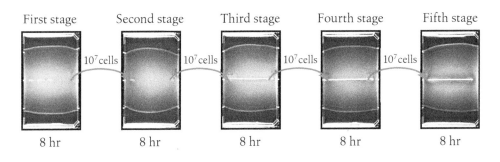

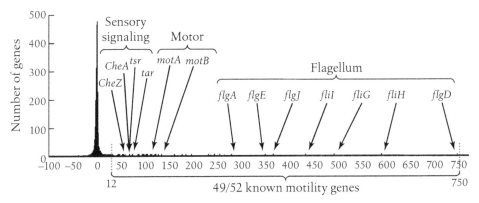

cheA, cheB, cheR, cheW, cheY, cheZ, dnaJ, dnaK, flgA, flgB, flgC,
flgD, flgE, flgF, flgG, flgH, flgI, flgJ, flgK, flgL, flgM, flgN, flhA,
flhB, flhC, flhD, fliA, fliC, fliD, fliE, fliF, fliG, fliH, fliI, fliK, fliL,
fliM, fliN, fliO, fliP, fliQ, fliR, fliS, fliZ, motA, motB, tap, tar, tsr

FIGURE 4.18

A modern version of experiments to search for chemotactic mutants. Cells are mutagenized by
exposure to transposable elements, which insert themselves randomly into the genome and
disrupt the function of individual genes. The population is then purified, as described in the
text, so that cells defective in chemotaxis are left behind in the center of the colony. Extracting
the DNA from these cells, one can make more copies of the DNA that include the sequences
of the transposable elements and hence are actually defective. The abundance of the different
genes is then measured by looking for binding (in the natural base-pairing way!) of this DNA
to templates formed from each gene in the bacterial genome. Results are reported as z-scores
(a measure of statistical significance) for enrichment of particular genes in the population.
Redrawn from Girgis et al. (2007).

experiments, shown in Fig. 4.18, can be expressed in terms of the statistical significance
of enrichment of the population for individual genes. It recovers, in one experiment,
decades of work isolating individual mutants.

When we make a list of proteins that, when lost to mutation, influence chemotaxis,
the list is not too long (Figs. 4.18 and 4.19):

- There are proteins involved in making the flagellar motor, which really is quite a
  complicated structure.
- There are the receptor molecules, which sit in the cell membrane. The small
  molecules that are actually being sensed are outside the cell, and they bind
  to these receptors. Sometimes there are additional adaptor proteins that help

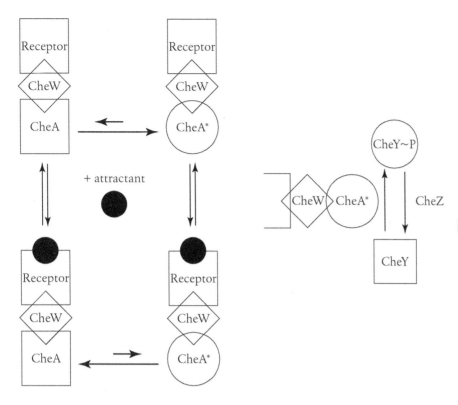

FIGURE 4.19

Biochemical amplification in the chemotactic response. At left, binding of chemoattractants to their receptors shifts the equilibrium between active (indicated by *) and inactive forms of the kinase CheA. At right, the active kinase phosphorylates CheY; this reaction is balanced by the action of the phosphatase CheZ. CheY~P is the phosphorylated form of CheY.

transport the small molecules in the space between the cell membrane and the surrounding outer wall.

■ There is a protein called CheY that can bind to the motor.

■ There is a protein called CheA that acts as an enzyme. It catalyzes a reaction in which a phosphate group is attached to the protein CheY. Such reactions are called protein phosphorylation, and many proteins change their structure and activity when phosphorylated—recall the multiple phosphorylation steps involved in the shut-off of rhodopsin (Section 2.3). In particular, CheY binds to the motor only in its phosphorylated form, which we write as CheY~P. Enzymes that phosphorylate proteins are called protein kinases or sometimes just kinases.

■ There is a separate protein CheZ that acts as an enzyme, catalyzing the removal of the phosphate groups from CheY~P. Thus, CheZ undoes the action of CheA. Enzymes that remove phosphate groups from proteins are called phosphatases.[13]

---

13. This might seem puzzling. Catalysts, after all, don't just increase the rate of a reaction—they also have to increase the rate of the reverse reaction, so as not to shift the (thermodynamic) equilibrium between reactants and products. When kinases phosphorylate proteins, they remove the phosphate group from an ATP molecule, and ATP is a common internal energy storage medium for cells; when the covalent bond that attaches the second or third phosphate group is broken, a large amount of free energy is released. Thus, although the kinase also catalyzes the reaction in which a phosphate group is stripped off the protein and added back to ADP (adenosine diphosphate), the free-energy difference opposing this reaction is huge, and it almost never happens. When a phosphatase removes a phosphate group from a protein, it just ejects this group into the surrounding solution, rather than putting it back in the place where the kinase found it. Thus, kinases and phosphatases catalyze opposing reactions, which are exactly opposite from the point of view of the target protein (here, CheY) but are not microscopic reverse reactions.

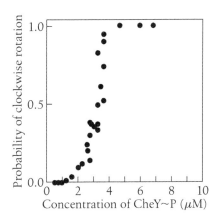

FIGURE 4.20

Phsophorylated CheY (CheY~P) binds to the flagellar motor and promotes clockwise rotation, which drives tumbling. The motor is extremely sensitive to small changes in CheY~P concentration. Data redrawn from Cluzel et al. (2000).

- There is a protein CheW that provides a link between the receptors and CheA.
- There are enzymes CheR and CheB, which add and remove methyl groups from the receptors. These methylation reactions are needed for the adaptation of the chemotactic response, that is, for the balance between running and tumbling (or clockwise and counterclockwise rotation of the motor in Fig. 4.17) to return to its initial value after exposure to a step change in the concentration of attractants or repellents.

So, there are many components, but not too many. To understand how these things fit together, let's work backward, trying to understand how the rotational bias of the motor depends on the concentration of CheY~P.

To measure the bias versus CheY~P, one has to do many tricks. It is relatively easy to measure the bias of the motor, either in experiments where the cell is tethered or where it is lying on a slide and one motor stump is sticking up with a bead attached. To know the concentration of a protein in a single cell, we need to make the protein visible. This is done by genetic engineering, replacing the normal CheY with a fusion between this protein and the green fluorescent protein, and arranging for the expression of this fusion protein to be controlled by signals that can be applied externally.[14] Finally, we need to know the concentration of the phosphorylated form of the protein, which is difficult to determine. But once phosphate groups are attached to a protein, they stay there until removed by another enzyme (the phosphatase). So, if we genetically engineer the bacterium to remove the phosphatase, we will surely disrupt the overall chemotactic response, but we then can be sure that all the CheY will be in its phosphorylated state. The result is shown in Fig. 4.20.

What we see most clearly from Fig. 4.20 is that the motor is remarkably sensitive to small changes in concentration of CheY~P. One can fit a function of the form

$$P_{\text{cw}} = \frac{c^n}{c^n + K^n},\tag{4.70}$$

with $K \sim 3\,\mu\text{M}$ and $n \sim 10$, although the data are almost within errors of being a

---

14. For more on the use of fluorescent proteins, see Section 4.3.

step function. If we take this function seriously, we can be more quantitative about sensitivity:

$$\frac{dP_{\text{cw}}}{dc} = n\frac{c^{n-1}}{c^n + K^n} - nc^{n-1}\frac{c^n}{(c^n + K^n)^2} \tag{4.71}$$

$$= \frac{1}{c}nP_{\text{cw}}(1 - P_{\text{cw}}), \tag{4.72}$$

$$\Rightarrow \frac{d\ln P_{\text{cw}}}{d\ln c} = n(1 - P_{\text{cw}}). \tag{4.73}$$

Thus, at small concentrations of CheY~P, where P_{cw} is small, a given fractional change in the concentration produces an n-fold larger fractional change in the probability of clockwise rotation. Even near the midpoint of the response, where $P_{\text{cw}} = 0.5$, with $n = 10$ we see that a small fractional change in concentration of CheY~P produces a fivefold larger fractional change in probability.

Problem 55: Absolute concentration measurements. In this problem you should try to understand how Cluzel et al. were able to put the CheY~P concentration on an absolute scale. Bacteria can be engineered to make a fluorescent version of many naturally occurring proteins. Although the resultant fluorescence signal seen under a microscope is proportional to the number of molecules under illumination, it can be difficult to measure the proportionality constant in an independent experiment. One can circumvent this problem by watching small numbers of molecules diffusing randomly in and out of an illuminated volume inside an individual cell and using the variance in the fluorescence intensity, along with its mean value, to make an absolute measurement of the concentration of the molecules.[15]

(a) Explain (qualitatively) how this measurement might work. What do you gain by using both the variance and the mean of this signal? How can the fluctuating fluorescence signal be analyzed further to give an estimate of the protein diffusion constant?

(b) Now let's convert the above intuition into a quantitative framework for analysis of the data. Consider the concentration $c(\mathbf{x}, t)$ of fluorescent molecules at different points in space and time. It fluctuates, and the deviation δc of the concentration from its average value $\bar{c}$ is uncorrelated between different points in space (but the same instant of time). Show that the analytic statement

$$\langle \delta c(\mathbf{x}, t)\delta c(\mathbf{x}', t)\rangle = \bar{c}\delta(\mathbf{x} - \mathbf{x}') \tag{4.74}$$

is equivalent to the intuitive remark that the variance of the number of molecules in a volume is equal to the mean number.

(c) If the system starts with some fluctuation in the concentration $c(\mathbf{x}, 0) = \bar{c} + \delta c(\mathbf{x}, 0)$, this profile will relax according to the diffusion equation,

$$\frac{\partial c(\mathbf{x}, t)}{\partial t} = D\nabla^2 c(\mathbf{x}, t). \tag{4.75}$$

---

15. Some of what gets done in this problem could be done more rigorously after reading Appendix A.6. I think the essential ideas do not really need the full technical apparatus, though perhaps this attempt to understand how an experiment works will provide more motivation to explore the theory fully. This problem was originally written together with Curt Callan.

Because the diffusion equation is linear, the profile of fluctuations at time t, $\delta c(\mathbf{x}, t)$, can be written as a linear operator acting on the initial condition $\delta c(\mathbf{x}, 0)$. Show that this linear relationship is

$$\delta c(\mathbf{x}, t) = \int d^3 y \left(\frac{1}{\sqrt{4\pi Dt}} \right)^3 \exp(-|\mathbf{x} - \mathbf{y}|^2/(4Dt))\, \delta c(\mathbf{y}, 0), \tag{4.76}$$

where D is the diffusion constant.

(d) When we bring light to a focus under the microscope, we effectively weight the points around the focus with a Gaussian function, so that the light intensity collected from the fluorescent molecules will be proportional to

$$s(t) = \int d^3 x \, c(\mathbf{x}, t) \exp(-|\mathbf{x}|^2/\ell^2), \tag{4.77}$$

where ℓ is the size of the focal region (roughly the size of the wavelength of light). Using Eq. (4.76) and Eq. (4.77), show that the temporal correlation function of this signal is given by

$$\langle \delta s(t) \delta s(0) \rangle \propto (|t| + \tau)^{-3/2}, \tag{4.78}$$

and relate the correlation time τ to the diffusion constant D and the size of the focal region ℓ. Hint: note that in doing the multidimensional Gaussian convolution integrals that show up in the last step of this computation, it is a good idea to do them Cartesian coordinate by Cartesian coordinate. This gives a precise method for extracting the diffusion constant from the fluctuating fluorescence signal.

Problem 56: The motor as a sensor of CheY~P. Although we usually think of *E. coli* as changing the concentration of CheY~P to modulate the direction of the flagellar motor, we can think of it the other way—that the motor serves as an internal sensor for the concentration of CheY~P. If we take a snapshot of one motor at time t, it is either going clockwise or counterclockwise, and thus is a binary variable $\sigma(t)$, with the convention that $\sigma = 1$ corresponds to clockwise rotation, and $\sigma = 0$ corresponds to counterclockwise.

(a) Show that the mean and variance of σ are given by $\langle \sigma \rangle = P_{cw}$ and $\langle (\delta\sigma)^2 \rangle = P_{cw}(1 - P_{cw})$.

(b) From Eq. (4.72), a change in concentration Δc produces a change in P_{cw}, $\Delta P_{cw} \equiv (dP_{cw}/dc)\Delta c$. Compute the signal-to-noise ratio for detecting the change Δc by observing σ, $S \equiv (\Delta \langle \sigma \rangle)^2/\langle (\delta\sigma)^2 \rangle$. Show that this quantity is maximal when $P_{cw} = 0.5$.

(c) We should be able to detect smaller changes in concentration by averaging over time. Specifically, if we average for a time τ_{avg}, then the signal-to-noise ratio should improve by a factor τ_{avg}/τ_c, where τ_c is the correlation time for the random fluctuations in σ. Experimentally, near $P_{cw} = 0.5$, switching between clockwise and counterclockwise states looks like a random process with a rate $r \sim 1.6 \text{ s}^{-1}$. Combine this with your results so far to show that one can reach reliable detection ($S = 1$) when

$$\frac{\Delta c}{c} \sim \sqrt{\frac{1\,\text{s}}{40\tau_{avg}}}. \tag{4.79}$$

(d) Berg and Purcell argued that if we try to measure concentrations with a sensor that has linear dimension a, then we will reach $S = 1$ when the fractional change in concentration is

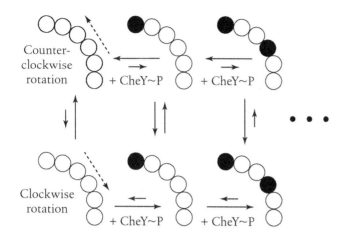

FIGURE 4.21

A model for the modulation of rotor bias by binding of phosphorylated CheY (CheY~P). These molecules bind independently to multiple sites around a ring. When all sites are empty (open circles), the equilibrium favors the counterclockwise rotating state. Binding is stronger to the clockwise state, however, so that as more sites are occupied (filled circles), the equilibrium shifts.

given by Eq. (4.63). The flagellar motor presents a target of roughly $a \sim 35$ nm to the cytoplasm, and we expect that such proteins as CheY~P diffuse with $D \sim 1$–$3\ \mu m^2/s$. With $c \sim 3\ \mu M$ near the midpoint of the response in Fig. 4.20, what is your prediction for the Berg-Purcell limit on the sensitivity of the motor as a detector of changes in CheY~P concentration? How does this compare with the (quasi-)experimental result in Eq. (4.79)?

The cooperativity with which CheY~P modulates the flagellar motor has, as discussed in Section 2.3, two very different interpretations. In one view, there are direct interactions among the different binding events. In the other view, the cooperativity is mediated by the switching of the motor between two states, as with R and T in hemoglobin or open and closed in the cGMP-modulated channels in rod cells. In the flagellar motor, the two-state model seems natural, because we know that there is both clockwise and counterclockwise rotation. Thus, we want to give a description in the spirit of the MWC model.

The flagellar motor is a rotary engine and has a ring structure, so let us imagine the multiple binding sites for CheY~ P as being arrayed around a ring, as in Fig. 4.21. CheY~P molecules bind independently to each site, but the strength of the binding depends on whether the whole structure is rotating clockwise or counterclockwise. Qualitatively, if binding is stronger in the clockwise state, then increasing the concentration of CheY~P will shift the equilibrium toward that state.

Quantitatively, we can work out the predictions of the model in Fig. 4.21 using statistical mechanics, on the hypothesis that all binding events and structural transitions of the motor between clockwise (subscript cw) and counterclockwise (subscript ccw) states come to equilibrium. One might worry about the latter assumption—after all, if the motor were truly at equilibrium, it would not be rotating and generating force— but let's proceed. Consider one possible state of the system, say, clockwise rotation with m out of the n sites filled by CheY~P molecules. We need to assign this state a weight in the Boltzmann distribution. We can assume that the clockwise state has an intrinsic (free) energy E_{cw}. With m molecules bound, the energy is lowered by $m F_{cw}$, where F_{cw} is the binding energy in the clockwise state, but we also have to take these m molecules out of solution, and this shifts the free energy by m times the chemical potential, $m\mu = mk_B T \ln(c/c_0)$, where c is the concentration of CheY~P, and c_0 is a reference concentration, as in Eq. (2.121). Finally, because the m occupied sites could chosen out of the n possibilities in many ways, there is a combinatorial factor. Putting these terms together, we have

$$\binom{n}{m} \exp\left[-\frac{1}{k_B T}\left(E_{cw} - mF_{cw} - mk_B T \ln(c/c_0)\right)\right]$$

$$= \binom{n}{m}\left(\frac{c}{K_{cw}}\right)^m e^{-E_{cw}/k_B T},$$

where $K_{cw} = c_0 e^{-F_{cw}/k_B T}$. To compute the probability of being in the clockwise state, we have to sum over all the different occupancies and normalize by the partition function, which includes a sum over the counterclockwise states:

$$P_{cw} = \frac{1}{Z} \sum_{m=0}^{n} \binom{n}{m}\left(\frac{c}{K_{cw}}\right)^m e^{-E_{cw}/k_B T} \tag{4.80}$$

$$= \frac{1}{Z} e^{-E_{cw}/k_B T} (1 + c/K_{cw})^n, \tag{4.81}$$

where

$$Z = e^{-E_{cw}/k_B T} (1 + c/K_{cw})^n + e^{-E_{ccw}/k_B T} (1 + c/K_{ccw})^n. \tag{4.82}$$

We can put this result in a more compact form,

$$P_{cw} = \frac{1}{1 + \exp\left[\theta - g(c)\right]}, \tag{4.83}$$

$$\theta = (E_{cw} - E_{ccw})/(k_B T), \tag{4.84}$$

$$g(c) = n \ln\left(\frac{1 + c/K_{cw}}{1 + c/K_{ccw}}\right). \tag{4.85}$$

Notice that if $K_{cw} \ll c \ll K_{ccw}$, then P_{cw} becomes the Hill function in Eq. (4.70). This derivation should be familiar from the discussion leading to Eq. (2.145) in Section 2.3.

Problem 57: MWC model of rotor bias.   Explore the parameter space of the model we have just described. Are there regimes, other than $K_{cw} \ll c \ll K_{ccw}$, where one can reproduce the steep dependence of P_{cw} on c observed by Cluzel et al. (2000)? Keep in mind that the actual number of binding sites n could be very large.

There are two really nontrivial ideas in the MWC model. First, even if the system we are thinking about is a protein composed of many subunits—four subunits in hemoglobin, and many more in the flagellar motor—we assume that the states are properties of the whole collection of subunits and not of the individuals. Thus, we do not imagine that small pieces of the motor are trying to rotate in opposite directions; we assume that the interactions among subunits are strong enough to enforce unanimity. The second idea, less relevant for the flagellar motor but important in general, is that the MWC model gives us a framework for including even more binding events, with different kinds of molecules. The rule is always the same: each binding event is independent, but the binding energy depends on the overall state of the system, so that (by detailed balance) binding will shift the equilibrium.

In many systems, it is not just a single class of ligands that binds. For hemoglobin itself, changes in pH, which presumably result in the binding and unbinding of protons, change the way in which oxygen binds.[16] For enzymes—proteins that catalyze a chemical reaction—it is not just the substrate that binds and is chemically altered, but other molecules bind as well and alter the activity of the enzyme. It is important that these other molecules are binding at other sites and are not directly interfering with substrate binding in enzymes or oxygen binding in hemoglobin. From the Greek for "other site," these effects are called allosteric, and the MWC model gives a general framework for thinking about allostery.

Coming back to chemotaxis, it is clear that part of the answer to how small changes in the concentrations of attractants or repellents can generate reliable changes in the swimming of bacteria is that the motor itself is very sensitive to small changes in the concentration of the internal signal, CheY~P. The extreme sensitivity of the motor also causes a problem, however. Figure 4.20 shows that extreme sensitivity must coexist with a very tight regulation, because if the concentration of CheY~P drifts far away from $c \sim K$, the cell loses all sensitivity to changes. More quantitatively, a corollary of the fact that $\sim \pm 10\%$ changes in the concentration of CheY~P are enough to modulate the probability of clockwise rotation by a factor of three is that a change of $\sim 30\%$ in the background concentration of CheY~P is enough to reduce the sensitivity to small changes by an order of magnitude. Can the cell actually hold the mean concentration of CheY~P in this narrow window where it maintains sensitivity? Or does the concentration, and hence the sensitivity, drift wildly?

We still need to understand the biochemical processes that lead from single molecular events, where attractants or repellents bind to receptors, to quasi-macroscopic changes in molecule number: at $c \sim 3\,\mu$M, a cell with volume $\sim 1\,\mu$m^3 has ~ 2000 molecules of CheY~P, so even a 10% change in concentration involves hundreds of molecules. Thus, there are at least two stages of gain in the system: between the CheY~P concentration and the probability of clockwise rotation, and between the ligand binding to receptors and the change in CheY~P concentration.

We know that the CheY~P concentration is increased by the activity of CheA, which is linked to the receptors through CheW (see Fig. 4.19). Is it possible that the gain from receptor occupancy to CheY~P concentration is the result solely of the catalytic activity of CheA? One active enzyme can catalyze the phosphorylation of many CheY molecules, but this is limited by the rate at which CheY molecules encounter the enzyme. If the total concentrations of CheY are in the range of $c \sim 10\,\mu$M, and this molecule has a diffusion constant similar to other cytoplasmic proteins, $D \sim 1\,\mu$m^2/s, then if the CheA presents a target of less than ~ 1 nm in size, we expect the maximal rate of these encounters to be $Dac \sim 10\,$s^{-1}. So, a single active CheA could not catalyze the production of several hundred CheY~P molecules within the 1 s rise time of the chemotactic response. Thus, there must be an additional stage of gain, in which the binding of a single ligand to its receptor influences multiple CheA proteins.

Experiments show that the receptor molecules are packed into a fairly dense lattice, and the CheW and CheA molecules are tightly associated with this membrane-bound

---

16. This is the Bohr effect, named after Christian Bohr, Niels's father.

structure. This physical proximity suggests models for the emergence of amplification in the activation of CheA that have the same flavor as the MWC models for the flagellar motor. In one version, clusters of receptor-CheW-CheA complexes act as a single large system with just two states, active and inactive, and the equilibrium between these states is shifted by the binding of ligands. Another version of this idea allows each receptor and associated CheA to switch independently between active and inactive states, but neighboring molecules have interactions that favor being in the same state. In this case, the lattice of receptors and kinases becomes analogous to an Ising ferromagnet, and the changing concentration of ligands acts as a magnetic field. Even if interactions are local, if they are tuned to be close to the critical point, then correlation lengths become long, and many receptors effectively respond together. Indeed, precisely at the critical point the sensitivity of the system to small changes in concentration grows with total number of receptors.

In the rod cell, amplification has a small contribution from cooperativity in the opening of channels by cGMP, but most of the gain comes from iterating the process in which one enzyme molecule can catalyze the conversion of many molecules of substrate. In chemotaxis, there are enzymatic steps, but cooperativity seems much more important. I don't think we yet have as complete an analysis as for the rod cell of the different sources of internal noise. Nor do we have a convincing explanation of how these noise sources are suppressed to the point that the whole system operates near the Berg-Purcell limit. One of the startling things about the system is that sensitivity to small changes in concentration is maintained across a wide range of background concentrations, and we return to this problem of adaptation in Section 5.3. We can see that mechanisms that achieve gain through cooperativity have the problem that they must trade sensitivity for dynamic range in a very severe way. Thus, even though we have the ingredients for understanding the extreme sensitivity of this system, these ingredients raise new questions of how precisely cells can control their own parameters, or whether there are other mechanisms at work that obviate the need for such tuning.

4.3   Molecule Counting, More Generally

Many of the crucial signals in biological systems—signals that are internal to cells, signals that cells use to communicate with one another, even signals that organisms exchange—are carried by changes in the concentration of specific molecules. The molecules range in size from single ions (e.g., calcium) to whole proteins. Such chemical signals act by binding to specific targets, whose synthesis and accessibility can also be controlled by the cell. A key point is that individual molecules move randomly, and so the arrival of signals at their targets has some minimum level of noise. As we shall see, several different systems operate with a reliability close to this physical limit: in essence, these systems are counting every molecule and making every molecule count.

In what follows we will see examples of chemical signaling in the decisions that cells make about whether to read out the information encoded in particular genes, in the trajectories that axons take toward their targets in the developing brain, and in the construction of spatial patterns in a developing embryo. Much of our thinking about precision, reliability, and noise in chemical signaling has been shaped by the phenomena of chemotaxis in bacteria, as described in the previous section. The last of the major questions left open by the Berg-Purcell analysis is whether we can do a full calculation that leads to their limit on the precision of concentration sensing. What Berg and Purcell wrote down makes absolutely no reference to the messy details of what actually happens to molecules as they are counted. This absence of detail could be wonderful, because it would mean that we can say something about the limits to precision in *all* biochemical signaling systems, regardless of details. Alternatively, the absence of details might be a disaster, a clue that we have simply missed the point.

To see what is at stake in the analysis of molecule counting, let's think about the regulation of gene expression (Fig. 4.22). As mentioned already, every cell in our bodies has the same DNA, and what makes a liver cell different from a neuron in your brain is that it reads out (expresses) different genes, making different proteins. Importantly, this is not just a discrete choice made once in your lifetime. Given that certain proteins are being made, the numbers of these molecules are constantly adjusted to match the needs of the cell. This also happens in bacteria, which adjust, for example, the concentrations of the enzymes needed to metabolize different nutrients that might or might not be present in the environment; much of what we know about the regulation of gene expression has its roots in work of this sort on metabolic control in bacteria.

There are many ways in which gene expression is controlled. To regulate the number of proteins in the cell, the cell can change either the rate at which they are made or the rate at which they are degraded, and both of these things happen. The synthesis of a protein involves two very different steps, transcription from DNA to messenger RNA (mRNA) and translation from mRNA to protein, and again there is regulation of both processes. We will focus on the regulation of transcription, that is, the reading of the DNA template to make mRNA.[17]

To make mRNA, a complex of proteins (including the RNA polymerase) must bind to the DNA and walk along it, spewing out the mRNA polymer as it walks. For this to happen, the RNA polymerase must bind to the right starting point along the DNA. One can imagine that this binding event can be inhibited simply by having other proteins bind to nearby sites along the DNA. Alternatively, the binding of proteins to slightly different positions near the starting point could help the RNA polymerase to find its way. Both of these things happen: proteins called transcription factors can act both as repressors and as activators of mRNA synthesis. The key step in this regulation is thought to be the binding of the transcription factors to specific sites near the RNA polymerase start site, as schematized in Fig. 4.22; the whole segment of DNA involved in the control and initiation of transcription is called the promoter. In higher organisms, the regions involved in regulation can be very large indeed and usually are called enhancers to avoid conjuring the simplified image in Fig. 4.22, which is more literally applicable in bacteria. Binding sites are specific, because the transcription factor protein is selective for particular DNA sequences. Much can be said about the nature of this

17. For a bit about the basics of DNA structure, see Appendix A.3.

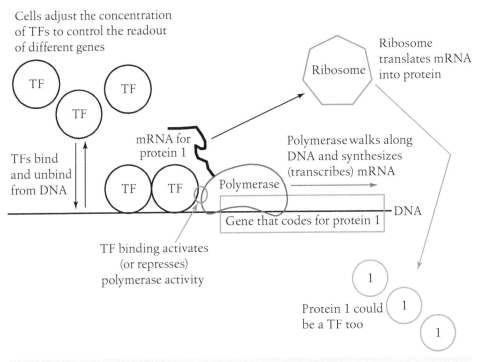

FIGURE 4.22

Control of gene expression by transcription factors (TFs). Synthesis of a protein involves transcription of the DNA coding for that particular protein and translation of the resulting messenger RNA (mRNA). An important component of control in these systems is the binding of transcription factors to the DNA at specific sites near the start of transcription, in the promoter or enhancer region. Transcription factors are themselves proteins, so this regulatory process naturally leads to a network of interactions; here we focus, for simplicity, on one input (the concentration of the transcription factor) and one output (the concentration of protein 1). Note that in bacteria all this happens in one compartment, whereas in eukaryotic cells the DNA is in the nucleus and mRNA is transported out to the cytoplasm, where translation occurs. Nothing in this figure is to scale.

specificity. For now the important point is that such regulatory systems are, in effect, sensors of the transcription factor concentration.

Problem 58: Autoregulation. Perhaps the simplest model of transcriptional regulation is one in which a gene regulates its own expression. Let the concentration (or number of molecules) of the protein be g, and assume that n of these molecules bind cooperatively to the promoter region of the gene. If the binding activates expression, and proteins are degraded in a simple first-order process with lifetime τ, then it is plausible that the dynamics of g is given by

$$\frac{dg}{dt} = r_{\max} \frac{g^n}{g^n + g_{1/2}^n} - \frac{g}{\tau}. \tag{4.86}$$

(a) Explain the significance of the parameters $r_{\max}$ and $g_{1/2}$. Show that there is a range of these parameters in which the system is bistable. More precisely, show that you can find three steady states, and that two of these are stable and one is unstable. What are the time constants for relaxation to these steady states? How do these times compare with the lifetime τ of the protein?

(b) The protein binding actually regulates the synthesis of mRNA, which in turn is translated by the ribosomes into protein. If m is the mRNA concentration (or number of molecules), then a plausible set of equations is

$$\frac{dm}{dt} = e_{\max} \frac{g^n}{g^n + g^n_{1/2}} - \frac{m}{\tau_m}, \tag{4.87}$$

$$\frac{dg}{dt} = r_{\text{trans}} m - \frac{g}{\tau_p}, \tag{4.88}$$

where $e_{\max}$ is the maximal transcription (expression) rate; r_{trans} is the rate at which mRNA molecules are translated into protein; and the lifetimes of protein and mRNA are τ_p and τ_m, respectively. Under what conditions will this more complete model be well approximated by the simpler model above? Are the steady states of the two models actually different? What about their stability?

(c) Suppose that instead of activating its own expression, the protein acts as a repressor of its own expression. Find the analog of Eq. (4.86) in this case, and show that there is only one steady state and that this state is stable.

(d) Expand your discussion of the autorepressor to include the mRNA concentration, as in Eq. (4.87) and Eq. (4.88). Find the steady state and linearize the equations around this point. Do you find exponential relaxation toward the steady state for all values of the parameters? Is it possible for the steady state to become unstable? Explain qualitatively what is happening, and go as far as you can analytically toward a complete description of the linearized dynamics.

The binding sites along DNA for the transcription factors have linear dimensions measured in nanometers, perhaps $a \sim 3\,\text{nm}$. The diffusion constants of proteins in the interior of cells is in the range of $D \sim 1\,\mu\text{m}^2/\text{s}$. Many transcription factors act at nanomolar concentrations, and it is useful to note that $1\,\text{nM} = 0.6\,\text{molecules}/\mu\text{m}^3$. Putting these numbers together, we have $Dac \sim 1.8 \times 10^{-3}\,\text{s}^{-1}$. Thus, the Berg-Purcell limit predicts that the smallest changes in transcription factor concentration that can be reliably detected are

$$\frac{\delta c}{c} \sim \frac{1}{\sqrt{Dac\tau_{\text{avg}}}} \sim \sqrt{\frac{10\,\text{min}}{\tau_{\text{avg}}}}. \tag{4.89}$$

Taken at face value, Eq. (4.89) suggests that truly quantitative responses—say, to 10% changes in transcription factor concentration—would require hours of integration. This time scale is seldom plausible.[18]

One should not take this rough estimate too literally. I think the message is not the exact value of the limiting precision, but rather that once concentrations fall to the nanomolar range, small changes will be very hard to detect. If cells do detect these small

---

18. If we want to make this rough estimate more precise, we also need to think more carefully about the kinetics by which transcription factors find their targets on the DNA. This is the subject of a large literature. For some pointers, see the references for this section in the Annotated Bibliography at the end of the volume.

changes, then almost certainly they will be bumping up against the physical limits set by counting molecules,[19] assuming that Berg and Purcell give us a good estimate of these limits. So, we need to check this assumption.

In Appendix A.6, we look in detail at how to make the Berg-Purcell limit more rigorous. The key idea is that fluctuations in concentration, and in many examples of binding to receptor sites, represent fluctuations in thermal equilibrium and thus are susceptible to the same analyses as Brownian motion, Johnson noise, and other examples of thermal noise. These analyses show how one can separate the limiting noise level from the extra noise that is associated with all the biochemical complexities that Berg and Purcell ignored. The result, then, is that the Berg-Purcell argument can be made rigorous, both for single receptors and for arrays of receptors, and their simple formula gives us a lower bound on the noise in biochemical signaling. This is important because, as noted at the start of this discussion, the Berg-Purcell limit does not make reference to any of the detailed biochemistry of what happens when the signaling molecules bind to their targets. Rather, the limit depends on the physical nature of the signal itself. The fact that we can make the Berg-Purcell argument rigorous encourages us to look more broadly and see whether there are other cases in which biological systems approach these physical limits to their signaling performance.

With confidence in the Berg-Purcell limit, we can apply it more widely. An important example of chemotaxis occurs during the development of the brain. Individual neurons start as relatively compact cells and then extend their axons to find the other cells with which they must make synapses. This process is guided by gradients in a variety of signaling molecules. Although there are many beautiful observations on these phenomena in vivo, it is not so easy to do a controlled experiment where one allows cells to migrate in well-defined gradients. One approach is to grow cells in a collagen matrix that is "printed" with droplets of growth factor at varying densities. Relatively quickly, diffusion acts to smear the rows of drops into a continuous gradient, which can be directly observed when the molecules are labeled with fluorophores. These measurements also allow an inference of the diffusion constant in this medium, $D \sim 8 \times 10^{-7}$ cm^2/s. The growth cones that guide the axon have linear dimensions $a \sim 10 \, \mu$m, and these experiments found that sensitivity to gradients is actually maximal in a concentration range near $c \sim 1$ nM. Under these conditions, then, we have $Dac \sim 500$ s^{-1}. Quite astonishingly, however, the cells seem to grow differentially in the direction of gradients that correspond to concentration differences of order one part in one thousand across the diameter of the growth cone. For this signal to be above the Berg-Purcell limit on the noise level, the cell must integrate for $\tau_{\mathrm{avg}} \sim 2000$ s, a reasonable fraction of an hour.

In truth, we don't know the time scale over which growth cones are integrating as they decide which way to turn, even in the more controlled in vitro experiments. We do know that the pace of neural development is slow—hours to days rather than minutes. Qualitative aspects of axonal behavior are consistent with the idea that the time scales of their movements are determined by the need to integrate long enough to generate reliable directional signals, from the rapid "exploration" by cellular appendages to the

---

19. It is worth emphasizing that "counting molecules" means counting molecules that arrive at their target. Thus, the limits to measurement are set not by the total number of molecules in the cell, but by their concentration.

dramatic slowing down near critical decision points. One such point is the optic chiasm, where the axons of ganglion cells emerging from the retina must decide whether to go toward the right or left half of the brain. It is attractive to think that the reliability with which cells in our brain find their targets is set by such basic physical principles, but we don't quite have enough data to say this with certainty.

Let us return to the problem that motivated our search for generality, the transcriptional regulation of gene expression. Until the past decade, there were essentially no direct measurements on the reliability of such regulatory mechanisms. Before we look at the new data, though, we need one more set of theoretical ideas.

Proteins are synthesized and degraded, and the simplest assumption is that these are single kinetic steps. Suppose we start just with synthesis, at some rate s molecules per second. We have seen that rate constants should be interpreted as the probability per unit time for individual molecular events. Thus, if we ask about the probability of finding exactly N molecules in the system at time t, this probability $P(N; t)$ changes as a result of two processes. Synthesis of a new molecule causes a transition $N - 1 \rightarrow N$, with probability s per unit time, and this increases $P(N; t)$. But synthesis also causes a transition $N \rightarrow N + 1$, at the same rate, and this reduces $P(N; t)$. Putting these terms together, we have the "master equation"

$$\frac{\partial P(N; t)}{\partial t} = s P(N - 1; t) - s P(N; t), \tag{4.90}$$

except at $N = 0$, where we have

$$\frac{\partial P(0; t)}{\partial t} = -s P(0; t). \tag{4.91}$$

We can solve these equations iteratively. We start with no molecules, so $P(0; 0) = 1$, while $P(N \neq 0; 0) = 0$. Then Eq. (4.91) tells us that

$$P(0; t) = e^{-st}. \tag{4.92}$$

If we substitute into Eq. (4.90) for $P(1; t)$, we have

$$\frac{\partial P(1; t)}{\partial t} = -s P(1; t) + s P(0; t) \tag{4.93}$$

$$\Rightarrow P(1; t) = \int_0^t dt' e^{-s(t-t')} s P(0; t') \tag{4.94}$$

$$= \int_0^t dt' e^{-s(t-t')} s e^{-st'} \tag{4.95}$$

$$= s e^{-st} \int_0^t dt' = e^{-st}(st). \tag{4.96}$$

We can go through the same calculation for $P(2; t)$:

$$P(2; t) = \int_0^t dt' e^{-s(t-t')} s P(1; t') \tag{4.97}$$

$$= e^{-st} \int_0^t dt' s^2 t' \tag{4.98}$$

$$= e^{-st} \frac{(st)^2}{2}. \tag{4.99}$$

This equation suggests that, for all N,

$$P(N; t) = e^{-st} \frac{(st)^N}{N!}. \tag{4.100}$$

Problem 59: Checking the Poisson solution.   Verify that Eq. (4.100) solves the master equation describing a single synthesis reaction at rate s, Eq. (4.90).

Equation (4.100) states that, as the synthesis reaction proceeds, the number of molecules that has been synthesized obeys the Poisson distribution. From what we have said about the Poisson distribution in the discussion of photon counting (Section 2.1 and Appendix A.1), we can compute the mean number of molecules, and we find

$$\langle N \rangle \equiv \sum_{N=0}^{\infty} N P(N; t) = st, \tag{4.101}$$

which makes perfect sense. Further, at all times the variance in the number of molecules is equal to the mean.

What happens when we add degradation to this picture? Now the state of the system can change in several ways, all of which will modify the probability that there are exactly N molecules. First, synthesis can cause the N molecules to become $N + 1$, reducing $P(N; t)$. Second, we can have the transition from $N - 1$ to N molecules, which increases $P(N; t)$. Note that these first two terms were already present in our simpler model. The third process is where degradation takes N molecules and eliminates one, resulting in $N - 1$ molecules. Because each molecule makes its transitions independently, the rate of this process must be proportional to N, and this reduces $P(N; t)$. Finally, if there were $N + 1$ molecules, degradation results in N, increasing $P(N; t)$; again because each molecule is independent, the rate of this process must be proportional to $N + 1$. Putting the terms together, we have

$$\frac{\partial P(N; t)}{\partial t} = -s P(N; t) + s P(N - 1; t) - k N P(N; t) + k(N + 1) P(N + 1; t), \tag{4.102}$$

where k is the probability per unit time for the degradation of one molecule.

The synthesis and degradation reactions can balance, generating a steady state. In this steady state the distribution of the number of molecules must obey

$$0 = s P(N - 1) - (s + k N) P(N) + k(N + 1) P(N + 1). \tag{4.103}$$

To solve this equation, it is useful to regroup the terms,

$$-sP(N-1) + kNP(N) = -sP(N) + k(N+1)P(N+1), \qquad (4.104)$$

where the left-hand side refers to the forward and backward rates between states with $N-1$ and N molecules, while the right-hand side refers to the transitions between N and $N+1$. All that we require is that the two sides be equal, but suppose we try to set each side separately to zero, which corresponds to "detailed balance" among the transitions into and out of each state. Then from the left-hand side we have

$$\frac{P(N)}{P(N-1)} = \frac{s}{kN}, \qquad (4.105)$$

while from the right we have

$$\frac{P(N+1)}{P(N)} = \frac{s}{k(N+1)}. \qquad (4.106)$$

But except for $N \rightarrow N+1$, these are the same equation. Thus, the steady state of this system does obey detailed balance, and we can solve by iterating Eq. (4.105):

$$P(1) = \frac{s}{k}P(0), \qquad (4.107)$$

$$P(2) = \frac{s}{2k}P(1) = \frac{(s/k)^2}{2}P(0), \qquad (4.108)$$

$$P(3) = \frac{s}{3k}P(2) = \frac{(s/k)^3}{3!}P(0), \qquad (4.109)$$

and, in general,

$$P(N) = \frac{(s/k)^N}{N!}P(0). \qquad (4.110)$$

Finally, we can fix the value of $P(0)$ by insisting that the distribution be normalized, and we find

$$P(N) = e^{-M}\frac{M^N}{N!}, \qquad (4.111)$$

which again is the Poisson distribution, with mean $M = s/k$.

Problem 60: The diffusion approximation. If N is not too small we expect that $P(N;t)$ and $P(N \pm 1;t)$ are not too different. Thus, we should be able to approximate using a Taylor series,

$$P(N \pm 1;t) \approx P(N;t) \pm \frac{\partial P(N;t)}{\partial N} + \frac{1}{2}\frac{\partial^2 P(N;t)}{\partial N^2}. \qquad (4.112)$$

(a) Show that this approximation turns the master equation in Eq. (4.102) into something that looks more like the diffusion equation. What is the effective potential in which the "coordinate" N is diffusing?

(b) Why does it make sense to stop the Taylor series after two derivatives? What happens if we stop after one?

(c) How does the steady state solution that you obtain in the diffusion approximation compare with the exact solution (the Poisson distribution)?

Problem 61: Langevin equations for chemical kinetics.   We know, as reviewed in Section 4.1, that we can describe Brownian motion by either a diffusion equation or a Langevin equation. In more detail, we started with kinetics that, in the macroscopic limit, correspond to the dynamics

$$\frac{dN(t)}{dt} = s - kN(t). \tag{4.113}$$

We would like to describe the noisy version of these dynamics as

$$\frac{dN(t)}{dt} = s - kN(t) + \zeta(t), \tag{4.114}$$

where—inspired by the Brownian motion example—we expect that the noise $\zeta(t)$ is white, but the strength might depend on the state of the system, so that

$$\langle \zeta(t)\zeta(t')\rangle = T_{\text{eff}}[N(t)]\delta(t - t'), \tag{4.115}$$

where to remind us of the analogy to Brownian motion, we can refer to the noise strength as an effective temperature T_{eff}. You should connect your work on this problem to your results from Problem 25 in Section 2.3.

(a) Find the effective temperature that will reproduce the diffusion equation that you derived in the preceding problem.

(b) If we integrate Eq. (4.114) over a very small time interval $\Delta\tau$, we obtain

$$\Delta N \equiv N(t + \Delta\tau) - N(t) \tag{4.116}$$

$$= [s - kN(t)]\,\Delta\tau + \int_0^{\Delta\tau} dt'\zeta(t + t'). \tag{4.117}$$

But if $\Delta\tau$ is small enough, we know that the changes in the number of molecules should be $\Delta N = 0$ or $\Delta N = \pm 1$. Going back to the master equation Eq. (4.102), identify these transition probabilities. From these probabilities, show that the mean change in the number of molecules is the first term in Eq. (4.117), $\langle\Delta N\rangle = [s - kN(t)]\Delta\tau$. Then show that the variance in ΔN is given by $\langle(\delta\Delta N)^2\rangle = [s + kN(t)]\Delta\tau$.

(c) To reproduce the variance in ΔN, we must have

$$\left\langle \left(\int_0^{\Delta\tau} dt'\zeta(t + t')\right)^2 \right\rangle = [s + kN(t)]\Delta\tau. \tag{4.118}$$

Use this expression, together with Eq. (4.115), to show that

$$T_{\text{eff}}[N(t)] = s + kN(t). \tag{4.119}$$

Does Eq. (4.119) agree with your result in part (a)?

So, these simplest of kinetic schemes for the synthesis and degradation of molecules predict that the distribution of the number of molecules ("copy numbers") should be Poisson. Certainly we can imagine kinetic schemes for which the fluctuations in copy number will be larger than Poisson. For example, if the simple picture of synthesis and degradation were correct for messenger RNA, but each mRNA leads to the synthesis of b proteins, then the mean number of proteins will be larger than the mean number of mRNA molecules by this factor b, $\langle N_{\mathrm{p}} \rangle = b \langle N_{\mathrm{mRNA}} \rangle$, but the variance will be larger by a factor of b^2, $\langle (\delta N_{\mathrm{p}})^2 \rangle = b^2 \langle (\delta N_{\mathrm{mRNA}})^2 \rangle$. Thus, if we count protein molecules, the variance will be larger than the mean, $\langle (\delta N_{\mathrm{p}})^2 \rangle = b \langle N_{\mathrm{p}} \rangle$, and hence the protein copy numbers are more variable than expected from the Poisson distribution. Notice that this is true even though we have assumed that the translation from mRNA to protein is completely noiseless, with each mRNA making exactly b proteins. Variance beyond the Poisson expectation here arises simply from amplification. This argument is exactly the same one made about photons and spikes from ganglion cells in the retina in Section 2.4.

With this background, what can we measure? Counting protein molecules is not easy. Over the past decades, we have seen a huge improvement in the methods of optical microscopy, to the point where we can literally see the light emitted from a single fluorescent molecule. But most biological molecules, and most proteins in particular, are not fluorescent. Indeed, until relatively recently the only proteins with interesting spectroscopic signatures in the visible part of the spectrum (e.g., the visual pigments and the heme proteins) involved a smaller molecular cofactor bound to the protein (retinal, heme). These cofactors are synthesized by separate, often complex, pathways. Thus, although it might be possible to engineer a cell to make a pigment protein just by splicing the relevant gene into its genome, it would be almost impossible to introduce the entire synthetic machinery for the cofactor. This is why the discovery of the green fluorescent protein (GFP) in a species of jellyfish turned out to be so important. In contrast to the proteins that require cofactors for their fluorescence, these molecules are intrinsically fluorescent. Since the isolation of the original GFP, many variants have been synthesized, in a variety of colors.

The simplest experiment to probe noise in the expression of a gene is to introduce the gene for GFP into a bacterium and just look at the levels of fluorescence. The brightness will be proportional to the number of molecules, and with luck we can even calibrate the proportionality factor. But expression levels could vary for uninteresting reasons. Cells vary in size as they grow and divide. There can be variations in the number of ribosomes, which will change the efficiency of translation, but it probably does not make sense to call these variations noise. How do we separate all these different sources of variation from genuine stochasticity in the processes of transcription and translation?

If we go back to Fig. 4.22, we see that the transcription of a gene into RNA is controlled by the binding of transcription factor proteins to a segment of DNA called the promoter or (in higher organisms) enhancer region. Suppose that we make two copies of the same promoter, but put one next to the gene for a GFP and one next to the gene for a red fluorescent protein, and then reinsert both of these into the genome. Now all variations in the state of the cell that affect the overall efficiency of transcription and translation will change the levels of green and red proteins equally. If the regulatory signals were noiseless, and the independent processes of transcription and translation of the two proteins were similarly deterministic, then every cell would be perfectly yellow,

FIGURE 4.23

Noise in the regulation of gene expression. A population of *Escherichia coli* express two fluorescent proteins of different colors, CFP and YFP, both under the control of the *lac* repressor. At left, expression is repressed, copy numbers are low, and color variations are substantial. Thus, although the two genes see the same regulatory signals, there is intrinsic variation in the output. At right, repression is relieved, expression levels are higher, and color variations are substantially smaller. From Elowitz et al. (2002). Reprinted with permission from AAAS.

having made equal amounts of green and red protein. Cells might differ in their total brightness, but the balance of red and green would be perfect. In contrast, if there really is noise in transcription and translation, or their regulation, then the balance of red and green will be imperfect, and if we look a population of genetically identical cells, they will vary in color as well as in brightness.

Figure 4.23 shows that our qualitative expectations for a two-color experiment are borne out in real experiments on *E. coli*, although "red" and "green" are actually yellow and cyan. In this experiment, the two fluorescent proteins are under the control of the *lac* promoter. In the native bacterium, this promoter controls the expression of enzymes needed for the metabolism of lactose, and if there is a better source of carbon available (or if lactose itself is absent) the bacteria do not want to make these enzymes. There is a transcription factor protein called lac repressor that binds to the *lac* promoter and blocks transcription. By changing environmental conditions, one can tap into the signals that normally tell the bacterium that it is time to turn on the *lac*-related enzymes and turn off the repression by inactivating the repressor proteins. Thus, not only can we get *E. coli* to make two colors of fluorescent protein, we can also arrange things so that we have control over the mean number of proteins that will be made. Everything that we have said thus far about noise in synthesis and degradation reactions predicts that if the cell makes more protein on average, then the fractional variance in how much protein is made should be reduced, and this is exactly what we see in Fig. 4.23.

More quantitatively, in Fig. 4.24 we see the decomposition of the variations into an extrinsic part that changes the two colors equally and an intrinsic part that corresponds to relative variations in the expression of the two proteins that are under nominally identical control. If synthesis and degradation of proteins were a Poisson process, then we expect that the variance would be equal to the mean; amplification of Poisson fluctuations in mRNA count would leave the variance proportional to the mean. Even if the Poisson model were exact, if we can't calibrate the fluorescence intensity to literally count the molecules, again all we could say is that the variance of what we measure will be proportional to the mean. In fact, the data are described well by

$$\frac{\langle(\delta F)^2\rangle}{\langle F\rangle^2} = \frac{A}{\langle F\rangle} + B, \tag{4.120}$$

where the fluorescence F is normalized, so that the mean under conditions of maximal expression is one, $A = 7 \times 10^{-4}$, and $B = 3 \times 10^{-3}$. If $B \to 0$, Eq. (4.120) is exactly the prediction of the Poisson model, and indeed B is small. Importantly, we can see

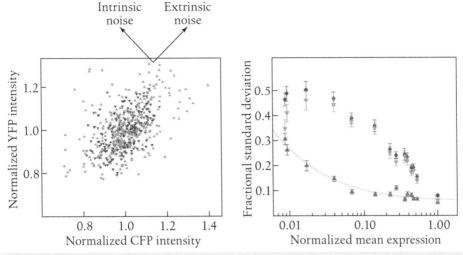

FIGURE 4.24

Separating intrinsic and extrinsic noise. At left, a scatter plot of the fluorescence from two different proteins (YFP and CFP) show the decomposition into variations in the overall efficiency of transcription and translation (extrinsic noise) and fluctuations that change the two expression levels independently (intrinsic noise). Blue and green data points are from two different strains of bacteria that are mssing different components of the relevant transcriptional regulatory apparatus. Adding these components back to the bacterium makes it possible to control the mean expression levels with external signals. At right, in such experiments the total variance has no simple dependence on the mean expression level, but the intrinsic noise (in red) goes down systematically as the mean expression level goes up. Quantitatively, we plot the standard deviation σ in fluorescence level, divided by the mean m, as a function of the mean. The dotted line is from Eq. (4.120). Redrawn from Elowitz et al. (2002).

the decrease in the fractional noise level with the increase in the mean. The absolute numbers also are interesting, because they tell us that cells can—at least under some conditions—set the expression level of a protein to an accuracy of better than 10%.

It has been appreciated for decades that the initial steps in the development of embryos provides an excellent laboratory in which to study the regulation of gene expression. As mentioned several times, what makes the different cells in our body different is, fundamentally, that they express different proteins. These differences in expression have a multitude of consequences, but the first step in making a cell commit to being one type or another is to turn on (and off) the expression of the correct set of genes. At the start, an embryo is just one cell, and through the first several rounds of cell division, it is plausible that the daughter cells remain identical. At some point, however, differences arise, and these are the first steps on the path to differentiation, or specialization of the cells for different tasks in the adult organism.

A much studied example of embryonic development is the fruit fly *Drosophila melanogaster*. We will learn much more about this system in Section 5.3, but for now the key point is that in making the egg, the mother sets the initial conditions for development in part by placing the mRNA for key proteins—referred to as the primary morphogens—at cardinal points in the embryo. As these messages are translated, the resulting proteins move through the embryo and act as transcription factors, activating

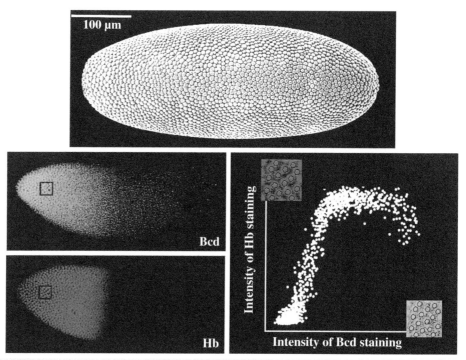

FIGURE 4.25

Bicoid (Bcd) and Hunchback (Hb) in the early *Drosophila melanogaster* embryo. At top, an electron micrograph of the embryo in cell cycle 14, with thousands of cells in a single layer at the surface. Image from E. F. Wieschaus, with thanks. At the bottom left, the embryo has been exposed to antibodies against the proteins Bcd and Hb, and these antibodies in turn have been labeled by green and red fluorophores, respectively; the fluorescence intensity should be proportional to the protein concentration, perhaps with some background. Bicoid is a transcription factor that activates the expression of Hb, and at the bottom right is a scatter plot of the output (Hb) versus input (Bcd), where each point represents the state of one nucleus from the images at left. From Gregor et al. (2007b).

the expression of other genes. An example is Bicoid (Bcd), for which the mRNA is localized at the (eventual) head; the diffusion and degradation of the Bcd protein leads to a spatial gradient in its concentration, and we can visualize this by fixing and staining the embryo with fluorescent antibodies, as shown in Fig. 4.25. A more modern approach is to fuse the gene for Bcd with a fluorescent protein and substitute it for the original gene; if we can verify quantitatively that the fusion protein replaces the function of the original, then we can measure the spatial profile of Bcd in a live embryo. Among other things, this approach makes it possible to demonstrate that the fluorescence signal from antibody staining really is proportional to the protein concentration, so we can interpret quantitatively the data from such images as those in Fig. 4.25.

From our point of view, in constructing the embryo, the mother has created an ideal experimental chamber. After just a few hours, there are thousands of cells in a controlled environment, exposed to a range of input transcription factor concentrations that we can literally read out along the embryo. We can also measure the response to these inputs, for example, the expression of the protein Hunchback (Hb) shown

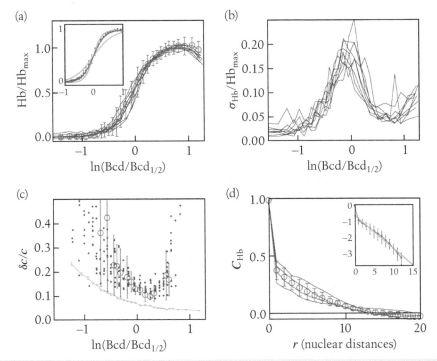

FIGURE 4.26

The transformation from Bicoid (Bcd) to Hunchback (Hb). Redrawn from Gregor et al. (2007b). (a) The input-output relation can be obtained starting from the scatter plot in Fig. 4.25, normalizing the fluorescence intensities as relative concentrations, and then averaging the output Hb expression level across all nuclei that have essentially the same input Bcd level. Blue curves show results for several individual embryos, and red circles with error bars show the mean and standard deviation of Hb expression level versus Bcd input for a single embryo. The inset shows that these data are well fit by a Hill relation (see the discussion around Eq. (4.70)) with $n = 5$ (in red), and substantially less well fit by $n = 3$ or $n = 7$ (in green). (b) Standard deviation σ of Hb output across the multiple nuclei with the same Bcd input in single embryos; different curves correspond to different individual embryos. (c) Combining the input-output relation and noise levels, we obtain the effective noise level referred to the input, as in Eq. (4.121). Blue points are raw data, green line is an estimate of measurement noise, and red circles are the results of subtracting the measurement noise variance, with error bars computed across nine embryos. (d) Correlations in Hb expression noise in different nuclei as a function of distance, measured in units of the mean distance between neighboring nuclei.

in Fig. 4.25. The targets of Bcd are themselves transcription factors, so conveniently they localize back to the nucleus, and hence each nucleus gives us one data point for characterizing the input-output relation. Taking seriously the linearity of antibody staining, we can plot the input-output relation between Bcd and Hb in appropriately normalized coordinates, as in Fig. 4.26. We can also measure the noise in expression by computing the variance across the many nuclei that experience essentially the same input Bcd level.

The first thing we see from Fig. 4.26 is that, consistent with the results from bacteria in Fig. 4.24, the embryo can regulate the expression of Hb to ~10% accuracy or better across much of the relevant dynamic range. How does this value compare with the physical limits? To measure the reliability of Hb's response to Bcd, we should refer the noise in expression back to the input—if we want to change the output by an amount equal to one standard deviation in the noise, how much do we have to change the input? The answer is given by propagating the variance backward through the input-output relation,

$$\langle(\delta\text{Hb})^2\rangle = \left|\frac{d\langle\text{Hb}\rangle}{d\ln c}\right|^2\left(\frac{\delta c}{c}\right)_{\text{eff}}^2, \tag{4.121}$$

where c is the concentration of Bcd, and $(\delta c/c)_{\text{eff}}$ defined in this way should be comparable to the Berg-Purcell limit. In Fig. 4.26 this effective noise level drops down to $(\delta c/c)_{\text{eff}} \sim 0.1$, so the system seems able to respond reliably to $\sim 10\%$ differences in concentration of the input transcription factor.

We have seen, in Eq. (4.89) and the surrounding discussion, that responding reliably to 10% differences in transcription factor concentrations would be very difficult, requiring hours of integration to push the noise level down to manageable levels. This time requirement seems generally implausible, but in the fly embryo it is impossible, because the whole process from laying the egg to the establishment of the basic body plan (several steps beyond the expression of Hb) is complete in 3 hr or less. This apparent paradox depends on estimating some key parameters, but in the Bcd-Hb system, they can be measured, and the solution to the problem of sensitivity versus noise does not seem to lie here.

Problem 62: Effective diffusion constants. Many molecules have their motion through cells impeded by interactions with large fixed structures, such as the scaffolding proteins that support the internal structure of the cell. There is a simple model for this in which the molecules we are watching, whose free concentration in solution is $c_{\text{free}}(\mathbf{x}, t)$ can bind and unbind from fixed sites that are at some density ρ. Then if a fraction $f(\mathbf{x}, t)$ of the sites at $\mathbf{x}$ are filled, the dynamics are described by

$$\frac{\partial c_{\text{free}}(\mathbf{x}, t)}{\partial t} = D\nabla^2 c_{\text{free}}(\mathbf{x}, t) - \rho k_{\text{on}} c_{\text{free}}(\mathbf{x}, t)[1 - f(\mathbf{x}, t)] + \rho k_{\text{off}} f(\mathbf{x}, t) \tag{4.122}$$

$$\frac{\partial f(\mathbf{x}, t)}{\partial t} = k_{\text{on}} c_{\text{free}}(\mathbf{x}, t)[1 - f(\mathbf{x}, t)] - k_{\text{off}} f(\mathbf{x}, t). \tag{4.123}$$

(a) Explain what the different terms in these equations mean and why they correspond to the words above. Find spatially homogeneous stationary solutions (this is not difficult). Linearize around these steady states.

(b) For the usual diffusion equation

$$\frac{\partial c(\mathbf{x}, t)}{\partial t} = D\nabla^2 c(\mathbf{x}, t), \tag{4.124}$$

if we Fourier transform in space and time, so that

$$c(\mathbf{x}, t) = \int \frac{d^3 k}{(2\pi)^3} \int \frac{d\omega}{2\pi} e^{i\mathbf{k}\cdot\mathbf{x} - i\omega t}\tilde{c}(\mathbf{k}, \omega), \tag{4.125}$$

then we have

$$-i\omega\tilde{c}(\mathbf{k}, \omega) = -Dk^2\tilde{c}(\mathbf{k}, \omega). \tag{4.126}$$

This equation is solved by a dispersion relation $\omega = -iDk^2$, which means that modes with spatial frequency k decay with a rate Dk^2. Find the dispersion relation for the modes of the linearized equations you derived in part (a). Show that, in some limits, there is still a mode with $\omega = -iD_{\text{eff}}k^2$. What is the effective diffusion constant D_{eff} in terms of the other parameters in the problem?

(c) Having found the diffusive mode in part (b), can you take limits of the original equations to show that they approximate diffusion but at a slower rate? Can you show that the total observable concentration $c(\mathbf{x}, t) = c_{\text{free}}(\mathbf{x}, t) + \rho f(\mathbf{x}, t)$ obeys this diffusion equation?

(d) When we look at this system, for example, by making the molecules fluorescent, we observe a diffusion constant D_{eff} that is smaller than the real diffusion constant D and a concentration c that is larger than c_{free}. What happens to the product $D_{\text{eff}}c$ versus Dc_{free}? How does this relate to the Berg-Purcell limit, or to the diffusion limit on reaction rates?

The fly embryo is unusual in that, for much of its early development, there are no membranes between the cells. Thus, Hb mRNA synthesized in one nucleus will be exported to the neighboring cytoplasm, and the translated protein should be free to diffuse to other nuclei. As a result, the Hb level in one nucleus should reflect an average over the Bcd signals from many cells in the neighborhood. If Hb has a diffusion constant similar to that of Bcd, then in a few minutes the molecules can cover a region that includes ~50 nuclei, and averaging over 50 independent Bcd signals is enough to convert the required integration time from hours to minutes. If this scenario is correct, there should be correlations among the Hb expression noise in nearby nuclei, and this is what we see in Fig. 4.26d. Indeed, the correlation length of the fluctuations is just what is needed to span the minutes/hours discrepancy. These results suggest strongly that the reliability of the Hb response to Bcd is barely consistent with the physical limits, but only because of spatial averaging.

Can we give a fuller analysis of noise in the Bcd-Hb system? In particular, we see from Fig. 4.26b that the noise level has a very characteristic dependence on the input concentration, which we can also replot versus the mean output, as in Fig. 4.27. This is an interesting way to look at the data, because in the limit where the Poisson noise of synthesis and degradation is dominant, we should have

$$\langle (\delta \text{Hb})^2 \rangle_{\text{Poisson}} = \alpha \langle \text{Hb} \rangle, \tag{4.127}$$

where the constant α depends on the units in which we measure expression but reflects the absolute number of independent molecules being made. In contrast, if the random arrival of transcription factors at their target is dominant, we should have Eq. (4.121) with the effective noise given by the Berg-Purcell limit, so that

$$\langle (\delta \text{Hb})^2 \rangle_{\text{BP}} = \left| \frac{d\langle \text{Hb} \rangle}{d \ln c} \right|^2 \cdot \frac{1}{N_{\text{cells}} D a c \tau_{\text{avg}}}, \tag{4.128}$$

where we have added a factor to include, as above, the idea that Hb expression levels at one cell depend on an average over N_{cells} nearby cells. Empirically, the mean expression level is well approximated by a Hill function,

$$\langle \text{Hb} \rangle = \frac{c^n}{c_{1/2}^n + c^n}, \tag{4.129}$$

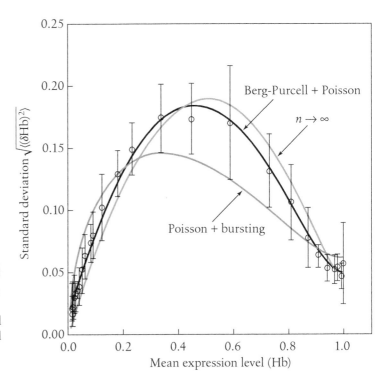

FIGURE 4.27

Noise in Hunchback (Hb) expression as a function of the mean expression level. This graph replots the data from Fig. 4.26 and compares it with several models described in the text. Error bars are standard deviations across multiple individual embryos. Redrawn from Tkačik et al. (2008).

where we choose units in which the maximum mean expression level is one, and the data are fit best by $n = 5$. Then we have

$$\frac{d\langle\text{Hb}\rangle}{d\ln c} = n\langle\text{Hb}\rangle\left(1 - \langle\text{Hb}\rangle\right),\tag{4.130}$$

and hence, after some algebra,

$$\langle(\delta\text{Hb})^2\rangle_{\text{BP}} = \beta\langle\text{Hb}\rangle^{2-1/n}\left(1 - \langle\text{Hb}\rangle\right)^{2+1/n},\tag{4.131}$$

$$\beta = \frac{n^2}{N_{\text{cells}}Dac_{1/2}\tau_{\text{avg}}}.\tag{4.132}$$

If we have both the Berg-Purcell noise at the input to transcriptional control and the Poisson noise at the output, then we expect the variances to add, so that

$$\langle(\delta\text{Hb})^2\rangle = \langle(\delta\text{Hb})^2\rangle_{\text{BP}} + \langle(\delta\text{Hb})^2\rangle_{\text{Poisson}}\tag{4.133}$$

$$= \beta\langle\text{Hb}\rangle^{2-1/n}\left(1 - \langle\text{Hb}\rangle\right)^{2+1/n} + \alpha\langle\text{Hb}\rangle.\tag{4.134}$$

Figure 4.27 shows how this prediction compares with experiment. Because $n = 5$ is known from the input-output relation, we have to set the parameters α and β. At maximal mean expression, $\langle\text{Hb}\rangle = 1$, and Eq. (4.134) predicts $\langle(\delta\text{Hb})^2\rangle = \alpha$, so we can read this parameter directly from the behavior at the right-hand edge of the graph ($\sqrt{\alpha} \sim 0.05$). We have just one parameter β left to fit, but it will determine the height, shape, and position of the peak in the noise level versus mean, so it is not at all guaranteed that we will get a reasonable fit. In fact the fit is very good, and we find

$\beta \sim 0.5$. It is interesting that the dependence of the variance on the mean seems very sensitive, because if we let the Hill coefficient become large, even the best fit of Eq. (4.134) systematically misses the data, as shown by the $n \to \infty$ curve in Fig. 4.27. Other subtly different models also fail, as you can see in Problem 64.

Problem 63: Details of Hb noise. Discuss the meaning of the parameters α and β. Can you relate them to meaningful physical quantities? Do we have independent data to see whether these numbers make sense?

Problem 64: Transcriptional bursting? The key point about noise in synthesis and degradation is that we expect the variance to be monotonic as a function of the mean (as in the Poisson model), but this is not what we see in Fig. 4.27. An alternative model that could explain the peak of noise at intermediate expression levels is that the transcription site switches between active and inactive states, generating a burst of mRNA molecules while in the active state. You should be able to refer to the discussion of noise in binding and unbinding without diffusion (leading up to Eq. (A.346)) and build up the predictions of this model.

(a) Suppose that switching into the active state occurs at a rate k_{on}, and the switch back to the inactive state occurs at a rate k_{off}. These rates must vary with the concentration of the input transcription factor, because it is only by switching between active and inactive states that the system can modulate the mean output. It seems plausible that the mean output is proportional to the probability of being in the active state. Are there any conditions under which this would not be true?

(b) Show that if the mean output is proportional to the probability of being in the active state, then the random switching will contribute to the output variance a term

$$\langle (\delta Hb)^2 \rangle_{burst} = \langle Hb \rangle \left(1 - \langle Hb \rangle \right) \cdot \frac{\tau_c}{\tau_{avg}}, \tag{4.135}$$

where the correlation time is $\tau_c = 1/(k_{on} + k_{off})$, the output is measured in units such that the maximal mean value is $\langle Hb \rangle = 1$, as above, and we assume that the averaging time is long compared with τ_c.

(c) Switching into the active state is associated with transcription factor binding. In contrast, switching back to the inactive state does not require any additional binding events. Thus, it is plausible that the rate k_{off} is independent of the input concentration c. What is the dependence of k_{on} required to reproduce the mean input-output relation in Eq. (4.129)? Is there a mechanistic interpretation of this dependence?

(d) As an aside, can you give an alternative description based on the MWC model, as in our discussion of the bacterial rotary motor (Section 4.2)? Notice that now you need to think about the kinetics of the transitions between the two states, not just the free energies. This question is deliberately open ended.

(e) Combine your results in parts (b) and (c) to show that the analog of Eq. (4.134) in this model is

$$\langle (\delta Hb)^2 \rangle = \langle (\delta Hb)^2 \rangle_{burst} + \langle (\delta Hb)^2 \rangle_{Poisson} \tag{4.136}$$

$$= \gamma \langle Hb \rangle \left(1 - \langle Hb \rangle \right)^2 + \alpha \langle Hb \rangle. \tag{4.137}$$

Give an expression for γ in terms of the original parameters of the model. Explain why the steepness of the Hill function (that is, the parameter n) does not appear directly in determining the shape of the relation between variance and mean.

(f) Fig. 4.27 shows the best fit of Eq. (4.137) to the data, which is not very good. Without doing a fit, check to see whether there is any choice of parameter that can predict a peak in $\langle(\delta\mathrm{Hb})^2\rangle$ at $\langle\mathrm{Hb}\rangle \sim 0.4$, as observed in the data. More generally, the model predicts a relation between the point at which the noise is maximal and the height of this maximum. Show that this relation is inconsistent with the data.

To summarize, we can now observe directly the noise in gene expression. Although one could emphasize that these fluctuations are, under some conditions, quite large, it seems more surprising that there are conditions where they are quite small. Cells can set the output of their genetic control machinery with a precision of $\sim 10\%$ or better, thus doing much more than switching genes on and off—intermediate levels of expression are meaningful. Thus, in particular, we must make measurements with an accuracy of better than 10%, which is not always easy to do. More fundamentally, the precision with which cells can control expression levels is not far from the limits set by the random arrival of the relevant signaling molecules (transcription factors) at their targets. We could imagine cells that use more copies of all the transcription factors, and thus could achieve greater precision—or be sloppier, and reach the same precision—but this does not seem to be what happens. I don't think we understand why evolution has pushed cells into this particular corner.

So far we have discussed noise as a small fluctuation around the mean. It is also possible that, in the same way that thermal noise can result in a nonzero rate for chemical reactions, noise in chemical kinetics can generate spontaneous switching among otherwise stable states. There is some elegant physics here, so I encourage you to explore the relevant references for this section in the Annotated Bibliography.

Problem 65: Extreme sensitivity, but slowly.   A newly discovered bacterium responds to a chemical signal by emitting light. The bacteria are roughly spherical, with diameter $d \sim 2\ \mu\mathrm{m}$, and hence are clearly visible under the microscope. The chemical signal is shown to be a small protein, presumably secreted by other bacteria; the protein diffuses through the extracellular medium with a diffusion constant $D \sim 10\ \mu\mathrm{m}^2/\mathrm{s}$. Very careful experiments establish that each individual bacterium either emits light at full intensity or is essentially dark and that changing the concentration c of the signaling protein changes the probability of being in the two states. Larger values of c correspond to higher probabilities of being in the light-emitting state, so that $p_{\mathrm{light}}(c)$ is monotonically increasing. There is a specific concentration $c = c_{1/2}$ of the signaling protein such that $p_{\mathrm{light}}(c_{1/2}) = p_{\mathrm{dark}}(c_{1/2}) = 0.5$. When poised at $c = c_{1/2}$, the system switches back and forth between the two states spontaneously at a rate of $\sim 1\ \mathrm{hr}^{-1}$. Remarkably, a change in c by just 10% is sufficient to shift the probabilities from $p_{\mathrm{light}} = 0.5$ to $p_{\mathrm{light}} = 0.9$ or $p_{\mathrm{light}} = 0.1$ when the concentration is increased or decreased, respectively.

(a) After some confusion in early experiments, it is found that everything said above is true, but the half-maximal concentration $c_{1/2} = 10^{-12}$ M. Is this possible? Justify your answer clearly and quantitatively.

(b) One group proposes that this extreme sensitivity is not at all surprising, because after all, proteins can bind to other proteins with dissociation constants as small as $K_D \sim 10^{-15}$ M. Does this observation of very tight binding have anything to do with the physical limits on sensitivity? Why or why not?

(c) Another group notes that 10^{-12} M corresponds to $\sim 10^{-3}$ molecules in the volume of the bacterium. They argue that this number provides evidence for homeopathy, in which drugs are claimed to retain their effectiveness at extreme dilution, perhaps even to the point where the doses contain less than one molecule on average. Can you resolve their confusion?

Problem 66: How simple can it be? Further studies of the new light-emitting bacterium described in Problem 65 aim at identifying the molecules involved. The first such experiment shows that if you block protein synthesis, the system cannot switch between the dark and light states, indicating that the switch involves a change in gene expression rather than (for example) a change in phosphorylation or methylation states of existing proteins, as in chemotaxis. A systematic search that knocks out individual genes, looking for effects on behavior, finds only one gene that codes for a DNA-binding protein. When this gene is knocked out, all bacteria are permanently dark. More detailed experiments show that these bacteria not only are dark, they also do not express the proteins required for generating light.

(a) Draw the simplest schematic model suggested by these results. Be sure that your model explains why there are two relatively stable states (light and dark) rather than a continuum of intermediates and that your model is consistent with the knock-out experiments.

(b) Assume that the signaling protein binds to some receptor on the surface of the cell, which triggers a cascade of biochemical events. For simplicity you can imagine that the output of this cascade is some molecule, the concentration of which is proportional to the average occupancy of the receptors over some window of time. Explain how this molecule can couple to your model in part (a) to influence the probability of the cell being in the dark or light states.

(c) Formalize your models from parts (a) and (b) by writing differential equations for the concentrations of all relevant species. Show how these equations imply the existence of discrete light and dark states. Can you see directly from the equations why changing the receptor occupancy will shift the balance between these states? It might be hard to explain the behavior near the midpoint ($c = c_{1/2}$), but it should be possible to explain the dominance of the dark state as $c \to 0$ and the light state as $c \to \infty$.

(d) Describe qualitatively all sources of noise that could enter your model. Do you have any guidance from experiment about which sources are dominant?

(e) Consider the point where $c = c_{1/2}$. Explain qualitatively what features of your model are responsible for determining the ~ 1 hr time scale for jumping back and forth between the light and dark states.

(f) See how far you can go in turning your remarks in part (e) into an honest calculation!

This section has several themes. First, bacterial chemotaxis provides us with an example of chemical sensing that is interesting, not just in itself but also as an example of a vastly more general phenomenon. Importantly, experiments on chemotaxis set a quantitative standard that should be emulated in the exploration of other chemical signaling systems, from the embryo to the brain. Second, as explained in Appendix A.6, the intuitive argument of Berg and Purcell can be made rigorous. What they identified is a limit to chemical signaling that is very much analogous to the photon shot noise limit in vision or imaging more generally. Molecules do many complicated things, but they have to reach their targets to do them, and this process is random, which sets a limit to the precision of almost everything that cells do.[20] Finally, real cells operate close to this limit, not just in specialized tasks, such as chemotaxis, but in the everyday business of regulating gene expression. Although other noise sources are clearly present, the noise floor that results from the Berg-Purcell limit never seems far away, and in some cases cells may push all the way to the point where this is the dominant noise source.

4.4   More about Noise in Perception

We have already said a bit about noise in visual perception, in the case where perception amounts to counting photons. But this phenomenon is just one corner of our perceptual experience, and we'd like to know whether some of the same principles are relevant outside this limit. In this section we look at a few instances, sampled from different organisms and different sensory modalities. I think one of the important ideas here is that considerations of noise—and processing strategies for reaching reliable conclusions in the presence of noise, perhaps even optimizing performance—cut across these many different systems, which often are the subjects of quite isolated literatures.

It has been known for some time that bats navigate by generating ultrasonic calls and listening for the echoes, forming an image of their world much as in modern sonar. To get a feeling for the precision of this behavior, there is a simple, qualitative experiment that is best explained with a certain amount of (literal) hand waving. Some bats will happily eat mealworms if you toss them into the air. Before tossing them, however, you can dip them into a little bit of flour. To eat the worm, the bat must "see" it and then maneuver its own body into position, finally sweeping the worm up in its wing and bringing it to its mouth (Fig. 4.28). But if the worm has been dusted with flour, this will leave a mark on the wing. Now repeat the experiment many times with same bat (but, of course, different worms). If you look at the bat's wing, you might expect to see many spots of flour, but in fact all the spots are on top of one another. This suggests that the entire process—not just identifying the location of the worm in the air, but the acrobatic movements required to scoop it up—have a precision of ~1 cm. In echolocation, position estimates are based on the time delays of the echoes, and with a sound speed of ~340 m/s, this spatial precision corresponds to a timing precision of $\delta t \sim 30 \, \mu s$. This rough estimate already is interesting, although maybe not too shocking: we can detect a few microseconds of difference in the arrival times of

---

20. It is possible to produce light that does not obey Poisson statistics for the photon counts, raising the question of whether we could generate comparable noise reductions in chemical processes. I think the answer is "yes"—for example, one could transport molecules to their targets by an active process that is more orderly than diffusion—but this seems enormously costly, as first emphasized by Berg and Purcell themselves. It is, however, worth thinking about. More subtly, some chemical reactions involve enormous numbers of steps, so that the fractional variance in the time required for completion of the reaction by one molecule becomes very small, as in the discussion of rhodopsin shut-off in Section 2.3. Indeed, transcription itself can be seen as an example, where it is possible for the time required to synthesize a single mRNA molecule—once transcription has been initiated—to be nearly deterministic, so that this process does not contribute a significant amount of noise.

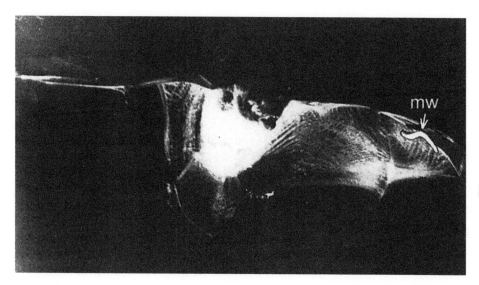

FIGURE 4.28

A big brown bat swooping in to capture a mealworm (mw). Photograph by F. A. Webster. Reprinted from Simmons (1989), with permission from Elsevier.

sounds between our two ears, which is how we can localize the source of low-frequency sounds. Barn owls do even better, detecting $\delta\tau \sim 1\,\mu\text{sec}$ between their ears.

As an aside, it was Rayleigh who understood that our brains need to use different cues for localization in different frequency ranges, just because of the physics of sound waves. At high frequencies (short wavelengths) our head casts an acoustic shadow, and there is a difference in intensity between our ears—the sound comes from the side that receives the louder signal. But at low frequencies, the wavelength is comparable to or larger than the size of the head, and there is no shadow. There is, however, a time or phase difference, but it is small. To demonstrate our sensitivity to these small time differences directly, he sat Lady Rayleigh in the gazebo behind their home and arranged for tubes of slightly different length to lead from a sound source to her two ears. A fabulous image.

Problem 67: Time differences and binaural hearing. Show that when a sound source is far away, the difference in propagation time to your two ears is independent of the distance to the source. What does determine this time difference? For your own head, what is the time difference for a source at an angle of $\sim 10°$ to the right of the direction your nose is pointing?

To be more quantitative, one would like to get the bats to report more directly on their estimates of echo delay, as in Fig. 4.29. In one class of experiments, bats stand at the base of a "Y"-shaped apparatus with loudspeakers on the two arms. Their ultrasonic calls are monitored by microphones and returned through the loudspeakers with programmable delays. In a typical experiment, the artificial echoes produced by one side of the Y are at a fixed delay τ, while the other side alternately produces delays of $\tau \pm \delta\tau$. The bat is trained to take a step toward the side that alternates, and the question is how small we can make $\delta\tau$ and still have the bat make reliable decisions. Early experiments suggested that delay differences of $\delta\tau \sim 1\,\mu\text{sec}$ were detectable, and perhaps more surprisingly that delays of $\sim 35\,\mu\text{sec}$ were less detectable, as shown in

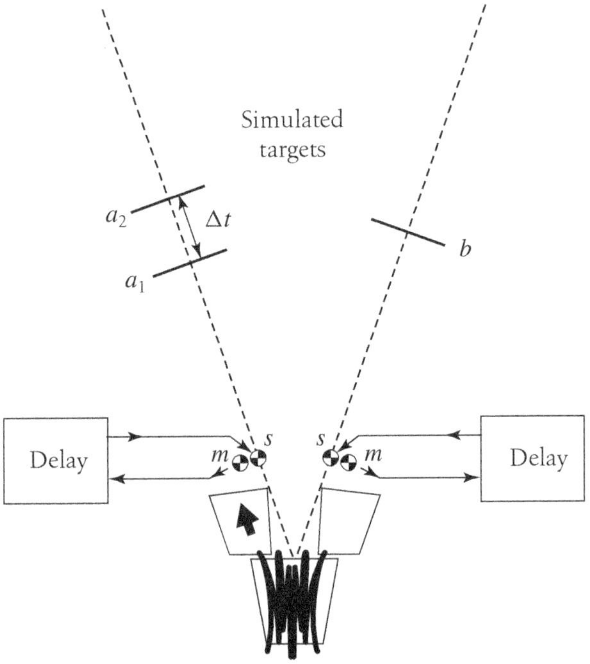

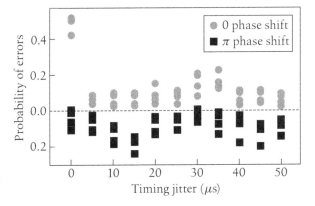

FIGURE 4.29

Schematic of the "Y" apparatus for testing echo timing-discrimination performance in bats. When the bat calls to the right, a microphone (m) picks up the signal and replays it through a speaker (s), introducing a delay to simulate a real target at position b. When the bat calls to the left, the delay alternates from call to call by an amount Δt, simulating a real target that moves back and forth between positions a_1 and a_2. Redrawn from Simmons et al. (1990).

FIGURE 4.30

Performance of four different bats at echo jitter discrimination. Echoes can be returned with no phase shift (circles), or with a phase shift of π (squares); errors for the phase-shifted echoes are plotted downward. The phase shift itself is detectable with almost no errors, and there is confusion around $\delta\tau \sim 35\ \mu$s. This "confusion peak" shifts and splits with the introduction of a phase shift. Redrawn from Simmons et al. (1990).

Fig. 4.30. The latter result might make sense if the bat were trying to measure delays by matching the detailed waveforms of the call and echo, because these sounds have most of their power at frequencies near $f \sim 1/(35\ \mu\text{sec})$—the bat can be confused by delay differences corresponding to an integer number of periods in the acoustic waveform. One can even see the $n = 2$ "confusion resonance" if one is careful. A phase shift introduced into the artificial echo shifts the confusion peak, as expected.

Let's think about this result more formally. Suppose that we are expecting a sound (or any signal) that has a time dependence $s_0(t)$, but we do not know when it will arrive, so what we actually observe will be $s_0(t - \tau)$ embedded in some background of noise. That is,

$$s(t) = s_0(t - \tau) + \eta(t), \tag{4.138}$$

where $\eta(t)$ is the noise. Let's assume, for simplicity, that the noise is white, with some spectral density $\mathcal{N}$. Then, as explained in Appendix A.2, the probability density for the function $s(t)$ becomes

$$P[s(t)|\tau] = \frac{1}{Z} \exp\left[-\frac{1}{2\mathcal{N}} \int dt \, \left(s(t) - s_0(t-\tau)\right)^2\right], \qquad (4.139)$$

where Z is a normalization constant and the notation reminds us that this is the distribution if we know the delay τ. If instead the delay is $\tau + \delta\tau$, we have

$$P[s(t)|\tau + \delta\tau] = \frac{1}{Z} \exp\left[-\frac{1}{2\mathcal{N}} \int dt \, \left(s(t) - s_0(t-\tau-\delta\tau)\right)^2\right]. \qquad (4.140)$$

As in our previous discussions of discrimination between two alternatives in Section 2.1, when we are faced with a particular signal $s(t)$ and have to decide whether the delay was τ or $\tau + \delta\tau$, the relevant quantity is the (log) likelihood ratio:

$$\lambda[s(t)] \equiv \ln\left(\frac{P[s(t)|\tau + \delta\tau]}{P[s(t)|\tau]}\right) \qquad (4.141)$$

$$= -\frac{1}{2\mathcal{N}} \int dt \, \left(s(t) - s_0(t-\tau-\delta\tau)\right)^2 + \frac{1}{2\mathcal{N}} \int dt \, \left(s(t) - s_0(t-\tau)\right)^2 \qquad (4.142)$$

$$= \frac{1}{\mathcal{N}} \int dt \, s(t) \left[s_0(t-\tau-\delta\tau) - s_0(t-\tau)\right]. \qquad (4.143)$$

If the delay really is τ, then the mean log likelihood ratio is defined by

$$\langle\lambda[s(t)]\rangle_\tau \equiv \left\langle \frac{1}{\mathcal{N}} \int dt \, s(t) \left[s_0(t-\tau-\delta\tau) - s_0(t-\tau)\right] \right\rangle_\tau, \qquad (4.144)$$

where $\langle\cdot\rangle_\tau$ denotes an average over the distribution $P[s(t)|\tau]$. But from Eq. (4.138), the waveforms $s(t)$ drawn from this distribution can be decomposed into signal plus noise, so that

$$\langle\lambda[s(t)]\rangle_\tau = \left\langle \frac{1}{\mathcal{N}} \int dt \, \left[s_0(t-\tau) + \eta(t)\right]\left[s_0(t-\tau-\delta\tau) - s_0(t-\tau)\right] \right\rangle, \qquad (4.145)$$

where now averaging just involves averaging over the noise $\eta(t)$. But the noise appears only linearly, and it has zero mean, so it drops out, giving us

$$\langle\lambda[s(t)]\rangle_\tau = \frac{1}{\mathcal{N}} \int dt \, s_0(t-\tau)\left[s_0(t-\tau-\delta\tau) - s_0(t-\tau)\right] \qquad (4.146)$$

$$= \frac{1}{\mathcal{N}}[C(\delta\tau) - C(0)], \qquad (4.147)$$

where

$$C(t) = \int dt' \, s_0(t')s_0(t'-t) \qquad (4.148)$$

is the autocorrelation function of the expected signal. Similar calculations yield

$$\langle \lambda[s(t)]\rangle_{\tau+\delta\tau} = \frac{1}{\mathcal{N}}[C(0) - C(\delta\tau)], \tag{4.149}$$

$$\langle (\delta\lambda[s(t)])^2\rangle_\tau = \langle (\delta\lambda[s(t)])^2\rangle_{\tau+\delta\tau} \tag{4.150}$$

$$= \frac{2}{\mathcal{N}}[C(0) - C(\delta\tau)]. \tag{4.151}$$

The log-likelihood ratio $\lambda[s(t)]$ is a Gaussian random variable (inherited from the Gaussian statistics of the noise η), so these few moments provide a complete description of the problem of discriminating between delays τ and $\tau + \delta\tau$. The end result is that the discrimination problem is exactly that of a single Gaussian variable (λ), with signal-to-noise ratio

$$SNR = \frac{(\langle \lambda[s(t)]\rangle_{\tau+\delta\tau} - \langle \lambda[s(t)]\rangle_\tau)^2}{\langle (\delta\lambda[s(t)])^2\rangle} = \frac{2}{\mathcal{N}}[C(0) - C(\delta\tau)]. \tag{4.152}$$

Thus, we see that SNR is large as soon as the jitter $\delta\tau$ is big enough to break the correlations in the waveform. Conversely, SNR falls if shifting by $\delta\tau$ brings the waveform back into correlation with itself, as will happen for an approximately periodic signal, such as the echolocation pulse.

Problem 68: Details of the signal-to-noise ratio for detecting jitter in echolocation.   Fill in the details leading to Eq. (4.152).

(a) How does this result change if the discrimination involves not just a time shift $\delta\tau$ but also a sign flip or π phase shift?

(b) Recall the relationship between error probability and the signal-to-noise ratio from Problem 10. Is it practical to try and estimate the correlation function $C(\tau)$ by measuring the error probability as a function of $\delta\tau$? What if you also have access to experiments with a sign flip, as in part (a)? If you have errors in the measurement of the error probability, how do these propagate back to estimates of the underlying $C(\tau)$?

(c) Compare your results in part (b) with the construction of "compound jitter discrimination curves" by Simmons et al. (1990). Can you suggest improvements in their data analysis methods?

This argument about discriminability assumes that the bat's brain actually can compute using the entire acoustic waveform $s(t)$, rather than some more limited features; in this sense we are describing the best that the bat could possibly do. It is interesting that such a calculation predicts confusion at delays where the autocorrelation function of the bat's call has a peak and that such confusions are observed. However, this calculation seems hopelessly optimistic: access to the acoustic waveform means, in particular, access to features that are varying on the microsecond time scale. If we record the activity of single neurons emerging from the ear as they respond to pure tones, then we can see the action potentials phase lock to the tone, but this effect is significant only up to some maximum frequency. Beyond this high-frequency cut-off, the overall rate of spikes increases with the intensity of the tone, but the timing of the spikes seems unrelated to the details of the acoustic waveform. Although there is controversy about the precise value of the cutoff frequency for phase locking, there seems to be no hint in the literature that it could be as high at 30 kHz. Taking all this at face value, it seems implausible that the auditory nerve actually transmits to the brain anything like a complete replica of the echo waveforms.

There is a second problem with this seemingly simple calculation. If we expand the SNR for small $\delta\tau$, we have

$$SNR = \frac{2}{\mathcal{N}}[C(0) - C(\delta\tau)] \approx \frac{C(0)}{\mathcal{N}} \cdot \left[\frac{C''(0)}{C(0)}\right](\delta\tau)^2. \qquad (4.153)$$

We expect that the term in brackets, which has the units of $1/(\text{time})^2$, is determined by the time scale on which the echolocation pulse is varying, something like $\sim 35 \, \mu\text{sec}$. In contrast, the first term, $C(0)/\mathcal{N}$, measures how loud the echo is relative to the background noise and is dimensionless. In acoustics it is conventional to measure in decibels, where 10 dB represents a factor of ten difference in acoustic power or energy. A typical quiet conversation produces sounds ~ 30 dB above our threshold of hearing and hence above the limiting internal noise sources in the ear, whatever these may be. The bat's echolocation pulses are enormously loud, and although the echoes may be weak, it still is plausible that (at least in the laboratory setting) they are ~ 60 dB above the background noise. Thus, our calculation predicts a signal-to-noise ratio of one when the differences in delay $\delta\tau$ are measured in tens of *nanoseconds*, not microseconds. I think this result was viewed as so obviously absurd that it was grounds for throwing out the whole idea that the bat uses detailed waveform information, even without reference to data on what the auditory nerve can encode.

In an absolutely stunning development, however, Simmons and colleagues went back to their experiments, produced delays in the appropriate range—convincing yourself that you have control of acoustic and electronic delays with nanosecond precision is not so simple—and found that the bats could do what they should be able to do as ideal detectors: they detect 10 *nano*second differences in echo delay, as shown in Fig. 4.31. Further, they added noise in the background of the echoes and showed that performance of the bats tracked the ideal performance over a range of noise levels. This is a wonderful example with which to start this section of our discussion, because we have no idea how the bat manages this amazing feat of signal processing.

The problem of echo delay discrimination has just enough structure to emphasize an important point: when we make perceptual decisions, we are not identifying signals, we are identifying the distribution out of which these signals have been drawn. This distinction becomes even more important as we move toward more complex tasks, where the randomness is intrinsic to the signal rather than being just a result of added noise. As an example, a single spoken word can generate a wide variety of sounds, all the more varied when embedded in a sentence. Identifying the word really means saying that the particular sound we have heard comes from this distribution and not another. Importantly, probability distributions can overlap, and hence there are limits on the reliability of discrimination.

Some years ago, Barlow and colleagues launched an effort to use these ideas of discrimination among distributions to study progressively more complex aspects of visual perception, in some cases reaching into the psychology literature for examples

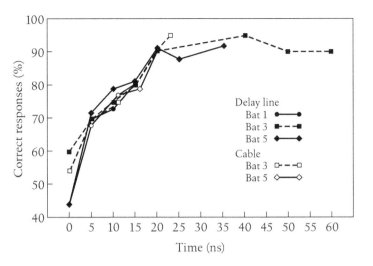

FIGURE 4.31

Bat echo discrimination performance at very small delays. As a control, the experimenters employed different methods of introducing delays (delay line and cable), and we can see reproducibility across several individual bats. Similar experiments in the presence of background noise show that the minimum time delay for reliable discrimination varies as expected from Eq. (4.153). Redrawn from Simmons et al. (1990).

of gestalt phenomena—where our perception is of the whole rather than its parts. One such example is the recognition of symmetry in otherwise random patterns. Suppose that we want to make a random texture pattern. One way to do this is to draw the contrast $C(\mathbf{x})$ at each point $\mathbf{x}$ in the image from some simple probability distribution that we can write down. An example is to make a Gaussian random texture, which corresponds to

$$P[C(\mathbf{x})] \propto \exp\left[-\frac{1}{2}\int d^2x \int d^2x' C(\mathbf{x})K(\mathbf{x}-\mathbf{x}')C(\mathbf{x}')\right], \qquad (4.154)$$

where $K(\mathbf{x}-\mathbf{x}')$ is the kernel or propagator that describes the texture. By writing K as a function of the difference between coordinates, we guarantee that the texture is homogeneous; if we want the texture to be isotropic, we take $K(\mathbf{x}-\mathbf{x}') = K(|\mathbf{x}-\mathbf{x}'|)$. Using this scheme, how do we make a texture with symmetry, say with respect to reflection about an axis?

Problem 69: Texture discrimination.   Show that Eq. (4.154) can be rewritten as

$$P[C(\mathbf{x})] \propto \exp\left[-\frac{1}{2}\int \frac{d^2k}{(2\pi)^2}\frac{|\tilde{C}(\mathbf{k})|^2}{S_C(\mathbf{k})}\right], \qquad (4.155)$$

where $S_C(\mathbf{k})$ is the (now two-dimensional) power spectrum, connected as usual to the correlation function

$$\langle C(\mathbf{x})C(\mathbf{x}')\rangle = \int \frac{d^2k}{(2\pi)^2} S_C(\mathbf{k})e^{i\mathbf{k}\cdot(\mathbf{x}-\mathbf{x}')}. \qquad (4.156)$$

Suppose that you have the task of discrimination between images drawn from distributions characterized by two different power spectra, $S_C(\mathbf{k})$ and $S_C(\mathbf{k}) + \Delta S_C(\mathbf{k})$. Show that, assuming you have access to a large area of the image, the discrimination problem for small $\Delta S_C(\mathbf{k})$ is again like that of a single Gaussian variable. Explain what role is played by the assumption of a "large area," and what defines "large" in this context. How does the signal-to-noise ratio for discrimination depend on area?

The statement that texture has symmetry about an axis is that for each point $\mathbf{x}$, we can find the corresponding reflected point $\hat{\mathbf{R}} \cdot \mathbf{x}$ and that the contrasts at these two points are very similar. These conditions should be true for every point. This symmetry can be enforced by choosing

$$
P_\gamma[C(\mathbf{x})] \propto \exp\left[-\frac{1}{2} \int d^2x \int d^2x' \, C(\mathbf{x}) K(\mathbf{x} - \mathbf{x}') C(\mathbf{x}') \right.
$$

$$
\left. + \frac{\gamma}{2} \int d^2x \, |C(\mathbf{x}) - C(\hat{\mathbf{R}} \cdot \mathbf{x})|^2 \right], \qquad (4.157)
$$

where γ measures the strength of the tendency toward symmetry. Clearly as $\gamma \to \infty$ we have an exactly symmetric pattern, quenching half of the degrees of freedom in the original random texture. In contrast, as $\gamma \to 0$, the weakly symmetric textures drawn from P_γ become almost indistinguishable from a pure random texture ($\gamma = 0$). Given images of a certain size and a known kernel K, there is a limit to the smallest value of γ that can be distinguished reliably from zero, and we can compare this statistical limit to the performance of human observers. This is more or less what Barlow did, although he used blurred random dots rather than the Gaussian textures considered here; the idea is the same, and all the details become the same in the limit of many dots. The result is that human observers come within a factor of two of the statistical limit for detecting γ or its analog in the random dot patterns.

One can use similar sorts of visual stimuli to think about motion, where rather than having to recognize a match between two halves of a possibly symmetric image, we have to match successive frames of a movie. Here again human observers can approach the statistical limits, as long as we stay in the right regime: we seem not to make use of fine dot positioning (as would be generated if the kernel K only contained low-order derivatives), nor can we integrate efficiently over many frames. These results are interesting, because they show the potentialities and limitations of optimal visual computation, but also because the discrimination of motion in random movies is one of the places where people have tried to make close links between perception and neural activity in the (monkey) cortex.

Let us look in detail at the case of visual motion estimation, using not humans or monkeys, but the visual system of the fly, which we have met already in Section 2.1. If you watch a fly flying around in a room or outdoors, you will notice that flight paths tend to consist of rather straight segments interrupted by sharp turns and acrobatic interludes. These observations can be quantified through the measurement of trajectories during free flight and in experiments where the fly is suspended from a torsion balance or a fine tether. Given the aerodynamics for an object of the fly's dimensions, even flying straight is tricky. In the torsion balance one can demonstrate directly that motion across the visual field drives the generation of torque, and the sign is such as to stabilize flight against rigid body rotation of the fly. Indeed one can close the feedback loop by measuring the torque that the fly produces and using this torque to (counter)rotate the visual stimulus. The result is an imperfect "flight simulator" for the fly in which the only cues to guide the flight are visual; under natural conditions the fly's mechanical sensors play a crucial role. Despite the imperfections of the flight simulator, the tethered fly will fixate on small objects, thereby stabilizing the appearance of straight flight. Similarly, aspects of flight behavior under free flight conditions can be understood if flies generate torques in response to motion across the visual field, and this response is remarkably fast, with a latency of just ~30 msec. The combination of free flight and torsion-balance experiments strongly suggests that flies can estimate their angular velocity from visual input alone and then produce motor outputs based on this estimate.

Voltage signals from the receptor cells are processed by several layers of the brain, each layer having cells organized on a lattice that parallels the lattice of lenses visible from the outside of the fly. As shown in Fig. 4.32, after passing through the lamina, the medulla, and the lobula, signals arrive at the lobula plate. This plate contains a stack of about 50 cells that are sensitive to different components of motion. These cells have imaginative names, such as H1 and V1, which respond to horizontal and vertical components of motion, respectively. If one kills individual cells in the lobula plate, then the simple experiment of moving a stimulus and recording the flight torque no longer works, strongly suggesting that these cells are an obligatory link in the pathway from the retina to the flight motor. Taken together, these observations support a picture in which the fly's brain uses photoreceptor signals to estimate angular velocity and encodes this estimate in the activity of a few neurons.[21] What can we say about the physical limits to the precision of this computation?

Suppose that we look at a pattern of typical contrast C and it moves by an angle $\delta\theta$, as schematized in Fig. 4.33. A single photodetector element will see a change in contrast of roughly $\delta C \sim C \cdot (\delta\theta/\phi_0)$, where ϕ_0 is the angular scale of blurring due to diffraction. If we can measure for a time τ, we will count an average number of photons $R\tau$, with R the counting rate per detector, and hence the noise can be expressed as a fractional precision in intensity of $\sim 1/\sqrt{R\tau}$. But fractional intensity is what we mean by contrast, so $1/\sqrt{R\tau}$ is really the contrast noise in one photodetector. To get the signal-to-noise ratio, we should compare the signal and noise in each of the N_{cells} detectors, then add the squares if we assume (as for photon shot noise) that noise is independent in each detector while the signal is coherent:

$$SNR \sim N_{cells} \cdot \left(\frac{\delta\theta}{\phi_0}\right)^2 C^2 R\tau. \tag{4.158}$$

Motion discrimination is hard for flies, because they have small lenses and hence blurry images (ϕ_0 is large), and because they have to respond quickly (τ is small). For instance, typical photon counting rates in a laboratory experiment are $R \sim 10^4\,\mathrm{s}^{-1}$, and outside on a bright day one can get to $R \sim 10^6\,\mathrm{s}^{-1}$. Under reasonable laboratory conditions—and taking account of all the factors that go in front of our rough

---

21. You should be skeptical of any claim about what the brain computes, or more generally what problems an organism has to solve to explain some observed behavior. The fact that flies can stabilize their flight using visual cues, for example, does *not* mean that they compute motion in any precise sense—they could use a form of "bang-bang" control that needs knowledge only of the algebraic sign of the velocity, although I think that the torsion-balance experiments argue against such a model. It also is a bit mysterious why we find neurons with such understandable properties: one could imagine connecting photoreceptors to flight muscles by means of a network of neurons in which there is nothing that we could recognize as a motion-sensitive cell. Thus, it is not obvious that the fly must compute motion or that there must be motion-sensitive neurons. But this is what we find.

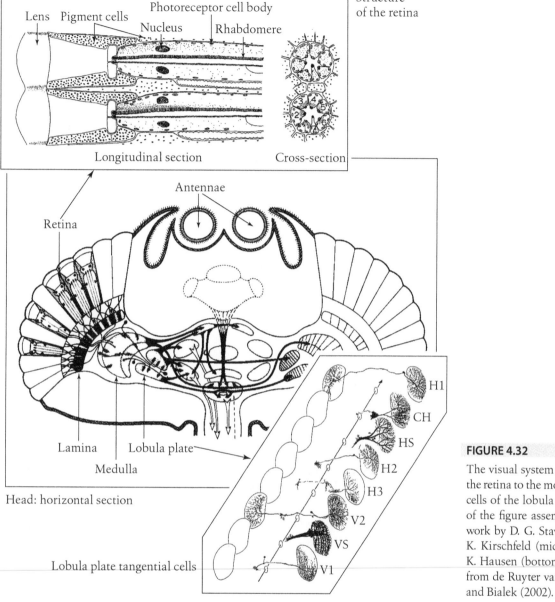

Structure
of the retina

Lens   Pigment cells   Photoreceptor cell body   Nucleus   Rhabdomere

Longitudinal section   Cross-section

Retina

Antennae

Lamina   Lobula plate

Medulla

Head: horizontal section

Lobula plate tangential cells

H1
CH
HS
H2
H3
V2
VS
V1

FIGURE 4.32

The visual system of a fly, from
the retina to the motion-sensitive
cells of the lobula plate. Pieces
of the figure assembled from
work by D. G. Stavenga (top),
K. Kirschfeld (middle), and
K. Hausen (bottom). Modified
from de Ruyter van Steveninck
and Bialek (2002).

Eq. (4.158) in a more careful calculation—the optimal estimator would reach $SNR = 1$
at an angular displacement of $\delta\theta \sim 0.05°$.

We can test the precision of motion estimation in two very different ways. One is
similar to the experiments we have discussed already, where we are forced to choose
between two alternatives and measure the reliability of this choice. A single neuron
responds to sudden steps of motion with a brief volley of action potentials, which we
can label as occurring at times $t_1, t_2, \cdots$. We as observers of the neuron can look at
these times and try to decide whether the motion had amplitude θ_+ or θ_-; the idea
is exactly the same as in earlier discussions of discrimination of signal versus noise,

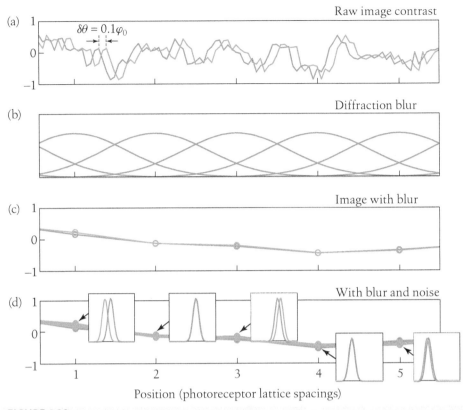

FIGURE 4.33

The limits to motion detection. (a) A possible pattern of contrast (normalized light intensity) versus position or angle in the visual world. Blue denotes the original pattern, and orange illustrates a shift by 1/10 of the spacing between photoreceptors. (b) Blurring and sampling of the image, with Gaussian apertures that provide a model for the optics of the fly's eye. Note that the spacing between photoreceptors is comparable to the width of the diffraction blur. (c) The signal arriving at each photoreceptor. The blurring reduces the contrast enormously. (d) The effect of adding noise, here with an amplitude expected if each snapshot involves counting an average of 10^3 photons. Insets show the distribution of signals plus noise in response to the original (blue) and shifted (orange) images. Despite the large differences between the two initial patterns, only one of the five receptor cells shown here would be able to come near to reliable detection. The experiments described in the text are done under conditions of even smaller signal-to-noise ratios.

but here we have to measure the relevant probability distributions rather than making assumptions about their form; see Fig. 4.34. Doing the integrals, one finds that looking at spikes generated in the first ∼30 msec after the step (as in the fly's behavior) we can reach the reliability expected for $SNR = 1$ at a displacement $\delta\theta = |\theta_+ - \theta_-| \sim 0.12°$, within a factor of two of the theoretical limit set by noise in the photodetectors.

It is worth noting a few more points that emerge from Fig. 4.34 and further analyses of this experiment. First, on the ∼30 msec time scale of relevance to behavior, there are only a handful of spikes. This is partly what makes it possible to do the analysis so completely, but it also is a lesson for how we think about the neural representation of

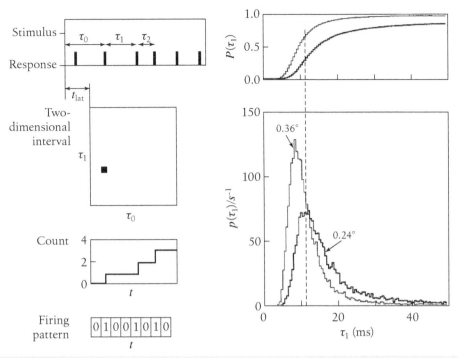

FIGURE 4.34

Motion discrimination with the fly's H1 neuron. At left is a schematic of the spikes in response to a transient stimulus, such as a step of motion. We can describe the response by the time until the first spike τ_0 after a latency t_{lat}, the time from the first spike to the second τ_1, ... Alternatively we can just count the spikes that have occurred up to a certain time after the stimulus, or we could, at some fixed time resolution, describe the whole pattern of spikes as a binary word. In each case we can analyze the discriminability of different stimuli by accumulating, over many repeated presentations of each stimulus, the distribution of responses. At right, is an example of this analysis, focusing on the single interspike interval τ_1 in response to steps that differ in size by 0.12°. Long intervals correspond to the weaker stimulus, and from the cumulative probability distributions in the top panel, we can read off the probabilities of correct identification of each stimulus. Redrawn from de Ruyter van Steveninck and Bialek (1995).

information in general. Second, we can dissect the contributions of individual spikes to show that each successive spike makes a nearly independent contribution to the signal-to-noise ratio for discrimination, so there is essentially no redundancy. Finally, the motions we are discussing—motions close to the physical limits of detectability, and motions that real neurons can represent reliably—are much smaller than the lattice spacing on the retina or the nominal diffraction limit of angular resolution of ~1°. Analogous phenomena have been known in human vision for more than a century, and are called hyperacuity.

The step-discrimination experiment gives us a very clear view of reliability in the neural response, but as with the other discrimination experiments discussed above, it is not a very natural task. An alternative is to ask what happens when the motion signal (angular velocity $\dot{\theta}(t)$) is a complex function of time. Then we can think of the signal-to-noise ratio in Eq.(4.158) as being equivalent to a spectral density of displacement noise $N_\theta^{\text{eff}} \sim \phi_0^2/(N_{\text{cells}}C^2R)$ or a generalization in which the photon counting rate is replaced by an effective frequency-dependent rate related to the noise characteristics of the photoreceptors, as in Fig. 2.15. It seems likely, as discussed above, that the fly's visual system really does make a continuous or running estimate of the angular velocity and that this estimate is encoded in the sequence of discrete spikes produced by neurons like H1. It is not clear that any piece of the brain ever decodes this signal in an explicit way, but if *we* could do such a decoding, we could test directly whether the accuracy of our decoding reaches the limiting noise level set by noise in the photodetectors.

An approach to decoding spike trains is shown in Fig. 4.35. The idea is that each spike contributes a small transient blip to our estimate of the signal versus time, and to obtain the full estimate, we add up all these small contributions. Thus, if the signal we are interested in is $s(t)$, our estimate is

$$s_{\text{est}}(t) = \sum_i f(t - t_i), \tag{4.159}$$

where $\{t_i\}$ are the spike arrival times as before, and we can choose the filter $f(t)$ to minimize the errors[22]

$$\chi^2 \equiv \int dt \left| s(t) - s_{\text{est}}(t) \right|^2. \tag{4.160}$$

Like most neurons, H1 has a sign preference for its inputs—motion in one direction generates more spikes, whereas motion in the opposite direction generates fewer spikes. Thus, large negative velocities cause H1 to go silent, and in these periods we would have no basis for inferring the detailed waveform of velocity versus time. Fortunately, the fly has two H1 neurons, one on each side of the head, with opposite direction preferences. We could record from both cells, or we could use the fact that the two cells see opposite motions relative to their own preference and look at the responses of one neuron to both a stimulus and the opposite motion. If the spikes in these two cases are $\{t_i^+\}$ and $\{t_i^-\}$, we can make a more symmetric reconstruction:

$$s_{\text{est}}(t) = \sum_i \left[f(t - t_i^+) - f(t - t_i^-) \right]. \tag{4.161}$$

Again, we choose the filter $f(t)$ to minimize χ^2.

In Figure 4.35 we see that the reconstruction of the velocity waveform in fact is quite accurate. More quantitatively, the power spectrum of the errors in the reconstructed signal approaches the limit set by noise in the photoreceptor cells, within a factor of two at high frequencies. Further, one can change, for example, the image contrast and show that the resulting error spectrum scales as expected from the theoretical limit.

To the extent that the fly's brain can estimate motion with a precision close to the theoretical limit, we know that the act of processing itself does not add too much noise. But being quiet is not enough: to make maximally reliable estimates of nontrivial stimulus features like motion, one must be sure to do the correct computation. Making

---

22. As in most such problems, we should be careful to avoid overfitting, and so we take a long experiment and, for example, learn the filter $f(t)$ by minimizing χ^2 computed from the first 90% of the data, and then test the quality of the reconstruction in the remaining 10%.

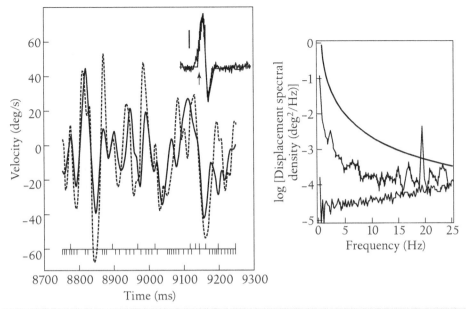

FIGURE 4.35

Decoding continuous motion signals from spikes generated by the H1 neuron. At left, the dashed curve indicates the true stimulus, angular velocity as a function of time. The solid line is the result of the decoding process, from Eq. (4.161). Tick marks below the stimulus indicate the spikes generated in a single presentation of this stimulus (upward ticks) or its negative (downward ticks). This consideration of a hypothetical neuron that sees the negative stimulus is meant to restore symmetry between positive and negative velocities and corresponds roughly to the response of the H1 neuron on the other side of the fly's head, which has the opposite direction selectivity. At right is the spectral density of errors in the reconstruction. The error is reported as a displacement error, so the spectrum grows as $1/\omega^2$ for low frequencies. Also shown is the spectrum of the stimulus (smooth line) and the limiting noise level computed from the actual noise levels measured in fly photoreceptors under the same conditions as for these experiments on H1. Reconstruction error and the physical limit to precision converge at high frequencies, so that the fly approaches optimal performance. Redrawn from Bialek et al. (1991).

this idea precise is in the same spirit as the discussion, in Section 2.4, of pooling single-photon signals from multiple rod cells at the level of bipolar cells. There we saw how the different orders of nonlinearity and summation result in very different final signal-to-noise ratios, even though all we are trying to do is add. Here the problem is more difficult, because the fly needs to estimate a feature of the visual world that is not directly reflected in the signals of any single receptor cell.

Problem 70: (Relatively) simple estimation problems. Suppose that someone draws a random number x from a probability distribution $P(x)$. Rather than seeing x itself, you see only a noisy version, $y = x + \eta$, where η is drawn from a Gaussian distribution with variance σ^2, so that

$$P(y|x) = \frac{1}{\sqrt{2\pi\sigma^2}} \exp\left[-\frac{1}{2\sigma^2}(y - x)^2\right]. \tag{4.162}$$

Having seen y, your job is to estimate x.

(a) Show that everything you know about x by virtue of observing y can be written in a way that suggests an analogy with statistical mechanics,

$$P(x|y) = \frac{1}{Z(y)} \exp\left[-\frac{V_{\text{eff}}(x)}{k_B T_{\text{eff}}} + \frac{F_{\text{eff}} x}{k_B T_{\text{eff}}}\right], \tag{4.163}$$

where

$$\frac{V_{\text{eff}}(x)}{k_B T_{\text{eff}}} = -\ln P(x) + \frac{x^2}{2\sigma^2}, \tag{4.164}$$

$$k_B T_{\text{eff}} = \sigma^2, \tag{4.165}$$

$$F_{\text{eff}} = y. \tag{4.166}$$

(b) From the discussion in Section 2.4, we know that if we define "best" to be the estimator that minimizes χ^2, then the best estimator is the conditional mean,

$$x_{\text{est}}(y) = \int dx \, x \, P(x|y). \tag{4.167}$$

Construct $x_{\text{est}}(y)$ in the case where $P(x)$ is a Gaussian with unit variance. Show that this estimate, although "best," is systematically wrong. That is, if we average $x_{\text{est}}(y)$ over the distribution $P(y|x)$, we do not recover x itself. Explain why this estimate can still be the best one.

(c) Now consider the case $P(x) = (1/2) \exp(-|x|)$. Show that, even though the transformation from what we are interested in (x) to what we measure (y) is linear, the optimal estimator is nonlinear. In particular, if rather than asking for an estimator that minimizes χ^2, we ask for the most probable value of x given y, show that the optimal estimator involves a threshold nonlinearity.

Motion estimation is an example of the more general problem of perceptual estimation. The data to which the brain has access are the responses of receptor cells, and the goal is to estimate some feature of the world. The first key step is to use Bayes' rule, combining the noisy data from the receptors with our prior knowledge that some things are more likely than others. Schematically, we have

$$P(\text{feature}|\text{receptor responses})$$
$$= \frac{P(\text{receptor responses}|\text{feature}) P(\text{feature})}{P(\text{receptor responses})}. \tag{4.168}$$

The second key step is to note that receptors typically do not respond directly to the features of interest but rather to raw sensory signals, such as light intensity, sound pressure in the auditory system, or the concentrations of specific molecular species in complex odors. Continuing schematically, let's denote the full spatiotemporal pattern of light intensities falling on the retina by $\mathcal{I}$. Receptor responses really depend on $\mathcal{I}$, which in turn is correlated with the feature that we want to estimate. Thus, we have

$$P(\text{receptor responses}|\text{feature})$$
$$= \int D\mathcal{I} \, P(\text{receptor responses}|\mathcal{I}) P(\mathcal{I}|\text{feature}), \tag{4.169}$$

and putting all the terms together, we have

$$P(\text{feature}|\text{receptor responses})$$

$$= \frac{1}{P(\text{receptor responses})} \int D\mathcal{I} \, P(\text{receptor responses}|\mathcal{I}) P(\mathcal{I}, \text{feature}). \quad (4.170)$$

If the lights are bright, and the noise level in the photoreceptors is low, it is plausible that knowing the pattern of receptor responses is almost equivalent to knowing the spatiotemporal pattern of light intensities $\mathcal{I}$. Hence, viewed as a function of $\mathcal{I}$, the distribution $P(\text{receptor responses}|\mathcal{I})$ is very sharply peaked. Then the entire structure of the optimal computation that maps receptor responses to the desired feature is controlled by $P(\mathcal{I}, \text{feature})$, which is a property of the world that we live in rather than of our eyes or brains. This is perhaps our most important qualitative conclusion: optimal estimates of sensory features involve computations determined by the structure of the world around us. To the extent that our brains, and those of other animals, make optimal estimates, this means that the way in which we process the world is set by the physics of our environment, not by peculiarities of our biological hardware.

For the case of motion estimation, what is the structure of $P(\mathcal{I}, \text{feature})$? Let's think about a one-dimensional version of the problem, so that spatiotemporal pattern of light intensity is $\mathcal{I} \equiv I(x, t)$. Then if a small piece of the visual world is moving rigidly relative to us with a velocity v, we should have $I(x, t) = I_0(x - vt)$. We can then take derivates in space and time:

$$\partial_x I \equiv \frac{\partial I(x, t)}{\partial x} = I_0'(x - vt), \quad (4.171)$$

$$\partial_t I \equiv \frac{\partial I(x, t)}{\partial t} = -v I_0'(x - vt). \quad (4.172)$$

Thus, we can compute the velocity as a ratio of spatial and temporal derivatives:

$$v_{\text{est}} = -\frac{\partial_t I}{\partial_x I}. \quad (4.173)$$

This expression is correct, but we have derived it by pushing to extremes. First we said that noise in the receptor responses is negligible, so we are effectively computing functions of the light intensity itself. Then we assumed that the dynamics of the light intensity is determined only by motion at the single velocity v. If either of these assumptions breaks down, our gradient-based estimator of velocity, Eq. (4.173), gets into serious trouble.

When dealing with noisy data, we develop several intuitions. First, the nature of our measurements is such that there usually is relatively more noise at higher frequencies, both in time and space. Thus, to suppress noise, we average. Conversely, if we differentiate, we expect that noise will be amplified, because differentiation enhances higher frequencies. Second, when we have a noisy measurement, it is dangerous to put the result of this measurement in the denominator of a ratio—there is a chance of division by zero because of a fluctuation. The gradient-based estimator compounds these sins, differentiating and then taking a ratio. We expect that this estimator will be a disaster if our low-noise assumptions are violated.

Problem 71: Ratios of noisy numbers.   Suppose there are two numbers that we are trying to measure, a and b. Our measurements, $\hat{a}$ and $\hat{b}$, give us the values of a and b but with some added Gaussian noise, so that

$$P(\hat{a}|a) = \frac{1}{\sqrt{2\pi\sigma^2}}e^{-(\hat{a}-a)^2/2\sigma^2}; \tag{4.174}$$

for simplicity we assume that the noise level is the same for the measurements of b, so that

$$P(\hat{b}|b) = \frac{1}{\sqrt{2\pi\sigma^2}}e^{-(\hat{b}-b)^2/2\sigma^2}. \tag{4.175}$$

What we would like to do is to estimate the ratio $r \equiv a/b$ from measurements $\hat{a}$ and $\hat{b}$.

(a) Suppose we form a naive estimate just by taking the ratio of measurements, $r_{est}^{naive} = \hat{a}/\hat{b}$. Do a small simulation to examine numerically the probability distribution of this estimate. In particular, consider the case where $a = b = 1$, so the correct answer is $r = 1$. If $\sigma = 0.1$, presumably r_{est}^{naive} stays close to this correct answer, but what happens at $\sigma = 0.2$ or 0.5? How does the variance of the estimator r_{est}^{naive} change as the noise level σ increases? Be sure to check in your simulation that you have enough samples to get a reliable measure of the variance. Is there anything suspicious in this computation, especially at larger σ?

(b) Look more closely at the right tail of the distribution of r_{est}^{naive}, that is, the behavior of $P(r_{est}^{naive} \gg 1)$ in the case where $a = b = 1$. Plot your numerical results on linear, semilog, and log-log plots to see whether you can recognize the shape of the tail. Is the shape changing with the noise level σ? Try to make a precise statement based on your simulations. I have left this part somewhat open ended.

(c) Try to derive analytically the regularities that you found in part (b).

(d) Although we think of $\hat{a}$ and $\hat{b}$ as measurements of the separate variables a and b, really all we want to know is the ratio $r \equiv a/b$. Show that the best estimate can be written, using Bayes' rule, as

$$r_{est}(\hat{a},\hat{b}) = \int dr \frac{r}{P(\hat{a},\hat{b})} \int da \int db\, \delta\left(r - \frac{a}{b}\right) P(\hat{a}|a)P(\hat{b}|b)P(a,b). \tag{4.176}$$

Make as much progress as you can evaluating these integrals on the hypothesis that the prior distribution $P(a,b)$ is broad and featureless. If you want to proceed analytically, you may find it useful to introduce a Fourier representation of the delta function and look for a saddle-point approximation. Numerically, you could assume, for example, that $P(a,b)$ is uniform over some region of the a–b plane, and just do the integrals for representative values of $\hat{a}$ and $\hat{b}$, mapping the function $r_{est}(\hat{a},\hat{b})$. Can you verify that r_{est}^{naive} is close to optimal at very small values of σ? What happens at larger values of σ? If σ is fixed, what happens as $b \to 0$?

The most obvious problem with the gradient-based motion estimator in Eq. (4.173) is simply that it is not well defined when the spatial derivative becomes small. This problem exists even if noise in the photoreceptors is small. To address the problem, we have to understand what the distribution $P(\mathcal{I}, \text{feature})$ looks like. Conceptually, what we want to do is simple. Imagine taking a walk on a very still day, so that motions of the world relative to our retina (or relative to the fly's retina) are dominated by our own motion. Using a camera, we can take a movie as we walk and can also put a gyroscope on the camera to monitor its motion. What emerges from such an experiment, then, is a set of samples drawn out of the distribution $P(\mathcal{I}, \text{feature})$. In particular, pixel by pixel and moment by moment, we can compute the spatial and temporal derivatives in the movie, and measure the velocity as well, so that we sample the distribution $P(\partial_t I, \partial_x I, v)$. Sinha and de Ruyter van Steveninck have done this experiment, using a specially constructed camera that matches the optics of the fly's eye and the speed of the photoreceptors.[23]

If the gradient-based estimate of motion were exact, then the distribution $P(\partial_t I, \partial_x I, v)$ would be sharply peaked along a ridge where $v = -\partial_t I / \partial_x I$. To see whether this prediction is right, we can compute directly the optimal estimator. We know that the best estimate in the sense of χ^2 is the conditional mean, so we should compute

$$v_{\text{est}}(\partial_t I, \partial_x I) = \int dv \, v P(v | \partial_t I, \partial_x I) \tag{4.177}$$

$$= \int dv \, v \frac{P(\partial_t I, \partial_x I, v)}{P(v)}. \tag{4.178}$$

The results of this computation, based on a walk in the woods, are shown in Fig. 4.36. We see that, when the spatial gradients are large, the contours of constant v_{est} really are straight lines, as expected from the gradient-based estimator. But when the spatial gradients are smaller, a new structure emerges, which is more closely approximated by a product of derivatives, $v_{\text{est}} \propto (\partial I / \partial t) \times (\partial I / \partial x)$, rather than a ratio. As you can see in the following problem, the same product structure emerges if we go back to the general formulation and take the limit of high noise levels.

Problem 72: Series expansion of the optimal estimator at low signal-to-noise ratios. We know from Section 2.1 that photoreceptors in the fly respond linearly to changes in light intensity or contrast. If the fly is rotating relative to the world along an angular trajectory $\theta(t)$, then the spatiotemporal pattern of contrast (again in a one-dimensional model) is $C(x - \theta(t), t)$. Individual cells respond with voltages $V_n(t)$ given by

$$V_n(t) = \int dt' \, T(t - t') \int dx \, M(x - x_n) C(x - \theta(t'), t') + \delta V_n(t), \tag{4.179}$$

where $T(\tau)$ is the temporal impulse response function, $M(x - x_n)$ is an aperture function centered on a lattice point x_n in the retina, and $\delta V_n(t)$ is the voltage noise.

(a) Show that the distribution of all the voltages, given the trajectory, can be written as

$$P[\{V_n(t)\} | \theta(t)]$$

$$\propto \int DC \, P[C] \exp\left[-\frac{1}{2} \sum_n \int \frac{d\omega}{2\pi} \frac{|\tilde{V}_n(\omega) - \langle \tilde{V}_n(\omega) \rangle|^2}{N_V(\omega)}\right], \tag{4.180}$$

---

23. Although conceptually simple, to generate Fig. 4.36 requires measuring light intensities with spatial and temporal resolutions matched to that of the retina, but collecting much more light so that photon shot noise in these measurements will be less than that in the retina and one can meaningfully claim to measure intensity at the input to the visual system. For details, as always, see the Annotated Bibliography for this section.

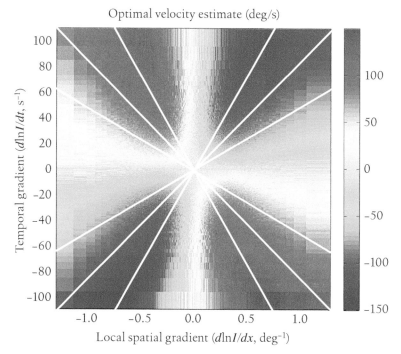

Optimal velocity estimate (deg/s)

FIGURE 4.36

Optimal estimates of angular velocity as a function of local spatial and temporal gradients of (log) light intensity. The estimates are computed from the theory described in the text, with the joint distribution of movies and motion sampled experimentally. Images are collected through an optical system that matches the fly's eye and are smoothed in time with a filter that optimizes estimation performance. At small signals, near the center of the plot, we see that moving along a line of constant physical velocity (in white; $\partial_t I + v \partial_x I = 0$) results in a changing estimate—a systematic error; only for large signals is the optimal estimate veridical. From experiments by S. R. Sinha and R. R. de Ruyter van Steveninck, with thanks.

where the mean voltages are, in the Fourier representation,

$$\langle \tilde{V}_n(\omega) \rangle = \tilde{T}(\omega) \int dx \, M(x - x_n) \int dt \, e^{+i\omega t} C(x - \theta(t), t), \qquad (4.181)$$

$N_V(\omega)$ is the power spectrum of the voltage noise, and $P[C]$ is the distribution of contrast that the fly would observe if held at $\theta = 0$.

(b) The optimal estimator is the conditional mean:

$$\dot{\theta}_{\text{est}}(t_0) = \int D\theta \, \dot{\theta}(t_0) P[\theta(t)|\{V_n(t)\}], \qquad (4.182)$$

$$P[\theta(t)|\{V_n(t)\}] = \frac{P[\{V_n(t)\}|\theta(t)] P[\theta(t)]}{P[\{V_n(t)\}]}. \qquad (4.183)$$

Evaluate all the integrals in a perturbation series, assuming that the average voltage responses are small compared with the noise level. You should find that the leading term is

$$\dot{\theta}_{\text{est}}(t) \approx \sum_{n,m} \int d\tau \int d\tau' V_n(t - \tau) K_{nm}(\tau, \tau') V_m(t - \tau'). \qquad (4.184)$$

Relate the kernel $K_{nm}(\tau, \tau')$ to expectation values in the distributions $P[C(x, t)]$ and $P[\theta(t)]$.

(c) Can you reformulate the expansion so that instead of expanding for small overall signal-to-noise ratio (small R), you expand for small instantaneous signals, that is, for small $V_n(t)$? What happens to the kernels in this case? It seems obvious that there should not be a linear term in this expansion. Can there be a third-order term? If such a term exists, what happens to the optimal estimate of velocity when we show the same movie but with inverted contrast (exchanging black for white)?

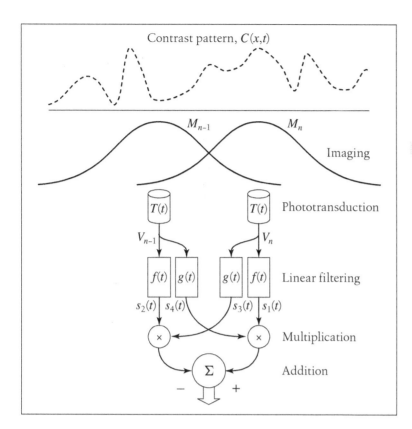

Contrast pattern, $C(x, t)$

M_{n-1} M_n

Imaging

$T(t)$ $T(t)$ Phototransduction

V_{n-1} V_n

$f(t)$ $g(t)$ $g(t)$ $f(t)$ Linear filtering

$s_2(t)$ $s_4(t)$ $s_3(t)$ $s_1(t)$

× × Multiplication

Σ Addition

− +

FIGURE 4.37

The correlator model of visual motion detection, adapted from Reichardt (1961). A spatiotemporal contrast pattern $C(x, t)$ is blurred by the photoreceptor point spread function, $M(x)$, and sampled by an array of photoreceptors, two of which (neighboring photoreceptors numbers $n - 1$ and n) are shown here. After phototransduction, the signals in each photoreceptor are filtered by two different linear filters, $f(t)$ and $g(t)$. The outputs of these filters from the different photoreceptors, $s_1(t)$ and $s_3(t)$ from photoreceptor n and $s_2(t)$ and $s_4(t)$ from photoreceptor $n - 1$, are multiplied, and one of these products is subtracted from the other by the addition unit Σ, yielding a direction-selective response. Redrawn from Bialek and de Ruyter van Steveninck (2005).

We can understand the low signal-to-noise ratio limit by realizing that when something moves, there are correlations between what we see at the two space-time points (x, t) and $(x + v\tau, t + \tau)$. These correlations extend to very high orders, but as the background noise level increases, the higher order correlations are corrupted first, until finally the only reliable thing left is the two-point function. In addition, closer examination shows that near-neighbor correlations are the most significant: we can be sure something is moving, because signals in neighboring photodetectors are correlated with a slight delay. This form of correlation-based motion computation, schematized in Fig. 4.37, was suggested long ago by Reichardt and Hassenstein based on behavioral experiments with beetles.

There are two clear signatures of the correlator model. First, because the receptor voltage is linear in response to image contrast, the correlator model confounds contrast with velocity: all things being equal, doubling the image contrast causes our estimate of the velocity to increase by a factor of four (!). This phenomenon is an observed property of the flight torque that flies generate in response to visual motion, at least at low contrasts, and the same quadratic behavior can be seen in the rate at which motion-sensitive neurons generate spikes, as shown in Fig. 4.38. Even humans experience the illusion of contrast-dependent motion perception at very low contrast. Although this behavior might seem strange, it has been known for decades.

The second signature of correlation computation is that we can produce movies with the right spatiotemporal correlations to generate a nonzero estimate $\dot{\theta}_{est}$ which

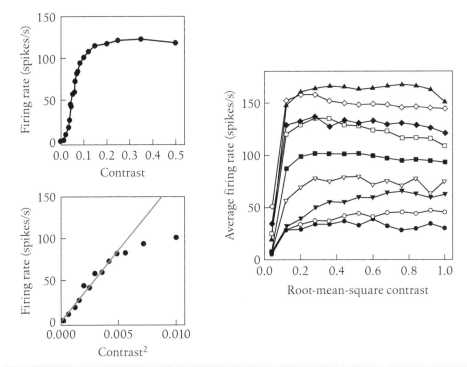

FIGURE 4.38

Responses of the H1 neuron to moving scenes with varying contrast. Scenes consist of bars with random intensities moving at constant velocity. At left, at one particular velocity we measure the rate at which H1 generates action potentials as a function of contrast. The lower panel expands the region at low contrast, emphasizing the quadratic behavior. At right are the responses at multiple velocities, showing that the saturated response at high contrast still is sensitive to the speed of movement. The different curves correspond to velocities increasing from 0.08 to 21°/s, with higher speeds generating monotonically higher firing rates in this range. Based on experiments described by de Ruyter van Steveninck et al. (1996). Thanks to Rob de Ruyter van Steveninck for this figure.

do not really have anything in them that we would describe as "moving." Consider a spatiotemporal white-noise movie $\psi(\mathbf{x}, t)$,

$$\langle \psi(\mathbf{x}, t)\psi(\mathbf{x}', t')\rangle = \delta(\mathbf{x} - \mathbf{x}')\delta(t - t'), \tag{4.185}$$

and then add the movie to itself with a weight and an offset:

$$C(\mathbf{x}, t) = \psi(\mathbf{x}, t) + a\psi(\mathbf{x} + \mathbf{\Delta x}, t + \Delta t). \tag{4.186}$$

Composed of pure noise, there is nothing really moving here. If you watch the movie, however, there is no question that you think something is moving in it, and the fly's neurons respond too (just like yours, presumably). Even more impressive is that if you change the sign of the weight a, then the direction of motion reverses, as predicted from the correlator model.

Problem 73: Motion from correlations alone.   Generate the image sequences described in the previous paragraph, and verify that you (and your friends) perceive them as moving.

(a) Play with the amplitude and sign of the weight a to see how it influences your perception. Can you find a regime in which the speed of motion seems to depend on $|a|$? Can you verify the reversal of motion when $a \to -a$?

(b) Compute the correlation function $\langle C(\mathbf{x}, t)C(\mathbf{x}', t')\rangle$; for simplicity you might want to confine your attention to a one-dimensional example. Consider also the correlation function for a genuine moving image, in which $C(\mathbf{x}, t) = C_0(\mathbf{x} - \mathbf{v}t)$. If $\mathbf{v} = \Delta\mathbf{x}/\Delta t$, how do the two correlation functions compare?

The optimal motion estimator illustrates the general trade-off between systematic and random errors. If we really are viewing an image that moves rigidly, so that $C(x, t) = C(x - vt)$, then there is no question that the "right answer" is to compute v as the ratio of temporal and spatial derivatives. Any departure from this choice involves making a systematic error. But, as discussed above, taking derivatives and ratios are both operations that are perilous in the presence of noise. To insulate the estimate from random errors driven by such noise (or, more generally, by aspects of the image dynamics that are not related to motion), we must calculate something that typically will not give the "right answer" even on average—we accept some systematic errors to reduce the impact of random errors. In the context of perception, systematic errors have a special name: illusions.

Could the theory of optimal estimation be a quantitative theory of illusions, grounded in physical principles? Colloquially, we say that to err is human, and it is conventional to assume that cases in which biological systems get the wrong answer to their signal-processing problems provide evidence regarding the inadequacies of the biological hardware. Is it possible that, rather than being uniquely human or biological, to err is the optimal response to the limits imposed by the physical world?

The long history of the correlator model provides ample testimony that insect visual systems make the kind of systematic errors expected from the optimal estimator, but precisely because of this long history, it is hard to view these observations as successful predictions. It would be more compelling if we could show that the same system that is well described by the correlator under some conditions crosses over to something more like the ratio of derivatives model at high signal-to-noise ratio, but demonstrating this has been elusive. The contrast dependence of the response in the motion-sensitive neurons saturates at high contrast, and this saturated response still varies with velocity (Fig. 4.38), as if the larger signals allow the system to disentangle ambiguities and recover a veridical estimate, but other experiments suggest that errors inherent in the correlator model persist even with strong signals. Humans easily see the illusion of motion with the noise movies of Eq. (4.186), as well as other motion illusions, but at high signal-to-noise ratios our visual systems recover estimates of velocity that are not systematically distorted, suggesting that in primates there also is some sort of crossover between different limits of the motion computation. Experiments under more natural, free flight conditions show that both flies and bees have access to veridical estimates of their translational velocity and can use them to control their flight speed, in contrast to what one would have expected from the correlator model. It is also worth noting that

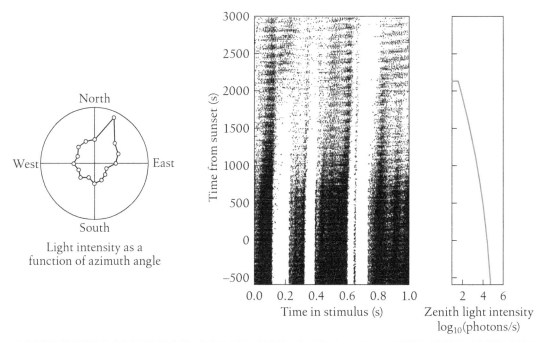

FIGURE 4.39

Responses of the H1 neuron as a fly is rotated, over a period surrounding sunset in mid-Fall; methods are as described in Lewen et al. (2001). The fly is rotated outside in a clearing surrounded by shrubs except in one direction (north-northeast) where the mean light intensity is larger (left). The rotational velocity repeats every 10 s, with 1 s of this period shown here. The falling light intensity (at right) is calibrated as photons per second per photoreceptor. The middle panel shows the spike trains from H1, which become more sparse and develop diagonal streaks—a response to the absolute angular position, or equivalently the spatial pattern, which varies because the integral of the velocity over one 10 s period is not quite zero. Thanks to Rob de Ruyter van Steveninck for this figure.

the responses of the motion-sensitive neurons are very different under more natural conditions.

Figure 4.39 shows responses of the H1 neuron to the rotation of a fly, outside under nearly natural conditions.[24] During the course of the experiment, the sun was going down, and so the mean light level varied by several orders of magnitude as the same trajectory of angular velocity versus time was repeated over and over. The integral of the trajectory was (accidentally) not quite zero, however, so that on each repetition the spatial pattern of light intensity was a bit different even if the angular velocity was the same.

At the start of the experiment, the responses are extremely vigorous and insensitive to variations in the spatial structure of the visual environment. As the light level falls, responses become weaker, but more dramatically, we see that there is a systematic variation from repetition to repetition, which appears as a diagonal pattern of spikes

---

24. If you want to study neural responses outside, as the sun is going down, you also have to be careful to stabilize the temperature of the fly, to be sure that what you see is a response to the changing light levels and not the resulting temperature changes.

across the upper part of Fig. 4.39. Thus, when signal-to-noise ratios are high in the natural environment, H1 responds to time-dependent velocities and largely ignores the spatial structure of its environment. In contrast, at lower signal-to-noise ratios the confounding of spatial structure and motion becomes more and more obvious. This pattern is in agreement with the expectations from optimal estimation theory, according to which such systematic errors arise only from the need to insulate the computation from random noise.

What we would really like is to have methods of dissecting the computation that has been done by a neuron simply by analyzing the relationship between visual inputs and spiking outputs under natural conditions. This is a huge challenge and obviously would be interesting in many other contexts. Approaches to this problem are discussed in the references for this section in the Annotated Bibliography, including results that come closest to a smoking gun for the crossover between correlator and gradient computations.

For visual signal processing, getting our hands on the true distribution of signals in the natural environment is a difficult experiment. For seemingly more complex cognitive judgments, the situation, perhaps surprisingly, is much simpler. To give an example, suppose that you are told that a member of the U.S. Congress has served for $t = 15$ years. What is your prediction for how long her total term will last? To keep things as simple as possible, let's assume you are not told anything about the politics of this congresswoman or her district; all you have to work with is $t = 15$ and your general knowledge of the turnover of elected officials. Your knowledge is probabilistic, so we use Bayes' rule to write

$$P(t_{\text{total}}|t) \propto P(t|t_{\text{total}})P(t_{\text{total}}). \tag{4.187}$$

If the moment at which the question is asked is not somehow synchronized to the length of congressional terms, then we have to assume that $P(t|t_{\text{total}})$ is uniform, $P(t|t_{\text{total}}) = 1/t_{\text{total}}$. Thus, our inference is controlled by the prior distribution $P(t_{\text{total}})$, and we can look this up in a database about the history of Congress. Finally, if you must pick one value of t_{total}, it makes sense in this context to choose the median, the point at which the actual value of t_{total} is equally likely to be longer or shorter than your estimate. As an example, if $P(t_{\text{total}})$ is a reasonably narrow Gaussian distribution, then for t much less than the mean $\langle t_{\text{total}} \rangle$, our best estimate of t_{total} is just $\langle t_{\text{total}} \rangle$ itself. In contrast, if the time t is much larger than the mean, then our best estimate is only slightly higher than t, which makes sense. Other priors can give qualitatively different results.

Problem 74: Estimating t_{total}. Derive the results just stated for the Gaussian prior. Consider also cases where $P(t_{\text{total}}) \propto t_{\text{total}}^{-\gamma}$ or $P(t_{\text{total}}) \propto t_{\text{total}}^{n} e^{-t_{\text{total}}/\tau}$.

The example of congressional terms is not unique. We could ask, as insurance companies do (albeit with more input data), about human lifespans: if you meet someone of age t, what is your best guess about his life expectancy? If you make a phone call and have been on hold for t minutes, what is your best guess about the total time you will have to wait? If you find yourself on line t of a poem, what is your

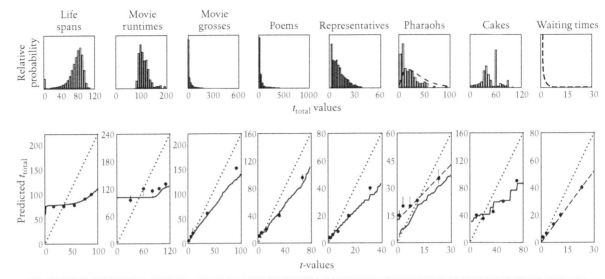

FIGURE 4.40

Estimation of totals based on one observation. The top row shows the priors $P(t_{total})$ measured from real-world data. The bottom panel compares people's predictions (points) based on one observation t with the optimal median estimator (solid lines) and a naive universal estimate $\hat{t}_{total} = 2t$. For the reigns of pharaohs and the telephone waiting times, dashed lines show optimal estimators for $P(t_{total}) \propto t_{total}e^{-t_{total}/\tau}$ ($\tau = 17.9$) and $P(t_{total}) \propto t_{total}^{-\gamma}$ ($\gamma = 2.43$), respectively. Redrawn from Griffiths and Tenenbaum (2006).

best guess about the total length of text? Nor is the structure of the problem bound to time, as such: suppose you learn that a movie has collected t dollars in gross receipts; what is your best guess about what its total earnings will be? All these problems have in common that we can look up the correct distribution $P(t_{total})$. Another important feature is that we can just go ask people what they think and see how they do relative to the predictions for optimal estimation based on the priors appropriate to the real world. The results from such an experiment are shown in Fig. 4.40.

I found the results of Fig. 4.40 quite astonishing when I first saw them. The time it takes to bake a cake comes from a very irregular distribution, but people seem to know this distribution and estimate accordingly. They are a bit confused about how long the pharoahs reigned, but their confusion is consistent: estimation of t_{total} behaves as if the subjects know the shape of $P(t_{total})$ but are off on the mean time scale, and if you ask another group of subjects to guess the mean reign of the pharoahs, they deviate from the right answer by the same factor. Important as the telephone problem may be, this is one case where there is no convenient data to which we can refer, so this case remains untested. In all the other cases, however, spanning widely disparate domains of knowledge and very different shapes for $P(t_{total})$, people are performing close to the optimum.

If we trace through the details of optimal estimation theory, it is evident that construction of the correct estimator involves knowing not only the distribution of signals but also the distribution of noise. Perhaps the simplest illustration of this is given by the problem of combining two measurements. Suppose that we are interested in x, but we observe

$$y_1 = x + \eta_1, \tag{4.188}$$
$$y_2 = x + \eta_2, \tag{4.189}$$

where the noise levels on the two measurements are generally different, $\langle \eta_1^2 \rangle = \sigma_1^2$ and $\langle \eta_2^2 \rangle = \sigma_2^2$; for simplicity we assume that the noise is Gaussian. Intuitively, we should be able to do better by combining the two observations than we would do by looking just at one of them, and we also expect that greater weight should be given to the more accurate measurement. Quantitatively, if the measurements are independent of one another, we have

$$P(x|y_1, y_2) = \frac{P(y_1, y_2|x)P(x)}{P(y_1, y_2)} \tag{4.190}$$

$$\propto P(x)P(y_1|x)P(y_2|x), \tag{4.191}$$

$$\propto P(x) \exp\left[-\frac{1}{2\sigma_1^2}(y_1 - x)^2 - \frac{1}{2\sigma_2^2}(y_2 - x)^2 \right]. \tag{4.192}$$

Then we can form the optimal estimator in the least squares sense,

$$x_{\text{est}}(y_1, y_2) \equiv \int dx\, x\, P(x|y_1, y_2) \tag{4.193}$$

$$= \frac{\sigma_2^2 y_1 + \sigma_1^2 y_2}{\sigma_1^2 + \sigma_2^2}, \tag{4.194}$$

where in the last step we assume that the prior $P(x)$ is broad compared with the noise levels in the data. Thus, as expected, the optimal estimate is a combination of the data, and the weights are inverse to their relative noise levels.

Problem 75: Cue combination.   Fill in the details leading to Eq. (4.194). Can you work out the same problem but with additional multiplicative noise, $y_n = e^{g_n}x + \eta_n$, where g_n is also Gaussian? In this case, it is possible to generate errors that are very large, so presumably, large disagreements between the data points y_1 and y_2 should not be resolved by simple averaging. See how much analytic progress you can make here, or do a simple simulation. This problem is deliberately open ended.

There are many situations in which we give strongly unequal weights to different data. A dramatic example is ventriloquism, in which we trust our eyes, not our ears, and assign the source of speech to the person (or the dummy) whose lips are visibly moving. To see whether we are giving weights in relation to noise levels, as would be optimal, we have to do an experiment in which we can manipulate the effective noise levels. This experiment was first done convincingly in tasks that require subjects to combine information from vision and touch. Although under normal conditions we give strong preference to our visual system, these data show convincingly that we do this only because our visual system provides much more accurate spatial information; if we can change their noise levels, people will change the weights given to different cues, as predicted by optimal estimation theory. There are even examples where the optimal estimates involve nonlinear combinations of the relevant cues, with the details of the nonlinearity dependent on the reliability of the individual cues, and in these cases as well we can see human subjects making the adjustments predicted by theory.

The examples of estimation that we have discussed thus far have in common that the distribution of the feature we are interested in estimating has a single well-defined peak given the input sensory data. In many cases, however, the data collected with our senses have multiple interpretations, perhaps even multiple interpretations that provide equally good explanations of what we have seen or heard. These ambiguous percepts arise in many contexts. When we experience these stimuli, our perceptions jump at random among the different possibilities. Could these random jumps originate from the same small noise sources that limit the reliability of our senses? The answer is not clear, but these corners of our perception have received renewed attention in recent years, in part because they provide a situation in which our perceptions fluctuate while the physical stimulus is constant, and in this situation we can start to distinguish between neurons that are responding to their inputs and those providing the signal that drives our subjective, conscious experience.

Perhaps a good way to summarize what we have learned from these many examples is that the case of photon counting in vision is a better guide to perception, in general, than one might have expected at the outset. Although it seems plausible that some creatures might be driven to the extremes of sensitivity as they try to make their way through the dark of night, it is less obvious that so many aspects of perception should be limited by fundamental noise sources, and that the brain should combine sense data with prior knowledge in ways that approximate optimal estimation. Importantly, optimal performance is not perfect performance, but rather the best possible compromise among different kinds of errors. But the structure of this compromise ultimately is driven by the irreducible noise sources in the physical signals that provide the input to our perceptual inferences.

4.5   Proofreading and Active Noise Reduction

Fluctuations are an essential part of being at thermal equilibrium. Thus, the fact that life operates in a relatively narrow range of temperature around 300 K means that some level of noise is inevitable. But being alive certainly is not being at thermal equilibrium. Can organisms use their nonequilibrium state to reduce the impact of nominally thermal noise? More generally, can we understand how to take a system in contact with an environment at temperature T and expend energy, driving it away from equilibrium, to reduce the effects of noise?

In his classic lectures *What Is Life?*, Schrödinger waxed eloquent about the fidelity with which genetic information is passed from one generation to the next, conjuring the image of a gallery with portraits of the Hapsburgs, their oddly shaped lips reproduced across centuries of descendants. Schrödinger was much impressed by the work of Timoféef-Ressovsky, Zimmer, and Delbrück, who had determined the cross-section for ionizing radiation to generate mutations, and used their results to argue that genes were of the dimensions of single molecules. Thus, the extreme stability of our genetic inheritance could not be based on averaging over many molecules, as a "naive classical physicist" might have thought, to use Schrödinger's evocative phrase. Now is a good time to set aside our modern insouciance and allow ourselves to be astonished,

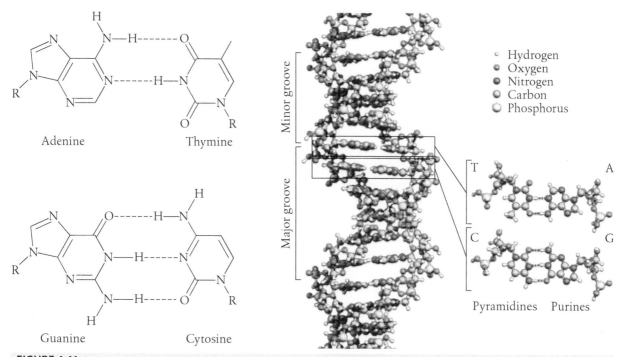

FIGURE 4.41

Base pairing in the Watson-Crick structure of DNA. At left is the hydrogen bonding between bases in the correct pairings, showing how they fit to satisfy the opportunities for hydrogen bonding, producing structures that are the same width and hence can fit into the double helix, as shown at right. The sugar and phosphate groups (R) are identical for all bases, which form the outer backbone(s) of the helix. From http://en.wikipedia.org/wiki/DNA, by Richard Wheeler.

as Schrödinger was, that so many of the phenomena of life are the macroscopic consequences of individual molecular events.

We now teach high school students that the key to the transmission of genetic information is the pairing of bases along the double helix—adenine (A) pairs with thymine (T), cytosine (C) pairs with guanine (G), as in Fig. 4.41. This is the triumph of the Watson and Crick theory of DNA structure.[25] The ideas of templates and structural complementarity that are at the heart of the double helix reappear many times—every time, in fact, that the organism needs to make reliable choices about which molecules to synthesize. But does structural complementarity solve the problem of reliability in biosynthesis?

That A pairs with T is really the statement that the (free) energy of a correct AT pair is much lower than that of an incorrect AC or AG pair. What matters here are the energy scales for chemical bonding. A genuinely covalent bond, such as the carbon-carbon or carbon-nitrogen bonds in the interior of the bases, results from the sharing of electrons

---

25. It would be almost silly to think you know something about "biophysics" (whatever you think the word means!) and not understand the interplay of theory and experiment that led to this revolution in the middle of the twentieth century. For a brief tour, see Appendix A.3.

between the atoms, and the energies are therefore on the scale of several electron volts.[26] Making the wrong base pairs would not require breaking any covalent bonds, so the energy cost will not be this large. If we tried to make an AG pair, it would be so big that it would not fit inside the backbone of the double helix; more precisely, we would have to make large distortions of the covalent bonds, and because these bonds are stiff, the energy cost would be huge. In contrast, if we try to make a CT pair, the backbone will hold the bases so far apart that they can't form hydrogen bonds. Thus, the minimal energy cost for a wrong base pair is the energy of two missing hydrogen bonds—on the order of $10\,k_BT$.

An energy difference of $\Delta F \sim 10\,k_BT$ means that the probability of an incorrect base pairing should be, according to the Boltzmann distribution, $e^{-\Delta F/k_BT} \sim 10^{-4}$. A typical protein is 300 amino acids long, which means that it is encoded by nearly 1000 bases; if the error probability is 10^{-4}, then replication of the DNA would introduce roughly one mutation in every tenth protein. For humans, with a billion base pairs in the genome, every child would be born with hundreds of thousands of bases different from his or her parents. If these predicted error rates seem large, they are—real error rates in DNA replication vary across organisms, but are in the range of 10^{-8}–10^{-12}, so that entire genomes can be copied almost without any mistakes.

The discrepancy between Boltzmann probabilities and observed error rates is much more widespread. When information encoded in the DNA is read out to make proteins, there are several steps where errors can occur. First is the synthesis of mRNA from the DNA template, a process not unlike the replication of the DNA itself, with the slight exception that the base A in DNA is matched to the base uracil (U) in mRNA rather than T in DNA. The genetic code is the codebook for translating from the language of bases along the mRNA into amino acids, as shown in Fig. 4.42. The cracking of this code is one of the great triumphs of twentieth-century science.

The rules for reading the genetic code are stored physically in the transfer RNA (tRNA) molecules, which at one end have a triplet of bases (the anticodon) that is complementary to a particular triplet of bases along the mRNA (the codon). At their other end is the amino acid that the codon represents, as shown schematically in Fig. 4.43. To make such molecules, there are specialized enzymes that recognize the tRNA and choose from the cellular soup the correct amino acid with which to "charge" the molecule. But some amino acids differ simply by the replacement of a CH_3 group with an H (as will be seen in more detail in Section 5.1); if we imagine the enzyme recognizing the first amino acid with a binding pocket that is complementary to the CH_3 group, then the second amino acid will also fit, and the binding energy will be weaker only by the loss of noncovalent contacts with the methyl group; it is difficult to see how this energy loss could be much more than $\sim 5\,k_BT$, corresponding to error rates of $\sim 10^{-2}$. If the error rates in tRNA charging were typically 10^{-2}, almost all proteins would have at least one wrong amino acid; in fact error rates are more like 10^{-4}, so that most proteins have no errors. There is one more step, at the ribosome, where tRNA molecules bind to their complementary sites along the mRNA and the amino acids

---

26. Chemists prefer to think per mole rather than per molecule, and they prefer joules to electron volts (I won't speak of calories). To have some numbers at your fingertips, remember that at room temperature, $k_BT = 1/40\,\text{eV} = 2.5\,\text{kJ/mole}$.

Second base
in mRNA
sequence

DNA bases
$$\begin{matrix} A & & U \\ G & \longrightarrow & C \\ T & & A \\ C & & G \end{matrix}$$
mRNA bases

		U		C		A		G
U	UUU	Phenylalanine	UCU	Serine	UAU	Tyrosine	UGU	Cysteine
	UUC	Phenylalanine	UCC	Serine	UAC	Tyrosine	UGC	Cysteine
	UUA	Leucine	UCA	Serine	UAA	**Stop**	UGA	**Stop**
	UUG	Leucine	UCG	Serine	UAG	**Stop**	UGG	Tryptophan
C	CUU	Leucine	CCU	Proline	CAU	Histidine	CGU	Arginine
	CUC	Leucine	CCC	Proline	CAC	Histidine	CGC	Arginine
	CUA	Leucine	CCA	Proline	CAA	Glutamine	CGA	Arginine
	CUG	Leucine	CCG	Proline	CAG	Glutamine	CGG	Arginine
A	AUU	Isoleucine	ACU	Threonine	AAU	Asparagine	AGU	Serine
	AUC	Isoleucine	ACC	Threonine	AAC	Asparagine	AGC	Serine
	AUA	Isoleucine	ACA	Threonine	AAA	Lysine	AGA	Arginine
	AUG	Methionine/start	ACG	Threonine	AAG	Lysine	AGG	Arginine
G	GUU	Valine	GCU	Alanine	GAU	Aspartic acid	GGU	Glycine
	GUC	Valine	GCC	Alanine	GAC	Aspartic acid	GGC	Glycine
	GUA	Valine	GCA	Alanine	GAA	Glutamic acid	GGA	Glycine
	GUG	Valine	GCG	Alanine	GAG	Glutamic acid	GGG	Glycine

First base in mRNA sequence

FIGURE 4.42

The genetic code. The four bases of DNA are transcribed into complementary bases of messenger RNA (mRNA), and then these bases, in groups of three, translate into amino acids. Note that with $4^3 = 64$ possibilities, there will be degeneracy in translating to the 20 amino acids; most of this degeneracy is in the choice of the third base. A, adenine; C, cytosine; G, guanine; T, thymine; U, uracil.

they carry are stitched together into proteins. Here, too, a discrepancy arises between thermodynamics and the observed error probabilities.[27]

Each of the events we have outlined—DNA replication, mRNA synthesis, tRNA charging, and protein synthesis on the ribosome—has its own bewildering array of biochemical details and is the subject of its own vast literature. As physicists we search for common theoretical principles that can organize this biological complexity, and I think that this problem of accuracy beyond the thermodynamic limit provides a wonderful model for this search. The key ideas go back to Hopfield and Ninio in the 1970s. Their classic papers usually are remembered for having contributed to the solution of the problem of accuracy, a solution termed "kinetic proofreading," which we will explore in a moment. But I think they should also be remembered for having recognized that there is a common physics problem that runs through this broad range of different biochemical processes.

To understand the essence of kinetic proofreading, it is useful to think about Maxwell's demon. Imagine a container partitioned into two chambers by a wall, with a small door in the wall. Maxwell conjured the image of a small demon who controls the door. If he[28] sees a molecule coming from the right at high speed, he opens the

---

27. A few notes. The tRNAs themselves are encoded in DNA, as are the charging enzymes. Thus, the genetic code itself should be subject to evolution. Also, tRNA pools can be controlled by the same mechanisms that control expression of proteins. These observations raise all sorts of interesting questions!

28. Why is it obvious that the demon is male?

GCGGAUUUUAGCUCAGDDGGGAGAGCGCCAGACUGAAYAuCUGGAGGUCCUGUGuuCGAUCCACAGAAUUCGCACCA

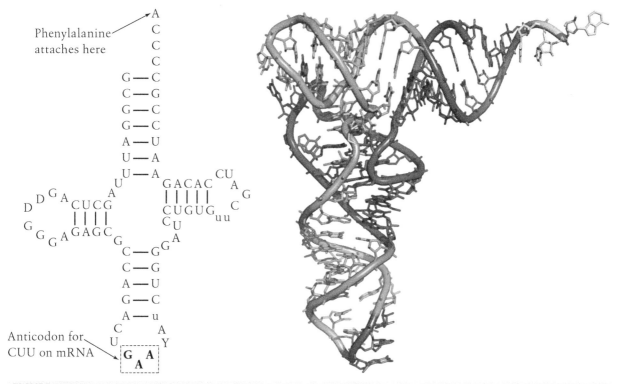

FIGURE 4.43

A transfer RNA (tRNA) molecule from yeast. This molecule connects the CUU codon on messenger RNA (mRNA) to the amino acid phenylanine, according to the genetic code in Fig. 4.42; D, Y, and u are chemically modified bases. The linear sequence is shown at the top, the "cloverleaf" structure indicating the opportunities for base pairing is shown at left, and the three-dimensional structure from X-ray crystallography is shown at right, from http://en.wikipedia.org/wiki/Transfer_RNA. See Fig. 4.42 for definitions of the bases.

door and allows it to go into the left chamber. Conversely, if he sees a molecule drifting slowly from the left, he opens the door and allows it to enter the right chamber. After some time, all the slow molecules are on the right, and all the fast molecules are on the left. But because the average kinetic energy of the molecules in a gas is proportional to the temperature, the demon has created a temperature difference: hot on the left, cold on the right. This temperature difference can be used to do useful work (e.g., running a heat engine), and thus, the demon appears to have created something out of nothing, violating the second law of thermodynamics.

There is nothing special about the demon's choice of molecular speed as the criterion for opening the door. It is a simple choice, because the result is a temperature difference, and we can imagine all sorts of appropriately nineteenth-century methods for extracting useful work from temperature differences. But if there are two kinds of molecules, *A* and *B*, and the demon arranges for the *A* molecules to accumulate in the left chamber and *B* molecules to accumulate in the right chamber, then there will be differences in chemical potential between the two chambers. There must be some way of using this difference to do work, even if as physicists we don't know enough chemistry to figure it out.

Problem 76: Pushing away from equilibrium. Consider a polymer made from A and B monomers. Suppose we start with pure poly-A, and use it as a template to construct a new polymer, much as in DNA replication (but simpler!). Template-directed synthesis works, because the A-A bond is stronger than the A-B bond by some free energy difference ΔG; we'll use the convention that $\Delta G > 0$. Then if we make a polymer of length N in which a fraction f of the monomers are incorrectly made to be B rather than A, the free energy of the system will have a contribution $Nf\Delta G$ relative to the perfectly copied poly-A. If the errors are made at random, however, then there is a contribution to the entropy of the polymer that comes from the sequence heterogeneity.

(a) Evaluate the entropy that comes from the random substitutions of A by B. What assumptions are you making in this calculation? Can you imagine these being violated by real molecules?

(b) Combine the entropy from part (a) with the "bonding" free energy $Nf\Delta G$ to give the total free energy of the polymer. Show that this quantity is minimized at $f_{eq} \propto \exp(-\Delta G/k_B T)$, as expected.

(c) How much free energy is stored in the polymer when $f < f_{eq}$? Can you give simple expressions when the difference $f_{eq} - f$ is small? What happens if (as we will see below) $f \approx f_{eq}^2$?

The demon's sin is to have generated a state of reduced entropy. We know that to enforce the second law, this nonequilibrium state must be paid for with enough energy to balance the books—to avoid building a perpetual motion machine, the demon must have dissipated an amount of energy equal to or greater than the amount of useful work that can be extracted from his reduction in the entropy of the system. The key insight of Hopfield and Ninio was that the problem of accuracy or low error rates was of this same kind: achieving low error rates, sorting molecular components with a precision beyond that predicted by the Boltzmann distribution, means that the cell is building and maintaining a nonequilibrium state, and it must spend energy in order to do so. Somewhere in the complexity of the biochemistry of these processes there must be steps that dissipate energy, and this dissipation has to be harnessed to improve the accuracy of synthesis.

To see how a mechanism to improve accuracy might work, let's look at the simplest model of a biochemical process catalyzed by an enzyme, as in Fig. 4.44. In essence, the chemical reaction of interest involves choosing among two (or more) substrate molecules, for example, the correct and incorrect base at a particular point along the strand of DNA that the cell is trying to replicate or transcribe into mRNA. To complete the reaction, the substrate has to bind to the enzyme, and this enzyme-substrate complex can be converted into the product. To have any possibility of correcting errors, it must be possible for the substrate to unbind from the enzyme before the conversion to product. This minimal scheme is the Michaelis-Menten model discussed in Section 2.3. The kinetics are described by

$$\frac{d[EA]}{dt} = k_+[A][E] - (k_- + V_{max})[EA], \tag{4.195}$$

$$\frac{d[EB]}{dt} = k'_+[B][E] - (k'_- + V'_{max})[EB], \tag{4.196}$$

$$[E]_{total} = [EA] + [EB] + [E], \tag{4.197}$$

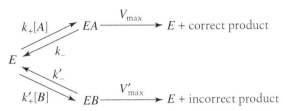

FIGURE 4.44

The simplest kinetic scheme in which an enzyme can choose correct or incorrect molecules out of solution, making correct or incorrect products.

where A is the correct substrate, B is the incorrect substrate, and $[E]_{\text{total}}$ is the (fixed) total concentration of enzyme molecules. The rate at which correct products are made is given by $V_{\text{max}}[EA]$, and the rate of making incorrect products is $V'_{\text{max}}[EB]$. If the overall rate of reactions is slow enough not to deplete the substrates, then we can compute these rates in the steady state approximation.

To compute the rate of errors, we don't even need to solve the entire problem. From Eq. (4.195) we can see that, in steady state,

$$[EA] = [E]\frac{k_+[A]}{k_- + V_{\text{max}}};\tag{4.198}$$

similarly, from Eq. (4.196), we have

$$[EB] = [E]\frac{k'_+[B]}{k'_- + V'_{\text{max}}}.\tag{4.199}$$

Thus, the error probability, or relative rate at which incorrect products are made, is given by

$$f \equiv \frac{\text{rate of making incorrect product}}{\text{rate of making correct product}}\tag{4.200}$$

$$= \frac{V'_{\text{max}}[EB]}{V_{\text{max}}[EA]}\tag{4.201}$$

$$= \left[\frac{k'_+[B]}{k'_- + V'_{\text{max}}}\right] \times \left[\frac{k_+[A]}{k_- + V_{\text{max}}}\right]^{-1} \times \left[\frac{V'_{\text{max}}}{V_{\text{max}}}\right].\tag{4.202}$$

All the reactions of interest share one important feature: the actual making and breaking of covalent bonds occurs on the other side of the molecule from the structure that defines correct versus incorrect, as is clear from the tRNA structure shown in Fig. 4.43. In the case of DNA replication, correctness has to do with the pattern of hydrogen bonding between the bases, on the inside of the helix, whereas the actual reaction required to incorporate one base into the growing polymer involves the phosphate backbone on the outside of the helix. These structural facts make it unlikely that the rate at which these bonds are formed is sensitive to the correctness of the substrate, $V'_{\text{max}} \approx V_{\text{max}}$, so this is not a source of selectivity. Then, from Eq. (4.202) it is clear that,

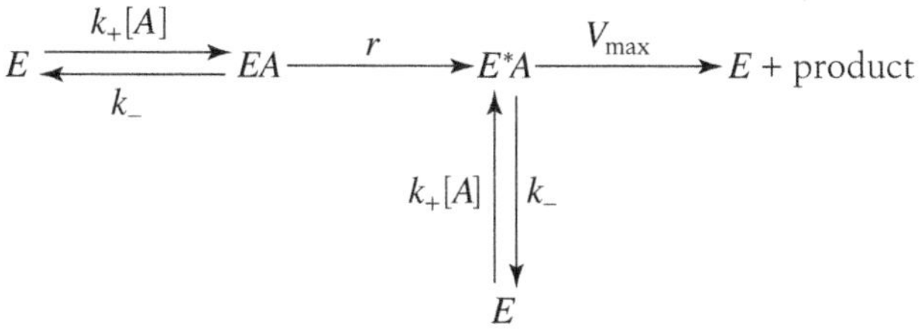

$$E \underset{k_-}{\overset{k_+[A]}{\rightleftharpoons}} EA \overset{r}{\longrightarrow} E^*A \overset{V_{\max}}{\longrightarrow} E + \text{product}$$

$$k_+[A] \Big\updownarrow k_-$$

$$E$$

FIGURE 4.45

The simplest scheme for kinetic proofreading. As described in the text, the key step is an irreversible transition from EA to E^*A, which gives a true second chance for equilibration with the free A molecules. E^* indicates a modified form of the enzyme E.

under these conditions, the error probability is minimized if the catalytic rate $V_{\max}$ is slow compared with the unbinding rates k_-, k'_-. This makes sense: if the catalytic step itself has no selectivity, then to maximize selectivity, one must give the wrong substrate a chance to fall off.

So, in this simplest kinetic scheme we have shown that the error probability is bounded:

$$f > \left(\frac{k'_+[B]}{k'_-}\right) \times \left(\frac{k_+[A]}{k_-}\right)^{-1}. \tag{4.203}$$

But this combination of rates and concentrations is exactly what determines the equilibrium binding of A versus B to the enzyme, and it hence can be written in terms of thermodynamic quantities,

$$f > \exp\left(-\frac{F_A - F_B}{k_B T}\right), \tag{4.204}$$

where F_A is the free energy for taking a single molecule of A out of solution and binding to the enzyme, and similarly for B; here binding energies are positive and are larger for tighter binding. Thus, we are back where we started, with an error probability determined by the Boltzmann distribution!

But the Michaelis-Menten scheme has a natural generalization. Suppose that, after binding, there is an irreversible transition to a new state, at a rate r, and that in this state the substrate can again be released from the enzyme, as in Fig. 4.45. In the simplest case, the events that determine binding and release of the (perhaps modified) substrate are the same as in the initial step, with the same rates. We can carry through the analysis of this kinetic scheme as before, using the same assumption that catalytic steps ($V_{\max}$ and r) have no selectivity, to find that

$$f > \exp\left(-\frac{F_A - F_B}{k_B T}\right) \exp\left(-\frac{F_{A^*} - F_{B^*}}{k_B T}\right). \tag{4.205}$$

But if the molecular interactions that select A over B are the same for A^* versus B^*, we expect $F_{A^*} - F_{B^*} \approx F_A - F_B$, and hence we have

$$f \rightarrow \left[\exp\left(-\frac{F_A - F_B}{k_B T}\right)\right]^2. \tag{4.206}$$

This is the essence of kinetic proofreading: by introducing an irreversible step into the kinetic scheme, a step that necessarily dissipates energy, it is possible to use the equilibrium selectivity twice and achieve an error probability that is the square of the nominal limit set by the Boltzmann distribution.

Problem 77: More on the basics of kinetic proofreading.   To begin, supply the details needed to derive Eq. (4.205). An even better exercise is to go through Hopfield's (1974) original paper, pen in hand, filling in all the missing steps. Then consider the following questions.

(a) In the simplest scheme, we saw that maximum selectivity occurs when V_{max} is slow compared with k_-. Is there a similar condition in the proofreading scheme? What does this tell us about the progress of the enzymatic cycle? More specifically, what is the fate of the typical substrate that binds to the enzyme? Is it converted to product or ejected from E^*?

(b) Consider a generalization of the kinetic scheme in Fig. 4.45 such that the nominally irreversible step with rate r is in fact reversible, with the reverse reaction at rate r'. To be general, imagine also the binding and unbinding from E^* can occur with rates that are different from the rates for E. Now there are detailed balance conditions that connect these different rates. Write down these conditions, and show how they effect the error probability. Can you say something general here? In particular, can you show how these conditions enforce the Boltzmann error rate in the absence of energy dissipation, no matter how many times the enzyme "looks" at the substrate?

How does this general idea of proofreading connect with the real biochemistry of these systems? In some sense the case of DNA replication (or transcription) is most obvious, as shown in Fig. 4.46. All the nucleotides incorporated into the growing strands of DNA or RNA start as nucleotide triphosphates, but once the final structure is formed only one phosphate is part of the backbone. Thus, at some point in the process, the high-energy phosphate bond must be cleaved, releasing roughly $20k_BT$ of free energy. If this is the irreversible step in proofreading, then is must be possible for the enzyme catalyzing the growth of the polymer to release the nucleotide after this cleavage, which means after it has been attached to the backbone of the growing chain. Thus, the enzyme must be not only a polymerase (catalyzing the polymerization reaction), it must also be an exonuclease (catalyzing the removal of nucleotides from the polymer). It had been known almost since the discovery of the polymerase that it also had exonuclease activity, but the idea of kinetic proofreading was needed to explain how this activity was connected, through energy dissipation, to proofreading and error correction. In the charging of tRNA, the process actually starts with an ATP molecule being cleaved, leaving an AMP attached to the amino acid before it reacts with the tRNA. In protein synthesis, the sequence of reactions is much more complex, but again there is an obligatory cleavage of a nucleotide triphosphate (in this case GTP → guanosine diphosphate (GDP)). All these examples are qualitatively consistent with the proofreading scenario,[29] and especially in the case of tRNA charging it has been

---

29. Hopfield has also emphasized that there are kinetic schemes in which proofreading still proceeds through energy-dissipating steps, but if the enzymes have some memory for past events, then the synthesis and dissipation can be separated in time, erasing some of the more obvious signatures from the simpler scheme. This idea may be especially important in thinking about more complex examples, such as protein synthesis on the ribosome or DNA replication in higher eukaryotes.

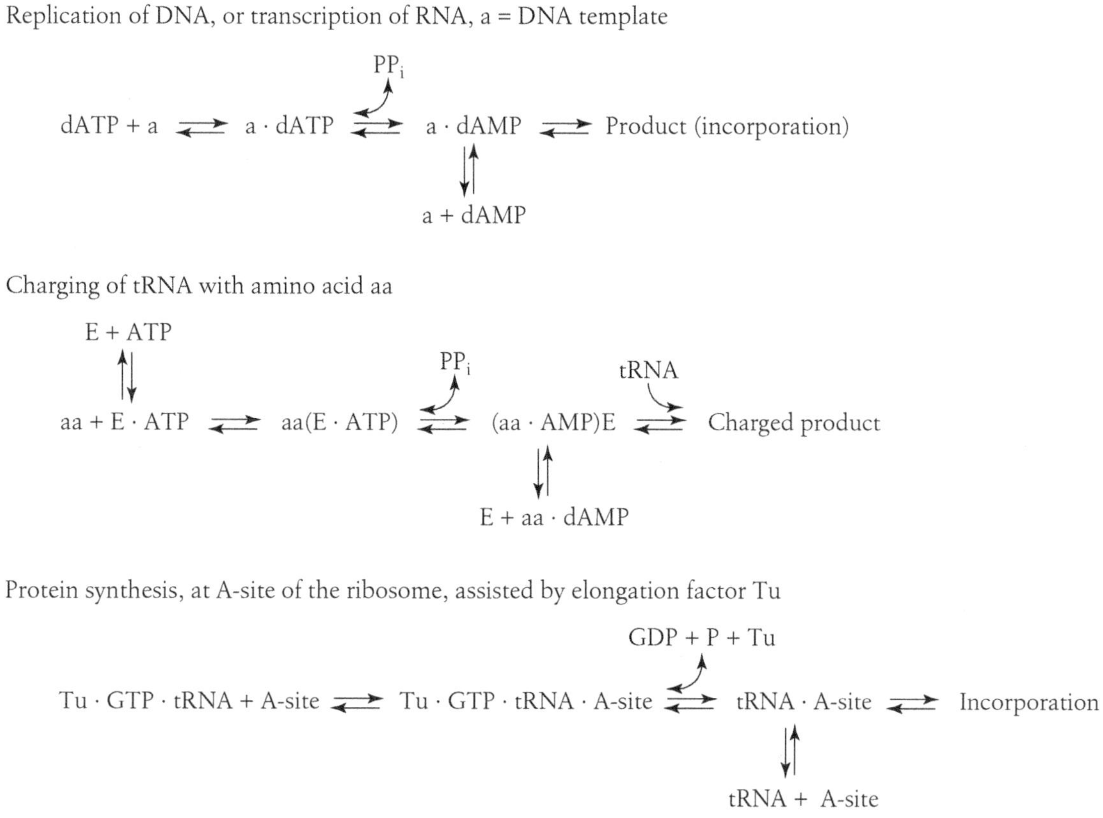

Replication of DNA, or transcription of RNA, a = DNA template

$$
dATP + a \; \xrightleftharpoons{} \; a \cdot dATP \; \overset{PP_i}{\xrightleftharpoons{}} \; a \cdot dAMP \; \xrightleftharpoons{} \; \text{Product (incorporation)}
$$

$$
\updownarrow
$$

$$
a + dAMP
$$

Charging of tRNA with amino acid aa

$$
E + ATP
$$

$$
\updownarrow
$$

$$
aa + E \cdot ATP \; \xrightleftharpoons{} \; aa(E \cdot ATP) \; \overset{PP_i}{\xrightleftharpoons{}} \; (aa \cdot AMP)E \; \overset{tRNA}{\xrightleftharpoons{}} \; \text{Charged product}
$$

$$
\updownarrow
$$

$$
E + aa \cdot dAMP
$$

Protein synthesis, at A-site of the ribosome, assisted by elongation factor Tu

$$
GDP + P + Tu
$$

$$
Tu \cdot GTP \cdot tRNA + A\text{-site} \; \xrightleftharpoons{} \; Tu \cdot GTP \cdot tRNA \cdot A\text{-site} \; \overset{}{\xrightleftharpoons{}} \; tRNA \cdot A\text{-site} \; \xrightleftharpoons{} \; \text{Incorporation}
$$

$$
\updownarrow
$$

$$
tRNA + A\text{-site}
$$

FIGURE 4.46

Connecting the proofreading scheme to specific biochemical processes. At the top, nucleotide triphosphates (e.g., adenosine triphosphate (ATP)) are incorporated as monophosphates (AMP) in DNA replication or the transcription to messenger RNA (mRNA), ejecting a double phosphate (PP_i). In the middle panel, the charging of transfer RNA (tRNA) molecules with amino acids (by the enzyme denoted E), involves an extra ATP. At bottom, a very simplified view of protein synthesis, in which the hydrolysis of guanosine triphosphate (GTP) to guansoine diphosphate (GDP) by the protein Tu provides the energy for proofreading at the ribosome. Redrawn from Hopfield (1974).

possible to pursue a more quantitative connection between theory and experiment, as discussed in the references for this section in the Annotated Bibliography.

Kinetic proofreading not only solves a fundamental problem—the problem Schrödinger confronted in the Hapsburg portraits—it also has been a source of new questions and ideas. If the accuracy of DNA replication depends not only on intrinsic properties of the DNA, but also on the detailed kinetics of the enzymes involved in replication, then the rate of mutations itself can be changed by mutations. It has long been known that there are mutator strains of bacteria that have unusually high error rates, and we now know that these strains have aspects of the proofreading apparatus disabled. One could imagine subtler changes, so that the mutation rate would become a quantitative trait. In this case the dynamics of evolution would be very different, because fluctuations in one direction in the space of genomes would change the rate of movement in all directions. Also, because accuracy depends on energy dissipation, in an environment with limited nutrients a trade-off takes place between the speed of growth and the fidelity with which genetic information is passed to the next generations; there is an optimization problem to be solved here. In protein synthesis, accuracy and even the overall kinetics will be affected by the availability of the different charged tRNAs, and this availability is under physiological control, so again there is the possibility that, especially for fast-growing bacteria where the problems are most serious, there is some tuning or optimization to be done.

Problem 78: Controlling the pace of evolution?   Imagine that we can represent the genotype of an organism by a single number x. This variable changes at random because of mutations from generation to generation, and it moves deterministically in response to selection in some "fitness landscape." Very schematically, let's describe all of this as Brownian motion in an effective potential V_{eff},

$$\frac{dx}{dt} = -\frac{\partial V_{\text{eff}}(x)}{\partial x} + \sqrt{2T(y)}\zeta_1(t), \qquad (4.207)$$

where $\langle \zeta_1(t)\zeta_1(t')\rangle = \delta(t-t')$, and we explicitly include the effective temperature of the Brownian motion $T(y)$. We write $T(y)$ to encode the fact that the rate of mutations, and hence the rate of these random variations, depends on the properties of proteins involved in the proofreading mechanisms, and these are also subject to mutations. So we should also write

$$\frac{dy}{dt} = -f(y) + \sqrt{2T(y)}\zeta_2(t), \qquad (4.208)$$

where we include the same noise in the dynamics of y that we had in x.

(a) Show that the deterministic parts of these dynamics are equivalent to motion in some combined potential $U(x, y)$. Given that the temperature depends on y, does the system really come to "equilibrium"?

(b) Show that there is an equilibrium distribution for y, and hence for the temperature itself. To do this, it will be useful to start with the diffusion equation that is equivalent to the Langevin Eq. (4.208), following the discussion in Appendix A.5.

(c) Suppose that $V_{\text{eff}}(x)$ has two wells, so that in the absence of mutations there are two stable genotypes. Presumably one is more favorable than the other, having smaller V_{eff}, and hence it will dominate the distribution of x at long times. But if conditions change, the relative fitness in the two genotypes could change, and to survive the system would have to move from one well to the other, surmounting the barrier between them. From Section 4.1, we expect this to take an exponentially long time if the temperature T is constant. What happens if the temperature can fluctuate, as it does here? How is this connected to the problem of varying barriers, discussed around Eq. (4.14)? If the variations in y are fast compared with those in x, how do these fluctuations change the expected rate of escape over the barrier?

(d) The deterministic dynamics do not couple x and y. Continuing with the double-well problem from part (c), show that the dominant trajectories for escape over a barrier separating two wells along x does, nonetheless, involve coupled motions of x and y.

Note that I have left the formulation of this problem a little vague, for example, not being too specific about the forms of $V_{\text{eff}}(x)$, $f(y)$, and $T(y)$. You should see how far you can go without making more specific assumptions, and then think about what forms would make sense. Knowing that proofreading changes a mutation rate of $\sim 10^{-4}$ into $\sim 10^{-8}$ or even smaller should give you a sense for the range of variations in $T(y)$. Can you use this information to get an estimate of how these extended dynamics can change the rate of escape over a fitness barrier?

Problem 79: Optimizing tRNA pools. There is a separate tRNA complementary to each of the 60 codons that code for amino acids (the remaining four codons stand for "start" and "stop"). The frequency with which these codons are used in the genome varies widely, both because proteins do not use all 20 amino acids equally and because different organisms use different synonymous codons (i.e., those coding for the same amino acid) with different frequencies. But, when it comes time to make protein, the cell needs access to the appropriate population of charged tRNAs. Naively one might expect that, if the supply of tRNA is limiting the rate at which a bacterium can make proteins and grow, then it would be good to have a supply of tRNA in proportion to how often the corresponding codon gets used. Let's see whether this is right. Suppose that protein synthesis is limited by arrival of the tRNA at the ribosome. Then the time required to incorporate one amino acid coded by codon i is $t_i \sim 1/(k[\text{tRNA}_i])$, where k is a second-order rate constant.

(a) The ribosome can incorporate roughly 20 amino acids per second. Estimate the concentration of tRNAs needed to be sure that this rate does not run afoul of the diffusion limit. Look up the real concentrations, and see how they compare.

(b) The average time required to incorporate one amino acid is $\bar{t} = \sum_i p_i/(k[\text{tRNA}_i])$, where p_i is the probability of codon i appearing in the cell's mRNA. If the cell can only afford a limited amount of tRNA, the natural constraint is on the total $\sum_i[\text{tRNA}_i]$. How should the individual concentrations be arranged to minimize the mean incorporation time $\bar{t}$? Is this result surprising?

(c) You might be tempted to say that, if the goal is to synthesize proteins as rapidly as possible, and the rates are limited by the arrival of tRNAs, then we should maximize the mean rate $\sum_i p_i k[\text{tRNA}_i]$. Why is this wrong?

The ideas of kinetic proofreading may be even more generally applicable than envisioned by Hopfield and Ninio. Many proteins have activity that is regulated by phosphorylation, as with the examples of CheY in bacterial chemotaxis and rhodopsin in the rod cell. Among these proteins are many kinases, enzymes that catalyze the phosphorylation of other proteins. This creates the possibility of a cascade—one kinase is phosphorylated, and it becomes active, phosphorylating a second kinase, which becomes active, and so on. At each step, one active molecule can hit many target molecules, and so there is amplification. Such kinase cascades are widely used in biochemical signal processing. Importantly, the phosphate groups that get attached to proteins are pulled from ATP, so phosphorylation is a prototypically irreversible energy-consuming reaction. In the immune system it has been suggested that this property can provide multiple stages of proofreading, contributing to self/nonself discrimination. More generally, as shown in Fig. 4.47, if activation of an enzyme requires two steps of phosphorylation, then these steps can be arranged in a proofreading scheme. Because there are many such pathways in the cell, proofreading in this case could increase specificity and reduce crosstalk.

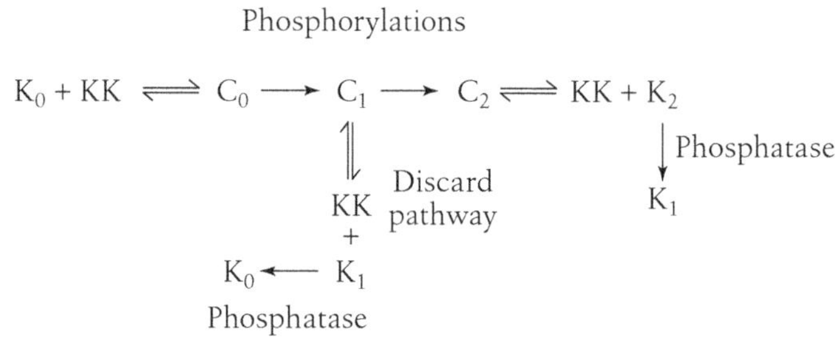

Phosphorylations

$$K_0 + KK \; \rightleftharpoons \; C_0 \longrightarrow C_1 \longrightarrow C_2 \rightleftharpoons KK + K_2$$

Discard pathway

$$KK + K_0 \longleftarrow K_1$$

Phosphatase

Phosphatase K_1

FIGURE 4.47

Kinetic proofreading in the phosphorylation of a kinase (K) by a kinase-kinase (KK). Activation of the kinase requires two steps of phosphorylation, and in this scheme the kinase-kinase can dissociate from its substrate after having transferred just one phosphate group. K_0, K_1, and K_2 denote the kinase with zero, one, and two attached phosphate groups, respectively; C_0, C_1, and C_2 are complexes of these species with KK. Note that the phosphorylation transitions $C_0 \rightarrow C_1$ and $C_1 \rightarrow C_0$ are irreversible. Redrawn from Swain and Siggia (2002).

Watson and Crick understood that the double helical structure of DNA, with its complementary strands, suggested a mechanism for the copying of genetic information from one generation to the next. But they also realized that the helical structure creates a problem, because the strands are entangled. The problem is most obvious in bacteria, where the chromosomes close into circles, but with very long molecules it is impossible to rely on spontaneous untangling, even if there is no formal topological obstruction. Eventually it was discovered that there is a remarkable set of enzymes that catalyze changes in the topology of circular DNA molecules, allowing the strands to pass through one another. In the process of relieving entanglement, these topoisomerases also reduce the energy stored in the supercoiling of these polymers. The problem is that being truly unlinked is a global property of the molecules, whereas the enzymes act locally. In the simplest models, then, topoisomerases would remove the obstacles to changing topology, but they could not shift the probability of being unlinked from its equilibrium value. Because making links or knots restricts the entropy of the molecule, there is an equilibrium bias in favor of unlinking, but this bias seems insufficient for cellular function. Indeed, as shown in Fig. 4.48, topoisomerases seem to leave fewer links than expected from the Boltzmann distribution, even in test tube experiments. In addition, if we look at the details of the biochemical steps involved, we can identify a series of steps that is equivalent to proofreading by the topoisomerases.

The ideas of proofreading have recently been revitalized by the opportunity to observe, more or less directly, the individual molecular events responsible for error correction. The key to this new generation of experiments is the realization that such molecules as RNA polymerase are molecular motors that move along the DNA strand as they function. Each step in this movement is presumably on the scale of the distance between bases along the DNA, $d \sim 3.4$ Å. The energy to drive this motion comes from breaking the phosphate bonds of the input nucleotides and is on the scale of $\sim 10 k_B T$. Thus, the forces involved are $F \sim 10 k_B T / d \sim 100$ pN.

When a dielectric sphere sits in an electric field, it polarizes, and the direction of the polarization is such that it lowers the energy. Thus, the energy of the sphere is lower in regions of high electric field. Because the energy is proportional to the square of the

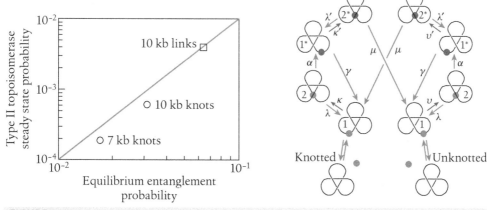

FIGURE 4.48

Kinetic proofreading in DNA unlinking. At left are experimental results redrawn from Rybenkov et al. (1997), showing that topoisomerases reach a linking probability roughly equal to the square of the expected equilibrium probability, suggesting a proofreading scheme. At right is a kinetic scheme illustrating the possibility of proofreading. Active topoisomerase molecules are shown in red, inactive in blue; green arrows denote transitions that are insensitive to the topology. All sensitivity is contained in the red arrows. This kinetic scheme is essentially a "folded" version of Hopfield's original scheme, shown in Fig. 4.45. Reprinted by permission from Macmillan Publishers, Ltd.: Yan et al. (1999).

field, this is true even if the field is oscillating in time. In particular, if we focus a light beam in a microscope, then the light intensity is higher in the focus. Light intensity is just the square of the electric field, so we expect that small dielectric objects will be attracted to focal spots, a phenomenon called optical trapping. Importantly, with realistic light intensities, the forces on micron-sized particles as they move in an optical trap indeed are on the scale of piconewtons, so it is possible to hold a molecular motor in place.

Problem 80: Optical trapping. The key to the experiments shown in Figs. 4.49 and 4.50 is the fact that small neutral particles can be trapped at the focus of a laser beam and that the forces generated in this way are on the same scale as those generated by individual biological motor molecules, such as the RNA polymerase.

(a) A good microscope can focus a laser beam to a spot of a bit less than $\sim 1\,\mu$m in diameter. Estimate the energy density of the electromagnetic field at the center of this spot if the total laser power is ~ 1 W.

(b) A dielectric object polarizes in response to an electric field $\mathbf{E}$. The polarization $\mathbf{P} = \chi \mathbf{E}$, where χ is the electric susceptibility. Show that this is equivalent to saying that the energy of the system is

$$E = \frac{1}{2\chi} \int d^3x\, |\mathbf{P}|^2 - \int d^3x\, \mathbf{P}{\cdot}\mathbf{E}, \qquad (4.209)$$

and use this equation to explain why the energy of a dielectric object is lower in regions of larger electric fields.

FIGURE 4.49

Schematic of an experiment to observe the function of RNA polymerase with single base-pair resolution. A laser beam is split, and the two resulting beams are focused to make optical traps for two micron-sized beads. Attached to one bead is a double-stranded DNA molecule, and attached to the other is an RNA polymerase molecule. As the polymerase synthesizes mRNA, it walks along the DNA, and the tether between the two beads is shortened. The intensities of the two beams are set so that the left-hand trap is stiffer, ensuring that most of the motion appears as a displacement of the right-hand bead, which is measured by projecting scattered light onto a position-sensitive detector. Redrawn from Shaevitz et al. (2003).

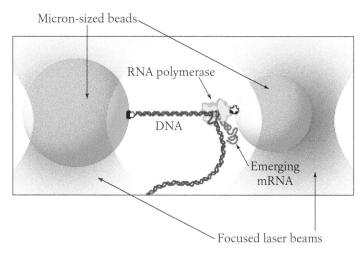

FIGURE 4.50

Motion of the RNA polymerase along DNA. (a) The top panel shows the position of the right-hand bead from Fig. 4.49 as the trap is moved in 1 Å steps, showing that these positions can be resolved. The bottom panel plots the active motion of the bead as the RNA polymerase synthesizes messenger RNA, showing the expected steps of 3.4 Å. Reprinted by permission from Macmillan Publishers, Ltd.: Abbondanzieri et al. (2005). (b) The top panel shows the average trajectory of the RNA polymerase aligned on the start and end of long pauses. The bottom panel displays the mean duration of pauses under different conditions, notably the addition of the "wrong" nucleotide inosine triphosphate (ITP). Redrawn from Shaevitz et al. (2003).

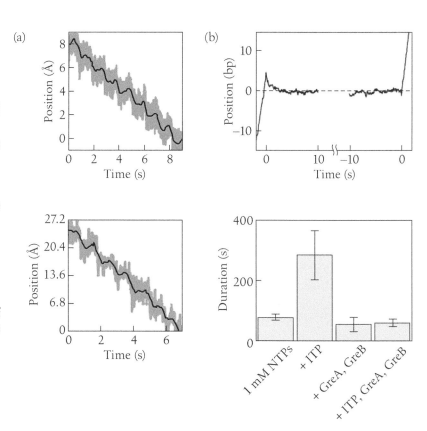

(c) In the units we have chosen, χ is dimensionless and tends to be of order one for real materials. Use this fact to relate the energy of a dielectric object in a field to the energy density of the field itself.

(d) Putting together all the above pieces, estimate the forces that a focused laser beam can apply to the ~1 μm sphere. The goal here is orders of magnitude, not factors of two.

Figure 4.49 shows the schematic of an optical trapping experiment on the RNA polymerase. Successive generations of technical improvements in these experiments have made it possible to track the motion of the polymerase with a resolution fine enough to see it step from one base pair to the next, as in Fig. 4.50. Importantly, in these experiments one can bathe the sample in a solution containing different nucleotides. If we add inosine triphosphate (ITP), which is not one of the standard four bases, it will sometimes be incorporated into the growing mRNA strand, but this is always a mistake. Under these conditions we can observe an increased frequency of pauses in the motion of the polymerase, followed by backtracking of 1–10 base pairs along a relatively stereotyped trajectory. If we remove from RNA polymerase the subunits thought to be involved in proofreading (GreA and GreB in Fig. 4.50b), then these error-induced pauses become very long.

To summarize, we can now see proofreading happening, molecule by molecule and base pair by base pair. The crucial point, however, is not the details of the mechanism in any particular case, but the enormous range of biological phenomena in which we see the same physics problem appearing. Solutions may differ in detail, but the outline always is the same—the need to achieve accuracy beyond that allowed by the Boltzmann distribution and the need for substantial energy dissipation to make this possible.

There is another broad class of examples in which there seems to be a discrepancy between the noise expected at thermal equilibrium and the performance of biological systems: the measurement of small displacements. In our inner ear, and in the ears of all other vertebrate animals, motions are sensed by hair cells, so named because of the tuft of "hairs" (more properly, stereocilia) that project from their top surface, as in Fig. 4.51. Although we usually think of ears as responding to airborne sounds, there are multiple chambers in the ear, some of which respond to sound, and others of

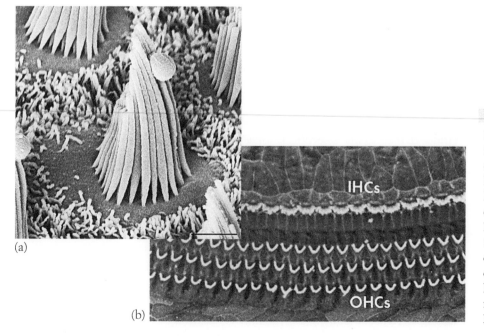

(a)

IHCs

OHCs

(b)

FIGURE 4.51

Hair cells of the vertebrate inner ear. (a) Bundle of stereocilia projecting upward from a hair cell in the bullfrog saccuᆞlus. Scale bar is 2 μm. Scanning electron micrograph from A. J. Hudspeth, with my thanks. (b) Cells in the mammalian cochlea: three rows of outer hair cells (OHCs) and one of inner hair cells (IHCs) at top. Scanning electron micrograph by E. M. Keithley, with my thanks.

which respond to lower frequency motions generated by rotation of the head, the largely constant force of gravity, or ground-borne vibrations. The core of all these systems, however, is the hair cell. When the stereocilia are bent, channels in the cell membrane open and close, which modulates an ionic current, as in other receptor cells that we have seen before. In a variety of systems it has been possible to open these organs, or even dissect out the hair cells, and to make direct mechanical measurements on the stereocilia. Typically, the bundle of hairs moves as a unit, and the stiffness is in the range of $\kappa \sim 10^{-3}$ N/m or less. Thus, the Brownian motion of the bundle should have an amplitude $\delta x_{\text{rms}} = \sqrt{k_B T / \kappa} \sim 2$ nm. This value seems small—stereocilia have lengths measured in microns.

There is a particular species of neotropical frog that exhibits clear behavioral responses to vibrations of the ground that have an amplitude of ~ 1 Å. Individual neurons that carry signals from the hair cells in the sacculus to the brain actually saturate in response to vibrations of just ~ 10 Å $= 1$ nm. Although there are controversies about the precise numbers, the motions of our eardrums in response to sounds we can barely hear are similarly on the atomic scale. Invertebrates don't use hair cells, but they also have mechanical sensors, and many of these respond reliably to motions in the angstrom or even subangstrom range.

By itself, the order-of-magnitude (or more) discrepancy between the amplitude of Brownian motion and the threshold of sensation might or might not be a problem. But surely it motivates us to ask whether, by analogy with kinetic proofreading, it is possible to lower the effective noise level by pushing the system away from thermal equilibrium. This also is an interesting physics problem, independent of its connection to biology.

Consider a mass hanging from a spring, subject to drag as it moves through the surrounding fluid, as in Fig. 4.52. By itself, the dynamics of this system are described by the Langevin equation, which by now we have seen several times:

$$m \frac{d^2 x(t)}{dt^2} + \gamma \frac{dx(t)}{dt} + \kappa x(t) = F_{\text{ext}}(t) + \zeta(t). \tag{4.210}$$

Here F_{ext} denotes external forces acting on the system, and the Langevin force, which summarizes the effects of random collisions with the molecules of the surrounding fluid, obeys

$$\langle \zeta(t)\zeta(t') \rangle = 2\gamma k_B T \delta(t - t'). \tag{4.211}$$

But suppose that we measure the position of the mass, differentiate to obtain the velocity, and then apply a feedback force proportional to this velocity, $F_{\text{feedback}} = -\eta dx(t)/dt$; then we have

$$m \frac{d^2 x(t)}{dt^2} + \gamma \frac{dx(t)}{dt} + \kappa x(t) = F_{\text{ext}}(t) + \zeta(t) + F_{\text{feedback}}(t) \tag{4.212}$$

$$= F_{\text{ext}}(t) + \zeta(t) - \eta \frac{dx(t)}{dt}, \tag{4.213}$$

$$m \frac{d^2 x(t)}{dt^2} + (\gamma + \eta) \frac{dx(t)}{dt} + \kappa x(t) = F_{\text{ext}}(t) + \zeta(t). \tag{4.214}$$

This system is equivalent to one with a new drag coefficient $\gamma' = \gamma + \eta$. But the fluctuating force has not changed—the molecules of the fluid do not "know" that we

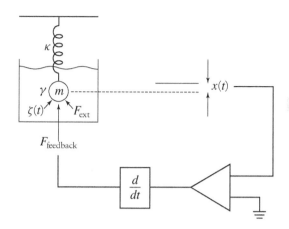

FIGURE 4.52

A schematic of active feedback, in which we observe the position $x(t)$ of a mass m on a spring with constant κ and apply a force proportional to the velocity. This force can serve to enhance or compensate the intrinsic drag γ, but because it is generated by an active mechanism (symbolically, through the amplifer) there need not be an associated change in the magnitude of the Langevin force $\zeta(t)$, as there would be at thermal equilibrium.

are applying feedback—so we can write

$$m\frac{d^2x(t)}{dt^2} + \gamma'\frac{dx(t)}{dt} + \kappa x(t) = F_{\text{ext}}(t) + \zeta(t),\tag{4.215}$$

$$\langle \zeta(t)\zeta(t')\rangle = 2\gamma k_B T \delta(t - t') = 2\gamma' k_B T_{\text{eff}} \delta(t - t'),\tag{4.216}$$

where $T_{\text{eff}} = T\gamma/\gamma' = T\gamma/(\gamma + \eta)$. Thus, by observing the system and applying a feedback force, we synthesize a system that is, effectively, colder and thus has (in some obvious sense, but we will need to be careful) less thermal noise.

This idea of active cooling is very old, but it has received new attention in the attempt to build very sensitive displacement detectors, for example, for the detection of gravitational waves. A recent example placed a 1 g mass in a laser interferometer and used the change in radiation pressure on the mass as a function of its position to generate the feedback force; this setup is different in detail from the model above but similar in spirit. The result was that the effective temperature could be brought down from $\sim$300 K to $\sim 7 \times 10^{-3}$ K, a reduction of roughly 40,000×, and it seems to be limited by noise in the laser itself.

It is important to be clear about exactly which measures of noise are reduced and which are not. The mean-square displacement of the oscillator—and hence, by equipartition, the apparent temperature—has been reduced. But when we try to drive the system with a force at the resonant frequency, the added damping means that it is more resistant, and hence the response to a given force is smaller. Thus, if we ask for the minimum force that we must apply (on resonance) to displace the oscillator by one standard deviation, this threshold force actually goes up, as if the system had more noise, not less. Finally, if we imagine that we can observe the position of the oscillator over a very long time, then what matters for detecting a small applied force at the resonant frequency is the spectral density of force noise, and this has not changed at all.

Problem 81: Effective noise levels. Do the real calculations required to verify the statements in the previous paragraph.

As an alternative to actively damping the oscillator, we can try to actively undamp it, using feedback of opposite sign:

$$m\frac{d^2x(t)}{dt^2} + \gamma\frac{dx(t)}{dt} + \kappa x(t) = F_{\text{ext}}(t) + \zeta(t) + F_{\text{feedback}}(t) \qquad (4.217)$$

$$= F_{\text{ext}}(t) + \zeta(t) + \eta\frac{dx(t)}{dt}, \qquad (4.218)$$

$$m\frac{d^2x(t)}{dt^2} + (\gamma - \eta)\frac{dx(t)}{dt} + \kappa x(t) = F_{\text{ext}}(t) + \zeta(t). \qquad (4.219)$$

Now the variance of the displacement is larger,

$$\langle(\delta x)^2\rangle = \frac{k_B T_{\text{eff}}}{\kappa} = \frac{k_B T}{\kappa} \cdot \frac{\gamma}{\gamma - \eta}, \qquad (4.220)$$

but the sensitivity to forces applied on resonance is also enhanced. If we have $F_{\text{ext}}(t) = F_0\cos(\omega_0 t)$, with $\omega_0 = \sqrt{\kappa/m}$, then the displacement will be $x(t) = x_0\sin(\omega_0 t)$, with $x_0 = F_0/[(\gamma - \eta)\omega_0]$. Thus, the signal-to-noise ratio in a snapshot of the motion becomes

$$\frac{x_0^2}{\langle(\delta x)^2\rangle} = \frac{F_0^2}{(\gamma - \eta)^2\omega_0^2} \cdot \frac{\gamma - \eta}{\gamma}\frac{\kappa}{k_B T} = \left[\frac{\kappa F_0^2}{(\gamma\omega_0^2)^2}\frac{1}{k_B T}\right] \cdot \frac{\gamma}{\gamma - \eta}. \qquad (4.221)$$

In this case the signal-to-noise ratio goes up in proportion to the amount of active undamping.

We can understand the impact of active undamping as a narrowing of the system bandwidth or a sharpening of the resonance around ω_0. Both the external force and the Langevin force drive the system in the same way. The difference is that we are considering an external force at the resonant frequency, whereas the Langevin force is white noise with equal power at all frequencies. By sharpening the resonance, active undamping reduces the total impact of this noise; because the bandwidth of the resonance is proportional to $\gamma - \eta$, the enhancement of the signal-to-noise ratio is also in proportion to this factor.

Taken at face value, it seems that we can increase the signal-to-noise ratio by an arbitrarily large factor—if we increase η so that $\gamma - \eta \to 0$, the resonance becomes infinitely sharp, and it becomes possible to detect arbitrarily small forces from just an instantaneous look at the position x. Any recipe for detecting arbitrarily small signals should be suspect, but what actually limits the growth of the signal-to-noise ratio in this case?

First, it should be clear that the increased signal-to-noise ratio comes at a cost. In a system with a sharp resonance, the time scale for response becomes long in inverse proportion to the bandwidth. Thus, as we let $\gamma - \eta \to 0$, the current position $x(t)$ becomes dependent on the forces $F_{\text{ext}}(t)$ in the distant past. This issue is serious, but it does not really set a limit to the smallest force we can detect.

Problem 82: A reminder about Green functions. The solution to the equation

$$m\frac{d^2x(t)}{dt^2} + (\gamma - \eta)\frac{dx(t)}{dt} + \kappa x(t) = F_{\text{ext}}(t) \qquad (4.222)$$

can be written in the form

$$x(t) = \int dt' \, G(t - t') F_{\text{ext}}(t'),$$

$$(4.223)$$

where $G(\tau)$ is the Green function or (time-domain) linear response function. Find $G(\tau)$, and verify that as $\gamma - \eta \to 0$, this function acquires weight at large τ, corresponding to a long memory or strongly nonlocal responses.

A second limit to the signal-to-noise ratio is set by noise in the amplifier itself. This certainly is a practical problem, and there may even be a fundamental problem, because linear amplifiers have a minimum level of noise set by quantum mechanics. There is some highly interesting physics here, and (confession time) there was a time when I worked very hard to convince myself that these quantum limits to measurement could be relevant to biological systems. This project failed, and I would rather not revisit old failures, so let's skip this one.

The third consideration that limits the narrowing of the bandwidth is the finite power output of any real amplifier. As we let $\gamma - \eta \to 0$, the amplitude of motion in response to a force at resonance grows as $1/(\gamma - \eta)$, and because there is a real drag force $-\gamma (dx/dt)$, the amplifier must dissipate power to drive these ever-larger motions. At some point this power requirement will become overwhelming, and the simple model $F_{\text{feedback}} = +\eta(dx/dt)$ has to break down. Intuitively, we expect that as x becomes larger, the strength of the feedback will decrease, so we can describe at least the beginning of this power limitation:

$$\eta \to \eta(x) \approx \eta_0[1 - (x/x_s)^2 + \cdots],$$

$$(4.224)$$

where x_s is the scale on which the amplifier loses linearity. Then we have

$$m\frac{d^2x(t)}{dt^2} + (\gamma - \eta_0)\frac{dx(t)}{dt} + \frac{\eta_0}{x_s^2}x^2(t)\frac{dx(t)}{dt} + \kappa x(t) = F_{\text{ext}}(t) + \zeta(t).$$

$$(4.225)$$

This equation has several important features. First, $\gamma = \eta_0$ is a bifurcation point. If $\gamma > \eta_0$, then in the absence of forces any small displacement from $x = 0$ will decay with time. In contrast, for $\gamma < \eta_0$, small displacements will oscillate and grow until the nonlinear term $\sim x^2(dx/dt)$ becomes significant; this is an example of a Hopf bifurcation. Second, if we poise the system precisely at the bifurcation point and drive it with a resonant force, then neglecting noise, we have

$$m\frac{d^2x(t)}{dt^2} + \frac{\gamma}{x_s^2}x^2(t)\frac{dx(t)}{dt} + \kappa x(t) = F_0 \cos(\omega_0 t).$$

$$(4.226)$$

Guessing that the solution is of the form $x(t) \approx x_0 \sin(\omega_0 t)$, we note that

$$x^2(t)\frac{dx(t)}{dt} \approx \omega_0 x_0^3 \sin^2(\omega_0 t) \cos(\omega_0 t)$$

$$(4.227)$$

$$= \frac{1}{4}\omega_0 x_0^3 \left[\cos(\omega_0 t) - \cos(3\omega_0 t)\right];$$

$$(4.228)$$

in the limit that the resonance is sharp, we know that the term at frequency $3\omega_0$ can't really drive the system, so we neglect it. With this approximation, substituting our guess for $x(t)$ into Eq. (4.226) yields

$$\frac{\gamma \omega_0}{4 x_s^2} x_0^3 = F_0,$$ (4.229)

or

$$x(t) = \left[\frac{4 F_0 x_s^2}{\gamma \omega_0} \right]^{1/3} \sin(\omega_0 t).$$ (4.230)

Thus, the response to applied forces is nonanalytic (at least in the absence of noise); the slope of the response at $F_0 = 0$ is infinite, as one expects from the linear equation above, but the response to any finite force is finite.

The fractional power behavior in Eq. (4.230) connects to a well-known but puzzling fact about the auditory system. As with any nonlinear system, if we stimulate the ear with sine waves at frequencies f_1 and f_2, we can hear combination tones built out of these fundamentals: $f_1 \pm f_2, 2f_1 - f_2$, and so on. In the human ear, the term $2f_1 - f_2$ (with $f_1 < f_2$) is especially prominent. What is surprising is that the subjective intensity of this combination tone is proportional to the intensity of the fundamental tones. If we imagine that combination tones arise from a weak nonlinearity that could be treated in perturbation theory, we would predict that if the input tones have amplitudes A_1 and A_2, then the amplitude of the combination tone should be $A_{2f_1-f_2} \propto A_1^2 A_2$. In contrast, the model poised precisely at the bifurcation point predicts $A_{2f_1-f_2} \propto (A_1^2 A_2)^{1/3}$, so that if we double the intensity of the input sounds, we also double the intensity of the combination tone, as observed.

Problem 83: Combination tones.   Do honest calculations to verify the statements about combination tones in the previous paragraph. Contrast the predictions far from the bifurcation point, where perturbation theory is applicable, with the predictions at the bifurcation point.

What happens to the nominally infinite signal-to-noise ratio in the linear model? As we increase the feedback η, the mean-square displacement increases, but Eq. (4.224) tells us that at larger x the effective strength of the feedback term decreases. We can try to see what will happen by asking for self-consistency. Suppose we replace the x-dependent value of the feedback term by an effective feedback strength that is given by the average:

$$\eta_{\text{eff}} \equiv \langle \eta(x) \rangle = \eta_0 [1 - \langle x^2 \rangle / x_s^2].$$ (4.231)

But if we have an effective feedback term, we can go back to the linear problem, and then Eq. (4.220) tells us that

$$\langle x^2 \rangle = \frac{k_B T}{\kappa} \cdot \frac{\gamma}{\gamma - \eta_{\text{eff}}}.$$ (4.232)

Combining these equations gives us a self-consistent equation for the position variance $\langle x^2 \rangle$:

$$\frac{\eta_0}{\gamma x_s^2} \langle x^2 \rangle^2 + \left(1 - \frac{\eta_0}{\gamma} \right) \langle x^2 \rangle = \frac{k_B T}{\kappa}.$$ (4.233)

Even if we let the strength of the bare feedback η_0 become infinitely large, this equation predicts that the effective feedback term will remain finite, and in particular we always have $\eta_{\text{eff}} < \gamma$, so we can never cross the bifurcation, at least in this approximation. Concretely, solving Eq. (4.233) and substituting back into Eq. (4.231) for the effective feedback, we find

$$\lim_{\eta_0 \to \infty} \frac{\gamma - \eta_{\text{eff}}}{\gamma} = \frac{k_B T}{\kappa x_s^2}. \tag{4.234}$$

Thus, the system can narrow its bandwidth to an extent that is limited by the dynamic range of the feedback amplifier, which in turn is related to its power output. Because active narrowing of the bandwidth reduces the effective noise level below the expected thermal noise, we have a situation very much analogous to kinetic proofreading: we can do better than Boltzmann, but it costs energy, and the more energy the system expends, the better it can do.

Problem 84: Noise levels in nonlinear feedback. Start by verifying Eq. (4.234). In the same approximation, calculate the response to applied forces, and show that the smallest force that can be detected above the noise in a single snapshot has been reduced by a factor of $\sim \sqrt{\kappa x_s^2 / k_B T}$ relative to what we would have without feedback. Then, there are several issues to consider.

(a) We have given two analyses. In the first, leading to Eq. (4.230), we neglect noise and take the nonlinearities seriously, finding that the response to small forces is nonanalytic. In the second, leading to Eq. (4.234) we treat the crucial nonlinear terms as a self-consistently determined linear feedback, and noise is included. In the second approach, the response to applied forces is linear. Can you reconcile these approaches? Presumably the first approach is valid if the applied forces produce displacements much larger than the noise level. Does this mean that the noise serves to "round" the nonanalytic behavior near $F = 0$?

(b) How do your results in part (a) affect your estimates of the smallest force that can be detected above the noise?

(c) You might be worried that our self-consistent approximation is a bit crude. An alternative is to simulate Eq. (4.225) numerically, bearing in mind the discussion in Section 4.1 about how to treat the Langevin force. Compare the results of your simulation with the predictions of the self-consistent approximation, for example, Eq. (4.233).

(d) You could also try an alternative analytic approach. If we rewrite Eq. (4.225) in the absence of external forces as

$$\frac{dx(t)}{dt} = v(t), \tag{4.235}$$

$$m\frac{dv(t)}{dt} = -\left[\gamma - \eta_0\left(1 - \frac{x^2(t)}{x_s^2}\right)\right]v(t) - \kappa x(t) + \zeta(t), \tag{4.236}$$

then you should be able to derive a Fokker-Planck or diffusion-like equation for the probability $P(x, v)$ of finding the system with instantaneous position x and velocity v. Can you find the steady state solution? How does this analytical result compare with your numerical results?

What have we learned from all this? Although there are limits, active feedback (with either sign) makes it possible to detect smaller signals than might otherwise be possible given the level of thermal noise. Pushing the system away from equilibrium, we spend energy to improve performance. This sounds like the sort of behavior biological systems might exploit.

If thermal noise is important, then it is useful to think about the bandwidth the system is using as it "listens" (in this case, literally) to its input, and the resulting exchange of energy. We recall that in a resonator, the time scale on which oscillations decay is $\tau \sim 1/\Delta f$, where Δf is the range of frequencies under the resonant peak. Thus, if we excite the resonator to an amplitude such that it stores energy E, this energy also decays on a time scale of $\sim \tau$. But in thermal equilibrium we know that the average energy is not zero, but rather $k_B T$, so the surrounding heat bath must provide a flux of power of $\sim k_B T / \tau \sim k_B T \Delta f$ to balance the dissipation. If we want to detect incoming signals above the background of thermal noise, then these signals have to deliver a comparable amount of power. A more careful calculation shows that this thermal noise power is $P = 4 k_B T \Delta f$.

Estimates of the power entering the inner ear at the threshold of hearing are $P \sim 4 \times 10^{-19}$ W. This value suggests that, to be sure the signals are above thermal noise, the ear must operate with a bandwidth of less than $\Delta f \sim 100$ Hz. There are several ways of seeing that this bandwidth is about right. If we record the responses of individual neurons emerging from the cochlea of animals like us, we can see that these responses are tuned. More quantitatively, as in Fig. 4.53, we can measure the sound pressure required to keep the neuron generating spikes at some fixed rate and see how this varies with the frequency of pure-tone inputs. This input required for constant output is minimal at one characteristic frequency of the neuron and rises steeply away from this minimum; for neurons with characteristic frequencies in the range of 1 kHz, the bandwidths are indeed $\Delta f \sim 100$ Hz. One can also try to measure the effective bandwidth in human observers, either by asking listeners to detect a tone in a noisy background and seeing how detection performance varies with the width of the noise, or by testing when one tone impairs the detection of another. More recently it has been possible to record the responses from individual receptor cells. All these bandwidth estimates are in rough agreement, and they also agree with the estimate based on comparing thermal noise with the power entering the ear at threshold, suggesting that filtering—in addition to its role in decomposing sounds into their constituent tones—really is essential in limiting the impact of noise. It is important that the resonance or filter that defines this bandwidth actually be in a part of the system where it can act to reject the dominant source of thermal noise. For example, placing the vibration-sensitive frog on a resonant table would mean that the whole system had a narrower bandwidth, but this would do nothing to reduce the impact of random motions of the stereocilia. It is extremely implausible that the passive mechanics of the stereocilia themselves can generate this narrow bandwidth.

Problem 85: Stereocilium mechanics.   Use the image of the hair bundle in Fig. 4.51 to estimate the mass and drag coefficient of the bundle as it moves through the surrounding fluid, which you can assume is water. Is the system naturally resonant? Overdamped or under-

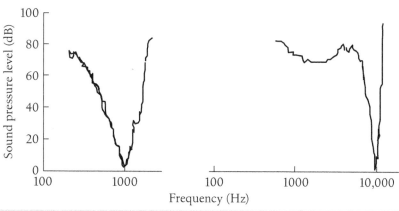

FIGURE 4.53

Tuning curves of cells in the cat auditory nerve. At left, a cell tuned to relatively low frequencies; at right, a cell tuned to higher frequencies. Curves are traced by sweeping frequency and adjusting the intensity, so that the rate of generating spikes is held constant. Because the spike generation is somewhat stochastic, the curves are a bit noisy. Note that the tuning at low frequencies is more symmetric, whereas at higher frequencies there is a pronounced asymmetry. Redrawn from Kiang et al. (1965).

damped? What bandwidth of filtering would be needed to be sure that fluid displacements of ∼1 Å are detectable above the thermal noise of the bundle? Is this roughly consistent with the observed threshold power?

In mammalian ears, the hair cells sit on top of a structure called the basilar membrane; the tips of the stereocilia are in contact with another structure, the tectorial membrane; and the entire organ, called the cochlea, is wrapped into a spiral and embedded in bone. Sound waves impinging on the eardrum are coupled to the cochlea to produce a pressure difference across the basilar membrane, which then vibrates, ultimately causing motions of the stereocilia. Because it is surrounded by fluid, motions of neighboring pieces of the basilar membrane are coupled, and the result is a wave that travels along the membrane. Because of gradations in the mechanical properties of the system, high-frequency waves have their peak amplitude near the entrance to the cochlea, and low-frequency waves have their peak near its end or apex. Helmholtz knew about the structure of the inner ear, and because he saw fibrous components in the various membranes, he imagined that these might be taut, resonant strings. Because the strings were of different lengths and thicknesses, varying smoothly along the length of the cochlea, the resonant frequency would also vary. Thus, Helmholtz had the basic picture of the cochlea as a mechanical system that analyzes incoming sounds into component frequencies, sorting them to different locations along the basilar membrane. It is not clear how seriously he took the details of the mechanics, but the picture of the ear as a frequency analyzer or bank of filters was taken very seriously, and indeed this picture accounts for many perceptual phenomena. The first direct measurements of basilar membrane motion were made by Békésy, who opened the cochleas of various animals, sprinkled reflecting flakes on the membrane, and observed its motion stroboscopically under the microscope.[30] Békésy saw the traveling wave of vibrations along the basilar membrane, and he saw the mechanical sorting of frequencies, which Helmholtz had predicted.

Problem 86: Cochlear mechanics.   The mammalian cochlea consists of two fluid-filled chambers separated by a structure that includes the hair cells and the basilar membrane. Let x be the coordinate along the length of the basilar membrane. Consider a simple model in which the fluid is described by a pressure difference p across the membrane, with resulting flows, neglecting details of the profiles perpendicular to the membrane. It is convenient to work in the (temporal) frequency domain, so all variables will be a function of x and ω, and assume that the fluid in the chambers (water) is effectively incompressible and inviscid on the relevant scales. Then if $v_{\mathrm{fl}}(x, \omega)$ is the average fluid velocity in the chambers at point x, Newton's equation for the fluid becomes

$$2\rho(-i\omega)v_{\mathrm{fl}}(x, \omega) = \frac{dp(x, \omega)}{dx}, \tag{4.237}$$

where the factor of two arises because there are two chambers. Conservation of the fluid, assuming incompressibility, would normally require $dv_{\mathrm{fl}}(x, \omega)/dx = 0$, but because the chambers are bounded by a membrane, fluid accumulation can be compensated by membrane motion, so that

$$S\frac{dv_{\mathrm{fl}}(x, \omega)}{dx} + bv_{\mathrm{BM}}(x, \omega) = 0, \tag{4.238}$$

where S is the cross-sectional area of the chambers, v_{BM} is the velocity of the membrane, and b is the width of the membrane. Finally, the motion of the membrane and the pressure difference across the membrane must be connected through the mechanical properties of the membrane itself. In the simplest model, we imagine that the membrane has mass, drag, and stiffness per unit area given by m, γ, and κ, respectively, so that

$$\left[(-i\omega)m + \gamma + \frac{1}{-i\omega}\kappa\right]v_{\mathrm{BM}}(x, \omega) = p(x, \omega). \tag{4.239}$$

(a) Show that these equations are all dimensionally correct, and be sure you understand the meaning of all terms.

(b) Assuming that all parameters vary only slowly with x, show that these equations can be combined into a single second-order equation of the form

$$\frac{d^2 p(x, \omega)}{dx^2} + k^2(x, \omega)p(x, \omega) = 0. \tag{4.240}$$

Give the relation between $k(x, \omega)$ and the other parameters in the problem.

(c) The basilar membrane has a gradient of mechanical properties, so that it is stiffest at the entrance to the cochlea and gradually becomes less stiff deeper inside; a good approximation

---

30. Many of Békésy's key contributions are collected in a volume published late in his life, along with reminiscences and quasi-philosophical remarks. As an example, he notes that in science good enemies are much more valuable than good friends, because enemies will take the time to find all your mistakes. Unfortunately, in the process of this dialogue, some of the enemies become friends and hence, by Békésy's criteria, their usefulness is lost.

is $\kappa = \kappa_0 e^{-\alpha x}$. Thus, near the entrance we can always find frequencies low enough that the term $\propto \kappa$ dominates Eq. (4.239). Show that in this case Eq. (4.240) really does describe propagating waves, with a real value of $k(x, \omega)$. In particular, in this limit, you should find that $k(x, \omega) = (\omega/c)e^{\alpha x/2}$, where c has units of a velocity.

(d) If k is a real constant, independent of x, then Eq. (4.240) is solved by $p(x, \omega) \propto e^{\pm ikx}$. In the more general case we can write $p(x, \omega) = e^{i\phi(x, \omega)}$; show that the phase obeys

$$\left[\frac{d\phi(x, \omega)}{dx}\right]^2 = k^2(x, \omega) + i\frac{d^2\phi(x, \omega)}{dx^2}. \tag{4.241}$$

Develop an expansion for $\phi(x, \omega)$ in the limit that $dk(x, \omega)/dx$ is small but nonzero, and explain the relation of this expansion to the WKB approximation in quantum mechanics. What is the condition for the validity of this approximation? How does the spatial variation in κ influence the amplitude (rather than the phase) of basilar membrane motion?

(e) If there is a well-defined resonance in the mechanics of the membrane, then at high frequencies the term $\propto m$ should dominate Eq. (4.239). Show that now $k(x, \omega)$ is imaginary, so the wave decays rather than propagating. Can you connect this decay qualitatively to the steep high-frequency slopes of the tuning curves in Fig. 4.53? Summarize all your results by sketching the pattern of basilar membrane motion versus x at some fixed frequency. Alternatively, sketch the behavior versus frequency at one point x.

(f) In quantum mechanics the breakdown of the WKB approximation typically is connected to reflection of the wave function off features in the underlying potential. Reflections of the wave traveling along the basilar membrane seem like a bad thing, and are (mostly) avoided. If the parameters m and γ are relatively constant, so all the variations are in κ, where is the WKB approximation most in danger of breaking down? Can you give a condition for it not to break down? How does this constrain the sharpness of the resonance in the basilar membrane?

Békésy was immediately impressed with the scale of motions in the inner ear. To make the basilar membrane vibrate by $\sim 1\,\mu$m and hence be easily visible under the light microscope, he had to deliver sounds at what would be the threshold of pain, ~ 120 dB SPL.[31] If we just extrapolate linearly, $1\,\mu$m at 120 dB SPL corresponds to 10^{-12} m at 0 dB SPL, or ~ 0.01 Å (!). This is an astonishingly small displacement.

Békésy also observed that the frequency selectivity of the basilar membrane motion was quite modest. The peak of the vibrations in response to a single frequency spreads over a distance along the cochlea that corresponds to more than ten times the apparent bandwidth over which we integrate. This discrepancy seems to have caused more concern than the extrapolated displacement. On the one hand, if it is correct, it suggests that there are mechanisms to sharpen frequency selectivity that come after the mechanics of the inner ear, perhaps at the level of neural circuitry. Békésy was

---

31. SPL stands for sound pressure level. It is conventional in acoustics to measure the intensity of sounds logarithmically relative to some standard. Ten decibels (dB) corresponds to a power ratio of $10\times$, so 20 dB corresponds to a factor of $10\times$ higher sound pressure variations. For human hearing the standard reference (0 dB SPL) is a pressure of 2×10^{-5} N/m^2, which is close to the threshold of hearing at frequencies near 2 kHz.

very much taken with the ideas of lateral inhibition in the retina, and suggested that this might be a much more general concept for neural signal processing. On the other hand, Békésy studied dissected cochleae that were, not to put too fine a point on it, dead. By the 1970s, it became clear that individual neurons emerging from the cochlea had frequency selectivity that was sharper than suggested by Békésy's measurements. In addition, it was found that (especially in mammals) this selectivity was extremely fragile, dependent on the health of the cochlea—so much so that the tuning properties of individual neurons could be changed within minutes by blocking blood flow to the ear, recovering just as quickly when the block was relieved.

Observations on the fragility of cochlear tuning emphasized the challenge of making direct mechanical measurements on more intact preparations, and presumably at more comfortable sound levels. To make measurements of smaller displacements, tools from experimental physics were brought to bear: the Mössbauer effect, laser interferometry, and Doppler velocimetry. At the same time, several groups turned to nonmammalian systems, which seemed like they would be more robust, such as the frog sacculus and the turtle cochlea. Especially in these systems it proved possible to make much more quantitative measurements on the electrical responses of the hair cells and eventually on their mechanical properties. In the midst of all this progress came the most astonishing evidence for active mechanical filtering in the inner ear.

If we build an active filter via feedback and try to narrow the bandwidth as much as possible, we are pushing the system to the edge of instability, as emphasized above. It is not difficult to imagine that, with active feedback provided by biological mechanisms, some sort of pathology could result in an error that pushes the gain past the bifurcation, turning a narrow bandwidth filter into an oscillator. If incoming sounds are efficiently coupled to motions of the active elements in the inner ear, then spontaneous oscillations of these elements will couple back, and the ear will emit sound. Strange as it may seem, careful surveys show that almost half of all ears have a spontaneous otoacoustic emission; a rather quiet, narrowband sound that can be detected by placing a microphone in the ear canal, as shown in Fig. 4.54. Importantly, the statistics of the sounds being emitted are not those of filtered noise, but rather those expected from a true oscillator—the distribution of instantaneous sound pressures has a minimum at zero, as expected if the quiet state is unstable.

The emission of sound provides dramatic qualitative evidence that there are active elements in the mechanics of the inner ear. If we look carefully at the mechanics of individual bundles of stereocilia, we again see spontaneous motions, and these are larger than expected from Brownian motion—there is some amplification visible even at the level of single cells. Although there are controversies about the details and possibilities for variations across species, there seems to be no more disagreement that some aspects of auditory mechanics indeed are active in the sense that we have been describing them here: our ears have mechanical filters, and this is the basis for our ability to discriminate frequencies and thus to perceive harmony, as Helmholtz imagined, but these filters are active, not passive. Less clear is whether these active filters function to reduce noise. Do real ears push all their filters to the edge of instability, operating at or very near the Hopf bifurcation? Does this provide the maximum possible noise reduction consistent with the available power to drive the amplifiers? I think these questions still are open.

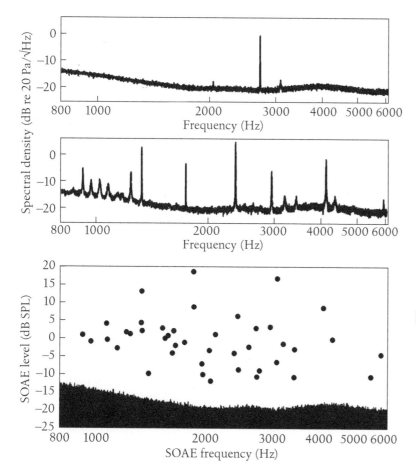

FIGURE 4.54

Spontaneous emission of sounds from the human ear. Top panels show the spectral density of sounds in the ear canals of two subjects. Bottom panel shows the intensities and frequencies of 41 spectral peaks found in eight subjects, compared with the noise background; SOAE denotes a spontaneous oto-acoustic emission. Redrawn from van Dijk et al. (2011).

4.6 Perspectives

Many of life's phenomena exhibit a startling degree of reliability and precision. Organisms reproduce and develop with surprising predictability, and our own perceptual experience of the world feels certain and solid. In contrast, when we look inside a single cell, or even at the activity of single neurons in brain, things look very noisy. Are the building blocks of biological behavior really so noisy? If so, how can we understand the emergence of reliability and certainty from all this mess?

Many of the problems faced by living organisms can be phrased in terms of sensing, processing, and responding to signals. If we look at a part of a system involved in such sensory tasks, we have to be careful in assessing noise levels. As a simple example, if we build a system in the lab that measures a small signal, and somewhere in this system there is an amplifier with high gain, then surely we will find places in the circuitry where the voltage fluctuations are large. Alternatively, there might be no gain, just a lot of noise. Thus, the variance of the noise at one point in the system, by itself, tells us nothing about its true degree of noisiness.

When we build sensors in the lab, we measure their noise performance by referring the noise to the input—estimating the noise level that would have to be added to the signals that we are trying to sense so as to account for the noise that we see at the output.

This effective noise level is also the noise that limits the detectability of small signals or the discriminability of signals that are very similar to one another. Importantly, for many sensors there are physical limits on this effective noise at the input, which allows us to put the noise performance on an absolute scale.

What we have done in this chapter is to look at several systems in which it has been possible to carry out the program of referring noise to the input. This is by no means a closed subject; a minority of systems has been analyzed in this way. Nonetheless, it is striking that, in so many disparate instances, the noise performance of biological systems indeed is close to the relevant physical limits. Of course, this observation is in the spirit of what we learned from the case of photon counting in vision (Chapter 2), but it seems much more general. I continue to find this extreme performance of biological systems quite astonishing, from bacteria to bats to flies to our own brains.

Perhaps the most surprising results are those in which organisms actually have control over all relevant signals, yet still operate in a regime where pushing to the limits of what physics allows them to detect is actually useful. Why don't cells just use higher concentrations of transcription factors, so that comparably precise modulations of gene expression would be possible even without suppressing spurious noise sources? Are the molecules so expensive that using more would be costly? More generally, we have been discussing optimization of performance given some constraints (e.g., the overall concentrations of the relevant molecules), but what sets these constraints? Is there some more general formulation in which we could imagine a family of possible organisms, moving along some optimal trade-off between resources and performance? We see hints of this more general approach in the case of kinetic proofreading, where the more accurate synthesis of proteins, for example, costs energy, and so there must be a balance between the cost of errors and that of correcting them. Although such ideas are not new, progress in experimental methods means that now we can imagine addressing these problems in quantitative measurements on single cells.

Although the subject of this chapter is one that I have thought about for a long time, I still think we are missing the most general and compelling formulation of the problems. Thus, active filtering and kinetic proofreading both are ways of reducing noise levels by spending energy to maintain a nonequilibrium state, but somehow our descriptions of these problems look very different. We can think about the dynamics of circuits that make decisions in neural networks or in biochemical networks using the same mathematics, but it seems harder to design experiments on biochemical networks where we can manipulate the accumulation of evidence for one choice or another the way we do in perceptual experiments. It is natural to discuss neural computations as combining incoming data with prior knowledge and to assess whether this combination is done optimally, but even though the same combination must be relevant in biochemical and genetic signaling, it has not been explored as thoroughly. Finally, it should be emphasized that the number of cases in which we can say precisely what we mean by "optimal performance" still is very small. These ideas are clearest in simple contexts (e.g., estimating features drawn from well-behaved distributions in backgrounds of Gaussian noise), yet organisms have to solve much more complex problems in much more complex environments. Our analytic skills just aren't yet strong enough to think about understanding speech with the same precision that we think about detecting a quiet tone or estimating the timing of an echo. The fact that real-world problems are complex and multifaceted is not an argument against theorizing along the lines explored here but rather is a challenge that needs to be met more directly.

No Fine Tuning

Imagine making a model of all the chemical reactions that occur inside a cell. Surely this model will have many thousands of variables, described by thousands of differential equations. If we write down these many differential equations with the right general form but choose the parameters at random, it seems reasonable to guess the resulting dynamics will be chaotic. Although there are periodic spurts of interest in the possibility of chaos in biological systems, this sort of generic behavior of large dynamical systems is not what characterizes life. However, it is not acceptable to claim that everything works because every parameter has been set to just the right value—in particular, these parameters depend on details that might not be under the cell's control, such as the temperature or concentration of nutrients in the environment. More specifically, the dynamics of a cell depend on how many copies of each protein the cell makes, and one either has to believe that everything works no matter how many copies are made (within reason), or that the cell has ways of exerting precise control over this number. Either answer would be interesting. This problem—the balance between robustness and fine tuning—arises at many different levels of biological organization. The goal of this chapter is to look at several examples, from single molecules to brains, hoping to see the common themes.

Physics, especially theoretical physics, is the search for concise mathematical descriptions of Nature, and to a remarkable extent this search has been successful. The dirty laundry of this enterprise is that our mathematical descriptions of the world have parameters. In a sense, one mathematical structure describes several possible worlds, which would be somewhat different if the parameters were chosen differently. Sometimes this variety is a good thing—in condensed matter physics, for example, the different parameter values might correspond to genuinely different materials, all of which are experimentally realizable. However, if the predictions of the model are too sensitive to the exact values of the parameters, there is something vaguely unsatisfying about our claim to have explained things. Such strongly parameter-dependent explanations are often called finely tuned, and we have grown to be suspicious of fine tuning. Experience suggests that if parameters need to be set to precise (or somehow unnatural) values, then we are missing something in our mathematical description of Nature.

One needs, of course, to be cautious in identifying instances of fine tuning. As an example, many beautiful phenomena associated with solar eclipses depend on the fact that, seen from our vantage point on the earth, the angular size of the moon is almost exactly equal to the angular size of the sun. As far as we know, this is a coincidence, and is not connected to anything else. There are many planets with moons—even more if we count the planets orbiting other stars—and we happen to live on one of them. Thus, we are sampling one out of many possibilities, and so rare things will happen. What we need to worry about are cases in which fine tuning seems essential to making things work, and where we see this in representative examples, or in all examples. We'll see plenty of these problematic cases.

In biological systems, there may be different reasons to be suspicious of fine tuning. For many processes, what we call parameters are certainly dynamical variables on longer time scales (e.g., the number of copies of a protein), and there is widespread doubt that cells can regulate these dynamics precisely. More fundamentally, the parameters of biological systems are encoded in the genome, and for evolution to occur it seems necessary that, closely related to the genomes we see today, there must be genomes (and hence parameter values) that also generate functional organisms of reasonable fitness. These ideas have entered the literature as the need for robustness and evolvability. Note that although the physicist's suspicion of fine tuning is a statement about the kind of explanation that we find satisfying, any attempt to enshrine robustness and evolvability as specifically biological principles involves hypotheses, either about the ability of cells to control their internal states or about the dynamics of evolution.

In this chapter we look at several examples of the fine-tuning problem, starting at the level of single molecules and then moving up to the dynamics of single neurons, the internal states of single cells more generally, and networks of neurons. These different biological systems are the subjects of nonoverlapping literatures, and so part of what I hope to accomplish in this chapter is to highlight the commonality of the physics questions that have been raised in these very different biological contexts.

5.1   Sequence Ensembles

The qualitative ideas about robustness versus fine tuning can be made much more concrete by focusing on single protein molecules. We have met many proteins already—the rhodopsin molecule, the cGMP-gated channels in the rod cell membrane, the components of the biochemical amplification mechanisms in vision and chemotaxis, the transcription factors controlling gene expression, and so on—but we have not yet dug into their structure, so now is the time.

Proteins are heteropolymers, and each monomer along the polymer chain is chosen from 20 possible amino acids (Fig. 5.1). When we look at the proteins made by one particular organism, each has some specific sequence coded in DNA and translated by means of tRNAs, as shown in Fig. 4.43. If a typical protein is 200 amino acids long, then there are $20^{200} \sim 10^{260}$ possible sequences, out of which a bacterium might choose a few thousand, and more complex creatures like us choose a few tens of thousands. Although different organisms do make slightly different choices, even if we sum over all life forms on the earth, we will find that real proteins occupy a tiny fraction of the available volume in sequence space.

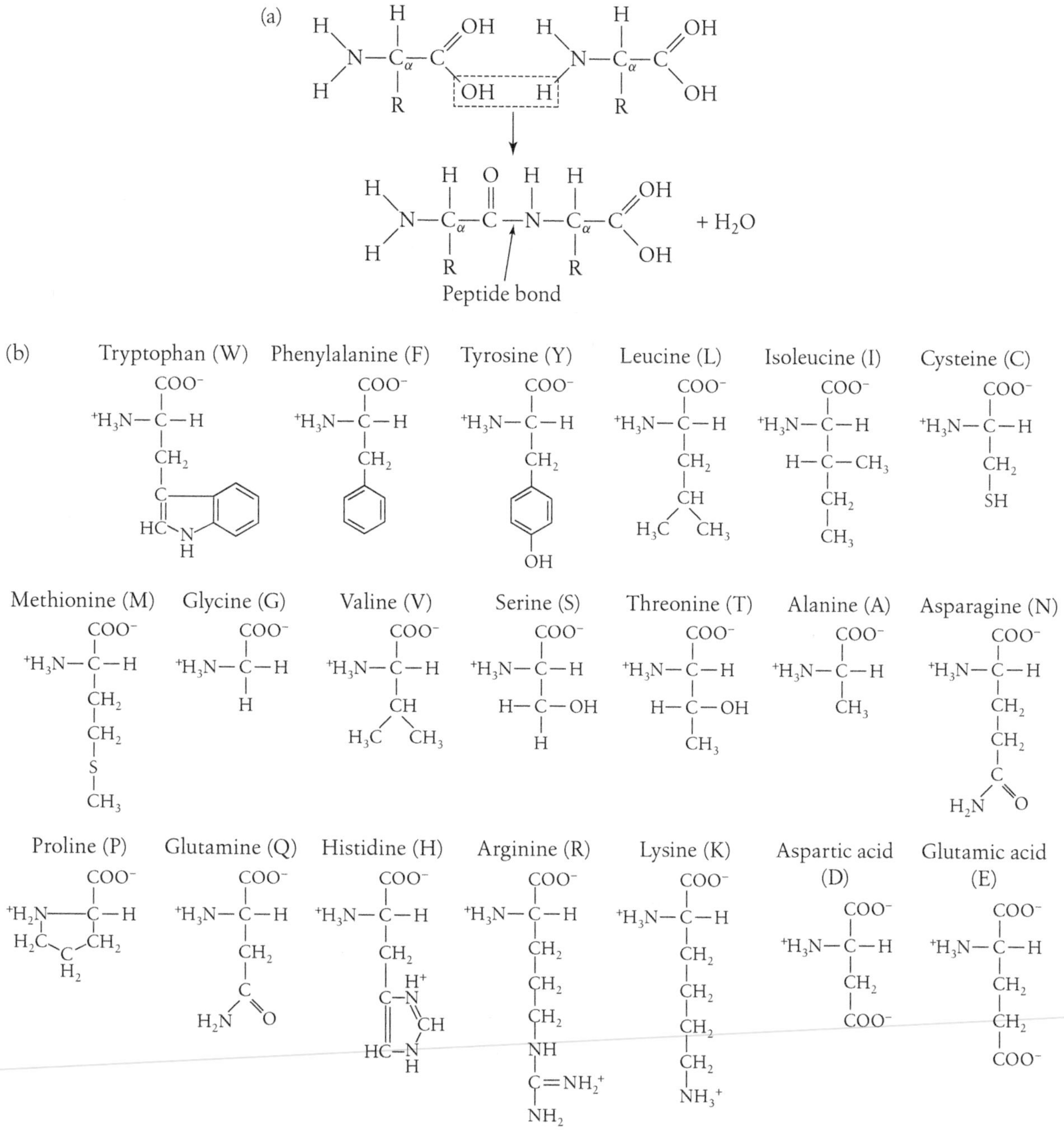

FIGURE 5.1

The basic structure of amino acids and the peptide bond. (a) Two amino acids. Different amino acids are distinguished by different groups R attached to the α-carbon. Proteins are polymers of amino acids, and the chemical step in polymerization is the formation of the peptide bond by removal of a water molecule. (b) The 20 different amino acids, arranged from most hydrophobic (top left) to most hydrophilic (bottom right).

Proteins with different sequences fold up into different structures and carry out different functions. Unlike most polymers, these structures are compact and (within some flexibility) unique. Although things can be more complicated, it probably is fair to say that the building blocks of protein structures are the secondary structural elements, α-helices and β-sheets. In the 1950s, Pauling and Corey realized that the amino acids, as they are connected along the chain of the protein (Fig. 5.1), could form repetitive hydrogen bonds if the chain folded in the right way. This observation led to the suggestion of helical and sheet structures. Evidence for helices and sheets appeared almost immediately in X-ray diffraction data from proteins, and things became even clearer as structures were solved from protein crystals. Figure 5.2 shows the comparison between the original theoretical proposal of Pauling and Corey and a relatively modern high-resolution structure of a small piece of the largely α-helical protein myoglobin.

Whether a given piece of protein folds into a helix or sheet depends on the intrinsic propensities of the individual amino acids along the chain, but also on the environment provided by the rest of the protein, because the structure is quite densely packed. Thus, the sequence obviously matters. Yet, it can't be that the *exact* sequence matters—this can be checked experimentally. Although some changes are disastrous (e.g., trying to bury a charged amino acid deep in the interior of the protein), many amino acid substitutions leave the structure and function of a protein almost completely unchanged, and many more generate quantitative modulations of function that could be useful in different environments or for closely related organisms.

Although protein function is tolerant to a wide range of sequence changes, not all sequences really make functional proteins; we can synthesize proteins by choosing amino acids at random, and almost none of them will fold. As shown below, we can even bias our choices at each site, trying to emulate a known family of proteins, but it still is true that if we choose each amino acid independently, most proteins don't fold.

As a crude theoretical model of a protein, we can coarse grain to keep track of the positions $\mathbf{r}_i$ of each α-carbon atom (as defined in Fig. 5.1) along the chain, not worrying about the detailed configuration of the side chains that project from the backbone, giving us the schematic in Fig. 5.3. Successive amino acids are bonded to each other, with a relatively fixed bond length ℓ, and when the chain folds to bring two amino acids near each other, they have an interaction that depends on their identity S_i, plus an excluded volume interaction that is independent of identity. So the total energy looks something like

$$E(\{\mathbf{r}_i\}) = \frac{\kappa}{2} \sum_i \left(|\mathbf{r}_{i+1} - \mathbf{r}_i| - \ell \right)^2 + \frac{1}{2} \sum_{ij} V(S_i, S_j) u(\mathbf{r}_i - \mathbf{r}_j)$$

$$+ \frac{1}{2} \sum_{ij} V_0(\mathbf{r}_i - \mathbf{r}_j), \tag{5.1}$$

where the stiffness κ should be large, the function $u(\mathbf{r})$ needs a shape to express the fact that amino acids have their optimal interaction at finite separation of their centers, and $V_0(\mathbf{r})$ should be relatively short ranged to express the excluded volume effect. We could try to be a little more realistic and have an extra variable for each amino acid, to keep track of the configuration of the side chain that projects from the position $\mathbf{r}_i$.

Problem 87: Screening.   We are assuming, in Eq. (5.1), that all interactions extend only over short distances, but we also know that there are charged groups. In this problem you show that

(a)   (b)

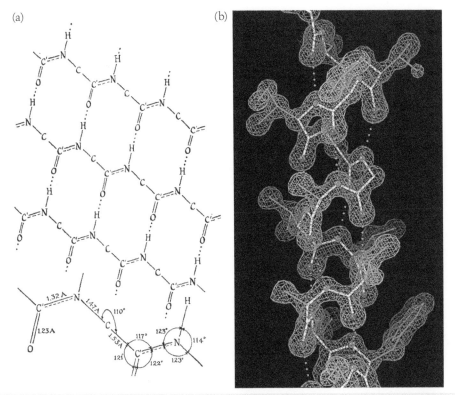

FIGURE 5.2

The α-helix. (a) The original proposal from Pauling and Corey (1951). The key idea is that, with reasonable lengths and angles for all the chemical bonds, the polymer of amino acids presents a periodic array of hydrogen bond acceptors (oxygen atoms) on one side and hydrogen bond donors (H atoms singly bonded to nitrogens) on the other; by wrapping the chain around a cylinder, to form a helix, these donors and acceptors can be aligned. Not shown in this figure are the R groups of Fig. 5.1 that distinguish the different amino acids; they should project out of the plane or outward from the cylindrical-helical surface. Whether this structure will be favorable then depends on the ability of these side chains to pack with neighboring pieces of the protein structure. (b) A short segment of the myoglobin molecule, with the electron density reconstructed from X-ray crystallography (http://en.wikipedia.org/wiki/Alpha_helix). The backbone is shown in white and the side chains in cyan; the oxygen atoms are red, and the hydrogen bonds are dotted.

the long-ranged Coulomb interaction is screened. For simplicity, let's imagine that everything is happening in an aqueous solution with only two types of ions, one positive and one negative (e.g., a simple salt solution, where the ions are Na^+ and Cl^-). Let the density of the two ions be $\rho_+(\mathbf{x})$ and $\rho_-(\mathbf{x})$, respectively. If the local electrical potential is $\phi(\mathbf{x})$, then in equilibrium the charge densities must obey

$$\rho_\pm(\mathbf{x}) = \rho_0 \exp\left[\pm\frac{q_e\phi(\mathbf{x})}{k_B T}\right], \tag{5.2}$$

where q_e is the charge on the electron, and ρ_0 is the density or concentration of ions in the absence of fields. Suppose that we introduce an extra charge Z at the origin. Convince yourself that the potential then obeys

$$\nabla^2\phi(\mathbf{x}) = -\frac{1}{\epsilon}\left[Zq_e\delta(\mathbf{x}) + q_e[\rho_+(\mathbf{x}) - \rho_-(\mathbf{x})]\right], \tag{5.3}$$

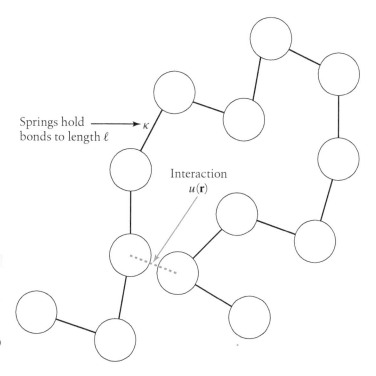

Springs hold bonds to length ℓ — κ

Interaction $u(\mathbf{r})$

FIGURE 5.3

A model for proteins based on Eq. (5.1). Bonds with stiff springs connect neighboring amino acids, which interact through a potential $u(\mathbf{r})$ when they get close. The strength of the interaction is determined by the identity of the amino acids through the term $V(S_i, S_j)$ in Eq. (5.1).

where ϵ is the dielectric constant. The combination of Eq. (5.2) and Eq. (5.3) is often called the Poisson-Boltzmann model, because Eq. (5.2) is the Boltzmann distribution and Eq. (5.3) is the Poisson equation of electrostatics.

(a) Show that, if the spatial variations in potential are small, Eq. (5.2) and Eq. (5.3) can be combined to give

$$\nabla^2 \phi(\mathbf{x}) + \frac{1}{\lambda^2}\phi(\mathbf{x}) = \frac{1}{\epsilon}Zq_e\delta(\mathbf{x}). \tag{5.4}$$

What is the length λ in terms of the other parameters in the problem?

(b) You may remember that Eq. (5.4) has solutions that decay exponentially far from the origin; this behavior is the same as for a force mediated by the exchange of a massive particle as opposed to the electromagnetic force, mediated by the massless photon.[1] In this context, Eq. (5.4) is called the Debye-Hückel equation. Solve Eq. (5.4) to give this result explicitly. If the typical concentration of ions in solution is $\rho_0 \sim 100$ mM, what is the value of λ?

(c) With only two univalent ion species, their relative concentrations are fixed by neutrality, and thus, there is only one parameter ρ_0 that enters the discussion. Generalize the derivation of the linearized Eq. (5.4) to the case where there are many species of ions, and compute λ.

---

1. Historically, this idea goes back to Yukawa, who imagined the strong force between protons and neutrons being mediated by the exchange of a heavy particle. We now know that this model was on the right track, but there were more layers of the strong interaction to be uncovered; solutions to Eq. (5.4) are still called Yukawa potentials. A more direct connection to the standard model of particle physics is in the case of the weak interaction, where the large mass of the W$^{\pm}$ and Z bosons determines the short range over which the weak interaction is effective.

(d) Going back to the two-species case in Eq. (5.3), can you solve the problem without making the linearizing approximation that leads to Eq. (5.4)? With spherical symmetry it is a one-dimensional problem, so you should be able to solve it numerically. With ρ_0 in the range of 100 mM as in part (b), how good is the linearized theory?

Having worked through this problem, does it seem reasonable that even electrostatic interactions are effectively local?

If we set the interaction $V = 0$, then Eq. (5.1) describes a polymer that takes a self-avoiding random walk. If $V = -U$, then there is a net attraction that causes collapse of the polymer into a more compact phase at low temperature, but this state is still disordered, because there is nothing to prefer one compact configuration over another. If V depends on the amino acid identities, then if we choose the sequence at random, the effective interaction between monomers i and j will also be random. This problem is complicated, but we know a great deal about the behavior of systems where the Hamiltonian contains terms chosen at random.

The prototype of a system with random interactions is the spin glass. Imagine a solid in which, at every site, there is a magnetic dipole that can point up or down, and hence can be described by an Ising spin $\sigma_\mu = \pm 1$ at site μ. If neighboring spins tend to be parallel, then we can write the Hamiltonian as

$$H = -J \sum_{\langle i, j \rangle} \sigma_i \sigma_j, \tag{5.5}$$

where $\langle i, j \rangle$ denotes neighboring sites, and J is the interaction energy between spins. In the classic spin glass materials, magnetic impurities are dissolved in a metal, so the distances between neighbors are random. Further, when the conduction electrons in the metal respond to the magnetic impurity, they polarize, but in a metal all electronic states involved in responses to small perturbations are near the Fermi surface. Hence they have a very limited range of momenta or wavevectors in their wave functions. This limitation in momentum space corresponds to an oscillation in real space, so the polarization surrounding a single magnetic impurity oscillates with distance. A neighboring impurity will be influenced by this polarization, and so the effective interaction between the two impurities can be positive or negative, at random, depending on the distance between them. This suggests a Hamiltonian of the form

$$H = -\sum_{ij} J_{ij} \sigma_i \sigma_j, \tag{5.6}$$

where J_{ij} is a random number. In a real system these interactions would be nonzero only for nearby spins, but there is a natural mean-field limit in which we allow all spins to interact; this is the Sherrington-Kirkpatrick model.

The key qualitative idea in spin glass theory is *frustration*, schematized in Fig. 5.4. In the case of the "ferromagnetic" Ising model in Eq. (5.5), each term in the Hamiltonian can be made as negative as possible by having all the spins point in the same direction, either up or down. But in the spin glass case, we may find (for example) that spin 1 is coupled to spins 2 and 3 with ferromagnetic interactions $J_{12} > 0$ and $J_{13} > 0$, but spins 2 and 3 are coupled to each other with an antiferromagnetic interaction, $J_{23} < 0$. In such a triangle, no configuration of the spins can optimize all terms in the energy function

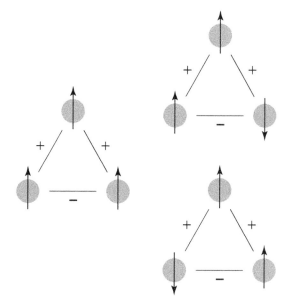

FIGURE 5.4

Three frustrated spins. Signs on the bonds indicate the signs of J_{ij} in Eq. (5.6). No matter what configuration of spins we choose, one of the bonds is always unsatisfied.

simultaneously—the interactions compete. As you can see in this simple problem with three spins, a consequence of this competition is that many states of the system with low energy are nearly degenerate. Importantly, in systems with many spins these low-lying states correspond to very different spin configurations.

Problem 88: Simulating (small) spin glasses.   Consider a mean-field spin glass, as in Eq. (5.6), in which the couplings J_{ij} are drawn at random from a Gaussian distribution; for simplicity start with the assumption that the mean of this distribution is zero and the variance is one. Notice that with N spins there are exactly 2^N states of the system as a whole, so that up to $N = 20$ (or even a bit more), you can enumerate all these states without taxing the memory of your computer.

(a) Write a simple program (e.g., in MATLAB) which, starting from a particular random matrix J_{ij}, gives the energies of all states in an N-spin system.

(b) Find the ground-state energy of an N-spin system, and do this many times for independent choices of the random interactions J_{ij}. Show that, if the distribution out of which the J_{ij} are drawn is held fixed, then the ground-state energy does not seem to be extensive (i.e., proportional to N) as N varies. In contrast, if the variance of J scales $\propto 1/N$, show that the average ground-state energy does seem to be proportional to the number of spins. Can you give an analytic argument for why this scaling should work?

(c) The exact ground-state energy depends on the particular choice of the interactions J_{ij}. One might hope that, as the system becomes large, there is self-averaging, so that the energy per spin becomes independent of these details in the limit $N \to \infty$. Do you see any signs of this behavior?

(d) Having normalized the variance of the couplings $\langle J^2 \rangle = 1/N$, so that the ground-state energy is on the order of -1 per spin, compute the gap Δ between the ground state and the first excited state of the system, again for many realizations of the matrix J_{ij}. How does the probability distribution of this gap behave at small values of the gap? In particular, is there a finite probability density as $\Delta \to 0$? How does this behavior of the gap compare with what you expect in a ferromagnet?

(e) Show that at least some of the low-lying states have spin configurations that are very different from the ground state. Again, contrast this with the case of a ferromagnet.

The statistical mechanics of spin glasses is a very beautiful subject, and we could spend a whole semester on it. What we need for the moment, however, is an intuition, something of the sort one can get from the numerical simulation in Problem 88. In systems with substantial frustration, we expect many locally stable low-energy states that are very far apart in the relevant state space. Thus, rather than having a well-defined ground state, with small fluctuations around this state, there are many inequivalent near-ground states, often with large barriers between them. If we think of the dynamics of the system as motion on an energy surface, then this surface will be rough, with many valleys separated by high passes; indeed, the Sherrington-Kirkpatrick model has valleys within valleys, hierarchically.

What does all this teach us about the protein-folding problem? To the extent that we can make analogies between spin glasses and heteropolymers with random sequences, we expect that these randomly chosen proteins will not, in general, have unique ground-state structures. Instead, there will be many inequivalent structures with nearly the same low energy, separated by large barriers. Several groups have used modern tools from the statistical mechanics of disordered systems to make this intuition precise, and indeed the random heteropolymer is a kind of glass—the polymer has compact, locally stable structures, but there are many of these, and the system tends to get stuck in one or another such local minimum at random. This contrasts sharply with the ability of real proteins to fold into particular, compact conformations that are (at some level of coarse graining) uniquely determined by the sequence. The real problem is even worse, because we have only considered the statistical mechanics of one polymer in solution; in practice the folded state of proteins competes not only with the higher entropy unfolded state but also with states in which multiple protein molecules aggregate and precipitate out of solution.

The conclusion is that the proteins occuring in Nature cannot be typical of sequences chosen at random. At the same time, not every detail of the amino acid sequence can be important. This is perhaps the most fundamental example of the general question we are exploring in this chapter—our description of life cannot depend on fine tuning, but neither are the phenomena of life generic. Concretely, we can ask how to describe the ensemble of sequences that we see in real proteins. One possibility is that this ensemble is profoundly shaped by history, and surely at some level this is true—we can trace evolutionary relationships through sequence data. Another possibility is that the ensemble of *possible* sequences is enormously constrained by physical principles: ensuring that a protein will fold into some compact, reproducible structure is extremely difficult, perhaps hard enough to explain the dramatically restricted range of sequences and even structures that we observe in real proteins.

The problem we are formulating here is related to, but different from, a much more widely discussed problem. The general question of how protein structure emerges from the underlying amino acid sequence is referred to as the the protein-folding problem. As a practical matter, one would like to predict the three-dimensional structure of the folded state, starting with only the sequence. Many approaches to this problem are based not on a physical model for the interactions but on attempts to generalize from many known examples of sequence-structure pairs. Faced with a particular sequence from Nature, this approach can be an extraordinarily effective. But it does not tell us why some heteropolymers fold into compact, reproducible states, while others do not, and why (presumably) some sequences will never be seen in real organisms. It is this more general version of the question that concerns us here.

In a typical sequence chosen at random, interactions among the different amino acids will be frustrated, blocking the system from finding a single well-isolated folded structure of minimum energy. A candidate principle for selecting functional sequences is thus the minimization of this frustration. If frustration is absent, few if any major energetic barriers will lie on the path from an unfolded state to the compact, native conformation, although the need for local structural rearrangements along the path may imply an irreducible roughness to the energy surface that, in a coarse-grained picture, will limit the mobility of the system along its path. This scenario has come to be called a folding funnel, emphasizing that the energy landscape is dominated by a single valley, into which all initial configurations of the system will be drawn, as shown schematically in Fig. 5.5.

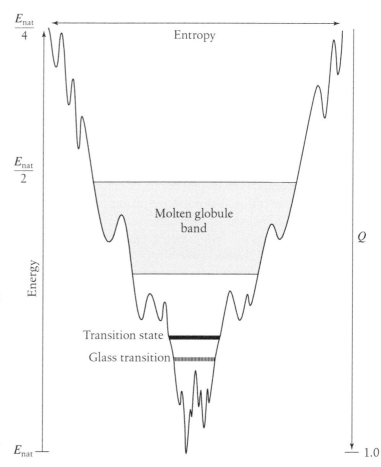

FIGURE 5.5

A schematic energy landscape for protein folding. An order parameter Q counts the fraction of correct contacts between amino acids that have been formed, and it increases as we move downward. Schematically, the funnel narrows, because this increase in the number of contacts reduces the entropy of the chain. The landscape has a small amount of roughness, and bands of energy mark stages in the folding process. Redrawn from Onuchic et al. (1995).

At a technical level, if frustration is absent, then we can look at the ground state or native structure and read off an approximation to the interactions. Thus, in a ferromagnet, all spins are parallel in the ground state, and if we simply look at each neighboring pair, we would guess that there is a ferromagnetic interaction between them; absent any other data, we should assume that all these interactions have the same strength. Although this might not be exactly right, the Hamiltonian obtained in this way will have the correct ground state. In contrast, this does not work with spin glasses, because the (near-)ground states necessarily leave some fraction of the interactions unsatisfied because of frustration. In this spirit, if we look at a small protein, we can generate a potential energy function that ties neighboring amino acids together along the chain and, in addition, has bonds between amino acids that are in contact in the folded state. We should choose the scale of the potential to have more or less the correct distance between amino acids, and the right order of magnitude for the free-energy difference between folded and unfolded states.

Models that bond together amino acids that should form contacts and neglect all other interactions have a long history and are referred to as Gō models. This approach involves an energy function of the form

$$E = \frac{1}{2} \sum_{\text{bonds}} \kappa_r (r - r_0)^2 + \frac{1}{2} \sum_{\text{angles}} \kappa_\theta (\theta - \theta_0)^2$$

$$+ \frac{1}{2} \sum_{\text{dihedrals}} \sum_n \kappa_\phi^{(n)} \left[1 + \cos(n(\phi - \phi_0)) \right]$$

$$+ \epsilon \sum_{i < j-3} \left[5 \left(\frac{\sigma_{ij}}{r_{ij}} \right)^{12} - C_{ij}^{\text{native}} 6 \left(\frac{\sigma_{ij}}{r_{ij}} \right)^{10} \right], \tag{5.7}$$

where the various κs are stiffnesses that hold bond lengths r and angles θ, ϕ to their native values along the chain. The crucial terms are those with scale ϵ, which serve to bond together pairs of residues ij forming a contact in the native, folded state ($C_{ij}^{\text{native}} = 1$) while pushing apart those that do not ($C_{ij}^{\text{native}} = 0$). In principle the different bonds can have specific lengths σ_{ij}, but this is not so important qualitatively.

What are the dynamics predicted by the energy function in Eq. (5.7)? We can simulate these dynamics, and along the trajectory we can measure the fraction Q of the contacts that should form in the folded state that have actually been made; by construction, as this order parameter increases, the energy of the system decreases. But making contacts lowers the entropy of the polymer, and exactly how much the entropy is lowered depends on which contacts are made. The detailed simulations show that the free energy as a function of Q has roughly a double-well structure, as in Fig. 5.6. Importantly, one can also sample the configurations in the transition state between the wells and ask which contacts have been made by the time the molecule finds its way to the top of the barrier. Because there are no competing interactions, the prediction is that the ensemble of transition-state configurations must reflect only the geometry of the target folded state.

Can we test the predictions of such simulations? We expect, from the general arguments in Section 4.1, that the rate of folding will have an approximately Arrhenius temperature dependence, $k \propto \exp(-\Delta F / k_B T)$, where ΔF is the free-energy difference

FIGURE 5.6

Gō models for two particular proteins, dihydrofolate reductase (DHFR, at left) and interleukin 1β (IL-1β, at right). Along the x-axis in all figures is a parameter Q measuring the fraction of native contacts that have formed. The top panels show the root-mean-square difference between the structures and the ground state, with colors denoting the energy. Note that, because there are no competing interactions, the energy decreases linearly as more of the native contacts are formed. But different values of Q can be achieved by different numbers of configurations, until at $Q = 1$ there is only one possible structure. Thus, the entropy generally declines with Q, although there is also some structure along the way determined by the geometry of the native fold. The result, shown in the bottom panels, is that the free energy has two distinct minima, corresponding to folded ($Q \approx 1$) and unfolded ($Q \approx 0$) states. Different curves correspond to different temperatures, as indicated. Redrawn from Clementi et al. (2000a). Copyright © 2000 by the National Academy of Sciences USA.

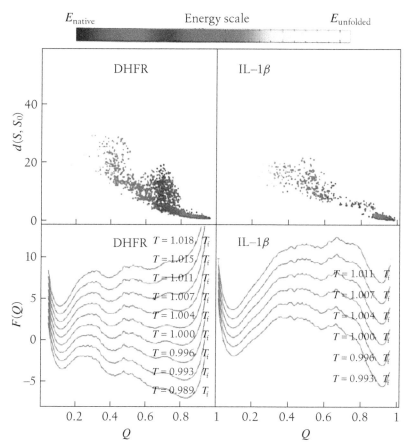

between the unfolded state and the transition state at the top of the barrier. Imagine that we mutate the protein to change amino acid i. This has some effect on the free-energy of every contact between i and j, and we can measure at least the sum of these effects by measuring the change in the free energy difference between the folded and unfolded states. But if along the reaction coordinate Q in Fig. 5.6 these contacts are made (on average) only once $Q > Q_c$, where the Q_c is the position of the transition state, then changing their energy does not change the activation free energy for the folding reaction. In contrast, if these contacts are made at $Q < Q_c$, they contribute to the free energy of the transition state and should change the rate of folding. Roughly speaking, the ratio between changes in the (kinetic) free energy of activation and the (thermodynamic) free energy of folding tells us the fraction of contacts involving residue i that are formed in the transition state. This quantity is something we can get directly from the computations summarized in Fig. 5.6; it is also something one can measure experimentally. Theory and experiment are in surprisingly good agreement, which strongly suggests that, at least for small proteins, frustration really has been minimized.

Problem 89: The location of transition states.   Suppose that the dynamics of a chemical reaction are described, as in Fig. 4.1, by motion of a coordinate x in a potential $V(x)$ that has

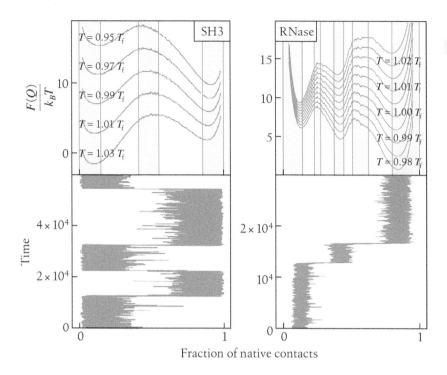

FIGURE 5.7

Simulations of folding for two proteins, using Gō models. At each instant of time (in arbitrary units) in the simulation we can count the fraction Q of native contacts, as in Fig. 5.6; sampling the probability distribution of Q, we infer the free energy $F(Q)$. At left are simulations of an SH3 domain, which is known to fold rapidly with no obvious intermediate states between folded and unfolded. At right are simulations of the enzyme RNase, which folds more slowly and occupies a well-defined intermediate state. These differences are captured by the Gō models, suggesting that frustration does not play a role in slowing the folding of the larger molecules. Reprinted from Clementi et al. (2000b), with permission from Elsevier.

two minima separated by a barrier. Let the locations of the two minima be at x_1 and x_2, while the peak of the barrier is at a position x_t. Assume that rate constants for transitions between the two wells are governed by the Arrhenius law. Now imagine that we apply a small force f directly to the coordinate x. How does this change the equilibrium between the two states? How does it change the rate of transition, say, from the states near x_1 to the states near x_2? Notice that these are measurable quantities. Can you combine them to infer the location of x_t along the line from x_1 to x_2? In particular, can you say something without knowing any additional parameters?

Some proteins are known to fold slowly, pausing in well-defined intermediate states. Does this behavior represent a failure to relieve all the frustration, or is it somehow intrinsic to the size and structure of these molecules? One can make Gō models of these slower proteins and compare them with the smaller two state folders. Results of such a comparison are shown in Fig. 5.7. Perhaps surprisingly, intermediates emerge in the folding of the larger protein even in a model where there is no intrinsic frustration from the interactions among different kinds of amino acids.

To summarize, the statistical mechanics of random heteropolymers alerts us to a severe problem in building functional proteins. A natural hypothesis is that this problem is solved completely, or optimally, by selecting sequences that minimize frustration. Although this approach seems abstract, it suggests a simplified but concrete model for the dynamics of folding that makes detailed predictions in good agreement with experiment.

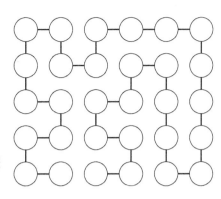

FIGURE 5.8

Compact folded structure of an $N = 30$ polymer on a square lattice.

A second approach to our problem looks more explicitly at the mapping between sequences and structures. The observation that changes in amino acid sequence (mutations) do not necessarily change protein structure tells us that many sequences map into the same structure. But what about the other direction of the mapping? If we imagine some compact structure of a hypothetical protein, can we find a sequence that will fold into this structure? This is the inverse folding problem, or the problem of protein design.

To address the inverse folding problem, it is helpful to step back and work on a simpler version of the problem. Imagine that there are just two kinds of amino acids, hydrophobic (H) and polar (P). Polar residues are happy to be next to one another, but they are equally happy to be on the outside surface of the protein, interacting with water. Hydrophobic residues are much happier to be next to one another, and this includes the effect of not being near water. Finally, for hydrophobic residues, it is likely that having a polar neighbor is marginally better than having water as a neighbor. Thus, there are three interaction energies, $E_{PP} > E_{HP} > E_{HH}$, where lower energy is (as usual) more favorable. To simplify yet further, let us assume that the structure of the protein lives on a lattice, as in Fig. 5.8. Now it is clear what we mean by compact structures—if the protein is $N = 27$ amino acids long, for example, a compact structure is one that fills a $3 \times 3 \times 3$ cube—and similarly the definition of neighbor is unambiguous.

Once we have simplified the problem, it is possible to attack it by exhaustive enumeration. On the $3 \times 3 \times 3$ cube, for example, there are $\sim$50,000 inequivalent compact structures, and there are $2^{27} \sim 10^8$ sequences of this length in the HP model. These numbers are large, but hardly astronomical, so one can explore these sequences and structures completely (as is true for two-dimensional models with $N = 30$ and 36). To begin, out of 2^{27} sequences, less than 5% have a unique compact structure with minimum energy; the majority of sequences have multiple degenerate ground states with inequivalent structures. Conversely, there are nearly 10% of compact structures for which no sequence finds that structure as its ground state, and the vast majority of structures are connected to just a handful of sequences. But if we ask how many sequences map into a given structure (N_s), there is a long tail to the distribution of this number (Fig. 5.9, at left), and some structures have thousands of sequences that all reach that structure as their ground state. We can say that these structures are highly designable. Structures with large N_s also have a large energy gap between the compact ground state and the next highest energy conformation, so that highly designable structures are also thermodynamically stable.

(a)

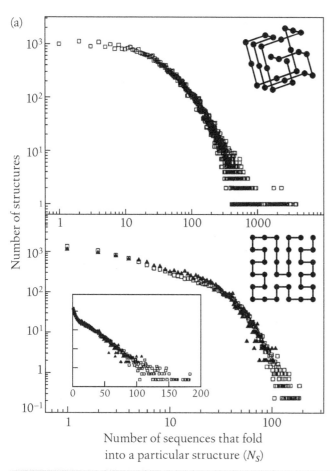

(b)

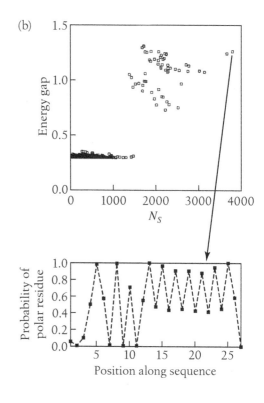

Number of structures

Number of sequences that fold
into a particular structure (N_S)

FIGURE 5.9

Exhaustive simulations of compact structures on a lattice. Redrawn from Li et al. (1996). (a) The number of structures that are the ground state for exactly N_s distinct HP sequences, plotted versus N_s for 3 × 3 × 3 (top) and 6 × 6 (bottom) lattices. Note the small number of structures that are the ground states for huge numbers of sequences. (b) The energy gap between the ground state and the first excited state, showing that stability correlates with N_s. The most highly designable structure has a distinctive pattern of hydrophobic and polar residues alternating with residues that are free to be either type with nearly equal probability.

What are these highly designable structures? It is hard to extrapolate from such small systems, but certainly the structures with largest N_s have more symmetry and show hints of extended elements, such as helices and sheets, which echo real secondary structural elements, as seen in the insets to Fig. 5.9. Can we understand why designability is so variable, and why these particular structures are highly designable?

Finding sequences that stabilize certain structures can be done in two ways. What we really want are sequences such that the desired structure is actually the ground state, which means we have to check all other possible competing structures. A weaker notion is to ask for a sequence that assigns a low energy to the desired structure, perhaps even the lowest possible energy across all sequences. If we are just trying the lower the energy, then the problem of choosing sequences is relatively simple—we should try to put the polar residues on the outside and the hydrophobic residues on the inside. This version of the inverse problem seems at most weakly frustrated, so there are downhill paths to find good sequences.

Analytic approaches to designability describe protein structure not in terms of the positions of all the amino acids, but in terms of the contact matrix C_{ij} defined in the discussion of Eq. (5.7). Assuming that all long-ranged interactions are screened, we can approximate the energy of the molecule as having contributions only from amino acids that are in contact,

$$E = \sum_{ij} C_{ij} \sum_{\mu\nu} s_i^\mu V_{\mu\nu} s_j^\nu, \tag{5.8}$$

where $s_i^\mu = 1$ if the amino acid at site i is of type μ, and $s_i^\mu = 0$ otherwise. The matrix $V_{\mu\nu}$ summarizes the interactions among the different types of amino acids. To approach the weaker notion of designability, we need to ask how many sequences give a particular structure a low energy. But asking about the numbers of sequences with a particular energy is just like doing statistical mechanics where we keep the structure fixed and instead allow the sequence $\{s_i^\mu\}$ to be the dynamical variable. This suggests that we compute the partition function in sequence space:

$$Z_{\text{seq}}(C) = \sum_{\{s_i^\mu\}} \exp\left[-\beta \sum_{ij} C_{ij} \sum_{\mu\nu} s_i^\mu V_{\mu\nu} s_j^\nu\right], \tag{5.9}$$

where $\beta = 1/(k_B T)$. Again, this computation is hard in general, but we can get some intuition by doing a high-temperature (small β) expansion.

Summing over all sequences is equivalent to averaging over a distribution in which all sequences are equally likely. Computing the average value of an exponential generates a series of cumulants, or connected correlations:

$$\langle e^{-x} \rangle = \exp\left[-\langle x \rangle + \frac{1}{2}\langle x^2 \rangle_c - \frac{1}{3!}\langle x^3 \rangle_c + \cdots\right], \tag{5.10}$$

$$\langle x^2 \rangle_c = \langle x^2 \rangle - \langle x \rangle^2 = \langle (x - \langle x \rangle)^2 \rangle, \tag{5.11}$$

$$\langle x^3 \rangle_c = \langle (x - \langle x \rangle)^3 \rangle, \tag{5.12}$$

and so on. To use this series in evaluating $Z_{\text{seq}}(C)$, we need to compute quantities of the form

$$\left\langle \sum_{\mu\nu} s_i^\mu V_{\mu\nu} s_j^\nu \right\rangle,$$

or

$$\left\langle \left(\sum_{\mu\nu} s_i^\mu V_{\mu\nu} s_j^\nu\right)^2 \right\rangle.$$

Because we are averaging over a distribution in which all sequences are equally likely, the vector $\mathbf{s}_i$ that specifies the choice of amino acid at site i is independent of the vectors $\mathbf{s}_j$ for any $j \neq i$. A slight inconvenience is that the averages of individual components are not zero, but rather $\langle s_i^\mu \rangle = 1/20$, since there are 20 amino acids. Thus, it is useful to write

$$E = \sum_{ij} C_{ij} \sum_{\mu\nu} s_i^\mu V_{\mu\nu} s_j^\nu$$

$$= \frac{1}{400} \left(\sum_{\mu\nu} V_{\mu\nu} \right) \sum_{ij} C_{ij} + \sum_{ij} C_{ij} \sum_{\mu\nu} \delta s_i^\mu V_{\mu\nu} \delta s_j^\nu, \tag{5.13}$$

where the δs_i now are independent random variables with zero mean. The constant term, which does not depend on the sequence, is proportional to $\sum_{ij} C_{ij}$, which counts the total number of contacts. Thus, if $\sum_{\mu\nu} V_{\mu\nu} < 0$, this term favors compactness, but in comparing different compact structures, as above, we can absorb this into the zero of energy. Then the free energy becomes

$$F_{seq}(C) \equiv -\frac{1}{\beta} \ln[Z_{seq}(C)] = A\,\mathrm{Tr}(C^2) + B\,\mathrm{Tr}(C^3) + \cdots, \tag{5.14}$$

where the coefficients depend on the details of the potential $V_{\mu\nu}$, and the term $\propto \mathrm{Tr}(C)$ is absent because $\mathrm{Tr}(C) = 0$.

Problem 90: Details of $F_{seq}(C)$. Derive Eq. (5.14), carrying the expansion out to at least one more order. Relate the coefficients in the expansion explictly to the properties of the potential $V_{\mu\nu}$.

Because C is a symmetric matrix, with elements that are either 1 or 0, $\mathrm{Tr}(C^2)$ just counts the number of contacts; $\mathrm{Tr}(C^3)$ counts the number of connected paths that lead from site i to site j to site k and back to site i. Similarly, the trace of higher powers counts the number of longer paths. But we can also take a less local view and note that $\mathrm{Tr}(C^n) = \sum_i \lambda_i^n$, where λ_i are the eigenvalues of the matrix C. As we consider higher powers in the expansion, the result is dominated more and more by the largest of these eigenvalues. Experimenting with small structures as in the discussion above, one can show that the designability of a structure really does correlate strongly with the largest eigenvalue of the contact matrix, and the most designable structures have the largest eigenvalues, as in Fig. 5.10. This is especially interesting, because the calculation we have outlined here does not depend on details of the assumptions about the interactions between amino acids—all that matters is locality.

As noted, computing $F_{seq}(C)$ gives us a weak notion of designability, counting the number of sequences for which a particular structure will have low energy. If we are willing to simplify our model of the interactions, then we can make progress on the stronger notion of designability, that many sequences have the same minimum-energy structure. Suppose we return to the model in which there are just two kinds of amino acids, hydrophobic and polar. Further, let's describe the structure in a similarly binary fashion, labeling each amino acid by whether it is on the surface of the molecule or in

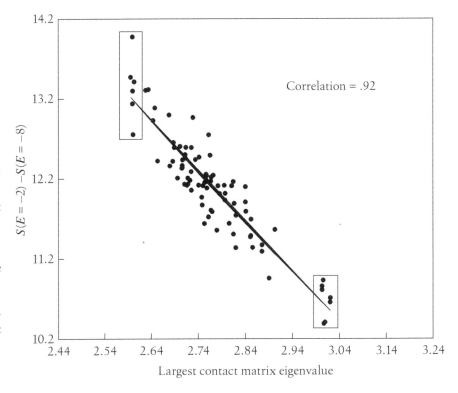

FIGURE 5.10

The connection between designability and the eigenvalues of the contact matrix. The energy $E = -2$ is typical, and $E = -8$ is in the tail of the distribution. The ordinate measures the difference in entropy between the ensemble of sequences consistent with these energies, as computed in a Monte Carlo simulation that gradually cooled this ensemble. Different points refer to different compact structures. Redrawn from England and Shakhnovich (2003).

the interior.[2] Now there is a plausible energy function—hydrophobic residues prefer interior sites, polar residues prefer the surface. Thus, the energy will be minimized when the binary description of the sequences ($s_i = +1$ for hydrophobic, $s_i = -1$ for polar) matches the binary description of the structure ($\sigma_i = +1$ for the interior, $\sigma_i = -1$ for the surface). Although we might not be able to calculate the exact energy function, ground-state structures should correspond to the minimum of a very simple energy function that just counts the violations of the hydrophobic/interior–polar/surface rule:

$$E \propto \sum_i (s_i - \sigma_i)^2. \tag{5.15}$$

An important point about this binary description of structures and sequences is that although all binary strings $\{s_i\}$ represent possible amino acid sequences, not all binary strings $\{\sigma_i\}$ are possible compact structures of a polymer. Thus, in the space of binary strings, and hence HP sequences, there are special points that correspond to realizable protein structures. The energy function in Eq. (5.15) tells us that the ground-state structure for any sequence is the nearest such point, where nearness is measured by a natural metric, the Hamming distance (counting the number of bits that disagree in the binary string). The set of sequences that will fold into one particular structure are those that fall within the Voronoi polygon surrounding the binary description of that structure, as shown in Fig. 5.11. In this picture, the sequence literally encodes the

---

2. On a lattice, with the protein folded into a compact structure, this categorization of sites is unambiguous, although one might worry a bit about the more general case.

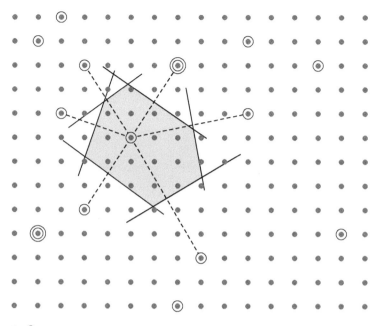

- • Sequence
- ◯ Nondegenerate structure
- ◎ Degenerate structure

FIGURE 5.11

Designability as seen in the binary description of sequences and structures. Each point represents a possible binary (hydrophobic-polar) sequence of amino acids, of which only a small subset correspond to sequences of inside/outside structures that are realizable as compact configurations of linear polymers. With the energy function from Eq. (5.15), sequences are predicted to fold into the nearest structure. To find this nearest structure we use the Voronoi construction, with dashed lines connecting one structure to its neighbors, then bisecting these lines to define an enclosed polygon, shown hatched. All the sequences in the hatched region will fold into the structure at its center. Redrawn from Li et al. (1998).

structure, and the folding process provides a kind of error correction in this code, mapping arbitrary binary strings back to the sparse set of realizable structures. By choosing structures that are far from other structures in this binary representation, one guarantees that many sequences will map to that one structure. Again this picture can be tested against simulations of the lattice models (as in the discussion above), and the results are consistent.

The lesson from this discussion is that not all structures are created equal, and selection of structures for their designability induces a nontrivial distribution on the space of sequences. This constraint restricts the set of allowed sequences but at the same time focuses precisely on those sequences for which not all details of the sequence have functional relevance.

There is yet another approach that tries to address the ensemble of allowed sequences, leaning on theory but also using a more direct experimental exploration. To appreciate this approach, you need to know that proteins form families. We have already met many examples of this: rhodopsin is one of the G-protein coupled receptors, and it interacts with transducin, which is a member of the G-protein family; in bacterial chemotaxis CheA is a kinase, attaching phosphate groups to CheY, and there is a kinase involved in turning off rhodospin as well; Bcd and Hb are transcription factor proteins, all of which bind to DNA, but they can be classified into subfamilies based on the mode of binding, and so on.

Beyond the kinases, there are other families of enzymes. To digest food, for example, we need to cut up the proteins that we ingest, and all cells also need to cut up old proteins that have been damaged or have outlived their usefulness in other ways. Cutting the

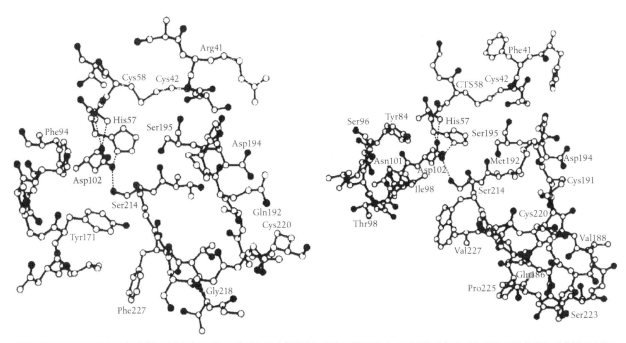

FIGURE 5.12

Comparison of the structure of the bacterial enzyme SGPA (right) and chymotrypsin (left), in the neighborhood of the active site. Note in particular the very similar geometrical relations among His57, Asp 102, and Ser 195, the triad of residues involved in the catalytic events. Reprinted from Brayer et al. (1978), with permission from Elsevier.

peptide bond quickly and efficiently requires a carefully engineered catalyst, but cells also need control over which sequences they are cutting. Thus, there are several families of protein-cutting proteins—proteases—and remarkable structural similarities exist among molecules separated by billions of years of evolutionary history. An example is shown in Fig. 5.12, comparing the structure of the bacterial enzyme SGPA and the mammalian enzyme chymotrypsin. These molecules have recognizably similar amino acids along only ∼25% of their sequences, yet the structures are very similar, especially in the active site where the crucial chemical events occur—the proteins fold to bring these key elements into a very specific geometrical arrangement, despite the sequence differences. Other interesting examples of protein families include smaller parts of proteins (domains) that can fold on their own and function as the interfaces between different molecules; there are hundreds of examples in some of these families.

If we line up the sequences for all the proteins in a family,[3] as in Fig. 5.13, we find that each site has some preferences for one amino acid over another. With enough members in the family, we get a decent estimate of the probability that an amino acid will be chosen in each position along the sequence. Perhaps the simplest hypothesis about the ensemble of allowed sequences is that amino acids are chosen independently at every site, with these probabilities. It should be emphasized that such one-body constraints

---

3. Sequence alignment is far from a simple problem, either in principle or in practice. For a discussion, see the references for this section in the Annotated Bibliography.

(a) $\sigma_1, \cdots$ σ_{30}

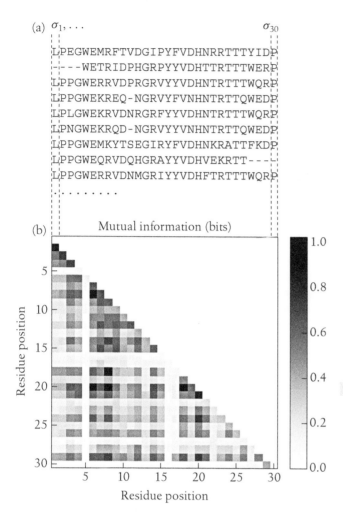

LPEGWEMRFTVDGIPYFVDHNRRTTTYIDP
---WETRIDPHGRPYYVDHTTRTTTWERP
LPPGWERRVDPRGRVYYVDHNTRTTTWQRP
LPPGWEKREQ-NGRVYFVNHNTRTTQWEDP
LPLGWEKRVDNRGRFYYVDHNTRTTTWQRP
LPNGWEKRQD-NGRVYYVNHNTRTTQWEDP
LPPGWEMKYTSEGIRYFVDHNKRATTFKDP
LPPGWEQRVDQHGRAYYVDHVEKRTT---L
LPPGWERRVDNMGRIYYVDHFTRTTTWQRP
·········

(b) Mutual information (bits)

FIGURE 5.13

Alignment of the WW domains. These small protein domains are named for their two highly conserved tryptophans (W). Redrawn from Mora and Bialek (2011). (a) Sequences in the family. (b) Correlations between amino acids at pairs of sites, measured by the mutual information (in gray; scale bar at right). The amino acids are indicated by the one-letter codes from Fig. 5.1, with − for gaps.

are strong, reducing the entropy of the allowed sequences from a nominal $\sim \log(20)$ per site down to $\sim \log(3)$ per site. But this reduction is not enough: if we synthesize proteins at random out of this distribution, it is almost impossible to find one that folds into something like the functional structure characteristic of the original family.

Given that one-body models do not work, the next logical step is to look at two-body effects—looking across the family of proteins, substitutions at one site tend to be correlated with substitutions at other sites. Can we sample an ensemble of sequences that captures these pairwise correlations? Let us imagine, for simplicity, that there are only two kinds of amino acid; the real case of twenty possibilities just needs more notation. Then we can use $\sigma_i = +1$ for one kind of amino acid at position i and $\sigma_i = -1$ for the other. The relative frequency of the two choices is measured by the "magnetization" $\langle \sigma_i \rangle_{\text{expt}}$, where the subscript reminds us that we measure this from data. Similarly, the correlations between amino acid substitutions at pairs of sites is measured by

$$C_{ij}^{\text{expt}} \equiv \langle \sigma_i \sigma_j \rangle_{\text{expt}} - \langle \sigma_i \rangle_{\text{expt}} \langle \sigma_j \rangle_{\text{expt}}. \tag{5.16}$$

Imagine creating an artificial family of M sequences $\{\sigma_i^\mu\}$, with $\mu = 1, 2, \cdots, M$. From this set of replica sequences we can compute the same expectation values that we computed from the real family of sequences:

$$\langle \sigma_i \rangle_{\text{model}} = \frac{1}{M} \sum_{\mu=1}^{M} \sigma_i^\mu, \tag{5.17}$$

$$C_{ij}^{\text{model}} = \frac{1}{M} \sum_{\mu=1}^{M} \sigma_i^\mu \sigma_j^\mu - \langle \sigma_i \rangle_{\text{model}} \langle \sigma_j \rangle_{\text{model}}. \tag{5.18}$$

We would like to arrange for the model family of sequences to have these quantities match the experimental ones. The first part ($\langle \sigma_i \rangle_{\text{model}} = \langle \sigma_i \rangle_{\text{expt}}$) is easy, because we can do this just by choosing the amino acids at every site independently with the same probabilities as in the experimental family. For the two-point correlations, we can form a measure of error between our model sequence ensemble and the real family,

$$\chi^2 = \sum_{ij} \left| C_{ij}^{\text{model}} - C_{ij}^{\text{expt}} \right|^2, \tag{5.19}$$

and then we can promote this mean-square error to an energy function, adjusting the M sequences according to a Monte Carlo simulation with slowly decreasing (effective) temperature. At low temperatures, this procedure should generate an ensemble of sequences that reproduce the pairwise correlations in the naturally occurring sequences. This procedure has been implemented for a real family of proteins, and novel sequences drawn out of the resulting ensemble have been synthesized. Remarkably, a finite fraction of these sequences fold into something close to the proper native structure, and these folded states are essentially as stable as are the natural proteins, as shown schematically in Fig. 5.14.

In the limit that we consider a very large family ($M \to \infty$) of artificial sequences, and as we take the effective temperature to zero, the Monte Carlo procedure draws samples out of a probability distribution that perfectly matches the measured one-point and two-point correlations but otherwise is as random or unstructured as possible. Statistical mechanics teaches us that entropy is a measure of randomness or disorder (which will be made more precise in Section 6.1), and so "as random as possible" should mean that the distribution of sequences has the maximum possible entropy consistent with the measured correlations. We will meet the maximum entropy idea again in Section 5.4, with more details in Appendix A.7. For now, note that the maximum entropy distribution of sequences takes the form

$$P(\{s_i\}) = \frac{1}{Z} \exp\left[\sum_{i=1}^{N} u_i(s_i) + \frac{1}{2} \sum_{i,j=1}^{N} \mathcal{V}_{ij}(s_i, s_j) \right], \tag{5.20}$$

where the fields u_i and the interactions $\mathcal{V}_{ij}$ must be chosen to reproduce the one-point and two-point correlations, where now we allow for the amino acid identity at each site to take on all 20 values, $s_i = 1, 2, \cdots, 20$. Actually finding these fields and interactions is the inverse of the usual problem in statistical mechanics and can be challenging. But if we can solve this problem, the maximum entropy method provides a potential answer to the question we posed at the outset—if random sequences do not fold, and the exact

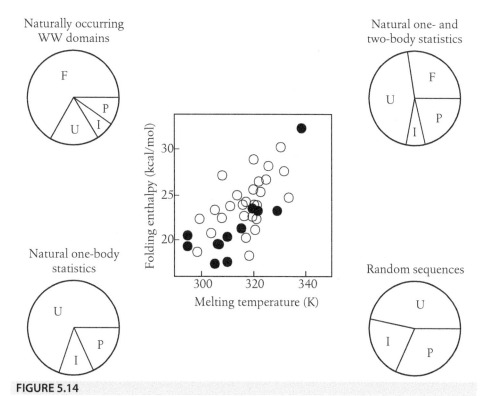

FIGURE 5.14

Artificial WW domains that respect pairwise sequence correlations approximate the properties of naturally occurring sequences. Pie charts show fractions of folded (F), unfolded (U), insoluble (I), and poorly expressing (P) sequences. Even some of the 42 natural domains fail to fold or express well when taken out of context, but none of the 19 random sequences or 43 sequences consistent with one-body statistics fold at all; the sequences consistent with both one- and two-body statistics fold about half as often as natural sequences. The center plot shows that those artificial sequences (filled symbols) that do fold do so with similar thermodynamics as the natural sequences (open symbols). Redrawn from Socolich et al. (2005).

sequence does not matter, how do we describe the ensemble of sequences consistent with a given protein structure or function? Equation (5.20) gives an explicit answer, a formula for the probability that a particular sequence will occur. Importantly, the form of the distribution is the same as the Boltzmann distribution, with the interactions and fields defining an effective energy surface in the space of sequences.

Problem 91: A small maximum entropy model. Consider a set of K Ising spins, $\sigma_1, \sigma_2, \ldots,$ σ_K, where each $\sigma_i = \pm 1$.

(a) For $K = 4$ suppose we observe that all spins have zero average magnetization, $\langle \sigma_i \rangle = 0$, and there are some pairwise correlations among all the pairs, $\langle \sigma_i \sigma_j \rangle = C_2$ for all $i \neq j$. The maximum entropy model consistent with these observations is

$$P(\{\sigma_i\}) = \frac{1}{Z(J)} \exp\left[\frac{J}{2} \sum_{i \neq j} \sigma_i \sigma_j \right]. \tag{5.21}$$

Find $Z(J)$ by summing over all the $2^K = 16$ states, and use this result to find the relationship between the measurable correlation C_2 and the value of J in the model. In addition to correlations among pairs, we can ask about the correlations among all four spins. It is conventional to subtract off what we might have expected from the pairs, defining

$$C_4 \equiv \langle \sigma_1 \sigma_2 \sigma_3 \sigma_4 \rangle - \left[\langle \sigma_1 \sigma_2 \rangle \langle \sigma_3 \sigma_4 \rangle + \langle \sigma_1 \sigma_3 \rangle \langle \sigma_2 \sigma_4 \rangle + \langle \sigma_1 \sigma_4 \rangle \langle \sigma_3 \sigma_2 \rangle \right]. \tag{5.22}$$

In physics we refer to these pairs as connected correlations, because of their relation to connected Feynman diagrams in field theory; statisticians might refer to such objects as cumulants (compared to moments). Show that your model in Eq. (5.21) makes a nontrivial prediction for C_4 as a function of C_2. The lesson here is that the least-structured model consistent with low-order correlations can predict higher order correlations.

(b) Let's take the Boltzmann form of our models seriously, so that, for example, Eq. (5.21) describes spins interacting in pairs, where the interaction energy is J, and $J > 0$ favors parallel spins. Suppose we have $K = 3$, and spins 2 and 3 do not interact with each other, so that the whole system is described by

$$P(\{\sigma_i\}) = \frac{1}{Z(J_{12}, J_{13})} \exp \left[J_{12} \sigma_1 \sigma_2 + J_{13} \sigma_1 \sigma_3 \right]. \tag{5.23}$$

Show that, despite the absence of direct interactions between spins 2 and 3, the correlation $\langle \sigma_2 \sigma_3 \rangle \neq 0$. How large can these correlations be?

One of the important lessons of statistical mechanics is that correlations can extend over much longer distances than the underlying interactions; indeed, near a critical point correlations extend over arbitrarily long distances, even in systems where the interactions are only among neighboring atoms or molecules. Thus, although we may detect significant correlations among the amino acid substitutions at many pairs of sites, it is possible that these correlations can be explained by Eq. (5.20) with the interactions $\mathcal{V}_{ij}$ being nonzero only for a very small fraction of pairs ij. Because the physical interactions between amino acids are short ranged, we expect that the joint choice of residues at sites i and j will have a direct effect on the probability that the sequence belongs to a given protein family only if the sites i and j are physically close to one another in the three-dimensional structure of the molecule. This idea was worked out in detail for pairs of receptors and associated signaling proteins in bacteria, and it was possible to identify, with high reliability, the amino acids that make up the region of contact between these molecules, as shown in Fig. 5.15. This success raises the tantalizing possibility that we could read off the physical contacts between amino acids—and hence infer the three-dimensional structure of proteins—from analysis of the covariations in amino acid substitutions across a large family.

The amino acid sequences of proteins are translations of the DNA sequences. But there are large parts of DNA that are not coding for proteins. Important parts of this "noncoding" DNA are involved in transcriptional regulation, as discussed in Section 4.3. The key steps of this regulatory process involve the binding of transcription factor proteins to DNA, and the architecture of the regulatory network depends on the specificity of these protein-DNA interactions. When we draw an arrow from one transcription factor to its target gene, then as schematized in Fig. 4.22, there must be a short sequence of DNA in or around the target gene to which the transcription factor can bind. That a given transcription factor activates or represses one gene but

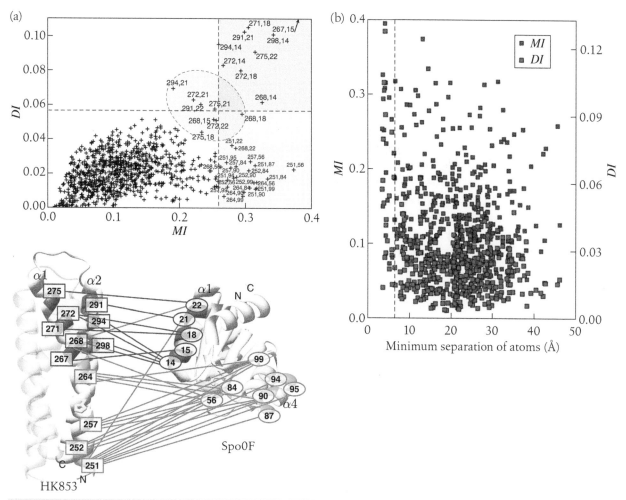

FIGURE 5.15

Interactions between residues in the ensemble of sequences predict spatial proximity. Redrawn from Weigt et al. (2009), with thanks to the authors. (a) The top plot is a scatter plot of the mutual information (MI) between amino acid substitutions at pairs of sites ij and the corresponding "direct information" (DI) between these sites, calculated by allowing for the interaction $\mathcal{V}_{ij}$ in Eq. (5.20) but eliminating the interactions with all other sites. Red and green shaded regions identify pairs of sites that have substantial direct interactions (red) or large mutual information without direct interactions (green), indicating that observed correlations are mediated through other sites; blue marks pairs on the border. The bottom maps these pairs to the known structure of this protein pair, showing that pairs with large DI indeed are in physical proximity, whereas those with large indirect MI are not. (b) A summary of the data in panel (a), showing how DI and MI correlate with the physical distance between the amino acids in the pair.

not another then is controlled by the presence or absence of the relevant sequences. But some transcription factors are quite promiscuous, and in higher organisms the relevant sequences often are quite short, so this specificity is not all-or-none. Rather we should think that every short sequence is a possible binding site, and a binding energy depends on the sequence.

A short piece of DNA sequence can be described by variables $\mathbf{s} \equiv \{s_i^\mu\}$, where $s_i^\mu = 1$ if the base at position i is of type μ; we have $\mu = 1, 2, 3, 4$ and $i = 1, 2, \cdots, L$, where L is the length of the possible binding site. Then if we look at one transcription factor, there is some binding energy of that factor to the DNA, $E(\mathbf{s})$, for every possible sequence. What does the function $E(\mathbf{s})$ look like? Obviously, if it is a constant, then there is no specificity at all—a given transcription factor will influence every gene in the genome—which can't be right. In contrast, if the binding is strong only for one specific sequence $\mathbf{s}_0$ (i.e., $E(\mathbf{s}_0) = -E_0$ with large $E_0 > 0$), whereas $E(\mathbf{s} \neq \mathbf{s}_0) \sim 0$, then the transcription factor can successfully target a small subset of genes. But then the landscape for evolutionary change becomes very rugged—changing a single base can completely eliminate one of the regulatory "arrows" in the network or create a new one of equal strength to all previous arrows—and this does not seem right either.

We can turn our question about the form of $E(\mathbf{s})$ around and ask about the set of sequences that will act as functional binding sites, presumably those sequences that have $E(\mathbf{s})$ in some range. In one limit, this ensemble would include all sequences; in the other limit, there would be just one sequence. Thus, the issue of specificity in protein-DNA interaction is the same problem we have been considering in the case of protein folding: where do real biological systems sit along the continuum between completely random sequences at one extreme and unique sequences at the other?

Many ideas for analyzing the nature of the sequence ensemble for binding sites involve the starting assumption that each base contributes linearly to the total binding energy, so that

$$E(\mathbf{s}) = \sum_{i=1}^{L} \sum_{\mu=1}^{4} W_{i\mu} s_i^{\mu}, \tag{5.24}$$

where $W_{i\mu}$ are the weights given to each position i. One of the first ideas was, in the language we have already used, a maximum entropy argument. If all we know is that functional binding sites must have some average binding energy $\langle E \rangle$, then the maximum entropy distribution consistent with this knowledge is

$$P(\mathbf{s}) = \frac{1}{Z} \exp\left[-\lambda E(\mathbf{s})\right], \tag{5.25}$$

which is the Boltzmann distribution at some effective temperature $\propto 1/\lambda$. Importantly, if the energy is additive, as in Eq. (5.24), then the probability of the entire sequence is a product of probabilities at the different sites:

$$P(\mathbf{s}) = \frac{1}{Z} \prod_{i=1}^{L} \exp\left[-\lambda \sum_{\mu=1}^{4} W_{i\mu} s_i^{\mu}\right]. \tag{5.26}$$

Thus, the expected frequency of occurrence of the different bases at each site—that is, the probability that $s_i^{\mu} = 1$—can be related directly to the weight matrix:

$$f_{i\mu} \propto \exp\left[-\lambda W_{i\mu}\right]. \tag{5.27}$$

Thus, if we could get a fair sampling of the ensemble of sequences we could just read off the matrix elements $W_{i\mu}$.

Problem 92: Random sequences.   Suppose that the genome is approximately random, so that at each site we have some probability of observing each base, but the actual choices are independent at every site. If the binding energy of a protein is given by Eq. (5.24), what can you say about the distribution of binding energies across all possible binding sites in the genome? If the range of the matrix elements $W_{i\mu}$ is bounded, can you say something about how the "best" binding site will compare with the bulk of the distribution? Even in a small bacterium, there are $\sim 10^6$ possible binding sites competing with the best binding site; are there plausible parameters such that the best site "wins," or will a large fraction of the proteins always be bound to the wrong sites? All of this is phrased with deliberate vagueness; part of the problem is for you to make the question more precise.

When these ideas first emerged in the 1980s, in work by Berg and von Hippel, there were few examples where one could point to multiple known binding sites for a single transcription factor. What was available to Berg and von Hippel were ~ 100 examples of the DNA sequences to which RNA polymerase binds when it begins transcribing. This is another example of protein-DNA interaction that is not a regulatory interaction but is an essential part of all gene expression.[4] Further, there had been in vitro kinetic measurements on transcription, so they knew something directly about the binding energies. If experiments are done in the regime where the binding sites are usually empty, then the observed transcription rates will be proportional to the concentration of polymerase and the equilibrium constant $K \propto \exp[-\beta E(\mathbf{s})]$. The comparison is shown in Fig. 5.16, including some estimates of errors in the measurements and predictions. The agreement is quite good. Thus, it really does seem that one can, at least roughly, estimate the energetics of binding events from the statistics of sequences, which is quite surprising.

The sequencing of whole genomes from many organisms created the opportunity for much more systematic exploration of sequence ensembles. That the number of transcription factors is very much smaller than the number of genes means that, generally, even in a single organism there must be many examples of binding sites for each transcription factor. Thus, it seems likely that similar sequences—sequences with good binding energies—will appear more frequently than would be expected at random, and these sequences should, in the simplest cases, be positioned near the start sites of transcription.

In written language, short sequences of letters that occur more frequently than expected by chance have a name—words. When we read, however, there are spaces and punctuation that mark the limits of the words, so we can recognize them. Interestingly, this is less true for spoken language, where the sounds of words often run together, and pauses or gaps are both less distinguishable and less reliable indicators of word boundaries. In fact, we really do not need these markers, even in the case of written text, as you can see by reading Fig. 5.17.

In the simplest view, words are independent, and all structure arises from the fact that not all combinations of letters form legal words. Then, if we know the boundaries

---

4. Even in this case the number of sequences is not very large, and we should remember that we are trying to estimate the frequencies of four different bases at each site. To improve their estimates, Berg and von Hippel (1987) used pseudocounts, a procedure explained in Eq. (A.462), Appendix A.8. To be completely faithful to the history, we should note that Berg and von Hippel never used the words "maximum entropy," and instead much discussion surrounds detailed hypotheses about the dynamics of sequence evolution that might lead to the Boltzmann distribution in Eq. (5.25).

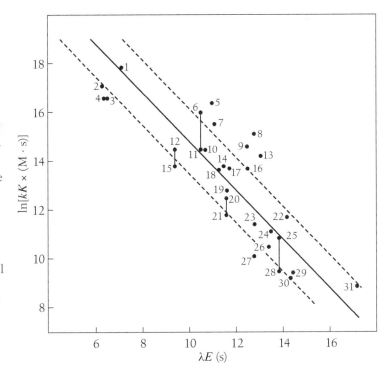

FIGURE 5.16
Sequence dependence of RNA polymerase activity compared with predictions from a maximum entropy model. The vertical axis measures effective second-order rate constants for the initiation of transcription by combination of RNA polymerase and different promoter sequences. The horizontal axis measures scaled binding energies predicted from a maximum entropy model based on ~100 sequences. Points refer to independent biochemical experiments, with lines connecting measurements on the same sequences, giving a sense of the error bars. A solid line with slope -1 is shown to guide the eye, with dashed lines indicating roughly the errors in the model arising from the finite sample size. Redrawn from Berg and von Hippel (1987).

FIGURE 5.17
A passage from Beckett's *Waiting for Godot*, spoken by Vladimir. All punctuation and spaces have been removed, but (hopefully) the text can still be understood.

theresmanalloverforyou
blamingonhisbootsthefa
ultsofhisfeethisisgetting
alarmingoneofthethieves
wassaveditsareasonable
percentagegogo

between words, the probability of observing a particular text becomes

$$P = \prod_w [P(w)]^{n_w}, \tag{5.28}$$

where n_w is the number of occurrences of the word w in the text, and $P(w)$ is the probability of this word. But we do not really know a priori the correct way of segmenting the text into words, and so we need to sum over all possible segmentations. Each segmentation S generates a different combination of words, so the count $n_w(S)$ depends on S. However, the probability that a word appears is a property of the language, not of our segmentation, and should be constant:

$$P = \sum_S \prod_w [P(w)]^{n_w(S)}. \tag{5.29}$$

If we think of this as a model for a long text, then given the vocabulary defined by the set of possible words $\{w\}$, maximizing the likelihood of the data amounts to setting the predicted probability of each word equal to the (normalized) mean number of occurrences of that word when averaged over all segmentations. Because the text is one dimensional, there are methods to sum over segmentations that are analogous to transfer matrix methods for one-dimensional models in statistical mechanics. The real challenge in looking at a genome is that we do not know the vocabulary.

One approach to learning the vocabulary is iterative: start with the assumption that words are single letters, then add two-letter words when the frequency of letter pairs is significantly higher than predicted by the model, and so on. To capture the functional behavior of real biological systems, one needs to include words with gaps, such as TTTCCNNNNNNGGAAA, in which "N" can be any nucleotide. Indeed, this example is one of the longer words that emerges from an analysis of possible regulatory regions of the yeast genome, and it corresponds to the binding site for MCM1, a protein involved in (among other things) control of the cell cycle. Globally, this approach to building a dictionary identifies hundreds words of more than four bases that pass reasonable tests of significance. At the time of the original work, $\sim$400 nonredundant binding sites were known whose function had been confirmed directly by experiment, and the dictionary reproduced one-quarter of these, a success rate 18 standard deviations outside what might have been expected by chance. One can do better by repeating the analysis using as input text only the regulatory regions of genes whose expression level is affected during particular processes or by the deletion or overexpression of other genes. Yet more power is added to the analysis by using the genomes of closely related organisms. Our ability to identify functional sequence elements simply from their statistics is prima facie evidence that, in the continuum from randomness to fine tuning, real biological systems are not at the random limit.

A very different approach to our problem involves exploring sequence space more systematically, as in the work described above for proteins. In a relatively short time, several different technologies have emerged for doing this, each with its own strengths and weaknesses. Much excitement was generated, in particular, by the development of genome-wide methods that allow, for example, probing the binding of a single transcription factor to all of the $\sim$6000 different promoter regions for genes in a single-celled organism, such as $E.$ $coli$ or yeast. In such "protein-binding microarrays," however, there seems to be no reliable calibration of the relation between the measurable fluorescence from labeled proteins and the actual binding probability. If we see a very bright spot, we can be sure that the protein is bound, but the actual distribution of fluorescence intensities has a long tail, as in Fig. 5.18. Where in this tail do we decide that we have a hit?

In the experiments shown in Fig. 5.18, fluorescence is a proxy for protein binding, and if the system comes to equilibrium, then the probability of protein binding depends on the DNA sequence through the binding energy $E(\mathbf{s})$. The space of sequences is huge, but the model of Eq. (5.24) says that the binding energy is a linear function of the sequence. Thus, fluorescence should depend on sequence only through a single linear projection. Finding this projection is an example of the dimensionality reduction problem that arose in thinking about neural computation, as discussed in the Annotated Bibliography for Section 4.4. The key idea is that, no matter how complicated or noisy the relationship that connects energy to binding to fluorescence, the

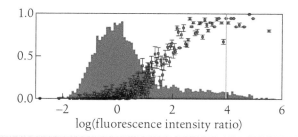

FIGURE 5.18

Protein-binding microarray data on the yeast transcription factor Abf1. In blue, a histogram of the fluorescence intensities across all ~6000 regulatory regions from the yeast genome, from Mukherjee et al. (2004). Flourescence intensity is measured relative to background, and the units of the histogram are arbitrary. The green line was drawn in the original experiments to define the threshold for binding. In red, with error bars, estimates of the probability that binding has occurred as a function of the fluorescence level, from the analysis described in the text. Redrawn from Kinney et al. (2007).

sequence can't provide more information about the output of the experiment than it does about the more fundamental quantity $E(\mathbf{s})$. Similarly, if we try to summarize the sequence by any reduced description, we will lose information unless our reduction corresponds to estimating $E(\mathbf{s})$ itself. Thus, if we search for a one-dimensional description—corresponding to a single linear projection of the sequence that preserves as much information[5] as possible about the experimental output—then the projection we find must be the best linear approximation to $E(\mathbf{s})$, up to a scale factor.[6]

Figure 5.19 shows examples of the weight matrices $W_{i\mu}$ obtained from the analysis of maximally informative dimensions in experiments on the yeast transcription factor Abf1, which is assumed to interact with a 20-base-long segment of the DNA. Individual matrix elements typically are determined with better than 10% accuracy, and the interaction of the protein with the DNA evidently is dominated by two approximately symmetric regions of five bases, separated by a gap of another five bases. Importantly, using this method, it is possible to analyze both in vitro (protein-binding microarray) and in vivo experiments, and get consistent answers. In contrast, if we just draw a conservative threshold on the signal strengths (e.g., the green line in Fig. 5.18), then these different sorts of experiments lead to divergent interpretations. Once we have confidence in the estimates of $E(\mathbf{s})$, we can go back and ask how the probability that the protein is bound is related to the fluorescence intensity; this relation is shown in Fig. 5.18. There is nothing about the analysis that forces this relationship to be smooth or monotonic, but it is.

Can we go further and relate these linear models of binding energy to the control of gene expression itself? Suppose that we put the expression of a fluorescent protein under the control of a known promoter and then randomly mutate the sequence. We can then generate an ensemble of bacteria with slightly different sequences, each of

---

5. "Information" here is used in the technical sense, in bits. See Section 6.1.

6. The actual computation is somewhat more involved because the possible regulatory regions are much larger than the binding sites, and so we have to test not all projections but all possible projections along the relevant ~500 base regions. For details, see Kinney et al. (2007).

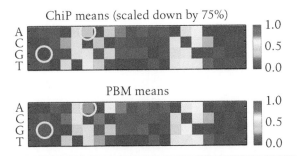

FIGURE 5.19

Weight matrices $W_{i\mu}$ for Abf1 in yeast, from analysis of chromatin immunoprecipitation (in vivo, top) and protein-binding microarray (in vitro, bottom) experiments. In these analyses the overall scale of $E(\mathbf{s})$ is not determined by the data, and so the two results have been scaled to maximize their similarity. Importantly, the two experiments are done in vivo and in vitro, respectively, but nonetheless generate very similar estimates of the underlying matrix governing protein-DNA interactions. Matrix elements with relatively poor agreement are circled in yellow, contrasting with a random element circled in gray; even the largest differences have little effect on the predicted binding energies.

which will express the fluorescent protein at different levels, presumably because the relevant transcription factor is binding more or less strongly. Experimentally, one can sort the cells by their fluorescence, sequence the promoter regions, and then search once more for a reduction of dimensionality that captures as much information as possible. If the mutations are sprinkled throughout the promoter region, we expect that there are at least two relevant dimensions, corresponding to the binding energy of the transcription factor and the binding energy of the RNA polymerase. The results of such an experiment and analysis are shown in Fig. 5.20.

As before, the search for maximally informative dimensions does not determine the scale of the energies. But if we take seriously that the quantities emerging from the analysis really are energies, then we can compute the probability that the RNA polymerase site is occupied, and it is this occupancy that presumably controls the initiation of transcription. If the energies for binding of the transcription factor (CRP) and RNA polymerase are ϵ_c and ϵ_r, respectively, then the probability of the polymerase site being occupied is

$$\tau = \frac{1}{Z} C_r e^{-\epsilon_r/k_B T} \left(1 + C_c e^{-\epsilon_c/k_B T} e^{-\epsilon_{\text{int}}/k_B T}\right), \tag{5.30}$$

where the partition function is

$$Z = 1 + C_c e^{-\epsilon_c/k_B T} + C_r e^{-\epsilon_r/k_B T}$$
$$+ C_r C_c e^{-\epsilon_r/k_B T} e^{-\epsilon_c/k_B T} e^{-\epsilon_{\text{int}}/k_B T}, \tag{5.31}$$

C_c and C_r are the concentrations of the transcription factor and the RNA polymerase, respectively, and ϵ_{int} is the interaction energy between the two proteins when they are both bound to the DNA. The two binding energies are quantities whose relation to the sequence should already have been determined by the search for maximally informative dimensions, except for the scale and zero of energy. To combine these energies, we need to set the scale ($k_B T$) and the zero (equivalently, the concentrations of the proteins),

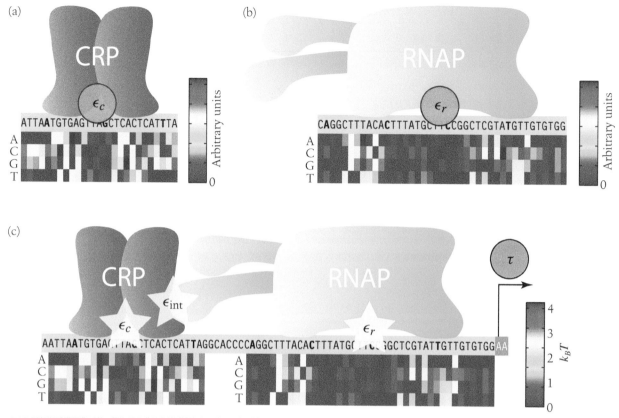

FIGURE 5.20

Analysis of experiments in which the expression of a fluorescent protein (measured by τ) is placed under the control of promoter sequences that are randomly mutated versions of the native sequence binding the transcription factor CRP. Redrawn from Kinney et al. (2010). (a,b) Separate analyses yield the weight matrices $W_{i\mu}$ for the CRP binding site (which determines the sequence dependence of the CRP binding energy ϵ_c) and for the RNA polymerase (RNAP) binding site (with binding energy ϵ_r), up to an arbitrary scale factor. (b) A combined analysis places these energies on an absolute scale and determines the interaction energy ϵ_{int}.

and we have to fit one more parameter: the interaction energy ϵ_{int}. The results of such an analysis are shown at the bottom of Fig. 5.20. For this particular system there are independent measurements of ϵ_{int}, and the agreement has $\sim 10\%$ accuracy. Even better, one can show that the single number τ in Eq. (5.30) captures as much information about the sequence dependence of the expression level as do the two numbers ϵ_c and ϵ_r. These results give us confidence that the use of statistical mechanics and linear energy models really does make sense here.

Problem 93: RNA polymerase occupancy.   Derive Eq. (5.30). Generalize to the case where there are two or more transcription factors, each of which can touch the RNA polymerase and contribute an interaction energy. Show that even if the binding of each transcription factor is independent (that is, there are no direct interactions among these factors), their mutual interactions with the RNA polymerase give rise to an effective cooperativity in the regulation of transcription. What is the relation of this picture to the MWC models of cooperativity?

Problem 94: Analysis of DNA sequences.   The file `seqs.txt` contains data from the experiments of Kinney et al. (2010); focus on the data labeled crp-wt. There are a large number of sequences for the promoter region of a gene, and each sequence is associated with a batch B0, B1, . . . , B9. B1 through B9 are those sequences that generate expression levels in some specified ranges, and B0 is a sample of all sequences used in the experiment. At each site along the sequence, you should have enough samples to estimate the joint distribution of the base at that site (four possibilities) and the batch number (nine possibilities). Using the ideas in Chapter 6, compute the amount of information that the base gives about the expression level (or its surrogate in this experiment, the batch number). Can you identify regions of the sequence that are particularly informative? Does the information ever go to zero, suggesting that some bases are completely irrelevant to the control of expression?

Now that we have some confidence in our description of the binding energies, we can go back and ask once more about the statistics of sequences and the problem of robustness versus fine tuning. The essential point, I think, is that although proteins bind to segments of DNA that are ~10 bases long, to first approximation all that matters is the binding energy, which is one number. Out of $\sim 4^{10}$ sequences, there are many ways of achieving the same value of the binding energy. The natural hypothesis is that evolution allows nearly random wandering among the isoenergetic sequences; comparing the genomes of closely related organisms gives results consistent with this hypothesis.

Our discussion has focused on the interaction of (mostly) single proteins with single DNA sequences. The real problem is in a larger context—there are more than 100 transcription factors, even in relatively small bacteria, and several thousands of functional binding sites. It is not enough that one protein bind to one site with a sensible energy; it must also be true that this protein does not bind to the wrong sites, nor that other proteins bind to its intended site. If we can take seriously the representation of a protein's sequence specificity by a linear model, then we can start to ask how different transcription factors distribute themselves in the space parameterized by the matrix elements $W_{i\mu}$, and whether this distribution provides maximum discriminability among the alternative regulatory signals. As explained in the references (see the Annotated Bibliography for this section), there are some interesting efforts in this direction, but much remains to be done.

5.2   Ion Channels and Neuronal Dynamics

The functional behavior of neurons involves the generation and processing of electrical signals, voltage differences, and current flows across the cell membrane. As noted in our discussion of the rod photoreceptor cell (Section 2.3), the membrane itself is insulating, and hence there would be no interesting electrical dynamics if not for the presence of specific conducting pores or channels. These pores are protein molecules that can change their structure in response to various signals, including the voltage across the membrane, which means that the system of channels interacting with the voltage constitutes a potentially complex, nonlinear dynamical system. We can also think of the ion channels in the cell membrane as a network of interacting protein molecules, with the interactions mediated through the transmembrane voltage. In contrast to many other such biochemical networks, we actually know the equations that describe the network dynamics, and as a result the questions of fine tuning versus robustness can be posed rather sharply.

When we move from thinking about individual neurons to thinking about circuits and networks of neurons, which really do the business of the brain, it is easy to imagine that the neurons are circuit elements with some fixed properties. We enhance this tendency by drawing circuit diagrams in which we keep track of whether neurons excite or inhibit one another, but nothing else about their dynamics is made explicit. Indeed, we will take this point of view in Section 5.4, just to keep things tractable. Despite our hopes for simplicity, our genome encodes $\sim 10^2$ different kinds of channels, each with its own kinetics. This range is expanded even further by the fact that many of these channels have multiple subunits, and cells can splice together the subunits in different combinations. On the one hand, this creates enormous flexibility and presumably adds to the computational power of the brain. On the other hand, this range of possibilities raises a problem of control. A typical neuron might have eight or nine different kinds of channels, and we will see that the dynamics of the cell depend rather sensitively on how many of each kind of channel is present. In keeping with the theme of this chapter, it might seem that cells need to tune their channel content very precisely, yet this needs to happen in a robust fashion.

To explore the trade-off between fine tuning and robustness in neurons, we need to understand the dynamics of the channels themselves.[7] For simplicity, let's neglect the spatial structure of the cell and assume we can talk about a single voltage difference V between the inside and the outside of the cell. Then because the membrane acts as a capacitor with capacitance C, we can write, quite generally,

$$C\frac{dV}{dt} = I_{\text{channels}} + I_{\text{ext}}, \tag{5.32}$$

where I_{ext} is any external current that is being injected (perhaps by us as experimenters) and I_{channels} is the current flowing through the channels. Each open channel acts more or less as an Ohmic conductance, and the structure of the channel endows it with specificity for particular ions. Because the cell works to keep the concentrations of ions different on the inside and outside of the cell, the thermodynamic driving force for the flow of current includes both the electrical voltage and a difference in chemical potential; it is conventional to summarize this by the reversal potential V_i for the currents flowing through channels of type i, which might involve a mix of ions. Because current only flows through open channels, we can write

$$I_{\text{channels}} = -\sum_i g_i N_i f_i \cdot (V - V_i), \tag{5.33}$$

where g_i is the conductance of one open channel of type i, N_i is the total number of these channels, f_i is the fraction that are open, and V_i is the reversal potential. If each

---

7. As with many of the topics discussed in this text, we could spend an entire semester on ion channels and not exhaust the subject. I admit that in some sections of the book I feel that I am providing a good guide to potentially complex matters, whereas in other sections I feel very strongly the weight of the things I am leaving out. As always, I encourage you to dig into the references in the Annotated Bibliography for this section.

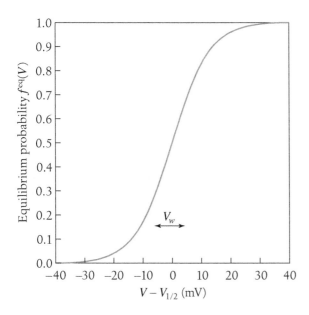

FIGURE 5.21

Activation curve for an ion channel, from Eq. (5.36), with $Q = 4$.

channel has just two states, open and closed, then the dynamics would be described by

$$\frac{df_i}{dt} = -\frac{1}{\tau_i(V)} \left[f_i - f_i^{eq}(V)\right] . \tag{5.34}$$

The equilibrium fraction of open channels as a function of voltage, $f_i^{eq}(V)$, often is called the activation curve, and $\tau_i(V)$ is the time constant for relaxation to this equilibrium.

What is a reasonable shape for the activation curve? We are describing a protein molecule that can exist in two states, and the equilibrium between these two states depends on voltage. This is possible only if the transition from closed to open rearranges the charges in the protein. In the simplest model, the opening of the channel effectively moves a charge Q across the membrane, and so the free-energy difference between open and closed states will be $\Delta F = F_0 - QeV$. Then the equilibrium probability of a channel being open will be given by

$$f^{eq}(V) = \frac{1}{1 + \exp\left[(F_0 - QeV)/k_B T\right]} \tag{5.35}$$

$$= \frac{1}{1 + \exp\left[-(V - V_{1/2})/V_w\right]}, \tag{5.36}$$

where the point of half maximal activation is $V_{1/2} = F_0/(Qe)$, and the width of the activation curve is $V_w = k_B T/Qe$, as shown in Fig. 5.21. The charge Q is referred to as the gating charge. At room temperature we have $k_B T/e = 25$ mV, so that even with relatively small values of Q we expect channels to make the transition from closed to open in a window of ~ 10 mV. The location of the midpoint $V_{1/2}$ depends on essentially all aspects of the protein structure in the open and closed states, so it is harder to develop intuition for this parameter.

In the absence of external inputs, all currents through the channels have to cancel. If only one kind of channel dominates, it will drive the voltage across the membrane to V_i, the reversal potential for that particular channel. Cells have pumps that maintain differences in the concentration of ions between the inside and outside of the cell, but these differences are not infinitely large. If, for example, there is a ratio of 100 between the internal and external concentrations of an ion with charge 1, then the reversal potential for the flow of this ion will be $\pm(k_BT/e)\ln(100) \sim 115\,\text{mV}$, where the sign depends on whether the higher concentration is inside or outside the cell. This estimate suggests that full dynamic range of voltage changes across the membrane will be limited to $\pm 100\,\text{mV}$, which is correct.

To get started on the dynamics, it is useful to identify the "resting potential" $V = V_0$, and study small perturbations around this steady state. The full dynamics are

$$C\frac{dV}{dt} = -\sum_i g_i N_i f_i \cdot (V - V_i) + I_{\text{ext}}, \tag{5.37}$$

$$\frac{df_i}{dt} = -\frac{1}{\tau_i(V)}\left[f_i - f_i^{\text{eq}}(V)\right], \tag{5.38}$$

and the linearization is

$$C\frac{d\delta V}{dt} = -\sum_i g_i N_i f_i^{\text{eq}}(V)\delta V - \sum_i g_i N_i(V_0 - V_i)\delta f_i + I_{\text{ext}}, \tag{5.39}$$

$$\frac{d\delta f_i}{dt} = -\frac{1}{\tau_i(V_0)}\left[\delta f_i - \frac{df_i^{\text{eq}}(V)}{dV}\bigg|_{V=V_0}\delta V\right]. \tag{5.40}$$

Fourier transforming, we can solve for the channel dynamics, then substitute and collect terms to find

$$\left[-i\omega C + \frac{1}{R_0} + \sum_i \frac{g_i N_i(V_0 - V_i)[df_i^{\text{eq}}(V)/dV]_0}{1 - i\omega\tau_i(V_0)}\right]\delta\tilde{V}(\omega) = \tilde{I}_{\text{ext}}(\omega). \tag{5.41}$$

The resting resistance of the membrane is defined by

$$\frac{1}{R_0} = \sum_i g_i N_i f_i^{\text{eq}}(V). \tag{5.42}$$

The term in brackets in Eq. (5.41) is the inverse impedance (or admittance) of the system.

Problem 95: Details of membrane impedance.   Fill in the steps leading to Eq. (5.41).

To understand what is going on here, it is helpful to think about channels with fast $(1/\tau_i \gg \omega)$ or slow $(1/\tau_i \ll \omega)$ responses. The fast channels renormalize the resistance:

$$\frac{1}{R_0} \to \frac{1}{R_0} + \sum_{i\in\text{fast}} g_i N_i(V_0 - V_i)\frac{df_i^{\text{eq}}(V)}{dV}\bigg|_{V=V_0}. \tag{5.43}$$

Importantly, the correction to the resistance can be either positive or negative. Suppose that, as in Fig. 5.21, the channels tend to open in response to increasing voltage, as most channels do. Then $[df_i^{eq}(V)/dV]_0 > 0$. But if this channel is specific for an ion with a reversal potential above the resting potential ($V_i > V_0$), then opening the channel creates a stronger tendency to pull the voltage toward this higher potential, which is a regenerative effect—a negative resistance. The power supply for this negative resistor is provided by the pumps that maintain the reversal potentials.

If the channels are slow, they make a contribution to the imaginary part of the admittance, along with the capacitance,

$$-i\omega C \rightarrow -i\omega C + \frac{1}{-i\omega} \sum_{i \in \text{slow}} \frac{g_i N_i}{\tau_i(V_0)} (V_0 - V_i) \frac{df_i^{eq}(V)}{dV}\bigg|_{V=V_0}. \tag{5.44}$$

Again the sign depends on details. If the channels are opened by increasing voltage and the reversal potential is below the resting potential, then their contribution is (almost) like an inductance and can generate a resonance by competing with the capacitance. This resonance is at a frequency

$$\omega_* = \left[\frac{1}{C} \sum_{i \in \text{slow}} \frac{g_i N_i}{\tau_i(V_0)} (V_0 - V_i) \frac{df_i^{eq}(V)}{dV}\bigg|_{V=V_0} \right]^{1/2}. \tag{5.45}$$

Problem 96: Equivalent circuits. Equation (5.41) shows that each type of channel contributes a parallel path for current flow through the membrane. The impedance of this path is defined by

$$\frac{1}{\tilde{Z}_i(\omega)} = g_i N_i f_i^{eq}(V) + \frac{g_i N_i (V_0 - V_i)[df_i^{eq}(V)/dV]_0}{1 - i\omega\tau_i(V_0)}. \tag{5.46}$$

Without resorting to the fast/slow approximations above, draw an equivalent circuit using the standard lumped elements (capacitance, resistance, inductance) that realizes this impedance. Show how the parameters of the lumped elements relate to the parameters of the channels.

So, we have seen that even in response to small signals, the dynamics of ion channels generate an interesting complement of electronic parts: resistors, inductors, and negative resistors. Certainly these elements together can make a filter, playing the effective inductance of the channels against the intrinsic capacitance of the membrane, as noted above. The negative resistor can sharpen the resonance and even generate an instability; on the other side of the instability is a genuine oscillator.

Problem 97: Oscillations. Construct a minimal model for ion channels in the cell membrane that supports a stable limit-cycle oscillation of the voltage.

The negative resistance alone means that we can have (without oscillations) an instability of the steady state around which we were expanding, presumably because

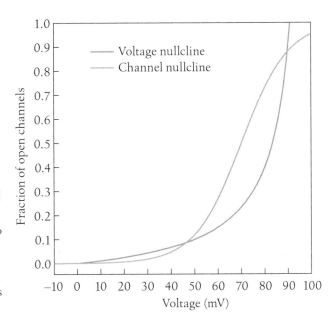

FIGURE 5.22

Bistability in a simple model of a neuron. The channel nullcline is Eq. (5.49), and the voltage nullcline is Eq. (5.50). To be explicit, we choose $f_{eq}(V)$ from Eq. (5.36), with $V_{1/2} = 70$ and $V_w = 10$, and $G_{leak}/gN = 0.1$. Note that there are three crossing points, corresponding to steady states. The low- and high-voltage states are stable; the intermediate-voltage state is unstable.

the real system is multistable. To see this behavior more clearly, consider just two types of channels—a leak channel, which has a total conductance G_{leak} independent of the voltage and has a reversal potential of zero, and some other channel, which opens in response to increasing voltage and has a reversal potential V_r. Then the dynamics are

$$C\frac{dV}{dt} = -G_{leak}V - gNf \cdot (V - V_r), \tag{5.47}$$

$$\frac{df}{dt} = -\frac{1}{\tau(V)}[f - f_{eq}(V)]. \tag{5.48}$$

The steady state solutions are determined by solving two simultaneous equations, usually called the nullclines, obtained by setting the time derivatives equal to zero:

$$f = f_{eq}(V), \tag{5.49}$$

$$V = V_r\frac{f}{f + G_{leak}/gN}; \tag{5.50}$$

these nullclines are shown schematically in Fig. 5.22 for some reasonable choice of parameters. We can see that there are three solutions to the two simultaneous equations; two are stable and one is unstable. The two stable states correspond, roughly, to one state in which all channels are closed and the voltage is zero (the reversal potential of the leak), and one state in which all channels are open and the voltage is near the reversal potential for these channels. The bistability means that, if the cell starts in the low-voltage state, injection of a relatively small, brief current can drive the system across a threshold (separatrix), so that it falls into the high-voltage state after the current pulse is complete. This behavior constitutes a form of memory (interesting, although not very realistic), but it also substantially amplifies the incoming signal, especially if the parameters are tuned so that the difference in voltage to the unstable state is small.

Problem 98: Bistability. Work through a concrete example of the ideas in the previous paragraphs, perhaps using the detailed model from Fig. 5.22. Verify analytically the claims about stability of the three different steady states. Explain how these analytic criteria can be converted into a test for stability of each steady state that can be read off directly from the plots in Fig. 5.22. Analyze the response to brief pulses of current, showing that there is a well-defined threshold for switching from one stable state to the other.

All the different kinds of dynamics we have seen thus far—filtering, oscillation, and bistability—can be generated by just one kind of channel with only two states. Real neurons are more complex. One important class of dynamics that we can't quite see in the simplest models is excitability. In this case, a small pulse again drives the system across a threshold, but what would have been a second stable state is destabilized by relaxation of some other degrees of freedom. The result is that the system takes a long, and often stereotyped, trajectory through its phase space before coming back to its original steady state after the input pulse is over. The action potential is an example of such excitable dynamics.

Our understanding of ion channels goes back to the classic work of Hodgkin and Huxley in the 1940s and 1950s. They studied the giant axon, a single cell, visible to the naked eye, which runs along the length of a squid's body and along which action potentials are propagated to trigger the squid's escape reflex. Passing a conducting wire through the interior of the long axon, they short-circuited the propagation, ensuring that the voltage across the membrane was spatially uniform, as in our idealization above.[8] They then studied the current that flowed in response to steps of voltage. If the picture of channels is correct, then with the voltage held constant, there should be an (Ohmic) flow of current through the open channels. If we step suddenly to a new value of the voltage, Ohm's law states that the current through the open channels will change immediately, but there will be a prolonged time dependence that results from the open or closing of channels as they equilibrate at the new voltage. In the simple model with two states, this changing current should relax exponentially to a new steady state; in particular, the initial slope of the current should be finite.

Hodgkin and Huxley found that the relaxation of the current at constant voltage has a gradual start, as if the channels had not one closed state but several, and the molecules had to go through these states in sequence before opening. They chose to describe these dynamics of the currents by imagining that, for the channel to be open, there were several independent molecular gates that all had to be open. Each gate could have only two states and would obey simple first-order kinetics, but the probability that the channel is open would be the product of the probabilities that the gates were open. In the simple case that the multiple gates are identical, the probability of the channel being open is just a power of the gating variable describing the probability that one gate is open. Hodgkin and Huxley also discovered that at least one important class of channels opens in response to increased voltage and then closes over time. They

---

8. There is a video of Hodgkin himself (along with a colleague) recreating some of these experiments in the 1970s: http://youtu.be/k48jXzFGMc8. You can also find the great anatomist J. Z. Young dissecting a squid, and taking us along the rather astonishing path to realizing that there are single nerve axons more than 1 mm in diameter: http://youtu.be/pw6_Si5jOpo.

described this mechanism by postulating that in addition to activation gates that were opened by increasing voltage, there were inactivation gates that closed in response to increasing voltage but had slower kinetics. Putting the pieces together, they described the fraction of open channels as

$$f_i = m_i^{\alpha_i} h_i^{\beta_i}, \tag{5.51}$$

where m and h are activation and inactivation gates, respectively, and the powers α and β count the number of these gates that contribute to the opening of one channel. The kinetics are then described by

$$\frac{dm_i}{dt} = -\frac{1}{\tau_i^{(m)}(V)} \left[m_i - m_i^{eq}(V) \right], \tag{5.52}$$

$$\frac{dh_i}{dt} = -\frac{1}{\tau_i^{(h)}(V)} \left[h_i - h_i^{eq}(V) \right], \tag{5.53}$$

and finally the voltage (again neglecting spatial variations) obeys

$$C\frac{dV}{dt} = -\sum_i g_i N_i m_i^{\alpha_i} h_i^{\beta_i} \cdot (V - V_i). \tag{5.54}$$

Problem 99: Two gates.   Suppose that each channel has two independent structural elements (gates), each of which has two states. Assuming that the two gates are independent of each other, fill in the steps showing that the dynamics of the channels are as described above. In particular, show that after a sudden change in voltage, the fraction of open channels starts to change as $\propto t^2$, not $\propto t$ as expected if the entire channel only has two states.

Problem 100: Hodgkin and Huxley revisited.   The original equations written by Hodgkin and Huxley are:[9]

$$C\frac{dV}{dt} = -\bar{g}_L(V - V_L) - \bar{g}_{Na}m^3h(V - V_{Na}) - \bar{g}_K n^4(V - V_K) + I(t), \tag{5.55}$$

$$\frac{dn}{dt} = \frac{0.01(-V + 10)}{e^{(-V+10)/10} - 1}(1 - n) - 0.125e^{-V/80}n, \tag{5.56}$$

$$\frac{dm}{dt} = \frac{0.1(-V + 25)}{e^{(-V+25)/10} - 1}(1 - m) - 4e^{-V/18}m, \tag{5.57}$$

$$\frac{dh}{dt} = 0.07e^{-V/20}(1 - h) - \frac{1}{e^{(-V+30)/10} + 1}h, \tag{5.58}$$

where Na and K refer to sodium and potassium channels, respectively; time is measured in milliseconds and V is measured in millivolts. These equations are intended to describe a small

---

9. The only difference from the original paper is that we use the modern sign convention for the voltage. Notice that this original formulation is in terms of a "maximal conductance" for each type of "current," whereas in modern language we could talk about the number of each type of channel. In fact, the more phenomenological description persists, because it corresponds more directly to what is measured, but it allows us to forget that such parameters as $\bar{g}_K$ actually measure the number of copies of a protein that have been inserted into the membrane.

patch of the membrane, and so many parameters are given per unit area: $C = 1\,\mu\text{F/cm}^2$, $\bar{g}_L = 0.3\,\text{mS/cm}^2$, $\bar{g}_{\text{Na}} = 120\,\text{mS/cm}^2$, and $\bar{g}_K = 36\,\text{mS/cm}^2$; the reversal potentials are $V_L = 10.613\,\text{mV}$, $V_{\text{Na}} = 115\,\text{mV}$, and $V_K = -12\,\text{mV}$, all measured from the resting potential.

(a) Rewrite these equations in terms of equilibrium values and relaxation times for the gating variables, for example,

$$\frac{dm}{dt} = -\frac{1}{\tau_m(V)}\left[m - m_{\text{eq}}(V)\right]. \tag{5.59}$$

Plot these quantities. Can you explain, intuitively, the form of the curves?

(b) Simulate the dynamics of the Hodgkin-Huxley equations in response to constant current inputs. Show that there is a threshold current above which the system generates periodic pulses. Explore the frequency of the pulses as a function of current.

(c) Suppose that the injected current consists of a mean (less than the threshold you identified in part (b)) plus a small component at frequency ω. By some appropriate combination of analytic and numerical methods, find the impedance $Z(\omega)$ for different values of the mean injected current. Show that the membrane has a resonance, and explore what happens to this resonance as the mean current is increased toward threshold. How do your results connect to the frequency of pulses above threshold?

(d) Real axons are essentially long thin cylinders. Show that, if we allow the voltage to vary along the length of the axon, a current per unit area should flow across the membrane of

$$I = \frac{a}{2R}\frac{\partial^2 V}{\partial z^2}, \tag{5.60}$$

where z is the coordinate along the cylinder, a is its radius, and R is the resistivity of the fluid filling the axon, assuming that resistance outside the axon is negligible. For the squid giant axon, $a \sim 250\,\mu\text{m}$ and $R \sim 35\,\text{ohm}\cdot\text{cm}$. Use this result to write equations for the voltage and gating variables along the axon. Note that only the equation for the voltage has spatial derivatives. Why?

(e) Simulate the response of a long segment of the axon to a current pulse injected at one end. Show that small pulses result in spatially restricted voltage responses, whereas larger inputs produce a propagating pulse. Confirm that these pulses become more stereotyped as they propagate and have a velocity that is independent of the input current. What is this velocity? How does it compare to the observed speed of action potentials, $v \sim 20\,\text{m/s}$?

Problem 101: Simplification.   It is difficult to make analytic progress in understanding the dynamics of a system with five variables. There is a history of trying to approximate the Hodgkin-Huxley model by exploiting the fact that the different variables have very different time scales. See how far you can go along this path. I have left this problem deliberately open ended. For some approaches, see the references in the Annotated Bibliography for this section.

It is good to pause here and review how we know that the Hodgkin-Huxley description of ion channels is correct. The initial triumph, which you are asked to reproduce in problem 100, is the prediction of the propagating action potential itself, as in Fig. 5.23, with the correct speed. The model also predicts that, as the action potential passes, there is a net flux of potassium and sodium across the membrane. On long time scales,

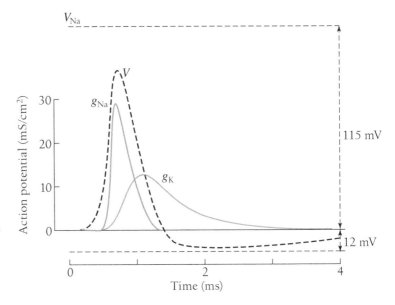

FIGURE 5.23

The action potential that emerges from the Hodgkin-Huxley model. The voltage trace is shown by the heavy dashed line, and the time dependence of the underlying conductances for sodium and potassium are shown by solid lines. Note that the rise of the action potential is associated with the opening of sodium channels, whereas the fall is a combination of these channels closing and the opening of potassium channels. The dynamic range of the electrical signal is bounded by the reversal potentials for these two ions (V_{Na} and V_K), as shown.

this flux must be balanced by the action of pumps that maintain the concentration differences between the inside and outside of the cell. But either by looking quickly or by poisoning the pumps, one should be able to detect the flux (e.g., using radioactive tracers); this works, quantitatively.

Nature provides a variety of toxins that block the action potential in different ways,[10] and we can also use artificial blockers, for example, by using ions with very large radius that can literally plug the hole in open channels. It is striking that these agents act selectively on different channels, and one can verify that this way of isolating the dynamics of sodium and potassium channels matches the Hodgkin-Huxley description. If we can arrange for the channels to open but be blocked, then the structural change of the channel molecule upon opening should still move the gating charge across the membrane. In addition, if we are careful, this movement should be measurable essentially as a delayed capacitive response to changes in the applied voltage. These gating currents have indeed been detected, and the magnitude of the gating current matches quantitatively what is predicted from the voltage dependence of the activation curve. In some cases this agreement can be extended to genetically engineered channels, where one can show that changes in the activation curve and gating currents track one another.

If individual channels are independent of one another, then their opening and closing events should be independent. If we look at a small patch of the membrane, there will not be that many channels present, and we might be able to see that the discrete events in the individual molecules do not quite average out—there should be noise from the random opening and closing of the single channels. This channel noise has been detected and has the spectral properties expected from the Hodgkin-Huxley model. Finally, if we look at even smaller patches of the membrane, and have

---

10. The most famous of these toxins might be tetrodotoxin, produced by the puffer fish. This molecule blocks the sodium channel and hence eliminates action potentials. It is worth remembering that these toxins serve a positive function for the organisms that produce them.

proportionately more sensitive amplifiers, we should be able to see the opening and closing of single channels. Again, this has been done. Most importantly, we can look at the distribution of times that individual channels spend in the open and closed states and connect this distribution to the kinetics predicted by the Hodgkin-Huxley model and its generalizations. Although these more-detailed measurements have revealed new features of channel kinetics even in well-studied examples, in outline the picture given to us by Hodgkin and Huxley has stood the test of time.

Now that we have confidence in our mathematical description of neurons, it is time to appreciate just how many parameters are involved. A typical cell expresses eight or nine different kinds of channels. Each channel is described by the dynamics of two gating variables. If activation and inactivation curves have a simple sigmoidal form as in Fig. 5.22, then there are two parameters for each such curve—the voltage at half activation and the slope or width—and at least one more parameter to set the time scale of the kinetics. Finally, there is a parameter to count the total number of channels, or equivalently the maximal conductance achieved if all channels are open. All together, then, this is ∼7 parameters per channel type, or roughly 50 parameters for the entire neuron, conservatively. Importantly, to a large extent the cell actually has control over these parameters and, in a meaningful sense, can adjust them almost continuously.

How do these adjustments occur? Most obviously, the total number of open channels is controlled in the same way that all other protein copy numbers are controlled. Sometimes, because of a clear connection to experiment, one speaks of the maximal conductance associated with a particular type of channel ($G_i^{\max} = g_i N_i$), but this terminology obscures the fact that this parameter really is the total number of copies of the protein that the cell has expressed and inserted into the membrane. The parameters of the activation curves and the time constants are intrinsic properties of the proteins, but they too can be adjusted in several ways. Like all proteins, ion channels can be covalently modified by phosphorylation or other actions. More importantly, the genome encodes a huge number of different ion-channel proteins; the human genome has 90 different potassium channels alone. Although they do form classes based on their dynamics, considerable variation exists within classes, and because many of these genes have multiple alternative splicings, there is the potential for almost continuous parameter variation. These different mechanisms of variation interact; as an example, different splicing variants can exhibit different sensitivities to phosphorylation.

Problem 102: Continuous adjustment of electrical dynamics.   To illustrate the possibility of nearly continuous adjustments in the electrical dynamics of neurons, consider the case of the hair cells in the turtle ear. In these cells (see Section 4.5), one contribution to frequency selectivity comes from a resonance in the electrical response of the hair cell itself. This resonance is driven by a combination of voltage-gated calcium channels and calcium-activated potassium channels. A detailed model of this system is described by Wu and Fettiplace (2001). Try to understand what they have done, and reproduce the essential theoretical results. In particular, what is the role of "details" (e.g., the building of channels out of combinations of different subunits) in generating the correct qualitative behavior?

One well-studied example of channel dynamics is in the stomatogastric ganglion of crabs and lobsters, schematized in Fig. 5.24. This network of ∼30 neurons generates a

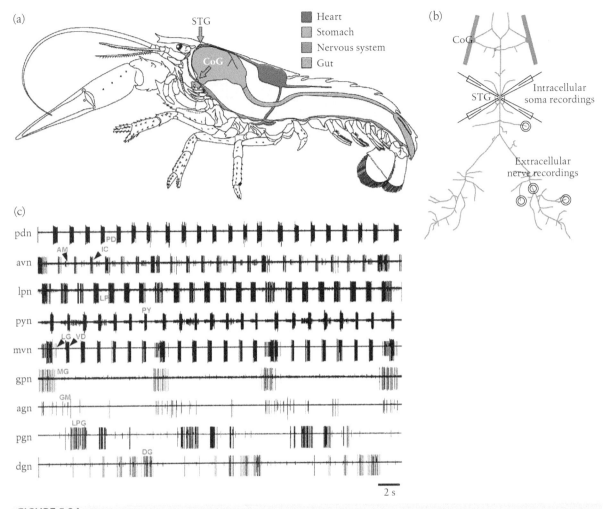

FIGURE 5.24

The stomatogastric ganglion (STG) in crustaceans. Reprinted with permission of Annual Reviews, Inc., from Marder and Bucher (2007); permission conveyed through Copyright Clearance Center, Inc. (a) The location of the STG and the commissural ganglion (CoG) in a lobster. (b) Schematic of the ganglion dissected out of the animal, and the opportunities for recording the activity of the neurons. (c) Simultaneous extracellular recordings from nine motor nerves at the output of this network. Names indicate particular neurons that can be identified in each individual (as with the named neurons in the fly visual system discussed in Section 4.4), and in some cases (e.g., avn, mvn) we can identify spikes from several individual neurons in the recording from one nerve. There are two main rhythms, the faster pyloric rhythm in cells PD, LP, PY, VD and IC, and the slower gastric-mill rhythm in cells MG, DG, GM, LPG, and LG.

rhythm, and this rhythm in turn drives muscles that actuate teeth in the crab stomach, grinding its food. Evidently getting the correct rhythm is important in the life of the organism. Records of the electrical signals from individual neurons show that several of the cells produce periodic bursts of action potentials, and a handful of cells are pacemakers that can generate this periodic pattern without input from the other cells. In one such cell (the lateral pyloric neuron), experiments show that there are seven different channel types. An important feature of this cell, shared by many other cells, is the presence of voltage-gated calcium channels. As action potentials occur, they trigger calcium flux into the cell. Because some channels are also directly affected by the calcium concentration, a complete model must include a description of the calcium buffering or pumping that counterbalances this flux.

It is worth being explicit about all these ingredients in the dynamics of the lateral pyloric neuron, not least to get a sense for the state of the art in such analyses. This will, however, take us to a level of detail that I have largely tried to avoid until this point in the text. It is essential, however, because this level of detail is where the problems become apparent. Once we have identified the problem, we can zoom back out to a more schematic view.

As before, we neglect the spatial structure of the cell, so there is just one relevant voltage difference V between the inside and outside of the cell, which obeys Eq. (5.54),

$$C\frac{dV}{dt} = -\sum_i g_i N_i m_i^{\alpha_i} h_i^{\beta_i}(V - E_i) + I_{\text{ext}}, \tag{5.61}$$

where I_{ext} is any externally injected current, and E_i is the reversal potential for channel type i. The kinetics of the gating variables m_i and h_i are governed by Eq. (5.52) and Eq. (5.53), respectively. For most channels, we can take the equilibrium values of the gating variables to be given by the generalization of Eq. (5.36),

$$m_i^{\text{eq}}(V) = \frac{1}{1 + \exp[-(V - V_{1/2}^{m_i})/V_w^{m_i}]}, \tag{5.62}$$

$$h_i^{\text{eq}}(V) = \frac{1}{1 + \exp[-(V - V_{1/2}^{h_i})/V_w^{h_i}]}, \tag{5.63}$$

and the time constants for relaxation of the gating variables are phenomenologically,

$$\frac{1}{\tau_i^{(m)}(V)} = \frac{k_i^{(m)}}{1 + \exp[-\gamma_i^{(m)}(V - V_i^{(m)})]}, \tag{5.64}$$

$$\frac{1}{\tau_i^{(h)}(V)} = \frac{k_i^{(h)}}{1 + \exp[-\gamma_i^{(h)}(V - V_i^{(h)})]}. \tag{5.65}$$

As shown in Table 5.1, this description works for several channel types, one selective for potassium (the delayed rectifier), two for calcium, and one mixed (the inward rectifier), plus a leak that exhibits no significant time or voltage dependence of its conductance.

Two of the important channel types allow calcium to flow into the cell. As we will see, this current is big enough to change the concentration of calcium inside the cell, which has a variety of effects on other processes, including one of the channels that does not fit the simple description we have given so far. So, we need to describe the dynamics of the calcium concentration itself. The simplest model is that the calcium relaxes back to some internally determined steady state, $[\text{Ca}]_0 = 0.05 \, \mu\text{M}$, with a rate $k_{\text{Ca}} = 360 \, \text{s}^{-1}$, and the current through the open calcium channels is driving an increase in the intracellular calcium concentration. In this case,

$$\frac{d[\text{Ca}]}{dt} = -k_{\text{Ca}} \left([\text{Ca}] - [\text{Ca}]_0\right) + AI_{\text{Ca}}, \tag{5.66}$$

TABLE 5.1

Subset of channels in the lateral pyloric neuron

Channel type	$g_i N_i$ (μS)	E_i (mV)	Midpoint (mV)	Width (mV)	Rate (s^{-1})
i = 1: delayed rectifier	0.35	$E_K = -80$			
Activation equilibrium ($\alpha_1 = 4$)			$V_{1/2}^{m_1} = -25$	$V_w^{m_1} = 17$	
Activation kinetics			$V_1^{(m)} = 10$	$1/\gamma_1^{(m)} = 22$	$k_1^{(m)} = 180$
i = 2: Ca^{++} current 1	0.21	E_{Ca}			
Activation ($\alpha_2 = 1$)			$V_{1/2}^{m_2} = -11$	$V_w^{m_2} = 7$	50
Inactivation ($\beta_2 = 1$)			$V_{1/2}^{h_2} = -50$	$V_w^{h_2} = -8$	16
i = 3: Ca^{++} current 2	0.047	E_{Ca}			
Activation ($\alpha_3 = 1$)			$V_{1/2}^{m_3} = -22$	$V_w^{m_3} = 7$	10
i = 4: inward rectifier	0.037	-10			
Activation equilibrium ($\alpha_4 = 1$)			$V_{1/2}^{m_4} = -70$	$V_w^{m_4} = -7$	
Activation kinetics			$V_4^{(m)} = -110$	$1/\gamma_1^{(m)} = 13$	$k_1^{(m)} = 0.33$
i = 5: leak	0.1	-50			
i = 6: A-current	2.2	$E_K = -80$			
Activation equilibrium ($\alpha_6 = 3$)			$V_{1/2}^{m_6} = -12$	$V_w^{m_6} = 26$	
Activation kinetics					$k_6^{(m)} = 140$
Inactivation equilibrium ($\beta_{6a} = 1$)			$V_{1/2}^{h_{6a}} = -62$	$V_w^{h_{6a}} = 6$	
Inactivation kinetics					$k_{6a}^{(h)} = 50$
Inactivation equilibrium ($\beta_{6b} = 1$)			$V_{1/2}^{h_{6b}} = -40$	$V_w^{h_{6b}} = -12$	
Inactivation kinetics					$k_{6b}^{(h)} = 3.6$

For the delayed rectifier and the second type of calcium channel, there is no evidence for inactivation. The negative value of $V_w^{(h_2)}$ means, from Eq. (5.36), that the probability of the inactivation gate being open decreases with increasing voltage. For calcium channels, the reversal potential varies, depending on the calcium concentration inside the cell, as in Eq. (5.67), and the relaxation times do not have a detectable voltage dependence. The voltage dependence of the inward rectifier kinetics is opposite to Eq. (5.64), that is, $1/\tau \propto 1 + \exp[-\gamma_i^{(m)}(V - V_i^{(m)})]$. The leak current, by convention, is the current that exhibits no voltage or time dependence of its conductance. From Buchholtz et al. (1992).

where I_{Ca} is the total calcium current ($I_{Ca} = I_2 + I_3$ from Table 5.1). The constant $A = 300 \ \mu$M/nC is inversely proportional to the volume into which the current flows, which experimentally comes out to be much smaller than the total volume of the cell body. As the concentration of calcium changes, the reversal potential for the calcium currents also changes,

$$E_{Ca} = \frac{k_B T}{2e} \ln\left(\frac{[Ca]_{out}}{[Ca]}\right), \tag{5.67}$$

where the calcium concentration outside the cell is $[Ca]_{out} = 13$ mM.

We are still missing three channel types in this cell. First, there is another potassium channel that is almost described by our standard model, but the inactivation seems to involve two processes that occur on different time scales. This behavior can be captured by replacing

$$h_6 \to x(V)h_{6a} + [1 - x(V)]h_{6b},\tag{5.68}$$

where the weighting function is

$$x(V) = \frac{1}{1 + \exp[-(V - 7)/15]},\tag{5.69}$$

with V measured in mV, as before.

Second, there is a fast sodium channel not unlike the ones that Hodgkin and Huxley found in the squid giant axon, with $\alpha_7 = 3$ and $\beta_7 = 1$. The activation is sufficiently fast that it can be approximated as instantaneous, so that m_7 is always at its equilibrium value, which varies with voltage in a slightly more complicated way than for the other channels,

$$m_7 = m_7^{eq}(V) = \frac{1}{1 + 136 \exp[-(V + 34)/13] \left(1 - \exp[-(V + 6)/20]\right)/(V + 6)},\tag{5.70}$$

where V again is measured in mV. The inactivation gates obey

$$\frac{dh_7}{dt} = a_7(V)(1 - h_7) - b_7(V)h_7,\tag{5.71}$$

where the rates,

$$a_7(V) = 40 \exp[-(V + 39)/8],\tag{5.72}$$

$$b_7(V) = \frac{500}{1 + \exp[-(V + 40)/5]},\tag{5.73}$$

are measured in s^{-1}. The total conductance contributed by these channels is large, $g_7 N_7 = 2300\ \mu$S, although they are only open briefly.

The last type of channel, like the first two in Table 5.1, is selective for potassium ions, but the probability of the channel being open is modulated by the intracellular calcium concentration. This channel has $\alpha_8 = \beta_8 = 1$, and the equilibrium state of the inactivation gate depends only on the calcium concentration:

$$h_8^{eq} = \frac{1}{1 + [Ca]/(0.6\ \mu M)}.\tag{5.74}$$

The equilibrium state of the activation gate, in contrast, depends both on voltage and on calcium,

$$m_8^{eq} = \frac{1}{1 + \exp[-(V + f[Ca])/23]} \cdot \frac{1}{1 + \exp[-(V + 16 + f[Ca])/5]} \cdot \frac{[Ca]}{2.5\ \mu M + [Ca]},\tag{5.75}$$

where $f = 0.6$ mV/μM. The relaxation rates, $k_8^{(m)} = 600$ s^{-1} and $k_8^{(h)} = 35$ s^{-1}, show little if any voltage dependence. This seems like a complicated model, but it fits the experimental results very well, as shown in Fig. 5.25.

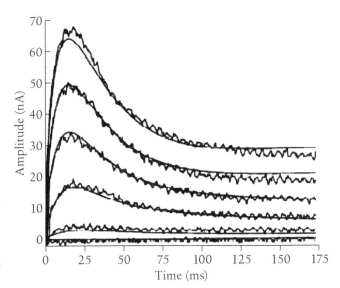

FIGURE 5.25

Dynamics of the calcium-dependent potassium current in response to voltage steps of varying size. Redrawn from Buchholtz et al. (1992). Experimental data (noisy traces) from Golowasch and Marder (1992); solid lines are from the model including Eq. (5.74) and Eq. (5.75).

Problem 103: Calcium-dependent potassium conductances.   Develop a microscopic picture to explain the combination of voltage and calcium dependences seen in Eq. (5.74) and Eq. (5.75). Remember that these equations describe the equilibrium fractions of molecules in particular states, so you need to relate these fractions back to the free energies of the different states. Connect your discussion with the MWC models introduced in the discussion of the cGMP gated channels in rod cells (Section 2.3).

The model of the lateral pyloric neuron described here represents the culmination of many years of effort, both in experiments on this particular system and in the exploration of these fully realistic generalizations of the Hodgkin-Huxley model to what seems the more typical case, with many different channel types functioning together. This model also represents a level of detail and complexity that I have tried to avoid so far, so some explanation is called for. First, the complexity consists largely of variations on a theme. Many channels are known to be described by the general picture of multiple activation and inactivation gates, so this picture provides a framework in which each new type of channel can be fit. Second, the complexity is justified by a large body of data. Independent experiments have been done on other systems, exploring quantitatively each type of channel seen in this neuron, and detailed experiments on this one cell have teased out the contributions of each channel type.

Problem 104: Justifying complexity.   Go through Buchholtz et al. (1992), Golowasch and Marder (1992), and Golowasch et al. (1992), and explain the justification for each channel type in the model discussed above.

Indeed, the program of describing the electrical dynamics of single neurons in terms of generalized Hodgkin-Huxley models, usually with many different channel types functioning together, became a small industry. It really worked. In some cases one could go so far as to characterize the kinetics of particular channel types through measure-

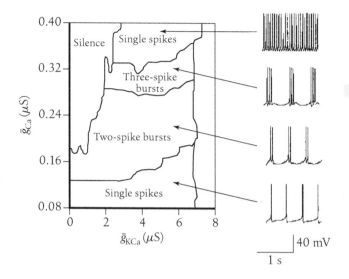

FIGURE 5.26

Simulations of a detailed model, with seven types of channel, for the lateral pyloric neuron in the stomatogastric ganglion of the crab. Changes in the pattern of activity are shown as a function of the numbers of two different kinds of channel, where channel number is expressed as the maximal conductance when all channels are open ($\bar{g}_{Ca}$ and $\bar{g}_{KCa}$). Note that relatively small changes in these parameters can result in both quantitative and qualitative changes in the pattern of electrical activity, running the full range from silence to single spike firing to bursting. Redrawn from Le Masson et al. (1993).

ments on single molecules and then put these single-molecule properties together to reproduce the functional behavior of the cell as a whole. This body of work is beautiful and implements what many people would like to do in other systems, building from measured properties of individual molecular events up to macroscopic biological function. As emphasized above, we can think of the ion channels in the cell membrane as a network of interacting proteins, where the interaction is mediated by the voltage across the membrane rather than direct protein-protein encounters, and where the equations for the dynamics of individual channels have a firm foundation. It is not unreasonable to claim that ion channels in the cell membrane are in fact the best-understood examples of biochemical networks, although the language typically used in describing these systems obscures this connection.

Despite their success, it came to be known, though not widely commented on, that these models of coupled ion-channel dynamics had a problem. Although experiments often characterize the activation curves and kinetics of the individual channels, it is hard to make independent measurements of the total number of channels (or equivalently, the maximum conductance when all channels are open). Thus, one is left adjusting these parameters, trying to fit the overall electrical dynamics of the neuron—for example, the rhythmic bursting of the pyloric neuron. This fitting turns out to be delicate; as one adjusts the (many) parameters, one finds bifurcations to qualitatively different behaviors in response to relatively small changes. An example of this behavior is shown in one two-dimensional slice through the seven-dimensional space of channel numbers in the pyloric model (Fig. 5.26).

From a physicist's point of view, this all seems a mess. There are many details one has to keep track of and many parameters to adjust.[11] One might be tempted just to walk away, and count this model as a part of biology we don't want to know about. But

---

11. As in the case of kinetic proofreading, I think there is a tendency to remember the original papers as having proposed mechanisms that solve problems. But in many ways, it was a much deeper contribution to formulate the problems. Even if the solutions turn out not to be precisely the ones chosen by Nature, the problems are important.

FIGURE 5.27

Mean calcium concentration follows the pattern of electrical activity. Main figure gives a coarse map of mean calcium concentration as a function of the same two variables shown in Fig. 5.26. Regions are labeled A, B, . . . , E in order of increasing mean concentration. Region A corresponds to near-zero concentration, region B to $c < 0.1\,\mu$M, up to region E in which $0.4 < c < 0.5\,\mu$M. Small figure at right shows that the region of bursting activity corresponds almost perfectly to the region of parameter space in which the mean calcium concentration is between 0.1 and 0.3 μM, so that holding the calcium level fixed will stabilize bursting. Redrawn from Le Masson et al. (1993).

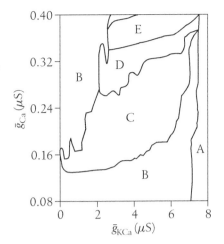

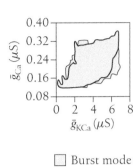

☐ Burst mode

there is a deep question here. If we have trouble adjusting the parameters of models to reproduce the observed functional behaviors of particular cells, how do the cells themselves adjust these parameters to achieve their correct functions? How does a cell choose the correct number of each type of channel to express? One could imagine that the cell has some sort of lookup table—I am a cell of type α, so I should express N_1^α molecules of channel type 1, N_{37}^α molecules of channel type 37, and so on. This is a bit implausible. More likely would be that the cell has some way of monitoring its activity, asking "how close am I to doing the right thing?", and generating an error signal that could be used to drive changes in the expression of the channels or perhaps their insertion into the membrane.

How can a neuron "know" whether it is exhibiting the desired pattern of electrical activity? It would need some signal that couples voltage changes across the membrane, which are quite fast, to the biochemical events regulating gene expression, which are quite slow. One idea is to use the intracellular calcium concentration as an intermediary. We know that many cellular processes are regulated by calcium, so one end of this mechanicsm is easy to imagine. But in the models described above the calcium concentration is an explicit part of the dynamics, so we can calculate, for example, the time-averaged calcium concentration as function of the parameters of the model. What we see in Fig. 5.27 is that [Ca^{++}] does an excellent job of tracing the pattern of electrical activity in this cell. Thus, if the system needs to stabilize a pattern of rhythmic bursting, it can do so by feedback mechanisms that try to hold the calcium concentration near a target value of $C_0 \sim 0.2\,\mu$M.

Let us suppose that the expression of each channel protein is regulated by calcium, so that

$$\tau_i \frac{dN_i}{dt} = N_i^{\max} f_i([\text{Ca}^{++}]/C_0) - N_i, \tag{5.76}$$

where $f_i(x)$ is a sigmoidal function, such as

$$f_i(x) = \frac{1}{1 + x^{\pm n}}. \tag{5.77}$$

These equations have their steady state at $N_i = N_i^{max} f_i([Ca^{++}]/C_0)$, but the calcium concentration must be determined self-consistently through the full dynamics of the channels and voltage. We should choose the signs of the calcium dependences to ensure stability: channels that allow excitatory currents to flow will tend to drive increases in $[Ca^{++}]$, and so they should be opposed by a decreasing function $f_i(x)$, whereas channels that allow inhibitory currents to flow should be controlled by an increasing function $f_i(x)$. Once the signs are chosen, if the regulation functions are steep (large value of n in Eq. (5.77)), and the maximum possible numbers of channels (N_i^{max}) are large, then the dynamics will always be pulled into regimes where $[Ca^{++}] \approx C_0$.

Problem 105: A simple example of a self-tuning neuron. Imagine a neuron with three types of channels. Two of these are always open, and they have different reversal potentials V_+ and V_-; there are N_+ and N_- copies of these two types of channel, and they have the same single-channel conductance. Further, there are channels that allow calcium to flow into the cell, and these channels have a probability of opening that depends on voltage. Calcium flow into the cell is opposed by a pump that would cause the internal calcium concentration to relax exponentially if there were no influx.

(a) Write out the equation for the dynamics of the voltage in the cell. In the approximation that calcium currents have a negligible impact on the voltage, show that there is a steady state in which the voltage is a weighted average of the two reversal potentials.

(b) Write out equations for the dynamics of the internal calcium concentration and the probability of the calcium channel being open. Feel free to hypothesize some simple activation curve for the channel, or look in the literature for inspiration. Introduce calcium-dependent dynamics for the $N_\pm$ as in Eq. (5.76) and Eq. (5.77). For simplicity, assume that the time constants for relaxation of the channel numbers are the same for the two channels, as are the maximal expression levels.

(c) What does your model predict? In particular, you have the choice of signs in Eq. (5.77); try the case where N_+ has the $+$ in this equation, and conversely for N_-. Is there a stable steady state voltage, even when the numbers of channels are allowed to fluctuate? How does this stability depend on the reversal potentials $V_\pm$?

This example, where the voltage is stabilized at a steady state, is quite simple and misses some crucial features of the dynamics of real neurons. Still, you can explore generalizations. Can you, for example, solve a similar model with many different kinds of channels, still neglecting their gating, and show that there is again a stable steady state voltage if the signs of the calcium regulation functions in Eq. (5.77) are chosen in correct relation to the reversal potential? How important are details, such as the assumption that the maximum numbers of the different channels are the same? This is a deliberately open-ended problem; you should try to see how far you can go without having to do simulations, which seem essential once we include realistic channel dynamics.

How can we tell whether something like this sort of self-tuning really is happening? If neurons "knew" how many of each kind of channel to make, then they would try to make this number no matter what the conditions were. For example, inputs from other neurons would drive changes in the electrical activity but not changes in channel expression. However, if the cell is maintaining some mean calcium concentration, or some other measure of activity, then changing the environment in which the neuron

FIGURE 5.28

Changing intrinsic properties of the stomato-gastric ganglion neurons. At left, an experiment in which one cell is ripped from the network and placed in isolation. At first (top) the electrical activity shifts from rhythmic bursts to repeated ("tonic") firing of single action potentials. After 2 days in culture, the cell is silent but responds to small positive currents with tonic firing; after 3 days the response consists of bursts not unlike those in the native network environment. At bottom, continuous recordings demonstrate that this switch from tonic firing to bursting can occur within an hour. At right, 1 hr of stimulation with negative current pulses drives a shift from bursting to tonic firing, which is reversed after 1 hr of no stimulation. All these changes in activity reflect changes in the numbers of different types of ion channels in the cell membrane, as predicted from the models discussed in the text. From Turrigiano et al. (1994). Reprinted with permission from AAAS.

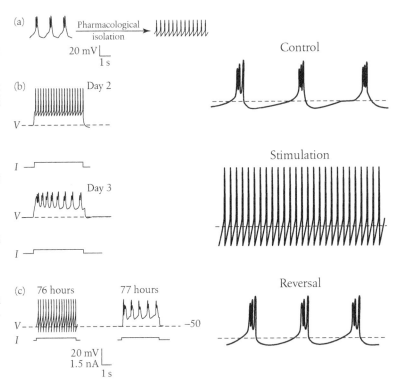

operates will change channel expression. As an extreme example, if we rip the neuron from its network and put it in a dish, the normal pattern of rhythmic bursting will go (wildly) wrong, but the calcium-sensitive dynamics of the channel expression levels will eventually bring the system back into something close to the original pattern. In this new state, the channels are playing different roles in the dynamics, because the driving forces for ionic current flow are different, but the final pattern of activity is the same. A literal version of this rather dramatic scenario actually works experimentally, as shown in Fig. 5.28.

We have noted already that, in invertebrates, such as flies and crabs, the structure and function of neurons in many circuits is sufficiently reproducible from individual to individual within a species that it makes sense to give these cells names and numbers. This discussion of stabilizing patterns of activity rather than expression levels suggests that such reproducibility of function can be achieved without exactly reproducing the number of copies of each channel protein. Further, although the slice through parameter space shown in Fig. 5.27 suggests that the region compatible with normal function is convex, this geometry in fact is not general, and real models often have banana-shaped volumes in parameter space that are consistent with particular patterns of electrical activity. Again, such results are consistent with experiments, most impressively in experiments that measure directly the number of copies of mRNA for several channel types in single cells.

You might worry that we have replaced the tuning of channel copy numbers with a fine tuning of the regulatory mechanisms on all channels. In fact, it is not plausible that calcium acts directly on expression of genes. It is more likely that calcium binds to some protein, and when its binding sites are occupied, the protein can act, directly or indirectly, as a transcription factor. Then the fact that all the genes have the same calcium dependence of their steady state values reflects their common regulation by the same calcium-binding protein. Exploring this scenario in more detail, we see that the kinetics of binding and unbinding of calcium to the sensitive protein can span the time scales of action potentials, bursts, and even the basic rhythm itself. By combining signals from calcium-binding proteins with different kinetics, the more subtle details in the pattern of electrical activity can be stabilized.

A model that explains the behavior of cells only when parameters are finely tuned provokes suspicion that we are missing something. One possibility—often the most plausible—is that the model simply is wrong. The models that we have for biological systems are not like the Navier-Stokes equations for fluids or the standard model of particle physics; we have many reasons to suspect that we are simply solving the wrong equations. But the electrical dynamics of neurons are a special case. Our mathematical models of channel dynamics emerged as accurate summaries of a huge body of data and are nearly exact on the time scales that are experimentally accessible; it is for this reason that we have gone into rather more detail here than in other sections of the text. Rather than rejecting the models, we therefore must conclude that we are missing something, presumably on time scales longer than the experiments that are used to characterize the channel kinetics. In particular, what look like constant parameters must become dynamical variables on long time scales. The simplest implementation of this idea seems to work, and it generates several dramatic experimental predictions that have subsequently been confirmed. Indeed, this theoretical work on the problem of parameter determination has launched a whole subfield of experimental neurobiology, investigating the activity-dependent regulation of the intrinsic electrical properties of neurons.

5.3 The States of Cells

Cells have internal states. Sometimes these states are expressed in a very obvious way, even to external observers, as when we see the alternating black and white stripes of a zebra. In other cases, the states are hidden, as when a neuron stops responding to a constant external stimulus but then rebounds when the stimulus is removed; the amplitude of the rebound reflects the initial amplitude of the stimulus, which must have been stored in some internal state, separate from the output. In these two examples, we also see that these internal states can be (approximately) discrete or continuous. In many cases, the states of cells are known to be encoded by the concentrations of particular, identifiable molecules, and these concentrations in turn reflect a balance of multiple kinetic processes. If we try to transcribe these qualitative ideas simply into quantitative models, we will find that the states of cells depend on parameters. Most obviously, these states will depend on absolute concentrations, and there is a widespread suspicion that absolute concentrations are highly variable, making them poor candidates for the markers of cellular state. More generally, it would seem that, unless we are careful, states will depend sensitively on parameters, providing another example of the problem of fine tuning versus robustness that we have been discussing.

When you tie your shoes in the morning, you can feel the pressure against the skin of your foot, but very quickly this sensation dissipates. When you step outside on a

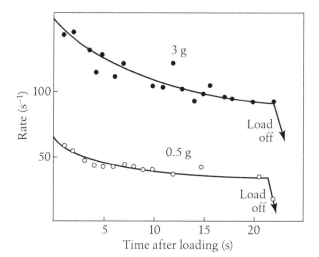

FIGURE 5.29

The original experiments demonstrating adaptation in the response of single sensory neurons (here from the muscle spindle) exposed to constant stimuli (weights). Weights are added at time $t = 0$, and we follow the rate at which this single neuron generates spikes. Redrawn from Adrian and Zotterman (1926a).

bright summer morning, you are aware of the light, but soon everything looks normal, and you would have trouble reporting accurately the absolute light level. These are examples of sensory or perceptual adaptation, in which we gradually become unaware of constant stimuli while maintaining sensitivity to small changes in these incoming signals. One of the first things discovered when it became possible to record the signals propagating along individual nerve fibers is that this adaptation occurs, at least in part, in the response of the single cells that first convert sensory inputs into electrical signals, as shown in Fig. 5.29. Further, as we have seen in the discussion of bacterial chemotaxis (Section 4.2), adaptation occurs even in the sensory systems of single-celled organisms. As we will discuss in connection with the problems of information transmission in neural coding (Section 6.4), adaptation can be a rich and complex phenomenon, being driven not just by constant background signals but also by the statistical structure of fluctuations around this background.

In the simplest case, where adaptation consists of reducing the response to constant signals while maintaining sensitivity to small transient changes, there is a natural schematic model (Fig. 5.30) in which a rapid positive response to the sensory input is canceled by a slower negative response. In several systems we can identify the molecular or cellular components that correspond to these different branches, and we will discuss the example of bacterial chemotaxis in detail. For the moment, however, our concern is more general. If adaptation is accomplished through some pathway that is independent of the basic response to incoming stimuli, then the "gains" of the two pathways are set by independent parameters. If we want the responses to constant inputs to be small, then these two gains must be very similar, so that they nearly cancel. In particular, if we want truly zero response to constants—zero net gain at zero frequency—then the signals passing through the two branches need to cancel exactly, which seems to require fine tuning of the parameters.

Before saying that we have found a problem, we should examine the precision of cancellation that is actually required. In the example of the fly photoreceptors, discussed in Section 2.1, we saw that the system acts as a nearly ideal photon counter up to rates of $\sim 10^5$ photons/s. If the response to a single photon lasts (at its shortest) ~ 10 ms, then the

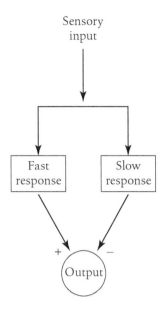

FIGURE 5.30

A schematic of the mechanisms underlying sensory adaptation. The branch that generates fast responses ensures that sudden changes in input will be transduced faithfully. The branch with the slower response causes a gradual decay of the output in response to constant inputs. To have truly zero response to constant input requires that the two branches be perfectly balanced.

cell is effectively counting up to ~1000. But, as we noted, single-photon responses are on the order of a few millivolts, so if these responses just add up, the voltage across the cell membrane would have to change by several volts, which is not going to happen—something like 90–99% of this response needs to be canceled to fit into the available dynamic range. In fact the process is more subtle, because adaptation results not in the subtraction of a constant but rather in a rescaling of the responses to single-photon events, but this gives us an idea of the scale of effect.

In the case of bacterial chemotaxis, we have seen in Section 4.2 that adaptation is essential for function (see especially Problems 49–51). Because the cell makes decisions based on the time course of concentrations along its trajectory, having a response to constant stimuli would mean that the cell effectively confuses "things are good" for "things are getting better," which would impede progress up the gradient of desirable chemicals. Direct measurements of the clockwise versus counterclockwise rotation of the flagellar motor, as in Fig. 4.17, show that the response to a small step in the concentration of attractant molecules decays to zero, so that adaptation is nearly perfect. Another way of seeing this is if one exposes the cells to concentrations that are exponentially increasing in time, the fraction of time the motor spends running clockwise becomes constant, depending on the rate of exponential increase, rather than rising up to saturation; an example is shown in Fig. 5.31.

Problem 106: Exponential ramps. If a cell has receptors to which molecules bind independently, then the fraction of occupied receptors will be given by $f = c/(c + K)$, where c is the concentration, and K is the binding constant. Show that if the concentration c changes slowly, then

$$\frac{df}{dt} = \frac{d \ln c}{dt} f(1 - f).$$
(5.78)

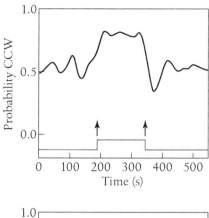

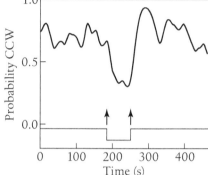

FIGURE 5.31

Response of *Escherichia coli* to exponentially increasing (top) or decreasing (bottom) concentrations of an attractant. Response is measured as the probability of counterclockwise rotation (CCW), averaged over a small window of time; stimulus onsets and offsets are marked by arrows. Redrawn from Block et al. (1983).

Show further that, with $c \approx K$, an exponentially increasing concentration produces an approximately constant rate of increase in f. Go back to Block et al. (1983) and convince yourself that they really are in this regime, relating your argument to the results in Fig. 5.31. Can you use these measurements to get an estimate of the absolute sensitivity of the system? How much does the motor bias shift when one extra receptor molecule is occupied per second?

If we observe freely swimming bacteria, then we can count the rate at which they initiate tumbles and see that this also adapts to constant stimuli. Figure 5.32 shows an unnatural but dramatic example, in which a population of bacteria is suddenly exposed to millimolar concentrations of aspartate, starting from zero background concentration. Tumbling is almost completely suppressed for nearly 10 minutes, but eventually recovers to within ∼10% of its initial rate, even though the initially saturating stimulus continues to be present.

To understand how it is possible to achieve near-perfect adaptation without fine tuning of parameters (as one might have thought from Fig. 5.30), we have to dig into the details of the molecular mechanisms involved. In Section 4.2 we outlined the fast events involved in the positive part of the chemotactic response (Fig. 4.19). To review briefly, receptor molecules on the cell surface form a complex with the enzyme CheA (a kinase), held together by a scaffolding molecule CheW. The complex is in equilibrium between the active (CheA*) and inactive (CheA) states, and this equilibrium is shifted by binding of attractant or repellent molecules to their receptors; for attractants, binding shifts

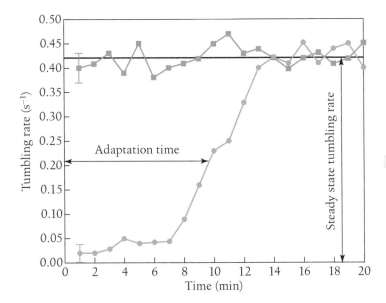

FIGURE 5.32

Experiments on adaptation in a large population of *Escherichia coli*. At time $t = 0$, the population is exposed to a high concentration of an attractive chemical, and as a result the bacteria almost stop tumbling. Over time, they adapt (blue), and the average rate of tumbling approaches the steady state value observed in the absence of stimuli (orange). Redrawn from Alon et al. (1999).

the equilibrium toward the inactive state. The active kinase CheA* phosphorylates the protein CheY, which can diffuse through the cell from the receptor complex to the flagellar motor, where it binds and favors clockwise rotation, driving the tumbling motion of the cell; the action of the kinase is opposed by a phosphatase, CheZ. Thus, an increase in the attractant concentration drives the kinase toward its inactive state, reducing the rate of phosphorylation of CheY; the continued action of the phosphatase results in a reduction of the CheY~P concentration, which reduces the probability of tumbling. This whole pathway is extraordinarily sensitive, responding reliably to individual molecules as they bind to their receptors.

How does the extremely sensitive response of the chemotactic system get canceled when stimuli are maintained at constant levels? In addition to binding the chemoattractant or repellent molecules, the receptors can be modified by covalent attachment of methyl groups. Much as with ligand binding, these modifications shift the equilibrium between active and inactive conformations of the kinase CheA—binding of attractants favors the inactive state, addition of methyl groups favors the active state. The active kinase not only phosphorylates CheY, leading to clockwise rotation of the motor, it also phosphorylates CheB, and then CheB~P removes methyl groups from the receptor. Thus, when an attractant lowers the activity of the kinase, it also allows more methyl groups to be attached, driving the activity back toward its original level—adapting, as shown in Fig. 5.33.

Although the methylation system provides a pathway to cancel the effect of the immediate response to sensory inputs, it is not clear that this cancellation should be anywhere near exact. In general, one would need to tune the activity of the methylation and demethylation enzymes to make sure that their effects exactly balance the direct response to sensory input. So, this system provides an example of our general problem of fine tuning, as emphasized by Barkai and Leibler. In addition to identifying the problem, they proposed that one can evade this need for fine tuning by assuming that the demethylation enzyme CheB only recognizes the active state of the receptor-kinase

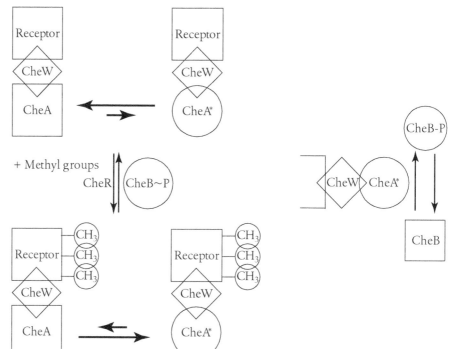

FIGURE 5.33

Methylation of the receptors allows for adaptation of the chemotactic response. At left, addition of methyl (CH_3) groups by CheR acts, similarly to ligand binding, as an allosteric effector, shifting the equilibrium between the active (CheA*) and inactive (CheA) states of the kinase; the schematic is meant to indicate that there are multiple methylation sites. At right, the feedback loop is closed by having the active kinase CheA* trigger activation of the demethylation enzyme CheB.

complex and ignores the inactive conformation. If this is true, the phosphorylation of CheB is not needed to close the feedback loop.

To see how the Barkai-Leibler scheme works, let's imagine that the receptor complex, which might include a cluster of several receptor molecules, switches as a whole between active and inactive states, in the spirit of the MWC model that we have discussed in several other contexts (e.g., Section 2.3). There is some free-energy difference ΔF between these states, and there are two contributions to this difference—one from the binding of attractants and one from methylation. Assume that the contribution of the methyl groups is additive and that the contribution from ligand binding has some arbitrary dependence on ligand concentration $F_L(c)$, which we could work out from a model like that in Fig. 5.33 (see Problem 109). Then the number of active enzymes A^* is given by

$$A^* = \frac{A_{\text{total}}}{1 + \exp\left[F_L(c) - n_{\text{M}}\Delta_{\text{M}}\right]}, \tag{5.79}$$

where n_M is the number of methyl groups per receptor complex, A_{total} is the total number of CheA molecules, and Δ_{M} is the contribution of each methyl group to the free energy difference between the two states. This number reflects a balance between the activities of the enzymes CheR and CheB, so we can write schematically

$$\frac{dn_M}{dt} = V_R - V_B, \tag{5.80}$$

where V_R and V_B are the "velocities" of the methylation and demethylation enzymes, respectively.

The key assumptions suggested by Barkai and Leibler are that CheR is running at some maximal rate, limited by its internal dynamics and not by the availability of substrate, while the velocity of CheB does depend on the availability of its substrate A^* according to some function $f(A^*)$ that we don't need to specify. Then we have

$$\frac{dn_M}{dt} = V_R^{\max} - V_B^{\max} f(A^*). \tag{5.81}$$

To reach steady state ($dn_M/dt = 0$), we must have

$$A^* = A_0^* = f^{-1}(V_R^{\max}/V_B^{\max}), \tag{5.82}$$

independent of the ligand concentration c. Thus, all steady states in the system must have the same level of activation of the kinase, hence the same level of phosphorylation of CheY and the same rate of tumbling. These steady states at varying c are not identical—they involve different levels of methylation—but they have the same functional output.

Problem 107: Allosteric model for chemotactic receptors. The schematic in Fig. 4.19 is equivalent to an MWC model in which the whole complex of the receptor, CheW and CheA, has two states, and the equilibrium is shifted by binding of the attractant molecule. In Fig. 5.33, attachment of methyl groups also shifts this equilibrium, but the binding and unbinding of these groups is part of an energy-yielding reaction and so does not have to obey detailed balance. Show that, nonetheless, these schematics generate Eq. (5.79), which has a decidedly Boltzmann form. Why does this work? What would change if groups or clusters of N receptor complexes were tied together and forced to all be in the same activation state?

If the scenario sketched here is correct, then we should be able to test it by manipulating the activity of the methylation and demethylation enzymes, using the modern tricks of molecular biology to modify the genome of *E. coli*. To begin, one can replace CheB with a mutant form that cannot be phosphorylated; adaptation still works and still is nearly perfect, suggesting that phosphorylation is not the key step in closing the feedback loop. Then the normal CheR gene can be deleted and replaced with a plasmid that carries the CheR coding region under the control of a promoter responding to external signals. In this way ~100-fold variations in CheR expression levels can be generated, from half the normal level to 50× overexpression, as shown in Fig. 5.34. Throughout this range, adaptation to large inputs (as in Fig. 5.32) is within ~10% of being exact. Although the mean rate of tumbling to which the system adapts (as well as the time scale of this adaptation) depends on the amount of CheR in the cell, changing the level of CheR does not change the fact that this steady mean tumbling rate is independent of input ligand concentrations.

Some of the earliest experiments on chemotaxis foreshadowed the dissociation between the precision of adaptation and other aspects of the dynamics. Figure 5.35 shows that in a single cell, the mean duration of runs and tumbles (more precisely, periods of clockwise and counterclockwise rotation of the motor) drift systematically

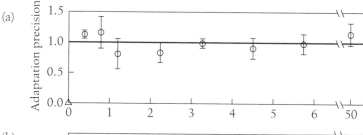

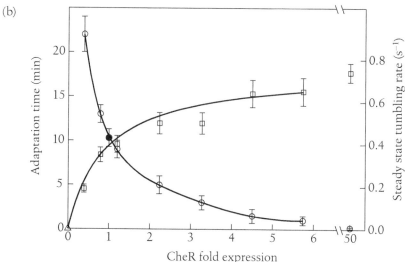

FIGURE 5.34

Chemotactic responses in the presence of varying amounts of CheR. Redrawn from Alon et al. (1999). (a) Adaptation precision is measured as the ratio of the mean tumbling rates in the presence and absence of 1 mM aspartate (as in the experiments of Fig. 5.32). (b) The actual tumbling rates and the time required to reach steady state after sudden exposure to 1 mM aspartate are shown.

over time, changing by nearly a factor of four over the course of an afternoon's experiment. But the balance between these periods—the probability that the motor is running counterclockwise—remains almost constant. This experiment wasn't aimed at testing the precision of adaptation, but it seems clear that the motor bias and hence, from Fig. 4.20, the concentration of CheY~P in the cell, is being held relatively constant even as other things change.

There is a lot of evidence that the methylation level of the receptors really is the molecular representation of the cell's adaptation state. As such, we might have expected that over- or under-expressing the enzyme which carries out the methylation reaction would shift the actual state of the system, and this would show up as a change in the output. In the model considered here, however, this last expectation is violated. The absolute level of kinase activity, and hence the absolute tumbling rate, does indeed change when we change the expression level of CheR. But Fig. 5.34 shows us that the average steady state response to an applied step in attractant concentration remains zero, independent of the CheR level: the precision of the balance between the processes responsible for excitation and adaptation does not depend on fine tuning of the underlying kinetics.

Problem 108: Calcium-driven adaptation in neurons.   Consider a neuron that generates spikes at rate r. Let's assume the response to external inputs I involves this rate relaxing toward some steady state,

$$\tau \frac{dr}{dt} = r_{max} f(I, [Ca]) - r,$$
(5.83)

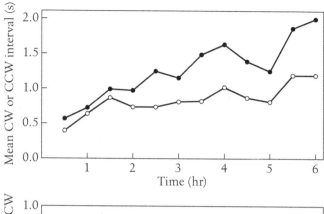

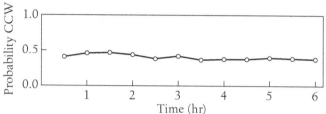

FIGURE 5.35

Observations on a single *Escherichia coli* over a very long time. Top panel shows the mean durations of clockwise (CW; open symbols) and counterclockwise (CCW; closed symbols) periods of motor rotation, while the bottom panel shows the probability of counterclockwise rotation. Redrawn from Block et al. (1983).

where the response depends both on the inputs and on the intracellular calcium concentration, through the function $f(I, [Ca])$. Write an equation for the dynamics of [Ca], assuming that each spike brings in a fixed number of calcium ions and that there is a pump extruding the ions at some opposing rate. The pumping rate must depend on the concentration, but for the moment take this dependence as some unknown function $V_{pump}([Ca])$. Find equations that describe the steady state of this system. Are there conditions on $V_{pump}([Ca])$ that lead to a steady state spike rate that is independent of the input I? If the input changes suddenly, does the spike rate still respond? Explain how this relates to the discussion of chemotaxis given here.

Are we done? I think there is still more to this problem. To begin, the fact that motor output is an extremely steep function of the CheY~P concentration (Fig. 4.20) means that successful adaptation requires more than just a constant level of CheY-P in steady state, independent of the input signal. This level actually has to fall into a very narrow range, or else the cells will be always running or tumbling. The parameters determining the steady state level of CheY~P are independent of the properties of the motor, which determine the functional operating range for this concentration. This seems like the same sort of balancing problem that Barkai and Leibler were worried about, but in a different part of the system, where their solution has no obvious analog.

You should also be a little suspicious about the simple equations introduced above. At best, they are some sort of mean field theory in a system where fluctuations could be important. In particular, one might worry that the rate of removing methyl groups depends on how many are there, especially if that number goes to zero. There must be some regime in which the simple argument is right, but we need a more honest calculation. Some of the references in the Annotated Bibliography for this section address this problem.

Finally, although one can manipulate the *E. coli* genome to change the expression levels of individual proteins by large factors, it is not clear whether such variations

(a)   (b)   (c)

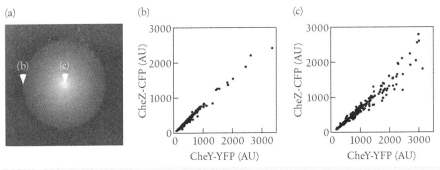

FIGURE 5.36

Better chemotactic performance is associated with correlated fluctuations in protein levels. Redrawn from Løvdok et al. (2009). (a) *Escherichia coli* swarm outward toward attractants. Cells have been engineered to express CheY and CheZ only under the control of a promoter induced by external signals. If we select cells from regions B or C of the swarm, we see that the cells that have been efficient (b) have tightly correlated variations in the two protein levels, whereas inefficient cells (c) have weaker correlations. Thus, selection for chemotactic efficiency will drive down the *relative* fluctuations in expression levels, even with substantial tolerance for variation in the absolute levels.

occur naturally. In higher organisms (like us), the length of DNA that codes for the amino acid sequence of a single protein may be broken into multiple segments (exons) that are separated by noncoding sequences (introns). When the mRNA molecule is transcribed, it has to be edited to bring the exons together before translation, which introduces a whole new layer of regulation. In bacteria, there are examples of the opposite situation—rather than having the gene for a single protein broken into pieces, the genes for multiple proteins are joined together, so that a single mRNA molecule contains the code for more than one protein. Such linked genes are called operons. Importantly this means that (leaving out a few details), fluctuations in the relative concentration of proteins in the same operon can arise only in the process of translation, not in transcription.

The many protein components of the chemotactic system are encoded on just two operons. The *mocha* operon encodes CheA and CheW, along with the flagellar motor proteins; the *meche* operon encodes CheR, CheB, CheY, and CheZ, along with two classes of receptor proteins. Recent experiments indicate that there is covariation even between the expression levels of CheA and CheY, suggesting that the cell can in fact control at least the relative concentrations of these proteins fairly precisely. Further, there is direct evidence that tight correlation between protein concentrations actually improves chemotactic performance, as shown in Fig. 5.36.

Problem 109: Balancing CheY and CheZ.   Make a simple model to explain why, as seen in Fig. 5.36, variations in the relative levels of CheY and CheZ are detrimental to chemotactic performance, whereas correlated variations are not. This problem is deliberately open ended.

It is interesting to compare the problem of robustness versus fine tuning in the case of chemotaxis with what we learned in the case of ion channels (Section 5.2). For ion channels, function really does depend sensitively on the number of copies of the

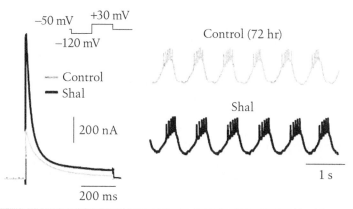

FIGURE 5.37

Responses to overexpression of a channel. At left, injecting mRNA for the A current channel (see Table 5.1) produces, after 72 hr, an increase in the current that flows when the voltage is stepped through the range expected to activate this channel. Thus, injecting the mRNA really does result in more channels being synthesized and inserted into the membrane. At right, a demonstration that this increased number of channels (in the "Shal" trace) does not perturb the basic pattern of activity (seen in the control). This behavior is possible only because the cell compensates by increasing the expression levels of other channels. Redrawn from MacLean et al. (2003).

different proteins in the network, and neurons have evolved control mechanisms that use their functional output (or a near surrogate) to control these copy numbers. Importantly, there are many ways to achieve the same function, so it is not the number of copies of each component that is tuned, but rather some possibly complex combinations of these quantities. For chemotaxis, the message of the experiments shown in Fig. 5.34 is that large variations in the copy number of just one component can be tolerated, pointing toward networks that are intrinsically insensitive to this parameter variation rather than toward any hidden control or tuning mechanisms. This suggests that one system is tuned, and the other is robust.

In contrast, Fig. 5.36 shows that, as with ion channels, variations in the relative copy numbers of the proteins in the chemotaxis network cause a degradation of function, and so presumably these variations are controlled under normal conditions, whether simply by grouping genes into operons or by more active mechanisms. Experiments more directly analogous to Fig. 5.34 have now been done in stomatogastric ganglion neurons, showing that control mechanisms can compensate for overexpression of particular channel types by changing the expression levels of other channels; see Fig. 5.37. Perhaps surprisingly, these compensation mechanisms are triggered even if the first channel is nonfunctional, and hence does not affect the electrical output, suggesting that there are signals internal to the transcriptional and translational networks that encode something about the correct functional operating point of the system. This phenomenon could be much more general.

Before moving on, there is also a somewhat philosophical point to be made about the mechanism of robustness in chemotaxis, or perhaps even about the idea of robustness itself. If we expect the function of a biochemical network to be robust against parameter variation, this robustness must be a property of the network topology—which nodes (molecules) are connected by arrows (reactions). In the specific model

considered by Barkai and Leibler, for example, it is essential that CheB acts on CheA*
as a substrate, but not on the inactive CheA. What is important in this case is the
absence of a link in the network connecting CheB with CheA.

The particular links that appear (or do not appear) in a biochemical network reflect
the specificity of the various enzymatic and protein-protein binding reactions. Substrate
specificity is a classical topic in biochemistry, and much of what we understand about
this topic was learned through painstaking experiments on purified samples of particular enzymes. The ideas of robustness emerged at a time when the community started
to wonder whether something was a bit hopeless about the overall project of this classical biochemical approach. Although one can study individual enzymes in detail, many
interesting biological functions emerge from networks with many interacting proteins
engaged in dozens (if not hundreds) of individual reactions. Further, the conditions
inside the living cell may be far from those we can reproduce in a test tube. How, then,
could we ever study every one of the relevant reactions under the right conditions?
Seen in this light, it just does not seem plausible that the accumulated biochemical
knowledge will add up to give us an understanding of how cells function as complete
systems. Robustness was one of several ideas offered as an alternative—if Nature has
selected networks that are robust to parameter variations, then the (already somewhat
hopeless) project of measuring all these parameters could safely be abandoned. But
because network topology is an encoding of substrate specificity, we can't really brush
aside all of classical biochemistry. Indeed, the example brought forward by Barkai and
Leibler is one in which the biochemistry is subtle, with one protein recognizing different
conformations of another. At the end of the day, then, biochemistry has its revenge—
robustness may be an emergent system-level property, driven by network topology, but
this topology is an expression of the underlying, detailed biochemistry.

Perhaps the most obvious example of cells having internal states is in the cycle of
events that allow cells to grow and divide. One of the triumphs of late-twentieth-century
molecular and cellular biology was the identification of the molecules that are at the
heart of the cell's clock, the essential components that rise and fall as the cell proceeds
through its cycle. The key protein components are called cyclins, and they are involved
in a network of phosphorylation and dephosphorylation reactions about which we
know many details. We can abstract from these details a network of 11 molecules that
activate or inhibit one another, as shown in Fig. 5.38. In the simplest view each node of
the network is either "on" ($S_i = 1$) or "off" ($S_i = 0$). The connections among the nodes
can be summarized by a matrix W_{ij}, where $W_{ij} = 1$ (-1) if an active node j activates
(inhibits) node i. The natural dynamics proceeds in discrete time steps,

$$S_i(t+1) = \Theta\left[\sum_j W_{ij}S_j(t)\right], \qquad (5.84)$$

where $\Theta[\cdot]$ is the unit step function,[12]

$$\Theta[x \geq 0] = 1, \qquad (5.85)$$
$$\Theta[x < 0] = 0. \qquad (5.86)$$

---

12. We have to be careful about the possibility that $\sum_j W_{ij}S_j(t) = 0$. In this case there is no "force" driving
a change in S_i, and so we assume it remains fixed, $S_i(t+1) = S_i(t)$.

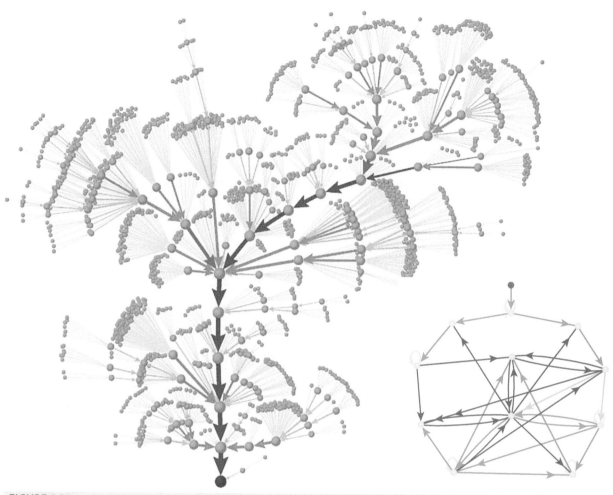

FIGURE 5.38

Dynamics in a simple model of the cell cycle. At bottom right is an abstraction of the known biochemical reactions into a reduced network; to emphasize the abstraction, the names of the molecular components have been suppressed. Green arrows denote activation, red arrows denote inhibition, and yellow arrows indicate self-inhibition or degradation. The dynamics, as explained in the text, are the simplest possible Boolean updating consistent with this pattern of connections. Of the 2^{11} states in this model, fully 86% of the states in the network (green dots) flow to a single fixed point (blue dot), corresponding to the resting state of the cell. The trajectories into this state organize as shown at left, so that there is a definite path (blue arrows), corresponding plausibly to the other states of the cell cycle. Redrawn from Li et al. (2004). Copyright © 2004 by the National Academy of Sciences USA.

The dynamics of this model have an extraordinary simplicity. Of the 2^{11} possible states, 86% flow to a single fixed point, corresponding to the resting state of the cell. If we imagine that a stimulus comes into the network—for example, a signal indicating that the cell has grown to a critical size—the system will be kicked out of this stationary state and then relax back. All the trajectories that lead back to the stationary state are organized into a tree, with a main trunk that is strongly attracting (Fig. 5.38). Thus, almost all the dynamics we actually see involves motion along a stereotyped path, and we can identify the steps along this path with the known states of the cell cycle.

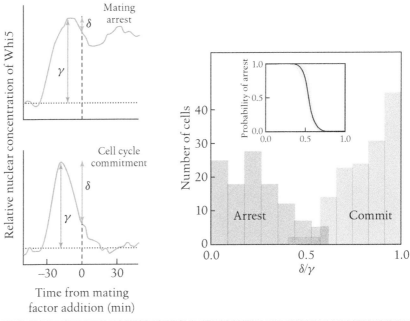

FIGURE 5.39

A single molecular species predicts cellular decisions. Budding yeast were genetically engineered to produce a fluorescent version of the protein Whi5, which moves out of the nucleus at a crucial step in the cell cycle. Cells can be exposed to a "mating factor," which can divert the cell from its regular progression through the cycle, depending on the time at which the signal is delivered. At top left, the time course of Whi5 nuclear concentration in a case where exposure to the mating factor stops the cycle; at bottom left, a case where the cell was already committed to progressing around the cycle. At right, the numbers of cells committed versus arrested as a function of the relative nuclear concentration at the moment when the mating signal is delivered; the inset summarizes these results as a probability of arrest, with the error band showing 95% confidence limits. Redrawn from Doncic et al. (2011).

That we can strip the full dynamics of a complex biochemical network down to roughly ten components (and then replace the dynamics by the simplest possible Boolean model) certainly suggests that the essential features of the cell's states and transitions are coded in the topology of the network and not in the detailed parameters. However, if we reach into the network and monitor the concentrations of individual components, we can correlate the cell's decision to transition from one state to the next with this concentration crossing a threshold with a precision on the order of $\sim 10\%$, as in Fig. 5.39. These views are not necessarily in conflict, but certainly they point in different directions. In one view, states and transitions are collective, and this collective behavior serves to suppress sensitivity to quantitative detail. In the other view, the system encodes information about state precisely in the quantitative variations of concentration.

Another example of robust output from a biological network is that almost everyone you know was born with five fingers on each hand. In insects, one can count even more instances in which discrete pieces of the body are arranged in a repetitive pattern, from the segments of the body itself (as in the beautiful caterpillar shown in Fig. 5.40) to the hairs or bristles on the body surface. Essentially every member of a particular

FIGURE 5.40

Insects provide many examples of repeated, reproducible structures visible outside the body. Image of tiger moth caterpillar by R. Tomlinson, Henderson State University, with my thanks.

species has the same number of body segments, the same number of hairs, and even the positions of the hairs are reproducible from individual to individual. It is not at all obvious how this level of reproducibility is achieved.

In the spirit of this discussion, we can distinguish two broad classes of explanations for the reproducibility of pattern formation (morphogenesis) in the embryo. In the first kind of explanation, the organism works to set the initial and boundary conditions very precisely, and each step in the process has been tuned to minimize noise. Alternatively, the second type of explanation posits that it is possible that noise and errors abound, but error-correction mechanisms pull the pattern back to its ideal structure. It also is possible that both scenarios are correct: Nature has selected for systems with minimal noise and taken care to control the conditions of development, but error-correction mechanisms are still needed to deal with the vagaries of a fluctuating environment.

To appreciate why the observations of reproducibility in morphogenesis are so puzzling, we need to review some basic mechanisms by which patterns form in the developing embryo. We also need to check our qualitative impressions of reproducibility against quantitative data. Let's start with the background. We have already met some of these mechanisms in our discussion of noise in gene expression (Section 4.3).

Embryos start as just one cell, the fertilized egg, and then this cell divides many times. Every one of these cells (as in our adult bodies) has the same DNA, assuming that nothing has gone wrong. What makes the different cells different is that they express different genes. The genes code for proteins, but not all proteins are made in all cells; the reading of the code to make the proteins is called the expression of the genes, as we have discussed before. Embryos come in all shapes and sizes throughout the animal kingdom, but the fly embryo is an interesting model system for many reasons, and as with many before us, we will focus our attention on this system. One reason is that there is a well-developed genetics for fruit flies (*Drosophila melanogaster*), made possible not least by their rapid growth and reproduction. Embryonic development itself is rapid as well, leading from a fertilized egg to the hatching of a fully functional maggot (the larvae of flies, analogous to caterpillars for butterflies) within 24 hr. All of this happens inside an egg shell, so there is no growth—pattern formation occurs at constant volume. The egg is ∼1/2 mm long, so one starts with one rather large cell, which has one nucleus. In the maggot there are ∼50,000 cells. For the first 3 hr of development, during which the blueprint for the body plan is laid out, something special happens: the nuclei multiply without building walls to make separated cells. Thus, for ∼3 hr the fly embryo is close

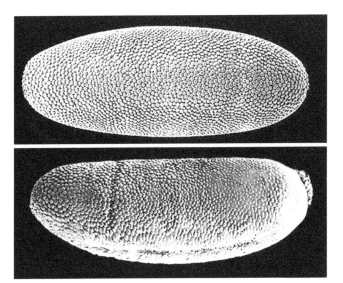

FIGURE 5.41

Electron micrographs of a *Drosophila* embryo in cycle 14, before (top) and after (bottom) gastrulation. Note, in particular, the cephalic furrow roughly one-third of the distance from the left in the bottom image. Thanks to E. F. Wieschaus for these images.

to the physicists' idealization of a box in which chemical reactions are occurring, with the different molecules free to move from one end of the box to the other (perhaps even by diffusion, although this is a subtle question).

The duplication of the nuclei is more or less synchronous for the first 13 mitotic divisions, or nuclear cycles, which is visually quite striking. During cycles 8–10, almost all the nuclei move to the surface of the egg, where they form a fairly regular two-dimensional lattice; conveniently, with all the nuclei at the surface of the egg, we have a much better chance to see what is going on (see Figs. 5.41 et seq.). With each subsequent cycle, this lattice dissolves and reforms. In cycle 14, the synchronous duplication of nuclei stops, and there is a pause while the embryo builds membranes between the nuclei to make separate cells. If you stop the action at this point and take an electron micrograph of the embryo, what you see is at the top in Fig. 5.41. There are ∼6000 cells on the surface. This is smaller than 2^{13}, but that's because not all the nuclei make it to the surface; some stay in the interior of the embryo, probably not by accident, because these become cells with special functions. All the cells look pretty much alike. If instead of stopping at this point, we wait just 15 minutes more, we see something very different, shown at the bottom in Fig. 5.41. Notice that there is a vertical cleft, about one-third of the way from the left edge of the embryo. This feature is the cephalic furrow, and it defines which part of the body will become the head. There is also a furrow along the bottom of the embryo, which is where the one layer of cells on the surface starts to fold in on itself so that there can be two "outside" surfaces, a process called gastrulation (think about the inside and outside of your cheek, both of which are outside the body from the topological point of view—we are not simply connected!).

The embryo does not just break into a head and a nonhead. In fact there are many different pieces to the body, usually called segments, as noted above. The question is how the cells at different points in the embryo "know" to become parts of different segments. The answer is quite striking, and its discovery is one of the great triumphs of modern biology. Long before cells start moving around and making the three-dimensional structures found in the fully developed organism, there is a blueprint that

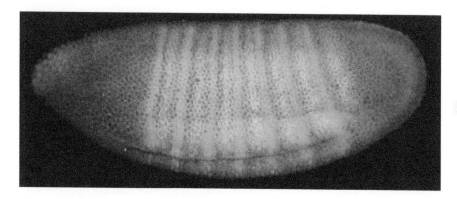

FIGURE 5.42

Expression patterns of two pair-rule genes in the early *Drosophila* embryo. The proteins are visualized by staining with fluorescent antibodies. Thanks again to E. F. Wieschaus for this image.

can be made visible by asking about the expression levels of particular genes. A now-classic set of genetic experiments showed that the number of genes that are relevant in these early patterning events is small, on the order of 100 out of the roughly 15,000 genes in the whole fly genome; if we focus on the pattern along the anterior-posterior axis of the embryo, the number of relevant genes is ~20. Most of these genes code for transcription factors that control the expression of other genes.

Suppose we stop the action in the embryo at cycle 14 and measure the concentration of two of these key proteins. One way to do this is to make antibodies against the protein we are interested in and then make antibodies against the antibodies; before using the secondary antibodies, we attach to them a fluorescent dye molecule. So if we expose the embryo first to one antibody (which should stick to the protein we are interested in, and not anywhere else, if we are lucky) and then to the other, we should have the effect of attaching fluorescent dyes to the protein we are looking for. Hence, under a microscope, the brightness of the fluorescence should indicate the concentration of the protein. One such experiment is shown in Fig. 5.42. Evidently the concentration of the proteins varies with position, and this variation corresponds to a striped pattern. The stripes should remind you of the segments in the fully developed animal, and this correspondence is quite precise. Mutations that move the stripes around, or delete particular stripes, have the expected correlates in the pattern of segmentation. To illustrate this point, we can blow up corresponding pieces of this image and the electron micrograph in Fig. 5.41, showing the cephalic furrow (Fig. 5.43); hopefully you can see how the furrow occurs at a place defined by the locations of the green and orange stripes. At the moment the names of these molecules do not really matter. What is important is to realize that the macroscopic structure of the fully developed organism largely follows a blueprint laid out within ~3 hr; this blueprint is written as variations in the expression level of different genes. Furthermore, we know which genes are the important ones, and there aren't too many of them.

We have pushed the problem of pattern formation in the embryo back to spatial variations in the pattern of gene expression, but how do these arise? You could imagine, as Turing did, that these patterns reflect a spontaneous breaking of symmetries in the egg. To understand the Turing scenario, imagine two genes (perhaps the two whose expression levels are pictured in Fig. 5.43) that code for proteins that both act as transcription factors, regulating the expression of these same genes g_1 and g_2. Then the rate of synthesis of protein 1, for example, will depend on the concentration of both

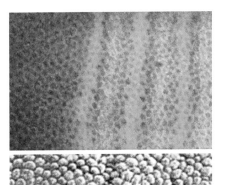

FIGURE 5.43

A combination of Figs. 5.41 and 5.42, emphasizing that the cephalic furrow occurs along a single line of cells that can be identified from the pattern of pair-rule gene expression.

proteins 1 and 2, and similarly for protein 2, so we can write locally

$$\frac{dg_1}{dt} = r_{\max} f_1(g_1, g_2) - \frac{g_1}{\tau}, \tag{5.87}$$

$$\frac{dg_2}{dt} = r_{\max} f_2(g_1, g_2) - \frac{g_2}{\tau}, \tag{5.88}$$

where $r_{\max}$ is the maximum rate of synthesis, $1/\tau$ is the rate at which proteins are degraded (we assume that these are the same for both genes), and f_1 and f_2 describe the dependence of the synthesis rates on the concentrations of the transcription factors. But the proteins can also diffuse, and we need to include this behavior in our description. For simplicity let's work in a continuum model, in which cells are sufficiently dense that they do not need to be considered as discrete elements, but let's stay in one dimension. Then we have

$$\frac{\partial g_1}{\partial t} = D_1 \frac{\partial^2 g_1}{\partial x^2} + r_{\max} f_1(g_1, g_2) - \frac{g_1}{\tau}, \tag{5.89}$$

$$\frac{\partial g_2}{\partial t} = D_2 \frac{\partial^2 g_2}{\partial x^2} + r_{\max} f_2(g_1, g_2) - \frac{g_2}{\tau}, \tag{5.90}$$

where the different proteins have different diffusion constants D. Can this simple system form spatial patterns?

Imagine that we search for a spatially uniform steady state of this system. This state corresponds to expression levels that solve the equations

$$g_1^s = r_{\max} \tau f_1(g_1^s, g_2^s), \tag{5.91}$$

$$g_2^s = r_{\max} \tau f_2(g_1^s, g_2^s), \tag{5.92}$$

and we can normalize the expression levels by the factor $r_{\max} \tau$, eliminating this quantity from the equations. Linearizing around this steady state, we find

$$\tau \frac{\partial}{\partial t} \begin{bmatrix} \delta g_1 \\ \delta g_2 \end{bmatrix} = \begin{bmatrix} \lambda_1^2 \frac{\partial^2}{\partial x^2} + \gamma_{11} - 1 & \gamma_{12} \\ \gamma_{21} & \lambda_2^2 \frac{\partial^2}{\partial x^2} + \gamma_{22} - 1 \end{bmatrix} \begin{bmatrix} \delta g_1 \\ \delta g_2 \end{bmatrix}, \tag{5.93}$$

where the length scales $\lambda_i = \sqrt{D_i \tau}$, and the constants γ_{ij} are the derivatives of the functions f_i at the steady state:

$$\gamma_{ij} = \frac{\partial f_i(g_1, g_2)}{\partial g_j} \bigg|_{g_1^s, g_2^s}. \tag{5.94}$$

In particular, if we look for variations that have a definite wavelength, $\delta g = \delta \tilde{g} e^{ikx}$, the time dependence of these spatial modes is governed by

$$\tau \frac{\partial}{\partial t} \begin{bmatrix} \delta \tilde{g}_1 \\ \delta \tilde{g}_2 \end{bmatrix} = \begin{bmatrix} -\lambda_1^2 k^2 + \gamma_{11} - 1 & \gamma_{12} \\ \gamma_{21} & -\lambda_2^2 k^2 + \gamma_{22} - 1 \end{bmatrix} \begin{bmatrix} \delta \tilde{g}_1 \\ \delta \tilde{g}_2 \end{bmatrix}. \tag{5.95}$$

If we search for solutions that vary as $e^{\Lambda t / \tau}$, then the values of Λ depend on k and are set by the eigenvalues of the matrix on the right-hand side of this equation.

In the Turing scenario, the mode at $k = 0$ is stable: the steady state we have identified really would be a steady state in the absence of diffusion. Solving for the eigenvalues of the matrix in Eq. (5.95) at $k = 0$, we find

$$\Lambda_\pm = \frac{1}{2} \left[\mathcal{T} \pm \sqrt{\mathcal{T}^2 - 4\mathcal{D}} \right], \tag{5.96}$$

where $\mathcal{T}$ and $\mathcal{D}$ are the trace and determinant, respectively, of the matrix:

$$\mathcal{T} = (\gamma_{11} + \gamma_{22}) - 2, \tag{5.97}$$

$$\mathcal{D} = (\gamma_{11} - 1)(\gamma_{22} - 1) - \gamma_{12}\gamma_{21}. \tag{5.98}$$

Stability ($\Lambda_\pm < 0$) is ensured if the trace ($\mathcal{T} = \Lambda_+ + \Lambda_-$) is negative, and the determinant ($\mathcal{D} = \Lambda_+ \Lambda_-$) is positive. This can be achieved, however, even if $\gamma_{11} > 1$, which corresponds to a situation in which the steady state would be unstable for g_1 alone but is stabilized by the feedback loop through g_2. Under these conditions, for appropriate combinations of the length constants $\lambda_{1,2}$, we can find $\Lambda(k \neq 0) > 0$, so that the steady state is unstable against the growth of spatially periodic patterns. Although this linear analysis is not sufficient to establish what actually happens once this instability sets in, intuitively we expect that the pattern with the largest value of Λ will grow the fastest and dominate.

Problem 110: Details of Turing. Fill in the details of the argument above. In particular, if you go back to the linearized equations (Eq. (5.95) at $k \neq 0$), show explicitly that there are solutions $\propto e^{\Lambda(k)t/\tau}$ with $\Lambda(k) > 0$, provided that $\gamma_{11} > 1$. Are there any other conditions that need to be met? In particular, what about the length constants λ_1 and λ_2?

This scenario, for better or worse, is not how it works. When the mother makes the egg, she places the mRNA for a handful of proteins at cardinal points. For example, the mRNA for the protein Bicoid (Bcd) is placed at the end of the embryo that will become the head; importantly, the mRNA is attached to the end of the egg and is not free to move. Once the egg is laid, translation of this mRNA begins, and the resulting

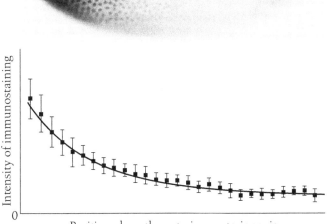

FIGURE 5.44

Antibody staining for the protein Bicoid (Bcd) in the early *Drosophila* embryo. The plot at the bottom represents means and 2× standard deviations from ten embryos; units of staining intensity are arbitrary. Reprinted from the original experiments by Driever and Nüsslein-Vollhard (1988a), with permission from Elsevier.

Bcd protein is free to move through the embryo. If we use the same trick as above and stop the action, labeling the embryo with antibodies against the protein, we see images like those in Fig. 5.44. Evidently there is a rather smooth gradient in the concentration of Bcd protein, high at one end and low at the other. A cell sitting at some point in the embryo thus can determine where it is along this long (anterior-posterior) axis by measuring the Bcd concentration. This is an example of the very general idea of positional information in the embryo.

Because Bcd is a transcription factor, it provides an input signal to a whole network of interacting genes, and this network can (if we speak colloquially) interpret the positional information, ultimately driving the emergence of the beautiful striped patterns as seen in Fig. 5.42. Measurements of the profile of Bcd concentration c show rather decent agreement with an exponential decay, as was already noted in the very first experiments (Fig. 5.44), so that

$$c(x) \approx c_0 e^{-x/\lambda}, \tag{5.99}$$

where x is the distance measured from the anterior end of the egg. Suppose, then, that the cephalic furrow is placed at the point where the Bcd concentration reaches some threshold value θ_{cf}. The position of the cephalic furrow is then

$$x_{cf} = \lambda \ln(c_0/\theta_{cf}). \tag{5.100}$$

Thus, if c_0 changes by $\sim$10%, the location of the furrow would shift by $\delta x_{cf} \sim 0.1\lambda$. Modern experiments show that $\lambda \sim 100 \ \mu$m, and the location of the cephalic furrow is reproducible with a standard deviation of $\sim$1% of the length of the embryo, or $\sim$5 μm in absolute length. In fact, one can look at other positional markers, such as the loca-

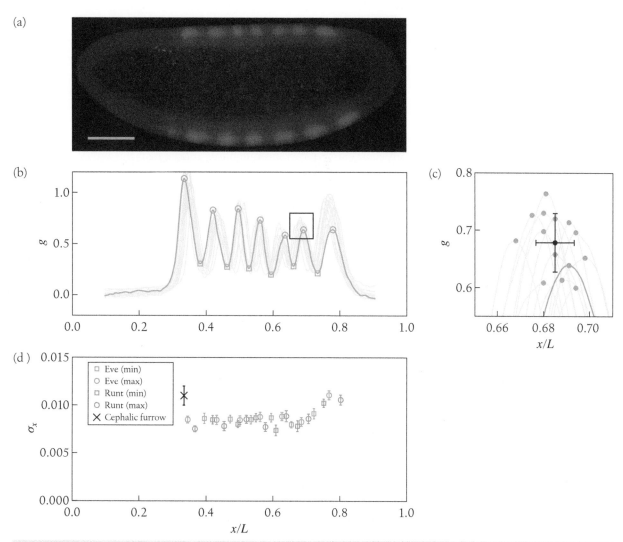

FIGURE 5.45

Reproducibility of various spatial markers along the anterior-posterior axis in the early *Drosophila* embryo. Redrawn from Dubuis et al. (2012). (a) Fluorescent antibody staining of the protein Eve; scale bar is 100 μm. (b) Normalized spatial profiles of the fluorescent intensity (g) in 14 embryos; the darker line is the embryo shown in panel (a). (c) A small region is blown up to show the variability of the peak; error bars show standard deviations of position and amplitude. (d) Standard deviations σ_x of position for several peaks and troughs of gene expression profiles, as well as for the position of the cephalic furrow measured in live embryos.

tions of peaks or troughs in the striped patterns of expression for the pair-rule genes in Fig. 5.42, and they are all reproducible at the ~1% level, as shown in Fig. 5.45. Thus, taken at face value, if the Bcd profile provides the basic map of position along the anterior-posterior axis, then the absolute concentration of Bcd, c_0, would have to be reproducible to better than ~10% from embryo to embryo to generate the observed reproducibility of these patterns. This problem exists even before we ask how to maintain constant proportions in the face of variations in the overall size of the embryo.

It may be dangerous to try and put ourselves into the minds of our fellow scientists, but I think that, when people started to think about this problem quantitatively, it seemed implausible that the reproducibility of embryonic development would depend on controlling absolute concentrations with 10% accuracy. To be more specific, I certainly know many thoughtful people who found this idea implausible, even if it was a deduction from well-known experiments; surely some step in the chain of reasoning must be wrong. However, the paper first characterizing the spatial profile of Bcd protein actually reported data on the variations across embryos (see Fig. 5.44), and the results are roughly consistent with ∼10% reproducibility, at least near the anterior end of the embryo.

Let's take seriously the simplest possible model for the spatial profile of Bcd, described above in words: the mRNA placed by the mother acts (as it is translated) as a source, the Bcd protein diffuses through the embryo, and the protein is also degraded by some first-order reaction. If we simplify and think of the system as just being one-dimensional (along the anterior-posterior axis), then the concentration $c(x, t)$ should obey

$$\frac{\partial c(x, t)}{\partial t} = D\frac{\partial^2 c(x, t)}{\partial x^2} - \frac{1}{\tau}c(x, t), \tag{5.101}$$

where τ is the lifetime of the protein against degradation. The boundary conditions are

$$-D\frac{\partial c(x, t)}{\partial x}\bigg|_{x=0} = R, \tag{5.102}$$

$$\frac{\partial c(x, t)}{\partial x}\bigg|_{x=L} = 0, \tag{5.103}$$

where R is the strength of the source at $x = 0$, and the last condition states that there is no flux out of the other end of the embryo. If development is slow enough for the system to come to steady state and the embryo is long, the concentration profile becomes

$$c_s(x) = \frac{R\tau}{\lambda}e^{-x/\lambda}, \qquad \lambda = \sqrt{D\tau}. \tag{5.104}$$

Problem 111: Details of the Bcd profile.

(a) What are the units of concentration in one dimension? Show that, with this proper choice of units, R is the number of Bcd molecules being translated per second.

(b) Derive the steady state solution Eq. (5.104). What is the precise criterion for the embryo to be long enough that this approximation is accurate?

(c) It is natural to ask whether the Bcd profile really reaches steady state during the early stages of development. Although this is an experimental question, we can ask what the simplest model predicts. Intuitively, there is some time scale t_* at which you expect the solution of Eq. (5.101) to reach steady state. How does this time scale relate to the other parameters of the problem? Answer this without doing any detailed calculations, and think about how your intuition might go astray.

(d) Try to do a more detailed calculation than you did in part (c) to address the approach to steady state. It is useful to assume from the beginning that L is large (in the sense of part (b)) and to replace the boundary condition at $x = 0$ with a source in the symmetrized version of the problem,

$$\frac{\partial c(x,t)}{\partial t} = D\frac{\partial^2 c(x,t)}{\partial x^2} - \frac{1}{\tau}c(x,t) + 2R\delta(x), \qquad (5.105)$$

where now $-\infty < x < \infty$; be sure you understand why we need a factor of 2 in front of the source term. At $t = 0$, before any protein has been translated, we must have $c = 0$ everywhere. By Fourier transforming in space, show that the exact time-dependent solution is

$$c(x,t) = 2R\int_{-\infty}^{\infty} \frac{dk}{2\pi} \frac{e^{ikx}}{Dk^2 + 1/\tau}\left[1 - e^{-(Dk^2+1/\tau)t}\right]. \qquad (5.106)$$

Verify that this approaches $c_s(x)$ from Eq. (5.104) as $t \to \infty$.

(e) Find a simple closed form for the time derivative of concentration at a point, $\partial_t c(x,t)$. Show that, expressed as a fraction of the local steady state concentration, this derivative peaks at a point $x_* = 2\lambda t/\tau$, and that at this peak $[\partial_t c(x_*,t)]/c_s(x_*) = 1/\sqrt{\pi\tau t}$.

(f) Suppose we could establish experimentally that, for example, after $t = 1\,\mathrm{hr}$, at each point x that we can see, $c(x,t)$ changes by less than $1\%/\mathrm{min}$ (or $\sim 10\%$ across the time required for the cell cycle). What can you conclude about the parameters of the system, taking the simple model seriously?

This simplest model recovers Eq. (5.99), which was suggested by the data. It gives us an explicit formula for the length constant λ and tells us (not surprisingly) that the absolute concentration scale c_0 is proportional to the strength of the source—that is, to the rate at which proteins can be translated from the mRNA bound to the anterior end of the embryo.

In this simple model, then, if we want c_0 to be reproducible with 10% accuracy, the source strength must also be reproducible. Is it plausible that the mother can count out mRNA molecules, with 10% accuracy, and create an environment in the embryo where the efficiency of translation is similarly well controlled? Alternatively, can we escape from these requirements of fine tuning by moving away from the simplest model?

Suppose that the processes that degrade the Bcd molecule act not on individual molecules, but on dimers, and these dimers are rare. We then expect that the concentration of dimers will be proportional to the square of the Bcd concentration, and instead of Eq. (5.101), the dynamics become

$$\frac{\partial c(x,t)}{\partial t} = D\frac{\partial^2 c(x,t)}{\partial x^2} - \frac{1}{\tau c_2}c^2(x,t), \qquad (5.107)$$

where c_2 is the concentration scale for dimer formation. Now the steady state solution must obey

$$\frac{d^2 c_s(x)}{dx^2} = \frac{1}{D\tau c_2}c_s^2(x). \qquad (5.108)$$

If we look for a solution of the form $c_s(x) = Ax^n$, we have

$$\frac{d^2(Ax^n)}{dx^2} = \frac{1}{D\tau c_2}(Ax^n)^2 \tag{5.109}$$

$$An(n-1)x^{n-2} = \frac{A^2}{D\tau c_2}x^{2n}, \tag{5.110}$$

which is solved by $n = -2$ and $A = 6D\tau c_2$. Thus, far from the source, the concentration profile is $c_s(x) = 6D\tau c_2/x^2$, independent of the strength of the source. More precisely, to match the boundary condition describing the source at $x = 0$, we must have

$$c_s(x) = \frac{6D\tau c_2}{(x+x_0)^2}, \qquad x_0 = (12D^2\tau c_2/R)^{1/3}. \tag{5.111}$$

The strength of the source appears only in x_0; for $x \gg x_0$, this term is negligible, and for large R this condition itself sets in at very small x. In this model, then, just making the source very strong—but not setting the strength precisely—is sufficient to ensure that almost the entire concentration profile will be independent of variations in this source strength.

Problem 112: Fill in the arguments leading to Eq. (5.111).

It is interesting that a relatively small change in molecular mechanism makes such a dramatic change in the robustness of the system to variations in parameters. Although this discussion has emphasized the role of nonlinearities in the degradation process, one can achieve similar effects through nonlinearities in diffusion, which might arise if the diffusing molecules are bound to receptors and "handed" from one cell to the next rather than moving freely in solution.

One might object that there is no free lunch. Even though Eq. (5.111) predicts that the Bcd profile is largely independent of the source strength, the concentration scale is now set by c_2, which has something to do with the dimerization of the molecules. The source strength R depends on how many copies of mRNA the mother places in the egg, but the scale c_2 is determined by more global physicochemical parameters of the cytoplasm, and perhaps these are easier to control. A complementary approach is to give up on making a single morphogen signal reproducible and to assume that the embryo makes use of multiple signals, hoping that the dominant sources of variation are in a common mode that can be rejected by the network processing these signals. Several such models have been suggested, as explained in the references in the Annotated Bibliography for this section.

With all this theoretical background, what can we say about the experimental situation? As noted at the outset, there are hints from the earliest literature that Bcd profiles in *Drosophila* might indeed be reproducible. We also know that the notion of robustness should not be exaggerated. The success of classical genetics in identifying the components of these networks immediately tells us that the system is not resistant to the elimination of single components. More subtly, one of the key experiments in establishing that Bcd is a primary source of positional information was to change the

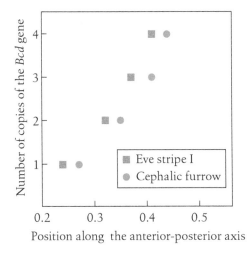

FIGURE 5.46

Variations in the position of spatial markers (as in Fig. 5.45) along the anterior-posterior axis in response to changes in the number of copies of the *bcd* gene. Redrawn from Driever and Nüsslein-Vollhard (1988b).

number of copies of the *bcd* gene; with more (or fewer) copies of the gene in its genome, the mother makes more (or less) mRNA and hence drives the strength of the Bcd source up (or down). In response to these changes, the patterns in the early embryo shift, with the cephalic furrow in particular moving: with higher concentrations of Bcd, the embryo tries to make a larger head, as shown in Fig. 5.46.[13] These results suggest that the embryo does not engage mechanisms that buffer the Bcd profile against variations in the strength of the source. For the morphogens whose concentration profile varies along the other axis of the embryo, however, there are signatures of the nonlinear degradation mechanism, which, as we have seen, can generate substantial robustness.

If there is no buffering, then it really does seem that reproducible outputs require reproducible inputs. Can we see this directly? As discussed in Section 4.3, one can genetically engineer flies to express a fusion of Bcd with GFP and show that this fusion protein quantitatively replaces the function of the native molecule. Figure 5.47 shows measurements of the concentration of Bcd in nuclei from 15 different embryos, using this Bcd-GFP fusion. The raw fluorescence intensity (or the inferred concentration) is plotted versus position along the anterior-posterior axis for each nucleus. Evidently the variability from embryo to embryo is small, with a standard deviation of less than 20%, and some of this variability can be traced to measurement errors, suggesting that the true variability is ~10% or even less. If the mother has only one copy of the gene for Bcd-GFP instead of the usual two, the fluorescence really is cut in half, so again there is no evidence of mechanisms that buffer the observable profile against variations in the strength of the source. This observation strongly suggests that the mother can place a reproducible number of mRNA molecules into the egg and that the apparatus for translation has a constant efficiency from embryo to embryo as well. It would be

---

13. It is not so easy to interpret these results quantitatively, because we do not really know if adding more copies of the gene produces *proportionately* higher concentrations of the protein. Still, Fig. 5.46 is prima facie evidence against complete robustness of the pattern to variations in the strength of the source. It is also interesting that, later in development, some of these errors can be corrected, although only a fraction of embryos survive to complete this error correction.

Measurements of the bicoid (Bcd) concentration in nuclei along the anterior-posterior axis of the *Drosophila* embryo. Each point corresponds to one nucleus in one embryo; points of the same color come from the same embryo, and error bars show the means and standard deviations across the 15 embryos in the experiment. The vertical axis shows the fluorescence signal in embryos engineered to make the Bcd–green fluorescent protein fusion, which can be calibrated to give the absolute concentration (at right). The horizontal axis shows the position of the nucleus x as a fraction of the overall length L of the embryo. Redrawn from Gregor et al. (2007b).

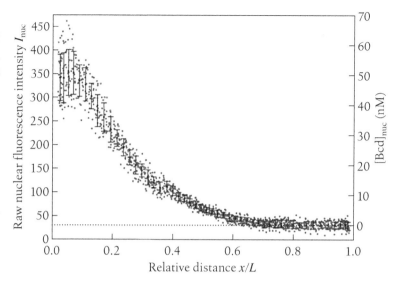

attractive to have direct measurements that confirm these conclusions. Of course, this also pushes the problem back. How does the mother count mRNA molecules with ∼10% accuracy? How does the embryo ensure that the efficiency of translation, which depends on myriad factors, is reproducible?

The problem we have been discussing thus far emerges as soon as we claim that position in the embryo is encoded by the concentration of specific molecules. In such a scheme, if we want neighboring cells to do different things reliably, then we will be driven to questions about how these cells can distinguish small differences in concentration, as discussed in Section 4.3.[14] Conversely, if we want two cells that occupy corresponding positions in different embryos to do the same thing, then we are driven to ask how the concentrations at these corresponding points can be made the same. These issues of precision and reproducibility arise even if the size of the embryo and the external conditions of development are identical. There is another problem, related to the variations in size of the embryo, and this is the problem of scaling.

To a remarkable extent, the proportions of organisms are constant, despite size variations. We all know people who have especially large heads, but certainly the proportions of the body vary much less than the overall size. Again, insects provide clear examples of this constancy, both within and across species. Different species of flies, for example, have embryos that span a factor of five or more in length, yet they have the same number of body segments, and individual segments have dimensions that scale with the overall size of the organism. You can see this scaling not just in the macroscopic patterns of the developed organisms but also in the patterns of gene expression, as shown in Fig. 5.48. Indeed, when we have looked at the problem of reproducibility above, we have implicitly used the idea of scaling, always plotting position as a fractional distance along the anterior-posterior axis.

---

14. See also the discussion of positional information, in bits, in Section 6.1.

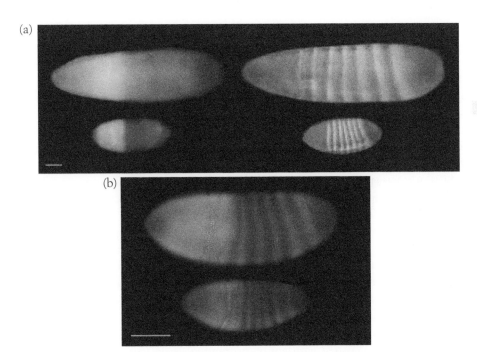

(a)

(b)

FIGURE 5.48

Immunofluorescence stainings for products of the gap and pair-rule genes in flies of different sizes. From Gregor et al. (2005). (a) Staining of *Lucilia sericata* (upper embryos) and *Drosophila melanogaster* (lower embryos) for Hunchback (green) and Giant (red) in the left column, and for Paired (green) and Runt (red) in the right column. (b) Staining of *D. melanogaster* (upper embryo) and *D. busckii* (lower embryo) for Hunchback (green) and Runt (red). Scale bars are 100 μm.

Scaling is deeply puzzling, perhaps especially so for physicists who have thought about pattern formation in nonbiological systems. To make this point, let's imagine making a model of the whole network of interactions that lead to, for example, the beautiful stripes of gene expression. In each nucleus there are chemical reactions corresponding to the transcription of the relevant genes, and the rates of these reactions are determined by the concentrations of the appropriate transcription factors. More equations will be needed to describe translation (although maybe one can simplify, if, for example, mRNA molecules degrade quickly and proteins live longer). Different points in space are coupled, presumably through diffusion of all these molecules, although we should worry about whether diffusion is the correct description. Even if you are not sure about the details, you can imagine the form of the equations: some sort of partial differential equations, in which the local time dependence of concentrations has contributions both from nonlinear terms describing the various chemical reactions and from spatial derivatives describing diffusion or other transport processes:

$$\frac{\partial g_i(x,t)}{\partial t} = D_i \nabla^2 g_i(x,t) + F_i(\{g_j\}). \tag{5.112}$$

This set of equations is some generalization of Eq. (5.89) and (Eq. 5.90) for the Turing scheme discussed above. But we have seen equations like these before in the study of nonbiological pattern forming systems such as Rayleigh-Bénard convection, directional solidification, and the like.

Many nonbiological pattern-forming systems generate periodic spatial patterns reminiscent of the segments in insects and the patterns of pair-rule gene expression. The scale of these patterns, however, is set by combinations of parameters in the equations. For example, we can combine a diffusion constant with a reaction rate or lifetime to get a length, as in the discussion of the Bcd profile above ($\lambda = \sqrt{D\tau}$). What happens if you put these equations in a larger box? Well, from Rayleigh-Bénard convection, we know the answer. In this system—a fluid layer heated from below—we see a collection of convective rolls, sometimes in the form of stripes and sometimes as two-dimensional cellular patterns (see Fig. 6.18 in the next chapter). Again, the length scale of the stripes is determined by the parameters of the equation(s). If you put the whole system in a bigger box, you get more stripes, not wider stripes.

The results in Fig. 5.48 come close to suggesting that we can put all the same equations into a bigger box, and the stripes come out wider in proportion to the length of the box. One might worry that these organisms are different, and so perhaps evolution has tuned the properties of the proteins involved so that the relevant combinations of parameters turn out to scale with embryo size. The differences can't be too large, because we can identify the same molecules as being involved through similarities of amino acid sequence, and because the same antibodies react with these molecules in different species. Still, it is possible that scaling across embryos in different species reflects an evolutionary adaptation.

If we look across related species of flies with embryos of very different sizes, then the Bcd profiles (as measured with antibody staining) seem to scale with the length of the egg. One can use the same experimental methods used in making the Bcd-GFP fusion more aggressively, extracting the sequences of Bcd from flies of different sizes and reinserting green versions of these different Bcds into the *Drosophila* genome. The striking result is that the resulting spatial profiles are those appropriate to the host embryo, not the source of the Bcd. Taken together, all these results suggest that, as with the problem of variability, the scaling problem (at least across species) is solved at the level of Bcd itself. It would appear that there is something about the environment or geometry of the embryo itself that couples the global changes in the size of the embryo to the local dynamics. It is not known whether such effects exist across the variations in egg size within a species, or if this effect only occurs, on average, across species.

Problem 113: Is scaling so mysterious?   Think of the (roughly ellipsoidal) embryo as a cylinder, with the source covering one end of this cylinder. Because most of the interior of the egg is yolk, we imagine that all degradation of proteins occurs near the surface of the egg. If the degradation reaction is rapid, then the surface of the embryo acts as a sink, and in the interior of the embryo the concentration obeys the diffusion equation with no additional terms.

(a) Assuming cylindrical symmetry, show that the steady state profile c_s obeys

$$\frac{\partial^2 c_s(x,r)}{\partial x^2} + \frac{1}{r}\frac{\partial}{\partial r}\left[r\frac{\partial c_s(x,r)}{\partial r}\right] = 0, \tag{5.113}$$

$$-D\frac{\partial c_s(x,r)}{\partial x}\bigg|_{x=0} = \frac{R}{\pi r_0^2}, \tag{5.114}$$

$$c(x, r = r_0) = 0, \tag{5.115}$$

where x measures position along the axis of the cylinder (the anterior-posterior axis of the embryo), r_0 is the radius of the cylinder, R is once again the number of molecules per second being injected by the source, and the last condition follows in the limit that degradation reactions at the surface are fast.

(b) Use the separation of variables method to look for solutions of the form $c_s(x, r) = e^{-x/\lambda} f(r)$, and derive the equation that determines $f(r)$. Hint: This is the equation for one of the well-studied special functions.

(c) Solve the equation you derived in part (b), being careful about the boundary conditions. Show that your solution implies a connection between the radius of the embryo r_0 and the length constant of the gradient, λ. Is this enough to solve the scaling problem? Perhaps there is something else we should measure, to see whether we are on the right track?

(d) Looking at images of the fly embryo displayed earlier in this section, estimate the radius r_0 assuming that the length of the embryo is $L \sim 0.5\,\text{mm}$. Does your prediction for λ in part (c) actually work quantitatively?

The discussion so far has taken seriously the idea that there are primary morphogens, placed by the mother, that provide the basic signal for position in the embryo. Position is a continuous variable, as is concentration. A profoundly different perspective emphasizes that, when development is finished, cells have adopted distinct types or fates, which define their function in the adult organism. These fates persist long after the primary morphogen signals have disappeared, and so they must represent stable states of the cells, thus bringing us back to the theme of this section. Cells even maintain their identities and states when separated from their neighbors, which suggests that the biochemical and genetic networks in each cell have multiple attractors. A minimal model of the networks relevant for development, then, would have the right number of attractors but a limited number of dynamical variables, perhaps many fewer than the number of genes involved in the entire network. As with the attractors in the Hopfield model for neural networks discussed in the next section, there is a plausible path to robustness, because changing the qualitative behavior of the system would actually require changing the number of attractors—the development of cells into types becomes a matter of topology rather than geometry in the model and hence is invariant to a finite range of parameter variation.

Problem 114: Two cell fates. Consider two genes, each of which encodes a transcription factor that represses the other. If g_1 (g_2) is the concentration of the protein encoded by the first (second) gene, then a plausible model is

$$\frac{dg_1}{dt} = r_{\max} f(g_2) - \frac{g_1}{\tau}, \tag{5.116}$$

$$\frac{dg_2}{dt} = r_{\max} f(g_1) - \frac{g_2}{\tau}, \tag{5.117}$$

where, as before, $r_{\max}$ is the maximum rate of synthesis, τ is the mean lifetime of these molecules against degradation, and the function $f(g)$ describes the repression of transcription.

To be concrete, we can choose

$$f(g) = \frac{K^n}{K^n + g^n}. \tag{5.118}$$

(a) Show that by proper choice of units for g and t, there is really only one free parameter in this model other than the exponent n.

(b) Find the steady state solutions. In particular, what are the conditions for having two stable solutions?

(c) Suppose the system is in a regime where there are two stable solutions, and we change the value of the free parameter by a small amount. Do the solutions persist? How robust are the two states? More seriously, what if we change the form of $f(g)$, for example,

$$f(g) = \frac{K^n}{K^n + \epsilon K^{n/2} g^{n/2} + g^n}, \tag{5.119}$$

so that the cooperativity of repression is not quite as sharp? Do the two stable states persist for finite ϵ?

The usual approach to models that have multiple stable states, as in Problem 114, is to give a microscopic formulation in which the variables are identifiable with particular molecular components and then try to show that these models have the desired properties. An alternative approach is more abstract. We know that as parameters in models are varied, stable states can appear and disappear at critical points, or bifurcations. Close to these critical points, the dynamics take on universal or "normal" forms, and in the right coordinate system, they become simple. If we imagine that, during the course of embryonic development, slow changes in the effective parameters drive the relevant networks through the bifurcations that create the multiple stable states, then with a little luck these universal forms of dynamics will describe the process by which cells make decisions. Further, if we think about the minimal models that implement these universal dynamics, we can classify them.

Figure 5.49 shows an example of a system that has two degrees of freedom and supports three stable states, corresponding to three cell fates, corresponding to what happens in development of the vulva in the roundworm *Caenorhabditis elegans*. The three fixed points of the dynamics can be arranged in three different ways, one of which corresponds roughly to a model in which cells can be pushed sequentially from one fate to the next by a graded morphogen signal, one of which depends on fine tuning of the decision processes, and one of which allows cells to make all possible transitions among pairs of states if driven by noise during the decision.

Once the correct topology is established, we can ask how changing the concentration of different molecular components (e.g., in genetic manipulations) maps into pushing the dynamics in one direction or another in the plane. Thus, rather than trying to use variables that are faithful to (often incompletely known) details, we can use coarse-grained variables that are guaranteed, in certain limits, to capture the essence of the dynamics, and then try to map these variables back to the particular genes involved, relating perturbations to quantities that can be varied or measured experimentally. Im-

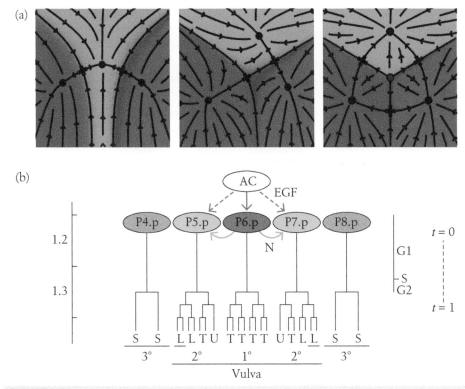

FIGURE 5.49

Three cell fates. From Corson and Siggia (2012), with thanks. (a) A dynamical system with two degrees of freedom can have three fixed points with different arrangements. At left, the stable states are ordered, so that graded perturbations can push the system from blue to green to red, or vice versa, but there is no way to jump directly between red and blue states. At right, in contrast, all transitions are possible, whereas in the middle the choice involved in a transition from blue to green versus red depends very sensitively on the initial conditions and noise along the trajectory. (b) The development of fates for a handful of cells in the worm *Caenorhabditis elegans* as they respond to growth factor (EGF) and interact with one another (N). The coloring of the cells denotes a commitment to fates at an early time (L2 and L3 denote larval stages; G1, S, and G2 denote stages in the cell cycle), which is then expressed by their offspring. 1°, 2° and 3° are names given to these lineages; L, T, U, and S mark the geometry of the cell divisions.

portantly, this approach makes clear from the start that robustness against parameter variation arises because meaningful cellular states are encoded in topology.

5.4   Long Time Scales in Neural Networks

The basic time scales of electrical dynamics in neurons are measured in milliseconds, yet the time scales of our mental experience are much longer. From the fraction of a second that we need to integrate sounds as we identify words or phrases, to the minutes of memory required for a phone number, to the decades over which our recollections of childhood experiences can stretch, the brain has access to time scales far beyond those describing the elementary events of action potential generation and synaptic transmission. If we write a set of dynamical equations, and the time scales that emerge to describe the whole system are much longer than the time scales appearing as parameters in the equations, then something special has happened. How does this work in the brain? How does the system ensure that this seemingly special separation of time scales occurs without the fine tuning of parameters?

One possible solution to the wide range of relevant time scales is to invoke a correspondingly wide range of mechanisms, and surely this is part of the answer. Thus, it seems unlikely that memories of things long past are stored as continuing patterns of electrical activity in the brain, which somehow last $\sim 10^{10} \times$ longer than their natural time scale and are always present to be examined as we reminisce. In contrast, for working memory—holding the words of a sentence in our minds, or doing mental arithmetic—the time scales involved seem at once long compared with natural time scales for electrical activity, yet too short to engage biochemical mechanisms, such as the regulation of gene expression, which could have more stable, semipermanent effects.

In fact, we know a whole class of examples in which long time scales emerge naturally. When a ball rolls down a hill, the time scale of the rolling may be short, but once at the bottom the ball can stay there (more or less) forever. So, perhaps we can arrange for the dynamics of neurons in an interconnected network to be like the motion of a particle on a (multidimensional) landscape, with nice deep valleys corresponding to patterns of activity that can persist for a long time once the system finds itself in the right neighborhood. In two hugely influential papers in the early 1980s, Hopfield showed how to do exactly this.

Neurons receive inputs from many other neurons. In the retina, we have seen single bipolar cells that receive input from tens to hundreds of rod photoreceptors (Section 2.4). In the central nervous system, the numbers are even more impressive: in the cortex a single cell collects signals from several thousand other cells, and in the extreme case of the cerebellum "many" actually means $\sim 10^5$. Conversely, although each cell has only one axon along which its output action potentials are sent, this axon can branch to contact thousands of other cells. Let's focus on one cell i, which receives inputs from many other cells j, as shown in Fig. 5.50. Schematically, we can imagine that each cell is either active or inactive, on or off, and hence the state of one cell can be represented by a binary variable $\sigma_i = \pm 1$; for the moment we will leave this model as a schematic and not try to interpret σ_i too closely in terms of action potentials or membrane voltage. In the simplest view, each cell j sends its output to cell i, and as these inputs are collected from the synapses, they are summed with some weights W_{ij}, which we can think of as the "strengths" of the synapses from cell j to cell i. Having summed its inputs, cell i must then decide whether to be on or off, comparing the total input to a threshold θ_i. This description is equivalent to saying that the state of cell i is set according to the equation

$$\sigma_i \rightarrow \text{sgn} \left[\sum_j W_{ij}\sigma_j - \theta_i \right]. \tag{5.120}$$

Models of this flavor go back at least to the 1940s, when McCulloch and Pitts explored the idea that the on/off states of neurons could implement a kind of logical

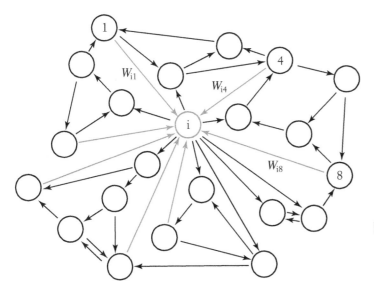

FIGURE 5.50

Schematic network of neurons, focusing on one cell i that receives inputs from many other cells $j = 1, 4, 8, \cdots$.

calculus.[15] Precisely because they can perform such general operations, these sorts of discrete dynamics can be almost arbitrarily complicated. Thus, without making some simplifying assumptions, it is hard to say anything about the dynamics generated by Eq. (5.120).

Suppose, however, that if neuron j synapses onto neuron i with strength W_{ij}, then neuron i synapses onto neuron j with the same strength, so that the matrix of synaptic strengths W_{ij} is symmetric. Then the updating of the state of neuron i in Eq. (5.120) serves to reduce an "energy" function defined by

$$E = -\frac{1}{2} \sum_{i,j} \sigma_i W_{ij} \sigma_j + \sum_i \theta_i \sigma_i. \tag{5.121}$$

This is the energy of the Ising model: the "spins" σ_i experience "magnetic fields" θ_i and interact with one another through couplings W_{ij}. Indeed, we can think of Eq. (5.120) as being the dynamics of a zero-temperature Monte Carlo simulation of an Ising model with energy defined by Eq. (5.121). Now we can make progress.

Problem 115: Energy in the Hopfield model. Show explicitly that the dynamics in Eq. (5.120) serves to decrease the energy function in Eq. (5.121).

If we can map the dynamics of a neural network onto the Ising model, then we can apply an enormous amount of our intuition (and mathematical tools) from statistical mechanics. Because the dynamics we have defined are at zero temperature—we are neglecting, for the moment, any noise in the neurons or synapses—it is possible to

15. This model should also remind you of the Boolean model for the biochemical network in the cell cycle, Eq. (5.84).

have collective states of the whole system that are stable forever. The simplest example is that with all thresholds equal to zero and all synaptic strengths equal and positive. Then the energy function becomes

$$E = -\frac{W}{2} \sum_{i,j} \sigma_i \sigma_j = -\frac{W}{2} \left(\sum_i \sigma_i \right)^2 . \tag{5.122}$$

This equation describes the mean-field ferromagnet: a ferromagnet because the energy is lowered by aligning all the spins, and mean-field because each spin interacts with all other spins.[16] In this model there are two stable ground states—all neurons on ($\sigma_i = +1$ for all i) and all neurons off ($\sigma_i = -1$ for all i). Two states aren't many, and these states seem especially odd, but maybe we are on the right track.

If instead of making all W_{ij} equal, we choose them at random, then the Ising model we have constructed is the mean-field or Sherrington-Kirkpatrick spin glass. We know that this system has many locally stable states, with an energy landscape that has valleys within valleys, as discussed in Section 5.1. This is probably too much, because the structure of these exponentially large number of states depends very sensitively on the precise form of the couplings W_{ij}. More generally, with N neurons, we only have $\sim N^2$ parameters at our disposal when we adjust the W_{ij}, so it is difficult to imagine how we could program the network to store exponentially many independent patterns.

To find a compromise between the ferromagnet and the spin glass, we make use of a trick borrowed from models for magnetism. Suppose that

$$W_{ij} = W \xi_i \xi_j, \tag{5.123}$$

where $\boldsymbol{\xi}$ is an arbitrary binary vector, $\xi_i = \pm 1$, and for simplicity let the thresholds $\theta_i = 0$. Then the energy becomes

$$E = -\frac{1}{2} \sum_{i,j} \sigma_i W_{ij} \sigma_j$$

$$= -\frac{W}{2} \sum_{i,j} \sigma_i \xi_i \xi_j \sigma_j \tag{5.124}$$

$$= -\frac{W}{2} \sum_{i,j} \left(\xi_i \sigma_i \right) \left(\xi_j \sigma_j \right) \tag{5.125}$$

$$= -\frac{W}{2} \sum_{i,j} \tilde{\sigma}_i \tilde{\sigma}_j, \tag{5.126}$$

where $\tilde{\sigma}_i = \xi_i \sigma_i$ is again a binary variable: $\tilde{\sigma}_i = \pm 1$. The transformation $\sigma_i \to \tilde{\sigma}_i$ is a discrete gauge transformation, so we say that the model with weights in Eq. (5.123) is gauge equivalent to a ferromagnet. Rather than the stable states of the system being

---

16. The attentive reader will notice a slight glitch in Eq. (5.122). Presumably the initial sum over pairs ij should exclude i = j, because neurons (in this picture) do not interact with themselves, and in any case $\sigma_i^2 = 1$, so this self-interaction makes only a constant contribution to the energy. But we need this term to complete the square in the last step. So, we should really add and subtract the relevant constant, shifting the zero of energy. But this doesn't change anything, so I let it slip by here.

$\sigma_i = +1$ for all i and $\sigma_i = -1$ for all i, the stable states are $\sigma_i = +\xi_i$ and $\sigma_i = -\xi_i$. Importantly, this construction can be generalized.

Rather than Eq. (5.123), consider what happens if we have two binary patterns $\boldsymbol{\xi}^{(1)}$ and $\boldsymbol{\xi}^{(2)}$, and we form the matrix W_{ij} as

$$W_{ij} = W \left(\xi_i^{(1)} \xi_j^{(1)} + \xi_i^{(2)} \xi_j^{(2)} \right). \tag{5.127}$$

Now we have

$$E = -\frac{1}{2} \sum_{i,j} \sigma_i W_{ij} \sigma_j$$

$$= -\frac{W}{2} \left[\sum_{i,j} \sigma_i \xi_i^{(1)} \xi_j^{(1)} \sigma_j \right] - \frac{W}{2} \left[\sum_{i,j} \sigma_i \xi_i^{(2)} \xi_j^{(2)} \sigma_j \right] \tag{5.128}$$

$$= -\frac{W}{2} \left[\left(\boldsymbol{\xi}^{(1)} \cdot \boldsymbol{\sigma} \right)^2 + \left(\boldsymbol{\xi}^{(2)} \cdot \boldsymbol{\sigma} \right)^2 \right]. \tag{5.129}$$

Clearly the energy will be low if the pattern of neural activity $\boldsymbol{\sigma}$ is parallel to the vector $\boldsymbol{\xi}^{(1)}$ or to the vector $\boldsymbol{\xi}^{(2)}$. But in a high-dimensional space, two randomly chosen vectors are, with high probability, nearly orthogonal. Thus, as $N \to \infty$, the two terms in the Hamiltonian can't both be important at once. Thus, the energy function will have a minimum near $\boldsymbol{\sigma} = \boldsymbol{\xi}^{(1)}$ and a separate minimum near $\boldsymbol{\sigma} = \boldsymbol{\xi}^{(2)}$, as well as the flipped versions of these states, $\boldsymbol{\sigma} = -\boldsymbol{\xi}^{(1)}$ and $\boldsymbol{\sigma} = -\boldsymbol{\xi}^{(2)}$.

Problem 116: Random vectors in high dimensions. Consider random binary vectors $\mathbf{v}$ in an N-dimensional space: $\mathbf{v} \equiv \{v_1, v_2, \ldots, v_N\}$, where each $v_i = \pm 1$ is chosen independently and at random. The angle ϕ between two such vectors is defined in the usual way by normalizing the dot product:

$$\cos \phi \equiv \frac{1}{N} \mathbf{v}^{(1)} \cdot \mathbf{v}^{(2)}. \tag{5.130}$$

Before calculating anything, explain why, if $\mathbf{v}^{(1)}$ and $\mathbf{v}^{(2)}$ are chosen independently, it must be that $\langle \cos \phi \rangle = 0$. Calculate the variance $\langle \cos^2 \phi \rangle$ to show that the typical values of $|\cos \phi| \sim 1/\sqrt{N}$, which vanishes as $N \to \infty$. Can you use the central limit theorem to say something about the whole probability distribution $P(\cos \phi)$ in this limit? Show that the distribution can be written exactly as

$$P(z = \cos \phi) = \int \frac{dk}{2\pi} e^{-ikz} \left[\cos(k/N) \right]^N. \tag{5.131}$$

Connect this result to the predictions of the central limit theorem. Develop a saddle point approximation, so that you can calculate, at large N, $P(z)$ for values of $|z| \gg 1/\sqrt{N}$. Verify your approximations with a simulation.

The key idea now is to go further, with not just two patterns but K patterns, writing the weights as

$$W_{ij} = W \sum_{\mu=1}^{K} \xi_i^{(\mu)} \xi_j^{(\mu)}. \tag{5.132}$$

Then the energy becomes

$$E = -\frac{W}{2} \sum_{ij} \sigma_i \left[\sum_{\mu=1}^{K} \xi_i^{(\mu)} \xi_j^{(\mu)} \right] \sigma_j = -\frac{W}{2} \sum_{\mu=1}^{K} \left(\boldsymbol{\xi}^{(\mu)} \cdot \boldsymbol{\sigma} \right)^2 . \tag{5.133}$$

Certainly if $K \ll N$, our intuition from the case of two patterns should carry over, because almost all the vectors $\boldsymbol{\xi}^{(\mu)}$ will be nearly orthogonal, and we should find that the energy function has $2K$ minima, near the vectors $\pm\boldsymbol{\xi}^{(\mu)}$. However, if we let K itself become large, we must get back to the spin glass model in which there are many locally stable states, but they do not have any connection to the patterns $\boldsymbol{\xi}^{(\mu)}$ that we have programmed into the system. Somewhere in between these regimes—that is, between $K \ll N$ and $K \sim N$—there must be a transition, and it is plausible that this transition is sharp in the limit of large N. Thus, we expect that for large networks, as we try to store more and more memories, there is a genuine phase transition between states corresponding to effectively recallable memories and some sort of confused or disordered state.

Problem 117: Simulating the Hopfield model.   Given a matrix W_{ij}, it is straightforward to simulate the dynamics of the Hopfield model, as defined by Eq. (5.120); try the simplest case, with $\theta_i = 0$. To run the simulation, you can go through these steps:

1. Start a collection of N spins in some randomly chosen state.
2. Choose one spin i at random.
3. Set $\sigma_i = \text{sgn} \left[\sum_j W_{ij}\sigma_j \right]$.
4. Choose another spin and repeat the update, again and again.

Produce a series of simulations to convince yourself that, with W_{ij} chosen as in Eq. (5.132) and a small value of K, the dynamics always stop in the neighborhood of one of the vectors $\boldsymbol{\xi}^{(\mu)}$ that you have used in sculpting the energy landscape. Explore what happens as K becomes larger. If you jump to $K \sim N/2$, can you see the emergence of more random stopping points for the dynamics? Perhaps even if you start at one of the vectors $\boldsymbol{\xi}^{(\mu)}$, does the interference from the other vectors destabilize this state? If the dynamics stops at a state $\boldsymbol{\sigma}_s$, define an order parameter by finding the nearest vector $\boldsymbol{\xi}^{(\mu)}$ and measuring the normalized dot product:

$$m_s = \max_{\mu} \frac{1}{N} \left| \boldsymbol{\xi}^{(\mu)} \cdot \boldsymbol{\sigma}_s \right| . \tag{5.134}$$

From many random starting points, what is the mean value of m_s as a function of K and N? As N gets larger, do you see the emergence of a thermodynamic limit, where the (intensive) order parameter $\langle m_s \rangle$ depends only on the ratio K/N? Are there signs of a phase transition at some critical value of K/N?

To get a sense of the transition between recallable memories and the disordered state, we can think of the energy function in Eq. (5.133) as telling us that there is an effective field on each spin:

$$h_i^{(\text{eff})} = \sum_{j\neq i} \sum_{\mu=1}^{K} \xi_i^{(\mu)} \xi_j^{(\mu)} \sigma_j. \tag{5.135}$$

Suppose the system is in one of the stored memory states, so that $\sigma_i = \xi_i^{(\nu)}$. Then the field on spin i becomes

$$h_i^{(\text{eff})} = \sum_{j\neq i} \sum_{\mu=1}^{K} \xi_i^{(\mu)} \xi_j^{(\mu)} \xi_j^{(\nu)} \tag{5.136}$$

$$= \xi_i^{(\nu)} \sum_{j\neq i} \xi_j^{(\nu)} \xi_j^{(\nu)} + \sum_{\mu\neq\nu} \xi_i^{(\mu)} \sum_{j\neq i} \xi_j^{(\mu)} \xi_j^{(\nu)}. \tag{5.137}$$

Because the patterns being stored are binary, the first term simplifies, so that

$$h_i^{(\text{eff})} = (N-1)\xi_i^{(\nu)} + \delta h_i, \tag{5.138}$$

where

$$\delta h_i = \sum_{\mu\neq\nu} \xi_i^{(\mu)} \sum_{j\neq i} \xi_j^{(\mu)} \xi_j^{(\nu)}. \tag{5.139}$$

If the stored patterns are chosen at random, then δh_i is on average equal to zero. This field is the sum of many terms, so it should come from a Gaussian distribution in the limit of large N, and the variance is given by

$$\langle (\delta h_i)^2 \rangle = \left\langle \left(\sum_{\mu\neq\nu} \xi_i^{(\mu)} \sum_{j\neq i} \xi_j^{(\mu)} \xi_j^{(\nu)} \right)^2 \right\rangle \tag{5.140}$$

$$= \sum_{\mu_1,\mu_2\neq\nu} \sum_{j,k\neq i} \langle \xi_i^{(\mu_1)} \xi_j^{(\mu_1)} \xi_j^{(\nu)} \xi_i^{(\mu_2)} \xi_k^{(\mu_2)} \xi_k^{(\nu)} \rangle \tag{5.141}$$

$$= \sum_{\mu_1,\mu_2\neq\nu} \sum_{j,k\neq i} \langle \xi_i^{(\mu_1)} \xi_j^{(\mu_1)} \xi_i^{(\mu_2)} \xi_k^{(\mu_2)} \rangle \langle \xi_j^{(\nu)} \xi_k^{(\nu)} \rangle, \tag{5.142}$$

where in the last step we make use of the fact that the pattern labeled by ν is independent of all other patterns $\mu \neq \nu$. But because each element of the pattern also is chosen at random as ± 1, we must also have $\langle \xi_j^{(\nu)} \xi_k^{(\nu)} \rangle = \delta_{jk}$, so that

$$\langle (\delta h_i)^2 \rangle = \sum_{\mu_1,\mu_2\neq\nu} \sum_{j\neq i} \langle \xi_i^{(\mu_1)} \xi_j^{(\mu_1)} \xi_i^{(\mu_2)} \xi_j^{(\mu_2)} \rangle \tag{5.143}$$

$$= \sum_{\mu_1,\mu_2\neq\nu} \sum_{j\neq i} \langle \xi_i^{(\mu_1)} \xi_i^{(\mu_2)} \rangle \langle \xi_j^{(\mu_1)} \xi_j^{(\mu_2)} \rangle \tag{5.144}$$

$$= \sum_{\mu_1,\mu_2\neq\nu} \sum_{j\neq i} \delta_{\mu_1\mu_2} \delta_{\mu_1\mu_2} \tag{5.145}$$

$$= (K-1)(N-1). \tag{5.146}$$

Thus, the field acting on spin i has a strength $\sim N$ causing that spin to align with the stored pattern labeled by ν, and a strength $\sim\sqrt{NK}$ causing it to align at random,

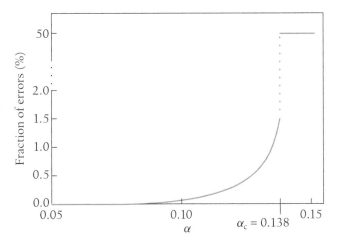

FIGURE 5.51

Recall of memories in the Hopfield model. The number of patterns K stored in the network is proportional to the number of neurons, so that $\alpha = K/N$ is the natural parameter. As α increases, the fraction of errors in the recalled patterns increases, until at some critical value α_c recall fails catastrophically, and the states of the network are not correlated with the stored patterns at all. Redrawn from Amit et al. (1985).

assuming that both K and N are large. As long as $\sqrt{NK} \ll N$, or equivalently $K \ll N$, the mean field dominates the random field, and the stored memory is stable. For some value of K/N, we expect that the random field will win: even if all $N-1$ neurons are in the state corresponding to a stored memory, this is not enough to force the last spin into the correct state, and hence the stored pattern becomes unstable.

It is not so easy to improve on this rough calculation. To do better, we really have to understand what happens in systems where the parameters of the energy function are chosen at random, which is technically demanding. As explained in the references in the Annotated Bibliography for this section, the result of these much more difficult calculations, summarized in Fig. 5.51, is not so far from our rough estimate: there are stable stored memories in large networks so long as $\alpha = K/N$ is smaller than a critical value $\alpha_c \approx 0.14$. The subtlety is that once the number of stored patterns is proportional to the number of neurons, the recall is not quite perfect—the patterns σ_i that are stable are ~1% different from the patterns ξ_i that were stored. This is fine, perhaps, so long as the patterns are distinguishable from one another. Importantly, in large networks ($N \to \infty$) the transition between recalling memories for $\alpha < \alpha_c$ and failure for $\alpha > \alpha_c$ is sharp, and failure is catastrophic.

Let's think about how much progress we have made toward solving our original problem. The Hopfield model shows how the dynamics of a neural network can correspond to "downhill" motion on an energy landscape, much like a ball rolling down a hill. Thus, the system as a whole has collective, macroscopic states that will persist for times arbitrarily long compared with the basic time scales of the system—the time scales on which the individual neurons update their microscopic states according to Eq. (5.120). Importantly, there are not just a few of these stable states, but many of them, in proportion to the number of neurons. Unlike the case of the ball coming to a stop at the bottom of the hill, the stability of these states is the result of activity, each neuron receiving continuous input from other neurons in the network; in effect the stable states are patterns of electrical activity that can reinforce themselves as they propagate through the network, embodying old ideas about the reverberation of activity patterns through the extensive feedback loops found in the brain.

It is tempting to think of the stable patterns of activity, $\sigma \approx \xi^{(\mu)}$ as *being* memories. When we set the synaptic connection matrix to the form shown in Eq. (5.132), we "store" the memories, and as the dynamics settles into one of its locally stable states, one of these memories is "recalled." Each of the stored memories has a large basin of attraction, so the network will recall the memory given only a relatively weak "hint" that the memory is somewhere in the neighborhood of the current state. I use quotation marks extensively here to highlight the fact that we are sliding from properties of the equations into the everyday language that we use in describing our internal mental experiences. This is dangerous. But, of course, it is also great fun.

A crucial property of the model is that a particular memory (e.g., $\mu = 42$) is not stored in any particular place. There is no single neuron or synapse that has responsibility for remembering this single recallable item. Instead, the memory is distributed over essentially all elements in the system. Correspondingly, if we eliminate one neuron or one synapse, there is no catastrophic loss of one memory, but at worst a gradual degradation of all memories; in the limit $K \ll N$, we might even imagine that, as $N \to \infty$, deletion of anything less than a finite fraction of cells or synapses would have a vanishingly small effect. This fault tolerance is a highly attractive property.

Problem 118: Fault tolerance. Develop a small simulation to illustrate the idea of fault tolerance in the Hopfield model.

You might worry that all this reasoning depends on a very particular form of the synaptic weight matrix, Eq. (5.132). But this form is both natural and, perhaps surprisingly, well connected to experiment. Suppose that the current state of activity in the network, $\sigma(t)$, represents something that we would like to store and be able to recall later. If every synaptic strength is changed by the rule

$$W_{ij} \rightarrow W_{ij} + W\sigma_i(t)\sigma_j(t), \qquad (5.147)$$

then, assuming that we have not already tried to store too many patterns in the network, the current state $\sigma(t)$ will act as one more pattern that can be recalled—one more stable state in the energy landscape; that is, the network will have "learned" the state $\sigma(t)$. The change in strength of the synapse from neuron j to neuron i depends only on the states of neurons i and j. Thus, although the memory is distributed throughout the network, the rule for storing the memory is completely local.

The rule for modification of synaptic strengths in Eq. (5.147), sometimes called a learning rule, means that, over time, the strength of the synapse from neuron j to neuron i will be proportional to the correlation between the activities of these two cells. Learning based on correlations is an idea that goes back at least to Hebb in the 1940s, although there are clear precursors in the writing of William James 50 years earlier. Both James and Hebb were making an intuitive leap between the macroscopic phenomena of human and animal learning and what they imagined could be the underlying neural mechanisms. Although their words admit some breadth of interpretation, to a remarkable extent they were right, and many synapses are found to exhibit "Hebbian plasticity."

If we record simultaneously from two neurons that are connected by a synapse, then an action potential in one neuron triggers a small current or voltage change in the other; the voltage changes are called (excitatory or inhibitory) postsynaptic potentials (EPSPs or IPSPs, respectively), and the amplitude of the EPSP measures the strength of the synapse. Repeated stimulation of one cell or the other, alone, typically is not

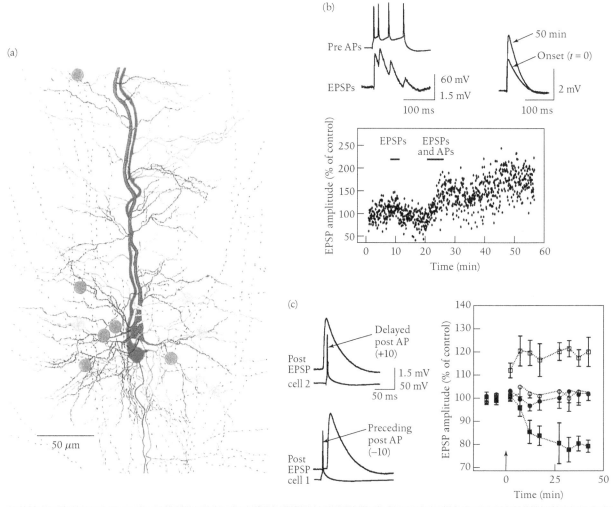

FIGURE 5.52

Spike timing-dependent plasticity in the cortex. From Markram et al. (1997). Reprinted with permission from AAAS. (a) Two neurons in a slice of rat cortex, drawn in red and black; green dots mark black → red synapses, and blue dots mark red → black synapses. (b) Measurement of excitatory postsynaptic potentials (EPSPs), demonstrating that these are stable over long times but increase when pre- and postsynaptic cells are jointly excited (EPSPs and action potentials APs) over a few minutes; these changes then persist for the duration of the experiment. (c) The sign of this effect is sensitive to the relative timing of the presynaptic spike (as monitored by the resulting EPSP) and the postsynaptic spike. "Pre before post" (upper traces, open squares) results in an increase of synaptic strength, whereas "post before pre" (lower traces, closed squares) results in a decrease.

sufficient to induce long-term changes in synaptic strength, but pairing these events—correlating the pre- and postsynaptic activity—does produce changes, and we can even see these in a slice of brain tissue kept alive in a dish, as in Fig. 5.52 where changes persist for ~1 hr. Some experiments demonstrate even longer lasting changes, and there is no reason to doubt that in an intact brain these changes would persist for days or even longer, perhaps being consolidated in subsequent processes that last for years.

The time scales of synaptic plasticity already pose important questions. Changes can be triggered by events on a time scale of milliseconds, or perhaps tens of milliseconds, corresponding to the occurrence of individual action potentials, but the consequences of these brief events persist for much longer. Thus, the molecular events at the synapse that are responsible for plasticity must have very short intrinsic time scales, so that they can respond to individual spikes, but the overall dynamics of these events must generate much longer time scales. Thus, as we try to understand how networks of neurons produce a nongeneric spread of time scales from the microscopic to the macroscopic, we uncover the same general problem in networks of molecules at each individual synapse. It seems likely that the solution is the same: the synapses can hold memories of their responses to brief events for a very long time, because they are driven into stable states. Such states could even be stable against the continuous replacement of all molecular components, which is necessary to explain how information can be stored for weeks, months, or years. Although much remains to be learned, we do have candidate molecular mechanisms to implement these ideas.

Problem 119: Something like CaM Kinase. Kinase molecules are enzymes that attach phosphate groups to other proteins, regulating their activity. Some kinases can themselves be regulated by these phosphorylation reactions. Let's see what happens if kinases can phosphorylate themselves, and if only phosphorylated kinases are active.

(a) Let the concentration of unphosphorylated kinase molecules be denoted by $[K]$, and the concentration of phosphorylated kinase molecules by $[KP]$. If the kinase operates on K as a substrate according to Michaelis-Menten kinetics (Section 2.3), explain why the rate of phosphorylation will be

$$\left.\frac{d[KP]}{dt}\right|_{\text{kinase}} = V_K[KP]\frac{[K]}{[K]+\alpha}, \tag{5.148}$$

and explain the significance of the parameters V_K and α; note that we have a proliferation of things called "K," so be careful about notation.

(b) Phosphate groups are removed from proteins by the action of a phosphatase (see the discussion in the case of bacterial chemotaxis, Section 4.2). Let's assume that the relevant phosphatase also acts via Michaelis-Menten kinetics, but all the phosphatase is active and has concentration [Ph]. Then show that the action of the phosphatase contributes a term

$$\left.\frac{d[KP]}{dt}\right|_{\text{phosphatase}} = V_{\text{Ph}}[\text{Ph}]\frac{[KP]}{[KP]+\beta}, \tag{5.149}$$

and again explain the meaning of the parameters.

(c) Combine the results in Eq. (5.148) and Eq. (5.149) to give a model for the dynamics of $[KP]$; remember that all kinase molecules are either phosphorylated or unphosphorylated, so that $[K] + [KP] = [K_t]$ is constant. Show that this model can have two stable states, so that once the active state has been turned on, it will stay on indefinitely (at least in this deterministic picture).

(d) Extend the model to include the destruction of both K and KP with some lifetime τ, and the synthesis of K at some fixed rate r. Do the stable states of concentration persist, even though the actual molecules are constantly being destroyed and replaced?

FIGURE 5.53

Action potentials recorded from single neurons in primate prefrontal cortex during a short-term memory experiment. A rhesus monkey is trained to open one of two doors when he receives a cue that they are unlocked (marked as response cues). Some time before the response cue, the subject has been allowed to see which of the doors has a piece of apple behind it; the times during which the apple is visible are marked as stimulus cues. All five neurons being monitored here seem to be active during the delay period, between the two cues, and this persistent activity plausibly is part of the memory that the subject holds. Redrawn from Fuster and Alexander (1971).

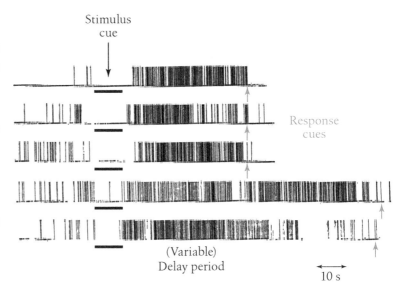

Looking closely at experiments, such as the one in Fig. 5.52, we can see that the detailed timing of the activity in the two neurons matters. In particular, if we record from both neurons, we can induce action potentials such that the presynaptic cell spikes just before the postsynaptic cell, or vice versa. When the presynaptic cell spikes first, the strength of the synapse increases, and when the postsynaptic cell spikes first, the strength of the synapse decreases; the transition between these regimes is less than ∼10 ms wide. Thus, changes in synaptic strength seem sensitive not just to the correlations between activity in the two connected cells but also to the (plausible) causal relation between the two cells—the synapse from i to j strengthens when the timing is such that spikes in cell i plausibly *cause* cell j to fire. Interestingly, this is what James suggested, a century ago.

What is the evidence that something like the Hopfield model is actually a correct description of real neural networks? The essence of the model, shorn of the analogies to magnetism, is that a recalled memory is a stable state of neural activity, one that persists in the absence of external stimuli by reverberating in the network. Such persistent activity of neurons has been observed. The canonical example occurs when an animal has to remember a sensory stimulus for a brief time (a few seconds to a minute) to compare it with another image or more simply because an immediate response would be impossible. In the period between the stimulus and the cue for the response, where the subject has to remember what has been seen or heard, these neurons continue to generate action potentials at a rate very different from the resting rate before exposure to the initial stimulus, as shown in Fig. 5.53. Although the behavior of each cell is different in detail, in many cases the activity during this delay period is steady, as if the system were simply locked into a new state. But the state into which the system falls differs depending on the image being remembered. Persistent activity is not just a feature of our cortex, but appears also in many other systems, from the primate spinal cord to the goldfish brainstem.

There is a very different way of connecting the Hopfield model to experiment. Imagine that we divide time into small bins of duration $\Delta\tau$. If $\Delta\tau$ is sufficiently small, then each neuron either generates an action potential in this bin, or it does not, so that the neural response is naturally binary: $\sigma_i = +1$ for a spike, $\sigma_i = -1$ for silence. For a large network it is impossible to measure the probability distribution of all the network states, $P(\boldsymbol{\sigma})$. But even recording from neurons one by one, it is possible to measure the mean rate at which each cell generates spikes, which is equivalent to the expectation value $\langle \sigma_i \rangle$. In addition, it is becoming increasingly common to record at least from pairs of cells, which makes it possible to estimate the correlations $C_{ij} \equiv \langle \sigma_i \sigma_j \rangle - \langle \sigma_i \rangle \langle \sigma_j \rangle$. One could ask, as a purely practical question, what do these measurements tell us about the full distribution $P(\boldsymbol{\sigma})$? In general, of course, there are infinitely many distributions (over the 2^N states of N cells) that are consistent with these $N(N + 1)/2$ measurements. Out of all these possible distributions, there is one that reproduces the measurements but otherwise describes a system that is as random or unstructured as possible. This is the maximum entropy distribution, as discussed in Section 5.1. I give a brief account of the maximum entropy idea here, but see also Appendix A.7.

The maximum entropy distribution consistent with a certain mean energy for a system is the Boltzmann distribution, and this construction generalizes. Suppose that we are looking for the probability distribution $P(\boldsymbol{\sigma})$, and we know the expectation values of some functions on the state, $\langle f_\mu(\boldsymbol{\sigma}) \rangle = \bar{f}_\mu$. Then to maximize the entropy of the distribution subject to these constraints, we use Lagrange multipliers λ_μ as usual. Thus, our problem is to maximize

$$\mathcal{F} = -\sum_\sigma P(\boldsymbol{\sigma}) \ln P(\boldsymbol{\sigma}) - \sum_\mu \lambda_\mu \left[\sum_\sigma P(\boldsymbol{\sigma}) f_\mu(\boldsymbol{\sigma}) - \bar{f}_\mu \right]$$

$$- \Lambda \left[\sum_\sigma P(\boldsymbol{\sigma}) - 1 \right], \tag{5.150}$$

where the last term (with Λ) fixes the normalization of the distribution. Following through the steps, the optimum is defined by

$$0 = \frac{\delta \mathcal{F}}{\delta P(\boldsymbol{\sigma})} = - \left[\ln P(\boldsymbol{\sigma}) + 1 \right] - \sum_\mu \lambda_\mu f_\mu(\boldsymbol{\sigma}) - \Lambda, \tag{5.151}$$

$$\ln P(\boldsymbol{\sigma}) = -\sum_\mu \lambda_\mu f_\mu(\boldsymbol{\sigma}) - (\Lambda + 1), \tag{5.152}$$

$$P(\boldsymbol{\sigma}) = \frac{1}{Z} \exp\left[-\sum_\mu \lambda_\mu f_\mu(\boldsymbol{\sigma}) \right], \tag{5.153}$$

where the partition function is $Z = e^{-(\Lambda+1)}$, or

$$Z(\{\lambda_\mu\}) = \sum_\sigma \exp\left[-\sum_\mu \lambda_\mu f_\mu(\boldsymbol{\sigma}) \right]. \tag{5.154}$$

The multipliers λ_μ are determined by matching the expectation values in the distribution to those observed experimentally. If we take this seriously as a statistical mechanics problem, the log of the partition function Z is the free energy, and derivatives of the free energy give us the various averages that we want to match to experiment. More precisely, we have the identity

$$\langle f_\nu(\boldsymbol{\sigma}) \rangle = -\frac{\partial \ln Z(\{\lambda_\mu\})}{\partial \lambda_\nu}, \tag{5.155}$$

so we have to solve the equations

$$-\frac{\partial \ln Z(\{\lambda_\mu\})}{\partial \lambda_\nu} = \bar{f}_\nu \tag{5.156}$$

to complete the construction of the model; in general this is a hard task, the inverse of what we usually do in statistical mechanics.

If the expectation values that we measure are $\langle \sigma_i \rangle$ and $\langle \sigma_i \sigma_j \rangle$, then the corresponding maximum entropy distribution can be written as

$$P(\boldsymbol{\sigma}) = \frac{1}{Z} \exp\left[\sum_{i=1}^{N} h_i \sigma_i + \frac{1}{2} \sum_{i \neq j}^{N} J_{ij} \sigma_i \sigma_j \right], \tag{5.157}$$

where the "magnetic fields" $\{h_i\}$ and the "exchange couplings" $\{J_{ij}\}$ have to be set to reproduce the measured values of $\{\langle \sigma_i \rangle\}$ and $\{C_{ij}\}$. This is an Ising model with pairwise interactions among the spins. What is crucial is that this model emerges here not through hypotheses about the network dynamics, but rather as the least-structured model that is consistent with the measured expectation values. The mapping to the Ising model is a mathematical equivalence, not an analogy, and the details of the model are specified by the data.

The emergence of the Ising model is an attractive aspect of the maximum entropy construction. But there is no obvious reason for real biological networks to have this maximum entropy property. Indeed, one might guess that there are complicated, higher order correlations that are important for the function of the network, and these will be missed by a maximum entropy model built only from pairwise correlations. It thus came as a surprise when it was found that these models really do provide an accurate description of the full correlation structure in the vertebrate retina as it responds to naturalistic stimuli. This result has led to considerable interest in the use of these models for the description of real biological networks, as described in Appendix A.7.

Problem 120: Maximum entropy model for a simple neural network.   Imagine that we record from N neurons and find that all of them have the same mean rate of spiking, $\bar{r}$. Further, if we look at any pair of neurons, the probability of both spiking in the same small window of duration $\Delta\tau$ is $p_c = (\bar{r}\Delta\tau)^2(1 + \epsilon)$. We want to describe this network as above, with Ising variables $\sigma_i = +1$ for spiking and $\sigma_i = -1$ for silence.

(a) Show that

$$\langle \sigma_i \rangle = -1 + \bar{r}\Delta\tau, \tag{5.158}$$

$$C_{ij} \equiv \langle \sigma_i \sigma_j \rangle - \langle \sigma_i \rangle \langle \sigma_j \rangle = 4\epsilon(\bar{r}\Delta\tau)^2. \tag{5.159}$$

(b) Because all neurons and pairs are equivalent, the maximum entropy model consistent with pairwise correlations has a simple form, which is a special case of Eq. (5.157) with all $J_{ij} = J$ and all $h_i = h$,

$$P(\boldsymbol{\sigma}) = \frac{1}{Z} \exp\left[h \sum_{i=1}^{N} \sigma_i + \frac{J}{2} \sum_{i \neq j}^{N} \sigma_i \sigma_j \right], \tag{5.160}$$

which is just the mean-field ferromagnet (assuming that J is positive). If N is large, one might expect that there is a "thermodynamic limit" in which quantities like energy and entropy become extensive (i.e., proportional to N). Show that this requires scaling of the coupling, $J = J_0/N$. With this scaling, derive the relationship between the derivatives of $\ln Z$ and the expectation values $\langle \sigma_i \rangle$ and C_{ij}.

(c) You may be familiar with the substitution tricks that we're about to use, but perhaps not. To be sure, let me take you through the steps. Notice that the interactions are described by

$$\frac{J}{2} \sum_{i \neq j}^{N} \sigma_i \sigma_j = \frac{J}{2} \sum_{i,j=1}^{N} \sigma_i \sigma_j - \frac{NJ}{2} = \frac{J}{2} \left(\sum_{i=1}^{N} \sigma_i \right)^2 - \frac{NJ}{2}. \tag{5.161}$$

Thus the partition function can be written as

$$Z = \sum_{\sigma} \exp\left[h \sum_{i=1}^{N} \sigma_i + \frac{J}{2} \sum_{i \neq j}^{N} \sigma_i \sigma_j \right] \tag{5.162}$$

$$= e^{-NJ/2} \sum_{\sigma} \exp\left[h \sum_{i=1}^{N} \sigma_i \right] \exp\left[\frac{J}{2} \left(\sum_{i=1}^{N} \sigma_i \right)^2 \right]. \tag{5.163}$$

Then the key step is to realize that

$$\exp\left[\frac{A}{2} (x)^2 \right] = \int \frac{d\phi}{\sqrt{2\pi A}} \exp\left[-\frac{\phi^2}{2A} + \phi x \right]. \tag{5.164}$$

Applied to Eq. (5.163) this expression allows us to write

$$Z = e^{-NJ/2} \sum_{\sigma} \exp\left[h \sum_{i=1}^{N} \sigma_i \right] \exp\left[\frac{J}{2} \left(\sum_{i=1}^{N} \sigma_i \right)^2 \right]$$

$$= e^{-NJ/2} \sum_{\sigma} \exp\left[h \sum_{i=1}^{N} \sigma_i \right] \int \frac{d\phi}{\sqrt{2\pi J}} \exp\left[-\frac{\phi^2}{2J} + \phi \sum_{i=1}^{N} \sigma_i \right] \tag{5.165}$$

$$= e^{-NJ/2} \int \frac{d\phi}{\sqrt{2\pi J}} \exp\left[-\frac{\phi^2}{2J} \right] \sum_{\sigma} \exp\left[(h + \phi) \sum_{i=1}^{N} \sigma_i \right]. \tag{5.166}$$

Now the spins have decoupled, and you should be able to do the sum over states, $\sum_{\sigma}$, inside the integral. Show that, with the scaling from part (b),

$$Z = e^{-NJ/2} \int \frac{d\phi}{\sqrt{2\pi J}} \exp\left[-NF(\phi; h, J_0) \right], \tag{5.167}$$

where the effective free energy $F(\phi; h, J_0)$ has no explicit N dependence.

(d) Use the method of steepest descent to approximate Eq. (5.167) at large N. Derive an expression for $\ln Z$ that captures both the leading behavior ($\ln Z \propto N$) and the first two corrections.

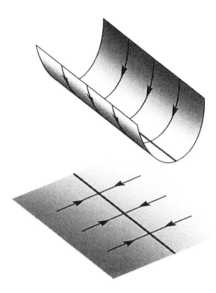

FIGURE 5.54

Energy landscape (top) and phase portrait (bottom) for system with a line of attractors. From Seung (1996). Copyright © 1996 by the National Academy of Sciences USA.

(e) To finish the construction of the model, we have to adjust h and J to match the measured means and pairwise correlations, Eq. (5.158) and Eq. (5.159). Using the scaling required for a thermodynamic limit, is there a prediction for the N dependence of the correlation strength ϵ? This should bother you—ϵ is a quantity that is *measured* from pairs of cells, and should not really depend on the number of cells in the network. Suppose we measure ϵ among more and more pairs of cells, so we have to describe larger and larger networks. Is it possible to have ϵ small and constant as $N \to \infty$? What conditions need to be met for this to happen?

The Hopfield model provides a scheme for stabilizing multiple, discrete patterns of activity. But there are situations in which the brain must hold a memory of a continuous variable. This problem is even less generic than the case of discrete attractors. To have a memory of a continuous variable, there must be (at least) a whole line or curve in state space along which the system can stop; if we think in terms of an energy landscape, then there must be one big valley, and the bottom of this valley must be precisely flat along one direction, as schematized in Fig. 5.54. Implausible as all this sounds, the brain really does hold memories of continuous variables, and it does so even in simple situations.

When you turn your head, cells in the semicircular canals, buried in the same bone as the cochlea, sense the rotational motion; this is called our vestibular sense. This angular motion input passes through the brain and drives a motor output that counter-rotates the eyes. This correction happens automatically and is called the vestibulo-ocular reflex. You can demonstrate it for yourself by shaking your head from side to side as you read this text. If you are holding the book at arm's length, then to read, you have to have your fovea—the $\sim 1°$ wide area of highest image quality—focused on the words as you read them. If you move your head from side to side and don't move your eyes to compensate, the text will blur. In fact, you (hopefully) have no trouble reading and shaking your head at the same time, suggesting that your eyes are being moved to compensate with an accuracy of better than $\sim 1°$. When you are reading, there are visual

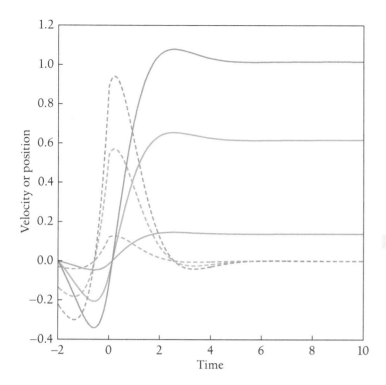

FIGURE 5.55

Integration as memory for a continuous variable. Dashed lines show possible velocity signals, and the solid lines show corresponding position signals, obtained by integrating the velocity. After the transient inputs die away, the output of the integrator is stable for all time (a memory) and can take on any real value. Units of position and time are arbitrary here.

cues to help guide your eye movements, but it turns out that even if you close your eyes or sit in a dark room, seeing nothing, your eyes still counterrotate to compensate for your head motions.

There is a subtlety of the vestibulo-ocular reflex, however. If you relax all the muscles to the eyes, then they rotate to a resting position in which you are looking more or less straight ahead (as defined by where the nose is pointing). Thus, if you turn your head to the right and stop, you need to keep tension on the eye muscles to be sure that they don't drift away from where you were looking before you turned. That is, to fully compensate for rotation of the head, you need a signal related to the desired angular displacement of the eyes. But the vestibular system is an inertial sensor, driven by angular accelerations; the mechanics of the semicircular canal turn this acceleration into a velocity signal over a wide range of frequencies, but the sensors really have zero response to constant displacements. Thus, the brain needs to take an input related (at best) to head velocity and generate an output related to head displacement—it has to integrate, where "integrate" has the literal meaning from calculus, rather than being a qualitative statement about the gathering of multiple signals. Although we do not usually think about it this way, an integrator is a device that, once the input signals die away, has a perfect memory for a continuous variable, as schematized in Fig. 5.55. Although these properties of the integral are obvious mathematically, it less obvious how to build a network of neurons that implements this mathematics.

Before continuing, it should be noted that the movement of our eyes is not the perfect integral of our head velocity. On a longer time scale, ~30 s, our eyes do drift back to a resting position if there is no further stimulus. But this time scale is very long compared with the natural time scales of individual neurons, perhaps by a factor of as much as $\sim 10^3$. Could this gap be closed by an emergent long time scale in the network, resulting from a line or curve of fixed points?

Suppose that the activity of each neuron is described by a coarse-grained continuous variable, such as the rate r at which it generates action potentials. If we inject a current I into the neuron directly, we find that the rate changes along some curve $r(I)$. Each spike arriving at a synapse onto cell i effectively injects current into that cell, but this current is smoothed by some dynamics, which we summarize by a time scale τ, and the spikes from cell j are weighted by the strength of the synapse W_{ij}. This suggests a simple model,

$$\tau \frac{dI_i}{dt} + I_i = \sum_j W_{ij} r(I_j) + I_i^{\text{ext}}, \tag{5.168}$$

where I_i^{ext} represents currents injected from outside the network, including from sensory inputs. Typical examples of the response function $r(I)$ are sigmoids, threshold linear relations, and other monotonic functions.

What would it mean for the dynamics of Eq. (5.168) to be an integrator? At the very least, the dynamics has to look like an integrator in its linear response to inputs, so let's see how this is possible. Assume that in the absence of inputs, there is some steady state at which $I_i = I_i^*$. Then if we linearize around this value, writing $I_i = I_i^* + u_i$, we have

$$\tau \frac{du_i}{dt} + u_i = \sum_j W_{ij} r'(I_j^*) u_j + I_i^{\text{ext}}. \tag{5.169}$$

As always with linear problems, we want to change coordinates so that matrices become diagonal. If we denote quantities in this new coordinate system by tildes, then we have

$$\tau \frac{d\tilde{u}_n}{dt} + \tilde{u}_n = \Lambda_n \tilde{u}_n + \tilde{I}_n^{\text{ext}}, \tag{5.170}$$

where the eigenvalues are defined as solutions to

$$\sum_j W_{ij} r'(I_j^*) \psi_j^{(n)} = \Lambda_n \psi_j^{(n)}. \tag{5.171}$$

If one of the $\Lambda_n \to 1$, then along this direction we have simply

$$\tau \frac{d\tilde{u}_n}{dt} = +\tilde{I}_n^{\text{ext}}, \tag{5.172}$$

$$\Rightarrow \tilde{u}_n(t) = \tilde{u}_n(0) + \frac{1}{\tau} \int_0^t dt' \, \tilde{I}_n^{\text{ext}}(t'), \tag{5.173}$$

so that $\tilde{u}_n$ is the time integral of its inputs. Thus, being an integrator means arranging the matrix of synaptic strength so that it (in appropriate units) has a unit eigenvalue, which means that, along this one mode, the signals being received from other cells in the network perfectly balance the decay processes within each cell. This is a critical point in the dynamics—if the eigenvalue is larger than one, the dynamics become unstable; if it is less than one, it is stable but an imperfect integrator. Only at the critical point is true integration achieved. If the system is within ϵ of the critical point, it will hold a memory for $\sim\tau/\epsilon$, so if we really need to span three orders of magnitude (or even two), then the adjustment to the critical point must be quite precise.

The language of eigenvalues and critical points makes precise our initial intuition that there is something highly nongeneric about memory for a continuous variable. Most valleys have a single lowest point, and balls keep rolling downhill until they find it. Only at the critical point is there one perfectly neutral direction in the valley, along which a ball has no force applied to it.

Problem 121: Details of the line attractor.   We can rewrite Eq. (5.169) as

$$\tau \frac{du_i}{dt} + u_i = \sum_j T_{ij} u_j + I_i^{\text{ext}}, \tag{5.174}$$

where T_{ij} are effective synaptic strengths.

(a) The inputs I_i^{ext} to the network are provided by other neurons, and especially in the vestibular system we know that these neurons are active even when the nominal signals (e.g., head velocity) are zero. Under what condition will this spontaneous activity not be integrated by the network? Why is this so important?

(b) If there is one mode of the network that serves as an integrator, while all other modes relax quickly, then the matrix T_{ij} has one eigenvalue equal to (or very close to) one, and all the other eigenvalues are small. Develop a perturbation theory to show how the eigenvalue near one will shift if the matrix elements $T_{ij} \to T_{ij} + \delta T_{ij}$, and use the gap in the spectrum of eigenvalues to (approximately) simplify the results.

(c) Can you use the perturbation theory in part (b) to estimate how accurately the matrix elements must be tuned to ensure that, for example, the memory of the integrator is ~ 30 s even when the microscopic time scales of the network are $\tau \sim 0.05$ s?

That the position of our eyes is the integral of the velocity signals from our semicircular canals, and that there is (apparently) a continuum of stable points where our eyes can sit, means that something like this description in terms of line attractors must be true for the system as a whole. Indeed, this model is more general: that we (and other animals) can stabilize a continuously variable set of postures means that the combined dynamics of our limbs, muscles, sensors, and brain must have a line or manifold of attractors. It is more challenging to point to a particular part of the system (e.g., a particular subnetwork of neurons in one part of the brain) and claim that the dynamics of this subsystem must have a line of attractors.

Seeing a model that explains things but only for particular choices of parameters is unsettling, as in our previous examples in this chapter. But in this case, we know that the relevant parameters—synaptic strengths—are adjustable, because this is how we learn. Also, we know that if we make errors, then under normal conditions (with the lights on) these errors are literally visible as slippage of the image on our retina as we turn our heads. There must be some way to use this error signal to adjust the synaptic weights and tune the network to its critical point. Does this happen?

To test the idea that the brain tunes the dynamics of the integrator circuit to its critical point, Major et al. did a seemingly simple but beautiful experiment using goldfish, which also exhibit oculomotor integration. Essentially they built a planetarium for the

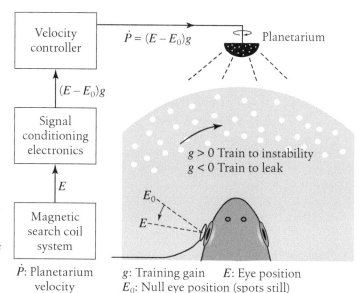

FIGURE 5.56

Schematic of the planetarium experiment. Eye movements are monitored and fed back to movements of the surrounding scene. Changing the gain drives learning, which can tune the network to be leaky or unstable. Redrawn from Major et al. (2004a).

goldfish and then coupled the rotation of this world to the goldfish eye movements, as shown in Fig. 5.56. Under normal conditions, when the eyes move by an angle θ, this is equivalent to the world moving the other way by the same angle. But if we give an additional rotation, we can create a situation in which the world slips on the retina even when the integrator network is set correctly. If the system in fact continuously uses slip signals to tune itself, this will drive a mistuning, either toward stability or instability. If we remove the feedback, we should then see that the fish can no longer stabilize its gaze, with the eyes either quickly relaxing to their resting position or exploding wildly away from rest, needing correction by frequent saccades.

Problem 122: Gain versus time constant.   Assume that a goldfish in the setup of Fig. 5.56 tunes the relevant neural network perfectly, so that there is a zero eigenvalue and hence precise integration of the input signals from the semicircular canals. If we take the animal out of the planetarium, the system will be either unstable, so that deviations from a stable eye position grow at a rate $k = 1/\tau$, or damped, so that deviations decay with a time constant τ. Derive the expected relationship between the gain g in the planetarium and the observable τ. How does this relationship compare with the data in the original papers?

The quick summary is that all of what we expect to see is observed experimentally, as summarized in Fig. 5.57. Importantly, one can record from neurons in the relevant circuit and demonstrate that the detuning of the behavioral integration is mirrored by changes in the dynamics of persistent neural firing. Although this does not prove that the line attractor scenario is correct, it does show that the long time scale of memory exhibited by the oculomotor integrator is the result of an active tuning process that uses visual feedback as a control signal. In this way, nongeneric behavior of the system is learned robustly.

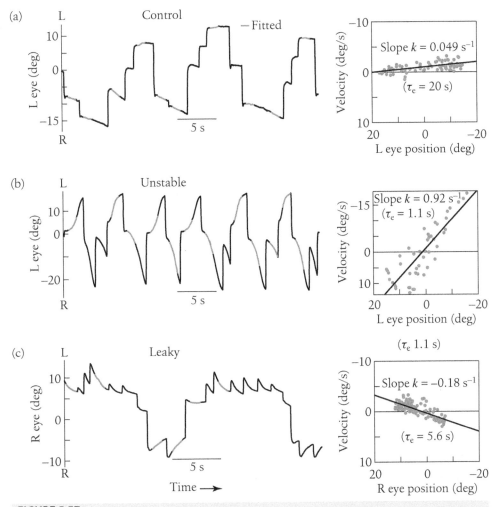

FIGURE 5.57

Results of the planetarium experiment. Redrawn from Major et al. (2004a). At left, eye position versus time. At right, eye velocity versus position along these trajectories. (a) Control experiments show the eye trajectories and position versus velocity plots before exposure to the feedback system in the planetarium. Note that the time constant of the system is ~20 s. After exposure to feedback, which should "teach" the system to be unstable (b) or leaky (c), trajectories and position versus velocity plots show the expected behaviors, with time constants for growth or decay on the order of 1 − 5 s.

5.5 Perspectives

The exploration of fine tuning versus robustness in biological systems encourages us to think beyond models for this or that particular system. To ask whether some function requires fine tuning of parameters, we have to imagine that the system we are looking at is just one member in a class of possible systems. Whatever the answer to our initial questions, this effort at generalization clearly is an important step on the path to a physicist's view of life.

When we think about individual proteins, generalization is easy—proteins are polymers, and there is a natural class of molecules that can be built from the same monomers but with different sequences. When we think about a biochemical or genetic network, with many interacting protein molecules, it seems natural to generalize to a class of networks that has the same topology but different parameters on each node or link. The ion channels in a single neuron provide an important example of a network of interacting proteins, where the interactions are mediated by the (global) transmembrane voltage and, importantly, experiments on single channel molecules serve to validate the equations describing what happens at each node. The natural generalization here is to consider classes of cells that express the same set of channels but in different numbers. Finally, for networks of neurons, the plasticity of the strengths of synaptic connections during learning makes it natural to think about classes of networks that have the same topology of connections among neurons but with different strengths of these connections. In all these cases, we can see that the generalization to a class of networks is not just a useful theoretical construct but is also something that has meaning in the life or evolution of the organism.

In the extreme, robustness would mean that functional behavior is largely invariant over the whole class of networks. If this really is the case, then we should be able to choose networks at random and have them function. This is essentially the strategy employed by many research groups searching for robustness in biochemical networks, and long before this effort there was a serious exploration of neural networks with randomly chosen strengths of synaptic connections among all the cells, using analytic methods borrowed from the dynamical theory of spin glasses. In the context of neural networks, the model with random connections indeed behaves chaotically, which seems odd, although it has been suggested that in the absence of other inputs this is the right answer—sensory inputs serve to drive the network out of the chaotic phase into an ordered state. For biochemical and genetic networks, chaos seems less generic, but to obtain functional behavior without adjusting parameters, there is general agreement that the topology of the network must be chosen carefully. There are several open questions here. Why is chaos not more common in large networks of biochemical reactions? What is the boundary between changing parameters (e.g., making the rate of one particular chemical reaction smaller) and changing topology (setting that rate exactly to zero)? To speak precisely about what will be typical of a randomly chosen network, we need a measure on the space of parameters. Is there a natural choice for this measure?

In most systems we have studied, the randomly chosen parameters do not correspond to functional behavior. Random amino acid sequences do not fold into functional proteins; randomly chosen numbers of ion channels will not generate the correct rhythms of electrical activity; and although random neural networks may perform some functions, they certainly do not provide for stable storage and recall of memories. In each of these cases there are mechanisms for tuning or selecting the functional regions of parameter space. In single neurons, adjusting the numbers of copies of different channels is a form of physiological adaptation, connecting electrical activity, intracellular messengers, and the control of gene expression. In neural networks, the strengths of synapses are adjusted during learning, and for some key processes this learning happens all the time—as perhaps is necessary if the behavior that the system is trying to stabilize is very far from typical in the space of possible networks. Finally, for amino acid and DNA sequences, the adjustment to functional behavior occurs on evolutionary time scales.[17] In this context, we can think of adaption, learning, and evolution as different mechanisms for accomplishing the same task, albeit on different time scales.

There is a sophisticated mathematical theory of learning, combining ideas from mathematics, computer science, and statistical physics. In particular, in different contexts this theoretical approach places bounds on what can be learned and how quickly. If we see adaptation, learning, and evolution as different approaches to the same problem, should there be a comparable theoretical framework limiting the speed of evolution or the effectiveness of adaptation? For evolution there is, in the long run, an obvious external definition of correct functional behavior (successful reproduction), and for learning there are often external signals (as in the case of the oculomotor integrator) that define the goal of the learning process. In adaptation, how do cells "know" the correct behavior that they are trying to stabilize? In the models for regulation of ion channel densities discussed in Section 5.2, this ability is (weakly) programmed into the cell by the parameters that define a target calcium concentration. Is there a more general definition of when cells are getting things right? Are there, as with learning, limits on how precisely the system can be tuned if it needs to adjust quickly?

To return to the opening remarks in this chapter, we wanted to distinguish between the usual physicist's mistrust of explanations that rest on fine tuning of parameters and some specifically biological notions of robustness or evolvability. Part of the motivation for robustness as a biological principle is the intuition that living organisms simply can't adjust parameters accurately enough to guarantee reliable, reproducible functions. I think this intuition turns out to be wrong—cells can and do exercise precise control over the numbers of molecules that they make, so that the absolute concentrations of relevant molecules *can be* reproducible from cell to cell (or, in the discussion of Section 5.3, embryo to embryo) with high precision. I emphasize "can be," because one clearly cannot conclude that all concentrations or molecule counts will be reproducible in this way. Indeed, the example of ion channels makes clear that, in the natural parameter space for the cell, there are many different ways of achieving essentially the same function, and so there is no reason for the cell to control the number of copies of any particular molecule very precisely. This is not evidence of sloppiness, but rather of a many-to-one mapping from microscopic parameters onto macroscopic functional properties.

That cells can exert precise control over the concentrations, or combinations of concentrations, of certain molecules does not solve all the organism's problems. Most fundamentally, life as a cold-blooded organism[18] means having to function across a range of conditions where all chemical reaction rates vary, often by an order of magnitude or more, with no guarantee that the different rates in a given network will scale together. For an example of this problem one need look no further than the familiar circadian rhythms, which have long been known to be invariant to temperature changes. At the same time, diversity of environments is one of the driving forces for speciation, so that (for example) fruit flies that live at different latitudes, and hence

---

17. It is worth emphasizing that, in the immune system, there is a kind of accelerated evolution within individual organisms, which serves to select a nontrivial distribution of sequences for the antibody molecules.

18. Most of the biomass on our planet is cold blooded, so this is a very general problem.

different temperature ranges, are genetically distinguishable. Natural history abounds with stories of animals that seek out very special environments in which to lay their eggs, casting doubt on any glib statement that embryonic development is robust against environmental perturbations. Still, simple laboratory experiments demonstrate that many aspects of life are nearly invariant over a wide range of environmental conditions, much wider than we might expect from simple models.

Locating life on the spectrum between precisely controlled (rather than finely tuned) dynamics and some more generic or robust behavior is an incredibly important question. It touches, as we have seen, on phenomena ranging from the states of single cells to the nature of our memories. It connects to theoretical ideas that have the potential to reach deeply into statistical physics and dynamical systems. At the risk of making clear the limits of my own understanding, however, I would say that we are still searching for the best formulations of these questions. We need more experimental guidance about what features of behavior are robust against which variations, and we need evidence that organisms actually face these variations in their natural environment. On the theoretical side, we need more anchor points like the random heteropolymer and the random neural network, where we have a complete analytic understanding of what is expected in the truly generic case. Most of all, we need a statistical mechanics of systems with random parameters that allows us to deal with the case where these parameters have nontrivial distributions. These are substantial challenges.

Efficient Representation

The generation of physicists who turned to biological phenomena in the wake of quantum mechanics noted that, to understand life, one has to understand not just the flow of energy (as in inanimate systems) but also the flow of information. There is some difficulty in translating the colloquial notion of information into something mathematically precise. Almost all statistical mechanics textbooks tell us that the entropy of a gas measures our lack of information about the microscopic state of the molecules, but often this connection is left a bit vague or qualitative. In 1948, Shannon proved a theorem that makes the connection precise: entropy is the unique measure of available information consistent with certain simple and plausible requirements. Further, entropy also answers the practical question of how much space we need to use in writing down a description of the signals or states that we observe. This leads to a notion of efficient representation, and in this chapter we explore the possibility that biological systems in fact form efficient representations, maximizing the amount of relevant information that they transmit and process, subject to fundamental physical constraints.

The idea that a mathematically precise notion of information would be useful in thinking about the representation of information in the brain came very quickly after Shannon's original work.[1] Therefore, a well-developed set of ideas exists about the how many bits are carried by the responses of neurons, in what sense the encoding of sensory signals into sequences of action potentials is efficient, and so on. More subtly, there is a body of work on the theory of learning that can be summarized by saying that the goal of learning is to build an efficient representation of what we have seen. In contrast, most discussions of signaling and control at the molecular level have left information as a colloquial concept. One of the goals of this chapter, then, is to bridge this gap. Hopefully, in the physics tradition, it will be clear how the same concepts can be used in thinking about the broadest possible range of phenomena. We begin, however, with the foundations.

---

1. There is also an interesting bit of history regarding the role of information theory, both formal and informal, in sharpening hypotheses about the nature of the genetic code. For details, see the references in the Annotated Bibliography for Section 4.5.

6.1   Entropy and Information

Two friends, Max and Allan, are having a conversation. In the course of the conversation, Max asks Allan what he thinks of the headline story in this morning's newspaper. We have the clear intuitive notion that Max will gain information by hearing the answer to his question, and we would like to quantify this intuition. Let us start by assuming that Max knows Allan very well. Allan speaks very proper English, being careful to follow the grammatical rules even in casual conversation. Because they have had many political discussions, Max has a rather good idea about how Allan will react to the latest news. Thus, Max can make a list of Allan's possible responses to his question, and he can assign probabilities to each of the answers. From this list of possibilities and probabilities we can compute an entropy S, which is done in exactly the same way as we compute the entropy of a gas in statistical mechanics.

Our intuition from statistical mechanics suggests that the entropy S measures Max's uncertainty about what Allan will say in response to his question, in the same way that the entropy of a gas measures our lack of knowledge about the microstates of all the constituent molecules. Once Allan gives his answer, all this uncertainty is removed—one of the responses occurred, corresponding to probability $p = 1$, and all the others did not, corresponding to $p = 0$—so the entropy is reduced to zero. It is appealing to equate this reduction in our uncertainty with the information we gain by hearing Allan's answer. Shannon proved that this is not just an interesting analogy; it is the *only* definition of information that conforms to some simple constraints.

If we want a general measure of how much information is gained on hearing the answer to a question, we have to put aside the details of the questions and the answers (although this might make us uncomfortable). We should understand answering a question to mean that we give a unique pointer to the answer, not necessarily that we give all the details. Thus, if you are taking a course on Shakespeare and someone asks what you are reading this week, it is not necessary to recite an entire sonnet; it is sufficient to give the title of the work, or even more compactly the page number on which it appears in some standard edition of the Bard's collected works. If we leave out the text of the questions and answers themselves, then all that remains are the probabilities p_n of hearing the different answers, indexed by $n = 1, 2, \cdots, N$, and so Shannon assumes that the information gained must be a function of these probabilities, $I(\{p_n\})$. The challenge is to determine this function.[2]

The first constraint is that, if all N possible answers are equally likely, then the information gained should be a monotonically increasing function of N—we learn more by asking questions that have a wider range of possible answers. The next constraint is that if our question consists of two parts, and if these two parts are entirely independent of each other, then we should be able to write the total information gained as the sum of the information gained in response to each of the two subquestions. Finally, more general multipart questions can be thought of as branching trees, as in Fig. 6.1, where the answer to each successive part of the question provides some

---

2. Notice that Shannon's "zeroth" assumption—that the information gained is a function of the probability distribution over the answers to our question—means that we must take seriously the notion of enumerating the possible answers. In this framework we cannot quantify the information that would be gained upon hearing a literally unimaginable answer to our question. It is interesting to think about whether this is a real restriction.

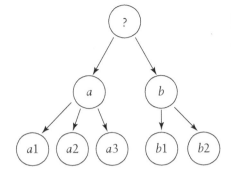

FIGURE 6.1

The branching postulate in Shannon's proof. The idea is to break a big question into multiple parts, as in the familiar game of twenty questions. We start with some initial question, at the top (?). Depending on the answer to this question (*a* or *b*), we ask a new question. This second question in turn has multiple possible answers (*a*1, *a*2, *a*3 or *b*1, *b*2). In this tree structure, the various subquestions live at branch points, with the answers emerging along the branches; finding our way to the full answer means following one path through the tree. The average information that we gain along this path should be additive—the weighted sum of information gained at every branch point.

further refinement of the probabilities; in this case we should be able to write the total information gained as the weighted sum of the information gained at each branch point. Shannon proved that the only function of the $\{p_n\}$ consistent with these three postulates—monotonicity, independence, and branching—is the entropy S, up to a multiplicative constant. It is worth going through the details, not least to be sure we understand how little is required to derive such a powerful result.

To prove Shannon's theorem, we start with the case where all N possible answers are equally likely. Then the information must be a function of N. Let this function be $I(\{p_n\}) = f(N)$. Consider the special case $N = k^m$. Then we can think of our answer— one out of N possibilities—as being given in m independent parts, and in each part we must be told one of k equally likely possibilities. But we have assumed that information from independent questions and answers must add, so the function $f(N)$ must obey the condition

$$f(k^m) = mf(k). \tag{6.1}$$

Notice that although we are focusing on cases where $N = k^m$, we have a condition that involves $f(k)$ for arbitrary k. Certainly $f(N) \propto \log N$ satisfies this equation. To show that this solution is unique, consider another pair of integers ℓ and n such that

$$k^m \leq \ell^n \leq k^{m+1}, \tag{6.2}$$

or, taking logarithms, we have

$$\frac{m}{n} \leq \frac{\log \ell}{\log k} \leq \frac{m}{n} + \frac{1}{n}. \tag{6.3}$$

Now because the information measure $f(N)$ is monotonically increasing with N, the ordering in Eq. (6.2) means that

$$f(k^m) \leq f(\ell^n) \leq f(k^{m+1}), \tag{6.4}$$

and hence from Eq. (6.1) we obtain

$$mf(k) \leq nf(\ell) \leq (m+1)f(k). \tag{6.5}$$

Dividing through by $nf(k)$, we have

$$\frac{m}{n} \leq \frac{f(\ell)}{f(k)} \leq \frac{m}{n} + \frac{1}{n}, \tag{6.6}$$

which is very similar to Eq. (6.3). The trick is now that with k and ℓ fixed, we can choose an arbitrarily large value for n, so that $1/n = \epsilon$ is as small as we like. Then Eq. (6.3) is telling us that

$$\left| \frac{m}{n} - \frac{\log \ell}{\log k} \right| < \epsilon, \tag{6.7}$$

and Eq. (6.6) for the function $f(N)$ can similarly be rewritten as

$$\left| \frac{m}{n} - \frac{f(\ell)}{f(k)} \right| < \epsilon. \tag{6.8}$$

Putting these expressions together, we have

$$\left| \frac{f(\ell)}{f(k)} - \frac{\log \ell}{\log k} \right| \leq 2\epsilon, \tag{6.9}$$

so that $f(N) \propto \log N$, as promised. Note that if we were allowed to consider $f(N)$ as a continuous function, then we could have made a much simpler argument. But, strictly speaking, $f(N)$ is defined only at integer arguments.

We are not quite finished, even with the simple case of N equally likely alternatives, because we still have an arbitrary constant of proportionality. The same issue arises in statistical mechanics: What are the units of entropy? In a chemistry course you might learn that entropy is measured in "entropy units," with the property that if you multiply by the absolute temperature (in kelvin), you obtain an energy in units of calories per mole; this happens because the constant of proportionality is chosen to be the gas constant R, which refers to Avogadro's number of molecules.[3] In physics courses entropy often is defined with a factor of Boltzmann's constant k_B, so that if we multiply by the absolute temperature we again obtain an energy (in joules) but now per molecule (or per degree of freedom), not per mole. Alternatively, many statistical mechanics texts take the sensible view that temperature itself should be measured in energy units—that is, we should always talk about the quantity $k_B T$, not T alone—so that the entropy, which after all measures the number of possible states of the system, is dimensionless. Any dimensionless proportionality constant can be absorbed by choosing the base that we use for taking logarithms. When measuring information, it is conventional to choose base two, because the most basic of questions have yes/no answers. Finally, then, we have $f(N) = \log_2 N$. The units of this measure are called bits, and one bit is the information contained in the choice between two equally likely alternatives.

Ultimately we need to know the information conveyed in the general case where our N possible answers all have unequal probabilities. Consider first the situation where all probabilities are rational, that is

$$p_n = \frac{k_n}{\sum_m k_m}, \tag{6.10}$$

---

3. I have to admit that when I read about "entropy units" (or calories, for that matter) I imagine that there was some great congress on units at which all such things were supposed to be standardized. Of course every group has its own favorite nonstandard units. Perhaps at the end of some long negotiations the chemists were allowed to keep entropy units in exchange for physicists continuing to use electron volts.

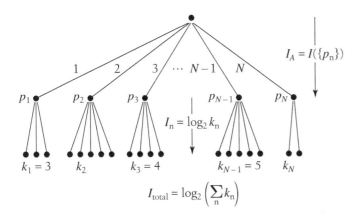

FIGURE 6.2

Grouping. To determine the information gained with unequal probabilities, we consider a "big question" with answers that fall into N groups. By hypothesis, in each group the k_n answers are equally likely.

where all the k_n are integers. If we can find the correct information measure for rational $\{p_n\}$, then by continuity we can extrapolate to the general case; the trick is that we can reduce the case of rational probabilities to the case of equal probabilities. To do this, imagine that we have a total of $N_{total} = \sum_m k_m$ possible answers, but that we have organized these into N groups, each of which contains k_n possibilities, as in Fig. 6.2. To specify the full answer, we would first find which group it is in, then find which of the k_n possibilities is realized. In this two-step process, the first step is to get the information we are really looking for—which of the N groups are we in—and so the information in the first step is our unknown function,

$$I_A = I(\{p_n\}). \tag{6.11}$$

In the second step, if we are in group n, then we will gain $I_n = \log_2 k_n$ bits, because this is just the problem of choosing from k_n equally likely possibilities. Because group n occurs with probability p_n, the *average* information we gain in the second step is

$$I_B = \sum_n p_n I_n = \sum_n p_n \log_2 k_n. \tag{6.12}$$

But this two-step process is not the only way to compute the information in the enlarged problem, because, by construction, the enlarged problem is just the problem of choosing from N_{total} equally likely possibilities. The two calculations have to give the same answer, so that

$$I_A + I_B = \log_2 (N_{total}), \tag{6.13}$$

$$I(\{p_n\}) + \sum_n p_n \log_2 k_n = \log_2 \left(\sum_m k_m \right). \tag{6.14}$$

Rearranging the terms, we find

$$I(\{p_n\}) = - \sum_n p_n \log_2 \left(\frac{k_n}{\sum_m k_m} \right) \tag{6.15}$$

$$= - \sum_n p_n \log_2 p_n. \tag{6.16}$$

Again, although the argument is worked out explicitly for the case where the p_n are rational, it must be the general answer if the information measure is a continuous function of the probabilities. So we are done: the average information gained on hearing the answer to a question is measured uniquely by the entropy of the distribution of possible answers.

It is worth pausing here to note that what Shannon did is very different from our conventional experience in using mathematics to describe the natural world. In most of physics, we have some set of observations (e.g., the motion of the planets in the night sky) that can be made quantitative (as Brahe did), and we search for mathematical structures that can explain and unify these data (as did Kepler and Newton). In contrast, Shannon considered an everyday phenomenon for which we have a colloquial language and asked whether this language itself could be made mathematically precise, without reference to quantitative data. It is remarkable that this actually worked, and that Shannon's construction has, as we will see, so many consequences.

When we try to quantify the information gained from hearing the answer to a question, it seems natural to think about a discrete set of possible answers. However, we also gain information from making measurements that have continuous outcomes. Naively, we are tempted to write

$$S_{continuum} = -\int dx\, P(x) \log_2 P(x),\qquad(6.17)$$

or some multidimensional generalization. The difficulty is that probability distributions for continuous variables have units—$P(x)$ has units inverse to the units of x—and we should be worried about taking logs of objects that have dimensions. Notice that if we wanted to compute a difference in entropy between two distributions, this problem would go away.

Problem 123: Dimensionality and the scaling of the entropy. As written, Eq. (6.17) does not really make sense, because we are taking the log of something with units. Suppose we try to clean this up and make bins along the x axis, each bin of width Δx and the nth bin centered at x_n. Then if the bins are reasonably small, the probability of falling in the nth bin is $p_n = P(x_n)\Delta x$.

(a) Show that if you calculate the entropy in the usual way, you find

$$S = -\sum_n p_n \log_2 p_n = S_{continuum} - \log_2(\Delta x)\qquad(6.18)$$

in the limit $\Delta x \to 0$. More generally, show that in D dimensions the entropy is

$$S = -\sum_n p_n \log_2 p_n = S_{continuum} - D \log_2(\Delta x).\qquad(6.19)$$

The result in Eq. (6.19) suggests that the scaling of the entropy with bin size provides a measure of the dimensionality D of the underlying space. This is especially interesting if the intrinsic dimensionality is different from the dimensionality we happen to be using to describe the system. As an example, if we describe a system by its position in a two-dimensional space (x, y), but really the points fall on a curve, then the right answer is that the system is one dimensional, not two dimensional.

(b) Write a small program in MATLAB to generate 10^6 points in the (x, y) plane that fall on the circle $x^2 + y^2 = 1$. Then divide the plane (you can confine your attention to the region $-2 < x < 2$, and similarly for y) into boxes of size $(\Delta x) \times (\Delta x)$, and estimate the fraction of points that fall in each box. From this estimate, compute the entropy, and see how it varies as a function of Δx. Can you identify the signature of the reduced dimensionality?

(c) Suppose that you take the 10^6 points from part (b) and add, to each point, a bit of noise in the x and y directions, for example, Gaussian noise with a standard deviation of $\sigma = 0.05$. Repeat the calculation of the entropy versus box size. If you look closely enough ($\Delta x \ll \sigma$), the underlying probability distribution really is two dimensional, because there is independent noise along x and y. But if your resolution is more coarse ($\Delta x \gg \sigma$), you won't be able to see the noise, and the points will appear to fall on a circle, corresponding to a one-dimensional distribution. Can you see this transition in the plot of $S(\Delta x)$?

The problem of defining the entropy for continuous variables is familiar in statistical mechanics.[4] In the simple example of an ideal gas in a finite box, the quantum version of the problem has a discrete set of states, so that we can compute the entropy of the gas as a sum over these states. In the limit that the box is large, sums can be approximated as integrals, and if the temperature is high, quantum effects are negligible. One might naively suppose that Planck's constant should disappear from the results in this limit, but this is not quite the case. Planck's constant has units of momentum times length, and so is an elementary area for each pair of conjugate position and momentum variables in the classical phase space; in the classical limit the entropy becomes (roughly) the logarithm of the occupied volume in phase space, but this volume is measured in units of Planck's constant. If we start with a classical formulation (as did Boltzmann and Gibbs, of course), then we would find ourselves facing the problems of Eq. (6.17); namely, that we are trying to take the logarithm of a quantity with dimensions. If we measure phase-space volumes in units of Planck's constant, then all is well. The important point is that the problems with defining a purely classical entropy do not stop us from calculating entropy differences, which are observable directly as heat flows. We shall find a similar situation for the information content of continuous variables.

In the simple case where we ask a question and there are exactly $N = 2^m$ possible answers, all with equal probability, the entropy is just m bits. But if we make a list of all possible answers, we can label each of them with a distinct m-bit binary number: to specify the answer, all we need to do is write down this number. Note that the answers themselves can be very complex—different possible answers could correspond to lengthy essays, but the number of pages required to write these essays is irrelevant. If we agree in advance on the set of possible answers, all we have to do in answering the question is to provide a unique label, as in the example of Shakespeare's collected works. If we think of the label as a code word for the answer, then in this simple case the length of the code word that represents the nth possible answer is given by $\ell_n = - \log_2 p_n$, and the average length of a code word is given by the entropy.

---

4. Indeed, this problem is so troublesome that it has led to a serious shift in our teaching. It is simpler to define everything in the case where states are discrete, which has led many people to argue that we should not teach statistical physics until after students have learned quantum mechanics. Whatever advantages this approach might have, it guarantees that many U.S. students never see anything statistical (beyond a few lectures on the kinetic theory of gases) until their third year of university, which is quite late.

The equality of the entropy and the average length of code words is much more general than our simple example. Before proceeding, however, it is important to realize that the entropy is emerging as the answer to two very different questions. In the first case we wanted to quantify our intuitive notion of gaining information by hearing the answer to a question. In the second case, we are interested in the problem of representing this answer in the smallest possible space. It is quite remarkable that the only way of quantifying how much we learn is to measure how much space is required to write it down.

In statistical mechanics we have the choice of working with a microcanonical ensemble (in which an ensemble of systems is distributed uniformly over states of fixed energy) or with a canonical ensemble (in which an ensemble of systems is distributed across states of different energies according to the Boltzmann distribution). The microcanonical ensemble is like our simple example with all answers having equal probability: entropy really is just the log of the number of possible states, as is carved on Boltzmann's tombstone. But we know that in the thermodynamic limit there is not much difference between the two ensembles. This suggests that we can recover a simple notion of representing answers with code words of length $\ell_n = -\log_2 p_n$ provided that we can find a suitable analog of the thermodynamic limit.

Imagine that instead of asking a question once, we ask it many times. As an example, every day we can ask a meteorologist for an estimate of the temperature at noon the next day. Now instead of trying to represent the answer to one question we can try to represent the whole stream of answers collected over a long period of time. Let us label the sequences of answers $n_1 n_2 \cdots n_N$, and these sequences have probabilites $P(n_1 n_2 \cdots n_N)$.[5] From these probabilities we can compute an entropy that must depend on the length of the sequence:

$$S(N) = -\sum_{n_1} \sum_{n_2} \cdots \sum_{n_N} P(n_1 n_2 \cdots n_N) \log_2 P(n_1 n_2 \cdots n_N). \qquad (6.20)$$

Now we can draw on our intuition from statistical mechanics. The entropy is an extensive quantity, which means that as N becomes large the entropy should be proportional to N; more precisely, we should have

$$\lim_{N \to \infty} \frac{S(N)}{N} = \mathcal{S}, \qquad (6.21)$$

where $\mathcal{S}$ is the entropy density for our sequence in the same sense that a large volume of material has a well-defined entropy per unit volume.

The equivalence of ensembles in the thermodynamic limit means that having unequal probabilities in the Boltzmann distribution has almost no effect on anything we want to calculate, so long as we ask only about systems with many degrees of freedom. In particular, for the Boltzmann distribution we know that, state by state, the log of the probability is the energy and that this energy is itself an extensive quantity. Further, we know that (relative) fluctuations in energy are small. But if energy is the log of probability, and relative fluctuations in energy are small, then almost all states we actually observe have log probabilities that are the same. By analogy, all the long sequences of answers must fall into two groups: those with $-\log_2 P \approx N\mathcal{S}$, and those

---

5. Notice that, at this point, we do not need to assume that successive questions have independent answers.

with $P \approx 0$. Now this argument is a bit sloppy, but it is the right idea: if we are willing to think about long sequences or streams of data, then the equivalence of ensembles tells us that "typical" sequences are uniformly distributed over $\mathcal{N} \approx 2^{NS}$ possibilities, and that this appproximation becomes more and more accurate as the length N of the sequences becomes large.

Problem 124: Probabilities and the equivalence of ensembles.[6] Consider an ideal monatomic gas in three dimensions, for which the energy is

$$E = \frac{1}{2m} \sum_{i=1}^{3N} p_i^2, \tag{6.22}$$

where m is the atomic mass, and p_i is the momentum of one atom in one direction. Note that the conventional symbols for momentum and probability are the same, unfortunately. We define the classical sum over states to be an integral over positions and velocities, normalized by appropriate powers of Planck's constant h.

(a) The partition function in the microcanonical ensemble is

$$Z_{\text{micro}}(E) \equiv \frac{1}{h^{3N}} \int d^N x \int d^N p \; \delta \left(E - \frac{1}{2m} \sum_{i=1}^{3N} p_i^2 \right) \tag{6.23}$$

$$= \left(\frac{V}{h^3} \right)^N \int d^N p \; \delta \left(E - \frac{1}{2m} \sum_{i=1}^{3N} p_i^2 \right). \tag{6.24}$$

If the energy is fixed with precision ϵ, then $Z_{\text{micro}}(E)\epsilon$ is the number of accessible states, all occurring with equal probability, and so the microcanonical entropy is $S_{\text{micro}}(E) = \log_2[Z_{\text{micro}}(E)\epsilon]$. Use the Fourier representation of the delta function and the method of steepest descent to derive the asymptotic behavior of $S_{\text{micro}}(E)$ at large N.

(b) In the canonical ensemble, at inverse temperature β, the probability of being in any state is given by the Boltzmann distribution,

$$P = \frac{1}{Z(\beta)} e^{-\beta E}, \tag{6.25}$$

where

$$Z(\beta) = \frac{1}{h^{3N}} \int d^N x \int d^N p \; \exp \left(-\frac{\beta}{2m} \sum_{i=1}^{3N} p_i^2 \right). \tag{6.26}$$

Evaluate $Z(\beta)$ and the entropy $S(\beta)$. Review what is meant when we say that the entropy is the same in the canonical and microcanonical ensembles at large N.

(c) The typical probability of a state in the canonical ensemble is $P_{\text{typical}} = 2^{-S(\beta)}$. Define the deviation from this typical probability, per atom, as $\Delta = (1/N) \log_2(P/P_{\text{typical}})$. What can you say about the distribution of Δ over all states? Can you make a precise version of the statement that most states have either almost the typical probability or zero probability? For example, can you put a bound on the fraction f of states that have $|\Delta| > \delta_c$? How does the relation between f and δ_c change with N?

---

6. Some of this problem may be familiar from a statistical mechanics class, though perhaps in slightly different language.

Problem 125: More about typicality.   Consider drawing N samples of a variable that can take on K different values, with probabilities $p_1, p_2, \cdots, p_K$. Let the sequence of samples that you observe be called $i_1, i_2, \cdots, i_N$, which has probability

$$P = \prod_{n=1}^{N} p_{i_n}. \tag{6.27}$$

Show that the average of $L = -(1/N) \log_2 P$ is the entropy of the underlying distribution, $S = -\sum_i p_i \log_2 p_i$. Say as much as you can about the distribution of L as N becomes large.

The idea of typical sequences, which is the information-theoretic version of a thermodynamic limit, is enough to tell us that our simple arguments about representing answers by binary numbers ought to work on average for long sequences of answers. An important if obvious consequence is that if we have many unlikely answers (rather than fewer more likely answers) then we need more space to write the answers down. More profoundly, this is true answer by answer: to be sure that long sequences of answers take up as little space as possible, we need to use $\ell_n \approx -\log_2 p_n$ bits to represent each individual answer n. Thus, even individual answers that are more surprising require more space to write down.

As a simple example, imagine that we have four answers, with probabilities $p_1 = 1/2$, $p_2 = 1/4$, and $p_3 = p_4 = 1/8$. Naively, if we use a binary representation, we will need two bits to represent the four possibilities. But the entropy is

$$S \equiv -\sum_{i=1}^{4} p_i \log_2 p_i = \frac{1}{2} \log_2 2 + \frac{1}{4} \log_2 4 + \frac{2}{8} \log_2 8 = \frac{7}{4}, \tag{6.28}$$

which is less than two bits (as it must be). Suppose that we represent the four possibilities by binary sequences:

$$1 \rightarrow 0, \tag{6.29}$$
$$2 \rightarrow 10, \tag{6.30}$$
$$3 \rightarrow 110, \tag{6.31}$$
$$4 \rightarrow 111. \tag{6.32}$$

Notice that the length of each code word obeys $\ell_i = -\log_2 p_i$, so we know that, on average, the number of binary digits used per answer will be equal to the entropy. This behavior illustrates the idea that, by using code words of different lengths, we can reduce the average amount of space we need to write things down.

Problem 126: Do we need commas?   When we represent a sequence of answers, we must be able to find the boundaries between the code words. If all the words have the same length, we can just count, but this does not work if we use unequal lengths. At worst, we could add an extra symbol to punctuate the stream of words, but this solution takes extra space and surely is inefficient. Convince yourself that the code defined by Eqs. (6.29)–(6.32) does not need any extra symbols—all sequences of code words can be parsed uniquely.

To complete the picture, we have to put the idea of typicality together with code words of varying length. Suppose that we look at a block of N answers, $n_1, n_2, \cdots, n_N$ as before; let's label this block (or state, to reinforce the analogy with statistical physics) by s, which occurs with probability p_s. We choose the labels so that all states are numbered in order of their probability, that is, $p_1 \geq p_2 \geq \cdots \geq p_K$, where K is the number of possible sequences of length N. For each state s we can compute the cumulative probability of lower energy (higher probability) states, $P_s \equiv \sum_{i=1}^{s-1} p_i$. Now take this cumulative probability and expand it as a binary number. If we stop after m_s digits, where m_s is chosen so that

$$-\log_2 p_s \leq m_s < -\log_2 p_s + 1, \tag{6.33}$$

then we guarantee that this binary representation of P_s will be different from any subsequent number with larger s, so it is a unique encoding of the state s, as shown schematically in Fig. 6.3. But now we can see that the average number of binary digits used to encode the blocks of length N will be

$$L(N) \equiv \sum_s p_s m_s, \tag{6.34}$$

and we can bound this value from both sides,

$$\sum_s p_s \left(-\log_2 p_s \right) \leq L(N) < \sum_s p_s \left(-\log_2 p_s + 1 \right), \tag{6.35}$$

$$S(N) \leq L(N) < S(N) + 1, \tag{6.36}$$

where S is the entropy of the N-answer blocks, $S(N) = -\sum_s p_s \log_2 p_s$.

If we count the length of the code per answer, then

$$\frac{S(N)}{N} \leq \frac{L(N)}{N} < \frac{S(N)}{N} + \frac{1}{N}. \tag{6.37}$$

But, as before, we know that the entropy per degree of freedom should approach a finite entropy density, as in Eq. (6.21), and now we see that the average code length per answer is within $1/N$ of this entropy density. Thus, as $N \to \infty$, the entropy and the minimum code length are equal.

To summarize, if we need to write down answers many times, then the minimum space required to write them down is, per answer, the entropy of the distribution out of which the answers are drawn. Notice that our choice of alphabet in which to write is arbitrary, but we also had an arbitrariness in choosing the units of entropy; this is the same arbitrariness, and we are being consistent in choosing logarithms to the base 2 and a binary alphabet. Thus, the statement that entropy is both the amount of information gained and the amount of space needed to write down what we have learned is not arbitrary, and there are no constants floating around to spoil the exact equality. To reach this maximally compact representation, we must at least implicitly use the structure of the probability distribution out of which the answers are drawn, adjusting the lengths of individual code words in relation to the probability of the answer.

Problem 127: Uniqueness of code words. Show that the choice in Eq. (6.33) really does guarantee the uniqueness of code words.

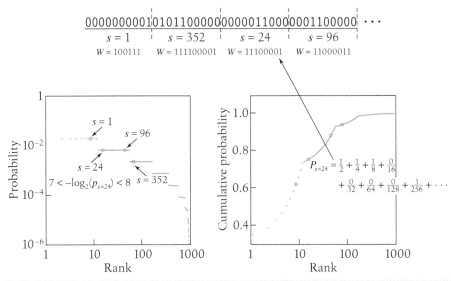

FIGURE 6.3

Coding of sequences with variable word length. In a stream where 0 and 1 occur independently, but with unequal probabilities, we can compress our description by coding N-bit blocks; here $N = 10$. Each block can be labeled by s, the number equivalent to the binary string (top). These states have widely varying probability p_s (lower left). We can compute the cumulative probability of states with higher probability (lower right), as described in the text, and use the binary expansion of this cumulative probability as the code word W. We stop the expansion at a number of digits given by rounding up from $-\log_2(p_s)$.

Problem 128: Coding rare events. Suppose that we have two possible answers, A and B, which occur with very unequal probabilities, $p_A \ll p_B$. Show that the entropy of the distribution of answers is approximately $S \approx p_A \log_2(e/p_A)$, where e is the base of natural logarithms. If we have a long sequence of answers, most are B with a sprinkling of As. Try to encode such a sequence in binary form, using a code in which some symbol (e.g., 1111) is reserved for A, and the blocks of B are encoded by writing the number of consecutive Bs as a binary number. To make this representation work—that is, to be sure that your encoding can be uniquely decoded—you obviously have to be careful in the special case where the number of Bs is equal to 15 (1111 in binary form). Are there any other problems? Does this code come close to the lower bound on code length set by the entropy?

The idea that there is a minimum amount of space required to write down a description of a system is incredibly important. At a practical level, we pay for the resources needed to write things down, or to transmit information from one place to another, and so there is a premium on using as little space as possible. The search for ways of packing information in the minimal amount of space is the problem of data compression. This result is the first indication of a general notion of efficiency in representing data, and we will see how this notion becomes relevant to biological systems.

The argument we have just given shows that once we know the probability distribution for the states s, we have a code that can be used to represent these states, and asymptotically this code is of minimum length. Suppose that states really are chosen out of a distribution $\mathbf{p} \equiv \{p_s\}$, but we don't know this; instead, we think that the distribution is $\mathbf{q}$. Then (neglecting terms that are unimportant in the large N limit), we assign a code word of length $\ell_s = - \log_2 q_s$ to each state, and so the mean code length is

$$L = - \sum_s p_s \log_2 q_s. \tag{6.38}$$

This expression is different than the entropy of the distribution $\mathbf{p}$, and the difference is

$$L - L_{\min} = L - S = - \sum_s p_s \log_2 q_s - \left[- \sum_s p_s \log_2 p_s \right] = \sum_s p_s \log_2 \left(\frac{p_s}{q_s} \right). \tag{6.39}$$

This quantity is zero if the two distributions are the same and is positive for any pair of distributions $\mathbf{p}$ and $\mathbf{q}$; it is called the Kullback-Leibler divergence between the two distributions and usually is written as

$$D_{\mathrm{KL}}(\mathbf{p}||\mathbf{q}) = \sum_s p_s \log_2 \left(\frac{p_s}{q_s} \right). \tag{6.40}$$

Notice that it is not a symmetric quantity and hence is not a metric on the space of distributions, although it does say something about the degree of similarity or difference between $\mathbf{p}$ and $\mathbf{q}$. The measure D_{KL} also is sometimes called the relative entropy of the distribution $\mathbf{p}$ with respect to $\mathbf{q}$.

To emphasize the role of D_{KL} as a measure of difference between distributions, suppose that we are given N samples and have to decide whether they came from $\mathbf{p}$ or $\mathbf{q}$. Out of the N samples, n_1 come from state 1, n_2 come from state 2, and so on. So the probability that the distribution $\mathbf{p}$ generated these samples is

$$P(\text{samples}|\mathbf{p}) = A \prod_s p_s^{n_s}, \tag{6.41}$$

where A is a combinatorial factor, and similarly

$$P(\text{samples}|\mathbf{q}) = A \prod_s q_s^{n_s}. \tag{6.42}$$

What we want to know is, given the samples, what is the probability P that they came from the distribution $\mathbf{p}$ as opposed to $\mathbf{q}$? Let us say that, a priori, the two possibilities are equally likely. Then, by Bayes' rule,

$$P = \frac{P(\text{samples}|\mathbf{p}) P(\mathbf{p})}{P(\text{samples})} \tag{6.43}$$

$$= \frac{P(\text{samples}|\mathbf{p})}{P(\text{samples}|\mathbf{p}) + P(\text{samples}|\mathbf{q})} \tag{6.44}$$

$$= \frac{1}{1 + 2^{-\Lambda}}, \tag{6.45}$$

where

$$\Lambda = \log_2 \left[\frac{P(\text{samples}|\mathbf{p})}{P(\text{samples}|\mathbf{q})} \right] = \sum_s n_s \log_2 \left(\frac{p_s}{q_s} \right). \tag{6.46}$$

As discussed previously, Λ is called the log likelihood ratio. Because it is proportional to all the n_s, it must also be proportional to N, and hence grows (on average) linearly with the number of samples. We can think of it as the accumulation of evidence for $\mathbf{p}$ versus $\mathbf{q}$, and the rate at which this evidence accumulates is, asymptotically,

$$\lim_{N \to \infty} \frac{1}{N} \Lambda = \sum_s \left[\lim_{N \to \infty} \frac{n_s}{N} \right] \log_2 \left(\frac{p_s}{q_s} \right) \tag{6.47}$$

$$= \sum_s p_s \log_2 \left(\frac{p_s}{q_s} \right) \tag{6.48}$$

$$= D_{\text{KL}}(\mathbf{p}||\mathbf{q}). \tag{6.49}$$

Thus, the Kullback-Leibler divergence is, like the entropy itself, the answer to two very different questions: the cost of coding data using codes based on the wrong distribution, and the ease of discriminating the distributions from each other based on samples.

Problem 129: A little more about the Kullback-Leibler divergence.

(a) Show that $D_{\text{KL}}(\mathbf{p}||\mathbf{q})$ is positive (semi-)definite and is minimized when $\mathbf{p} = \mathbf{q}$.

(b) $D_{\text{KL}}(\mathbf{p}||\mathbf{q})$ is unbounded, so some probability distributions are infinitely different from each other. Explain, using the connection to the accumulation of evidence, how to make sense of this divergence.

(c) If we have a family of distributions that depend on a parameter $\mathbf{p}_\theta$, show that $D_{\text{KL}}(\mathbf{p}_\theta||\mathbf{p}_{\theta'})$ behaves as $F(\theta) \times (\theta - \theta')^2$ when the parameters θ and θ' are close. Give an explicit formula for $F(\theta)$.

(d) Imagine that we draw N samples out of the distribution p_{θ_0}, but all we know is that the distribution is in the family p_θ. Use Bayes's rule to construct $P(\theta|\text{samples})$, and show that as N becomes large, this becomes peaked around the right answer, $\theta = \theta_0$. Show that the variance around this peak is related to $F(\theta_0)$.

(e) If the two distributions $\mathbf{p}$ and $\mathbf{q}$ are Gaussian, evaluate $D_{\text{KL}}(\mathbf{p}||\mathbf{q})$. Suppose that the two Gaussian distributions differ in either their means or their variances, but not both. You should find that the choice of changing mean versus variance makes a difference to the (a)symmetry of D_{KL}. Make this difference explicit, and use what we have shown about D_{KL} as a measure of discrimination to explain the origin of this difference.

Problem 130: Accumulating evidence.   Imagine that we can observe samples drawn out of one of two distributions, $p(x)$ or $q(x)$. At each tick of the clock, we see a new, independent sample x_t. After time T, the log likelihood ratio is

$$\Lambda(T) = \sum_t \log_2 \left[\frac{p(x_t)}{q(x_t)} \right], \tag{6.50}$$

which is equivalent to

$$\Lambda(T) = \Lambda(T - \Delta t) + \log_2 \left[\frac{p(x_t)}{q(x_t)} \right], \tag{6.51}$$

where Δt is the time between samples.

(a) Show that, as $\Delta t \to 0$, this expression becomes equivalent to

$$\frac{d\Lambda(T)}{dT} = v + \eta(T), \tag{6.52}$$

where $\eta(T)$ is white noise, $\langle \eta(T)\eta(T') \rangle = D\delta(T - T')$. Relate the drift velocity v and the diffusion constant D to the properties of the distributions $p(x)$ and $q(x)$.

(b) Consider the special case where $p(x)$ and $q(x)$ are Gaussian distributions with different means but the same variance. How are v and D related to the natural measures of signal-to-noise ratio in this case?

(c) As time progresses, the opportunity to collect more samples means that we can make increasingly reliable decisions about whether $p(x)$ or $q(x)$ is the correct distribution. If we decide as soon as $\Lambda(T)$ crosses a threshold θ, then we guarantee that we will make errors with only some known probability. What is this probability $P_e(\theta)$?

(d) Equation (6.52) is the Langevin equation for a particle that diffuses and drifts. Thus, we can write the diffusion equation for the probability distribution of Λ at one time:

$$\frac{\partial P(\Lambda, T)}{\partial T} = D \frac{\partial^2 P(\Lambda, T)}{\partial \Lambda^2} - v \frac{\partial P(\Lambda, T)}{\partial \Lambda}. \tag{6.53}$$

Show that at $T = 0$ the distribution $P(\Lambda, 0)$ is localized at $\Lambda = 0$. If we make a decision as soon as $\Lambda = \theta$, then we must have $P(\Lambda \geq \theta, T) = 0$, which is equivalent to an absorbing boundary at $\Lambda = \theta$. Find $P(\Lambda, T)$ subject to these two boundary conditions. Before going through the details, make some sketches of what you expect at different times.

(e) Although the absorbing boundary in part (d) imposes $P(\Lambda = \theta, T) = 0$, the derivative of the distribution is not zero at $\Lambda = \theta$, and hence there is a flux $J(T) = -D(\partial P(\Lambda, T)/\partial \Lambda)$ leaking through the boundary. Show that $J(T)$ is actually the probability per unit time that Λ reaches θ at time T (i.e., it is the distribution of first passage times). Explain why, as the problem has been set up here, this is equivalent to the distribution of times at which our observer will decide whether the samples come from $p(x)$ or $q(x)$. What is the mean of this decision time, $\bar{T}(\theta)$? Make a plot of the error probability versus the mean decision time, $P_e(\theta)$ versus $\bar{T}(\theta)$.

We all have experienced the sense that we are trading speed for accuracy as we make decisions. Your result in part (e) provides a concrete theory for this trade, in a context where evidence is accumulating over time. Such models, and their generalizations, have been used to analyze the performance of human observers in a variety of tasks, and even to motivate experiments on the responses of neurons plausibly involved in the decision making processes for nonhuman primates. For more, see the references in the Annotated Bibliography for this section.

The connection between entropy and information has (at least) one more very important consequence: correlations or order reduce the capacity to transmit information. Perhaps the most familiar example is in spelling. If all possible combinations of letters were legal words, then there would be $(26)^4 = 456{,}976$ four-letter words. But if you look through a large, reasonably coherent body of English text—the collected works of a prolific author, or the last year of newspaper articles—you will find that there at most a few hundred four-letter words being used. Most of this restriction of vocabulary comes from correlations among the letters in the word: once we have put a t in the first position, it is much more likely that we will put a vowel in the second position; if we want to put a consonant then it has a high probability of being an h, and so on. Although, correlations have signs—we speak both of correlation and anticorrelation—with respect to the entropy, all correlations have the same effect, namely, of reducing the entropy. Indeed, as explained in Appendix A.7, we can construct models for the probability distribution of the states in a system that are consistent with some measured correlations but otherwise have the maximum possible entropy. In addition, we can build a hierarchy of these models with ever-smaller entropies as we take into account more correlations; once we capture all relevant correlations, the entropy converges to its true value.

For four-letter words, as an example, the entropy for random letters would be $S_{rand} = 4 \log_2(26) = 18.8$ bits. In the collected works of Jane Austen, the one-body correlations, which measure unequal frequencies with which letters are used, reduces this value to $S_{ind} = 14$ bits. Taking into account the two-body correlations between pairs of letters cuts this entropy nearly in half, to $S_2 = 7.48$ bits; the true entropy of the distribution of four-letter words in these texts in only slightly less: $S = 6.92$ bits. Thus, the entropy is reduced nearly by a factor of three from the case of completely random letters, and most of this reduction is explained by one- and two-body correlations. Again, the important point is that these correlations, which may have many advantages, certainly have the consequence of reducing our vocabulary and hence our capacity to transmit information.

This seems an appropriate moment to reflect on the fact that entropy is a very old idea. It arises in thermodynamics first as a way of keeping track of heat flows, so that a small amount of heat dQ transferred at absolute temperature T generates a change in entropy $dS = dQ/T$. Although there is no function Q that measures the heat content of a system, there is a function S that characterizes the (macroscopic) state of a system independent of the path to that state. Now we know that the entropy of a probability distribution also measures the amount of space needed to write down a description of the (microscopic) states drawn out of that distribution.

Let us imagine, then, a thought experiment in which we measure (with some fixed resolution) the positions and velocities of all gas molecules in a small box and type these numbers into a file on a computer. There are relatively efficient programs (gzip, or "compress" on a UNIX machine) that compress such files to nearly their shortest possible length. If these programs really work as well as they can, then the length of the file tells us the entropy of the distribution out of which the numbers in the file are being drawn, but this is the entropy of the gas. Thus, if we heat up the room by $10°$ and repeat the process, we will find that the resulting data file is longer. More profoundly, if we measure the increase in the length of the file, we know the entropy change of the gas and hence the amount of heat that must be added to the room to increase the temperature. This connection between a rather abstract quantity (the length in bits of a computer file) and a very tangible physical quantity (the amount of heat added to a room) has long struck me as one of the more dramatic, if elementary, examples of the power of mathematics to unify descriptions of very disparate phenomena.

6.2 Noise and Information Flow

Returning to the conversation between Max and Allan (Section 6.1), we assumed that Max would receive a complete answer to his question, and hence that all his uncertainty would be removed. A more realistic description is that, for example, the world can take on many states w, and by observing data d we learn something, but not everything, about w. Before making observations, we know only that states of the world are chosen from some distribution[7] $P_W(w)$, and this distribution has an entropy $S[P_W(w)]$. Once we observe some particular datum d, our (hopefully improved) knowledge of w is described by the conditional distribution $P(w|d)$, which has an entropy $S[P(w|d)]$ that is smaller than $S[P_W(w)]$ if we have reduced our uncertainty about the state of the world by virtue of observations. We identify this reduction in entropy as the information that we have gained about w:

$$I(d \rightarrow w) \equiv S[P_W(w)] - S[P(w|d)]. \tag{6.54}$$

Notice that this quantity depends on exactly what datum d we have observed.

Strictly speaking, entropy is a property of the probability distribution out of which the states of a system are drawn. Thus, we write $S[P_W(w)]$ to mean the entropy of the states of the world when these are drawn out of $P_W(w)$. Similarly, we should write $S[P(w|d)]$ for the entropy of states of the world conditional on having observed the data d. Notice that $S[\cdot]$ is the same functional in both cases. But this notation is slightly cumbersome. Indeed, in statistical mechanics and thermodynamics we seldom talk about "the entropy of the distribution out of which the states of the gas have been drawn" (although we should); instead, we just say "the entropy of the gas." In this spirit, sometimes I will use the shorthand $S(w) \equiv S[P_W(w)]$ and $S(w|d) \equiv S[P(w|d)]$. I hope this does not cause any confusion.

With notational issues settled, let's go back to our problem. Having defined the information gained in Eq. (6.54), we should appreciate that this value is not guaranteed to be positive. Consider, for instance, data that tell us that all our previous measurements have larger error bars than we thought: clearly such data, at an intuitive level, reduce our knowledge about the world and should be associated with negative information. In other words, some data points d will increase our uncertainty about the state w of the world, and hence for these particular data the conditional distribution $P(w|d)$ has a larger entropy than the prior distribution $P_W(w)$, so that $I(d \rightarrow w)$ will be negative. However, we hope that, on average, gathering data corresponds to gaining information: although single data points can increase our uncertainty, the average over all data points does not.

If we average over all possible data—weighted, of course, by their probability of occurrence $P_D(d)$—we obtain the average information that d provides about w:

$$\langle I(d \rightarrow w) \rangle = S(w) - \sum_d P_D(d) S(w|D). \tag{6.55}$$

---

7. When we talk about the states w of the world, it is natural to say that these states are drawn from the distribution $P(w)$. Similarly, when talking about the data that we will collect, it is natural to write that particular observations d are drawn from the distribution $P(d)$. The problem is that $P(\cdot)$ refers to different functions in these two cases. To solve this notational problem, note that the states of the world w come from a set of possible states, $w \in W$, and so the distribution over these states can be written $P_W(w)$. Similarly, individual observations come from a set of possible observations, $d \in D$, and the distribution of these data should be written $P_D(d)$.

This expression can be rearranged and simplified, and the result is so important that it is worth being very explicit about the algebra:

$$\langle I(d \to w) \rangle = - \sum_w P_W(w) \log_2 P_W(w)$$

$$- \sum_d P_D(d) \left[- \sum_w P(w|d) \log_2 P(w|d) \right] \tag{6.56}$$

$$= - \sum_w \sum_d P(w, d) \log_2 P_W(w)$$

$$+ \sum_w \sum_d P(w|d) P_D(d) \log_2 P(w|d) \tag{6.57}$$

$$= - \sum_w \sum_d P(w, d) \log_2 P_W(w)$$

$$+ \sum_w \sum_d P(w, d) \log_2 P(w|d) \tag{6.58}$$

$$= \sum_w \sum_d P(w, d) \log_2 \left[\frac{P(w|d)}{P_W(w)} \right] \tag{6.59}$$

$$= \sum_w \sum_d P(w, d) \log_2 \left[\frac{P(w|d) P_D(d)}{P_W(w) P_D(d)} \right] \tag{6.60}$$

$$= \sum_w \sum_d P(w, d) \log_2 \left[\frac{P(w, d)}{P_W(w) P_D(d)} \right], \tag{6.61}$$

where we identify the joint distribution of states of the world and data: $P(w, d) = P(w|d) P_D(d)$.

It is a nontrivial result that, from Eq. (6.61), the average information that d provides about w is symmetric in d and w. Thus, we can also view the state of the world as providing information about the data we will observe, and this information is, on average, the same as the information that the data will provide about the state of the world. This "average information provided" therefore is often called the mutual information, and this symmetry will be very important in subsequent discussions; to remind ourselves of this symmetry we write $I(d; w)$ rather than $\langle I(d \to w) \rangle$.

One consequence of the symmetry or mutuality of information is that we can write the mutual information as a difference of entropies in two different ways:

$$I(d; w) = S(w) - \sum_d P_D(d) S(w|d), \tag{6.62}$$

$$I(d; w) = S(d) - \sum_w P_W(w) S(d|w). \tag{6.63}$$

If we consider only discrete sets of possibilities, then entropies are positive (or zero), so that these equations imply

$$I(d; w) \leq S(w), \tag{6.64}$$

$$I(d; w) \leq S(d). \tag{6.65}$$

The first equation tells us that by observing d, we cannot learn more about the world than there is entropy in the world itself. This makes sense: entropy measures the number of possible states that the world can be in, and we cannot learn more than we would learn by reducing this set of possibilities down to one unique state. Although sensible (and, of course, true), this statement is not terribly powerful: seldom are we in the position that our ability to gain knowledge is limited by the lack of possibilities in the world around us. In contrast, there is a tradition of studying biological systems as they respond to highly simplified signals, and under these conditions the lack of possibilities can be a significant limitation, substantially confounding the interpretation of experiments.

Equation (6.65), however, is much more powerful. It says that, whatever may be happening in the world, we can never learn more than the entropy of the distribution that characterizes our data. Thus, if we ask how much we can learn about the world by taking readings from a wind detector on top of the roof, we can place a bound on the amount we learn just by taking a very long stream of data, using these data to estimate the distribution $P_D(d)$, and then computing the entropy of this distribution.

The entropy of our observations thus limits how much we can learn, no matter what question we were hoping to answer, and so we can think of the entropy as setting (in a slight abuse of terminology) the capacity of the data d to provide or to convey information. As an example, the entropy of neural responses sets a limit to how much information a neuron can provide about the world, and we can estimate this limit even if we do not yet understand what it is that the neuron is telling us (or the rest of the brain).

Problem 131: Maximally informative experiments. Imagine that we are trying to gain information about the correct theory T describing some set of phenomena. At some point, our relative confidence in one particular theory is very high; that is, $P(T = T_*) > F \cdot P(T \neq T_*)$ for some large F. However, there are many possible theories, so our absolute confidence in the theory T_* might nonetheless be quite low, $P(T = T_*) \ll 1$. Suppose we follow the "scientific method" and design an experiment that has a yes-or-no answer, and this answer is perfectly correlated with the correctness of theory T_* but is uncorrelated with the correctness of any other possible theory—our experiment is designed specifically to test or falsify the currently most likely theory. What can you say about how much information you expect to gain from such a measurement? Suppose instead that you are completely irrational and design an experiment that is irrelevant to testing T_* but has the potential to eliminate many (perhaps half) of the alternatives. Which experiment is expected to be more informative? Although this is a gross cartoon of the scientific process, it is not such a terrible model of a game like "twenty questions." It is interesting to ask whether people play such question games following strategies that might seem irrational but nonetheless serve to maximize information gain. Related but distinct criteria for optimal experimental design have been developed in the statistical literature.

To see how the ideas of entropy reduction and information work in a real example, let's consider the response of a neuron to sensory inputs. As we have discussed in previous sections, most neurons in the brain generate a sequence of brief ($\sim 1\,\text{ms}$) identical electrical pulses called action potentials or spikes. Because these events are

FIGURE 6.4

A schematic of how a train of action potentials is converted to discrete words at different time resolutions $\Delta\tau$. There is a minimum interspike interval, the "refractory period" (here, $\sim$2 ms), so that for sufficiently small $\Delta\tau$, the words are binary. Highlighted is the case where $\Delta\tau = 2$ ms and $T = 8$ ms, so this segment of the spike train becomes three successive four-bit words, 0101, 0100, and 0110.

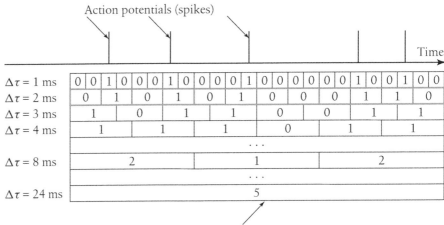

Words at different time resolutions

identical, we can think of them as marking points in time, and then we can build a discrete vocabulary of responses by fixing some limited time-resolution $\Delta\tau$, as in Fig. 6.4. More precisely, if $\Delta\tau$ is small, then in each small time window of duration $\Delta\tau$ we will see either one or zero spikes, and so the response is naturally discrete and binary. Then segments of the spike train of duration T can be thought of as $T/\Delta\tau$-letter binary words. In contrast, if $\Delta\tau$ is large, then we are representing the neural response by counting the number of spikes in these larger time windows, and again these counts can be strung together into words, though now these words are built from a more complex alphabet. Recording from a single neuron as the animal experiences some reasonably complex, dynamic sensory inputs, it is relatively easy to estimate the distribution of these words, $P(W)$, so long as we do not make the ratio $T/\Delta\tau$ too large. Then we can compute the entropy of this distribution, $S(T, \Delta\tau)$.

Figure 6.5 shows the results of experiments on the motion-sensitive neuron H1 in the fly visual system introduced in Section 4.4, when we discussed noise and the precision of visual motion estimation. In these experiments, the fly sees a randomly moving pattern, and H1 responds with a stream of spikes. If we fix $\Delta\tau = 3$ ms and look at $T = 30$ ms segments of the spike train, there are $2^{T/\Delta\tau} \sim 10^3$ possible words, but the distribution is strongly biased, and the entropy is only $S(T, \Delta\tau) \sim 5$ bits. This relatively low entropy means that we can still sample the distributions of words even out to $T \sim 50-60$ ms, which is interesting, because the fly can actually generate a flight correction in response to visual motion inputs within $\sim$30 ms.

The entropy $S(T, \Delta\tau)$ should be an extensive quantity, which means that, for large T, we should have $S(T, \Delta\tau) \propto T$. More strongly, if the correlations in the spike train are sufficiently short ranged, then we expect that at large T we will have

$$\frac{1}{T}S(T, \Delta\tau) = \mathcal{S}(\Delta\tau) + \frac{C(\Delta\tau)}{T} + \cdots, \tag{6.66}$$

where the omitted terms vanish more rapidly than $1/T$. In fact we see this behavior in the real data (Fig. 6.6), which suggests that we really can estimate the entropy rate $\mathcal{S}(\Delta\tau)$.[8]

---

8. For more on the problem of estimating entropy and information from real data, see Appendix A.8.

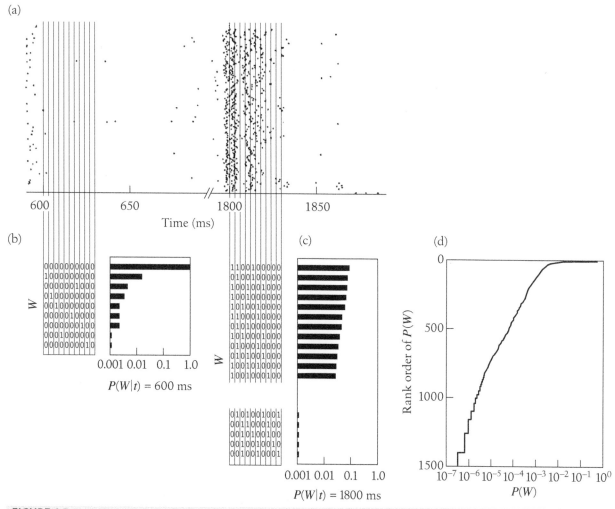

FIGURE 6.5

A neuron responds to dynamic stimuli with sequences of spikes. In this case, as described in the text, we look at the motion-sensitive neuron H1 in the fly's visual system. Redrawn from de Ruyter van Steveninck et al. (1997). (a) Each line across time is a single presentation of a movie, and dots mark the arrival times of spikes on each trial. (b,c) The spike trains are discretized into binary words with $\Delta\tau = 3$ ms resolution, and the distribution of words that occur at a particular moment in the movie, $P(W|t)$, is shown for $t = 600$ ms and $t = 1800$ ms. (d) The distribution of words averaged over all times, in rank order.

Connecting to the discussion above, the entropy rate $\mathcal{S}(\Delta\tau)$ sets a limit on the rate at which the spikes can provide information about the sensory input. When we make $\Delta\tau$ smaller, the entropy rate necessarily goes up, because previously indistinguishable responses map to different words at higher time resolution. Concretely, if we make $\Delta\tau$ smaller by a factor of two, then every 1 in the coarse words can become either a 01 or a 10 in the higher resolution words, and so we expect the entropy to increase by roughly one bit for every spike, as in Fig. 6.4.

Problem 132: Entropy and entropy rate in simple models. Going back to Chapter 2 and Appendix A.1, you know how to generate events drawn from a Poisson process with an

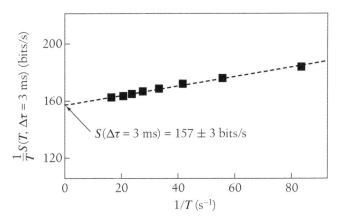

FIGURE 6.6

Entropy is extensive. From the experiments on the neuron H1 in Fig. 6.5, we compute the entropy of words at fixed time resolution $\Delta\tau = 3$ ms and variable length T, stopping when T is so large that we can no longer reliably sample the distribution $P(W)$. The data (error bars are smaller than the symbols) fall on the line predicted by Eq. (6.66), and we can thus extract an estimate of the entropy rate $\mathcal{S}(\Delta\tau)$. Redrawn from Strong et al. (1998a).

arbitrary time-dependent rate $r(t)$. Here you should take this (semi-)seriously as a model for spike trains, and use the resulting simulations to explore the entropy and entropy rate of neural responses.

(a) Start with $r = r_0$, a constant. Generate a long sequence of spikes (e.g., $\sim 10^4$). Choose a time resolution $\Delta\tau$ such that $r_0 \Delta\tau \ll 1$, and turn your simulated spike train into a binary sequence; for simplicity, ignore the (rare) occurrence of two spikes in one bin. Form words with $T/\Delta\tau$ bits, and estimate the distribution of these words from your simulated data. Compute the entropy of this distribution, and explore its dependence on T, r_0, and $\Delta\tau$. Do you see the emergence of an entropy rate, $S \sim \mathcal{S}T$?

(b) Explain why, for a Poisson process with a constant rate, $S = \mathcal{S}T$ should be exact. From this result, you can calculate $\mathcal{S}$ by thinking about just one bin of size $\Delta\tau$, and you should do this. How does your analytic result compare with the simulation results in part (a)?

(c) Suppose that $x(t)$ is a Gaussian stochastic process with correlation function

$$\langle x(t)x(t')\rangle = \sigma^2 e^{-|t-t'|/\tau_c}. \tag{6.67}$$

Samples of this process can be generated by simulating the Langevin equation,

$$\tau_c \frac{dx}{dt} = -x + \sqrt{2\sigma^2\tau_c}\,\eta(t), \tag{6.68}$$

where $\langle \eta(t)\eta(t')\rangle = \delta(t-t')$. Consider a Poisson process with rate $r(t) = r_0 e^{x(t)}$. Generate spike sequences for this process, and follow the procedures in part (a) to estimate the entropy in binary words of duration T at resolution $\Delta\tau$, with reasonable choices of parameters. Can you observe the emergence of extensive behavior, $S \sim \mathcal{S}T$? Does this (as seems plausible) require $T \gg \tau_c$? How do your results depend on σ?

A long-standing question in thinking about the brain has been whether the precise timing of individual spikes is important, or whether the brain is capable of counting spikes only in relatively coarse time bins, so that the rate of spikes over longer periods of time is all that matters. We now have the tools to give a more precise formulation of this question. As the time resolution increases, the entropy of the spike trains goes up, and hence so does the capacity of the neuron to convey information. The question is whether this capacity is used. Does the information about sensory inputs also rise as the time resolution is improved, or is the extra entropy just noise?

If the sensory inputs are called s, then the information that the spike sequences in some window T provide about these inputs can be written, as in Eq. (6.63), as a difference of entropies,

$$I(s; W) = S(W) - \langle S(W|s) \rangle_s, \tag{6.69}$$

where $\langle \cdot \rangle_s$ denotes an average over the distribution of inputs. We have already discussed the entropy of the neural vocabulary, $S(W)$; the problem is how to estimate $S(W|s)$, the entropy of the words given the sensory input s. To do this we need to sample the distribution $P(W|s)$, that is, the distribution of neural responses when the stimulus is fixed. At a minimum, this requires that we repeat the same stimuli many times. So, if the visual stimulus is a long movie, we have to show the movie over and over again. But how do we pick out a particular stimulus s from the continuous stream? One way is to realize that the flow of time in the movie provides an index into the stimuli, and all we need are averages over the distribution of stimuli. If the source of the stimuli is ergodic (which we can arrange to be true in the lab!), then an average over stimuli is equivalent to an average over time. So, if we repeat the movie many times and focus on events at time t relative to the start of the movie, we can sample, in repeats of the movie, the distribution $P(W|t)$, as in Fig. 6.5, and hence estimate $S(W|t)$. Finally, the information is obtained by explicitly replacing the ensemble average with a time average:

$$I(s; W) = S(W) - \langle S(W|t) \rangle_t. \tag{6.70}$$

Each of the entropy terms on the right should behave as in Eq. (6.66), and so we can extract an estimate of the information rate $R_{\text{info}}(\Delta\tau)$ as a function of time resolution. Results are shown in Fig. 6.7.

This figure shows that, as the time resolution is varied from 800 ms down to 2 ms, the information rate follows the entropy rate, with a nearly constant 50% efficiency. Although we should not generalize too much from one example, this result certainly suggests that neurons are making use of a significant fraction of their capacity in actually encoding sensory signals. Also, this is true even at millisecond time resolution. The idea that the entropy of the spike train sets a limit to neural information transmission emerged almost immediately after Shannon's work, but it was never clear whether these limits could be approached by real systems.

Problem 133: Information from single events. This section began by defining the information gained in a single observation. Here, we would like to find the parallel for individual neural responses, but there is a twist, because spikes are rare compared with silences. Thus, it makes sense to ask how much information is obtained per spike, or per nonsilent word W. Imagine that we look in a window of duration $\Delta\tau$ at time t, and we are looking for some event e—this event could be a single action potential, or some combination of multiple spikes with specific intervals between them. On average these events occur with some rate $\bar{r}_e$.

(a) In the small window $\Delta\tau$, either the event e occurs or it does not; for sufficiently small $\Delta\tau$, the probability of occurrence is $p_e = \bar{r}_e \Delta\tau$. What is the entropy of the binary variable marking the occurrence of the event? Can you simplify your result when $p_e \ll 1$? You will see

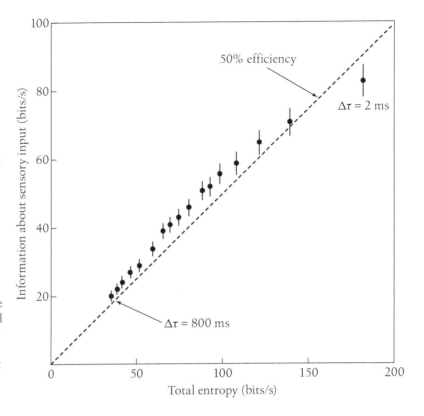

FIGURE 6.7

Entropy and information in a spike train. Experiments on the fly's motion-sensitive visual neuron were analyzed as described in the text (following Fig. 6.5) to estimate the total entropy and the information carried about the sensory input. As the time resolution of the analysis is varied from $\Delta\tau = 800$ ms down to $\Delta\tau = 2$ ms, we distinguish finer details of the neural response and expand the capacity of the putative neural code; this enhanced capacity is measured by the increasing entropy. Remarkably, across this huge range, capacity is used with almost constant efficiency. Redrawn from Strong et al. (1998a).

that the entropy in this limit is small, but so is the expected number of events. What is the entropy per event?

(b) If we know the sensory inputs to this neuron, then the probability of an event depends on time, locked to the time dependence of the sensory signal. Let's call the time-dependent rate $r_e(t)$. As in part (a), compute the entropy of the binary event/nonevent variable, but now conditional on knowledge of the sensory inputs.

(c) Combine your results in parts (a) and (b) to give an expression for the mean information that the occurrence or nonoccurrence of the event provides about the sensory input. Normalize by the expected number of events, to give bits per event. Is the limit $\Delta\tau \to 0$ well behaved? When the dust settles, you should find that the information per event is

$$I_e = \left\langle \frac{r_e(t)}{\bar{r}_e} \log_2\left[\frac{r_e(t)}{\bar{r}_e}\right] \right\rangle_t. \tag{6.71}$$

(d) As an alternative view of the same question, suppose that we observe a large window of time T. If T is sufficiently large, we can be sure that the event e will occur, but we don't know when. Thus, there is a probability $P(t) = 1/T$ for the time t at which this event occurs. In contrast, if we know the inputs, the probability that the event occurs at time t should be proportional to the time-dependent rate $r_e(t)$. Construct a normalized distribution $P(t|\text{inputs})$, and compute the mutual information between the inputs and the event time as the difference in entropy, $I_e = S[P(t)] - S[P(t|\text{inputs})]$. Does the result you obtain in this way agree with Eq. (6.71)? Why or why not?

Problem 134: Information from single spikes in a simple model. In Problem 132, you constructed a model spike train using a Poisson process with a time varying rate $r(t) = r_0 e^{x(t)}$, where $x(t)$ is a Gaussian stochastic process. Show that, for this model, the information carried by a single spike about $x(t)$ is linear in the variance of the signal $\langle x^2 \rangle$. This suggests that if the signal variance grows, the information carried by spikes grows with it, without bound. Explain what is wrong with this picture. Suppose instead that the spike rate $r(t)$ depends on x through some saturating function, for example

$$r(t) = \frac{r_0}{1 + \exp[-x(t) + \theta]}. \tag{6.72}$$

Reduce the formula for I_e in this model to a single integral that you can do numerically. Can you see how the results simplify as $\langle x^2 \rangle$ becomes large? Hint: Notice that this limit of Eq. (6.72) is equivalent to a model in which

$$r(t) = \frac{r_0}{1 + \exp[-\gamma(x(t) + \tilde{\theta})]}, \tag{6.73}$$

where $\gamma \to \infty$ while $\langle x^2 \rangle$ stays constant. Is there a setting of the threshold $\tilde{\theta}$ that maximizes I_e? Is there a cost to achieving this optimum?

You might worry that the high efficiency of coding seen in the fly's H1 neuron arises because the fly has relatively few neurons and thus is under greater pressure to be efficient. Although this may be true, it seems that high coding efficiencies are to be found even in animals like us and our primate cousins, who have huge numbers of neurons. In humans it is possible to record from individual receptor cells in our hands and fingertips, contacting the axons of these cells as they course along the arm to the spinal cord. Data are more limited than in the fly, so one has to be more careful to avoid systematic errors, but the lower bound on the efficiency of coding complex, dynamic variations in the indentation of the skin is more than 50%.

In the visual cortex of nonhuman primates, there is a classic series of experiments correlating the perception of motion with the activity of single neurons. The cortex can be divided into areas based on the response characteristics of the neurons in that area, and the medial temporal area (MT) has a preponderance of cells that respond to motion in small patches of the visual field, being excited to generate more spikes by motion in one direction and inhibited by motion in the opposite direction, as for the fly's neuron H1 discussed above. The experiments that link the activity of neurons in MT with perception are done with random dot patterns in which a fraction of the dots move coherently while the remaining dots are randomly deleted and replaced at new locations; the perception of motion direction becomes less reliable as the degree of coherence decreases. The evidence that single neurons are making a measurable contribution to the perceptual decision is strong, because one can correlate the number of spikes generated by a neuron with the animal's decision about leftward versus rightward motion even when the coherence is zero, and the animal is just guessing.

The experiments in MT focused on asking the animal to report a decision about motion direction across a 2 s window of stimulation. When we look at these random patterns, however, we see a certain amount of "jiggling," especially at low coherence. If we present exactly the same pattern of random dots versus time, we find that the neurons respond with a fair degree of reliability to the temporal details of the movie,

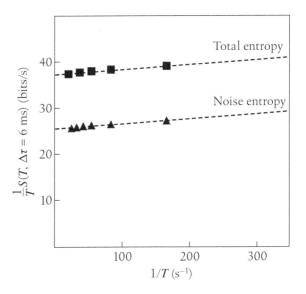

FIGURE 6.8

Entropy and information in spike trains from a motion-sensitive neuron in the primate visual cortex (the medial temporal area). Visual stimuli were random dot patterns moving with very little coherence among the dots. The experiments are by Britten et al. (1993) and the analysis by Strong et al. (1998b). From Eq. (6.70), the information is the difference between the total entropy and the noise entropy, each shown here.

certainly down to time scales of less than 10 ms. Figure 6.8 shows what this means in terms of the information carried by the spike trains about the time-varying details of the visual stimulus, rather than just the overall direction of motion. Here the efficiency, at $\Delta\tau = 6$ ms time resolution, is 25–30%. Experiments on the same neurons using stimuli that alternated between moving left and right[9] on the 30–100 ms time scale found information rates of 1–2.5 bits/spike, quite comparable to the results with H1. In summary, although there are differences in the details of the spike trains from motion-sensitive neurons in flies and monkeys, there is little difference in the amount of information they carry or in the efficiency with which this information is encoded with respect to the kinds of complex, dynamic stimuli that are relevant to the real world.

We now want to look at information transmission in the presence of noise, connecting back a bit to what we discussed in Chapters 2 and 4. Imagine that we are interested in some signal x, and we have a detector that generates data y, which is linearly related to the signal but corrupted by added noise:

$$y = gx + \xi. \tag{6.74}$$

It is reasonable in many systems to assume that the noise is Gaussian, either for fundamental physical reasons (as with thermal noise), or because it arises from a superposition of many independent, finite-variance sources, in which case the central limit theorem is valid. We also start with the assumption that x is drawn from a Gaussian distribution just because this is a simple place to start; we will see that we can use the maximum entropy property of Gaussians to make some more general statements based on this simple example. The question, then, is how much information observations on y provide about the signal x.

9. To be more precise, each neuron in MT has, as with H1 in the fly, a preferred direction of motion that drives the strongest responses. The stimuli in the experiments alternated between the preferred and antipreferred directions, chosen for each neuron.

The problem of information transmission with Gaussian signals and noise is sufficiently important that it is worth going through all the algebra quite explicitly; this is also one of those pleasing problems where, as we calculate, terms proliferate and then collapse into a much simpler result. So, onward. The statement that ξ is Gaussian noise means that

$$P(y|x) = \frac{1}{\sqrt{2\pi \langle \xi^2 \rangle}} \exp\left[-\frac{1}{2\langle \xi^2 \rangle}(y - gx)^2\right], \tag{6.75}$$

where we take the mean $\langle \xi \rangle$ to be zero, because it is just noise.[10] Our simplification is that the signal x also is drawn from a Gaussian distribution,

$$P_X(x) = \frac{1}{\sqrt{2\pi \langle x^2 \rangle}} \exp\left[-\frac{1}{2\langle x^2 \rangle}x^2\right], \tag{6.76}$$

and hence y itself is Gaussian:

$$P_Y(y) = \frac{1}{\sqrt{2\pi \langle y^2 \rangle}} \exp\left[-\frac{1}{2\langle y^2 \rangle}y^2\right], \tag{6.77}$$

$$\langle y^2 \rangle = g^2\langle x^2 \rangle + \langle \xi^2 \rangle. \tag{6.78}$$

To compute the information that y provides about x, we use Eq. (6.61), generalized to the case of continuous variables,

$$I(y;x) = \int dy \int dx\, P(x,y) \log_2\left[\frac{P(x,y)}{P_X(x)P_Y(y)}\right] \text{ bits} \tag{6.79}$$

$$= \frac{1}{\ln 2}\int dy \int dx\, P(x,y) \ln\left[\frac{P(y|x)}{P_Y(y)}\right] \tag{6.80}$$

$$= \frac{1}{\ln 2}\left\langle \ln\left[\frac{\sqrt{2\pi \langle y^2 \rangle}}{\sqrt{2\pi \langle \xi^2 \rangle}}\right] - \frac{1}{2\langle \xi^2 \rangle}(y - gx)^2 + \frac{1}{2\langle y^2 \rangle}y^2\right\rangle, \tag{6.81}$$

where by $\langle \cdot \rangle$ we understand an expectation value over the joint distribution $P(x,y)$. In Eq. (6.81), the first term is the expectation value of a constant. The third term involves the expectation value of y^2 divided by $\langle y^2 \rangle$, so we can cancel numerator and denominator. In the second term, we can take the expectation value first of y with x fixed, and then average over x, but because $y = gx + \xi$, the numerator is just the mean square fluctuation of y around its mean value, which again cancels with the $\langle \xi^2 \rangle$ in the

---

10. I don't intend to be flippant. There is a general rule, I think, that things we call noise have zero mean, essentially because we are always free to absorb any nonzero mean into the definition of "signal," or even into the choice of units. The clearest case is in the Langevin description of Brownian motion: *all* the forces on a Brownian particle result from collisions with the molecules of the surrounding fluid, and we call the average of this force the drag, leaving the random piece to be the Langevin force, which then has zero mean by definition.

denominator. So we have, putting the three terms together:

$$I(y; x) = \frac{1}{\ln 2} \left[\ln \sqrt{\frac{\langle y^2 \rangle}{\langle \xi^2 \rangle}} - \frac{1}{2} + \frac{1}{2} \right] \tag{6.82}$$

$$= \frac{1}{2} \log_2 \left(\frac{\langle y^2 \rangle}{\langle \xi^2 \rangle} \right) \tag{6.83}$$

$$= \frac{1}{2} \log_2 \left(1 + \frac{g^2 \langle x^2 \rangle}{\langle \xi^2 \rangle} \right) \text{ bits.} \tag{6.84}$$

Another way of arriving at this result is to remember that the information is a difference of entropies [Eq. (6.63)], but in this case the underlying distributions are all Gaussian. Thus, it is useful to know, in general, the entropy of a Gaussian distribution. Suppose that

$$P(z) = \frac{1}{\sqrt{2\pi \langle (\delta z)^2 \rangle}} \exp \left[-\frac{(z - \langle z \rangle)^2}{2 \langle (\delta z)^2 \rangle} \right]. \tag{6.85}$$

Now our task is to compute

$$S = -\int dz \, P(z) \log_2 P(z) = -\left\langle \log_2 P(z) \right\rangle. \tag{6.86}$$

But we have

$$\log_2 P(z) = \frac{1}{\ln 2} \left[\ln \left(\frac{1}{\sqrt{2\pi \langle (\delta z)^2 \rangle}} \right) - \frac{(z - \langle z \rangle)^2}{2 \langle (\delta z)^2 \rangle} \right], \tag{6.87}$$

and hence

$$S = -\left\langle \log_2 P(z) \right\rangle \tag{6.88}$$

$$= \frac{1}{\ln 2} \left[\ln \left(\sqrt{2\pi \langle (\delta z)^2 \rangle} \right) + \left\langle \frac{(z - \langle z \rangle)^2}{2 \langle (\delta z)^2 \rangle} \right\rangle \right] \tag{6.89}$$

$$= \frac{1}{\ln 2} \left[\frac{1}{2} \ln \left(2\pi \langle (\delta z)^2 \rangle \right) + \frac{1}{2} \right] \tag{6.90}$$

$$= \frac{1}{2} \log_2 \left[2\pi e \langle (\delta z)^2 \rangle \right]. \tag{6.91}$$

Notice that the entropy is independent of the mean, as we expect, because entropy measures variability or uncertainty.

Problem 135: Using the entropy of Gaussian distributions.   Use the general result on the entropy of Gaussian distributions, Eq. (6.91), to rederive Eq. (6.84) for the information transmission through the Gaussian channel.

We can gain some intuition by rewriting Eq. (6.84). Rather than thinking of our detector as adding noise after generating the signal gx, we can think of it as adding noise directly to the input and then transducing this corrupted input:

$$y = g(x + \eta_{\text{eff}}), \tag{6.92}$$

where $\eta_{\text{eff}} = \xi/g$. Note that the effective noise η_{eff} is in the same units as the input x; this is called referring the noise to the input and is a standard way of characterizing detectors, amplifiers, and other devices, as discussed in Chapter 4.[11] Written in terms of the effective noise level, the information transmission takes a simple form,

$$I(y; x) = \frac{1}{2} \log_2 \left(1 + \frac{\langle x^2 \rangle}{\langle \eta_{\text{eff}}^2 \rangle} \right) \text{ bits}, \tag{6.93}$$

or

$$I(y; x) = \frac{1}{2} \log_2 (1 + SNR), \tag{6.94}$$

where the signal-to-noise ratio is the ratio of the variance in the signal to the variance of the effective noise: $SNR = \langle x^2 \rangle / \langle \eta_{\text{eff}}^2 \rangle$.

The result in Eq. (6.94) is easy to picture. At the start, the signal is spread over a range $\delta x_0 \sim \langle x^2 \rangle^{1/2}$, but by observing the output of the detector, we can localize the signal to a small range $\delta x_1 \sim \langle \eta_{\text{eff}}^2 \rangle^{1/2}$, and the reduction in entropy is $\sim \log_2(\delta x_0/\delta x_1) \sim (1/2) \log_2(SNR)$, which is approximately the information gain.

Problem 136: A small point. Explain why the simple argument in the preceding paragraph, which seems sensible, does not give the exact answer for the information gain at small SNR.

To illustrate these ideas, consider the expression of the gap genes in the fly embryo (Sections 4.3 and 5.3). In response to the primary (maternally supplied) morphogens, these genes have varying levels of expression that provide a first step in building the blueprint for the fully developed organism. One of the basic ideas in developmental biology is that these expression levels carry positional information; that is, cells know where they are in the embryo, and hence their fate in the developed organism, as a result of knowing the concentrations of these molecules. It seems natural to ask whether we can quantify this positional information in bits. To do so, as in Fig. 6.9, we can look at many embryos and measure the concentration versus position in each one. If there is a perfect functional relationship with no noise, then the transmission of positional information is limited only by the number of samples that we take along the position

---

11. As a reminder, if we build a photodetector, it is not so useful to quote the noise level in volts at the output—we want to know how this noise limits our ability to detect dim lights. Similarly, when characterizing a neuron that uses a stream of pulses to encode a continuous signal, we do not really want to know the variance in the pulse rate (although this quantity is widely discussed); we want to know how noise in the neural response limits precision in estimating the real signal, which amounts to defining an effective noise level in the units of the signal itself. In the present case this is just a matter of dividing, but generally it is a more complex task, as when decoding spike trains (Fig. 4.35).

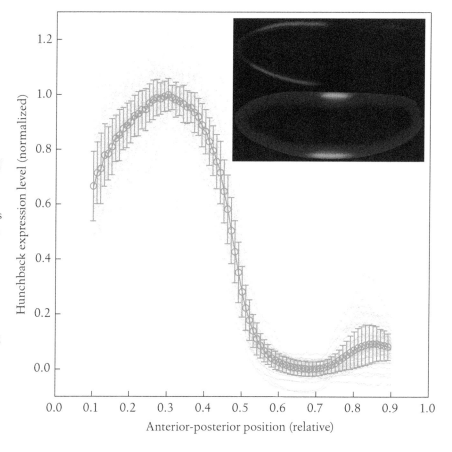

FIGURE 6.9

Spatial profiles of Hunchback (Hb) expression in the early *Drosophila* embryo. Small dots show experiments from individual embryos; circles with error bars are mean and standard deviation across 51 embryos. In the inset, the image in red shows fluorescent antibody staining for Hb, and green shows the corresponding measurement for Krüppel. These images are taken by optical sectioning along the midline of the embryo. The intensity is measured in a small area, roughly the size of a nucleus, that slides along the rim of the embryo where the nuclei are sitting. Redrawn from Dubuis et al. (2012).

axis; hence the information in bits will just be the log of the number of cells. But there is noise, and it sets a limit to the positional information.

The position along the embryo can be measured by $0 \leq x \leq 1$. If we assume that the cells acquiring positional information are distributed uniformly (which is approximately true), then $P_X(x) = 1$. The expression level of the gene we are looking at will be called g. What we need to know is the distribution of expression levels at one position, $P(g|x)$. Experiments supply samples from this distribution, but we may or may not have enough samples to characterize the whole distribution. What we can do more easily is to measure the mean $\bar{g}(x)$ and the variance $\sigma_g^2(x)$, and then approximate $P(g|x)$ as being Gaussian. One might worry that this approximation is uncontrolled, but in fact we can say more.

Suppose that all we know is the mean and variance of the distribution $P(g|x)$. The mutual information $I(g; x)$ is the difference between the entropy of the distribution $P(g)$ and the average entropy of the distribution $P(g|x)$:

$$I(g; x) = S[P(g)] - \langle S[P(g|x)] \rangle_x. \tag{6.95}$$

Thus, if we can put an upper bound on the entropy $S[P(g|x)]$, we can put a lower bound on the information. Suppose we search for a distribution $P(g|x)$ that maximizes the entropy while reproducing the measured mean and variance. This is an example of the maximum entropy problems that we have met before; they are discussed more fully in

Appendix A.7. The result that we need here is that the maximum entropy probability distribution consistent with given mean and variance is a Gaussian. Thus, in making the approximation that $P(g|x)$ is Gaussian, we place a lower bound on the mutual information $I(g;x)$.

In the example shown in Fig. 6.9, this variance at each position is relatively small, with $\sigma_g(x) \sim 0.1$ in units where the maximum mean expression level is 1. Following through the computation of entropies as outlined above, these data show that the expression level of Hb protein provides 2.26 ± 0.04 bits of information about position in the embryo. In the Gaussian approximation this is a lower bound on the information, but in fact the data sets are just large enough to make more direct estimates and to show that this bound is tight (see Problem 137). Classically, the gap genes have been described as specifying boundaries, dividing the embryo into patches of high (on) and low (off) expression. Evidently a simple on/off picture corresponds at most to one bit of positional information, and so a quantitative analysis teaches us that the focus on expression boundaries literally misses half of the story.

Problem 137: Details of positional information.   In the file Hb.mat you will find an example of data used to generate Fig. 6.9.[12] $Hb(x,i)$ is a matrix, with indices corresponding to $x = 1, 2, \cdots, L = 100$ points along the anterior-posterior axis and $i = 1, 2, \cdots, N = 20$ embryos; the entries in Hb correspond to Hb expression levels estimated by immunoflourescent staining. The data are normalized so that the mean expression level at each position has a maximum value of $\bar{g} = 1$ and a minimum value of $\bar{g} = 0$, as you should verify. As in Fig. 6.9, you should focus on the central 80% of the embryo.

(a) Pool the data from all positions so that you have many samples of the expression level g. Estimate the underlying distribution $P_G(g)$ by making bins along the g axis of some size Δg. Alternatively, you could make bins with varying size, adapted so that each bin contains the same number of samples. In both approaches, with the number of bins fixed you should compute the entropy $S[P_G(g)]$ of your estimated distribution and then see how this entropy depends on the number of embryos N that you include in the analysis. Can you see that the estimated entropy behaves as

$$S_{\text{est}} = S + \frac{A}{N} + \frac{B}{N^2} + \cdots,$$ (6.96)

as discussed in Appendix A.8?

(b) From Problem 123 we expect that varying the size of the bins Δg will produce a change in the estimated entropy such that $S_{\text{est}} = S_0 - \log_2(\Delta g)$. Can you see this change in the real data? What is the corresponding behavior that you expect to see with adaptive bins? Does it also show up in the data?

(c) If we have more bins, the sampling problem gets harder. The signature of this should be that the coefficients of the finite-size corrections A and B in Eq. (6.96) become larger as Δg becomes smaller. Show from the data that this is true. From Appendix A.8, do you have some analytic expectations for how this should work quantitatively? Are these expectations borne out in the data?

(d) With some fixed bin size Δg, make estimates of the conditional distributions $P(g|x)$ at each position x. Use these estimates to compute the entropy $S[P(g|x)]$ and hence, with your

---

12. Data are available through http://press.princeton.edu/titles/9911.html.

results from part (a), obtain an estimate of $I(g; x)$. Importantly, this estimate depends both on your choice of bin size and on the number of embryos included in your sample. Convince yourself that, by analogy with Eq. (6.96),

$$I_{est}(\Delta g, N) = I_{\infty}(\Delta g) + \frac{A'(\Delta g)}{N} + \frac{B'(\Delta g)}{N^2}. \tag{6.97}$$

Are the signs of A', B' different from those of A, B? Why or why not? At fixed N, $I_{est}(\Delta g, N)$ can have a strong dependence on Δg, but presumably the real answer does not depend on our arbitrary choice of bin size as $\Delta g \to 0$. Can you recover this behavior by extrapolating to $I_{\infty}(\Delta g)$? Can you give an analytic theory for the residual dependence on Δg? What happens to all of this in the case where we use adaptive bins?

(e) Develop a strategy for assigning error bars to the extrapolated $I_{\infty}(\Delta g)$. See also Appendix A.8.

(f) Suppose that instead of sampling the conditional probability distribution $P(g|x)$, we approximate it as Gaussian, following the discussion above. Use this approximation to compute the entropy $S[P(g|x)]$ and combine with your estimates of $S[P_G(g)]$ to estimate $I(g; x)$. Explain what happens to the bin size Δg in this calculation. How do your results from the Gaussian approximation compare with the more direct approach in the previous parts of the problem?

(g) Suppose that we randomize the data, replacing the position x by a random permutation of the numbers from 1 to L, chosen independently for each embryo. Now there should be no mutual information, but if we estimate $I_{rand}(g; x)$ by any naive method, the answer will be nonzero because of spurious correlations in our finite sample. Show that, so long as we do not have too many bins along the g axis, the extrapolation to $I_{\infty}(\Delta g)$ from Eq. (6.97) is zero within error bars for the randomized data. What happens as Δg gets smaller? How far along the path $\Delta g \to 0$ can you trust the extrapolation?

Thus far we have considered the information carried by a single variable. As a next step consider the case where we observe several variables $y_1, y_2, \cdots, y_K$ in the hopes of learning about the same number of underlying signals $x_1, x_2, \cdots, x_K$. The equations analogous to Eq. (6.74) are then

$$y_i = \sum_{j=1}^{K} g_{ij} x_j + \xi_i, \tag{6.98}$$

with the usual convention that we sum over repeated indices. The Gaussian assumptions are that each x_i and ξ_i has zero mean, but in general we have to think about arbitrary covariance matrices:

$$S_{ij} = \langle x_i x_j \rangle, \tag{6.99}$$

$$N_{ij} = \langle \xi_i \xi_j \rangle. \tag{6.100}$$

The relevant probability distributions are

$$P_X(\{x_i\}) = \frac{1}{\sqrt{(2\pi)^K \det S}} \exp\left[-\frac{1}{2} \sum_{i,j=1}^{K} x_i \cdot (S^{-1})_{ij} \cdot x_j \right], \tag{6.101}$$

$$P(\{y_i\}|\{x_i\}) = \frac{1}{\sqrt{(2\pi)^K \det N}} \exp\left[-\frac{1}{2} \sum_{i,j=1}^{K} \left(y_j - \sum_{k=1}^{K} g_{ik} x_k \right) \cdot (N^{-1})_{ij} \cdot \left(y_j - \sum_{m=1}^{K} g_{jm} x_m \right) \right], \tag{6.102}$$

where det S denotes the determinant of the matrix S, $(S^{-1})_{ij}$ is element ij in the inverse of the matrix S, and similarly for the matrix N.

To compute the mutual information, we proceed as before. First we find $P_Y(\{y_i\})$ by doing the integrals over the x_i,

$$P_Y(\{y_i\}) = \int d^K x\, P(\{y_i\}|\{x_i\}) P(\{x_i\}), \qquad (6.103)$$

and then we write the information as an expectation value,

$$I(\{y_i\}; \{x_i\}) = \left\langle \log_2\left[\frac{P(\{y_i\}|\{x_i\})}{P_Y(\{y_i\})}\right]\right\rangle, \qquad (6.104)$$

where $\langle\cdot\rangle$ denotes an average over the joint distribution $P(\{y_i\}, \{x_i\})$. As in Eq. (6.81), the logarithm can be broken into several terms such that the expectation value of each one is relatively easy to calculate. Two of three terms cancel, and the one that survives is related to the normalization factors that come in front of the exponentials. After the dust settles we find

$$I(\{y_i\}; \{x_i\}) = \frac{1}{2}\mathrm{Tr}\,\log_2[\mathbf{1} + N^{-1}\cdot(g\cdot S\cdot g^T)], \qquad (6.105)$$

where Tr denotes the trace of a matrix, $\mathbf{1}$ is the unit matrix, and g^T is the transpose of the matrix g.

Problem 138: The multidimensional Gaussian. Fill in the details leading to Eq. (6.105).

The matrix $g\cdot S\cdot g^T$ describes the covariance of those components of y that are contributed by the signal x. We can always rotate the coordinate system on the space of ys to make this matrix diagonal, which corresponds to finding the eigenvectors and eigenvalues of the covariance matrix; these eigenvectors also are called principal components. For a Gaussian distribution, the eigenvectors describe directions in the space of y that are fluctuating independently, and the eigenvalues are the variances along each of these directions. If the covariance of the noise is diagonal in the same coordinate system, then the matrix $N^{-1}\cdot(g\cdot S\cdot g^T)$ is diagonal and the elements along the diagonal are the signal-to-noise ratios along each independent direction. Taking the Tr log is equivalent to computing the information transmission along each direction using Eq. (6.94) and then summing the results.

An important case is when the different variables x_i represent a signal sampled at several different points in time. Then there is some underlying continuous function $x(t)$, and in place of the discrete Eq. (6.98) we have the continuous linear response of the detector to input signals:

$$y(t) = \int dt'\, M(t-t')x(t') + \xi(t). \qquad (6.106)$$

In this continuous case the analog of the covariance matrix $\langle x_i x_j\rangle$ is the correlation function $\langle x(t)x(t')\rangle$. We are usually interested in signals (and noise) that are stationary. This means, as discussed in Appendix A.2, that all statistical properties of the signal are invariant to translations in time: a particular pattern of wiggles in the function $x(t)$ is equally likely to occur at any time. Thus, the correlation function, which could in principle depend on two times t and t', depends only on the time difference:

$$\langle x(t)x(t')\rangle = C_x(t - t').\tag{6.107}$$

The correlation function generalizes the covariance matrix to continuous time, but we have seen that it can be useful to diagonalize the covariance matrix, thus finding a coordinate system in which fluctuations in the different directions are independent. From Appendix A.2 we know that this diagonalization is achieved in the Fourier representation, because (in the Gaussian case) different Fourier components are independent, and their variances are (up to normalization) the power spectra.

To complete the analysis of the continuous-time Gaussian channel described by Eq. (6.106), we again refer noise to the input by writing

$$y(t) = \int dt' M(t - t')[x(t') + \eta_{\text{eff}}(t')].\tag{6.108}$$

If both signal and effective noise are stationary, then each has a power spectrum; let us denote the power spectrum of the effective noise η_{eff} by $N_{\text{eff}}(\omega)$ and the power spectrum of x by $S_x(\omega)$ as usual. There is a signal-to-noise ratio at each frequency,

$$SNR(\omega) = \frac{S_x(\omega)}{N_{\text{eff}}(\omega)},\tag{6.109}$$

and as we have diagonalized the problem by Fourier transforming, we can compute the information just by adding the contributions from each frequency component, so that

$$I[y(t); x(t)] = \frac{1}{2}\sum_{\omega}\log_2[1 + SNR(\omega)].\tag{6.110}$$

Finally, to compute the frequency sum, we use the limiting behavior as T becomes large (see Eq. (A.170) in Appendix A.2):

$$\sum_{n} f(\omega_n) \to T\int \frac{d\omega}{2\pi} f(\omega).\tag{6.111}$$

Thus, the information conveyed by observations on a (large) window of time becomes

$$I[y(0 < t < T); x(0 < t < T)] \to \frac{T}{2}\int \frac{d\omega}{2\pi}\log_2[1 + SNR(\omega)]\,\text{bits.}\tag{6.112}$$

The information gained is proportional to the time of our observations, so it makes sense to define an information rate:

$$R_{\text{info}} \equiv \lim_{T\to\infty}\frac{1}{T}\cdot I[y(0 < t < T) \to x(0 < t < T)]\tag{6.113}$$

$$= \frac{1}{2}\int \frac{d\omega}{2\pi}\log_2[1 + SNR(\omega)]\,\text{bits/s.}\tag{6.114}$$

Note that in all these equations, integrals run over both positive and negative frequencies; if the signals are sampled at points in time spaced by τ_0, then the maximum (Nyquist) frequency is $|\omega|_{\text{max}} = \pi/\tau_0$.

Problem 139: How long to look? Integrating for longer times can suppress the effects of noise and hence presumably gain more information. Usually we would say that the benefits of integration are cut off by the fact that the signals we are looking at will change. But once we think about information transmission there is another possibility—perhaps we would learn more by using the same time to look at something new, rather than getting a more accurate view of something already seen. To address this possibility, let's consider the following simple model. We look at one thing for a time τ, and then jump to something completely new. Given that we integrate for τ, we achieve some signal-to-noise ratio $S(\tau)$.

(a) Explain why, in this simple model, if the noise is Gaussian, then the rate at which we gain information is at most

$$R_{\text{info}}(\tau) = \frac{1}{\tau} \log_2 \left[1 + S(\tau)\right] . \tag{6.115}$$

How does the assumption that we "jump to something completely new" enter into the justification of this formula?

(b) To make progress, we need a model for $S(\tau)$. Because this is the signal-to-noise ratio, let's start with the signal. Suppose that inputs are given by x, and the output is y. At $t = 0$, the value of y is set to zero, and after that our sensory receptor responds to its inputs according to a simple differential equation:

$$\tau_0 \frac{dy}{dt} = -y + x . \tag{6.116}$$

Show that $y(\tau) = x[1 - \exp(-\tau/\tau_0)]$. Now for the noise, suppose that $\eta_{\text{eff}}(t)$ has a correlation function

$$\langle \eta_{\text{eff}}(t)\eta_{\text{eff}}(t') \rangle = \sigma_0^2 e^{-|t-t'|/\tau_c} . \tag{6.117}$$

Show that if we average the noise over a window of duration τ, then the variance is

$$\sigma^2(\tau) \equiv \left\langle \left[\frac{1}{\tau} \int_0^\tau dt \, \eta_{\text{eff}}(t) \right]^2 \right\rangle \approx \sigma_0^2 \quad (\tau \ll \tau_0) \tag{6.118}$$

$$\approx \frac{2\sigma_0^2 \tau_c}{\tau} \quad (\tau \gg \tau_0). \tag{6.119}$$

Give a more general analytic expression for $\sigma^2(\tau)$. Put these factors together to get an expression for $S(\tau) = y^2(\tau)/\sigma^2(\tau)$. To keep things simple, you can assume that the time scale that determines the response to inputs is the same as that which determines the correlations in the noise, so that $\tau_c = \tau_0$.

(c) Hopefully you can show from your results in part (b) that $S(\tau \gg \tau_0) \propto \tau$. This behavior corresponds to our intuition that signal-to-noise ratios grow with averaging time, because we beat down the noise, not worrying about the possibility that the signal itself will change. What happens for $\tau \ll \tau_0$?

(d) Suppose that τ_0 is very small, so that all "reasonable" values of $\tau \gg \tau_0$. Then, from part (c), $S(\tau) = A\tau$, with A a constant. Using this assumption, plot $R_{\text{info}}(\tau)$; show that with proper choice of units, you do not need to know the value of A. What value of τ maximizes the information rate? Is this consistent with the assumption that $\tau \gg \tau_0$?

(e) In general, the maximum information is found at the point where $dR_{\text{info}}/d\tau = 0$. Show that this condition can be rewritten as a relationship between the signal-to-noise ratio and its logarithmic derivative, $z = d \ln S(\tau)/d \ln \tau$. From your previous results, what can you say about the possible values of z as τ is varied? Use this to bound $S(\tau)$ at the point of maximum R_{info}. What does this result say about the compromise between looking carefully at one thing and jumping to something new?

(f) How general can you make the conclusions that you draw in part (e)?

In the same way that we used the Gaussian approximation to put bounds on the positional information carried by the gap genes, we can put bounds on the information carried by sensory neurons. As discussed in Section 4.4, we can reconstruct continuous sensory input signals from the discrete sequences of action potentials, sometimes quite accurately. Concretely, the sensory stimulus $s(t)$ could be light intensity as a function of time in a small region of the visual field, sound pressure as a function of time at the ear canal, the amplitude of mechanical vibrations in such sensors as the frog sacculus, and so forth. We can estimate the signal from the spike times $\{t_i\}$ in a single neuron as

$$s_{\text{est}}(t) = \sum_i f(t - t_i), \tag{6.120}$$

where the filter $f(\tau)$ is chosen to minimize $\chi^2 = \langle |s_{\text{est}}(t) - s(t)|^2 \rangle$. Then the quality of the reconstructions can be evaluated by measuring the power spectrum of errors in the reconstruction and referring these errors to the input, frequency component by frequency component:

$$\tilde{s}_{\text{est}}(\omega) = g(\omega) \left[\tilde{s}(\omega) + \tilde{\eta}_{\text{eff}}(\omega) \right]. \tag{6.121}$$

Although the errors in the reconstruction might not be exactly Gaussian, the maximum entropy argument tells us that we can put a lower bound on the information that the spike train provides about the stimulus $s(t)$ by measuring the power spectrum of the effective noise η_{eff}. An example is shown in Fig. 6.10, from experiments on the mechanical sensors in the cricket and frog. Importantly, we can also put upper bounds on the entropy of the spike train, first by assuming that spikes occur independently, then by assuming that the intervals between spikes are independent and finally by allowing for correlations between successive intervals. With a lower bound on the information and an upper bound on the entropy, we have a lower bound on their ratio, the coding efficiency. In these systems, as with the case of the fly's motion-sensitive neuron (H1) in Fig. 6.7, efficiencies reach $\sim$50% with timing precision in the millisecond range.

By now both the direct and the reconstruction methods have been used to measure information rates and coding efficiencies in a wide range of neurons responding to sensory stimuli, from the first steps of sensory coding in invertebrates, such as the cricket cercal system in Fig. 6.10, to cells deep in primate visual cortex as in Fig. 6.8. The result that single neurons use 30–50% of their spike-train entropy to encode sensory information, even down to millisecond resolution, has been confirmed in many systems, as described in the references in the Annotated Bibliography for this section. An important thread running through this work is that information rates and coding efficiencies are higher, and the high coding efficiency extends to higher time resolution, when sensory inputs are more like those occuring in nature—complex, dynamic, and with enormous dynamic range. An example from the frog auditory system is shown in Fig. 6.11. In some cases, reaching such conclusions depends on using more sophisticated meth-

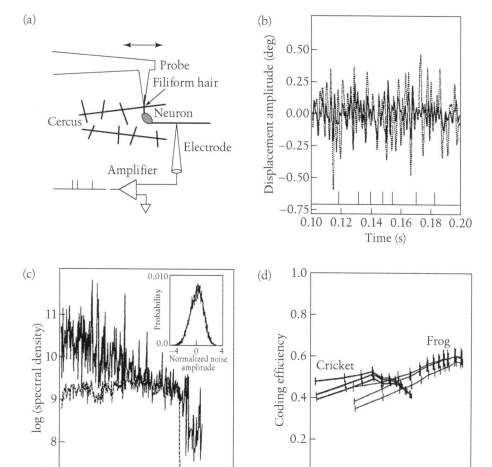

FIGURE 6.10

Coding efficiency in cricket and frog vibration sensors. Redrawn from Rieke et al. (1993). (a) A schematic of experiments on the cricket cercal sensors, with direct stimulation of the sensory (filiform) hairs and recording from the primary sensory neurons. (b) Stimulus (dashed line), reconstruction (solid line), and spikes in experiments on the cercal neurons. (c) Power spectral density of the signal (dashed), and the noise η_{eff} (solid) in the reconstructions, from Eq. (6.121). (d) Coding efficiency for single neurons in the cricket cercus and the frog sacculus, using successively higher order approximations to the spike train entropy. Variable timing precision is implemented by providing the reconstruction algorithm in Eq. (6.120) with spike times t_i at limited resolution.

ods for estimating the mutual information between spikes and stimuli, as described in Appendix A.8. These results suggest not only that the brain is capable of efficient coding, but also that this efficiency is achieved by matching neural coding strategies to the structure of natural sensory inputs. We will return to this idea in Section 6.4.

The results thus far give us the opportunity to explore the way in which noise limits information transmission. Imagine that we have measured the spectrum of the effective noise, $N_{eff}(\omega)$. By changing the spectrum of input signals, $S(\omega)$, we can change the rate of information transmission. Can we maximize this information rate? Clearly this problem is not well posed without some constraints: if we are allowed just to increase the amplitude of the signal—multiply the spectrum by a large constant—then we can always increase information transmission. We need to study the optimization of information rate with some fixed dynamic range for the signals. A simple example, considered by Shannon at the outset, is to fix the total variance of the signal, which is the same as fixing the integral of the spectrum. We can motivate this constraint by noting that if the signal is a voltage and we have to drive this signal through a resistive

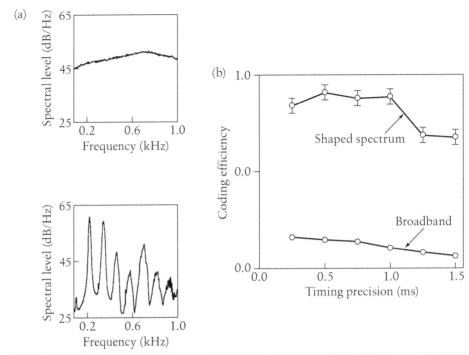

FIGURE 6.11

Coding efficiency in frog auditory neurons. Redrawn from Rieke et al. (1995). (a) The power spectrum of a broadband, artificial stimulus (top) and a stimulus shaped to have the same spectrum as bullfrog calls (bottom). These stimuli were played to the bullfrog while recording from individual auditory neurons emerging from the amphibian papilla. Reconstructing the sound pressure as a function of time allows us to bound the information transmission rate, as explained in the text, and from this we estimate the coding efficiency—the ratio of the information rate to the entropy rate. (b) In this example, the coding efficiency is substantially higher for the more naturalistic stimuli, approaching 90%.

element, then the variance is proportional to the mean power dissipation. Alternatively, it might be easy to measure the variance of the signals that we are interested in, and then the constraint is empirical.

So the problem we want to solve is maximizing R_{info} while holding $\langle x^2 \rangle$ fixed. As before, we introduce a Lagrange multiplier and maximize a new function:

$$\tilde{R} = R_{\text{info}} - \lambda \langle x^2 \rangle \tag{6.122}$$

$$= \frac{1}{2} \int \frac{d\omega}{2\pi} \log_2 \left[1 + \frac{S_x(\omega)}{N_{\text{eff}}(\omega)} \right] - \lambda \int \frac{d\omega}{2\pi} S_x(\omega). \tag{6.123}$$

The value of the function $S_x(\omega)$ at each frequency contributes independently, so it is easy to compute the functional derivatives,

$$\frac{\delta \tilde{R}}{\delta S_x(\omega)} = \frac{1}{2 \ln 2} \cdot \frac{1}{1 + S_x(\omega)/N_{\text{eff}}(\omega)} \cdot \frac{1}{N_{\text{eff}}(\omega)} - \lambda, \tag{6.124}$$

and the optimization condition is $\delta \tilde{R}/\delta S_x(\omega) = 0$. The result is that

$$S_x(\omega) + N_{\text{eff}}(\omega) = \frac{1}{2\lambda \ln 2}. \tag{6.125}$$

Thus, the optimal choice of the signal spectrum is one that makes the sum of signal and (effective) noise equal to white noise! This result, like the fact that information is maximized by a Gaussian signal, is telling us that efficient information transmission occurs when the received signals are as random as possible given the constraints. Thus, an attempt to look for structure in an optimally encoded signal (deep in the brain, perhaps) could be frustrating.

In general, complete whitening as suggested by Eq. (6.125) can't be achieved at all frequencies, because if the system has finite time resolution (for example), the effective noise grows without bound at high frequencies. Thus, the full solution is to have the spectrum determined by Eq. (6.125) everywhere that the spectrum is positive, and then to set the spectrum equal to zero outside this range. If we think of the effective noise spectrum as a landscape with valleys, the condition for optimizing information transmission corresponds to filling the valleys with water; the total volume of water is the variance of the signal, as we will see in the discussion leading to Fig. 6.12.

Problem 140: Whitening. Consider a system that responds linearly to a signal $s(t)$, with added noise $\eta(t)$:

$$x(t) = \int d\tau \, F(\tau)s(t - \tau) + \eta(t). \tag{6.126}$$

Assume that the noise is Gaussian and white, with power spectrum $\mathcal{N}_0$, so that

$$\langle \eta(t)\eta(t') \rangle = \mathcal{N}_0 \delta(t - t'). \tag{6.127}$$

For simplicity, assume that the signal $s(t)$ is Gaussian, with a power spectrum $S(\omega)$:

$$\langle s(t)s(t') \rangle = \int \frac{d\omega}{2\pi} S(\omega) \exp[-i\omega(t - t')]. \tag{6.128}$$

(a) Write an expression for the rate R_{info} at which the observable $x(t)$ provides information about the signal $s(t)$.

(b) The variance of the variable $x(t)$ is not well defined. Why? Consider just the component of $x(t)$ that comes from the signal $s(t)$, that is, Eq. (6.126) but with $\eta = 0$. Find an expression for the variance of this ouput signal.

(c) Consider the problem of maximizing R_{info} by adjusting the filter $F(\tau)$. Obviously the information transmission is larger if F is larger, so to make the problem well posed, assume that the variance of the output signal (from part (b)) is fixed. Show that this variational problem can be solved explicitly for $|\tilde{F}(\omega)|^2$, where $\tilde{F}(\omega)$ is the Fourier transform of the filter $F(\tau)$. Can you explain intuitively why only the modulus, and not the phase, of $\tilde{F}(\omega)$ is relevant here?

(d) Find the limiting form of the optimal filter as the noise becomes small. What does this filter do to the input signal? Explain why this makes sense. Saying that "noise is small" is slightly strange, because $\mathcal{N}_0$ has units. Give a more precise criterion for your small noise limit to be valid.

(e) Consider the case of an input with exponentially decaying correlations, so that

$$S(\omega) = \frac{2\langle s^2\rangle \tau_c}{1 + (\omega\tau_c)^2},$$ (6.129)

where τ_c is the correlation time. Find the optimal filter in this case, and use it to evaluate the maximum value of R_{info} as a function of the output signal variance. You should check that your results for R_{info}, which should be in bits per second, are independent of the units used for the output variance and the noise power spectrum. Contrast your result with what would happen if $|\tilde{F}(\omega)|$ were flat as a function of frequency, so that there was no real filtering (just a multiplication, so that the output signal variance comes out right). How much can you gain by building the right filter?

Problem 141: Finite time resolution.   Suppose that signals experience a delay as they are detected, but this delay itself can fluctuate. Then the output of the system $x(t)$ is related to the input $s(t)$ as

$$x(t) = s(t - \tau(t)),$$ (6.130)

where $\tau(t)$ is the fluctuating delay.

(a) Assume that the fluctuations are small, and show that this is equivalent to adding noise:

$$x(t) = s(t - \bar{\tau}) + \eta(t),$$ (6.131)

where $\bar{\tau}$ is the average delay. Find the correlation function and power spectrum of the equivalent noise $\eta(t)$ in terms of the correlation function for the delay fluctuations, $\langle \delta\tau(t)\delta\tau(t')\rangle$.

(b) We expect that fluctuations in the delay are slow, so that the correlations $\langle \delta\tau(t)\delta\tau(t')\rangle$ extend over times longer than correlations in the signal $s(t)$. Show that, in this limit, the power spectrum of the equivalent noise η simplifies and is proportional to the variance of the time delay, $\langle (\delta\tau)^2\rangle$.

(c) Show that the signal-to-noise ratio falls at high frequencies, no matter what the spectrum of the input signals is.

These ideas have been used to characterize information transmission across the first synapse in the fly's visual system. We have seen these data before, in thinking about how the precision of photon counting changes as the background light intensity increases (Section 2.1). Over a reasonable dynamic range of intensity variations, the average voltage response of the photoreceptor cell is related linearly to the intensity or contrast $C(t)$ in the movie, and the noise or variability $\delta V(t)$ is governed by a Gaussian distribution of voltage fluctuations around the average:

$$V(t) = V_{DC} + \int dt'\, T(t - t')C(t') + \delta V(t),$$ (6.132)

where V_{DC} is the mean voltage at zero contrast. This (happily) is the problem we have just analyzed.

As before, we think of the noise in the response as being equivalent to noise $\delta C_{\text{eff}}(t)$ that is added to the movie itself:

$$V(t) = V_{\text{DC}} + \int dt'\, T(t - t')[C(t') + \delta C_{\text{eff}}(t)]. \tag{6.133}$$

Because the fluctuations have a Gaussian distribution, they can be characterized completely by their power spectrum $N_C^{\text{eff}}(\omega)$, which measures the variance of the fluctuations that occur at different frequencies:

$$\langle \delta C_{\text{eff}}(t)\delta C_{\text{eff}}(t')\rangle = \int \frac{d\omega}{2\pi} N_C^{\text{eff}}(\omega)\exp[-i\omega(t - t')]. \tag{6.134}$$

There is a minimum level of this effective noise set by the random arrival of photons (shot noise). The photon noise is white if expressed as $N_C^{\text{eff}}(\omega)$, although it makes a nonwhite contribution to the voltage noise. As we have discussed, over a wide range of background light intensities and frequencies, the fly photoreceptors have effective noise levels that reach the limit set by photon statistics. At high frequencies there is excess noise beyond the physical limit, and this excess noise sets the time resolution of the system.[13]

The power spectrum of the effective noise tells us, ultimately, what signals the photoreceptor can and cannot transmit. How do we turn these measurements into bits? One approach is to assume that the fly lives in some particular environment and then calculate how much information the receptor cell can provide about this particular environment. But to characterize the cell itself, we might ask a different question: how much information can the cell transmit, in principle, even if we are free to adjust the signals in the environment to match the properties of the cell? To answer this question, we are allowed to shape the statistical structure of the environment so as to make the best use of the receptor (the opposite, presumably, of what happens in evolution!). This is just the optimization discussed above, so it is possible to turn the measurements on signals and noise into estimates of the information capacity of these cells. This was done both for the photoreceptor cells and for the large monopolar cells that receive direct synaptic input from a group of six receptors. From measurements on natural scenes the mean-square contrast signal was fixed at $\langle C^2\rangle = 0.1$. Results are shown in Fig. 6.12.

The first interesting feature of the results is the scale: individual neurons are capable of transmitting well above 1000 bits/s. This finding does not mean that this capacity is used under natural conditions, but rather speaks to the precision of the mechanisms underlying the detection and transmission of signals in this system—on a time scale of ~ 10 ms, the output of these cells can point to $2^{10} \sim 10^3$ reliably distinguishable visual signals. Second, information capacity continues to increase as the level of background light increases: noise due to photon statistics is relatively smaller in brighter lights, and this reduction of the physical limit actually improves the performance of the system even up to very high photon-counting rates, indicating once more that the physical limit is relevant to the real performance. Third, we see that the information capacity as a function of photon-counting rate is shifted along the counting-rate axis as we go from

---

13. The effective contrast noise $N_C^{\text{eff}}(\omega)$ is the inverse of what we plotted in Fig. 2.15 to compare with the actual photon counting rate.

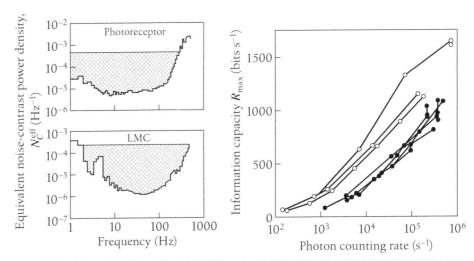

FIGURE 6.12

At left, the effective contrast noise levels in a single photoreceptor cell and a single monopolar cell (LMC, a second-order neuron in the fly's retina). The hatching shows the signal spectra required to whiten the total output over the largest possible range while maintaining the input contrast variance $\langle C^2 \rangle = 0.1$, as discussed in the text. At right, the resulting information capacities as a function of the photon counting rates in the photoreceptors, for photoreceptors (solid) and LMCs (open). Redrawn from de Ruyter van Steveninck and Laughlin (1996).

photoreceptors to the large monopolar cells, which corresponds (quite accurately!) to the fact that large monopolar cells integrate signals from six photoreceptors and thus act as if they captured photons at a rate six times higher than individual photoreceptors do. Finally, in the large monopolar cells information has been transmitted across a synapse, and in the process it is converted from a continuous voltage signal to discrete events corresponding to the release of neurotransmitter vesicles at the synapse. As a result, there is a new limit to information transmission that comes from viewing the large monopolar cell as a vesicle counter.

If every vesicle makes a measurable, deterministic contribution to the cell's response (a generous assumption), then the large monopolar cell's response is equivalent to reporting how many vesicles are counted in a small window of time corresponding to the photoreceptor time resolution. We do not know the distribution of these counts, but we can estimate (from other experiments, with uncertainty) the mean count. What can we say from the mean count alone? The vesicle count is a discrete variable, so the maximum amount of information that can be transmitted by a vesicle counter is the entropy of the count distribution. Formally, if n is the number of vesicles counted, and $P(n)$ is the distribution of this count, then we have

$$I(n; \text{stimulus}) \leq S[P(n)]. \tag{6.135}$$

If we know the mean vesicle count,

$$\langle n \rangle = \sum_{n=0}^{\infty} P(n)n, \tag{6.136}$$

then there is a maximum possible entropy, $S_{\max}(\langle n \rangle)$, which is computed in Appendix A.7, Eq. (A.413):

$$S_{\max}(\langle n \rangle) = \log_2(1 + \langle n \rangle) + \langle n \rangle \log_2(1 + 1/\langle n \rangle) \text{ bits.} \qquad (6.137)$$

Thus, although there is much we do not know about the details of synaptic transmission, we can say that the information transferred across the synapse is bounded:

$$I(n; \text{stimulus}) \leq S[P(n)] \leq S_{\max}(\langle n \rangle) = \log_2(1 + \langle n \rangle) + \langle n \rangle \log_2(1 + 1/\langle n \rangle) \text{ bits.}$$

$$(6.138)$$

No mechanism at the synapse can transmit more information than this limit.

Individual synaptic contacts are unlikely to transmit more than 10^2 vesicles/s in steady state, but the synapse between the photoreceptor and the large monopolar cell is special, because there are $\sim 10^3$ active zones or separate contact points between the two cells. Thus, within the $\Delta t \sim 5 \text{ ms}$ time resolution of the photoreceptors, the mean number of vesicles released could be as large as ~ 500. Then the limiting rate of information transmission is $R \sim S_{\max}(\langle n \rangle)/\Delta t \sim 1600 \text{ bits/s}$, which is remarkably close to the highest actual rates observed in Fig. 6.12.

To summarize, we have seen that there are limits to information transmission that are set by noise and by discretization. In biological systems we already know from Chapter 4 that noise is not negligible and discretization abounds—there are finite numbers of molecules, vesicles, spikes, and so forth. Biological systems thus face limits on how much information they can transmit, and we have seen a couple of hints that these limits are relevant—that the actual systems we find in Nature transmit nearly as much information as they can, given the resources they have devoted to the task. This observation motivates the hypothesis that the efficiency of information transmission, or the efficiency of representation, is one of those Xs that we discussed in Chapter 3, a candidate variational principle that may help us understand why biological mechanisms have been selected to have the parameters that they do. We will pursue this idea, but first, we need to address a more pressing question.

6.3 Does Biology Care about Bits?

The question for this section has been with us almost since Shannon's original work. On the one hand, the few examples we have seen in the last section certainly suggest that organisms are squeezing more bits out of their hardware than we might naively have expected, perhaps even coming close to physical limits on information transmission. On the other hand, bits seem very abstract compared with what really matters in life. Here we review old ideas about the connection of information to gambling and see how closely related ideas have reappeared in thinking about the life strategies of bacterial populations. Then we step back and try to look more generally at the connections among information, biological function and evolutionary fitness, and argue that evolution really can select for biological mechanisms that are efficient in an information-theoretic sense.

To start, let us consider a simple game; this may seem like a strange topic for a physics course, but please bear with me! I will flip a coin, and you bet on whether it will come up heads or tails. If you get it right, I double your money. If you're wrong, you lose what you bet. If the coin is fair, so that heads and tails each come up half the time, there really isn't anything to analyze, what happens is just chance. But if you know, for example, that the coin is biased, and that the probability of heads really is 60%, you might be tempted to put all your money on heads. On average, if you bet 1 dollar, you will receive $2 \times (0.6) = 1.2$ dollars in return, which sounds good. Indeed, if we play only once, then this is what you should do, because it will maximize your expected return.

But what happens if we are going to play repeatedly, which you might think is a better metaphor for life? Now if you put all your money on heads, there is a 40% chance that, in one flip, you'll lose it all. Suppose that instead you put a fraction f of your money on heads and a fraction $1 - f$ on tails. If we introduce a binary variable $n = 1$ for heads and $n = 0$ for tails, then on the ith flip your winnings will change by a factor

$$G_i = 2 \times \left[f n_i + (1 - f)(1 - n_i) \right], \tag{6.139}$$

where n_i marks what happens on the ith flip. After N successive flips you will have a gain

$$G_{\text{total}}(N) = 2^N \prod_{i=1}^{N} \left[f n_i + (1 - f)(1 - n_i) \right], \tag{6.140}$$

where we are assuming that you consistently put a fraction f of your accumulated winnings down as a bet on heads and the remainder on tails.

To continue, we want to write the product in Eq. (6.140) as the exponential of a sum; because n_i is either 0 or 1, we have

$$f n_i + (1 - f)(1 - n_i) = \exp \left[n_i \ln(f) + (1 - n_i) \ln(1 - f) \right]. \tag{6.141}$$

Thus, we can write the total gain as

$$G_{\text{total}}(N) = 2^N \prod_{i=1}^{N} \left[f n_i + (1 - f)(1 - n_i) \right] \tag{6.142}$$

$$= \exp \left[N \Lambda(f; \{n_i\}) \right], \tag{6.143}$$

where

$$\Lambda(f; \{n_i\}) = \ln 2 + \frac{1}{N} \sum_{i=1}^{N} \left[n_i \ln(f) + (1 - n_i) \ln(1 - f) \right]. \tag{6.144}$$

Written this way, $\Lambda(f; \{n_i\})$ defines a rate of exponential growth for your winnings. But $\Lambda(f; \{n_i\})$ depends not only on your betting strategy, summarized by the fraction f that you put on heads, but also on the sequence of heads and tails that come up in the game, denoted by $\{n_i\}$. The key point is that, if we play many times, so we can think about the limit $N \to \infty$, this dependence on the details of the flips goes away.

For any well-behaved random variable, the average over N observations must approach the mean computed from the probability distribution as N becomes large. In the present case, if n_i is a binary variable that takes the value $n_i = 1$ with probability p and $n_i = 0$ with probability $1 - p$, then as N becomes large we should have

$$\frac{1}{N} \sum_{i=1}^{N} n_i \to p, \tag{6.145}$$

and similarly

$$\frac{1}{N}\sum_{i=1}^{N}(1 - n_i) \to 1 - p. \tag{6.146}$$

We can use these equations to evaluate the long-term growth of your winnings, simplifying the results of Eq. (6.144):

$$\Lambda(f;\{n_i\}) = \ln 2 + \frac{1}{N}\sum_{i=1}^{N}\left[n_i\ln(f) + (1 - n_i)\ln(1 - f)\right] \tag{6.147}$$

$$= \ln 2 + \left(\frac{1}{N}\sum_{i=1}^{N}n_i\right)\ln(f) + \left(\frac{1}{N}\sum_{i=1}^{N}(1 - n_i)\right)\ln(1 - f)$$

$$\to \ln 2 + p\ln(f) + (1 - p)\ln(1 - f), \tag{6.148}$$

where again p is the probability of heads. To maximize the growth rate $\Lambda(f)$, as usual we differentiate and set the result to zero:

$$0 = \left.\frac{d\Lambda(f)}{df}\right|_{f=f_{opt}} = p\frac{1}{f_{opt}} + (1 - p)(-1)\frac{1}{1 - f_{opt}}; \tag{6.149}$$

$$\Rightarrow \frac{1 - p}{1 - f_{opt}} = \frac{p}{f_{opt}}, \tag{6.150}$$

or more simply, $f_{opt} = p$. This is an interesting result: you maximize the rate at which your winnings will grow by matching the fraction of your resources that you bet on heads to the probability that the coin will come up heads, and similarly for tails.

Problem 142: Check that $f_{opt} = p$ is a maximum, and not a minimum, of $\Lambda(f)$.

Problem 143: If we bet only once, then in this simple game the maximum mean payoff is obtained by betting on the most likely outcome. However, as we play many times—more precisely, in the limit that we play infinitely many times—a sort of matching strategy, or "proportional gambling" maximizes the growth rate. Explore the crossover between these limits. You might start with some simple simulations, and then see whether you can make analytic progress, perhaps saying something about the leading $1/N$ corrections at large N. I am leaving this deliberately vague and open ended, hoping that you will play around with these ideas.

Something even more interesting happens when we evaluate the optimal growth rate, that is, $\Lambda_{opt} = \Lambda(f_{opt})$:

$$\Lambda_{opt} = \Lambda(f = p) \tag{6.151}$$

$$= \ln 2 + p\ln(p) + (1 - p)\ln(1 - p) \tag{6.152}$$

$$= \ln 2 - \left[-p\ln(p) - (1 - p)\ln(1 - p)\right]. \tag{6.153}$$

These terms should be starting to look familiar. The term $\ln 2$ is the entropy for a binary variable (heads/tails) if you do not know anything about what to expect, and hence the two alternatives are equally likely. In contrast, the term in brackets, $-p \ln(p) - (1 - p) \ln(1 - p)$, is the entropy of a binary variable if you know that the two alternatives come up with probabilities p and $1 - p$. Thus, the optimal growth rate is the difference in entropy between what might happen with an arbitrary coin and what you know will happen with this coin. In other words, *the maximum rate at which your winnings can grow in a simple gambling game is equal to the information that you have about the outcome of a single coin flip.*

This connection between information theory and gambling was discovered in the 1950s by Kelly, who was searching for some interpretation of Shannon's work that did not refer to the process of communication. Obviously what we have worked out here is a very simple and special case, and we need to do much more to claim that the connection is general. But before going further, let me emphasize something about Kelly's result. At some intuitive level, we can all agree that if we know more about the outcome of the coin flip (or the horse race, or the stock market, or whatever), then we should be able to make more money. In a very general context, Shannon proved that "know more" should be quantified by various entropy-like quantities, but it is not obvious that the knowledge measured by Shannon's bits is actually useful knowledge when it comes time to make a bet. Kelly showed that the maximum rate at which your winnings can grow in his game *is* the information, and his proof is constructive, so we actually know how to achieve this maximum. This result really is quite astonishing.

Let's try to generalize what we have done. Suppose that on each trial i, there are many possible outcomes, $\mu = 1, 2, \cdots, K$; write $n_i^{(\mu)} = 1$ if on the ith trial the outcome is μ, and $n_i^{(\mu)} = 0$ otherwise. Further, let's say that you bet a fraction of your assets f_μ on each of the possible outcomes μ, and if μ actually happens, then each dollar bet on this outcome becomes g_μ dollars; all money bet on events that do not happen is lost. If you need an example of this sort of game, think of a horse race in which you get money back only if you pick the winner. We assume that the different outcomes occur with probability p_μ, but we do not assume anything about the relationship between these odds and the payoffs g_μ.

Having defined all the factors, the analog of Eq. (6.140) is

$$G_{\text{total}}(N) = \prod_{i=1}^{N} \left[\sum_{\mu=1}^{K} f_\mu g_\mu n_i^{(\mu)} \right]. \tag{6.154}$$

Now we can follow the same steps as before. We start by rewriting the total gain,

$$\ln G_{\text{total}}(N) = \sum_{i=1}^{N} \ln \left[\sum_{\mu=1}^{K} f_\mu g_\mu n_i^{(\mu)} \right] \tag{6.155}$$

$$= \sum_{i=1}^{N} \sum_{\mu=1}^{K} n_i^{(\mu)} \ln(f_\mu g_\mu), \tag{6.156}$$

and then we look for the behavior at large N:

$$\frac{1}{N}\ln G_{\text{total}}(N) = \sum_{\mu=1}^{K}\left[\frac{1}{N}\sum_{i=1}^{N}n_i^{(\mu)}\right]\ln(f_\mu g_\mu) \tag{6.157}$$

$$\rightarrow \Lambda(\{f_\mu\}) = \sum_{\mu=1}^{K}p_\mu\ln(f_\mu g_\mu). \tag{6.158}$$

We want to maximize the growth rate Λ, subject to the normalization condition that the fractions of our assets placed on all the options add up ($\sum_\mu f_\mu = 1$), so we introduce a Lagrange multiplier α and find the maximum of the function

$$\tilde{\Lambda}(\{f_\mu\}) = \sum_{\mu=1}^{K}p_\mu\ln(f_\mu g_\mu) - \alpha\left[\sum_{\mu=1}^{K}f_\mu - 1\right]. \tag{6.159}$$

The equations for the maximum are, again,

$$0 = \left.\frac{\partial\tilde{\Lambda}(\{f_\mu\})}{\partial f_\mu}\right|_{\{f_\mu\}=\{f_\mu^{\text{opt}}\}} = \frac{p_\mu}{f_\mu^{\text{opt}}} - \alpha, \tag{6.160}$$

$$\Rightarrow f_\mu^{\text{opt}} = \frac{p_\mu}{\alpha}; \tag{6.161}$$

because $\sum_\mu f_\mu = \sum_\mu p_\mu = 1$, we must have $\alpha = 1$, so that

$$f_\mu^{\text{opt}} = p_\mu. \tag{6.162}$$

Substituting, we find the maximum growth rate:

$$\Lambda_{\text{opt}} = \sum_{\mu=1}^{K}p_\mu\ln(p_\mu g_\mu). \tag{6.163}$$

The first interesting thing is that we recover from the simpler heads/tails problem the idea of proportional gambling (Eq. (6.162)): you maximize the rate at which your winnings will grow by matching the fraction of your resources bet on each horse in the race to the probability that this horse will win. Strangely, perhaps, this result is independent of the rewards or gains as expressed in the parameters $\{g_\mu\}$.

The second point is that we can see what it means for the odds to be truly fair. If our opponent in this game (the track operator) sets the returns in inverse proportion to the probability that each horse wins, $g_\mu = 1/p_\mu$, then the maximum growth rate of our winnings, Λ_{opt}, is exactly zero.

This notion of fairness leads to an information-theoretic interpretation of Λ_{opt}. Notice that we have done the calculation on the assumption that we have perfect knowledge of the distribution $\{p_\mu\}$. Perhaps the track operators have less knowledge, and so they set the odds *as if* the distribution were something else, which we can call $\{q_\mu\}$. More generally, we can define

$$q_\mu = \frac{1}{Z}\frac{1}{g_\mu}, \tag{6.164}$$

with Z chosen so that $\sum_\mu q_\mu = 1$. If $Z = 1$, then the payoffs $\{g_\mu\}$ are fair in the distribution $\{q_\mu\}$, whereas if $Z < 1$, the track operators are keeping something for themselves (as they are wont to do). Then we can see that

$$\Lambda_{\text{opt}} = -\ln Z + \sum_{\mu=1}^{K} p_\mu \ln\left(\frac{p_\mu}{q_\mu}\right). \tag{6.165}$$

The second term is the Kullback-Leibler divergence between the probability distributions $\mathbf{p} \equiv \{p_\mu\}$ and $\mathbf{q} \equiv \{q_\mu\}$, from Eq. (6.40):

$$D_{\text{KL}}(\mathbf{p}||\mathbf{q}) \equiv \sum_{\mu=1}^{K} p_\mu \ln\left(\frac{p_\mu}{q_\mu}\right). \tag{6.166}$$

As explained in the discussion leading up to Eq. (6.40), the Kullback-Leibler divergence measures the cost of coding signals with the wrong distribution. Equation (6.165) now shows us that better knowledge of the probability distribution does not just allow us to make shorter codes. The amount by which we can compress the data describing the sequence of winners in the horse race is exactly the amount by which our winnings can grow. More precisely, if we can build a shorter code than the one built implicitly by the track operators, then we will gain exactly in proportion to this shortening. Thus, in this context, we literally get paid for constructing more efficient representations of the data!

We have connected the growth rate of winnings to the efficiency with which we can represent data, but this is not quite as compelling as a direct connection to how much information we have about the outcome of the game, which is where we started in the case of coin flips; let's see if we can do better. Imagine that, on each trial i, we have access to some signal x_i that tells us something about the likely outcome. More precisely, when we observe x_i, the probability that the outcome will be μ on trial i is not p_μ but rather some conditional probability $p(\mu|x_i)$; if the signals x are themselves chosen from some distribution $P(x)$, then for consistency, we must have

$$p_\mu = \int dx\, P(x) p(\mu|x). \tag{6.167}$$

To use the extra information provided by the signal x, you will adjust your strategy to bet a fraction $f_\mu(x_i)$ on the outcome μ given that you have "heard" x_i. How does the extra information provided by x improve your winnings?

To compute the growth of winnings in the presence of extra information, we proceed along the same lines as before to find the analog of Eq. (6.158):

$$\Lambda[\{f_\mu(x)\}] = \int dx\, P(x) \sum_{\mu=1}^{K} p(\mu|x) \ln[f_\mu(x)g_\mu]. \tag{6.168}$$

Now we need to maximize this quantity, choosing strategies that are defined by the functions $f_\mu(x)$, where for each x we have the constraint that $\sum_\mu f_\mu(x) = 1$. Once again the solution to this optimization problem is proportional gambling, but now the proportions are conditional on your knowledge, so that the analog of Eq. (6.162) is

$$f_\mu^{\text{opt}}(x) = p(\mu|x). \tag{6.169}$$

This determines the optimal growth rate:

$$\Lambda_{\text{opt}} = \int dx\, P(x) \sum_{\mu=1}^{K} p(\mu|x) \ln[p(\mu|x)g_{\mu}]. \tag{6.170}$$

Problem 144: Fill in the steps leading to the derivation of $\Lambda[\{f_{\mu}(x)\}]$ in Eq. (6.168) and the consequences of optimizing this functional, Eq. (6.169) and Eq. (6.170).

What interests us here is not the details, but the gain in growth rate that is possible by virtue of having access to the signal x, that is, the difference between Λ_{opt} in Eq. (6.170) and Eq. (6.163):

$$\Delta\Lambda_{\text{opt}} = \int dx\, P(x) \sum_{\mu=1}^{K} p(\mu|x) \ln[p(\mu|x)g_{\mu}] - \sum_{\mu=1}^{K} p_{\mu} \ln[p_{\mu}g_{\mu}] \tag{6.171}$$

$$= \int dx\, P(x) \sum_{\mu=1}^{K} p(\mu|x) \ln[p(\mu|x)g_{\mu}] - \int dx\, P(x) \sum_{\mu=1}^{K} p(\mu|x) \ln[p_{\mu}g_{\mu}] \tag{6.172}$$

$$= \int dx\, P(x) \sum_{\mu=1}^{K} p(\mu|x) \ln\left[\frac{p(\mu|x)}{p_{\mu}}\right]. \tag{6.173}$$

The details of the payoffs g_{μ} drop out, and *the gain in growth rate is exactly the mutual information between the signal x and the outcomes μ.*

Once again information translates directly into the (increased) rate at which capital can grow. Thus, the abstract measure of information has a clear impact on very down-to-earth measures of performance in a real-world task. But, beyond metaphor,[14] what does this have to do with life?

The most direct connection between life and gambling is through the phenomenon of persistence. Many bacteria have two distinct lifestyles. In one (for example), they grow quickly in most environments but are very susceptible to being killed by antibiotics. In the other, they grow very slowly but survive the antibiotics. This scenario is almost exactly a horse race—if the bacterium bets correctly, it grows; but if it bets incorrectly, it dies (or grows at rates far below what is possible). Absent any direct measurements on the environment, a population of genetically identical bacteria will maximize its growth rate by a form of proportional gambling, so that even in a healthy person not taking antibiotics, some of the resident bacteria should persist in a state of slow growth and (eventual) antibiotic resistance.[15] The fraction of bacteria in this state reflects the population's estimate of the probability that they will encounter the hostile environment of antibiotics. We also see that gaining information about the environment opens the possibility of faster growth in precise proportion to the information gained.

In a world of two alternatives, there is not much information to gain. There are ex-amples of bacteria that choose among a wider variety of lifestyles, and these phenomena

---

14. Life is a gamble (etc.).

15. Here "resistance" is used colloquially. Technically, antibiotic resistance refers to a trait that is encoded genetically, and hence inheritable, rather than a lifestyle choice. The (choosable) state in which bacteria grow slowly but are not killed by antibiotics is called persistent.

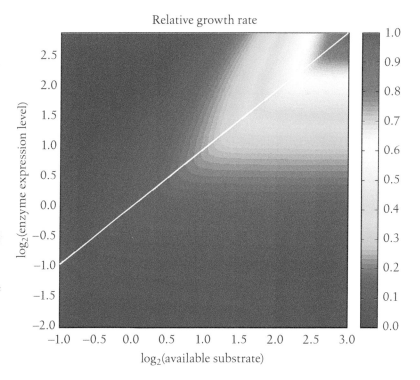

FIGURE 6.13

A schematic of bacterial growth rate as a function of available substrate concentration and enzyme expression level. The growth rate is a compromise between metabolizing the substrate and the cost of making the enzyme. The thin white line traces the optimal setting of expression level as a function of substrate availability.

(including the simple example of two alternatives) are called phenotypic switching. In the approximation that for each environment there is only one phenotype that grows, phenotypic switching is exactly the horse racing problem.

The example of phenotypic switching makes a nice map back to the early work about gambling but is perhaps still a bit too simple. Let's try to be more general. Imagine a bacterium that lives in an environment in which the concentrations of nutrients are fluctuating (slowly, so we do not have to worry about dynamics). To make use of the currently available nutrients, the bacterium must express the relevant enzymes involved in metabolism. Let's simplify and assume that there is one nutrient or substrate at concentration s and one relevant gene at expression level g. The bacterium will then grow at some rate $r(s, g)$ that depends both on the state of the world, s, and on its internal state g.

The growth rate of the bacterium is a compromise between two effects. On one hand, growth requires metabolism of the available nutrient, and so growth should be faster if there is either more nutrient or more enzyme. On the other hand, making the enzyme itself takes resources, which should slow the growth; in the limit of low nutrient concentrations, this cost can become dominant, and growth would stop if the cell tried to make too much enzyme. This scenario is shown schematically in Fig. 6.13.

Problem 145: A simple fitness landscape.   The schematic in Fig. 6.13 is based on a simple model. Suppose that growth is precisely proportional to the rate at which the enzyme degrades the substrate. In a Michaelis-Menten kinetic scheme for the enzyme (see the discussion following Eq. (2.146)), then the rate of degradation (in molecules per second) will be

$$V = V_{max} g \frac{s_{free}}{K + s_{free}}, \qquad (6.174)$$

where g is the number of copies of the enzyme molecule, V_{max} is the maximum rate at which the enzyme can run, s_{free} is the concentration of the substrate free in solution, and K is the Michaelis constant that sets the scale for half-saturation of the enzyme. The total substrate concentration is the sum of that free in solution and bound to the enzyme,

$$s = s_{free} + \frac{1}{\Omega} g \frac{s_{free}}{K + s_{free}}, \qquad (6.175)$$

where Ω is the cell volume. If the growth rate is proportional to the metabolic rate less a correction for the cost of making the enzymes, we should have

$$r(s, g) = \alpha g \frac{s_{free}}{K + s_{free}} - \beta g. \qquad (6.176)$$

Solve for s_{free} to rewrite $r(s, g)$ explicitly in terms of s. Then show that by proper choice of units, there is only one arbitrary parameter. What is the meaning of this remaining parameter? Make some reasonable choices, and plot your own version of Fig. 6.13.

Imagine a bacterium whose life is governed by Fig. 6.13. As the available substrate concentration fluctuates, one possibility is that all bacteria carefully adjust their enzyme expression levels to achieve optimal growth rate under each condition. An extreme alternative is that different bacteria in the population choose their expression levels at random out of some distribution, in the hope that some of them by chance have made good choices, much as in the proportional gambling scenario. In the first case, the expression level carries an enormous amount of information about the concentration of available substrate—indeed, if we imagine that the optimum is traced perfectly, then knowing the expression level would tell us the exact substrate concentration, which represents an infinite amount of information. In contrast, the gambling strategy involves no correlation of the internal and external states, and hence no information is conveyed. Evidently, the average growth rate across an ensemble of environments will be larger if the bacteria can adjust their expression levels perfectly, but maybe this is so obvious as not to be interesting. We know that there is some average growth rate that can be achieved with no information about the outside world, and that an infinite amount of information would allow the population to grow faster. What happens in between?

The mutual information between the internal state g and the external world s can be written as

$$I(g; s) = \int ds \, P(s) \int dg \, P(g|s) \log_2 \left[\frac{P(g|s)}{P_G(g)} \right]. \qquad (6.177)$$

We can make $I(g; s)$ as small as we like by letting $P(g|s)$ approach $P_G(g)$. But suppose that we want to maintain some average growth rate in the ensemble of environments defined by $P_S(s)$. This average growth rate is

$$\langle r \rangle = \int ds \, P_S(s) \int dg \, P(g|s) r(s, g). \qquad (6.178)$$

Now it seems clear that not all conditional distributions $P(g|s)$ are consistent with a given $\langle r \rangle$. What we would like to show is that there is a minimum value of $I(g;s)$ consistent with $\langle r \rangle$.

This problem is one of constrained minimization, so as usual we introduce a Lagrange multiplier and minimize

$$\mathcal{F}[P(g|s)] \equiv I(g;s) - \lambda \langle r \rangle - \int ds \, \mu(s) \int dg \, P(g|s), \tag{6.179}$$

where the second set of Lagrange multipliers $\mu(s)$ enforces normalization of the distributions $P(g|s)$ at each value of s. The key step in minimizing $\mathcal{F}$ is to evaluate the derivative of the information with respect to the conditional distribution:

$$\frac{\delta I(g;s)}{\delta P(g|s)} = \frac{\delta}{\delta P(g|s)} \int ds \, P_S(s) \int dg \, P(g|s) \log_2 \left[\frac{P(g|s)}{P_G(g)} \right] \tag{6.180}$$

$$= P_S(s) \log_2 \left[\frac{P(g|s)}{P_G(g)} \right] + \frac{1}{\ln 2} P_S(s) P(g|s) \cdot \frac{1}{P(g|s)}$$

$$- \frac{1}{\ln 2} \int ds' \, P_S(s') P(g|s') \frac{1}{P_G(g)} \frac{\delta P_G(g)}{\delta P(g|s)} \tag{6.181}$$

$$= P_S(s) \log_2 \left[\frac{P(g|s)}{P_G(g)} \right] + \frac{1}{\ln 2} P_S(s) - \frac{1}{\ln 2} P_G(g) \frac{1}{P_G(g)} P_S(s) \tag{6.182}$$

$$= P_S(s) \log_2 \left[\frac{P(g|s)}{P_G(g)} \right], \tag{6.183}$$

which is nice, because all the messy bits cancel out. Now we can solve the full problem:

$$0 = \frac{\delta \mathcal{F}[P(g|s)]}{\delta P(g|s)} \tag{6.184}$$

$$= \frac{\delta}{\delta P(g|s)} \left[I(g;s) - \lambda \int ds \, P_S(s) \int dg \, P(g|s) r(s,g) \right.$$

$$\left. - \int ds \, \mu(s) \int dg \, P(g|s) \right] \tag{6.185}$$

$$= P_S(s) \log_2 \left[\frac{P(g|s)}{P_G(g)} \right] - \lambda P_S(s) r(s,g) - \mu(s), \tag{6.186}$$

$$\log_2 \left[\frac{P(g|s)}{P_G(g)} \right] = \lambda r(s,g) + \frac{\mu(s)}{P_S(s)} \tag{6.187}$$

$$P(g|s) = \frac{1}{Z(s)} P_G(g) \exp \left[\beta r(s,g) \right], \tag{6.188}$$

where $\beta = \lambda \ln 2$; $Z(s) = \exp[\ln 2 \mu(s)/P_S(s)]$ is a normalization constant,

$$Z(s) = \int dg \, P_G(g) \exp \left[\beta r(s,g) \right]; \tag{6.189}$$

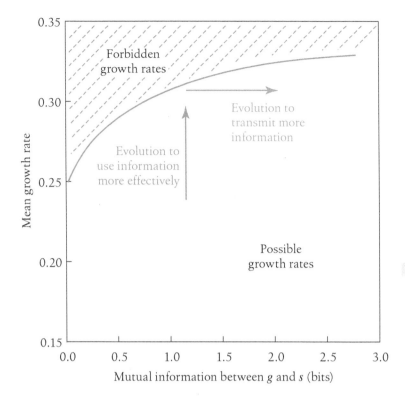

FIGURE 6.14

Mean growth rate as a function of the mutual information between expression levels and substrate availability for the system in Fig. 6.13. We assume that the (log) substrate is chosen from a distribution that is uniform over the 16-fold range shown in Fig. 6.13 and then solve for the optimal $P(g|s)$ using Eq. (6.188) and Eq. (6.189).

and we must obey

$$P_G(g) = \int ds\ P_S(s) P(g|s). \tag{6.190}$$

Notice that our solution for $P(g|s)$ is (roughly) a Boltzmann distribution, where $-r(g, s)$ plays the role of the energy, and β is the inverse temperature. As expected from this analogy, we can write the information and average growth rate as derivatives:

$$I(g; s) = \lambda \langle r \rangle - \int ds\ P_S(s) \log_2 Z(s), \tag{6.191}$$

$$\langle r \rangle = \int ds\ P_S(s) \frac{d \ln Z(s)}{d\beta}. \tag{6.192}$$

The Boltzmann form of the optimal solution in Eq. (6.188) helps our intuition. At small β, the distribution $P(g|s)$ is almost the same as $P_G(g)$, so that little information is conveyed between internal and external states. In contrast, as β becomes large, the distribution $P(g|s)$ becomes sharply peaked around the expression level $g_{opt}(s)$ that maximizes the growth rate. Varying β should trace out a curve of mean growth rate versus information, as shown in Fig. 6.14. We see from the derivation that this curve represents the maximum mean growth rate achievable given a certain amount of mutual information, or alternatively the minimum amount of information required to achieve a certain mean growth rate, $I_{min}(\langle r \rangle)$.

Problem 146: Asymptotics of growth rate versus information.   The precise form of the relationship between the mean growth rate and the minimum information depends, of course, on details of the function $r(s, g)$. Show that the behavior at large values of the minimum information is more nearly universal. To do this, develop an asymptotic expansion at large values of β,

$$P(g|s) = \frac{1}{Z(s)} P(g) \exp\left[\beta r(s, g)\right]$$

$$\approx \frac{1}{Z(s)} P(g) \exp\left[\beta r(s, g_{\text{opt}}(s)) + \frac{\beta}{2} A(g - g_{\text{opt}}(s))^2\right], \qquad (6.193)$$

$$A = \left.\frac{\partial^2 r(s, g)}{\partial g^2}\right|_{g=g_{\text{opt}}(s)}, \qquad (6.194)$$

and use this expansion to evaluate $Z(s)$, from which you can calculate $I_{\min}(\langle r \rangle)$. Can you generalize your discussion to the case where there are many substrates and many genes to control?

It is important to take seriously the scales in Fig. 6.14. It could have been that the full growth advantage derived from controlling expression levels was achievable with only a small fraction of a bit, or conversely, that it required many tens of bits. In fact, for this simple problem the answer is that cells can make use of more than one bit, but not too many more. Thus, (near-)optimal growth requires more than just turning a gene on and off, and presumably this is even clearer if we think about more realistic situations where there are multiple substrates and multiple genes. As we will see in the next section, the noise levels measured for the control of gene expression set a limit of $\sim 1-3$ bits to the information that can be transmitted through these control elements. Thus, the amount of information that cells need to optimize their growth in varying environments is plausibly close to the maximum they can transmit, and this limit in turn is set by the number of molecules that the cell is devoting to these tasks.

Just to be clear, it is useful to think about the alternatives. If information is cheap, so that it is easy for cells to transmit many bits, then evolution selects for mechanisms that drive the system upward in the information-fitness plane of Fig. 6.14. But if information itself is hard to come by, evolutionary pressure (which really only acts to increase growth rates) must necessarily drive cells outward along the information axis.

Sometimes the fact that organisms have to be flexible and survive in a fluctuating environment is offered as a qualitative argument against the possibility of optimization. Indeed, if the environment fluctuates, it may not be advantageous for organisms to drive toward perfect performance under any one set of conditions. But the argument we have given here shows that strategies for dealing with varied environments are themselves subject to optimization, making the most of a limited amount of information and eventually being pushed by selection to gather more bits.

In the problem of horse races, or phenotypic switching, information translated directly into a growth rate. Here we see that, more generally, there is a minimum amount of information needed to achieve a given average growth rate. In both cases, information is necessary but not sufficient. Thus, organisms *can* grow faster if they gather and represent more information, but this is not guaranteed—they might make poor use of the information and fail to reach the bound on their growth rate. We have focused here on achieving a certain average growth rate, but the whole discussion can be transposed to other domains. For example, if I ask you to point at a target that appears at random in your visual field and reward you in proportion to how close you come to the exact position of the target, then to collect a certain level of average reward, your brain must represent some minimum amount of information about the target location. Quite generally, we can imagine plotting some biological measure of performance (probability of catching a mate, nutritional value extracted from picking fruit, growth rate, happiness, etc.) versus the amount of information that the organism has about the relevant variables. This information-fitness plane will be divided by a curve that separates the possible from the impossible, because without a certain minimum level of information, higher fitness is impossible.

Problem 147: Information and motor control. One important task in motor control is tracking. We do this in many situations, but when objects move smoothly across our visual field, we track their motion by moving our eyes. This behavior is called smooth pursuit.

(a) The simplest version of the tracking problem is that we have an observable variable x and we try to generate a variable y that is as close as possible to x; notice that we have not yet introduced any dynamics. To make these ideas precise, assume that x is chosen out of a probability distribution $P_X(x)$, and that "as close as possible" will be measured by the mean-square error $\epsilon = \langle (y - x)^2 \rangle$. Develop, by analogy with the argument above, a variational principle for the choice of y—that is, for the distribution $P(y|x)$—in which you can find the minimum amount of mutual information $I(x; y)$ needed to reach a criterion value for ϵ.

(b) Show that the variational problem you formulated in part (a) is solved by

$$P(y|x) = \frac{1}{Z(x, \beta)} P_Y(y) e^{-\beta(y-x)^2}, \tag{6.195}$$

where β is the parameter that trades information for accuracy, and consistency requires that

$$Z(x, \beta) = \int dy \, P_Y(y) e^{-\beta(y-x)^2}, \tag{6.196}$$

$$P_Y(y) = \int dx \, P(y|x) P_X(x). \tag{6.197}$$

(c) Consider the special case where x is Gaussian; we can choose units so that

$$P_X(x) = \frac{1}{\sqrt{2\pi}} e^{-x^2/2}. \tag{6.198}$$

It seems plausible that, in this case, y is also Gaussian:

$$P_Y(y) = \frac{1}{\sqrt{2\pi\sigma_y^2}} e^{-y^2/(2\sigma_y^2)}. \tag{6.199}$$

Show that, with this ansatz, you can compute $Z(x, \beta)$ in Eq. (6.196) analytically, and then impose the consistency condition on Eq. (6.197). Does this lead to a solvable equation for the variance σ_y^2? Push as far as you can toward a complete solution, ultimately plotting $I(x; y)$ versus the mean-square error ϵ.

(d) Explore what happens if you change the cost function, and measure errors as $\epsilon = \langle u(x - y) \rangle$, using something other than quadratic u. In particular, consider the case where large errors have only finite cost, such as

$$u(z) = 1 - e^{-(z/z_0)^2}. \tag{6.200}$$

Most likely you will need to find a numerical approach to solve the consistency conditions. This problem is deliberately open ended.

In the information theory literature, the sort of bounds we are computing here go by the name of rate-distortion curves. For example, if we measure image quality by some complicated perceptual metric, then to have images of a certain quality, on average we will need to transmit a minimum number of bits. In this spirit, we can think about more complicated situations, such as organisms foraging or acting in response to sensory stimuli and collecting rewards. Although one is not rewarded specifically for bits, rate-distortion theory tells us that to collect rewards at some desired rate always requires a minimum number of bits of information.

When constructing a rate-distortion curve, we implicitly define some bits as being more relevant than others. Thus, if I need to match my state to that of the environment, presumably some environmental variables need to be tracked more accurately than others; because the rate-distortion curve gives the minimum number of bits, I need to put the precision (extra bits) in the right place. This is important, because it means that we have a framework for assigning value to bits. To be concrete, in Fig. 6.14 it is possible to imagine an infinite variety of mechanisms that gather the same number of bits but fail to achieve the maximum mean growth rate, either because they use the bits incorrectly or because they have gathered the wrong bits. Bits in and of themselves are not guaranteed to be useful, but to do useful things, there is a minimum number of bits required.

A different, perhaps simpler, view of these ideas is provided by the problem of search. There are many situations in which organisms need to find things, and there is a cost to taking too long. Suppose that we are looking for something that could be in one of N places. A priori we may know that some places are more likely than others, and so we can label the possibilities in order of their probabilities: location 1 has probability p_1, location 2 has probability p_2, and so on, with $p_1 \geq p_2 \geq p_3 \geq \cdots \geq p_N$. Because it is the most likely possibility, we should look first in location 1, and there is a probability p_1 that our search will be complete after this one time step. If the thing we are looking for is not at location 1, we look next at location 2, and with probability p_2 our search ends here. Continuing, we can see that the average number of steps in our search is

$$\bar{T} = \sum_{t=1}^{N} t p_t, \tag{6.201}$$

where again it is important that we have ordered the possible locations correctly. Notice that we have been very optimistic, not including any constraints that might prevent us from jumping between successive locations; to be more realistic, then, we should say that $\bar{T}$ is a lower bound on the mean time required to complete the search.

If we know the value of $\bar{T}$, then we can say that the probability distribution p_t has a maximum possible value for its entropy; this is the same argument as mentioned in relation to counting vesicles (near the end of Section 6.2). From the discussion of these maximum entropy problems in Appendix A.7, we have (adapting Eq. (A.413)),

$$S[\{p_t\}] \le \log_2(1 + \bar{T}) + \bar{T}\log_2(1 + 1/\bar{T}) \sim \log_2(\bar{T}e), \qquad (6.202)$$

where we make the approximation that $\bar{T} \gg 1$. Previously we used this to say something about the entropy (Section 6.2), but now we can turn things around to say something about the mean search time:

$$\bar{T} \ge \frac{1}{e}2^{S[\{p_t\}]}. \qquad (6.203)$$

Thus, the time it takes to find something is bounded by (the exponential of) our uncertainty about its location, where uncertainty is measured quantitatively by the entropy, as Shannon taught us. In the same way that bits are necessary but not sufficient for metabolic control, real searches might take longer than this entropic bound. But if we are searching efficiently, the only way to speed things up is to lower the entropy or, equivalently, to gain information.

When we search, we gain information. If we find our target, then we know exactly where it is, and the entropy falls to zero. Even if we do not find what we are looking for, we can cross one possibility off the list, which reduces the entropy of the distribution. If we go through the possibilities in order of their likelihood, as above, then the expected value of the entropy reduction is maximized at every step. Thus, in this case, minimizing the time required to find the target is equivalent to maximizing the rate at which we gain information about its location. It has been suggested that this observation can be made much more general.

When searching, we often have incomplete information and are constrained such that the next place we look must be close to where we looked last, in some metric. An extreme example is a moth searching for a mate, guided by the smell of pheromones. The moth flies continuously and thus cannot jump from one possible location to another; the data he collects are sparse, as the turbulent airflow converts even the steady emission of odorants into an intermittent sequence of plumes. In addition, locations of the mate can be identified only to the extent that they are upwind. At each moment, the moth has the choice to continue flying upwind or to cast left or right, hoping to pick up another plume. The problem usually is formulated by asking for the sequence of decisions (upwind versus cast), guided by sensory inputs (encounters with plumes) that would bring the moth as close as possible to the source. But such local greediness often is not effective in the long run, and more global computations are difficult. An alternative is to ask for the local decisions that maximize the gain of information about the location of the odor source, and it has been shown that this "infotaxis" algorithm actually outperforms other methods, in part because it provides a natural trade-off between exploring to gain new knowledge and exploiting the knowledge already acquired; further, this exploration-exploitation trade-off is implemented without having to adjust parameters. An example of infotaxis is shown in Fig. 6.15.

The infotaxis idea is interesting in and of itself as a search algorithm that can be used even when data are sparse and it is not possible to steer the search by local computations of spatial gradients. But in the context of our present discussion, the idea is important because it shows how a biologically grounded, goal-directed behavior can

(a)   (b)

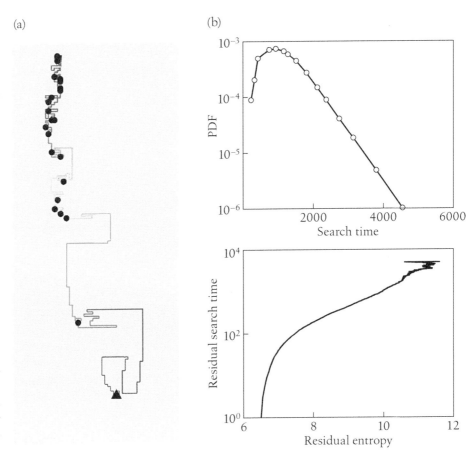

FIGURE 6.15

Infotaxis and the search for an odor source. (a) An example of a simulated search trajectory in a turbulent flow. Segments of the trajectory are color coded by their duration, and black dots mark the moments when the simulated moth encounters a plume. Note the alternations between straight upwind flight (upward in the figure) and casting perpendicular to the wind direction. (b) The distribution of time required for the search does not have a long tail, despite the massive fluctuations in the input odor signals along any given path. (c) Along trajectories the residual search time tracks the residual entropy, as in Eq. (6.202). Reprinted by permission from Macmillan Publishers, Ltd.: Vergassola et al. (2007).

be replaced with an abstract information-theoretic optimization principle. Not only does the information-theoretic formulation include the goal, it also provides us with a practical strategy for achieving the goal that seems to be more effective than anything else that has been proposed in this context. We do not know whether real organisms implement the infotaxis algorithm explicitly, but this is convincing evidence against the notion that optimizing information is somehow incompatible with the achievement of more concrete goals.

Problem 148: Using infotaxis to find the minimum of a function.   Imagine that we are trying to find the minimum of a function along a line, and we can make noisy measurements of both the function and its gradient at each point where we look. To keep things simple, suppose we are already in a window of x such that we can approximate the function as quadratic, $f(x) = \frac{1}{2}a(x - x_0)^2$; the problem is that we do not know a or x_0.

(a) Suppose that we can measure $y = f(x)$ and $z = f'(x)$ with some Gaussian errors δy and δz. From these measurements at one value of x, we can estimate a and x_0. Construct the matrix that describes the propagation of the errors, that is, J such that

$$\begin{bmatrix} \delta a \\ \delta x_0 \end{bmatrix} = \begin{bmatrix} J_{ay} & J_{az} \\ J_{x_0 y} & J_{x_0 z} \end{bmatrix} \begin{bmatrix} \delta y \\ \delta z \end{bmatrix}. \tag{6.204}$$

(b) Show that the entropy of a multidimensional Gaussian distribution is determined by the determinant of the covariance matrix. If the covariance matrix of the errors δy and δz is given, what can you say about the entropy associated with the uncertainty in the parameters (a, x_0) after this one measurement?

(c) If we start with some knowledge of the parameters expressed as a prior distribution $P(a, x_0)$, use Bayes' rule to show how this distribution is changed as the result of one measurement. Can you give an approximate expression for how much the entropy is reduced? Given $P(a, x_0)$, at what value of x is the expected entropy reduction, and hence the information gain, largest? How does the problem of maximizing this information gain relate to the problem of finding x_0 itself?

An interesting if incomplete connection of rate-distortion theory to biological systems is the case of protein structure. If I want to describe protein structures with high precision, I need to specify where every atom is located. But if sequence determines structure, then to some accuracy I just need to specify the amino acid sequence, which is at most $\log_2(20)$ bits per amino acid, and many fewer per atom. In fact, as discussed in Section 5.1, many different sequences generate essentially the same structure, so there must be an even shorter description. Thus, if we imagine taking the ensemble of real protein structures, there must be a description in very few bits that nonetheless generates rather small errors in predicting the positions of the atoms. Finding the optimally compact description (i.e., along the true rate-distortion curve) would be a huge help in understanding protein folding, because the joint table of sequences and (compactly described) structures would be much smaller. There is even an intuition that such a compact representation with high accuracy must exist to ensure rapid folding, essentially because the number of states needed for an accurate description should be connected to the number of states that the protein must search through as it folds. I am not sure how to make this notion rigorous, but it is interesting.

We can search for compact descriptions of protein structure by approximating the local path of the α-carbon backbone as steps on a discrete lattice, making the lattice progressively more complex. We can do better by moving off the lattice to cluster the natural dihedral angles describing the path from one amino acid to the next; results are shown in Fig. 6.16. Indeed, by the time we have assigned 10 or 20 states per amino acid, we can reconstruct structures with $1-2$ Å root-mean-square accuracy.

Another very specific connection between biology and bits is in the case of embryonic development. In the simplest model of morphogen gradients, each cell independently reads out the local concentration of the morphogen(s), and makes decisions—most importantly, about the regulation of gene expression—based on this local measurement, as in Fig. 6.17. In this model, the only information that a cell has about its position in the embryo is the morphogen concentration, and so the information that cells have about position can be no larger than the information that they extract about this concentration. In effect there is a communication channel from the morphogen to the expression levels of the genes that define the blueprint for development, and the information that can be transmitted along this channel sets a bound on the complexity and reliability of the blueprint. As an example, if we have N rows of cells along one axis of the embryo, and each row reliably adopts a distinct fate that we can determine by looking at the expression levels of a handful of genes, then (again, in the simplest

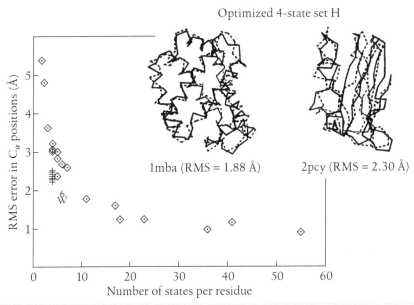

FIGURE 6.16

Rate-distortion curve (or its moral equivalent) for protein structures. The path of the α-carbon backbone is approximated by a discrete set of local moves along the chain, which is roughly equivalent to forcing the structure to live on a lattice. Diamonds correspond to lattices with different structures (e.g., three possible moves on a tetrahedral lattice); + and W correspond to discrete approximations obtained by clustering known structures. The y-axis measures the root-mean-square (RMS) error in the positions of all α-carbons along the chain. Inset shows two examples of protein structures (solid lines) compared with optimized four-state approximations (dashed lines). Redrawn from Park and Levitt (1995), with permission from Elsevier.

model) there must be $\log_2 N$ bits of information transmitted through the regulatory network that takes the morphogens as input and gives the gene expression levels as output. As in the above discussion of growth rates, this becomes interesting because, as we shall see, the information capacity of gene regulatory elements is quite limited. Rough estimates of the relevant quantities in the *Drosophila* embryo, using the kinds of experiments discussed in Sections 4.3 and 5.3, suggest that the embryo might indeed be forming patterns near the limits set by the information capacity of gene regulation.

What happens if the situation is more complicated than in Fig. 6.17? In particular, we know about plenty of systems that form patterns spontaneously, without any analog of the maternal signal to break the translational symmetry. It is important to realize that although patterns can form spontaneously, information can't really be created, only transmitted. In a crystal, for example, once we know that one atom is in a particular position, we can predict the position of other atoms, but only because of the bonds that connect the atoms. Because all atoms undergo Brownian motion, the transmission of information is not perfect, and knowledge of one atomic position provides only a limited number of bits about the position of another atom. This limit on information transmission becomes tighter as the temperature—and hence the noise level in the "communication channel" connecting distant atoms—becomes higher, until the crystal melts, and there is no information transmitted over long distances.

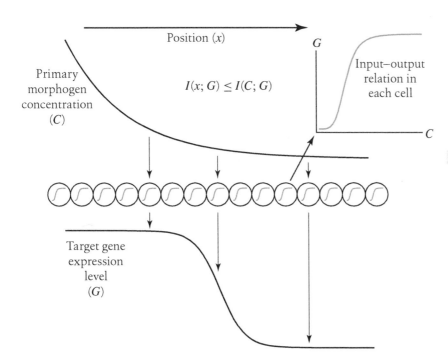

FIGURE 6.17

Information flow in a feed-forward model of genetic control in the early embryo. The concentration C of the primary morphogen depends on position x, and each cell responds independently by modulating the expression level G of some target gene (or genes). In this simple view, information about position only reaches the gene expression level through the intermediary of the primary morphogen concentration, and hence we have $I(x; G) \leq I(C; G)$.

Problem 149: Transmitting positional information along a chain. Imagine a collection of N atoms at positions x_i along a line, connected by springs with equilibrium length a, so that the potential energy of the collection is

$$V(\{x_i\}) = \frac{\kappa}{2} \sum_{i=1}^{N-1} (x_{i+1} - x_i - a)^2; \tag{6.205}$$

notice that we are not making the usual assumption that the line loops back on itself into a ring. The Boltzmann distribution is then a (slightly complicated) multidimensional Gaussian:

$$P(\{x_i\}) \equiv \frac{1}{Z} e^{-V(\{x_i\})/k_B T} = \frac{1}{Z} \exp\left[-\frac{\kappa}{2k_B T} \sum_{i=1}^{N-1} (x_{i+1} - x_i - a)^2 \right]. \tag{6.206}$$

Let's choose coordinates so that the center of the chain is exactly at $x = 0$, which we can achieve by taking $P(\{x_i\}) \rightarrow P(\{x_i\})\delta(\sum_i x_i)$.

(a) Starting with the case $N = 3$, show that you evaluate $P_i(x_i)$ directly by integrating over all $x_{j \neq i}$:

$$P_i(x_i) = \int d^{N-1} x_{j \neq i} P(\{x_i\}). \tag{6.207}$$

Hint: Start by integrating over x_1, and proceed forward toward x_i; repeat by starting with x_N and proceeding backward toward x_i.

(b) Use your results from part (a) to approximate the overall distribution of positions:

$$P(x) = \frac{1}{N} \sum_{i=1}^{N} P_i(x_i). \tag{6.208}$$

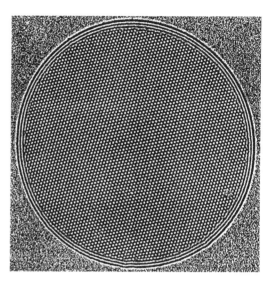

FIGURE 6.18

This looks like a perfect crystal of beads, but it actually is a small (~10 cm diameter) container filled with carbon dioxide at high pressure and heated from below. The image is formed by passing light through the gas, and is sometimes called a shadowgraph. The temperatures at the top and bottom of the container are held strictly constant (to within a few thousandths of a degree), so that the patterns will not be disrupted by variations in conditions. Similarly, the top and bottom of the container are extremely flat (smooth to within the wavelength of light), and the whole system is held horizontal with high precision, so that the direction of gravity is aligned with the axis of symmetry through the center of the circle. Reprinted, with permission, from Bodenschatz et al. (1991). Copyright © 1991 by the American Physical Society.

(c) Combine the results from parts (a) and (b) to compute how much information, in bits, the index i provides about the position x_i. How does this depend on N? How do your results compare with $I \sim \log N$ required to specify the position of every atom with some fixed precision?

In nonequilibrium systems, such as the Rayleigh-Bénard convection cell shown in Fig. 6.18, we see spatial patterns in which some local variable (e.g., temperature, fluid density, or velocity) at one position predicts the value of the corresponding variable at another position. Call this local variable $\phi(\mathbf{x})$. If we imagine a large ensemble of snapshots like the one in Fig. 6.18, we can build up the distribution functional $P[\phi(\mathbf{x})]$. The statement that we have a periodic pattern, for example, is equivalent to the statement that if we look at two points separated by an appropriately chosen vector $\mathbf{d}$, then $\phi(\mathbf{x}) \approx \phi(\mathbf{x} + \mathbf{d})$. But if we choose the first point $\mathbf{x}$ at random, we can get a broad range of values for $\phi_1 \equiv \phi(\mathbf{x})$, drawn from a distribution $P_1(\phi_1)$. Similarly, if we are choosing $\mathbf{x}$ at random, then $\phi_2 \equiv \phi(\mathbf{x} + \mathbf{d})$ is also broadly distributed; in fact, it must come from the same distribution as ϕ_1. But once we know ϕ_1, if there is a periodic pattern, then the distribution $P(\phi_2|\phi_1)$ must be sharply peaked around $\phi_1 = \phi_2$ and hence very different from the "prior" distribution of ϕ_2. But this is exactly the condition for there to be mutual information between ϕ_1 and ϕ_2. Thus, the existence of a spatial pattern is equivalent to the presence of mutual information between the local variables at distant points. Where does this information come from? As with the bonds connecting the atoms in the crystal, it must be transmitted through the dynamics of the system, which connect points only to their immediate neighbors.

In a strict interpretation of the concept of positional information, we actually require more than mutual information between local variables at distant points. We require that the value of some local variable(s), typically the expression levels of several genes, tell us about the location of the point where we have observed them. In this way, cells would "know" their position in the embryo by virtue of their expression levels, and these signals could drive further processes in a way that is appropriate to the cell's location—not just relative to other cells, but in absolute terms.[16] If we call the local variables $\{g_i\}$, for gene expression levels, then the positional information is $I(\mathbf{x}; \{g_i\})$. But the local variables at point $\mathbf{x}$ are controlled by a set of inputs that may include external, maternally supplied morphogens, the expression levels $\{g_i\}$ in neighboring cells, and perhaps other variables as well. We can always write the distribution of expression levels at one point in terms of these inputs:

$$P(\{g_i\}|\mathbf{x}) = \int d(\text{inputs})\, P(\{g_i\}|\text{inputs})\, P(\text{inputs}|\mathbf{x}). \qquad (6.209)$$

Noise in the control of gene expression corresponds to the fact that the distribution $P(\{g_i\}|\text{inputs})$ is not infinitely narrow. Now because, at any one point, information flows from $\mathbf{x}$ to the inputs to the $\{g_i\}$, we must have $I(\mathbf{x}; \{g_i\}) \leq I(\text{inputs}; \{g_i\})$, which is true no matter how complicated the inputs might be. More importantly, as hinted at in the analysis of the first synapse in fly vision (following Fig. 6.12), any input-output device has a maximum amount of information it can transmit that is determined by its noise level. Thus, if we consider the whole network of interactions that result in the regulation of the gene expression levels $\{g_i\}$, the noise in this network determines a maximum value for $I(\text{inputs}; \{g_i\})$, which sets a limit to the amount of positional information that cells in the embryo can acquire and encode with these genes.

Problem 150: The data-processing inequality.   The theorem that we need to establish the above argument is that if we have three variables, x, y, and z, such that

$$P(z|x) = \int dy\, P(z|y) P(y|x), \qquad (6.210)$$

then $I(z; x) \leq I(z; y)$ and $I(z; x) \leq I(y; x)$. Show that this statement is true.

To summarize, the reliability and complexity of the patterns that can form during embryonic development are limited by the amount of positional information that cells can acquire and represent. This information in turn is limited by the capacity of the genetic or biochemical networks whose outputs encode the positional information. Therefore, if real networks operate in a regime where this capacity is small, the complexity of body plans will be limited by the ability of the organism to squeeze as much information as possible out of these systems.

Most examples we have considered thus far have the feature that the information is about something that has obvious relevance for the organism. Can we find some more general way of arriving at notions of relevance? It is useful to have in mind an organism collecting a stream of data, whether the organism is like us, with eyes and ears, or like a bacterium, sensing the concentrations of various molecules in its external and internal

---

16. This is certainly what "positional information" means in the usual descriptions of the concept; see the discussion of the information carried by Hb expression levels in the fly embryo, near Fig. 6.9 in Section 6.2. There are almost no measurements of this information, in bits, so it remains possible that a real cell can determine much more about its relative position than about its absolute position. This would not change the spirit of what I am saying here, but the details would matter. This question is one of many open ones about information flow in the embryo.

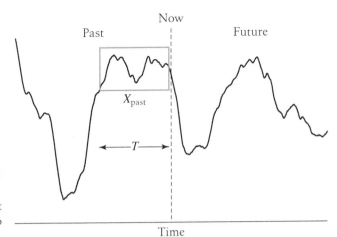

FIGURE 6.19

A schematic of the prediction problem. We observe a time series, and at some moment (now) we look back at a segment of the recent past with duration T, X_{past}. From this, we try to infer something about what will evolve in the future.

environments. Of all these data, the only part we can use to guide our actions (and eventually collect rewards, reproduce, etc.) is the part that has predictive power, because by the time we act, we are already in the future. Thus, we can ask how to squeeze out, of all the bits we collect, only those bits that are relevant for prediction.

More concretely, as in Fig. 6.19, if we observe a time series through a window of duration T (i.e., for times $-T < t \leq 0$), then to represent the data X_{past} we have collected requires $S(T)$ bits, where S is the entropy. But the information that these data provide about the future X_{future} (i.e., at times $t > 0$) is given by some $I(X_{past}; X_{future}) \equiv I_{pred}(T) \ll S(T)$. In particular, although for large T the entropy $S(T)$ is expected to become extensive, the predictive information $I_{pred}(T)$ is always subextensive. Thus, we expect that the data X_{past} can be compressed significantly into some internal representation X_{int} without losing too much of the relevant information about X_{future}. Formally, we can construct the optimal version of this mapping by solving

$$\max_{X_{past} \to X_{int}} \left[I(X_{int}; X_{future}) - \lambda I(X_{int}; X_{past}) \right], \tag{6.211}$$

where $X_{past} \to X_{int}$ is the rule for creating the internal representation and λ is a Lagrange multiplier. This sort of problem has been dubbed an "information bottleneck," because we try to preserve the relevant information while squeezing the input data through a narrow channel.

Problem 151: Predictive information is subextensive.   If we observe a stationary stochastic process $x(t)$ on the interval $t_1 < t \leq t_1 + T$, the entropy of the distribution $P[x(t)]$ depends only on T, not t_1; let's call this entropy $S(T)$.

(a) Use your intuition from statistical mechanics to explain why we expect $S(T)$ to grow extensively, that is, $S(T) \propto T$ at large T. More formally, show that at large T we have

$$S(T) \to \mathcal{S}T + S_1(T), \tag{6.212}$$

where

$$\lim_{T \to \infty} \frac{S_1(T)}{T} = 0. \tag{6.213}$$

Thus, although $S_1(T)$ can grow with T, it must grow more slowly than T itself—it is subextensive.

(b) Consider the case where time is discrete, and x is Markovian, so that $x(t + 1)$ depends on $x(t)$ but on no earlier history. Show that, in this case, $S_1(T)$ is just a constant.

(c) Consider the case where $X_{\text{past}} \equiv x(-T < t \leq 0)$ and $X_{\text{future}} \equiv x(0 < t < T')$. Show how the predictive information $I_{\text{pred}}(T, T') \equiv I(X_{\text{past}}; X_{\text{future}})$ is related to the function $S(T)$. You should be able to do this in general, without the Markov assumption. Show further that there is a finite limit as the duration of the future becomes infinite, and that this limit $I_{\text{pred}}(T)$ is subextensive in T.

In general, we should consider the mapping $X_{\text{past}} \to X_{\text{int}}$ to be probabilistic, so we can describe it by some conditional distribution $P(X_{\text{int}}|X_{\text{past}})$. Then the quantity we are trying to maximize becomes

$$-\mathcal{F} = \sum_{X_{\text{int}}, X_{\text{past}}} P(X_{\text{int}}|X_{\text{past}})P(X_{\text{past}}) \log_2 \left[\frac{P(X_{\text{int}}|X_{\text{past}})}{P(X_{\text{int}})} \right]$$

$$- \lambda \sum_{X_{\text{int}}, X_{\text{future}}} P(X_{\text{int}}|X_{\text{future}})P(X_{\text{future}}) \log_2 \left[\frac{P(X_{\text{int}}|X_{\text{future}})}{P(X_{\text{int}})} \right]. \quad (6.214)$$

This expression is written as if our choice of representation X_{int} depended directly on the future, but this can't be true. Any correlation between what we write down and what happens in the future is inherited from the data that we collected in the past, which means we can write

$$P(X_{\text{int}}|X_{\text{future}}) = \sum_{X_{\text{past}}} P(X_{\text{int}}|X_{\text{past}})P(X_{\text{past}}|X_{\text{future}}). \quad (6.215)$$

In addition, we have

$$P(X_{\text{int}}) = \sum_{X_{\text{past}}} P(X_{\text{int}}|X_{\text{past}})P(X_{\text{past}}). \quad (6.216)$$

As usual, we want to take the derivative of $\mathcal{F}$ with respect to the distribution $P(X_{\text{int}}|X_{\text{past}})$, being careful to add a Lagrange multiplier $\mu(X_{\text{past}})$ that fixes the normalization for each value of X_{past}, and then set the derivative to zero to find an extremum. The solution to this problem can be written as

$$P(X_{\text{int}}|X_{\text{past}}) = \frac{P(X_{\text{int}})}{Z(X_{\text{past}}; \lambda)} \exp\left(- \lambda D_{KL} \left[P(X_{\text{future}}|X_{\text{past}})||P(X_{\text{future}}|X_{\text{int}}) \right] \right),$$

$$(6.217)$$

where $D_{KL}(Q'||Q)$ denotes the Kullback-Leibler divergence between distributions Q' and Q, and $Z(X_{\text{past}}; \lambda)$ is a normalization constant. This is not a solution to our problem, but rather a self-consistent equation that the solution has to satisfy. The problem we are solving is an example of selective compression.

Problem 152: Give the full derivation leading to Eq. (6.217).

We should think of Eq. (6.217) as being like Eq. (6.188), but instead of adjusting an internal state in relation to the potential formed by the growth rate, here the effective potential is the (negative) Kullback-Leibler divergence, which measures the similarity between the distributions of futures given the actual past and given our compressed representation of the past. Thus, if two past histories lead to similar distributions of futures, they should be mapped into the same value of X_{int}, even if they are otherwise very different. This makes sense, because we are trying to throw away any information that does not have predictive power. When λ is very large, differences in the expected future need to be very small before we are willing to ignore them, whereas at small λ it is more important that our description be compact, so we are willing to make coarser categories. As in rate-distortion theory, there is no single right answer, but rather a curve that defines the maximum amount of predictive information we can capture given that we are willing to write down a certain number of bits about the past. Along this curve is a one-parameter family of strategies for mapping our observations on the past into some internal representation X_{int}.

Problem 153: Predictive information and optimal filtering.   Imagine that we observe a Gaussian stochastic process $x(t)$ that consists of a correlated signal $s(t)$ in a background of white noise $\eta(t)$, that is, $x(t) = s(t) + \eta(t)$, where

$$\langle s(t)s(t') \rangle = \sigma^2 \exp\left(-|t - t'|/\tau_c\right) \tag{6.218}$$

$$\langle \eta(t)\eta(t') \rangle = \mathcal{N}_0 \delta(t - t'). \tag{6.219}$$

As discussed in detail in Appendix A.2, the full probability distribution for the function $x(t)$ is

$$P[x(t)] = \frac{1}{Z} \exp\left[-\frac{1}{2} \int dt \int dt' \, x(t)K(t - t')x(t')\right], \tag{6.220}$$

where Z is a normalization constant.

(a) Construct the kernel $K(\tau)$ explicitly. Be careful about the behavior near $\tau = 0$.

(b) Break the data $x(t)$ into a past $X_{past} \equiv x(t < 0)$ and a future $X_{future} \equiv x(t > 0)$, relative to the time $t = 0$. Show that $P[x(t)]$ can be rewritten so that the only term that mixes past and future is of the form

$$\left[\int_{-\infty}^{0} dt \, g(-t)x(t)\right] \times \left[\int_{0}^{\infty} dt' \, g(t')x(t')\right], \tag{6.221}$$

where $g(t) = \exp(-t/\tau_0)$, with $\tau_0 = \tau_c(1 + \sigma^2\tau_c/\mathcal{N}_0)^{-1/2}$. More formally, if we define

$$z = \int_{-\infty}^{0} dt \, g(-t)x(t), \tag{6.222}$$

show that

$$P(X_{future}|X_{past}) = P(X_{future}|z). \tag{6.223}$$

Explain why the optimal internal representation of the predictive information, X_{int}, can only depend on z.

(c) Suppose you are given the past data $x(t \leq 0)$, and instead of being asked to predict the future, you are asked to make the best estimate of the underlying signal $s(t = 0)$. Show that this optimal estimate is proportional to z. Explore the connection of your results here to those in Problem 37, Section 2.4.

As you just showed in the last problem, the optimal representation of predictive information is equivalent, at least in simple cases, to the separation of signals from noise. In Section 6.5 we will see that extracting the predictive information from other kinds of time series is equivalent to learning the underlying parameters or rules that the data obey. In a somewhat more fanciful example, we can think of X_{past} as a word in a sentence, and X_{future} as the next word; then the mapping $X_{\text{past}} \rightarrow X_{\text{int}}$ is equivalent to making clusters of words. When λ is small, there are very few clusters, and they correspond very closely to parts of speech. As λ becomes larger, we start to discern categories of words that seem to have meaning. Indeed, the first exercise of this sort was to choose not two successive words as past and future, but rather the noun and verb in the same sentence, and then the impression (still subjective) that the resulting clusters of nouns have similar meanings is even stronger, as seen in Fig. 6.20. It is tempting to suggest that the optimal representation of predictive information is extracting meaning from the statistics of sentences.[17]

Let me pull the different arguments of this section together, even if imperfectly. What we really care about is how organisms can maximize some measure of performance—ultimately, their reproductive success—given access to some limited set of resources. In any broad class of possible biological mechanisms, there is an optimum that divides the fitness-resources plane into possible and impossible regions, as in the upper-right quadrant of Fig. 6.21; evolutionary pressure drives organisms toward this boundary. But for any measure of fitness or adaptive value, achieving some criterion level of performance always requires some minimum number of bits; this is the content of rate-distortion theory. Thus, there is a plane (in the upper-left quadrant of Fig. 6.21) of fitness versus information, and again a curve divides the possible from the impossible. Importantly, the information that an organism can use to gain a fitness advantage—even in the simple example of adjusting gene expression levels to match the availability of nutrients—is always predictive information, because the consequences of actions come after they are decided on.

Bits are not free. In simple examples, such as the Gaussian channel in Section 6.2, the information that can be transmitted depends on the signal-to-noise ratio, which in turn depends on the resources the organism can devote to the problem, whether these resources are naturally counted as action potentials or molecules. If we think about the bits that will be used to direct an action, then there are many costs—the cost of acquiring the information, of representing the information, and the more obvious physical costs of carrying out the resulting actions, but we always can assign these costs

---

17. Some people would be horrified by this suggestion. For some discussion of the pros and cons of statistical approaches to language (which is a huge subject!), see the references in the Annotated Bibliography for this section.

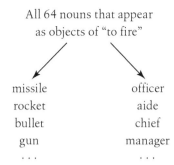

All 64 nouns that appear
as objects of "to fire"

missile   officer
rocket   aide
bullet   chief
gun   manager
. . .   . . .

FIGURE 6.20

A precursor of the information bottleneck problem. In one year of Associated Press news reports, there are 64 nouns (X_{noun}) that appear as the direct object of the verb "to fire," and these nouns also are paired with 2146 other verbs (X_{verb}). Following the ideas in the text, imagine compressing the description of the nouns, $X_{noun} \rightarrow X_{int}$, while trying to preserve the information that the compressed description conveys about the verb that appears with the noun. That is, maximize $I(X_{int}; X_{verb})$ while holding $I(X_{int}; X_{noun})$ fixed. The solution to the problem is shown when $I(X_{int}; X_{noun}) \approx 1$ bit supports two distinct values of X_{int}; the nouns that map to the two values of X_{int} with high probability are listed. This classifies the nouns by their meaning, separating weapons (firing a missile) from job titles (firing a manager). Importantly, this classification is based only on the co-occurrence of the nouns with verbs in sentences; there is no supervisory signal that distinguishes the different senses of the verb. From Pereira et al. (1993).

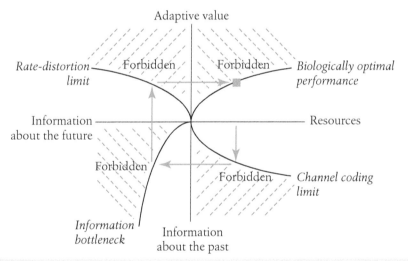

FIGURE 6.21

Connecting the different optimization principles. Lines indicate curves of optimal performance, separating allowed from forbidden (hashed) regions of each quadrant. In the upper-right quadrant is the biologically relevant notion of optimization, maximizing fitness or adaptive value at fixed resources. But actions that achieve a given level of adaptive value require a minimum number of bits, and because actions occur after plans, these are bits about the future (upper left). In contrast, the organism has to "pay" for bits, and hence there are minimum resource costs for any representation of information (lower right). Finally, given some bits (necessarily obtained from observations on the past), there is some maximum number of bits of predictive power (lower left). To find a point on the biological optimum, one can try to follow a path through the other three quadrants, as indicated by the arrows. Redrawn from Bialek et al. (2007).

to the symbols at the entrance to the communication channel. The channel capacity separates the information-resources plane into accessible and inaccessible regions, as in the lower-right quadrant of Fig. 6.21. Ideas about metabolically efficient neural codes, for example, can be seen as efforts to calculate this curve in specific models. The information we are talking about now is information that we actually collect: information about the past. To close the connections among the different quantities, we need the information bottleneck idea, which tells us that—given the structure of the world we live in—having a certain number of bits of information about the future requires capturing some minimum number of bits about the past.

To summarize, if an organism wants to achieve a certain mean fitness, it needs a minimum number of bits of predictive power, which requires collecting a minimum number of bits about the past; this in turn necessitates some minimum cost or available resources. Usually we think of evolution as operating in the trade-off between resources and fitness, but this balance has echoes in the other quadrants of Fig. 6.21, where information-theoretic bounds are at work. These connections provide a path whereby evolution can select for mechanisms that approach these bounds, even though evolution itself does not know about bits.

6.4   Optimizing Information Flow

We have seen that organisms should care about bits—for every criterion level of performance that a system wants to achieve, there is a minimum number of bits that it needs. If bits are cheap, or easy to acquire, then this need for a minimum number of bits is true but not much of a constraint. However, if the physical constraints under which organisms operate imply severe limits on information transmission, then the minimum number of bits may approach the maximum number available, and strategies that maximize efficiency in this sense may be critical to biological function.

One of the central ideas in thinking about the efficiency with which bits can be collected and transmitted is that what we mean by efficient (and, in the extreme, optimal) depends on context, as indicated schematically in Fig. 6.22. The top panel shows a typical sigmoidal input-output relation, which might describe the expression level of a gene versus the concentration of a transcription factor, the probability of spiking in a neuron as a function of the intensity of the sensory stimulus, or the like. In the bottom panel we see different possibilities for the distributions out of which the input signals might be drawn. For the two distributions in orange, the input signals are confined to the saturated regions of the input-output relation, leaving the output almost always in the fully "off" or "on" state. In these situations, the output is always the same and is unaffected by the changes in the input that actually occur with reasonable probability, and the system is essentially useless. More subtly, for the distribution in blue, input signals are in the middle of the input-output relation, where its slope is maximal, but the dynamic range of these variations is small, so that the variations in output are only a small fraction of what is possible, and these variations might well be obscured by any reasonable level of noise. Finally, for the distribution in green, the dynamic range of the likely inputs is just big enough to push the system through the full dynamic range of the input-output relation, generating large (maximal?) variations in the output.

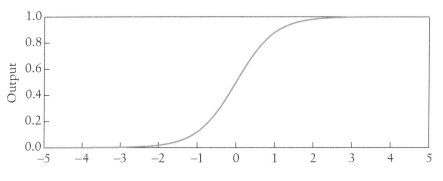

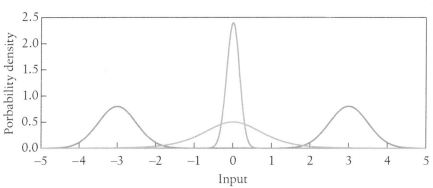

FIGURE 6.22

At top, an example of an input-output relation. At bottom, different possible probability distributions for the inputs. As described in the text, the orange and blue distributions are poorly matched to the input-output relation, whereas the green distribution seems to be a better match.

It is easy for everyone to agree that, in Fig. 6.22, the blue and orange distributions of inputs are poorly matched to the input-output relation, and the green distribution is well matched, but it takes a little more courage (and courts more controversy) to make a precise mathematical statement about what constitutes a good match, or the best match. What we will try out as a definition of "best" is that outputs should provide as much information as possible about the inputs.

Let's start with input x, chosen from a distribution $P_X(x)$, and assume that this is converted into one output y by a system that has an input-output relation $g(x)$ but also some added noise:

$$y = g(x) + \xi. \tag{6.224}$$

When we plot an input-output relation, as in Fig. 6.22, we (implicitly) are referring to the *average* behavior of the system, because realistically there must be some level of noise, and hence the input and output are related only probabilistically. We now make this assumption explicit by adding the noise ξ. To keep things simple, we assume that this noise is Gaussian, with some variance σ^2 and as usual zero mean. In principle, the variance of the output noise could depend on the value of the input. This dependence will be important below, so we'll write $\sigma_y^2(x)$ as a reminder that we are talking about the variance of the output (hence the subscript), but it may depend upon the input.

To compute the amount of information that y provides about x, we need various probability distributions. Specifically, we want to evaluate

$$I(y;x) = \int dx \int dy\, P(x,y) \log_2 \left[\frac{P(x,y)}{P_X(x)P_Y(y)} \right] \tag{6.225}$$

$$= \int dx \int dy\, P(x,y) \log_2 \left[\frac{P(y|x)}{P_Y(y)} \right]. \tag{6.226}$$

It is the conditional distribution $P(y|x)$ that describes, in the most general setting, the probabilistic relationship between input and output. The overall distribution of outputs is given by

$$P_Y(y) = \int dx\, P(y|x)P_X(x). \tag{6.227}$$

With the hypothesis that the noise ξ is Gaussian, Eq. (6.224) tells us that

$$P(y|x) = \frac{1}{\sqrt{2\pi\sigma_y^2(x)}} \exp\left[-\frac{(y-g(x))^2}{2\sigma_y^2(x)} \right]. \tag{6.228}$$

The information can be written (as usual) as the difference between two entropies:

$$\begin{aligned}
I(y;x) &= \int dx \int dy\, P(x,y) \log_2 \left[\frac{P(y|x)}{P_Y(y)} \right] \\
&= -\int dy\, P_Y(y) \log_2 P_Y(y) \\
&\quad - \int dx\, P_X(x) \left[-\int dy\, P(y|x) \log_2 P(y|x) \right].
\end{aligned} \tag{6.229}$$

But the conditional distribution $P(y|x)$ is Gaussian, with variance $\sigma_y^2(x)$, so we can substitute for the conditional entropy from Eq. (6.91) to give

$$I(y;x) = -\int dy\, P_Y(y) \log_2 P_Y(y) - \frac{1}{2\ln 2}\int dx\, P_X(x) \ln[2\pi e\sigma_y^2(x)]. \tag{6.230}$$

The distribution of outputs $P_Y(y)$ is broadened by two effects. First, as x varies, the mean value of y changes. Second, even with x fixed, noise causes variations in y. But if the noise is small, the first effect should dominate, which will simplify our problem:

$$P_Y(y) = \int dx\, P_X(x)P(y|x) \tag{6.231}$$

$$= \int dx\, P_X(x) \frac{1}{\sqrt{2\pi\sigma_y^2(x)}} \exp\left[-\frac{(y-g(x))^2}{2\sigma_y^2(x)} \right] \tag{6.232}$$

$$= \int dz \left| \frac{dz}{dx} \right|^{-1} P_X(x=g^{-1}(z)) \frac{1}{\sqrt{2\pi\sigma_y^2(z)}} \exp\left[-\frac{(y-z)^2}{2\sigma_y^2(z)} \right], \tag{6.233}$$

where we have changed variables to $z = g(x)$, which is allowed if the input-output relation is monotonic. But now we can view the integral as an average over a distribution of z, and if the noise is small, we can always write

$$\int dz \, F(z) \frac{1}{\sqrt{2\pi\sigma_y^2(z)}} \exp\left[-\frac{(y-z)^2}{2\sigma_y^2(z)}\right] \approx F(z=y) + \frac{1}{2}\sigma_y^2(z=y) \left. \frac{d^2 F(z)}{dz^2}\right|_{z=y} + \cdots,$$

$$(6.234)$$

for any function $F(z)$. Keeping just the leading term, at small noise levels we have

$$P_Y(y) \approx \left[\left|\frac{dz}{dx}\right|^{-1} P_X(x = g^{-1}(z))\right]_{z=y}.$$

$$(6.235)$$

This expression looks complicated, but it's not. In fact it is the same as ignoring the noise altogether and saying that there is some deterministic transformation from x to y, $y = g(x)$, in which case we must have

$$P_X(x)dx = P_Y(y) \, dy.$$

$$(6.236)$$

By the same reasoning, we can also view the variance $\sigma_y^2(x)$ as being a function not of the input x but rather of the output y, so we'll write $\sigma_y^2(y)$.

Problem 154: Details of the small noise approximation, part one.   Show that Eq. (6.236) really is the same as Eq. (6.235).

In the small noise approximation, then, the mutual information between x and y thus can be written as

$$I(y;x) \approx -\int dy \, P_Y(y) \log_2 P_Y(y) - \frac{1}{2\ln 2} \int dy \, P_Y(y) \ln[2\pi e\sigma_y^2(y)]. \quad (6.237)$$

This expression invites us to think of $P_Y(y)$ as the "free" distribution that we can vary as we try to optimize the information transmission. We started with the problem of varying the distribution of inputs, but now things are formulated in terms of the distribution of outputs; Eq. (6.236) tells us that these are equivalent in the low noise limit. To do the optimization correctly, we have to add a Lagrange multiplier that fixes the normalization of the distribution. Thus, we are interested in the functional

$$\tilde{I} \equiv I(y;x) - \mu \int dy \, P_Y(y).$$

$$(6.238)$$

As usual, to optimize we set the derivative equal to zero:

$$0 = \frac{\delta \tilde{I}}{\delta P_Y(y)}\bigg|_{P_Y(y)=P_{\text{opt}}(y)} = -\frac{1}{\ln 2}\left[\ln P_{\text{opt}}(y) + 1\right]$$

$$-\frac{1}{2\ln 2}\ln[2\pi e\sigma_y^2(y)] - \mu, \tag{6.239}$$

$$\ln P_{\text{opt}}(y) = -\frac{1}{2}\ln[2\pi e\sigma_y^2(y)] - (1 + \mu\ln 2), \tag{6.240}$$

$$P_{\text{opt}}(y) = \frac{1}{\sqrt{2\pi e\sigma_y^2(y)}}e^{-(1+\mu\ln 2)}. \tag{6.241}$$

We can write this result more simply by gathering together the various constants:

$$P_{\text{opt}}(y) = \frac{1}{Z}\frac{1}{\sigma_y}, \tag{6.242}$$

where Z must be chosen so that the distribution is normalized:

$$Z = \int \frac{dy}{\sigma_y}. \tag{6.243}$$

With this result for the optimal distribution, the mutual information is

$$I_{\text{opt}} = \log_2\left[\frac{Z}{\sqrt{2\pi e}}\right]. \tag{6.244}$$

Problem 155: Extrema of the mutual information.   Once again we need to check that we have found an optimum, rather than some other type of extremum, in the dependence of the mutual information on the distribution of outputs, Eq. (6.237). You can do this explicitly by computing second (functional) derivatives or by appealing to general convexity properties of the entropy. Notice that our ability to write the information so simply as a functional of the output distribution alone is a feature of the low noise approximation. More generally, we should view the mutual information as a functional of the input distribution $P_X(x)$ and the conditional distribution(s) $P(y|x)$. Show that, in this more general setting, once $P(y|x)$ is known, the mutual information has a well-defined maximum as a functional of $P_X(x)$.

Problem 156: Details of the small noise approximation, part two.   Carry out the small noise approximation to the next leading order in the noise level σ_y^2. Step by step, you should find $P_Y(y)$ and then an expression for the information $I(y;x)$. What can you say about the problem of optimizing $I(y;x)$ in this case?

The result for the optimal distribution of outputs, Eq. (6.242), is telling us something sensible: we should use the different outputs y in inverse proportion to how noisy they are. Suppose, however, that the noise level is constant. Then what we find is that the distribution of outputs should be uniform. How can the system do this? Recall that in the low noise limit, the relationship between input and output is nearly deterministic, so we have Eq. (6.236): $P_Y(y)dy = P_X(x)dx$. But we also have that $y = g(x)$ in this approximation. If $P_Y(y)$ is uniform, then we have

$$P_Y(y) = \frac{1}{y_{max} - y_{min}}, \tag{6.245}$$

and hence

$$\frac{dy}{dx} = \frac{dg(x)}{dx} = (y_{max} - y_{min})P_X(x), \tag{6.246}$$

$$g(x) = (y_{max} - y_{min})\int_{x_{min}}^{x} dx'\, P_X(x'). \tag{6.247}$$

Thus, in this simple limit, the optimal input-output relation is proportional to the cumulative probability distribution of the input signals.

Problem 157: How general is Eq. (6.247)?   We have derived Eq. (6.247) by assuming that the noise is additive, Gaussian, small, and has a variance that is constant across the range of inputs or outputs. Show that you can relax the assumption of Gaussianity (while keeping the noise small, additive, and independent of inputs) and still obtain the same result for the optimal input-output relation.

Equation (6.247) makes clear that any theory involving optimizing information transmission or efficiency of representation inevitably predicts that the input-output relation must be matched to the statistics of the inputs. Here the matching is simple: in the right units, we could just read off the distribution of inputs by looking at the (differentiated) input-output relation. Although this is obviously an over-simplified problem, it is tempting to test the predictions, which is exactly what Laughlin did in the context of the fly's visual system.

Laughlin built an electronic photodetector with aperture and spectral sensitivity matched to those of the fly retina and used it to scan natural scenes, sampling the distribution of input light intensities $P(\mathcal{I})$ as it would appear at the input to the photoreceptors. In parallel he characterized the second-order neurons of the fly visual system—the large monopolar cells that receive direct synaptic input from the photo-receptors, as seen in Sections 2.1 and 6.1—by measuring the peak voltage response to flashes of light. The agreement with Eq. (6.247) was remarkable, as shown in Fig. 6.23, especially considering that there are no free parameters. Although there are obvious open questions, this is a really beautiful result that inspires us to take these ideas more seriously.

This simple model automatically carries some predictions about adaptation to over-all light levels. If we live in a world with diffuse light sources that are not directly visible, then the intensity that reaches us at some point is the product of the effective brightness of the source and some local reflectances. As it gets dark outside the reflectances do not change—these are material properties—and so we expect that the distribution $P(\mathcal{I})$ will look the same except for scaling. Equivalently, if we view the input as the loga-rithm of the intensity, then to a good approximation $P(\log \mathcal{I})$ just shifts linearly along the $\log \mathcal{I}$ axis as mean light intensity goes up and down. But then the optimal input-

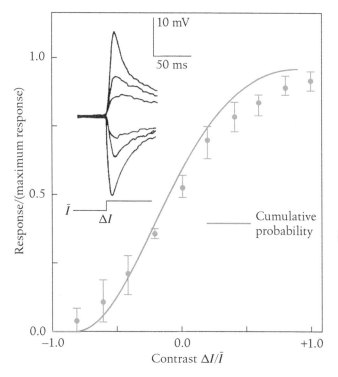

FIGURE 6.23

Input-output relations of large monopolar cells compared with the prediction of Eq. (6.247). Brief changes in light intensity relative to a mean background produce transient voltage changes in the large monopolar cells (inset), and the peaks of these responses are taken as the cell's output. Normalized responses are compared to the cumulative probability distribution of light intensities, as described in the text. Redrawn from Laughlin (1981).

output relation $g(\mathcal{I})$ would exhibit a similar invariant shape, with shifts along the input axis when expressed as a function of $\log \mathcal{I}$. This behavior is in rough agreement with experiments on light/dark adaptation in a wide variety of visual neurons.

As I have emphasized before, the problems of signals, noise, and information flow in the nervous system have analogs in the biochemical and genetic machinery of single cells. For the simple problem of one input and one output, we can move beyond analogy and actually use the same equations to describe these very different biological systems.

Suppose that we have a single transcription factor that controls the expression of one target gene. Now we can think of the input x as the concentration of the transcription factor and the output y as the expression level of the gene. As in Laughlin's discussion of the fly retina, we are (perhaps dangerously) ignoring dynamics. In the context of gene regulation this model probably is best seen as a quasi-steady state approximation, in which the changes in transcription factor concentration are either slow or infrequent, so that the resulting gene expression level has a chance to find its appropriate steady level in response.

We have discussed the problems of noise in the control of gene expression in Section 4.3, and a crucial feature of that discussion is that the noise levels cannot be constant. In the simplest case, we are counting molecules, and counting zero molecules allows for no variance, whereas counting the maximum number of molecules leaves lots of room for variation. For the problem at hand, this means—because of Eq. (6.242)— that the distribution of outputs maximizing information transmission can't be uniform.

In Section 4.3 we identified (at least) three noise sources in the regulation of gene expression. One term is the shot noise in the synthesis and degradation of the mRNA or protein (output noise). The second is the randomness in the arrival of transcription factor molecules at their target site (input noise), and the third is from the kinetics of the switching events that occur on binding of the transcription factors. We have argued that cells can reduce the impact of this last term by proper choice of parameters, leaving two fundamental sources of noise. The shot noise generates a variance at the output proportional to the mean, whereas the random arrivals are equivalent to a fluctuation in input concentration $(\delta c/c)^2 \propto 1/c$. Putting these sources together we have (from the discussion leading to Eq. (4.134)) the variance in the expression level:

$$\sigma_g^2(c) = \alpha \bar{g}(c) + \frac{B}{c} \cdot \left| \frac{d\bar{g}(c)}{d\ln c} \right|^2, \tag{6.248}$$

where α and B are constants, and $\bar{g}(c)$ is the mean expression level as a function of the input transcription factor concentration c. As usual we normalize the measurements of expression levels so that the maximum $\bar{g}(c) = 1$. Finally, if we can assume that the input-output relation is well approximated by a Hill function,

$$\bar{g}(c) = \frac{c^n}{c^n + K^n}, \tag{6.249}$$

then we can write the variance as a function of the mean, as in Eq. (4.134):

$$\sigma_g^2(\bar{g}) = \alpha \bar{g} + \beta \bar{g}^{2-1/n}(1 - \bar{g})^{2+1/n}. \tag{6.250}$$

The parameter $A = \beta/\alpha$ measures the relative importance of input and output noise; large A means that the input noise is dominant near the midpoint of the input-output relation.

Figure 6.24 shows the results for the optimal distributions of expression levels, derived using the general result of Eq. (6.242) with the noise variance from Eq. (6.250). We hold the cooperativity fixed ($n = 5$) and consider what happens as we change the relative importance of the input and output noise (A). As long as output noise is dominant, $A < 1$, the optimal distribution is monotonically decreasing. If we take the results seriously, the distribution has a singularity as we approach zero expression level. There is no physical reason why this can't happen, but we also can't trust our calculation here, because at some point the noise $\sigma \propto \bar{g}^{1/2}$ will become larger than the mean as $\bar{g} \to 0$. Nonetheless, it is clear that when output noise is dominant, the optimal distribution of expression levels is relatively featureless, biased toward low expression levels. The strength of this bias is considerable, so that the probability of having more than half-maximal activation, $\int_{1/2}^1 dg\, P(g)$, is a bit less than 30%.

At larger values of A, where input noise is more important, the optimal distribution of expression levels becomes bimodal. This behavior is especially interesting, because extreme bimodality corresponds to a simple on/off switch. True switchlike behavior runs counter to the idea that information transmission is being maximized: we might expect that maximizing information transmission involves making extensive use of intermediate expression levels, whereas building a reliable switch means exactly the opposite—avoiding intermediate levels. In fact, few of the classic examples of genetic switches are perfect on/off switches; maximizing information transmission can lead to relatively low probabilities of occupying intermediate levels, just depending on the

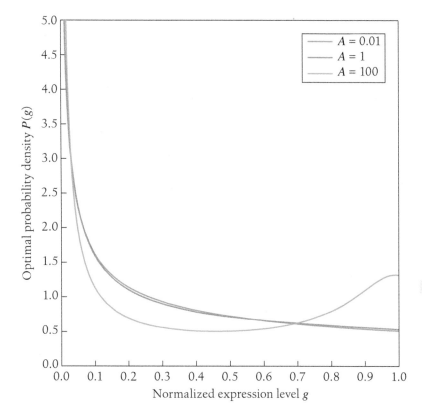

FIGURE 6.24

Optimal distributions of (output) gene expression levels. The transmission of information is maximized from a single transcription factor to a single target gene. Different curves correspond to different relative contributions from input and output noise (A), as in Eq. (6.250).

structure of the noise in the system. Thus, what looks like an imperfect switch may be optimizing information flow.

We can bring this theoretical discussion down to earth by considering a real system. As discussed in Section 4.3, measurements have been made on the input-output relation and noise level for the control of the *hunchback* gene by the transcription factor Bicoid (Bcd) in the early *Drosophila* embryo. If we take the formalism above seriously, we can use these measurements to predict, with no free parameters, the distribution of Hb expression levels, which can also be extracted from the experiments. To do this correctly, we should go beyond the small noise approximation and solve the full optimization problem numerically; the results are shown in Fig. 6.25.

Figure 6.25 is the direct analog of Laughlin's result for the fly retina. As in that case, the agreement of theory and experiment is excellent, and again it should be emphasized that there are no free parameters—these are not models we are fitting to data, but quantitative predictions from theory. Furthermore, we can show from the data that the actual amount of information[18] being transmitted from Bcd to Hb is 0.88 ± 0.09 of the limit set by the measured noise levels. Thus, we can see directly that the system is operating near its optimum. This optimum corresponds to significantly more than one

18. This is a good place to remember the technical difficulties involved in estimating information from finite samples of data. See Appendix A.8.

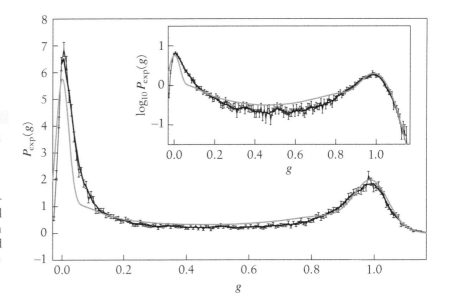

FIGURE 6.25

Distributions of Hunchback expression levels in the early fruit fly *Drosophila* embryo. In orange, the distribution predicted by optimizing information transmission given the measured input-output relation and noise in the control of Hunchback by Bicoid. In black, with error bars, the distribution as measured experimentally; black line connects the data points to guide the eye. Redrawn from Tkačik et al. (2008a).

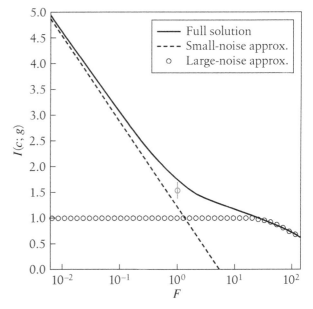

FIGURE 6.26

Changes in information transmission from Bicoid to Hunchback if all noise variances are scaled by a factor F. This is equivalent to scaling, by a factor $1/F$, the numbers of all relevant molecules—proteins, mRNAs, and so forth. At each noise level we compute the maximum information transmission. Limiting behaviors in the small and large noise approximations are shown for reference. The real system (in orange, with error bar) is in an intermediate regime, although close to the small noise limit. Redrawn from Tkačik et al. (2008a).

bit, which means that intermediate expression levels, beyond an on/off switch, are being used reliably. Finally, because we understand how the absolute numbers of molecules influence the noise level in the system (see, again, Section 4.3), we can compute that more bits would be very expensive—doubling the information would require 20 times as many molecules, as shown in Fig. 6.26.

The orders of magnitude here are important and are evident from much rougher calculations. Suppose that the only source of noise is shot noise in the synthesis and degradation of molecules, so that the variance in gene expression level is proportional to the mean: $\sigma_g^2 = \alpha \bar{g}$. But then from Eq. (6.243) and Eq. (6.244) we can calculate the maximum amount of information that this gene can carry:

$$I_{\text{opt}} = \log_2 \left[\frac{Z}{\sqrt{2\pi e}} \right], \tag{6.251}$$

$$Z = \int_0^1 \frac{dg}{\sigma_g} = \int_0^1 \frac{dg}{\sqrt{\alpha g}} = \frac{2}{\sqrt{\alpha}}, \tag{6.252}$$

$$\Rightarrow I_{\text{opt}} = \frac{1}{2} \log_2 \left(\frac{2}{\alpha \pi e} \right) \text{ bits}, \tag{6.253}$$

where again we have adopted the convention that mean expression levels range between zero and one. Surveying results on noise in gene expression (see the Annotated Bibliography for Section 4.3), we see that a precision of $\sim 10\%$ at maximal expression is typical, which presumably corresponds to the synthesis of ~ 100 independent molecules. But then $\alpha \sim 0.01$, and we find $I_{\text{opt}} \sim 2.3$ bits. Doubling this information capacity would require reducing α by a factor of nearly 25, which means increasing the number of independent molecules by the same factor.

Problem 158: Limits from input noise. Consider a system for controlling gene expression that is limited entirely by noise from the random arrival of transcription factors at their targets. Then from Berg and Purcell (Section 4.2) we know there is an equivalent input noise $\sigma_c \sim \sqrt{c/Da\tau}$, where D is the diffusion constant, a is the target size, and τ is the integration time. Show that, by analogy with Eq. (6.251) and Eq. (6.252), you can write the information capacity of the system as

$$I_{\text{opt}} = \log_2 \left[\frac{Z}{\sqrt{2\pi e}} \right], \tag{6.254}$$

$$Z = \int_0^{c_{\max}} \frac{dc}{\sigma_c}, \tag{6.255}$$

where $c_{\max}$ is the maximum concentration of input molecules. Notice that $N_{\text{in}} = c_{\max} V$ is the maximum number of (free) input molecules, where V is the relevant volume—the entire cell, in the case of a bacterium or the nucleus in the case of a eukaryote. In contrast, in our discussion above, $\alpha \sim 1/N_{\text{out}}$, with N_{out} the maximum number of independent output molecules. Make clear that both input and output noise set a limit on information transmission that has the same dependence on the maximum number of available molecules. Are there parameter choices that cells can make that cause these limits to be quantitatively similar? In particular, if the regulatory element is embedded in a network such that the output molecules are also transcription factors, it might make sense that increasing the numbers of input versus output molecules would make comparable contributions to information flow. Is this reasonable, given what you know about the parameters? How could you incorporate the fact that the "independent" molecules are probably the mRNAs, whereas the active input molecules are proteins?

The conclusion from these calculations is that bits are genuinely expensive for cells. It is possible to do more than turn genes on and off, pushing 2–3 bits of information through a single genetic regulatory element, but significantly more than this would require orders-of-magnitude larger investments in making the relevant molecules. This strongly suggests that, to the extent that cells need information—and we know from the arguments in the preceding section that they do—there will be pressure to squeeze as much information as possible out of a limited number of molecules. Because the dependence of the information capacity on the number of available molecules is at best logarithmic, even if the cell can afford twice as many molecules, it would be better to make two independent regulatory elements, potentially doubling the information rather just doubling the argument of the logarithm in such expressions as Eq. (6.253). The idea here is the same as in Problem 139, where you explored the trade-off between spending more time to increase the signal-to-noise ratio in one measurement and using the same time to look at multiple measurements of independent quantities in the world. In both cases, optimization in the presence of sensible resource constraints (time or number of molecules) would drive the system to find multiple independent paths for information flow, each of which functions at modest capacity. Although speculative, it is tempting to point out that, in the context of genetic regulatory circuits, such optimization amounts to a pressure for increased complexity. Importantly, this pressure is coming purely from physical considerations—trying to transmit more information at fixed cost.

Problem 159: Information flow through calcium-binding proteins.   Many biological processes are regulated by calcium. Typically the regulatory process begins with calcium binding to a protein. In almost all cases, there are multiple binding sites, and these sites interact cooperatively. We'd like to understand something about the signals, noise, and information flow in such regulatory systems. Not much has been done in this area, so this is a deliberately open-ended problem. Consider the relatively simple model shown in Fig. 6.27. This is a dimeric protein with four states, corresponding to empty and filled Ca^{++} binding sites on each of the two monomers. The sites interact, because the rate of unbinding from one site depends on the occupancy of the other site.

(a) Calculate the equilibrium probability of occupying each of the states in Fig. 6.27. Use these results to plot the fraction of occupied binding sites as a function of the calcium concentration c. You should be able to choose units that eliminate all parameters except for the dimensionless constant F. Show that for $F = 1$, your results are equivalent to having two independent binding sites, and the fraction of occupied sites becomes more strongly sigmoidal

FIGURE 6.27

Model of calcium binding to a dimeric protein. The rate at which calcium binds to each site, k_+, is assumed to be the same and independent of the occupancy of the other site. The unbinding rates, however, are different, depending on whether the other site is empty (k_-) or filled (k_-/F).

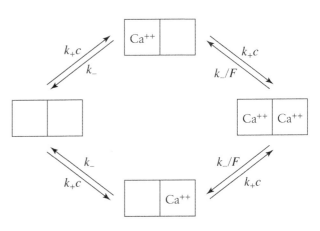

or switchlike as F becomes larger. Cooperativity means, in this context, that the free-energy change upon binding of a calcium ion to one site is increased by occupancy of the other site. Relate the parameter F to this free-energy difference or interaction energy. Can interaction energies of just a few times $k_B T$ make a difference in the shape of the plot of occupancy versus concentration?

(b) Suppose that we have N copies of this protein in the cell, all experiencing the same calcium concentration. Let the number of molecules with no bound calcium be n_0, the number with one bound calcium be n_1, and the number with two bound calcium be n_2; of course $\sum_j n_j = N$. Use your results from part (a) to calculate the mean values of each n_j and the covariance matrix $C_{jk} = \langle \delta n_j \delta n_k \rangle$. Verify that the determinant of the covariance matrix is zero in this formulation. Why is this true? Notice that we are only asking here about the fluctuations that would be seen in a single snapshot of the molecules, not about the dynamics or spectrum of this noise.

(c) It is widely assumed that in systems such as this, only the state with full occupancy of the binding sites is really active. In practice, then, the calcium-binding protein is associated with some other protein, such as an enzyme, and the enzyme becomes active only when both Ca^{++} ions are bound. Thus, the output of the system is something proportional to n_2. Calculate the change in the mean $\langle n_2 \rangle$ that results from a small change in calcium concentration $c \to c + \delta c$. Compare this result with the variance $\langle (\delta n_2)^2 \rangle$ to compute a signal-to-noise ratio, or the equivalent noise level δc_{rms} in the calcium concentration itself. Plot your results. Again, you should be able to put everything into unitless form, leaving only the parameter F. Does making the system more switchlike by increasing F make it more sensitive to small changes in concentration, as you might expect? Are there competing effects that could result in better performance at smaller F?

(d) Suppose that molecules with one bound calcium also are active. Then the output activity of the system is proportional to some mixture of n_1 and n_2, which we can write as $A = (1 - a)n_2 + an_1$; note that $a = 0$ is the case where only doubly bound states are active. Compute the sensitivity of the mean activity, $\partial \langle A \rangle / \partial c$, and the variance $\langle (\delta A)^2 \rangle$. If the system is operating at a particular calcium concentration c, can you lower the effective noise level

$$\sigma_c \equiv \sqrt{\langle (\delta A)^2 \rangle} \left| \frac{\partial \langle A \rangle}{\partial c} \right|^{-1} \tag{6.256}$$

by choosing $a \neq 0$? Can you lower the noise level at all calcium concentrations using the same value of a, or are there trade-offs?

(e) Use Eq. (6.254) and Eq. (6.255), together with your results for σ_c from part (d), to calculate the limit on information transmission through this system. Does thinking about the information transmitted, rather than just the noise level, help you to decide whether there is a uniquely best mixture of activity from the singly and doubly bound states? What is the impact of the cooperativity (here captured by the parameter F) on the information transmission?

(f) Your results from parts (a)–(e) suggest that, at least under some conditions, it would be useful if the system "reads out" some combination of the singly and doubly occupied states. Can you find hints in the literature of the predicted partial activation? For concreteness, focus on the case of calmodulin. Our discussion in this problem is for snapshots of the molecules, so "noise" just means the total variance. Suppose that the readout scheme effectively averages over a time longer than the times required for transitions among the different states. Then you need to compute the spectral density of the noise and follow the path discussed in the context of bacterial chemotaxis (Section 4.2 and Appendix A.6). Is there anything qualitatively new here, or just a change in details?

What we have been doing in the previous paragraphs is exploring the consequences of the general principle that Laughlin used (in a special case) 30 years ago—input-output relations should be matched to the distribution of inputs so as to maximize information transmission. One of many questions left open in Laughlin's original discussion is the time scale on which this matching should occur. If there is a well-defined distribution of input signals, stable on very long time scales, matching could occur through evolution. Another possibility is that the distribution is learned during the lifetime of the individual organism, but still learned only once. Finally, one could think about mechanisms of adaptation that would allow these systems to adjust their input-output relations in real time, tracking changes in the input distribution. Although I have emphasized that the questions of information flow and noise are the same in very different biological contexts, from the regulation of gene expression to the representation of sensory signals in the brain, it certainly is possible that the answers are different in these different cases. But the last possibility, real-time tracking of the input distribution, is interesting, because it opens the possibility for new experimental tests. These tests have actually been done in the context of neural coding.

We know that some level of real-time matching occurs, as in the example of light and dark adaptation in the visual system. We can think of this as neurons adjusting their input-output relations to match the mean of the input distribution. The real question, then, is whether there is adaptation to the distribution or just to the mean. Actually, there is also a question about the world we live in, which is whether there are other features of the distribution that change slowly enough to be worth tracking in this sense.

As an example, we know that many signals that reach our sensory systems come from distributions with long tails (for an example, see Fig. 6.28). In some cases (e.g., in olfaction, where the signal—odorant concentration—is a passive tracer of a turbulent flow) there are clear physical reasons for these tails, and indeed it has been an important theoretical physics problem to understand this behavior quantitatively. In most cases, the tails arise through some form of intermittency. Thus, we can think of the distribution of signals as being approximately Gaussian, but the variance of this Gaussian itself fluctuates; samples from the tail of the distribution arise in places where the variance is large. If this explanation is correct, then we should be able to tame the tails of the distribution by normalizing signals to their local variance, as shown in Fig. 6.29 for the sounds in Fig. 6.28. This scenario also holds for images of the natural world, so that there are regions of high variance and regions of low variance. The possibility of such variance normalization in images suggests that the visual system could code more efficiently by adapting to the local variance, in addition to the local mean (light and dark adaptation).

Problem 160: Growing tails.   Consider a random variable x chosen from a Gaussian distribution with variance v, but v itself fluctuates, so that

$$P_X(x) = \int_0^\infty dv \, P_V(v) \frac{1}{\sqrt{2\pi v}} e^{-x^2/2v}. \tag{6.257}$$

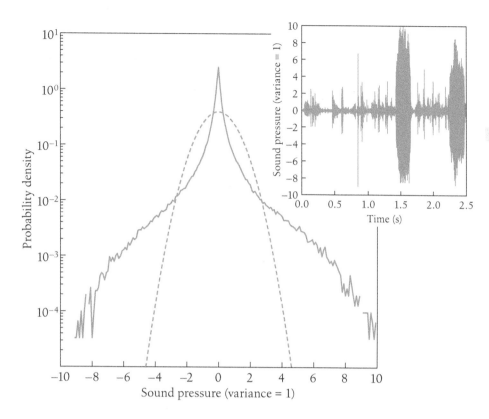

FIGURE 6.28

Intermittency in a natural sound ensemble. Sounds were recorded in a colony of zebra finches. The inset shows a sample of the sound pressure versus time, and the main figure shows the probability distribution of the instantaneous sound pressure. Note the alternating loud and soft periods, characteristic of natural sounds. These periods are reflected in the long tails of the distribution, which is far from Gaussian (dashed). From experiments by G. Kim, B. D. Wright, and A. J. Doupe, with my thanks.

In particular, suppose that the distribution of variances is $P_V(v) \propto \exp[-(v/v_0)^n]$. Derive an approximate expression for $P_X(x)$ at large x, and show that these tails of the distribution indeed are longer than Gaussian, as expected intuitively. What form is required for $P_V(v)$ so that the tails are exponential, $P_X(x) \sim \exp(-a|x|)$, for large $|x|$?

Adaptation to local variance, or more generally adaptation to input statistics beyond the mean, definitely happens at many stages of neural processing; two examples are shown in Fig. 6.30. The earliest experiments looked at the responses of retinal ganglion cells to sudden changes in the variance of their inputs and showed that there is a pattern similar to what one sees with sudden changes in mean. More ambitious experiments on the motion-sensitive neurons in the fly visual system mapped the input-output relation when inputs were drawn from different distributions. These experiments found that the input-output relation scales in proportion to the dynamic range of inputs, which is what we expect from the matching principle if noise levels are small; it was also checked that the precise proportionality constant in the scaling relation served to maximize information transmission. Further, by suddenly switching from one distribution to another, you can catch the system using the wrong code and transmitting less information, but the adaptation to the new distribution is fast, close to the limit set by the need to collect enough samples to verify there was a change. Related observations have been made for many systems, from low-level sensory neurons up to mammalian cortex, as described in the references in the Annotated Bibliography for this section. We can even see such effects in the response of single neurons to injected currents, suggesting

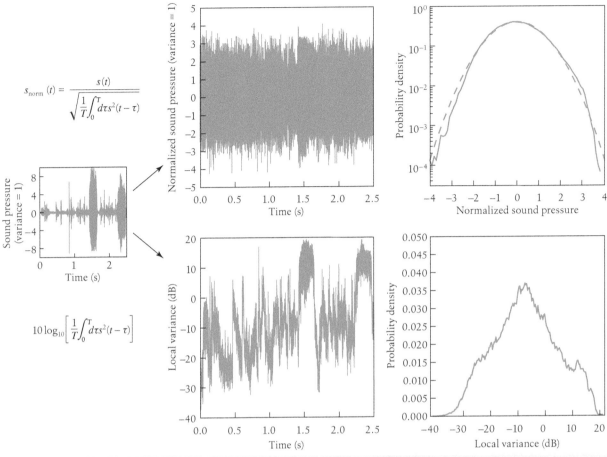

$$s_{norm}(t) = \frac{s(t)}{\sqrt{\frac{1}{T}\int_0^T d\tau s^2(t-\tau)}}$$

$$10\log_{10}\left[\frac{1}{T}\int_0^T d\tau s^2(t-\tau)\right]$$

FIGURE 6.29

Variance normalization of natural sounds. At each moment in time we can look at the sound pressure signal $s(t)$ and compute a local variance or power $\sigma^2(t) = (1/T)\int_0^T d\tau s^2(t-\tau)$. The upper panels show what happens when we construct the normalized signal $s_{norm}(t) = s(t)/\sigma(t)$; the distribution of s_{norm} becomes accurately Gaussian with proper choice of $T = 0.55$ ms; this form is in contrast to the strongly non-Gaussian distribution of s itself in Fig. 6.28. The local variance varies wildly, as shown in the lower panels, and is best expressed on a logarithmic or decibel scale.

that normalization can be a building block of neural computation. One issue in all this work is that some observed changes in apparent input-output relation are so rapid that it seems strained to describe them as adaptation, but the functional behavior is the same.

I think the adaptation experiments are important, because they give a whole new way of testing the ideas about matching between the input-output relation and the distribution of inputs—by changing the input distribution, if you believe the theory, we should drive changes in the input-output relation. It seems that this works. Can we imagine a similar experiment in the genetic or biochemical systems? In truth, there are few cases (aside from embryonic development) where we have quantitative measurements on the distributions of inputs under moderately natural conditions. If we change the distribution, then for the case of gene regulation it seems that input-

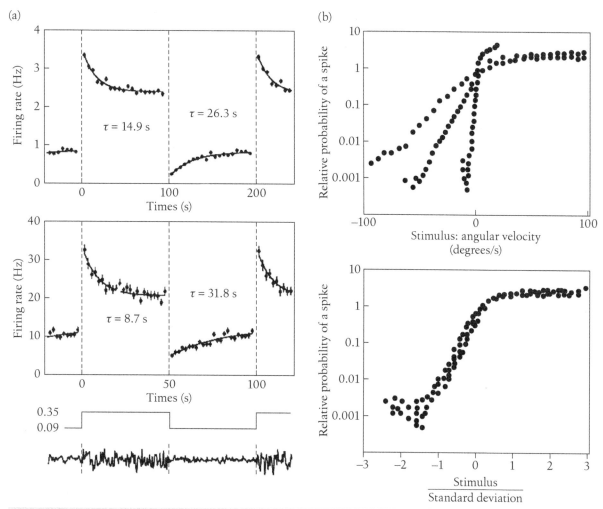

FIGURE 6.30

Adaptation to changing input distibutions seen in single neurons. (a) Adaptation of retinal ganglion cells to sudden changes in the variance of light intensity. Time dependence of spike rate in two cells (top panels) versus time during periodic switching of the variance (bottom). Redrawn from Smirnakis et al. (1997). (b) The top panel shows input-output relations for the fly motion-sensitive neuron H1 measured when inputs are drawn from different distributions. To be precise, one has to define the input as a filtered version of the velocity, and the methods for determining these filters are discussed in the references in the Annotated Bibliography for Section 4.4. The bottom panel shows that the input-output relations collapse when expressed a function of the stimulus in units of its standard deviation. Redrawn from Brenner et al. (2000).

output relations could change in response only on evolutionary time scales, but at least for bacteria, such evolutionary experiments are now quite feasible. There are models for network evolution that use information-theoretic quantities as a surrogate for fitness, in the spirit of the previous section. These models are generating interesting predictions, as described in the Annotated Bibliography for this section. It would be exciting to see laboratory evolution experiments that are the analog of the neural experiments in Fig. 6.30. In addition, it should be noted that the input-output relations of genetic regulatory elements are in fact plastic on shorter time scales, as transcription factors and the DNA itself can be covalently modified (phosphorylated, methylated, etc.) by enzymes whose activity is controlled through many other signals. I don't think we know whether there is an analog of adaptive coding in neurons hidden in these mechanisms.

So far the discussion has been about one input and one output in single genes or neurons. Almost all the really interesting systems, however, involve populations or networks of these elements. Indeed, one of the earliest ideas about optimizing information transmission in neural coding is that interactions among neighboring neurons in the retina serve to reduce the redundancy of the signals that they transmit, thus making better use of their capacity.

To get a feeling for how redundancy reduction works, consider a system in which there are N receptor cells that produce signals x_i, and these feed into a layer of N output neurons that take linear combinations of their inputs and add noise, so that the outputs of the system are

$$y_i = \sum_j W_{ij} x_j + \eta_i, \tag{6.258}$$

as shown schematically in Fig. 6.31. In the simplest case the noise will be Gaussian and independent in each output neuron: $\langle \eta_i \eta_j \rangle = \delta_{ij} \sigma^2$. Let's also assume, again for simplicity, that the distribution of the $\{x_i\}$ is also Gaussian, with zero mean and a covariance matrix $\langle x_i x_j \rangle = C_{ij}$. Then following the arguments in Section 6.2, the information that the outputs provide about the inputs is

$$I(\mathbf{y}; \mathbf{x}) = \frac{1}{2} \text{Tr} \log_2 \left(\mathbf{1} + \frac{1}{\sigma^2} W C W^T \right), \tag{6.259}$$

where W^T denotes the transpose of the matrix W. We can chose the matrix W, which defines the receptive fields of the output neurons, to maximize the information, but we need a constraint, because otherwise the answer is always to make W larger, to overwhelm the noise. A natural constraint, then, is to fix the overall dynamic range of the output signal:

$$\sum_i \langle y_i^2 \rangle = \text{Tr} \left(W C W^T \right) + N \sigma^2. \tag{6.260}$$

But if we transform to a basis where $W C W^T$ is diagonal, then the information becomes

$$I(\mathbf{y}; \mathbf{x}) = \frac{1}{2} \sum_\mu \log_2 \left(1 + \frac{\Lambda_\mu}{\sigma^2} \right), \tag{6.261}$$

where $\{\Lambda_\mu\}$ are the eigenvalues of the matrix $W C W^T$, and the constraint is that the sum of these eigenvalues is fixed. Then it is clear from the convexity of the logarithm that the best we can do is to have all the eigenvalues be equal, which means that $W C W^T$ is proportional to the unit matrix. But (leaving aside the contribution from the noise) $W C W^T$ is the correlation matrix of the output signals. Thus, in this simple model, we maximize information transmission by removing all correlations in the input and making the outputs independent of one another.

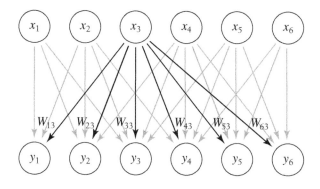

FIGURE 6.31

A schematic network, based on Eq. (6.258). The signal x_j provides input to y_i, with weight W_{ij}. All connections are present, but the connections from x_3 are highlighted.

Problem 161: Convexity and equalization. Show explicitly that if we want to maximize

$$I = \frac{1}{2} \sum_\mu \log_2(1 + \tilde{\Lambda}_\mu), \tag{6.262}$$

subject to the constraint

$$\sum_\mu \tilde{\Lambda}_\mu = C, \tag{6.263}$$

then the solution is to have all the $\{\tilde{\Lambda}_\mu\}$ be equal, that is, $\tilde{\Lambda}_\mu = \Lambda_0$.

In the retina, we expect that correlations, and perhaps also the transformations from input to output, are translation invariant. Indeed, the output (ganglion) cells of the retina can be classified, and if we look at a single class of cells their receptive fields in many cases tile the retina in a regular and fairly uniform array, as shown schematically in Fig. 6.32. Thus, if the receptor cell i is at position $\mathbf{r}_i$, perhaps on a lattice, and the output neurons are on the same lattice, we should have

$$C_{ij} = C(\mathbf{r}_i - \mathbf{r}_j), \tag{6.264}$$
$$W_{ij} = W(\mathbf{r}_i - \mathbf{r}_j). \tag{6.265}$$

Then the condition for independence at the outputs becomes

$$\delta_{ij} \propto \sum_{km} W_{ik} C_{km} W_{jm} \tag{6.266}$$

$$= \sum_{km} W(\mathbf{r}_i - \mathbf{r}_k) C(\mathbf{r}_k - \mathbf{r}_m) W(\mathbf{r}_j - \mathbf{r}_m). \tag{6.267}$$

We approximate the sums as integrals, so that

$$\delta_{ij} \approx \int d^2 r' \int d^2 r'' \, W(\mathbf{r}_i - \mathbf{r}') C(\mathbf{r}' - \mathbf{r}'') W(\mathbf{r}_j - \mathbf{r}'') \tag{6.268}$$

$$= \int \frac{d^2 k}{(2\pi)^2} |\tilde{W}(\mathbf{k})|^2 S(\mathbf{k}) e^{i\mathbf{k}\cdot(\mathbf{r}_i - \mathbf{r}_j)}, \tag{6.269}$$

On parasol   Off parasol

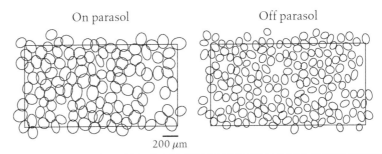

200 μm

FIGURE 6.32

Receptive fields of ganglion cells in the primate retina form an approximate lattice. By playing random movies to the retina, we can correlate the movies with the spikes generated by each neuron, and in this way estimate the small region in space to which one cell is responding. These experiments record from many neurons simultaneously and distinguish different types of cells, here two classes that generate more spikes in response to transient increases (on cells) or decreases (off cells) in light intensity. Approximating the sensitive regions, or receptive fields, of each cell by an ellipse, we see that the ellipses are nearly close packed, so that a single class of cells tiles the visual field. Redrawn from Shlens et al. (2006).

where $\tilde{W}(\mathbf{k})$ is the Fourier transform of $W(\mathbf{r})$, and we identify the Fourier transform of the correlation function $C(\mathbf{r})$ as the power spectrum $S(\mathbf{k})$. To satisfy this condition, $|\tilde{W}(\mathbf{k})|^2 S(\mathbf{k})$ must be constant, independent of $\mathbf{k}$, and if $W(\mathbf{r})$ is symmetric in space, this means that

$$\tilde{W}(\mathbf{k}) \propto \frac{1}{\sqrt{S(\mathbf{k})}}. \tag{6.270}$$

We expect that the power spectrum of the inputs to the retina will fall off at high frequencies, which means that the optimal weights W have the form of a filter that does the opposite, attenuating the low frequencies and enhancing high frequencies. In fact, experiments show that the power spectrum of contrast in natural scenes is scale invariant, so that $S(\mathbf{k}) \propto |\mathbf{k}|^{-\alpha}$, with the exponent α close to 2. Then the optimal weights W should vanish as $\mathbf{k} \to 0$, which means that the output of the retina should be insensitive to spatially uniform illumination; in contrast, the output should overemphasize gradients or edges. By including the effects of noise (as in Problem 162), we can see a crossover to spatial averaging at low signal-to-noise ratios; see Fig. 6.33. Qualitatively this prediction is correct: at high signal-to-noise ratios, we literally see this enhancement of edges not just in the responses of retinal ganglion cells but also in our perception, through the phenomenon of Mach bands (Fig. 6.34); this spatial differentiation gives way to integration as we lower the light levels and hence the signal-to-noise ratio.

Problem 162: Redundancy reduction versus noise reduction.   Equation (6.270) suggests that at large $\mathbf{k}$, where the power spectrum of input signals should be small, the weight in transferring these signals to the output should be large. This can't be completely right, because we expect that at very high (spatial) frequencies, signals will be lost in a background of noise. Go back to the start of this analysis and assume that the signals x_j already have a little bit of noise attached to them (as with photon shot noise in vision), so that

$$y_i = \sum_j W_{ij}(x_j + \xi_j) + \eta_i, \tag{6.271}$$

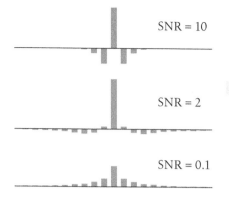

SNR = 10

SNR = 2

SNR = 0.1

FIGURE 6.33

Cross-sections through the optimal matrices W_{ij} in the problem of efficient coding with noise. The correlation function is assumed to be exponential, $C_{ij} \propto \exp(-|i - j|/\xi)$, with $\xi = 50$, much longer than the range of interactions shown here. At high signal-to-noise ratios (SNRs), the solution looks like a differentiator, which removes the second-order correlations in the signal, whereas at low SNR, the solution integrates to suppress noise. Redrawn from Atick and Redlich (1990).

(a) (b)

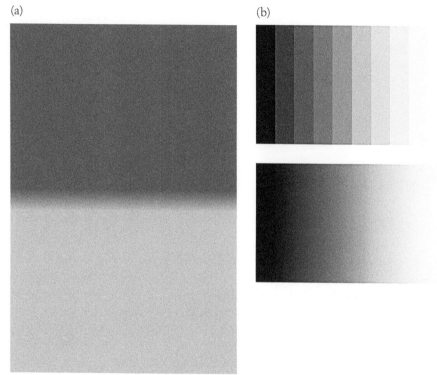

FIGURE 6.34

Mach bands. (a) A sudden transition between gray levels is perceived with an extra dark line on dark side of the transition, and an extra bright line on the bright side, as if the image were being filtered through a partially differentiating filter as in Fig. 6.33. From http://en.wikipedia.org/wiki/Mach_bands. (b) This filtering gives a sequence of discrete transitions from darker to lighter the appearance of nonmonotonicity, especially when compared with a continuous gradient in the bottom image. From J. Pomerantz, Rice University, with my thanks.

where everything is as before but $\langle \xi_i \xi_j \rangle = \delta_{ij} \sigma_0^2$. Follow the outline above and derive the form of the weights W_{ij} that optimize information transmission at fixed output variance. Verify that as $\sigma_0 \to 0$, you recover the simple picture in which the optimal W_{ij} serve to remove correlations. Show, in contrast, that as σ_0 becomes large, the optimal solution involves averaging over multiple inputs to beat down the noise.

Problem 163: Information available at the retina.   We can approximate cells in the retina as responding linearly to the contrast in the visual world, so that in one snapshot the signal in each cell can be written as

$$s_n = \int d^2x\, F(\mathbf{x} - \mathbf{x}_n) C(\mathbf{x}) + \xi_n, \tag{6.272}$$

where $F(\mathbf{x})$ is a filter that describes the effect of looking through the eye's lens, $\mathbf{x}_n$ is the position of the receptor cell, and ξ_n is a noise source, independent in each cell, that we can approximate as Gaussian with some variance σ^2. We can approximate the optical filtering as having a transfer function in Fourier space defined by that of an ideal lens,

$$|\tilde{F}(\mathbf{k})|^2 = (1 - |\mathbf{k}|/k_c)^2, \tag{6.273}$$

and $|\tilde{F}(\mathbf{k})| = 0$ for $|\mathbf{k}| > k_c$. We can approximate the positions of the cells as being on a lattice, and it is very close to being true that the lattice spacing matches the optics, so that there is no aliasing. In practice, then, if we work in Fourier space, we can just limit ourselves to $|\mathbf{k}| \leq k_c$. Finally, analysis of images collected on a walk through the woods shows that these natural scenes have a power spectrum

$$S(\mathbf{k}) \equiv \int d^2y\, \langle C(\mathbf{x})C(\mathbf{x} + \mathbf{y})\rangle e^{-i\mathbf{k}\cdot\mathbf{y}} = \frac{A}{|\mathbf{k}|^{2-\eta}}, \tag{6.274}$$

where the "anomalous dimension" $\eta = 0.19 \pm 0.01$.

(a) A natural quantity to set the scale of information transmission is the signal-to-noise ratio in a single receptor cell,

$$SNR_1 \equiv \frac{1}{\sigma^2}\left\langle \left[\int d^2x\, F(\mathbf{x} - \mathbf{x}_n)C(\mathbf{x})\right]^2 \right\rangle = \frac{1}{\sigma^2}\int \frac{d^2k}{(2\pi)^2}|\tilde{F}(\mathbf{k})|^2 S(\mathbf{k}), \tag{6.275}$$

where again we should limit the integral to $|\mathbf{k}| \leq k_c$. Evaluate SNR_1 in terms of the other parameters of the problem. You can approximate the allowed region of integration in the $\mathbf{k}$ plane (the Brillouin zone of the receptor lattice) as being circular, for simplicity.

(b) In Fourier space, the independent random noise ξ_n in each cell is equivalent to a white noise spectrum, $N(\mathbf{k}) = N_0$. The scale of the spectrum is set by the total noise variance,

$$\sigma^2 = \int \frac{d^2k}{(2\pi)^2} N(\mathbf{k}), \tag{6.276}$$

where the integration is over the Brillouin zone. Use this condition to evaluate N_0. Also, the area of the unit cell of the lattice, A_1, is inverse to the area of the Brillouin zone, so it is useful to have an expression for it:

$$A_1 = \left[\int \frac{d^2k}{(2\pi)^2}\right]^{-1}. \tag{6.277}$$

(c) Show that the information that receptor cells convey about the visual world is less than what is computed if we assume the contrast fluctuations are drawn from a Gaussian distribution with the measured power spectrum $S(\mathbf{k})$.

(d) If image contrast is drawn from a Gaussian distribution, then by analogy with the discussion leading up to Eq. (6.114) for signals in the time domain, we can define a signal-to-noise ratio at every spatial frequency $\mathbf{k}$ and then sum the information conveyed by every

Fourier component to give the information per unit area (or more usefully, the information per receptor cell):

$$I_1 = A_1 \int \frac{d^2k}{(2\pi)^2} \log_2 \left[1 + \frac{|\tilde{F}(\mathbf{k})|^2 S(\mathbf{k})}{N(\mathbf{k})} \right]. \tag{6.278}$$

First, show that this expression is correct. Then show that it can be rewritten to depend only on SNR_1, with all other parameters dropping out.

(e) Evaluate $I_1(SNR_1)$ numerically, over a plausible range $1 < SNR_1 < 10^3$ for human foveal cones in the daytime. What conclusions can you draw? How do your results compare with the common intuition that our visual systems are bombarded with huge amounts of information?

These arguments for whitening—removing spatial correlations—should also work in the time domain. If we are trying to transmit a time-dependent signal that has a power spectrum $S(\omega)$, then by analogy with Eq. (6.270) there should be a regime in which the optimal filter has the transfer function

$$\tilde{F}(\omega) \propto \frac{1}{\sqrt{S(\omega)}}. \tag{6.279}$$

If the spectrum falls off with frequency, then the filter will be high pass, suppressing low frequencies and enhancing higher frequencies. Indeed, many naturally occurring signals have power spectra that are $1/f$, so we would predict that $|\tilde{F}(\omega)|^2 \propto \omega$. This behavior seems sensible, roughly like a differentiator, but it's not: taking a derivative corresponds to $\tilde{F}(\omega) \propto -i\omega$, so that $|\tilde{F}(\omega)|^2 \propto \omega^2$. Thus, to provide the optimal representation of $1/f$ signals requires a filter that takes "half a derivative." There is no way to build fractional differentiators with a finite number of conventional linear components; for example, if the allowed components are resistors, capacitors, and inductors, one cannot build a circuit with an impedance $|Z(\omega)| \propto \sqrt{\omega}$ from a finite set of these components. Can biological systems build such filters? One of the places to look is in the first synapse of the fly's visual system, the same system where Laughlin analyzed the nonlinear input-output relation (Fig. 6.23). Measurements on the linear responses of photoreceptors and large monopolar cells to small variations in contrast can be combined to determine the transfer function of the synapse itself, as shown in Fig. 6.35. Remarkably, $|\tilde{F}(\omega)|^2 \propto \omega$ is an excellent approximation over nearly two decades in frequency.

Problem 164: Whitening and fractional differentiation. Do a full calculation, optimizing information transmission in the time domain, that leads to Eq. (6.279). Then show that no finite number of resistors, capacitors, and inductors can realize a fractional differentiator.

A simpler example of these ideas is perhaps provided by color processing. Roughly speaking, at one point in space our retina takes three samples, corresponding to the signals in the three different cones. These three signals are correlated, both because the absorption spectra of the pigments in the different cones overlap and because the

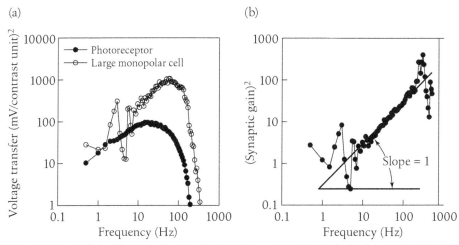

FIGURE 6.35

Voltage transfer across the first synapse in fly vision. Redrawn from Laughlin and de Ruyter van Steveninck (1996). (a) Recordings from photoreceptors and large monopolar cells, as in Fig. 6.12, measuring the transfer function from image contrast to cellular voltage versus frequency. (b) The ratio of these transfer functions defines the gain in transfer of voltage across the synapse. Note the unit slope on a log-log plot over nearly two decades.

reflectance spectra of the objects around us are rather smooth functions of wavelength.[19] By analogy with what we have seen thus far, if the retina is under pressure to maximize information transmission, then it should send these signals to the brain in some decorrelated form. Early guesses about the form of the correlations among the different cone signals suggested that the three decorrelated signals would correspond roughly to the sum of all the inputs (the total light intensity, ignoring color), an approximately red-minus-green signal and a blue-minus-yellow signal. In fact it is known that neurons throughout the visual system follow this pattern of "opponent" color processing.

To do a more quantitative analysis requires getting away from traditional color photography, because (for example) the three channels in a CCD camera do not have wavelength sensitivities that correspond exactly to that of our cones. Instead one can take hyperspectral images, essentially measuring the spectrum of light at each point in the scene, and then construct the expected signals that will be seen by each cone, given the absorption spectra of the three cone pigments. This analysis shows, quite remarkably, that the rough intuition about opponent processing is nearly exact, with the decorrelated signals being almost perfect integer combinations of the cone signals. If the three cone signals are $\mathcal{L}$, $\mathcal{M}$, and $\mathcal{S}$ for the long, medium, and short wavelengths, respectively, then the decorrelated signals are $\ell = \mathcal{L} + \mathcal{M} + \mathcal{S}$ (light intensity), $\alpha = \mathcal{L} + \mathcal{M} - 2\mathcal{S}$ (yellow minus blue), and $\beta = \mathcal{L} - \mathcal{M}$ (red minus green), where the coefficients are unity with an accuracy of $\sim 1\%$. Further, this linear transformation serves to generate truly independent signals, even though the underlying distributions

---

19. This is the same effect. The reflectance properties of most naturally occurring objects in our terrestrial environment are determined by the absorption spectra of organic pigments, and these tend to be broad; see Section 2.2 and Appendix A.4.

are not Gaussian. These very clean results come from a delicate interplay between the statistical structure of the world and the properties of our visual pigments. I don't know how accurately the coefficients in opponent color processing have been measured, but this is a striking prediction that certainly captures the qualitative behavior of the system and deserves to be tested more quantitatively.

In the example of Fig. 6.33, the optimal weights for transforming receptor signals into neural output correspond to a center-surround structure in which an output neuron at point $\mathbf{r}$ gives a positive weight to the receptor cell at point $\mathbf{r}$ and a negative weight to its neighbors. Alternatively, all weights from receptors to neurons can be thought of as positive, and the output neurons inhibit one another before sending their signals on to the brain. In our retinas things are complicated, because the transformation from photoreceptors to ganglion cells involves several intermediate cells, but in some simpler creatures, such as the horseshoe crab, the picture of lateral inhibition seems to be correct, and indeed the horseshoe crab was the first retina in which receptive fields were measured. Lateral inhibition is thought to be a general neural mechanism for sharpening the responses to stimuli that vary across an array of neurons, and we have seen that this sharpening is essential in decorrelating signals and enhancing the efficiency of information transmission. Could there be an analog of this for the transmission of information through genetic or biochemical networks? If we go back to the case of Bcd regulating the expression of Hb, as in Fig. 6.25, we know that this is just one piece of a larger network in which the primary morphogen Bcd feeds into a collection of gap genes, which in turn interact with one another. Because transcription factors tend to be either activators or repressors, in the absence of any other effects, all target genes would have correlated expression levels and hence provide redundant data about the concentration of the input. This redundancy can be removed by lateral inhibition, and that is what we see in the gap gene network (Fig. 6.36). The challenge is to take this quantitative analogy and turn it into a quantitative theory.

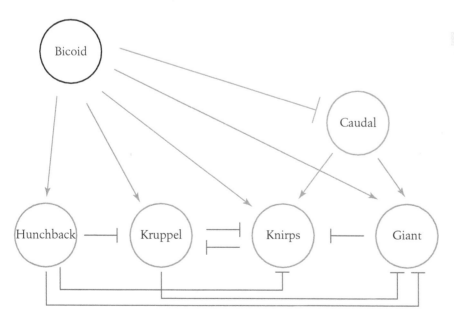

FIGURE 6.36

The gap gene network in the *Drosophila* embryo. Multiple genes are activated by the primary maternal morphogen, Bicoid, so that these different genes receive correlated signals. Efficient information transmission can be enhanced by having the different target genes activated at staggered thresholds, so that they fractionate the available dynamic range, but this does not solve the problem of correlations. Repressive interactions among the targets, however, could serve to make the final expression levels less redundant, much as lateral inhibition was proposed to reduce redundancy in the retina.

The representations of data constructed by the nervous system might be efficient in the sense we have considered here, but they have a more obvious feature—they are built from discrete action potentials or spikes. If we look with some reasonably fine time resolution, $\Delta\tau < 10$ ms, then because the average spike rates are less than 100 spikes/s, at any moment the typical neuron is silent. In this sense, the code is sparse. It is this sparseness that, among other things, makes it possible to decode spike trains using linear filters, as in Eq. (6.120). Spikes are expensive, requiring substantial energy expenditure, and perhaps it is this cost that drives the brain toward the construction of sparse representations.

To capture the role of spiking in a tractable form, let us see how far we can take the idea of linear reconstruction, as discussed in Section 4.4. If the sensory input is $s(t)$—for example, sound pressure as a function of time in the auditory system—we would like to have a family of neurons labeled by μ that spike at times $\{t_i^\mu\}$ such that

$$s_{\text{est}}(t) = \sum_\mu \sum_i f_\mu(t - t_i^\mu) \tag{6.280}$$

is as close as possible to the true signal. Notice that in this system the input is $s(t)$, and the output is the set of spikes $\{t_i^\mu\}$. If we imagine adjusting the input-output relations, the mapping $s(t) \to \{t_i^\mu\}$ will change, perhaps in complicated ways. But suppose we knew the functions $f_\mu(\tau)$. Then there would be best times t_i^μ for each spike, so that the match between $s_{\text{est}}(t)$ and $s(t)$ is as close as possible. We could imagine searching through some large space of input-output relations to find one that puts the spikes at these best times, or we could use the times themselves as our description of the input-output relation. Conversely, if we knew the spike times, we could adjust the filters $f_\mu(\tau)$, as in previous discussions. Can we solve both problems, subject to a constraint on the total number of spikes? This is hard, but by slightly softening the problem—allowing each term in Eq. (6.280) to have a varying amplitude—it becomes tractable.

Figure 6.37 shows the results of this approach, where a small population of neurons provides an efficient representation of natural sounds. There are several interesting features in these results. First, the filters $f_\mu(\tau)$ are localized in time; although they are tuned to particular frequencies, they are more like a wavelet than a Fourier representation, with support over a window of time that scales inversely with the characteristic frequency. Second, the filters also have a very asymmetric shape, with a sharp attack and a slower decay. Third, even at a particular characteristic frequency, there are multiple filters with noticeably different shapes. If we look through measurements of the impulse responses of neurons emerging from the mammalian ear, we see exactly these structures; there are even cells that overlay the predicted filters almost perfectly.[20] Importantly, these structures are lost if we try to build representations of very different sound ensembles.

By allowing for different total numbers of spikes, or limiting the time resolution with which the spikes are placed, codes of different qualities can be constructed. For

---

20. We need to be careful here. The impulse responses of the neurons are measured by the reverse correlation method, which, ideally, extracts a filter characteristic of the *encoding* of sounds into spikes; see references in the Annotated Bibliography for Section 4.4. In contrast, the filters $f_\mu(\tau)$ are characteristic of the *decoding* process. One can circumvent this problem by using reverse correlation to analyze the model code, with almost identical results. This suggests that the model, at least, is operating in a regime where the coding and decoding filters are similar, which happens exactly in the limit where all spikes are statistically independent from one another, so that there is no redundancy. Thus, the search for efficient codes may also drive the emergence of simplicity in decoding.

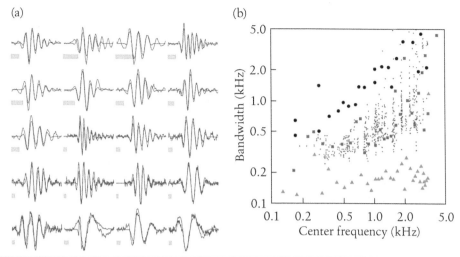

FIGURE 6.37

Ingredients for an efficient representation of natural sounds, as in Eq. (6.280). Redrawn from Smith and Lewicki (2006). (a) The functions $f_\mu(\tau)$ in red, compared with the impulse responses of single neurons in the cat auditory nerve in blue; gray scale bars are 5 ms long. All these filters are band pass, so that their Fourier transforms $\tilde{f}_\mu(\omega)$ have maximum magnitude at some characteristic frequency and fall to half maximal over some bandwidth. (b) A scatter plot of these bandwidths versus characteristic frequencies for the filters (red) and auditory neurons (small blue dots); filters trained on different ensembles (black circles and green triangles) exhibit very different behaviors.

these different codes, it is relatively easy to put an upper bound on the entropy of the spike trains and to measure the errors between $s_{\rm est}(t)$ and the true $s(t)$; putting these together, one obtains the rate–distortion curve for this family of codes. Applied to ensembles of human speech, the results are comparable to or better than conventional coding schemes. It does indeed seem that nature has found an efficient class of codes, not just in abstract terms.

In Section 6.3, we discussed the idea that organisms are not interested in information in general, but only in information that has predictive power. Given the statistics of the world in which we live, capturing a certain number of bits about the past can provide a maximum number of bits about the future, as in Fig. 6.21. Can we test the hypothesis that biological systems approach this limit, extracting just those bits from the past that carry maximum predictive information? The difficulty here is to measure the amount of information that neural responses, for example, carry about the future. We could do this by choosing particular features of the future to estimate from the neurons, much as we decoded spike trains to estimate stimuli (Section 4.4), but we might get the wrong answer if we choose the wrong features.

In the discussion leading up to Fig. 6.7, we saw how to measure the information carried by neural responses without choosing particular features of relevance. The key is that the information that responses carry about the stimulus is the difference between two entropies: the total entropy of the responses and the average entropy across multiple repetitions of the same stimulus. The challenge now is to observe responses to stimuli that vary while repeating their future. In fact this is not so hard, because

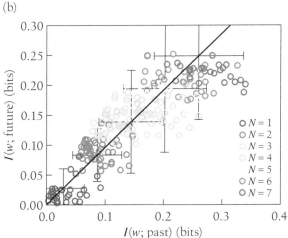

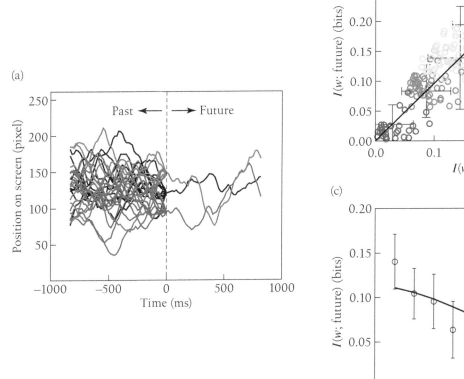

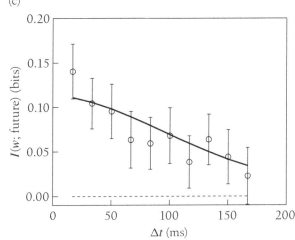

FIGURE 6.38

Near-optimal representation of predictive information in the retina. Redrawn from Palmer et al. (2012). (a) The common future experiment. The trajectory of a moving bar stimulus is shown as a function of time relative to the point at which sets of trials converge onto a common future; 10 trials for each of three futures are shown here. (b) Mutual information between population responses in retinal ganglion cells and the future of the stimulus versus information about the past of the stimulus, for groups of cells ranging from $N = 1$ to $N = 7$. For each of 53 individual cells in this experiment, the group with maximal information about the future is plotted. Each cell participates in at least one group that saturates the bound, shown as a solid line. (c) For one particular group of 5 cells from panel (b), the information about the future of the stimulus is plotted versus the time before the future onset, showing that these data (circles) track the bound (solid line) over a range of delays.

if the stimuli that we generate are drawn from a known stochastic process, we can draw samples of this process that are independent but guaranteed to converge on a common future at some moment in time, as shown in Fig. 6.38a. With such stimuli, we can ask whether the neural responses before the convergence time can be used to predict which of several common futures is coming, and in this way we can measure, without further assumptions, the information that these responses carry about the future, at least in a simple world described by a known stochastic process. Further, we can compare this information with the limit set by the amount of information that these responses carry about past stimuli. As shown in Fig. 6.38b,c, small populations of ganglion cells in the retina come within error bars of the limit as they encode a single moving object in their visual field. Although much remains to be done, this result certainly is suggestive.

6.5 Gathering Information and Making Models

The world around us, thankfully, is a rather structured place. Whether we are doing a careful experiment in the laboratory or taking a walk through the woods, the signals that arrive at our brains are far from random noise; there seem to be some underlying regularities or rules. Surely one task that all organisms must face is the learning or extraction of these rules and regularities, making models of the world, either explicitly or implicitly. In this section we explore how learning and making models is related to the general problem of efficient representation.

Perhaps the simplest example of learning a rule is fitting a function to data— we believe in advance that the rule belongs to a class of possible rules that can be parameterized, and as we collect data we learn the values of the parameters. This is something we all learned about in our physics lab classes (see Fig. 6.39 for a reminder), and even this simple example introduces many deep issues.

First, data usually come with some level of noise, and as a result any model really is (at least implicitly) a model of the probability distribution out of which the data are being drawn, rather than just a functional relationship. Indeed, one could argue that the general problem is always that of learning such distributions, and any rigid or deterministic rules emerge as a limit in which the noise becomes small or is beaten down by a large number of observations.

Second, we would like to compare different models, often with different numbers of parameters. We have an intuition that simpler models are better, and we want to make this intuition precise—is it just a subjective preference, or is the search for simplicity something we can ground in more basic principles? A related point is that where the classical curve-fitting exercises involve models with a limited number of parameters, we might want to go beyond this restriction and consider the possibility that the data are described by functions that are merely smooth to some degree.

Finally, we would like to quantify how much we are learning—and how much *can* be learned—about the underlying rules given a limited set of data. If there are limits to how much can be learned, is it possible that biology has constructed learning machines that are efficient in some absolute sense, pushing up against these limits? So, let's plunge in.

Imagine that we observe two streams of data x and y, or equivalently a stream of pairs $(x_1, y_1), (x_2, y_2), \cdots, (x_N, y_N)$. Assume that we know in advance that the xs are drawn independently and at random from a distribution $P_X(x)$, and the ys are noisy versions of some function acting on x,

$$y_n = f(x_n; \boldsymbol{\alpha}) + \eta_n, \tag{6.281}$$

where $f(x; \boldsymbol{\alpha})$ is one function from a class of functions parameterized by $\boldsymbol{\alpha} \equiv \{\alpha_1, \cdots, \alpha_K\}$ and η_n is noise, which for simplicity we assume is Gaussian with known variance

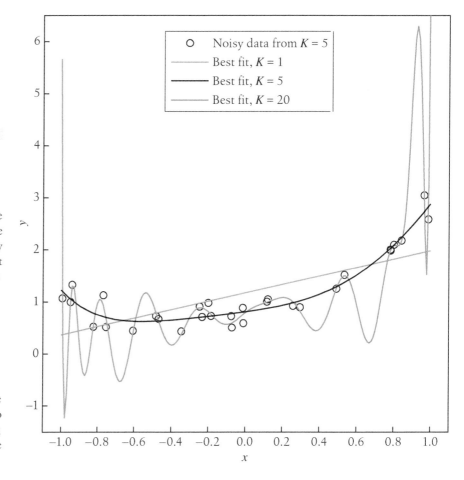

FIGURE 6.39

Fitting to polynomials. Given a collection of data points (black circles), $\{x_n, y_n\}$, we try to fit these data with polynomials of different degree K, with $K = 1, 5, 20$. As the degree of the polynomial—what we think of intuitively as the complexity of our model—increases, we can get closer to the data points, but at the same time we are introducing wild fluctuations, which seem unlikely to be correct. In fact, $K = 5$ is the correct answer, because the data points were generated by choosing the x_n at random, evaluating some fixed fifth-order polynomial, and then adding noise. To claim that we understand how to learn, we have to find a principled way of convincing ourselves that it is better to keep the poorer fit with the simpler model.

σ^2. We can even start with a very simple case, where the function class is just a linear combination of basis functions, so that

$$f(x; \boldsymbol{\alpha}) = \sum_{\mu=1}^{K} \alpha_\mu \phi_\mu(x). \tag{6.282}$$

The usual problem is to estimate, from N pairs $\{x_n, y_n\}$, the values of the parameters $\boldsymbol{\alpha}$; in favorable cases such as this we might even be able to find an effective regression formula. Probably you were taught that the way to do this is to compute χ^2,

$$\chi^2 = \frac{1}{\sigma^2} \sum_n \left| y_n - f(x_n; \boldsymbol{\alpha}) \right|^2, \tag{6.283}$$

and then minimize this expression to find the correct parameters $\boldsymbol{\alpha}$. You may or may not have been taught *why* this is the right thing to do, which is what we want to understand here.

If our model, Eq. (6.281), is correct, what is the probability that we observe the data points $\{x_n, y_n\}$? Let's start by asking about the locations of the points x_n where we get samples of the functional relationship between x and y. In the standard examples of curve fitting, the examples are given to us and there is nothing more to say; thus, we might as well assume that the points x_n are chosen randomly and independently out of some distribution $P_X(x)$, perhaps just the uniform distribution on some interval. One might ask whether there is a good choice for the next point x_{n+1}, perhaps a point that will give us the maximal information about the underlying parameters $\boldsymbol{\alpha}$. This is the problem faced in the design of experiments—how do we choose what to measure given what we already know?—but let's leave this aside for the moment.

The model in Eq. (6.281) is a statement about the conditional probability distribution of y_n given x_n. Specifically, y_n is a Gaussian random variable with a mean value of $f(x_n; \boldsymbol{\alpha})$ and a variance of σ^2, so that

$$P(y_n|x_n, \boldsymbol{\alpha}) = \frac{1}{\sqrt{2\pi\sigma^2}} \exp\left[-\frac{(y_n - f(x_n; \boldsymbol{\alpha}))^2}{2\sigma^2}\right]. \tag{6.284}$$

By hypothesis, the noise on every point is independent, which means that

$$P(\{y_n\}|\{x_n\}, \boldsymbol{\alpha}) = \prod_{n=1}^{N} P(y_n|x_n, \boldsymbol{\alpha}). \tag{6.285}$$

Now we can put things together to write the probability of the data given the parameters of the underlying model:

$$P(\{x_n, y_n\}|\boldsymbol{\alpha}) = \left[\prod_{n=1}^{N} P(y_n|x_n, \boldsymbol{\alpha})\right] \times \left[\prod_{n=1}^{N} P_X(x_n)\right] \tag{6.286}$$

$$= \left[\prod_{n=1}^{N} P_X(x_n)\right] \prod_{n=1}^{N} \frac{1}{\sqrt{2\pi\sigma^2}} \exp\left[-\frac{(y_n - f(x_n; \boldsymbol{\alpha}))^2}{2\sigma^2}\right] \tag{6.287}$$

$$= \exp\left[\sum_{n=1}^{N} \ln P_X(x_n) - \frac{N}{2}\ln(2\pi\sigma^2) - \frac{1}{2}\chi^2\right], \tag{6.288}$$

where we identify χ^2 from Eq. (6.283). Notice that the only place where the parameters appear is in χ^2, and $P \propto e^{-\chi^2/2}$. Thus, finding parameters minimizing χ^2 also serves to maximize the probability that the model could have given rise to the data. This sounds like a good thing to do, and certainly maximizing the probability of the data (usually called "maximum likelihood") feels more fundamental than minimizing χ^2. But what are we really accomplishing by maximizing P?

The discussion in Section 6.1 showed that the entropy is the expectation value of $-\log P$, and that it is possible to encode signals so that the amount of space required to specify each signal uniquely is on average equal to the entropy. In such optimal encodings, each possible signal s drawn from $P(s)$ is encoded in a space of approximately $-\log_2 P(s)$ bits, and this approximation becomes more accurate as the signal becomes a long sequence of independent measurements, as in the example at hand. Thus, any model probability distribution implicitly defines a scheme for coding signals that are drawn from that distribution, so if we make sure that the data have high probability in the distribution (small values of $-\log P$), then we also are making sure that our code or representation of these data is compact. Thus, good old-fashioned curve fitting really is finding efficient representations of data, which is the same principle that we discussed in the previous section in contexts ranging from the regulation of gene expression to neural coding. To be clear, in the earlier discussion we took for granted some physical or resource constraints (e.g., the noise level or limited numbers of molecules) and tried to transmit as much information as possible. Here we do the problem the other way around, searching for a representation of the data that will require the minimum set of resources.

It is important that the connection between curve fitting and efficient representation is not an analogy but rather a mathematical identity. If our measurements really are contaminated by Gaussian random noise, then minimizing χ^2 is the same as finding a model that generates the data with maximum likelihood, and implicitly this defines a "code" for the data that is as short as we can find. Intuitively, a good model gives us a compact representation, because most features of the data can be calculated from the model without looking at the data itself. Concretely, if $y = f(x)$ and we are given (x, y) pairs, then knowledge of the function f means that we don't really need to write down the values of y—we can calculate them from the given values of x. If there are fluctuations around the predictions of the model, then we have to encode these fluctuations as well, and this takes space, so that minimizing the fluctuations reduces the space needed to encode the data as a whole.

The equivalence of accurate models and compact representations formalizes a common belief, namely that if we really understand something, we should be able to express this understanding concisely.[21] But the preference for concision seems subjective. There ought to be something objective about our definition of understanding, or at least the notion of what makes a good model. The really hard question here is whether we can invert these arguments, and claim that the search for efficient representations of data will lead automatically to good models and perhaps even to understanding.

To take seriously the equivalence of good models and efficient representations, we have to dig a little more deeply. The claim that a model provides a code for the data is not complete, because at some point we have to represent our knowledge of the model itself. One idea is to do this explicitly—estimate how accurately each of the parameters is known, and then count how many bits will be needed to write down the parameters to that accuracy and add this to the length of our code. Another idea is more implicit—we do not really know the parameters, all we do is estimate them from the data, so it is not so obvious that we should separate coding the data from coding the parameters, although this might emerge as an approximation. In this view what we should do is to integrate over all possible values of the parameters, weighted by some prior knowledge, and thus compute the probability that the data could have arisen from the class of models being considered.

To carry out this program of computing the total probability of the data given the model class, we need to do the integral

---

21. There is some controversy about who first expressed this notion of concision as a marker of understanding. It goes back at least to Pascal, who wrote a letter in 1656 that included the apology "I would not have made this so long except that I do not have the leisure to make it shorter," suggesting that if he had thought more clearly about what he wanted to say, he could have said it in fewer words.

$$P(\{x_i, y_i\}|\text{class}) = \int d^K\alpha \, P(\alpha)P[\{x_i, y_i\}|\alpha] \tag{6.289}$$

$$= \int d^K\alpha \, P(\alpha) \exp\left[-\frac{N}{2}\ln(2\pi\sigma^2) - \frac{1}{2\sigma^2}\chi^2(\alpha; \{x_i, y_i\})\right]$$

$$\times \left[\prod_{n=1}^{N} P_X(x_n)\right], \tag{6.290}$$

where $P(\alpha)$ is the a priori distribution of parameters, maybe just a uniform distribution on some bounded region. But χ^2 is a sum over data points, which means it (typically) will be proportional to N. Thus, at large N we are doing an integral in which the exponential has terms proportional to N—and so we should use a saddle point approximation. To implement this approximation, let's write

$$P(\{x_i, y_i\}|\text{class}) = \exp\left[-\frac{N}{2}\ln(2\pi\sigma^2)\right]\left[\prod_n P_X(x_n)\right]\int d^K\alpha \, e^{-Ng(\alpha)}, \tag{6.291}$$

where the effective energy per data point is

$$g(\alpha) = \frac{1}{2N}\chi^2(\alpha; \{x_i, y_i\}) - \frac{1}{N}\ln P(\alpha). \tag{6.292}$$

The saddle point approximation is

$$\int d^K\alpha \, e^{-Ng(\alpha)} \approx e^{-Ng(\alpha^*)}(2\pi)^{K/2}\exp\left[-\frac{1}{2}\ln\det(N\mathcal{H})\right], \tag{6.293}$$

where α^* is the value of α at which $g(\alpha)$ minimized, and the Hessian $\mathcal{H}$ is the matrix of second derivatives of g at this point:

$$\mathcal{H}_{\mu\nu} = \frac{\partial^2 g(\alpha)}{\partial\alpha_\mu \partial\alpha_\nu}\bigg|_{\alpha=\alpha^*}. \tag{6.294}$$

At large N, $g(\alpha)$ is dominated by χ^2, so α^* must be close to the point where χ^2 is minimized. Putting the pieces together, we have

$$-\ln P(\{x_i, y_i\}|\text{class}) \approx \frac{N}{2}\ln(2\pi\sigma^2) - \sum_{n=1}^{N}\ln P_X(x_n) + \frac{1}{2}\chi^2_{\min}$$

$$+ \ln P(\alpha^*) - \frac{K}{2}\ln 2\pi + \frac{1}{2}\ln\det(N\mathcal{H}). \tag{6.295}$$

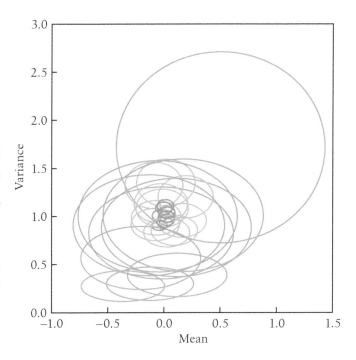

FIGURE 6.40

Confidence limits on the estimation of mean and variance for a Gaussian distribution. In several independent experiments, we choose $N = 10$ (blue), $N = 100$ (green), or $N = 1000$ (orange) points out of a Gaussian distribution with zero mean and unit variance. We estimate the mean and variance from the data in the usual way and draw error ellipses on the parameters that should contain 95% of the weight. The linear dimensions of these ellipses shrink by $\sim 1/\sqrt{10}$ as N increases by a factor of 10. The (log) area inside the ellipses measures the entropy of the uncertainty in parameters, and decreases in this area correspond to gains in information.

Note that $\mathcal{H}$ is a $K \times K$ matrix, and so $\det(N\mathcal{H}) = N^K \det(\mathcal{H})$. This representation allows us to group together terms based on their N dependence,

$$- \ln P(\{x_i, y_i\}|\text{class}) \approx - \sum_{i=1}^{N} \ln P(x_i) + \frac{1}{2}\chi^2_{\min} + \frac{N}{2}\ln(2\pi\sigma^2) + \frac{K}{2}\ln N + \cdots,$$

(6.296)

where the first three terms are $\propto N$, and the omitted terms (including what we have neglected in the saddle point approximation) are constant or decreasing as $N \to \infty$. Again, the negative log probability measures the length of the shortest code for $\{x_i, y_i\}$ that can be generated given the class of models.

In Eq. (6.296), the first term averages to N times the entropy of the distribution $P(x)$, which makes sense, because by hypothesis the xs are being chosen at random. The second and third terms are as before, the length of the code required to describe the deviations of the data from the predictions of the best-fit model; this length also grows in proportion to N. The fourth term must be related to coding our knowledge of the model itself, because it is proportional to the number of parameters. We can understand the $(1/2) \ln N$ because each parameter is determined to an accuracy of $\sim 1/\sqrt{N}$, as in Fig. 6.40, so if we start with a parameter space of size ~ 1, there is a reduction in volume by a factor of $\sqrt{N}$ and hence a decrease in entropy (gain in information) by $(1/2) \ln N$. Finally, the omitted terms do not grow with N.

Problem 165: Deriving the code length in a class of models. Fill in the details leading to Eq. (6.296). Find an explicit form for the omitted terms, and show that they do not grow with N. What assumptions do you need to make about the prior distribution $P(\boldsymbol{\alpha})$ to do this problem?

What is crucial about the term $(K/2) \ln N$ in Eq. (6.296) is that it depends explicitly on the number of parameters. In general we expect that by considering models with more parameters we can get a better fit to the data, which means that χ^2 can almost always be reduced by considering more complex model classes. But we know intuitively that this trend has to stop—we do not want to use arbitrarily complex models, even if they do provide a good fit to what is observed. It is attractive, then, that if we look for the shortest code that can be generated by a class of models, there is an implicit penalty or coding cost for increased complexity. It is interesting from a physicist's point of view that this term emerges essentially from consideration of phase space or volumes in model space. It is thus an entropy-like quantity in its own right, and the selection of the best model class can be thought of as a trade-off between this entropy and the "energy" measured by χ^2, a view to which we return below.

Thus, Eq. (6.296) tells us that there is a natural penalty for the complexity of our model. Although this term is linear in the number of parameters, it is only logarithmic in the number of data points. In contrast, $\chi^2_{\min}$ decreases with the number of parameters and is linear in the number of data points. In this way, the penalty for complexity becomes (relatively) less important the more data we gather: if we have only a few data points, then although we could lower χ^2 by fitting every wiggle, the phase space factor pushes us away from this solution toward simpler models; if, however, the wiggles are consistent as we collect more data, then this factor becomes less important, and we can move to the more complex models.

To see that this description really corresponds to a quantitative theory, we have to generate a data set and go through the process of fitting by means of minimization of the code length in Eq. (6.296). Let's consider polynomial functions. We can pick a polynomial by choosing coefficients a_μ at random, say, in the interval $-1 < a < 1$, where

$$f(x) = \sum_{\mu=0}^{K_{\text{true}}} a_\mu x^\mu. \tag{6.297}$$

We confine our attention to the range $-5 < x < 5$; in this range the function $f(x)$ has some overall dynamic range (measured, e.g., by its variance over this interval), and we assume the noise variance σ^2 is 1% of this signal variance. Then we can generate points according to

$$y_n = f(x_n) + \eta_n, \tag{6.298}$$

and try to fit them. Fitting to any polynomial of degree K by minimizing χ^2 is a standard exercise, and in this way we find $\chi^2_{\min}(K)$. Then we can find the value of K that minimizes the total code length in Eq. (6.296); this last step is just a competition between $\chi^2_{\min}(K)$ and $(K+1) \ln N$. The results of this exercise are shown in Fig. 6.41.

Figure 6.41 shows that our qualitative description of the competition between complexity and goodness of fit really works. First we note that with a large number of data points, minimizing the code length zeroes in on the correct order of the underlying polynomial ($K \to K_{\text{true}}$), despite the presence of noise that could be "fit" using more

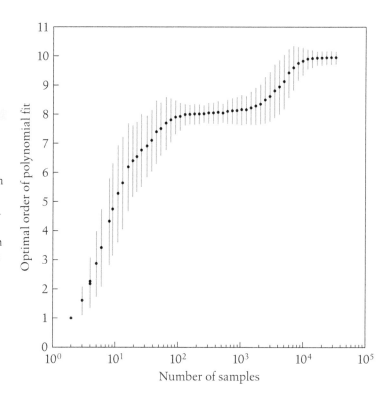

FIGURE 6.41

Fitting to polynomials, part two. Choose the coefficients of a polynomial with degree 10 at random, and then choose points at random in the interval $-5 < x < 5$; there is added noise (as in Eq. (6.298)) with a standard deviation set to 1/10 of the overall dynamic range of the function $f(x)$. We then try to fit polynomials of degree K and find the value of K that minimizes the code length in Eq. (6.296). Because the result depends both on the particular value of the polynomial coefficients and on the particular points x_n that we happen to sample, we choose 500 examples and look at the mean (points) and standard deviation (error bars) across this ensemble of examples. Although the optimal order of the polynomial in any given example is an integer, fractional values arise from averaging over many examples.

complex models. Next, we see that for smaller numbers of data points, the shortest code is biased toward simpler models. In the limit of only a handful of data points, the shortest code is often a straight line ($K = 1$). Put another way, we start with a bias toward simple models, and only as we uncover more data can the adding of greater complexity be supported.

Problem 166: Fitting and complexity. Generate a version of Fig. 6.41 for yourself, doing a simulation that follows the steps outlined in the text. If you do this using MATLAB, you will find the command `polyfit` to be useful.

(a) Start with a small version of the problem (e.g., fitting to $N = 20$ data points).

(b) Plot some of your intermediate results, just to get a feeling for what is going on. In particular, plot χ^2_{min} as a function of K, verifying that higher order polynomials always give better fits in the sense of smaller χ^2.

(c) Notice that χ^2_{min} is a function of K and N, but also a function of the particular points $\{x_i, y_i\}$ you have "observed" in the experiment and of the particular parameters $\{a_\mu\}$ that specify the real function you are trying to learn. When you choose a different set of parameters and test points $\{x_i\}$, from the same distribution, how different is the minimum energy per data point $\epsilon_{min} = \chi^2_{min}/N$ as a function of K? What happens to this variability as N gets larger?

(d) Perhaps the most important thing is to verify that minimizing the code length really does control the complexity of the fit, selecting a nontrivial optimum K. Convince yourself that, as in Fig. 6.41, the optimal K is smaller than K_{true} for small data sets and approaches K_{true} as you analyze larger data sets.

There are many reasons to prefer simpler models, and certainly the idea that we entertain more complex models only as we collect more data is in accord with our sense of how we understand the world. But all of this reasoning can seem a little soft and squishy. Indeed, given the evident complexity of life and the world around us, you might start to suspect that the preference for simple models is not an objective principle but rather a subjective choice made by humans—and more often by scientists than by humans in general.[22] Even some technical discussions leave this impression of subjectivity, suggesting that although there must be a trade-off between goodness of fit and complexity, the structure of this trade-off is something that we are free to choose, perhaps inventing a new "penalty for complexity" tuned to the details of each problem. As physicists we are raised to be suspicious of overly complex models, but again this preference for simplicity is often couched in (surprisingly) soft words about the elegance or brevity of the equations that describe the model. What we have seen here is that this notion can be made much more precise.

The power of information theory in this context is that, by consistently measuring code lengths in bits, we do not have to discuss our preference for simplicity as a separate principle from goodness of fit. Deviations from the model (badness of fit) and the complexity of the model both add bits to the overall code length, and the relative contributions are calculable with no adjustable constants. The absence of unknown constants is important, because if we had to specify weights for the different terms, we would once again inject subjectivity into the discussion of just how much we care about simplicity. Instead, we have one principle (search for the most compact description) and everything else follows. In particular, what follows is that limited experience (small N) biases us toward simpler models; as we accumulate more experiences (ultimately, as $N \to \infty$) we can admit more complex descriptions of the world.

This picture is satisfying, and I am inclined to say that we can declare victory—we understand what we are doing when making models, why simple models are preferable, and how the support for more complex models emerges. Nonetheless, there are several loose ends, and I am not sure that I know how to tie them all up.

The first and most obvious problem is that our discussion makes sense as long as we specify in advance a class of models, and more seriously, a hierarchy of such classes with increasing complexity. It is not at all obvious how to do this. Worse yet, plausible but wrong ways of selecting classes of models can lead to weird results, for example, if we have a function well described by a Fourier series with just a few terms, but we try fitting it to polynomials. Simplicity and complexity have meaning as code lengths only if we have a defined ensemble of possibilities to choose from, in much the same way that Shannon's original discussion of the information gained on hearing the answer to a question (Section 6.1) starts with the assumption that we know the distribution out of which answers will be drawn.

A second, and perhaps related, problem is that we are discussing models with a finite number of parameters. It might seem more natural, for example, to imagine that the relationship between x and y is just some smooth function, not necessarily describable with a finite number of parameters; that is, $f(x)$ should live in a function space and not in a finite-dimensional vector space. Now we have to specify a prior distribution not on the parameters, as with $P(\boldsymbol{\alpha})$ above, but on the functions themselves, $P[f(x)]$.

---

22. Even among scientists, physicists have a special affinity for simple models, often to the point of being the punchline in jokes, as in "consider the case of the spherical horse."

The simplest version of this problem is not with functional relations but just with probability distributions: suppose that we observe a set of points $x_1, x_2, \cdots, x_N$, which we assume are drawn randomly and independently out of a distribution $Q(x)$. How do we estimate Q? If the distribution we are looking for belongs to a family with a finite number of parameters, we proceed as before, but if all we know is that $Q(x)$ is a smooth function, then we have to specify a prior probability distribution on this space of distributions. From a physicist's point of view, probability distributions on such function spaces are just scalar field theories, and one can carry a fair bit of technology over to do real computations. The lesson from these computations is that, with some reasonable priors to implement what we mean by "smooth," everything works as it does in the case of finite parameters, but the prior does matter.

Problem 167: Taming the singularities.   The basic problem in trying to learn about a continuous probability distribution is to explain why, having observed a set of points $x_1, x_2, \cdots, x_N$, we shouldn't just guess that the distribution is of the form

$$Q(x) \sim \frac{1}{N} \sum_{i=1}^{N} \delta(x - x_i), \tag{6.299}$$

which of course generates precisely the data we have observed with maximal (infinite!) probability density. We all know that this is the wrong answer. The role of priors on the space of distributions is to express this knowledge. A very different approach to taming the singularities is sometimes called "kernel density estimation," in which we search for a probability distribution in the form

$$Q(x) = \frac{1}{K} \sum_{j=1}^{K} \frac{1}{\ell} F\left(\frac{x - y_j}{\ell}\right), \tag{6.300}$$

where ℓ is a characteristic length scale, $F(z)$ is some "bloblike" function, and the y_j are the centers of the blobs; F is normalized so that $\int dz \, F(z) = 1$. For concreteness let

$$F(z) = \frac{1}{\sqrt{2\pi}} e^{-z^2/2}. \tag{6.301}$$

If we let $K = N$ (generally not such a good idea), then it should be clear that the model that generates the data with the highest probability is one in which the kernel centers are on top of the data points, $y_i = x_i$ for all i. It should also be clear that this probability of the data increases for smaller ℓ, diverging as $\ell \to 0$. But we know that, to get control over complexity, we should compute the total probability of generating the data in this class of model. In this case the parameters of the model are the kernel centers $\{y_i\}$. Assume that everything happens in a box, so that $0 < x < L$, and similarly for $\{y_i\}$. By translation invariance the prior on the ys should be flat in this box. Calculate the total probability that this class of models generates the data in the limit $\ell \to 0$. Is the answer finite? If so, it means that the phase space factors are just strong enough to compensate for the goodness of fit and prevent anything from diverging in this limit. Can you find any other approximations that allow you to say anything about the optimal value of ℓ?

Quite generally, when we compute the total probability that a model can generate data, we are doing integrals like

$$P(\{x_i\}|\text{model class}) = \int DQ \, P[Q(x)] \prod_{i=1}^{N} Q(x_i), \qquad (6.302)$$

where $P[Q(x)]$ is the probability distribution function(al) on the space of distributions. It embodies all our prior knowledge, in whatever form—that the distribution can be described by a few parameters, or merely that it is smooth in some sense. To understand what is happening in this integral, it is useful to measure possible distributions $Q(x)$ relative to the true distribution $Q_{\text{true}}(x)$:

$$P(\{x_i\}|\text{model class}) = \left[\prod_{i=1}^{N} Q_{\text{true}}(x_i)\right] \int DQ \, P[Q(x)] \prod_{i=1}^{N} \left[\frac{Q(x_i)}{Q_{\text{true}}(x_i)}\right]. \qquad (6.303)$$

We can collect the product into an exponential,

$$P(\{x_i\}|\text{model class}) = \left[\prod_{i=1}^{N} Q_{\text{true}}(x_i)\right]$$
$$\times \int DQ \, P[Q(x)] \exp\left[N \frac{1}{N} \sum_{i=1}^{N} \ln\left(\frac{Q(x_i)}{Q_{\text{true}}(x_i)}\right)\right]. \qquad (6.304)$$

The average over data points x_i in the exponential approaches, at large N, an average over the true distribution:

$$P(\{x_i\}|\text{model class})$$
$$\rightarrow \left[\prod_{i=1}^{N} Q_{\text{true}}(x_i)\right] \int DQ \, P[Q(x)] \exp\left[N \int dx \, Q_{\text{true}}(x) \ln\left(\frac{Q(x)}{Q_{\text{true}}(x)}\right)\right]. \qquad (6.305)$$

In the exponential we have the Kullback-Leibler divergence between the true distribution $Q_{\text{true}}(x)$ and the candidate distribution $Q(x)$. It will be useful to denote this quantity by ϵ, because it will play the role of an effective energy in an analogy with statistical physics:

$$\epsilon = \int dx \, Q_{\text{true}}(x) \ln\left(\frac{Q_{\text{true}}(x)}{Q(x)}\right). \qquad (6.306)$$

But then in integrating over all possible distributions, if the only important quantity is ϵ, we should count the (weighted) volume of distributions that have a particular value of this energy:

$$\mathcal{N}(\epsilon) = \int DQ \, P[Q(x)]\delta\left[\epsilon - \int dx \, Q_{\text{true}}(x) \ln\left(\frac{Q_{\text{true}}(x)}{Q(x)}\right)\right]. \qquad (6.307)$$

Putting these definitions together, we can write

$$P(\{x_i\}|\text{model class}) \rightarrow \left[\prod_{i=1}^{N} Q_{\text{true}}(x_i)\right] \int d\epsilon \, \mathcal{N}(\epsilon) e^{-N\epsilon}. \qquad (6.308)$$

The remaining integral really does look like a statistical mechanics problem. In this view, ϵ is the energy (per data point), and $\mathcal{N}(\epsilon)$ is a density of states. The logarithm of the density of states is an entropy, so that $\mathcal{N}(\epsilon) = e^{S(\epsilon)}$.

Entropy is an extensive quantity. Thus, if the models we consider have K parameters, then there are effectively K degrees of freedom in our statistical mechanics problem, and so we expect $S(\epsilon) \propto K$. Thus, we can define an entropy density $s(\epsilon) = S(\epsilon)/K$, and then we have

$$P(\{x_i\}|\text{model class}) \propto \int d\epsilon \, \exp\left[-N\left(\epsilon - \frac{K}{N}s(\epsilon)\right)\right]. \tag{6.309}$$

At large N, the integral is dominated by the minimum of the free-energy density, $g(\epsilon) = \epsilon - Ts(\epsilon)$, where the role of temperature is played by $T = K/N$. This calculation makes explicit the idea that learning is statistical mechanics in the space of models, and that seeing more examples corresponds to lowering the temperature, cooling the system into an ordered state around the right answer. Depending on the space of possible models, and hence the function $s(\epsilon)$, there can be phase transitions—a sudden jump, as we collect more examples, from wandering around in model space to having a compelling fit to the data.

What would it mean to have a phase transition in learning? As we accumulate more examples, we are lowering the effective temperature in the equivalent statistical mechanics problem. At first this does not do much, in the same way that lowering the temperature of water from 80° C to 30° C does very little. But, at some point, a relatively small change in the number of examples we have seen produces a huge change in the distribution over models, freezing into a small volume surrounding the correct answer. This process would be something like the subjective "aha!" experience, where we suddenly seem to understand something or master a skill after a long period of exposure or training. Although we have all (I hope) experienced this phenomenon, it is not so easy to study quantitatively, and so I think we have no idea whether the statistical mechanics approach to learning provides a useful guide to understanding this effect.

It is interesting to look at the history of studies in animal learning in light of these results. Already in the 1920s and 1930s it was clear that, at certain tasks, animals could exhibit sudden rather than gradual learning. Although this phenomenon was observed well before Hebb, and decades before the observation of changes in synaptic strength driven by the correlation between pre- and postsynaptic neurons (see Section 5.4), there was already a general view that learning relied on statistical association and thus should be a continuous process. Thus, there was a question whether sudden learning represents a new mechanism beyond associative processes. The mapping of learning onto a statistical mechanics problem reminds us that when there are many degrees of freedom, continuous dynamics can have nearly discontinuous consequences.

Before leaving the image of energy-entropy competition, we should note a caveat. In getting to Eq. (6.309), we allowed N to become very large, so that averages over samples can be replaced by averages over the underlying distribution, and then used the resulting formulas with finite N to say something about how learning proceeds. Evidently this process is dangerous. It also was controversial when it first emerged, because the results seemed to conflict with an approach used by computer scientists that emphasized bounds on the learning curve. To explain how this controversy was resolved would take us far afield, so I will simply point to the references in the Annotated Bibliography for this section. The basic summary is that although I have cheated a bit, I have not lied: there is a well-defined approximation that leads to Eq. (6.309), and the resulting predictions can be made rigorous and shown to be consistent with known bounds. To be fair to the history, I should also note that the idea of phase transitions in learning did not emerge from general arguments but from concrete problems in learning by model neural networks. This is important, because it emphasizes that phase transitions are not just abstract possibilities but actually occur in systems that are not implausible models of real brains.

To what extent is our (or other animals') performance in situations where we learn understandable in terms of the theoretical structures we have been discussing? In the examples discussed above, what is being learned is a probability distribution, or some set of parameters describing the data that we observe. It is not so easy to ask even human subjects to report on their current estimates of these parameters, and it is completely unclear how we would do this in other organisms. In practice, subjects are usually asked to make a decision; in classical work on pigeons the decision is to peck or not to peck at a target, and humans are usually asked a yes-or-no question or asked to push one of a small set of buttons. Evidently the bandwidth of these experiments is limited— although we may be continuously updating an internal model with many parameters, what we report is on the order of one bit: yes or no.

One context that comes closer to the theoretical discussion, albeit in a simple form, concerns making decisions when the alternatives come with unequal probabilities. This scenario harkens back to our earliest topic, a human observer waiting for a dim flash of light in a dark room (see Section 2.1). As discussed in that context, optimal decisions— deciding that a signal is convincingly above the background of noise—are achieved by setting a threshold that depends on the probability that the signal is present. If this probability can change over time, then it must be learned. More prosaically, if we have to choose between two alternatives even in a limit where they are fully distinguishable, but the rewards for the different choices vary probabilistically, then we have to learn something about the underlying probabilities of reward to develop a sensible strategy. These experiments have attracted interest, because they might connect to our economic behavior, and because they provide settings in which we can search for the neural correlates of the subject's estimate of probability and value.[23]

There is a classical literature showing that human observers adjust their criteria for detecting signals to the probability that the signals occur. The question about learning is really how long it takes the subject to make this adjustment. In the simplest case, the probability changes suddenly, and we look for a change in behavior as a response. If the only behavioral output is a decision among two alternatives, we as observers also need to go through an inference process to decide when the first sign of a response occurs. In such an experiment, we have a complete probabilistic description of the trajectory taken by the sensory stimuli or rewards, so at any moment we can calculate the probability that the signals being shown are consistent with constant parameters

23. I think it is fair to say that the concept of value has attracted more attention in this context, because it seems more connected to economics. Indeed, there is now a whole field described as neuroeconomics. But perhaps the probabilistic nature of our inferences, even in the economic context, has been given less attention than it deserves.

or a recent, sudden change. Given the responses of the subject, we can also ask for the moment at which we see the first statistical sign of a change in behavior. In experiments where rats experience changing reward probabilities, the change in behavior occurs at times so soon after the changes in probability that the best evidence for the change is modest, corresponding to probabilities in the range 0.1–0.9; only rarely (in ~20% of trials) do rats wait to reach 99% certainty. On these very short time scales, the rate at which the rat collects rewards changes very little, suggesting that changes in strategy really are driven by learning the underlying probabilities rather than by tinkering until rewards accumulate.

In a similar spirit, we can do a longer experiment, as shown in Fig. 6.42, with the probabilities jumping among different levels, and track the dynamics of the behavior. These experiments were done in monkeys, with the goal of finding the neural correlates of the probability–value judgements being made by the observers, but our concern here is with the behavior itself. While making decisions, all the subject has access to is the discrete time series of rewards—if I push the left button, I may or may not be rewarded, but I have no way of knowing what would have happened if I had pushed the right button. Choice behavior should somehow be connected to the underlying probabilities of reward, and in the simplest model the observer makes estimates of these probabilities just by smoothing or filtering the time series of rewards themselves.

More directly, we can imagine that the behavioral choices themselves are based, probabilistically, on the outputs of such filters. If the time scale of smoothing is too short, then estimates will be noisy; whereas if the time scale is too long estimates will average together times when the probabilities are genuinely different. Analyzing the behavioral data, we can extract the filter that the monkey is actually using and then compare the time constants of this filter with the optimum. As shown in Fig. 6.42, the actual time constants are nearly equal to the optimal ones, and the differences correspond to a less than 1% difference in the rate at which the subjects collect rewards. Thus, the subjects in these experiments not only behave as if they were basing their behavior on running estimates of the underlying probabilities in their world, the algorithm they use for making these estimates is also near the optimum and hence is adapted to the statistics of that world.

Problem 168: Tracking probabilities.   A simplified version of the problem faced by the subjects in the experiments shown in Fig. 6.42 is as follows. Suppose that we observe a discrete, binary time series, $n_t = 0, 1$, at each time step t. There is an underlying probability p_t for observing $n_t = 1$ at each moment, and this probability varies in time. Faced with the time series n_t, one can try to estimate the probability as a running average over previous experience:

$$p_t^{\text{est}} = \sum_{\tau=1}^{\infty} f(\tau) n_{t-\tau}. \tag{6.310}$$

Find a formal expression for the filter $f(\tau)$ that minimizes the mean-square error $\langle |p_t - p_t^{\text{est}}|^2 \rangle$. Suppose that variations in the probability are correlated over some time τ_c, so that

$$\langle p_t p_{t'} \rangle = \frac{1}{4} + \langle (\delta p)^2 \rangle e^{-|t-t'|/\tau_c}. \tag{6.311}$$

(a)

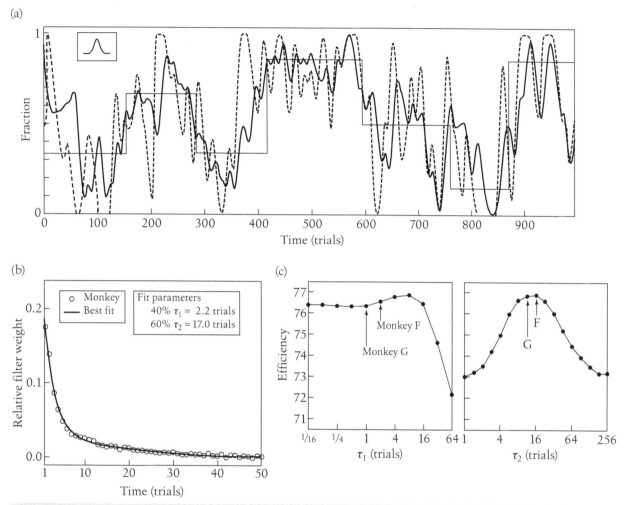

FIGURE 6.42

Tracking the changing probabilities of reward. Redrawn from Corrado et al. (2005). (a) The local frequencies of choosing one of two alternatives (heavy solid line) and being rewarded (dashed line), when the probability of this choice being rewarded jumps among different levels as shown (thin line). Frequencies are computed from discrete events by smoothing with the Gaussian kernel shown in the inset. (b) The filter inferred from the relationship between rewards and subsequent choices, together with a double exponential fit (solid line). (c) The efficiency of collecting rewards averaged over the whole session, assuming that the subject implements a filter with the time constants as shown in panel (b). The subjects' behaviors are best fit by parameters that generate efficiencies within 1% of the optimum.

What is the form of $f(\tau)$? How does the time over which this filter averages relate to the correlation time τ_c?

One approach to adding bandwidth to experiments on learning is to average over many subjects, so that the performance after N examples can be measured as a real number (e.g., the probability of subjects getting the right answer), even though the data from individuals is discrete (yes-or-no answers). But, as emphasized in Fig. 6.43, this

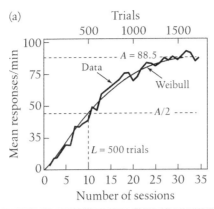
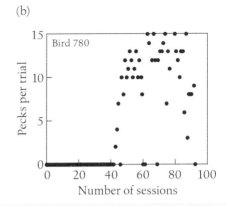

FIGURE 6.43

Learning curves in individuals versus groups. Redrawn from Gallistel et al. (2004). Pigeons experience a classical Pavlovian world, in which the illumination of a key on the wall of the experimental chamber is correlated with the appearance of a food reward, and they learn to peck at the key in anticipation of the food reward. Performance is measured by counting pecks, either per unit time or per trial, and smooth curve is a fit to the Weibull distribution. (a) Average performance in a large population of birds improves gradually and very slowly, requiring many hundreds of trials before reaching its half-maximal level. (b) Performance measured in one individual bird is noisy (because we have access only to the number of pecks as a behavioral output) but makes a relatively sudden transition to near saturating performance as the animal experiences ∼10 additional examples.

approach can be misleading. Individual subjects seem to learn simple tasks abruptly, but with transitions after different numbers of trials, so that average learning curves are smooth and gradual. This result is interesting, because more abrupt learning reminds us of performance as a function of signal-to-noise ratio in discrimination tasks, and because theory along the lines described above often predicts relatively rapid learning when the space of possibilities is small. The variations across individuals may then reflect differences in how the small problem posed by the particular experimental situation is weighted within the much larger set of possible behaviors available to the organism. But much needs to be done to make this notion precise.

Another approach to increasing the bandwidth of behavioral experiments is to look at continuous motor outputs rather than at decisions. An example is if we have to move an object through a medium that generates an unknown, anisotropic mobility tensor; as we practice, we learn more about the parameters of this environment and can move more accurately. Importantly, each trial of such an experiment generates an entire movement trajectory rather than just a single discrete decision. Analysis of these trajectories can reveal how the errors made in one trial influence the change of our internal model on the next trial. This experiment emphasizes learning of parameters that influence the movement itself. But more generally, that some movements are made in extraordinarily precise relation to sensory inputs (e.g., as we follow a moving target with our eyes) and that we can learn to anticipate the need for such movements (e.g., as targets follow predictable trajectories), suggest that analysis of continuous movements should provide us with a path to examine more details of the brain's internal model of the world. A simple version of this idea is that the latency for us to move our eyes toward one of two suddenly appearing targets depends on the relative probabilities of

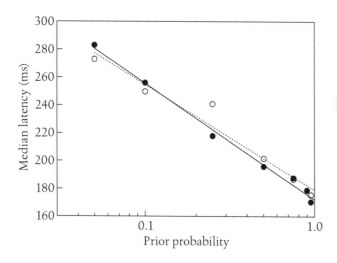

FIGURE 6.44

Latency for saccadic eye movements to targets of varying probability. Subjects are asked to move their eyes to a target that appears at a random time after they fixate on a small spot. The target is either to the left or right, with varying probabilities in blocks of trials. For two subjects (filled and empty symbols), the experimenter collects all trials in which the subjects move to a target of probability p and computes the mean latency of the eye movement. Redrawn from Carpenter and Williams (1995).

the targets—we move more quickly toward targets of higher probability, as shown in Fig. 6.44. It is tempting to think that the latency of movement gives us a readout of the brain's estimate of this probability. Again, there is much to do here.

Finally, a theoretical point. I have emphasized that learning a model amounts to building an efficient representation of the data observed, and hence the goal of learning is no different than the goals proposed in the previous section for the transmission of information through neural or genetic networks. This theoretical unity is attractive. But you might worry—why do we care about representing what we have observed in the past? What matters, to follow the discussion at the end of Section 6.3, is what is of use in guiding actions in the future. Thus, presumably we learn models that describe data collected in the past because we expect these models to still be true in the future, which allows us to make successful predictions. How does this logic connect to our ideas about efficient representation?

Recall from Section 6.3 that the predictive information in a time series (i.e., the information that observations on the past provide about the future) is equal to the subextensive component of the entropy. In the course of evaluating the probability of data given a class of models, in Eq. (6.296), we have implicitly calculated this subextensive entropy. Specifically, the negative log probability of a set of data at N time points has a term proportional to N (the extensive piece) and a term that grows only logarithmically with N (the leading subextensive piece). Thus, when we are observing a time series from which we can learn a model with K parameters, there is a subextensive entropy and hence predictive information of $\sim (K/2) \log_2 N$ bits. The "meaning" of this predictive information is precisely that we know something about the parameters underlying the data, and on the hypothesis that these parameters are constant, we can predict something about the future.

Problem 169: Predictive information in learning. Imagine that we observe a random variable x, drawn from a Gaussian distribution with zero mean. A priori the standard deviation itself is drawn from some distribution $Q(\sigma)$. We have observed N samples, $x_1, x_2, \cdots, x_N$, and we can expect to see a future in which there will be M more samples, $x_{N+1}, x_{N+2}, \cdots, x_{N+M}$.

We can think of the samples we have seen as our past, $X_{past} \equiv \{x_1, \cdots, x_N\}$, and the coming samples as our future, $X_{future} \equiv \{x_{N+1}, \cdots, x_{N+M}\}$. We are interested in the problem of extracting from X_{past} what is relevant for predicting X_{future}.

(a) Construct the joint distribution $P(X_{past}, X_{future})$ as an integral over σ. Without actually evaluating the integral, show that for fixed N, the conditional distribution $P(X_{future}|X_{past})$ obeys

$$P(X_{future}|X_{past}) = P(X_{future}|S), \tag{6.312}$$

$$S = \frac{1}{N} \sum_{i=1}^{N} x_i^2. \tag{6.313}$$

That is, optimal predictions of the future in this simple problem depend only on a running estimate of the variance. This dependence is described by saying that the running variance provides a sufficient statistic for prediction.

(b) What is the distribution of the running variance estimate, $P(S)$? Start by writing an exact expression, as an integral over all the observations $x_1, x_2, \cdots, x_N$, and then give an approximate result at large N.

(c) We would like to compress our description of the past into some internal representation, $X_{past} \to X_{int}$. As in Eq. (6.211) and the subsequent discussion, the goal is to maximize the information that X_{int} conveys about X_{future} ($I(X_{int}; X_{future})$), while constraining the amount of information that it captures about the past ($I(X_{int}; X_{past})$). Show that, from your results in part (a), the optimal mapping $X_{past} \to X_{int}$) obeys

$$P(X_{int}|X_{past}) = P(X_{int}|S). \tag{6.314}$$

You might guess that a good choice for X_{int} is a noisy and/or quantized version of S itself. What does this mean for the distribution $P(X_{int}|S)$? Explore this ansatz, and show that it provides a consistent solution to the optimization problem in Eq. (6.211).

(d) What you have shown in part (c) is that the optimal representation of predictive information in this problem is equivalent to learning the underlying parameter of the probability distribution that describes the data. Can you go further and show that this learning is in some sense optimal, providing the best possible estimate of the parameter given the data? Can you generalize to cases where the relevant distribution is described by more than one parameter? These are deliberately open-ended questions.

When we observe N data points, the total amount of information we have collected is a number of bits proportional to N. But in the case considered here, there are just $\sim(K/2) \log_2 N$ bits of information about the future. If we can separate these predictive bits from the nonpredictive background, we will have discovered the parameters of the underlying model. Thus, compressing the data while preserving the predictive information is exactly the same problem as learning. Interestingly, if we live in a world described by a complex model (large K), then the amount of predictive information is much larger than the information needed to describe the present.

6.6   Perspectives

Optimizing information transmission, or maximizing the efficiency with which information is represented, is the sort of abstract general principle that physicists find appealing. At the same time, this abstraction makes us suspicious about its relevance to the nitty-gritty of life. Thus, although information is essential for survival, surely much of what organisms do is bound up in the fact that some bits are more useful than others and in the challenges of acting rather than just collecting data. In this chapter we have seen both how interesting predictions flow from the abstract principles and how these principles connect, sometimes surprisingly, to the more quotidian facts of life. It is surely too soon, in this as in any other section of the book, to decide whether some of these candidate theoretical principles are on the right track, and in any case I am not a disinterested observer. What I emphasize here is that thinking about the optimization of information transmission has been productive, not least because it suggests genuinely new kinds of experiments. In many systems, these experiments have generated interesting results, independent of the theoretical motivation. In many other systems, even the first generation of experiments remains to be done.

Perhaps the most important point about information-theoretic optimization principles is that they force us to think about biological systems in context. Whereas classical biology routinely considered organisms in their natural settings, as biology has modernized and become more mechanistic, we see more and more work on systems shorn of their context. To give an example, it may be that the best-studied system for the regulation of gene expression is the *lac* operon in *Escherichia coli*. But how much do we know about the distribution of lactose concentrations encountered by these cells in their natural environments? We know that, under many conditions, the total number of *lac* repressor proteins in the cell is small, but what is the dynamic range of this number over the lifetime of the organism? Vastly more is known about the details of the DNA sequences that are targeted by transcription factors involved in the regulation of metabolic genes than is known about the real-world variations in nutrient conditions that create the need for metabolic regulation.

In the case of neural information processing, the ethologists—who often study systems specialized for the processing of particular sense data, such as bird song or bat echolocation—provided a persistent reminder about the importance of the natural context in understanding biological function. Perhaps our human ability to deal with a seemingly much wider range of data and tasks generates some resistance to thinking that lessons from a barn owl or an electric fish could be of relevance to how we explore higher brain function. The claim that at least some aspects of neural circuitry are arranged to generate efficient representations of incoming sense data provides a counterpoint, suggesting that even for a "general purpose" sensory system, context matters. By now there is a whole subfield of neuroscience focused on the structure and processing of natural signals, a field we might think of as a modern quantitative development of the early work in ethology. Because our sense organs are such high-quality devices, there are substantial experimental challenges in characterizing their natural inputs and delivering controlled versions of these natural signals in the laboratory. Precisely because natural signals are rich and complex, analyzing neural responses to these signals poses significant theoretical challenges. Progress on these experimental and theoretical problems is giving us more powerful tools with which to explore the brain, again independent of the sometimes distant motivation of optimization principles.

Thinking about information flow encourages us to ask about the structure of natural behavioral outputs as well as of natural sensory inputs. In the attempt to quantify animal (and human) behavior in the laboratory, there has been a tradition of constraining this behavior to a small discrete set of alternatives. This approach has been enormously powerful, not least because such constrained experiments are amenable to analyses in terms of signals and noise, as in our initial discussion of photon counting in vision (Chapter 2). Similarly, experiments on the control of gene expression in single-celled organisms often have focused on the switch in expression patterns associated with a sudden transition from one nutrient source to another. Even the ethologists tended to categorize, collapsing whole ranges of behavior onto a limited space of discrete choices. But behavior, in single cells or in entire humans, is vastly richer than choosing among discrete alternatives. As the technology for monitoring behavior improves, it becomes possible to ask whether the continuous variations in natural behaviors are just noise or are related systematically to goals and context. Even if behavior really is composed of choice among a small set of stereotyped possibilities—such as running and tumbling in *E. coli*—the timing of these choices can convey considerable information about the sensory inputs that drive them.

We have the impression that we are bombarded by complex data and that our behaviors are relatively limited. But the inputs to our sensory systems are highly structured, presumably because they derive from a limited set of causes and effects in the environment, and hence carry much less information about what is really out there than might be guessed from the available bandwidth; our receptors provide limited, noisy views of these inputs, reducing the information still further (see, e.g., Problem 163). At the other end, our motor outputs are quite rich, even if we tend to coarse grain and categorize these behaviors into limited classes. Could it be that motor outputs are so carefully shaped and timed in relation to sensory inputs from the environment that we (and other organisms) are making use of a large fraction of the information available about this environment? There is a huge experimental challenge in tracking information flow all the way from sensory input to motor output, even in simple cases; in more complex cases there is a substantial theoretical challenge in providing a framework for the analysis of such data.

One of the most important aspects of information theory is that bits have value. This is why, for example, there is a minimum number of bits needed to be sent over a telephone connection to be sure that speech is intelligible and speakers identifiable. For living organisms, the value of bits depends on many details, perhaps more detail than, as physicists, we would like to think about. What we can say, however, is that bits with no predictive power are valueless and that most bits we have collected over our lifetimes are in this valueless category. Thus, separating predictive information from the background of nonpredictive clutter is a formidable, and biologically relevant, challenge. Importantly, this general task seems to contain within it, as special cases, problems ranging from signal processing to learning, problems that we usually think of as belonging to different levels of biological organization with very different mechanisms. Perhaps this is, after all, a path to the sort of general principle we are seeking.

Outlook

Writing a book with the subtitle "Searching for Principles" invites an obvious question: did you find any? This seems an especially urgent question given that the book has grown so long, admittedly much longer than I thought it would be when I started the project. Let's go through the candidate principles one by one, to see what we have learned, and then look ahead. I'll be brief.

Noise Is Not Negligible

Starting with the case of photon counting, and continuing to many other examples, it is clear that real biological systems approach fundamental physical limits to their noise performance at well-defined detection, discrimination, and estimation tasks. I continue to find this fact amazing. In the context of sensory information processing, it is not so hard to imagine that evolution selects for organisms that suppress extraneous sources of noise and make more efficient use of the available signals from the outside world. Perhaps more surprising are the cases where the organism has control over the scale of the signal—in modulating the concentration of transcription factors, or the gradients of growth factors guiding the wiring of neurons—and yet there still is enough pressure to push these systems to the point where they approach the physical limits to signaling. All these examples point toward the idea that we can elevate the optimization of noise performance or reliability to a principle, and we have seen that this principle allows us to derive some nontrivial aspects of behavior and mechanism in many different systems. The difficulty is that implementing such optimization principles is easiest in cases where the tasks are simple and well defined, and surely this focus on simple tasks misses a great deal of what organisms really do to survive. This is not just a problem for theory: doing experiments in a natural context and providing a quantitative characterization of what organisms are doing in these more complex situations are huge challenges.

No Fine Tuning

Starting with our intuition that a reliance on fine tuning is suspicious, we have seen many examples where experiments demonstrate directly the robustness of biological function to variations in parameters. In some cases this flexibility really is quite astonishing. However, there are many cases in which we have equally clear experimental demonstrations of the limits to parameter variation, so that genuinely random choices do not lead to biological function. Perhaps as a result, the attempts to elevate robustness to a principle have not quite arrived at something completely general. In some cases, we can ask to maximize the volume in parameter space that is consistent with functional behavior, and this approach does seem to select for mechanisms that correspond to the ones chosen by real biological systems. In the limit, we can ensure parameter insensitivity by arranging for biological functions to be topological rather than geometrical features of the underlying dynamics. In other cases, apparent fragility or parameter sensitivity in one layer of mechanism is a hint that there is another layer of control (or learning or adaptation) that allows the whole system to find the relevant parameter regime robustly. Finally, in several cases we have explored models that are explicitly as random as possible while satisfying certain constraints, and these models seem to work, in that they give a good description of the distribution of states that the system can take on as it carries out its functions. All these ideas seem to be related, though not very precisely.

Efficient Representation

Starting with Shannon's theorem that the only way to quantify the intuitive notion of available information is through the entropy, we have seen that there are limits on how much information can be transmitted and on the minimum amount of information required for performance of certain tasks. Once again, several different real biological systems approach these limits, suggesting that we can elevate the optimization of information transmission to a principle. The consequences of this principle are many and varied, from the distribution of expression levels for a gene to the variety of filters found in the inner ear. Information-theoretic optimization can subsume more classical goal-directed behaviors, from search to learning a model of the world around us. There are connections to the problems of noise, but maximizing information transmission is much more than just minimizing noise. Although there are persistent doubts that the generality of information theory can be made consistent with the specificity of the tasks facing real organisms, we have seen that there are ways of attaching value to bits, and thus connecting bits to performance metrics for particular tasks. Most importantly, information-theoretic approaches place biological systems in context and thus focus attention on the matching of mechanisms to the environment in which the organism lives and functions. As with the ideas about optimizing noise performance, there are huge theoretical challenges in going beyond the classical examples and deriving the consequences of optimal information transmission in complex naturalistic environments.

Some Fond Hopes

Back in Chapter 1, I expressed the hope that we might find a theoretical physics of biological systems that is worthy of the name—something that reaches the level of compelling simplicity, generality, and quantitative predictive power that we have come to expect in our theories of the physical world, yet engages with the complexity and diversity of real biological systems. It is too much to claim that the candidate principles we have explored realize this dream, but I do think that our explorations illustrate the feasibility of the quest. There is much to be done along the lines outlined here, but it is also important to point to some grander possibilities that perhaps are now in better focus. We have seen that biological systems are good at many things, and we have been able to make this observation precise in several cases. Perhaps some of these examples of exceptional performance come close to being defining. Is it possible, for example, that being alive is essentially a matter of being able to make predictions, so that actions take on a meaning in relation to the future state of the environment? Can we imagine building an ensemble of chemical reaction networks, for example, that are as random as possible but capture a certain amount of predictive power about the fluctuations in available substrates? Might such an ensemble describe something like the original cells? In a very different direction, we have seen how physical principles can illuminate biological function at many levels, from single molecules to biochemical and genetic networks to the processing of raw sensory data for extracting useful features. But how far can this sort of explanation be taken? Can we imagine reaching toward higher cognitive phenomena? We have hints—from Bayesian estimation in cognitive problems (Fig. 4.40) or from the possibility that meaning can be extracted from the statistics of words (Fig. 6.20)—but there is a widespread sense that, somewhere along the path from the outside world to our subjective experience, physics stops, and some uniquely biological or even uniquely human considerations take over. As noted at the start, Helmholtz hoped otherwise. So do I.

Appendix

Some Further Topics

In these sections I collect topics that are off to the side, or in the background, of the main arguments made in the text. There is a mix of materials some students may have learned in previous courses and those pointing off in new directions. Thus, some readers may find the background essential, whereas others may wish I had pushed more things into the Appendix; yet others may find the asides more interesting than what I thought were the main points. I hope, however, that everyone finds something useful here. As with the main text, I try not to skip steps, and problems are embedded in the narrative.

A.1   Poisson Processes

The defining feature of the Poisson process is that each event (e.g., photon arrival) is independent of all the others, given that we know the rate $r(t)$ at which the events occur. Here we go through the detailed consequences of this simple assumption of independence; hopefully some of the results are familiar. Many textbook presentations make a big deal out of the distinction between a homogeneous Poisson process, in which the rate is a constant, $r(t) = \bar{r}$, and an inhomogeneous one, in which it can depend on time. The general case is not that hard, so I prefer to start with it.

Most conventional light sources are not exactly Poisson, but the approximation is very good. There are many more systems for which the Poisson model is a decent if not excellent approximation, and so we'll discuss these processes without further reference to photons: we are describing the statistics of arbitrary point events that occur at times $t_1, t_2, \cdots, t_N$.

The rate $r(t)$ can be thought of either as the mean rate of events that we would observe in the neighborhood of time t if we did the same experiment many times, or equivalently as the probability per unit time that we observe an event at t. There is the same dual definition for the concentration $c(\mathbf{x})$ of molecules—either the mean number of molecules per unit volume that we find in the neighborhood of a point $\mathbf{x}$, or the probability per unit volume that we observe a single molecule at $\mathbf{x}$.

Because the events are independent, the probability density for observing events at times $t_1, t_2, \cdots, t_N$ must be proportional to a product of the rates evaluated at these times:

$$P[\{t_i\}|r(\tau)] \propto r(t_1)r(t_2) \cdots r(t_N) \equiv \prod_{i=1}^{N} r(t_i). \tag{A.1}$$

But to get the exact form of the distribution, we must include a factor that measures the probability of *no* events occurring at any other times. The probability of an event occurring in a small bin of size $\Delta\tau$ surrounding time t is, by the original definition of the rate, $p(t) = r(t)\Delta\tau$, so the probability of no event must be $1 - p(t)$. Thus, we need to form a product of factors $1 - p(t)$ for all times not equal to the special t_i at which we observed events. Let's call this factor F:

$$F = \prod_{n \neq i}[1 - p(t_n)]. \tag{A.2}$$

Then the probability of observing events in bins surrounding the t_i is

$$P[\{t_i\}|r(\tau)](\Delta\tau)^N = \frac{1}{N!}F\prod_{i=1}^{N}\left[r(t_i)\Delta\tau\right], \tag{A.3}$$

where the $N!$ corrects for all the different ways of assigning labels $1, 2, \cdots, N$ to the events we observe.

To proceed, we pull out all factors related to the t_i and isolate the terms independent of these times:

$$\begin{aligned}
P[\{t_i\}|r(\tau)](\Delta\tau)^N &= \frac{1}{N!}F\prod_{i=1}^{N}\left[r(t_i)\Delta\tau\right] \\
&= \frac{1}{N!}\prod_{n \neq i}\left[1 - r(t_n)\Delta\tau\right]\prod_{i=1}^{N}\left[r(t_i)\Delta\tau\right] \tag{A.4} \\
&= \frac{1}{N!}\prod_{n}\left[1 - r(t_n)\Delta\tau\right]\prod_{i=1}^{N}\left[\frac{r(t_i)\Delta\tau}{1 - r(t_i)\Delta\tau}\right]; \tag{A.5}
\end{aligned}$$

here $\prod_n$ denotes a product over all possible times t_n.

We have two products to compute in Eq. (A.5). Let's work on them separately. First we have

$$A \equiv \prod_{n}\left[1 - r(t_n)\Delta\tau\right] = \exp\left(\sum_{n}\ln\left[1 - r(t_n)\Delta\tau\right]\right). \tag{A.6}$$

We are interested in the case where the time bin $\Delta\tau$ is very small (we introduced these artificially, above), which means that we need to take the logarithm of numbers that are almost equal to one. The Taylor expansion of the logarithm allows us to write

$$\ln\left[1 - r(t_n)\Delta\tau\right] = -r(t_n)\Delta\tau - \frac{1}{2}\left[-r(t_n)\Delta\tau\right]^2 + \cdots, \tag{A.7}$$

so that we have

$$A = \frac{1}{N!} \exp \left(\sum_n \ln \left[1 - r(t_n) \Delta \tau \right] \right) \tag{A.8}$$

$$= \frac{1}{N!} \exp \left(\sum_n [-r(t_n) \Delta \tau] - \frac{1}{2} \sum_n [-r(t_n) \Delta \tau]^2 + \cdots \right). \tag{A.9}$$

This expression involves a sum over bins in the exponential, with factors of the bin width $\Delta \tau$. The sum converges, as the bins become small, to an integral, because for any smooth function $f(t)$, we have

$$\lim_{\Delta \tau \to 0} \sum_n f(t_n) \Delta \tau = \int dt \, f(t). \tag{A.10}$$

In the present case this means that

$$\exp \left(\sum_n [-r(t_n) \Delta \tau] - \frac{1}{2} \sum_n [-r(t_n) \Delta \tau]^2 + \cdots \right)$$
$$\to \exp \left[-\int dt \, r(t) - \frac{1}{2} \Delta \tau \int dt \, r^2(t) + \cdots \right]. \tag{A.11}$$

The second integral in the exponential has an extra factor of $\Delta \tau$, which comes from the $(\Delta \tau)^2$ in the previous expression, but if we really let $\Delta \tau$ go to zero this must be negligible as long as the rate does not become infinite.

The second product that we need to evaluate in Eq. (A.5) is

$$B \equiv \prod_{i=1}^{N} \frac{r(t_i) \Delta \tau}{1 - r(t_i) \Delta \tau}. \tag{A.12}$$

Expanding in $\Delta \tau$ gives

$$B = (\Delta \tau)^N \left[\prod_{i=1}^{N} r(t_i) \right] \left[1 - \sum_{j=1}^{N} r(t_j) \Delta \tau + \cdots \right]. \tag{A.13}$$

Putting the two factors together and keeping only the leading term as $\Delta \tau \to 0$, we have

$$P[\{t_i\}|r(\tau)](\Delta \tau)^N \equiv \frac{1}{N!} A B \tag{A.14}$$

$$= \frac{1}{N!} \exp \left[-\int dt \, r(t) - \frac{1}{2} \Delta \tau \int dt \, r^2(t) + \cdots \right]$$
$$\times (\Delta \tau)^N \left[\prod_{i=1}^{N} r(t_i) \right] \left[1 - \sum_{j=1}^{N} r(t_j) \Delta \tau + \cdots \right] \tag{A.15}$$

$$\to \frac{(\Delta \tau)^N}{N!} \exp \left[-\int_0^T dt \, r(t) \right] \prod_{i=1}^{N} r(t_i), \tag{A.16}$$

so that

$$P[\{t_i\}|r(\tau)] = \frac{1}{N!} \exp\left[-\int_0^T dt\, r(t)\right] \prod_{i=1}^N r(t_i), \qquad (A.17)$$

where we have set the limits on the integral to refer to the whole duration of our observations, from $t = 0$ to $t = T$. Note that this expression is a probability density for the N arrival times $t_1, t_2, \cdots, t_N$ and hence has units of (time)$^{-N}$.

It is a useful exercise to check the normalization of the probability distribution in Eq. (A.17). We want to calculate the total probability, which involves taking the term with N events and integrating over all N arrival times, then summing on N. Let's call this sum Z:

$$Z \equiv \sum_{N=0}^\infty \int_0^T dt_1 \int_0^T dt_2 \cdots \int_0^T dt_N\, P[\{t_i\}|r(t)] \qquad (A.18)$$

$$= \sum_{N=0}^\infty \int_0^T dt_1 \int_0^T dt_2 \cdots \int_0^T dt_N \frac{1}{N!} \exp\left[-\int_0^T dt\, r(t)\right] \prod_{i=1}^N r(t_i). \quad (A.19)$$

The exponential does not depend on the $\{t_i\}$ or on N, so we can take it outside the sum and integral. Furthermore, although we have to integrate over all N different t_i together (an N-dimensional integral), the integrand is a product of terms that depend on each individual t_i. Thus, we really have a product of N one-dimensional integrals, and they turn out to be N copies of the same integral:

$$Z = \exp\left[-\int_0^T dt\, r(t)\right] \sum_{N=0}^\infty \frac{1}{N!} \int_0^T dt_1 \cdots \int_0^T dt_N\, r(t_1) \cdots r(t_N) \qquad (A.20)$$

$$= \exp\left[-\int_0^T dt\, r(t)\right] \sum_{N=0}^\infty \frac{1}{N!} \int_0^T dt_1\, r(t_1) \int_0^T dt_2\, r(t_2) \cdots \int_0^T dt_N\, r(t_N)$$

$$(A.21)$$

$$= \exp\left[-\int_0^T dt\, r(t)\right] \sum_{N=0}^\infty \frac{1}{N!} \left[\int_0^T dt\, r(t)\right]^N. \qquad (A.22)$$

The series expansion of the exponential function is

$$\exp(x) = \sum_{N=0}^\infty \frac{1}{N!} x^N, \qquad (A.23)$$

so we can actually do the sum in Eq. (A.22):

$$\sum_{N=0}^\infty \frac{1}{N!} \left[\int_0^T dt\, r(t)\right]^N = \exp\left[+\int_0^T dt\, r(t)\right]. \qquad (A.24)$$

Substituting, we have

$$Z = \exp\left[-\int_0^T dt\, r(t)\right] \sum_{N=0}^{\infty} \frac{1}{N!} \left[\int_0^T dt\, r(t)\right]^N$$

$$= \exp\left[-\int_0^T dt\, r(t)\right] \times \exp\left[+\int_0^T dt\, r(t)\right] \tag{A.25}$$

$$= 1, \tag{A.26}$$

which completes our check on the normalization of the distribution.

Next we would like to derive an expression for the distribution of counts, written as $P(N|\langle N\rangle)$ to remind us that the shape of the distribution depends (as we will see) only on its mean. To do this, we take the full probability distribution $P[\{t_i\}|r(\tau)]$, pick out the term involving N events, and then integrate over all possible arrival times of these events:

$$P(N|\langle N\rangle) = \int_0^T dt_1 \cdots \int_0^T dt_N\, P[\{t_i\}|r(\tau)] \tag{A.27}$$

$$= \int_0^T dt_1 \cdots \int_0^T dt_N \frac{1}{N!} \exp\left[-\int_0^T dt\, r(t)\right] \prod_{i=1}^{N} r(t_i). \tag{A.28}$$

As in the discussion leading to Eq. (A.22), the exponential factor can be taken outside the integral, and the N-dimensional integral reduces to a product of N one-dimensional integrals:

$$P(N|\langle N\rangle) = \int_0^T dt_1 \cdots \int_0^T dt_N \frac{1}{N!} \exp\left[-\int_0^T dt\, r(t)\right] \prod_{i=1}^{N} r(t_i)$$

$$= \frac{1}{N!} \exp\left[-\int_0^T dt\, r(t)\right] \int_0^T dt_1 \cdots \int_0^T dt_N \prod_{i=1}^{N} r(t_i)$$

$$= \frac{1}{N!} \exp\left[-\int_0^T dt\, r(t)\right] \left[\int_0^T dt\, r(t)\right]^N \tag{A.29}$$

$$= \frac{1}{N!} \exp(-Q) Q^N, \tag{A.30}$$

where

$$Q \equiv \int_0^T dt\, r(t). \tag{A.31}$$

In particular, the probability that no events occur in the time from $t = 0$ to $t = T$ is $P(0) = \exp(-Q)$, or

$$P(0|\langle N\rangle) = \exp\left[-\int_0^T dt\, r(t)\right]. \tag{A.32}$$

With the probability distribution of counts from Eq. (A.30), we can compute the mean and the variance of the count. To obtain the mean, we compute

$$\langle N \rangle \equiv \sum_{N=0}^{\infty} P(N)N \tag{A.33}$$

$$= \sum_{N=0}^{\infty} \frac{1}{N!} \exp(-Q) Q^N N \tag{A.34}$$

$$= \exp(-Q) \sum_{N=0}^{\infty} \frac{1}{N!} Q^N N. \tag{A.35}$$

We have already made use of the series expansion for the exponential, Eq. (A.23). To sum this last series, we notice that

$$Q^N N = Q \frac{\partial}{\partial Q} Q^N, \tag{A.36}$$

so that

$$\langle N \rangle = \exp(-Q) \sum_{N=0}^{\infty} \frac{1}{N!} Q^N N$$

$$= \exp(-Q) \sum_{N=0}^{\infty} \frac{1}{N!} Q \frac{\partial}{\partial Q} Q^N \tag{A.37}$$

$$= \exp(-Q) Q \frac{\partial}{\partial Q} \sum_{N=0}^{\infty} \frac{1}{N!} Q^N \tag{A.38}$$

$$= \exp(-Q) Q \frac{\partial}{\partial Q} \exp(+Q), \tag{A.39}$$

where in the last step we again identify the series for the exponential. The derivative of the exponential is just the exponential itself,

$$\frac{\partial}{\partial Q} \exp(+Q) = \exp(+Q), \tag{A.40}$$

so that

$$\langle N \rangle = \exp(-Q) Q \frac{\partial}{\partial Q} \exp(+Q)$$

$$= \exp(-Q) Q \exp(+Q) = Q. \tag{A.41}$$

Thus, the mean count is what we have called Q, the integral of the rate.

Now we can write the count distribution directly in terms of its mean:

$$P(N|\langle N \rangle) = \exp(-\langle N \rangle) \frac{\langle N \rangle^N}{N!}, \tag{A.42}$$

which is Eq. (2.1), at the start of our discussion of photon counting in vision (Section 2.1).

A similar calculation yields the variance of the count distribution. We start by computing the average of N^2:

$$\langle N^2 \rangle = \sum_{N=0}^{\infty} N^2 P(N). \tag{A.43}$$

Substituting for $P(N)$ from Eq. (A.30) and rearranging, we have

$$\langle N^2 \rangle = \sum_{N=0}^{\infty} N^2 P(N)$$

$$= \sum_{N=0}^{\infty} N^2 \exp(-Q) \frac{1}{N!} Q^N \tag{A.44}$$

$$= \exp(-Q) \sum_{N=0}^{\infty} \frac{1}{N!} N^2 Q^N. \tag{A.45}$$

The trick is once again to write the extra factors of N (here N^2) in terms of derivatives with respect to Q. Now we know that

$$\frac{\partial^2}{\partial Q^2} Q^N = N(N-1) Q^{N-2}, \tag{A.46}$$

so we can write

$$Q^2 \frac{\partial^2}{\partial Q^2} Q^N = (N^2 - N) Q^N, \tag{A.47}$$

which is almost what we want. But we can use the formula in Eq. (A.36) to finish the job, obtaining

$$N^2 Q^N = Q^2 \frac{\partial^2}{\partial Q^2} Q^N + Q \frac{\partial}{\partial Q} Q^N. \tag{A.48}$$

Now we can substitute into Eq. (A.45) and follow the steps corresponding to Eqs. (A.37)–(A.41):

$$\langle N^2 \rangle = \exp(-Q) \sum_{N=0}^{\infty} \frac{1}{N!} N^2 Q^N$$

$$= \exp(-Q) \sum_{N=0}^{\infty} \frac{1}{N!} \left[Q^2 \frac{\partial^2}{\partial Q^2} Q^N + Q \frac{\partial}{\partial Q} Q^N \right] \tag{A.49}$$

$$= \exp(-Q) Q^2 \frac{\partial^2}{\partial Q^2} \sum_{N=0}^{\infty} \frac{1}{N!} Q^N + \exp(-Q) Q \frac{\partial}{\partial Q} \sum_{N=0}^{\infty} \frac{1}{N!} Q^N \tag{A.50}$$

$$= \exp(-Q) Q^2 \frac{\partial^2}{\partial Q^2} \exp(+Q) + \exp(-Q) Q \frac{\partial}{\partial Q} \exp(+Q) \tag{A.51}$$

$$= \exp(-Q) Q^2 \exp(+Q) + \exp(-Q) Q \exp(+Q) \tag{A.52}$$

$$= Q^2 + Q. \tag{A.53}$$

We have already identified Q as equal to the mean count, which means that the mean-square count can be written as

$$\langle N^2 \rangle = \langle N \rangle^2 + \langle N \rangle. \tag{A.54}$$

The variance of the count is defined by

$$\langle (\delta N)^2 \rangle \equiv \langle N^2 \rangle - \langle N \rangle^2 \tag{A.55}$$

$$= [\langle N \rangle^2 + \langle N \rangle] - \langle N \rangle^2 = \langle N \rangle. \tag{A.56}$$

Thus, the variance of the count for a Poisson process is equal to the mean count. The standard deviation of the Poisson distribution is then the square root of the mean, and this square root of N law is one of the most important intuitions about the statistics of counting independent events—photons, molecules, ...

The next important characteristic of the Poisson process is the interval between events. The probability per unit time that we observe an event at time t is given by the rate $r(t)$. The probability that we observe no events in the open interval $[t, t + \tau)$ is given by

$$P(0) = \exp\left[-\int_t^{t+\tau} dt' \, r(t') \right]. \tag{A.57}$$

The probability per unit time that this interval is closed by an event is again the rate, now at time $t + \tau$. Thus, the probability per unit time that we see events at t and $t + \tau$, with no events in between, is given by

$$P(t, t + \tau) = r(t) \exp\left[-\int_t^{t+\tau} dt' \, r(t') \right] r(t + \tau). \tag{A.58}$$

In the simple case that the rate is constant, this expression is just $P(t, t + \tau) = r^2 e^{-r\tau}$. In contrast, if the rate varies, the average probability for observing two events separated by an empty interval of duration τ is

$$P_2(\tau) = \left\langle r(t) \exp\left[-\int_t^{t+\tau} dt' \, r(t') \right] r(t + \tau) \right\rangle, \tag{A.59}$$

where $\langle \cdot \rangle$ is an average over these variations in rate.

The probability density of intervals is really the conditional probability that the next event will be at $t + \tau$ given that there was an event at t. To form this conditional probability we need to divide by the probability of an event at t, which is just the average rate. Again, in the simple case of a constant rate, this yields the probability density of interevent intervals:

$$p(\tau) = r e^{-r\tau}. \tag{A.60}$$

This exponential form is one of the classic signatures of a Poission process. We can think of it as arising because the moment at which the interval closes has no memory of the moment at which it opened, and so the probability that there has not been an event must be a product of terms for the absence of an event in each small time slice $\Delta \tau$, as in the derivation above, and in the limit this product becomes an exponential.

Our last task is to evaluate averages over Poisson processes, such as the one in Eq. (2.41):

$$\left\langle \sum_i V_0(t - t_i) \right\rangle = \sum_{N=0}^{\infty} \int_0^T dt_1 \cdots \int_0^T dt_N \, P[\{t_i\}|r(t)] \sum_i V_0(t - t_i). \tag{A.61}$$

We proceed systematically, looking at one term in the sum and doing the integrals one at a time.

One term in the sum means that we choose, for example, i = 1 *and* one particular value of N. This term is

$$\int_0^T dt_1 \cdots \int_0^T dt_N P[\{t_i\}|r(t)]V_0(t - t_1)$$

$$= \int_0^T dt_1 \cdots \int_0^T dt_N \, \exp\left[-\int_0^T d\tau \, r(\tau)\right] \frac{1}{N!} r(t_1)r(t_2) \cdots r(t_N)V_0(t - t_1). \tag{A.62}$$

The exponential factor (along with the $1/N!$) is constant and again can be brought outside the integral. Now we rearrange the order of the integrals:

$$\int_0^T dt_1 \int_0^T dt_2 \cdots \int_0^T dt_N r(t_1)r(t_2) \cdots r(t_N)V_0(t - t_1)$$

$$= \int_0^T dt_1 \, r(t_1)V_0(t - t_1) \int_0^T dt_2 \, r(t_2) \cdots \int_0^T dt_N \, r(t_N) \tag{A.63}$$

$$= \left[\int_0^T dt_1 \, r(t_1)V_0(t - t_1)\right]\left[\int_0^T d\tau r(\tau)\right]^{N-1}. \tag{A.64}$$

But the fact that we chose i = 1 was arbitrary; we would have gotten the same answer for any i = 1, 2, $\cdots$, N. Thus, summing over i is the same as multiplying by N. This leaves us with the sum on N, so we put everything back together to find

$$\left\langle \sum_i V_0(t - t_i) \right\rangle = \exp\left[-\int_0^T d\tau \, r(\tau)\right]\left[\int_0^T dt_1 \, r(t_1)V_0(t - t_1)\right]$$

$$\times \sum_{N=0}^{\infty} \frac{N}{N!}\left[\int_0^T d\tau \, r(\tau)\right]^{N-1} \tag{A.65}$$

$$= \exp\left[-\int_0^T d\tau \, r(\tau)\right]\int_0^T dt_1 \, r(t_1)V_0(t - t_1)$$

$$\times \sum_{N=0}^{\infty} \frac{1}{N!}\left[\int_0^T d\tau \, r(\tau)\right]^{N} \tag{A.66}$$

$$= \exp\left[-\int_0^T d\tau \, r(\tau)\right]\int_0^T dt_1 \, r(t_1)V_0(t - t_1)$$

$$\times \exp\left[+\int_0^T d\tau \, r(\tau)\right] \tag{A.67}$$

$$= \int_0^T dt_1 \, r(t_1)V_0(t - t_1). \tag{A.68}$$

Thus, our simple model of summing pulses from single photons generates a voltage that, on average, responds linearly to the rate of photon arrivals,

$$\langle V(t) \rangle = V_{DC} + \int dt' V_0(t - t') r(t'), \tag{A.69}$$

which is Eq. (2.42) in the main text.

Actually, we have shown something more general, which will be useful below. The expectation value we have computed is of the form $\langle \sum_i f(t_i) \rangle$. What we have seen is that summing over arrival times is, on average, equivalent to integrating over the rate:

$$\left\langle \sum_i f(t_i) \right\rangle = \int d\tau \, r(\tau) f(\tau). \tag{A.70}$$

Intuitively, this makes sense: the sum over arrival times approximates a density along the time axis, and this density is the rate, with units of (events)/(time).

Now we need to do the same calculation, but for the correlation function of the voltage. Again we have

$$V(t) = \sum_i V_0(t - t_i), \tag{A.71}$$

and we want to compute $\langle V(t) V(t') \rangle$. The arrival times of photons are independent of one another—this is the essence of the Poisson process—and so we should have

$$\langle V(t) V(t') \rangle = \left\langle \sum_i V_0(t - t_i) \sum_j V_0(t' - t_j) \right\rangle \tag{A.72}$$

$$= \sum_{i,j} \langle V_0(t - t_i) V_0(t' - t_j) \rangle \tag{A.73}$$

$$= \sum_{i \neq j} \langle V_0(t - t_i) V_0(t' - t_j) \rangle + \sum_i \langle V_0(t - t_i) V_0(t' - t_i) \rangle \tag{A.74}$$

$$= \sum_{i \neq j} \langle V_0(t - t_i) \rangle \langle V_0(t' - t_j) \rangle + \sum_i \langle V_0(t - t_i) V_0(t' - t_i) \rangle, \tag{A.75}$$

where we use the independence of t_i and t_j for $i \neq j$ in the last step. It is useful to add and subtract the diagonal $i = j$ term from the sum, so that

$$\langle V(t) V(t') \rangle = \sum_{i \neq j} \langle V_0(t - t_i) \rangle \langle V_0(t' - t_j) \rangle + \sum_i \langle V_0(t - t_i) \rangle \langle V_0(t' - t_i) \rangle$$
$$+ \sum_i \langle V_0(t - t_i) V_0(t' - t_i) \rangle - \sum_i \langle V_0(t - t_i) \rangle \langle V_0(t' - t_i) \rangle \tag{A.76}$$

$$= \sum_{i,j} \langle V_0(t - t_i) \rangle \langle V_0(t' - t_j) \rangle$$
$$+ \sum_i \left[\langle V_0(t - t_i) V_0(t' - t_i) \rangle - \langle V_0(t - t_i) \rangle \langle V_0(t' - t_i) \rangle \right]. \tag{A.77}$$

The key step now is to notice that we can rearrange the sums and expectation values in the first term,

$$\sum_{i,j} \langle V_0(t - t_i)\rangle \langle V_0(t' - t_j)\rangle = \sum_i \langle V_0(t - t_i)\rangle \sum_j \langle V_0(t' - t_j)\rangle \tag{A.78}$$

$$= \left\langle \sum_i V_0(t - t_i)\right\rangle\left\langle \sum_j V_0(t' - t_j)\right\rangle \tag{A.79}$$

$$= \langle V(t)\rangle\langle V(t')\rangle, \tag{A.80}$$

where in the last step we recognize the voltage itself, from Eq. (A.71). Thus, Eq. (A.77) can be rewritten as an equation for the covariance of the voltage fluctuations:

$$\langle \delta V(t)\delta V(t')\rangle \equiv \langle V(t)V(t')\rangle - \langle V(t)\rangle\langle V(t')\rangle \tag{A.81}$$

$$= \sum_i \left[\langle V_0(t - t_i)V_0(t' - t_i)\rangle - \langle V_0(t - t_i)\rangle\langle V_0(t' - t_i)\rangle\right]. \tag{A.82}$$

The integrals simplify if we confine our attention to the case where the rate is constant, $r(t) = \bar{r}$. If we are looking in an interval $0 < t < T$, then any single arrival time t_i is uniformly distributed in this interval. Thus, for any function $f(t)$, we have

$$\langle f(t - t_i)\rangle = \frac{1}{T}\int_0^T d\tau f(t - \tau). \tag{A.83}$$

This average is independent of the index i, so if we sum over events, we just multiply by the expected number of events, which is $\bar{r}T$:

$$\sum_i \langle f(t - t_i)\rangle = (\bar{r}T)\frac{1}{T}\int_0^T dt' f(t - t') = \bar{r}\int_0^T dt' f(t - t'); \tag{A.84}$$

this is Eq. (A.70) specialized to $r(\tau) = \bar{r}$. But when using these results to evaluate the averages in Eq. (A.82), we have to be careful because, in one of the terms the product of averages is taken before summing. Thus, the equation becomes

$$\langle \delta V(t)\delta V(t')\rangle = \sum_i \left[\langle V_0(t - t_i)V_0(t' - t_i)\rangle - \langle V_0(t - t_i)\rangle\langle V_0(t' - t_i)\rangle\right]$$

$$= \sum_i \left[\frac{1}{T}\int_0^T d\tau V_0(t - \tau)V_0(t' - \tau)\right]$$

$$- \sum_i \left[\frac{1}{T}\int_0^T d\tau V_0(t - \tau)\frac{1}{T}\int_0^T d\tau' V_0(t' - \tau')\right] \tag{A.85}$$

$$= (\bar{r}T)\frac{1}{T}\int_0^T d\tau V_0(t - \tau)V_0(t' - \tau)$$

$$- (\bar{r}T)\frac{1}{T^2}\int_0^T d\tau V_0(t - \tau)\int_0^T d\tau' V_0(t' - \tau') \tag{A.86}$$

$$\to \bar{r}\int_0^T d\tau V_0(t - \tau)V_0(t' - \tau), \tag{A.87}$$

where in the last step we consider the limit $T \to \infty$.

It is especially useful to convert the correlation function of voltage fluctuations into the corresponding power spectrum. In general, the power spectrum can be written as

$$S_V(\omega) = \int d\tau \, e^{+i\omega\tau} \langle \delta V(t+\tau)\delta V(t)\rangle, \tag{A.88}$$

and so in this case we have

$$S_V(\omega) = \int d\tau \, e^{+i\omega\tau} \bar{r} \int d\tau' \, V_0(t+\tau-\tau')V_0(t-\tau') \tag{A.89}$$

$$= \bar{r} \int d\tau \int d\tau' \, e^{+i\omega\tau} e^{+i\omega(t-\tau')} V_0(t+\tau-\tau') e^{-i\omega(t-\tau')} V_0(t-\tau') \tag{A.90}$$

$$= \bar{r} \left[\int d\tau \, e^{+i\omega(\tau+t-\tau')} V_0(\tau+t-\tau') \right]$$
$$\times \left[\int d\tau' e^{-i\omega(t-\tau')} V_0(t-\tau') \right] \tag{A.91}$$

$$= \bar{r} |\tilde{V}_0(\omega)|^2, \tag{A.92}$$

where in the last step we recognize the Fourier transform of the pulse shape $V_0(t)$. This is what we need to establish Eq. (2.66) of the main text. This relationship between the pulse shape and power spectrum for Poisson processes is called Campbell's theorem.

Problem 170: Approach to Gaussianity.   If the mean rate of events is $\bar{r}$, and each pulse $V_0(t)$ has a duration τ, then the voltage at any moment in time typically involves contributions from $\sim \bar{r}\tau$ pulses. If this number becomes large, the fluctuations in voltage should become Gaussian, by the central limit theorem. Explore the approach to Gaussianity, either by analytically computing higher moments or by doing a numerical simulation. This problem is deliberately open ended.

A.2   Correlations, Power Spectra, and All That

Consider a function $x(t)$ that varies in time. We would like to describe a situation in which these variations are random, drawn out of some distribution. But now we need a distribution for a function rather than for a finite set of variables. This should not bother us, because such constructions are central to much of modern physics, for example, in the path integral approach to quantum mechanics. Distributions of functions are called "distribution functionals" when we need to be precise.

One strategy for constructing distribution functionals is to start by discretizing time, so that we have at most a countable infinity of variables:

$$x(t_1), x(t_2), x(t_3), \cdots.$$

Let's assume for simplicity that the mean value of x is zero. Then the first nontrivial characterization of the statistics of x is the covariance matrix:

$$C_{ij} = \langle x(t_i)x(t_j)\rangle. \tag{A.93}$$

If a single variable y is drawn from a Gaussian distribution with zero mean and variance σ^2, then we have

$$P(y) = \frac{1}{\sqrt{2\pi\sigma^2}} \exp\left[-\frac{y^2}{2\sigma^2}\right]. \tag{A.94}$$

The generalization to multiple variables is

$$P(\{x_i\}) = \frac{1}{\sqrt{(2\pi)^N \det C}} \exp\left[-\frac{1}{2}\sum_{i,j=1}^N x_i (C^{-1})_{ij} x_j\right], \tag{A.95}$$

where det is the determinant and $(C^{-1})_{ij}$ is element ij of the matrix inverse to C; if we think of the $\{x_i\}$ as a vector $\mathbf{x}$, then we can write, more compactly,

$$P(\{x_i\}) = \frac{1}{\sqrt{(2\pi)^N \det C}} \exp\left[-\frac{1}{2}\mathbf{x}^T \cdot C^{-1} \cdot \mathbf{x}\right], \tag{A.96}$$

where $\mathbf{x}^T$ is the transpose of the vector $\mathbf{x}$. Just to be clear, this expression describes a Gaussian distribution, and many interesting signals are non-Gaussian. Still, it's a good place to start.

Problem 171: Gaussian integrals. If you have not done this before, now is a good time to check that the probability distribution in Eq. (A.96) is normalized. This requires you to show that

$$\int d^N x \, \exp\left[-\frac{1}{2}\mathbf{x}^T \cdot C^{-1} \cdot \mathbf{x}\right] = \sqrt{(2\pi)^N \det C}. \tag{A.97}$$

While you're at it, you should also show that

$$\ln \det C = \text{Tr} \ln C, \tag{A.98}$$

where det is the determinant and Tr is the trace. This is simplest in the case which matters here, where C is a symmetric matrix with positive eigenvalues. As an aside, explain why covariance matrices must have positive eigenvalues.

The covariance matrix C_{ij} can have an arbitrary structure, constrained only by symmetry ($C_{ij} = C_{ji}$) and positivity of its eigenvalues. But when the index i refers to discrete time points, we have an extra constraint that comes from invariance under translations in time. Because there is no clock, we must have that

$$\langle x(t)x(t')\rangle = C_x(t - t'), \tag{A.99}$$

with no dependence on the absolute time t or t'. To evaluate the exponential in the Gaussian distribution, we need to compute the inverse of the matrix C. To start, let's look at an example where the correlation function is

$$C_x(t - t') = e^{-|t-t'|/\tau_c}, \tag{A.100}$$

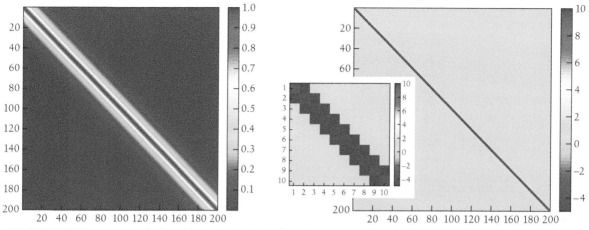

FIGURE A.1

Covariance matrix and its inverse. Elements of the matrix are indexed by the x- and y-axes, and shading measures the value of the matrix elements, according to the scale bars. At left, the covariance matrix in Eq. (A.101), with $\Delta t / \tau_c = 0.1$. At right, the inverse matrix, with inset showing a 10×10 submatrix surrounding the diagonal.

where τ_c is the correlation time, and hence with $t_n = n \Delta t$, we have

$$C_{ij} = \exp\left[-\left(\frac{\Delta t}{\tau_c} \right) |i - j| \right]. \tag{A.101}$$

This matrix and its inverse are shown in Fig. A.1 for $\Delta t / \tau_c = 0.1$. We see that, at least in this case, the inverse matrix C^{-1} consists almost entirely of zeros, except in the immediate neighborhood of the diagonal. If we make Δt smaller, the number of nonzero elements in C_{ij} increases, because what matters is the physical time scale τ_c. But in $(C^{-1})_{ij}$, the range of nonzero elements off the diagonal stays fixed as $\Delta t \to 0$. Thus, the inverse matrix actually is the discretization of a local differential operator.

Problem 172: Recognizing a differential operator.   Verify the statements just made about the $\Delta t \to 0$ limit.

Reflexively, seeing that we have to compute inverses and determinants of matrices, we should think about diagonalizing C. Recall from quantum mechanics that the eigenfunctions of an operator have to provide a representation of the underlying symmetries. In this case, the relevant symmetry is time translation, so we know to look at the Fourier functions $e^{-i\omega t}$. In fact, once we have the hint about using a Fourier representation, we don't need the crutch of discrete time points any more. Let's see how this works.

We define the Fourier transform with the conventions

$$\tilde{x}(\omega) = \int_{-\infty}^{\infty} dt\, e^{+i\omega t} x(t), \tag{A.102}$$

$$x(t) = \int_{-\infty}^{\infty} \frac{d\omega}{2\pi} e^{-i\omega t} \tilde{x}(\omega). \tag{A.103}$$

Now if we compute the covariance of two frequency components, we have

$$\langle \tilde{x}(\omega)\tilde{x}(\omega') \rangle = \left\langle \int_{-\infty}^{\infty} dt\, e^{+i\omega t} x(t) \int_{-\infty}^{\infty} dt'\, e^{+i\omega' t'} x(t') \right\rangle \tag{A.104}$$

$$= \int_{-\infty}^{\infty} dt\, e^{+i\omega t} \int_{-\infty}^{\infty} dt'\, e^{+i\omega' t'} \langle x(t) x(t') \rangle \tag{A.105}$$

$$= \int_{-\infty}^{\infty} dt\, e^{+i\omega t} \int_{-\infty}^{\infty} dt'\, e^{+i\omega' t'} \int_{-\infty}^{\infty} \frac{d\Omega}{2\pi} e^{-i\Omega(t-t')} S_x(\Omega), \tag{A.106}$$

where we introduce the Fourier transform of the correlation function:

$$S_x(\Omega) = \int_{-\infty}^{\infty} d\tau\, e^{+i\Omega\tau} C_x(\tau), \tag{A.107}$$

$$C_x(t - t') = \int_{-\infty}^{\infty} \frac{d\Omega}{2\pi} e^{-i\Omega(t-t')} S_x(\Omega). \tag{A.108}$$

Now we can rearrange the integrals in Eq. (A.106):

$$\langle \tilde{x}(\omega)\tilde{x}(\omega') \rangle = \int_{-\infty}^{\infty} dt\, e^{+i\omega t} \int_{-\infty}^{\infty} dt'\, e^{+i\omega' t'} \int_{-\infty}^{\infty} \frac{d\Omega}{2\pi} e^{-i\Omega(t-t')} S_x(\Omega),$$
$$= \int_{-\infty}^{\infty} \frac{d\Omega}{2\pi} S_x(\Omega) \left[\int_{-\infty}^{\infty} dt\, e^{i(\omega-\Omega)t} \right] \left[\int_{-\infty}^{\infty} dt'\, e^{i(\omega'+\Omega)t'} \right]. \tag{A.109}$$

This is moment to recall some properties of the Dirac delta function. The delta function has the property that

$$\delta(z) = 0 \quad \text{for } z \neq 0, \tag{A.110}$$

$$\int dz\, \delta(z) = 1, \tag{A.111}$$

if the domain of the integral includes $z = 0$. Then the Fourier transform of the delta function is

$$\int dz\, e^{+iqz} \delta(z) = 1, \tag{A.112}$$

so that

$$\delta(z) = \int_{-\infty}^{\infty} \frac{dq}{2\pi} e^{-iqz}. \tag{A.113}$$

Thus, we recognize, in Eq. (A.109), that

$$\int_{-\infty}^{\infty} dt\, e^{i(\omega-\Omega)t} = 2\pi\delta(\omega - \Omega), \tag{A.114}$$

$$\int_{-\infty}^{\infty} dt'\, e^{i(\omega'+\Omega)t'} = 2\pi\delta(\omega' + \Omega). \tag{A.115}$$

Substituting back into Eq. (A.109), we have

$$\langle \tilde{x}(\omega)\tilde{x}(\omega') \rangle = \int_{-\infty}^{\infty} \frac{d\Omega}{2\pi} S_x(\Omega) \left[\int_{-\infty}^{\infty} dt \, e^{i(\omega-\Omega)t} \right] \left[\int_{-\infty}^{\infty} dt' \, e^{i(\omega'+\Omega)t'} \right].$$

$$= \int_{-\infty}^{\infty} \frac{d\Omega}{2\pi} S_x(\Omega) 2\pi \delta(\omega - \Omega) 2\pi \delta(\omega' + \Omega) \tag{A.116}$$

$$= S_x(\omega) 2\pi \delta(\omega' + \omega). \tag{A.117}$$

Thus, although different time points can be correlated with one another in complicated ways, the covariance of frequency components has a much simpler structure: $\tilde{x}(\omega)$ is correlated only with $\tilde{x}(-\omega)$.

This covariance structure, which couples positive and negative frequency components, makes sense, because we are using a complex representation for real variables. To make a real variable $x(t)$, the Fourier transform must obey

$$\tilde{x}(-\omega) = \tilde{x}^*(\omega), \tag{A.118}$$

where $\tilde{x}^*$ is the complex conjugate of $\tilde{x}$, so positive and negative frequency components are not independent—in fact, they are redundant. It might be more natural to write Eq. (A.117) as

$$\langle \tilde{x}(\omega)\tilde{x}^*(\omega') \rangle = S_x(\omega) 2\pi \delta(\omega' - \omega), \tag{A.119}$$

making clear that frequency components are correlated with themselves, not with other frequencies.

We could instead think about the real and imaginary parts of the positive frequency components, which can be written as

$$\tilde{x}_{\text{Re}}(\omega) = \frac{1}{2} \left[\tilde{x}(\omega) + \tilde{x}(-\omega) \right], \tag{A.120}$$

and

$$\tilde{x}_{\text{Im}}(\omega) = \frac{1}{2i} \left[\tilde{x}(\omega) - \tilde{x}(-\omega) \right]. \tag{A.121}$$

With this representation, we can use the result in Eq. (A.117):

$$\langle \tilde{x}_{\text{Re}}(\omega)\tilde{x}_{\text{Re}}(\omega') \rangle = \left\langle \frac{1}{2} \left[\tilde{x}(\omega) + \tilde{x}(-\omega) \right] \frac{1}{2} \left[\tilde{x}(\omega') + \tilde{x}(-\omega') \right] \right\rangle \tag{A.122}$$

$$= \frac{1}{4} \left[\langle \tilde{x}(\omega)\tilde{x}(\omega') \rangle + \langle \tilde{x}(\omega)\tilde{x}(-\omega') \rangle \right]$$

$$+ \frac{1}{4} \left[\langle \tilde{x}(-\omega)\tilde{x}(\omega') \rangle + \langle \tilde{x}(-\omega)\tilde{x}(-\omega') \rangle \right] \tag{A.123}$$

$$= \frac{S_x(\omega)}{4} 2\pi \left[\delta(\omega + \omega') + \delta(\omega - \omega') \right.$$

$$\left. + \delta(-\omega + \omega') + \delta(-\omega - \omega') \right]. \tag{A.124}$$

Because we are looking only at positive frequencies, $\omega + \omega'$ can never be zero, and hence the first and last delta functions can be dropped. The remaining two are actually the same, so we have

$$\langle \tilde{x}_{\text{Re}}(\omega)\tilde{x}_{\text{Re}}(\omega')\rangle = \frac{1}{2}S_x(\omega)2\pi\delta(\omega-\omega'). \tag{A.125}$$

Similar calculations show that the imaginary parts of $\tilde{x}(\omega)$ have the same variance,

$$\langle \tilde{x}_{\text{Im}}(\omega)\tilde{x}_{\text{Im}}(\omega')\rangle = \langle \tilde{x}_{\text{Re}}(\omega)\tilde{x}_{\text{Re}}(\omega')\rangle \tag{A.126}$$

$$= \frac{1}{2}S_x(\omega)2\pi\delta(\omega-\omega'), \tag{A.127}$$

while real and imaginary parts are uncorrelated,

$$\langle \tilde{x}_{\text{Re}}(\omega)\tilde{x}_{\text{Im}}(\omega')\rangle = 0. \tag{A.128}$$

Problem 173: The other phase. Derive Eq. (A.127) and Eq. (A.128).

What does all this mean? We think of the random function of time $x(t)$ as being built out of frequency components, and each component has a real and an imaginary part. The structure of the covariance matrix is such that different frequency components do not covary,[1] which makes sense—if we have covariation of different frequency components, then we can beat them against one another to make a clock running at the difference frequency, which would violate time-translation invariance. Similarly, that real and imaginary components do not covary means that there is no preferred phase, which again is consistent with (indeed, required by) time-translation invariance.

We should be able to put these results on the covariance matrix together to describe the distribution functional for a Gaussian function of time. Because the real and imaginary parts are independent, let's start with just the real parts. We should have

$$P[\{\tilde{x}_{\text{Re}}(\omega)\}] \propto \exp\left[-\frac{1}{2}\int_0^\infty \frac{d\omega}{2\pi}\int_0^\infty \frac{d\omega'}{2\pi}\tilde{x}_{\text{Re}}(\omega)\mathcal{A}(\omega,\omega')\tilde{x}_{\text{Re}}(\omega')\right], \tag{A.129}$$

where $\mathcal{A}$ is the inverse of the covariance:

$$\int \frac{d\omega'}{2\pi}\mathcal{A}(\omega,\omega')\langle \tilde{x}_{\text{Re}}(\omega')\tilde{x}_{\text{Re}}(\omega'')\rangle = 2\pi\delta(\omega-\omega''). \tag{A.130}$$

We can find $\mathcal{A}$ by substituting the explicit expression for the covariance and doing the integrals:

$$2\pi\delta(\omega-\omega'') = \int \frac{d\omega'}{2\pi}\mathcal{A}(\omega,\omega')\langle \tilde{x}_{\text{Re}}(\omega')\tilde{x}_{\text{Re}}(\omega'')\rangle$$

$$= \int \frac{d\omega'}{2\pi}\mathcal{A}(\omega,\omega')\frac{1}{2}S_x(\omega')2\pi\delta(\omega'-\omega'') \tag{A.131}$$

$$= \frac{1}{2}\mathcal{A}(\omega,\omega'')S_x(\omega''). \tag{A.132}$$

---

1. Again, it should be emphasized here that "covary" means at second order; there can be higher order correlations among different frequency components, although these are constrained. See, for example, Problem 16.

Thus, we have

$$A(\omega, \omega'') = \frac{1}{S_x(\omega'')} 4\pi \delta(\omega - \omega'').$$
(A.133)

Substituting back into Eq. (A.129) for the probability distribution, we have

$$P[\{\tilde{x}_{\mathrm{Re}}(\omega)\}] \propto \exp\left[-\frac{1}{2}\int_0^\infty \frac{d\omega}{2\pi} \int_0^\infty \frac{d\omega'}{2\pi} \tilde{x}_{\mathrm{Re}}(\omega) \mathcal{A}(\omega, \omega') \tilde{x}_{\mathrm{Re}}(\omega')\right]$$
(A.134)

$$= \exp\left[-\frac{1}{2}\int_0^\infty \frac{d\omega}{2\pi} \int_0^\infty \frac{d\omega'}{2\pi} \tilde{x}_{\mathrm{Re}}(\omega) \frac{4\pi \delta(\omega - \omega')}{S_x(\omega')} \tilde{x}_{\mathrm{Re}}(\omega')\right]$$
(A.135)

$$= \exp\left[-\int_0^\infty \frac{d\omega}{2\pi} \frac{\tilde{x}_{\mathrm{Re}}^2(\omega)}{S_x(\omega)}\right].$$
(A.136)

The same argument applies to the imaginary parts of the Fourier components, and these are independent of the real parts, so we have

$$P[x(t)] = P[\{\tilde{x}_{\mathrm{Re}}(\omega), \tilde{x}_{\mathrm{Im}}(\omega)\}]$$
(A.137)

$$\propto \exp\left[-\int_0^\infty \frac{d\omega}{2\pi} \frac{\tilde{x}_{\mathrm{Re}}^2(\omega) + \tilde{x}_{\mathrm{Im}}^2(\omega)}{S_x(\omega)}\right]$$
(A.138)

$$= \frac{1}{Z}\exp\left[-\int_0^\infty \frac{d\omega}{2\pi} \frac{|\tilde{x}(\omega)|^2}{S_x(\omega)}\right]$$
(A.139)

$$= \frac{1}{Z}\exp\left[-\frac{1}{2}\int_{-\infty}^\infty \frac{d\omega}{2\pi} \frac{|\tilde{x}(\omega)|^2}{S_x(\omega)}\right],$$
(A.140)

where we have introduced the normalization constant Z.

It is useful to look at the example illustrated in Fig. A.1. In that case $C_x(\tau) = \exp(-|\tau|/\tau_c)$, so the power spectrum is

$$S_x(\omega) = \int_{-\infty}^\infty d\tau\, e^{+i\omega\tau} e^{-|\tau|/\tau_c}$$
(A.141)

$$= \int_{-\infty}^0 d\tau\, e^{(+i\omega + 1/\tau_c)\tau} + \int_0^\infty d\tau\, e^{(+i\omega - 1/\tau_c)\tau}$$
(A.142)

$$= \frac{1}{(+i\omega + 1/\tau_c)} + \frac{1}{-(+i\omega - 1/\tau_c)}$$
(A.143)

$$= \frac{\tau_c}{1 + i\omega\tau_c} + \frac{\tau_c}{1 - i\omega\tau_c}$$
(A.144)

$$= \frac{2\tau_c}{1 + (\omega\tau_c)^2}.$$
(A.145)

Thus, the probability distribution functional has the form

$$P[x(t)] = \frac{1}{Z}\exp\left[-\frac{1}{2}\int_{-\infty}^\infty \frac{d\omega}{2\pi} \frac{|\tilde{x}(\omega)|^2}{S_x(\omega)}\right]$$

$$= \frac{1}{Z}\exp\left[-\frac{1}{4\tau_c}\int_{-\infty}^\infty \frac{d\omega}{2\pi}[1 + (\omega\tau_c)^2]|\tilde{x}(\omega)|^2\right].$$
(A.146)

The Fourier transformation can be thought of as a rotation in the space of functions. In this view, functions are vectors, and proper rotations preserve the lengths of vectors. The corresponding statement for the Fourier transform is Parseval's theorem:

$$\int_{-\infty}^{\infty} \frac{d\omega}{2\pi} |\tilde{x}(\omega)|^2 = \int dt\, x^2(t).$$
(A.147)

We can use the same result a bit more subtly, to recognize that integrals with powers of frequency are equivalent to integrals over time derivatives:

$$\int_{-\infty}^{\infty} \frac{d\omega}{2\pi} (\omega\tau_c)^2 |\tilde{x}(\omega)|^2 = \tau_c^2 \int_{-\infty}^{\infty} \frac{d\omega}{2\pi} |-i\omega\tilde{x}(\omega)|^2$$
(A.148)

$$= \tau_c^2 \int dt \left[\frac{dx(t)}{dt} \right]^2,$$
(A.149)

where we recognize $-i\omega\tilde{x}(\omega)$ as the Fourier transform of $dx(t)/dt$. Thus, we can write

$$P[x(t)] = \frac{1}{Z} \exp\left[-\frac{1}{4\tau_c} \int_{-\infty}^{\infty} \frac{d\omega}{2\pi} [1 + (\omega\tau_c)^2] |\tilde{x}(\omega)|^2 \right]$$

$$= \frac{1}{Z} \exp\left[-\frac{1}{4\tau_c} \int dt \left(\tau_c^2 \dot{x}^2(t) + x^2(t) \right) \right].$$
(A.150)

This expression shows explicitly, as promised above, that inverting the covariance matrix gives rise to differential operators. This example also is nice because it produces a probability distribution functional for trajectories $x(t)$ that reminds us of a (Euclidean) path integral in quantum mechanics, in this case for the harmonic oscillator.

Problem 174: Parseval's theorem. Verify Eq. (A.147).

Let's push a little further and see whether we can evaluate the normalization constant Z. By definition, we have

$$Z = \int \mathcal{D}x \, \exp\left[-\frac{1}{4\tau_c} \int dt \left(\tau_c^2 \dot{x}^2(t) + x^2(t) \right) \right],$$
(A.151)

where $\int \mathcal{D}x$ denotes an integral over all the functions $x(t)$. We have the general result for an N-dimensional Gaussian integral (Problem 171):

$$\int d^N x \, \exp\left[-\frac{1}{2} \mathbf{x}^{\mathrm{T}} \cdot \hat{A} \cdot \mathbf{x} \right] = \sqrt{\frac{(2\pi)^N}{\det \hat{A}}}$$
(A.152)

$$= \sqrt{(2\pi)^N} \exp\left[-\frac{1}{2} \operatorname{Tr} \ln \hat{A} \right],$$
(A.153)

where $\hat{A}$ is a matrix. Here the number of dimensions must become infinite, because we are integrating over functions. As you may recall from discussions of the path integral in quantum mechanics, there is some arbitrariness about how this is done, or, more formally, in how we define the measure $\mathcal{D}x$. A fairly standard choice is to absorb the

$\sqrt{2\pi}$, so that, in the time window $0 < t < T$, we have

$$\mathcal{D}x = \lim_{dt \to 0} \prod_{n=0}^{T/dt} \frac{dx(t_n)}{\sqrt{2\pi}}, \quad t_n = n \cdot dt. \tag{A.154}$$

Notice that before letting $dt \to 0$, the integral is over a finite number of points, so we should be able to carry over the results we know and just interpret the limits correctly.

The Gaussian functional integrals that we want to do have the general form

$$\int \mathcal{D}x \, \exp\left[-\frac{1}{2} \int dt \int dt' x(t) \hat{K}(t, t') x(t')\right],$$

where $\hat{K}$ is an operator. Carrying over what we know from the case of finite matrices [Eq. (A.153)], we have

$$\int \mathcal{D}x \, \exp\left[-\frac{1}{2} \int dt \int dt' x(t) \hat{K}(t, t') x(t')\right] = \exp\left[-\frac{1}{2} \operatorname{Tr} \ln \hat{K}\right]. \tag{A.155}$$

The only problem is to define $\operatorname{Tr} \ln \hat{K}$. Because $\hat{K}$ is an operator, we can ask for its spectrum, that is, its eigenvalues and eigenfunctions. Thus, we need to solve the equations

$$\int_0^T dt' \hat{K}(t, t') u_\mu(t') = \Lambda_\mu u_\mu(t), \tag{A.156}$$

where we are careful to note that here, $0 < t < T$. In the basis formed by the eigenfunctions, $\hat{K}$ is diagonal. As with matrices, when an operator is diagonal, we can take the logarithm element by element, and then computing the trace requires us to sum over these diagonal elements. Traces and determinants are invariant, so we can use this convenient basis and not worry about generality. Thus, we have

$$\operatorname{Tr} \ln \hat{K} = \sum_\mu \ln \Lambda_\mu. \tag{A.157}$$

How does this work for our case? Comparing Eq. (A.150) and Eq. (A.155), we have

$$\int dt \int dt' x(t) \hat{K}(t, t') x(t') = \int dt \left[\frac{\tau_c}{2} \left(\frac{dx(t)}{dt}\right)^2 + \frac{1}{2\tau_c} x^2(t)\right]. \tag{A.158}$$

To put this equation into a more standard form, we need to integrate by parts:

$$\int dt \left[\frac{\tau_c}{2} \left(\frac{dx(t)}{dt}\right)^2 + \frac{1}{2\tau_c} x^2(t)\right] = \int dt \, x(t) \left[-\frac{\tau_c}{2} \frac{d^2}{dt^2} + \frac{1}{2\tau_c}\right] x(t). \tag{A.159}$$

This allows us to identify

$$\hat{K}(t', t) = \delta(t' - t) \left[-\frac{\tau_c}{2} \frac{d^2}{dt^2} + \frac{1}{2\tau_c}\right]. \tag{A.160}$$

This operator is linear, and is also time-translation invariant (again). So we know that the eigenfunctions are $e^{-i\omega t}$, and because we are in a finite window of duration T, we should use only those frequency components that fit into the window, $\omega_n = 2\pi n/T$ for integer n. Then we have

$$\int_0^T dt\, \delta(t' - t) \left[-\frac{\tau_c}{2} \frac{d^2}{dt^2} + \frac{1}{2\tau_c} \right] e^{-i\omega_n t} = \left(\frac{\tau_c \omega_n^2}{2} + \frac{1}{2\tau_c} \right) e^{-i\omega_n t'}, \quad (A.161)$$

so that the eigenvalues are

$$\Lambda(\omega_n) = \left(\frac{\tau_c \omega_n^2}{2} + \frac{1}{2\tau_c} \right) = \frac{1 + (\omega_n \tau_c)^2}{2\tau_c}. \quad (A.162)$$

Notice that they are just the inverses of the power spectrum:

$$\Lambda(\omega_n) = \frac{1}{S_x(\omega_n)}. \quad (A.163)$$

This result makes sense when we look back at Eq. (A.140).

To finish the calculation, we have

$$Z = \exp\left[-\frac{1}{2} \sum_\mu \Lambda_\mu \right] \quad (A.164)$$

$$= \exp\left[-\frac{1}{2} \sum_n \ln\left(\frac{1}{S_x(\omega_n)} \right) \right] \quad (A.165)$$

$$= \exp\left[\frac{1}{2} \sum_n \ln S_x(\omega_n) \right]. \quad (A.166)$$

Finally, we need to do the sum. As the time window T becomes large, the spacing between frequency components, $\Delta\omega = 2\pi/T$, become small, and the sum approaches an integral.[2] Thus, for any function of ω_n, we have

$$\sum_n f(\omega_n) = \frac{1}{\Delta\omega} \sum_n \Delta\omega f(\omega_n) \quad (A.168)$$

$$\rightarrow \frac{1}{\Delta\omega} \int d\omega\, f(\omega) \quad (A.169)$$

$$= T \int \frac{d\omega}{2\pi} f(\omega). \quad (A.170)$$

At last, this result gives

$$Z = \exp\left[\frac{T}{2} \int \frac{d\omega}{2\pi} \ln S_x(\omega) \right]. \quad (A.171)$$

---

2. There is an analogous result for summing over the states of particles in a box in quantum systems. In this case the states are labeled by their wavevector $\mathbf{k}$, and in three dimensions we have

$$\sum_{\mathbf{k}} \rightarrow V \int \frac{d^3 k}{(2\pi)^3}, \quad (A.167)$$

where V is the volume of the box.

Putting the pieces together, we have the probability distribution functional for a Gaussian $x(t)$:

$$P[x(t)] = \exp\left[+\frac{T}{2}\int_{-\infty}^{\infty}\frac{d\omega}{2\pi}\ln S_x(\omega) - \frac{1}{2}\int_{-\infty}^{\infty}\frac{d\omega}{2\pi}\frac{|\tilde{x}(\omega)|^2}{S_x(\omega)}\right]. \qquad (A.172)$$

Not every case we look at will be Gaussian, but this result helps get us started.

Problem 175: Generality.   We made an effort to evaluate Z in the specific case where $C_x(\tau) = e^{-|\tau|/\tau_c}$, but we wrote the final result in a very general form, Eq. (A.172). Show that this slide into generality is justified.

Problem 176: Nonzero means and signal-to-noise ratios.   We should be able to carry everything through in the case where the mean $x(t)$ is not zero. For example, for just background noise described by some spectrum $\mathcal{N}(\omega)$, we have

$$P_{\text{noise}}[x(t)] = \exp\left[+\frac{T}{2}\int_{-\infty}^{\infty}\frac{d\omega}{2\pi}\ln\mathcal{N}(\omega) - \frac{1}{2}\int_{-\infty}^{\infty}\frac{d\omega}{2\pi}\frac{|\tilde{x}(\omega)|^2}{\mathcal{N}(\omega)}\right]. \qquad (A.173)$$

If there is an added signal $x_0(t)$, the distribution functional becomes

$$P_{\text{signal}}[x(t)] = \exp\left[+\frac{T}{2}\int_{-\infty}^{\infty}\frac{d\omega}{2\pi}\ln\mathcal{N}(\omega) - \frac{1}{2}\int_{-\infty}^{\infty}\frac{d\omega}{2\pi}\frac{|\tilde{x}(\omega) - \tilde{x}_0(\omega)|^2}{\mathcal{N}(\omega)}\right]. \qquad (A.174)$$

Suppose that you observe some particular $x(t)$, and you have to decide whether this came from the signal or noise distribution; that is, you have to decide whether the signal was present. For simplicity, assume that the two possibilities are equally likely a priori. As discussed in Chapter 2, to make such decisions optimally, you should use the relative probabilities that the signal or noise could give rise to the data. In particular, consider computing the log likelihood ratio:

$$\lambda[x(t)] \equiv \ln\left(\frac{P_{\text{signal}}[x(t)]}{P_{\text{noise}}[x(t)]}\right). \qquad (A.175)$$

(a) Give a simple expression for $\lambda[x(t)]$. Show that it is a linear functional of $x(t)$.

(b) Show that, when the $x(t)$ are drawn at random out of either P_{signal} or P_{noise}, $\lambda[x(t)]$ is a Gaussian random variable. Find the means, $\langle\lambda\rangle_{\text{noise}}$ and $\langle\lambda\rangle_{\text{signal}}$, and the variances $\langle(\delta\lambda)^2\rangle_{\text{noise}}$ and $\langle(\delta\lambda)^2\rangle_{\text{signal}}$, in the two distributions. Hint: You should see that $\langle(\delta\lambda)^2\rangle_{\text{noise}} = \langle(\delta\lambda)^2\rangle_{\text{signal}}$.

(c) Sketch the distributions $P_{\text{noise}}(\lambda)$ and $P_{\text{signal}}(\lambda)$. Show that your ability to make reliable discriminations is determined only by the signal-to-noise ratio,

$$SNR = \frac{\left(\langle\lambda\rangle_{\text{signal}} - \langle\lambda\rangle_{\text{noise}}\right)^2}{\langle(\delta\lambda)^2\rangle}, \qquad (A.176)$$

and that we can write

$$SNR = \int_{-\infty}^{\infty}\frac{d\omega}{2\pi}\frac{|\tilde{x}_0(\omega)|^2}{\mathcal{N}(\omega)}. \qquad (A.177)$$

(d) In rod cells, a single photon produces a current pulse with the approximate form $x_0(t) = I_1(t/\tau)^3 e^{-t/\tau}$. The power spectrum of continuous background noise is approximately $\mathcal{N}(\omega) = A/[1 + (\omega\tau)^2]^2$, with the same value of τ. Evaluate the peak current I_{peak} and total variance of the background noise, σ_I^2. A naive estimate of the signal-to-noise ratio is just $SNR_{\text{naive}} = (I_{\text{peak}}/\sigma_I)^2$. Show that the optimal signal-to-noise ratio, computed from Eq. (A.177), is larger. Why?

A.3   Diffraction and Biomolecular Structure

Many of the molecules that carry out essential biological functions have structures that are now known to atomic resolution. This knowledge has made for a revolution, not just because we can sometimes literally see how things work, but also because the structures supply a framework for asking new questions. Most detailed experimental information about the structure of biological molecules has come from X-ray diffraction measurements, and so we focus on these.

My experience is that most students have more experience of scattering theory in the context of quantum mechanics than in classical electromagnetism, so let's use the quantum language to get started. X-ray photons are incident on a sample, and we consider scattering events that result in a shift of the photon's energy by $\hbar\omega$ and a shift in its momentum by $\hbar\mathbf{q}$. The amplitude for this scattering event must be proportional to the $(\mathbf{q}, \omega)$ spatiotemporal Fourier component of the relevant density in the sample. For an electromagnetic wave what matters is (roughly) the charge density. Thus, the cross–section for elastic ($\omega = 0$) scattering is

$$\sigma(\mathbf{q}) \propto \left| \int d^3x \, e^{i\mathbf{q}\cdot\mathbf{x}} \rho(\mathbf{x}) \right|^2. \tag{A.178}$$

It is useful to have in mind the geometry, as shown in Fig. A.2. If the X-ray photons approach the sample collimated along the x-axis, they have an initial wavevector $\mathbf{k}_0 = k\hat{\mathbf{x}}$, where $\hat{\mathbf{x}}$ is a unit vector pointing along the x-axis, $k \equiv |\mathbf{k}_0| = 2\pi/\lambda$, and λ is the wavelength. If they emerge with a final wavevector $\mathbf{k}_f$ at an angle θ relative to the $\hat{x}$-axis, then $\mathbf{q} \equiv \mathbf{k}_f - \mathbf{k}_0$, and the magnitude of the scattering vector (or, up to a factor $\hbar$, momentum transfer) is

$$|\mathbf{q}| = |\mathbf{k}_f - \mathbf{k}_0| \tag{A.179}$$

$$= \sqrt{|\mathbf{k}_f - \mathbf{k}_0|^2} \tag{A.180}$$

$$= \sqrt{|\mathbf{k}_f|^2 - 2\mathbf{k}_f\cdot\mathbf{k}_0 + |\mathbf{k}_0|^2} \tag{A.181}$$

$$= \sqrt{k^2 - 2k^2\cos\theta + k^2} \tag{A.182}$$

$$= \sqrt{2k^2(1 - \cos\theta)} = 2k\,\sin(\theta/2). \tag{A.183}$$

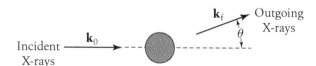

FIGURE A.2

The geometry for X-ray scattering. Photons are incident with wavevector $\mathbf{k}_0$, and emerge after scattering through angle θ with wavevector $\mathbf{k}_f$.

Thus, scattering by a small angle corresponds to a small momentum transfer. The classic results about X-ray diffraction concern the case where the density profile is periodic, as in a crystal. If the periodicity corresponds to displacement by d (let's think along one dimension, for the moment), then the density can be expressed as a discrete Fourier series, which means (from Eq. (A.178)) that $\sigma(\mathbf{q})$ will have delta functions at $|\mathbf{q}| = 2\pi n/d$, with n an integer. Combining this with Eq. (A.183), we find the angles that satisfy the Bragg condition:

$$2\pi n/d = (4\pi/\lambda)\sin(\theta/2) \Rightarrow \sin(\theta/2) = n\lambda/2d. \tag{A.184}$$

This expression is the Bragg condition, except that what we have identified as the scattering angle is twice the usual definition.

The first great triumph of X-ray diffraction in elucidating the structure of biological molecules came with the structure of DNA. This is an often told, and often distorted, piece of scientific history. Watson and Crick predicted the structure of DNA by arguing that a few key facts about the molecule, when combined with the rules of chemical bonding, were enough to suggest an interesting structure that would have consequences for the mechanisms of genetic inheritance. It was known that DNA was composed of four different kinds of nucleotide bases: adenine (A), thymine (T), guanine (G), and cytosine (C). Importantly, Chargaff had surveyed the DNA of many organisms and shown that although the ratios of A to G, for example, vary enormously, the ratios A/T and C/G do not. Watson and Crick realized that the molecular structures of the bases are such that A and T can form favorable hydrogen bonds, as can C and G. Further, the resulting hydrogen-bonded base pairs are the same size and thus could fit comfortably into a long polymer, as shown in Fig. A.3. Piling on top of one another, the base pairs would also experience a favorable stacking interaction among the π-bonded electrons in their rings. Finally, if one looks carefully at all the bond angles where the planar bases connect to the sugars and phosphate backbone, each successive base pair must rotate relative to its neighbor, and although there is some flexibility the favored angle was predicted to be $2\pi/10$ radians, or $36°$.

Quite independently of his collaboration with Watson, Crick had been interested in the structure of helical molecules and in the X-ray diffraction patterns that they should produce. Thus, when Watson and Crick realized that the structure of DNA might be a helix, they were in a position to calculate what the diffraction patterns should look like, and thus compare with the data emerging from the work of Franklin, Wilkins, and collaborators. So, let's look at the theory of diffraction from a helix.

It is best to describe a helix in cylindrical coordinates: z along the axis of the helix, r outward from its center, and an angle ϕ around the axis. Helical symmetry is the statement that translations along z are equivalent to rotations of the angle ϕ. Thus, a continuous helical structure would have the property that

$$\rho(z, r, \phi) = \rho(z + d, r, \phi + 2\pi d/\ell), \tag{A.185}$$

for any displacement d, where ℓ is the displacement corresponding to a complete rotation. For a discrete helical structure, the same equation is true, but only for values of d that are integer multiples of a fundamental spacing d_0.

(a)

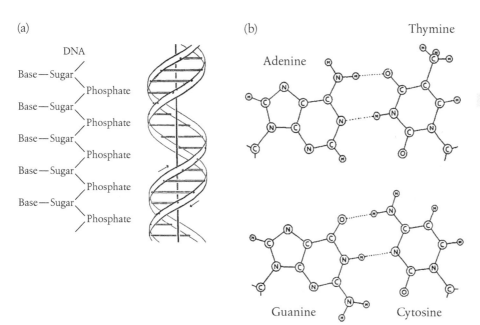

(b)

FIGURE A.3

The structure of DNA. (a) The polymeric pattern of bases, sugars, and phosphates, and the famous double helix. (b) The pairings A-T and G-C, illustrating the similar sizes of the correct pairs. Note that the donor-acceptor pattern of hydrogen bonds discriminates against the incorrect A-C and G-T pairings. Reprinted by permission from Macmillan Publishers, Ltd.: Watson and Crick (1953b).

For the continuous helix, what Eq. (A.185) is telling us is that the dependence on the two variables z and ϕ really collapses to a dependence on one combined variable:

$$\rho(z, r, \phi) = g(r, \phi - 2\pi z/\ell). \tag{A.186}$$

We know that any function of angle can be expanded as a discrete Fourier series,

$$f(\phi) = \sum_{n=-\infty}^{\infty} \tilde{f}_n e^{-in\phi}, \tag{A.187}$$

so in this case we have

$$\rho(z, r, \phi) = \sum_{n=-\infty}^{\infty} \tilde{g}_n(r) e^{-in(\phi - 2\pi z/\ell)}. \tag{A.188}$$

Our task is to compute

$$\int d^3x \, e^{i\mathbf{q}\cdot\mathbf{x}} \rho(\mathbf{x}). \tag{A.189}$$

In cylindrical coordinates, we can write $\mathbf{q} = (q_z\hat{\mathbf{z}}, \mathbf{q}_\perp)$, so that $\mathbf{q}\cdot\mathbf{x} = q_z z + q_\perp r \cos\phi$, where we choose the origin of the angle ϕ to make things simple and $q_\perp = |\mathbf{q}_\perp|$. Thus we have

$$e^{i\mathbf{q}\cdot\mathbf{x}} = e^{iq_z z} e^{iq_\perp r \cos\phi} \tag{A.190}$$

$$= e^{iq_z z} \sum_{n=-\infty}^{\infty} i^n J_n(q_\perp r) e^{in\phi}, \tag{A.191}$$

where

$$J_n(u) = \int_0^{2\pi} \frac{d\phi}{2\pi} e^{-in\phi} e^{iu \cos\phi} \tag{A.192}$$

are Bessel functions. Putting Eq. (A.191) together with the consequences of helical symmetry in Eq. (A.188), we have

$$
\int d^3x \, e^{i\mathbf{q}\cdot\mathbf{x}} \rho(\mathbf{x}) = \int_{-\infty}^{\infty} dz \int_0^{\infty} dr\, r \int_0^{2\pi} d\phi\, e^{iq_z z} \sum_{n=-\infty}^{\infty} i^n J_n(q_\perp r) e^{in\phi}
$$

$$
\times \sum_{m=-\infty}^{\infty} \tilde{g}_m(r) e^{-im(\phi - 2\pi z/\ell)} \tag{A.193}
$$

$$
= \sum_{n,m=-\infty}^{\infty} i^n \int_{-\infty}^{\infty} dz\, e^{iq_z z} e^{-i2\pi mz/\ell}
$$

$$
\times \int_0^{\infty} dr\, r\, J_n(q_\perp r) \tilde{g}_m(r) \int_0^{2\pi} e^{in\phi} e^{-im\phi}. \tag{A.194}
$$

The integral over ϕ forces $m = n$, and the integral over z generates delta functions at $q_z = 2\pi n/\ell$. Thus, for a continuous helix we expect that the X-ray scattering cross section will behave as

$$
\sigma(q_z, q_\perp) \propto \sum_{n=-\infty}^{\infty} \delta(q_z - 2\pi n/\ell) \left| \int_0^{\infty} dr\, r\, J_n(q_\perp r) \tilde{g}_n(r) \right|^2. \tag{A.195}
$$

In particular, if most of the density sits at a distance R from the center of the helix (which is not a bad approximation for DNA, because the phosphate groups have higher electron density than the rest of the molecule), then we have

$$
\sigma(q_z, q_\perp) \sim \sum_{n=-\infty}^{\infty} \delta(q_z - 2\pi n/\ell) |J_n(q_\perp R)|^2. \tag{A.196}
$$

Equation (A.196) states that diffraction from a helix generates a series of layer lines at $q_z = 2\pi n/\ell$, and from their spacing we should be able to read off the pitch of the helix (the distance ℓ along the z-axis corresponding to a complete turn). Further, if we look along a single layer line, we should see an intensity varying as $\sim |J_n(q_\perp R)|^2$. What is important here about the Bessel functions is that for small $q_\perp$ we have $J_n(q_\perp R) \propto (q_\perp R)^n$, and the first peak of the nth Bessel function occurs at a point roughly proportional to n. The resulting pattern is shown schematically in Fig. A.4.

Problem 177: Bessel functions.   Verify the statements about Bessel functions made above in enough detail to understand the diffraction patterns shown in Fig. A.4.

Let's see what happens when we move from the continuous to the discrete helix. To keep things simple, suppose that all the density indeed is concentrated at a distance R from the center of the helix, so that

$$
\rho(\mathbf{x}) = \frac{1}{R} \delta(r - R) \sum_n \delta(z - nd_0) \delta(\phi - n\phi_0), \tag{A.197}
$$

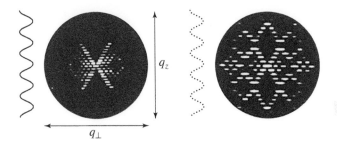

FIGURE A.4

Diffraction from continuous (left) and discrete (right) helices. From Holmes and Blow (1965), with thanks to K. C. Holmes.

where the rotation from one element to the next is $\phi_0 = 2\pi d_0/\ell$; notice that we do not really require ℓ/d_0 to be an integer. Now we have

$$
\int d^3x\, e^{i\mathbf{q}\cdot\mathbf{x}} \rho(\mathbf{x}) = \int_{-\infty}^{\infty} dz \int_0^{\infty} dr\, r \int_0^{2\pi} d\phi\, e^{iq_z z}
$$
$$
\times \sum_{n=-\infty}^{\infty} J_n(q_\perp r) e^{in\phi} \frac{1}{R} \delta(r-R)
$$
$$
\times \sum_{m=-\infty}^{\infty} \delta(z-md_0)\delta(\phi - m\phi_0). \qquad \text{(A.198)}
$$

Each of the integrals we need to do has a delta function, so integrating is just substituting:

$$
\int d^3x\, e^{i\mathbf{q}\cdot\mathbf{x}} \rho(\mathbf{x}) \propto \sum_{m=-\infty}^{\infty} \sum_{n=-\infty}^{\infty} J_n(q_\perp R)\delta(q_z + 2\pi n/\ell - 2\pi m/d_0). \qquad \text{(A.199)}
$$

Thus, the discrete helix involves a double sum of terms. For $m = 0$ we have the results for the continuous helix. But the sum over $m \neq 0$ causes the whole "X" pattern of the continuous helix to be repeated with centers at $(q_z = 2\pi m/d_0, q_\perp = 0)$; the line $q_\perp = 0$ is often called the "meridian," and so the extra peaks centered on $(q_z = 2\pi m/d_0, q_\perp = 0)$ are called "meridional reflections." All of this structure is shown in Fig. A.4. Just as the spacing of the layer lines allows us to measure the helical pitch ℓ, the spacing of the meridional reflections allows measurement of the spacing d_0 between discrete elements along the helix.

At this point you know what Watson and Crick knew. They had a theory of what the structure should be, and almost certainly they had already realized the implications of this structure, as they remarked in their first paper "It has not escaped our notice that the specific pairing we have postulated immediately suggests a possible copying mechanism for the genetic material" (Watson and Crick 1953a: 737). They also knew that if the structure was as they had theorized, then the diffraction pattern should display a number of key signatures—the regularly spaced layer lines, the "X" arrangement of their intensities, and the meridional reflections—that would provide both qualitative and quantitative confirmation of the theory. Thus, you should be able to imagine their excitement when they saw the clean X-ray diffraction pattern from

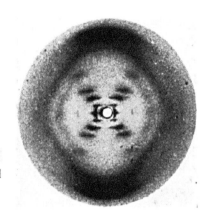

FIGURE A.5

The justly famous photograph 51, showing the diffraction from DNA molecules pulled into a fiber. Reprinted by permission from Macmillan Publishers, Ltd.: Franklin and Gosling (1953).

hydrated DNA, the famous photograph 51 taken by Rosalind Franklin (Fig. A.5). As far as anyone could tell, the proposed structure was right.

Problem 178: Discrete helices, more generally.   Show that most of what was said above can be generalized to an arbitrary discrete helix, without assuming that the density is concentrated at $r = R$. That is, use only the symmetry defined by Eq. (A.185) for $d = nd_0$.

Problem 179: Fibers versus crystals.   We have discussed the diffraction from a helix as if there were just one molecule and have not been very precise about the difference between amplitudes and intensities. Show that if there are many helices, all with their z-axes aligned but with random positions and orientations in the x-y plane, then the diffraction intensity from the ensemble of molecules depends only on the structure of the individual helices.

It is crucial to appreciate that, contrary to what is often said in textbooks, it was not possible to "determine" the structure of DNA by looking at diffraction patterns like those in Fig. A.5. However, if you thought you knew the structure, you could predict the diffraction pattern—in the regime where it could be measured—and see whether you got things right. This difference between experiments that support a theory (or find something that a theory tells us must exist) and experiments that discover something unexpected or genuinely unknown is an incredibly important distinction that is often elided.

So much has been written about this moment in scientific history that it would be irresponsible not to pause and reflect. However, I am not a historian, so I make just a few observations. Most importantly, I think, the story of the DNA structure combines so many themes in our understanding of science and society (separately and together) that it has an almost mythical quality, and as with the ancient myths everyone can find something in the story that connects to their own concerns. There is the enormous issue of gender inequality in the scientific community, something for which we hardly even had a vocabulary until decades after the event. There are the personalities of all the individuals, both as they were in 1953 and as they developed in response to the world-changing discovery in which they participated. There is the tragedy of Franklin's early death. There is the competition between Cambridge and London, and the interaction of an American collaborator with these very British social structures. Finally, there are issues that are more purely about the science, such as the interaction between theory and experiment, physics and biology. I encourage you to read more about this episode and its ramifications in the references listed in the Annotated Bibliography for this section.

To actually *determine* the structure of a large molecule by X-ray diffraction, we need to form crystals of those molecules. Crystals of a protein are not like crystals of salt or even small molecules. They are quite soft and contain a lot of water. The bonds between proteins, for example, in a crystal are much weaker than the bonds that hold each protein together. On one hand, this makes growing and handling the crystals quite difficult. On the other hand, it means that the internal structure of the protein in the crystal is more likely to be typical of its structure when free in solution.

A crystal in three dimensions is defined by the existence of three vectors $\mathbf{a}$, $\mathbf{b}$, and $\mathbf{c}$ such that the density is the same if we translate by integer combinations of these vectors:

$$\rho(\mathbf{x}) = \rho(\mathbf{x} + n\mathbf{a} + m\mathbf{b} + k\mathbf{c}). \tag{A.200}$$

Thus, the density can be expanded into a Fourier series,

$$\rho(\mathbf{x}) = \sum_{knm} \tilde{\rho}_{knm} \exp\left[i(k\mathbf{G}_a + n\mathbf{G}_b + m\mathbf{G}_c) \cdot \mathbf{x}\right], \tag{A.201}$$

where the $\mathbf{G}_i$ are the reciprocal lattice vectors. As a result, the X-ray scattering cross-section is a set of delta functions or Bragg peaks:

$$\sigma(\mathbf{q}) \propto \sum_{knm} |\tilde{\rho}_{knm}|^2 \delta(\mathbf{q} - k\mathbf{G}_a - n\mathbf{G}_b - m\mathbf{G}_c). \tag{A.202}$$

Problem 180: Details of diffraction.   Fill in the details leading to Eq. (A.202), including the relationship between the reciprocal lattice vectors $\mathbf{G}_i$ and the real lattice vectors $\mathbf{a}$, $\mathbf{b}$, and $\mathbf{c}$.

Even if we can make a perfect measurement of $\sigma(\mathbf{q})$, we only learn about the magnitudes of the Fourier coefficients, $|\tilde{\rho}_{knm}|^2$, which is not sufficient to reconstruct the density $\rho(\mathbf{x})$. This difficulty is called the "phase problem." For small structures it is not such a serious problem, because the constraint that $\rho(\mathbf{x})$ has to be built out of discrete atoms allows us to determine the positions of the atoms from the diffraction pattern. But for a protein, with thousands of atoms in each unit cell of the crystal, this is hopeless.

The phase problem was solved experimentally through the idea of isomorphous replacement. Suppose that we could attach to each molecule in the crystal one or more very heavy atoms, in well-defined (but unknown) positions. If we can do this without disrupting the packing of the molecules in the crystal, then the positions of the Bragg peaks will not change, but their intensities will. If we can approximate the density profiles of the heavy atoms as delta functions (which should be right, unless we look at very large $|\mathbf{q}|$), then we have

$$|\tilde{\rho}_{knm}|^2 \rightarrow \left| \tilde{\rho}_{knm} + \sum_{\mu} Z_{\mu} e^{i\mathbf{q}_{knm}\cdot\mathbf{x}_{\mu}} \right|^2, \tag{A.203}$$

where $\mathbf{q}_{knm} = k\mathbf{G}_a + n\mathbf{G}_b + m\mathbf{G}_c$, Z_{μ} is the charge of the μth heavy atom, and $\mathbf{x}_{\mu}$ is its position. In the simple case of one added heavy atom, we can choose coordinates so that its position is at the origin, and then it should be clear that the change in intensity on adding the heavy atom is directly sensitive to the value of $\cos\phi_{knm}$, where ϕ_{knm} is the phase of the complex number ρ_{knm}. This technique is almost enough, but to resolve the remaining twofold ambiguity $\cos\phi_{knm} = \cos(-\phi_{knm})$, one needs at least two different examples of adding heavy atoms to the structure.

As in many sections of the text, what we are discussing here could be expanded to a full course on its own; some references are collected in the Annotated Bibliography for this section. Although we do not explore fully, it is worth drawing attention to two points. First, there are other ways of solving the phase problem, notably by using tunable X-ray sources. As we tune the energy of the incoming X-rays, we can hit the edge of absorption bands, and at these energies the scattering from individual atoms has a (known!) phase shift. By measuring the change in the diffraction pattern that arises from this phase shift, we do something that is almost equivalent to adding a heavy atom. This technique works if there is a single atom (or a small number of them) whose absorption edges can be approached, and so it is especially well suited to proteins that contain metal atoms. The second point is the difference between accuracy and resolution. In practice, we can measure clean diffraction patterns only out to some maximal value of the scattering vector $|\mathbf{q}|$, which suggests that when we try to do the Fourier synthesis of the electron density, we will be limited to components whose spatial period is longer than $2\pi/|\mathbf{q}|$. This is true, and this shortest measurable period is referred to as the "resolution" of the diffraction pattern. But if we know that the underlying pattern of electron density is (1) positive, and (2) composed of blobs corresponding to discrete atoms, then we are not just trying to synthesize an arbitrary function from its Fourier components. We can use our general knowledge of the underlying structure to refine our estimate of the electron density pattern. In this way, for example, we can estimate the positions of individual atoms in the structure to much better than the nominal resolution of the diffraction pattern. Thus, one routinely reads about knowing the positions of atoms in proteins with an "accuracy" of $\sim$0.1–0.2 Å, even though the "resolution" of diffraction patterns is only 1–2 Å.

The density really consists of discrete blobs corresponding to atoms, and—if we can look at sufficiently high resolution—additional density in the bonds between atoms. For the moment let's think just about the atoms. Then the density has the form

$$\rho(\mathbf{x}) \approx \sum_{\mu} f_{\mu} \delta(\mathbf{x} - \mathbf{x}_{\mu}), \tag{A.204}$$

where $\mathbf{x}_{\mu}$ is the position of the μth atom, and f_{μ} is an effective charge or scattering density associated with that atom. Thus, the scattering cross-section behaves as

$$\sigma(\mathbf{q}) \sim \sum_{\mu\nu} f_{\mu} f_{\nu} e^{i\mathbf{q}\cdot(\mathbf{x}_{\mu}-\mathbf{x}_{\nu})}. \tag{A.205}$$

Importantly, the positions of atoms fluctuate. The time scale of these fluctuations typically is much shorter than the time scale of the experiment, so we will see an average:

$$\sigma(\mathbf{q}) \sim \left\langle \sum_{\mu\nu} f_\mu f_\nu e^{i\mathbf{q}\cdot(\mathbf{x}_\mu - \mathbf{x}_\nu)} \right\rangle. \tag{A.206}$$

Assuming that the fluctuations in position are Gaussian around some mean, then we have

$$\sigma(\mathbf{q}) \sim \left\langle \sum_{\mu\nu} f_\mu f_\nu e^{i\mathbf{q}\cdot(\mathbf{x}_\mu - \mathbf{x}_\nu)} \right\rangle$$

$$\equiv \sum_{\mu\nu} f_\mu f_\nu \left\langle e^{i\mathbf{q}\cdot\mathbf{r}_{\mu\nu}} \right\rangle \tag{A.207}$$

$$\sim \sum_{\mu\nu} f_\mu f_\nu e^{i\mathbf{q}\cdot\langle\mathbf{r}_{\mu\nu}\rangle} e^{-\frac{1}{2}|\mathbf{q}|^2\langle(\delta\mathbf{r}_{\mu\nu})^2\rangle}, \tag{A.208}$$

where $\mathbf{r}_{\mu\nu} = \mathbf{x}_\mu - \mathbf{x}_\nu$, and for simplicity we assume that the fluctuations are isotropic. Thus, the scattering intensity at $\mathbf{q}$ is attenuated relative to what we expect from a fixed structure by an amount $e^{-\frac{1}{2}|\mathbf{q}|^2\langle(\delta\mathbf{r}_{\mu\nu})^2\rangle}$. These quantities are called the Debye-Waller factors. Thus, although X-ray diffraction is a static method, it is sensitive to dynamical fluctuations in structure.

If we can fit the intensities of the diffraction spots carefully, even in a large molecule such as a protein, we can see that the Debye-Waller factors are essential and can be used to extract estimates of the sizes of the atomic motions. There are also techniques that allow us to look more directly at these motions, and to a large extent the results are consistent, although care must be exercised, because different methods are sensitive to motions on different time scales.

A.4 Electronic Transitions in Large Molecules

In this section I outline an honest calculation that reproduces the intuition of Figs. 2.22 and 2.24. These same ideas will be important in thinking about electron-transfer reaction in Section 4.1, so they are worth a detour.

The system has two electronic states, which we can represent as a spin one-half; let spin down be the ground state and spin up be the excited state. The Born-Oppenheimer approximation shows that we can think of the atoms in a molecule as moving in a potential determined by the electronic state,[3] which we denote by $V_\uparrow(\mathbf{q})$ and $V_\downarrow(\mathbf{q})$ in the excited and ground states, respectively; $\mathbf{q}$ stands for all the atomic coordinates (not just the one that we sketch!). Because we are observing photon absorption, there must be a matrix element that connects the two electronic states and couples to the electromagnetic field; assume that, absent symmetries, this coupling is dominated by an electric dipole term. In principle the dipole matrix element $\mathbf{d}$ could depend on the atomic coordinates, but we neglect this effect.[4] Putting the pieces together,

---

3. As in the main text, I use "atoms" and "nuclei" interchangeably.

4. In practice, this effect is small. You should think about why this is true.

we have the Hamiltonian for the molecule,

$$\mathbf{H} = \mathbf{K} + \frac{1}{2}(1 + \sigma_z)V_\uparrow(\mathbf{q}) + \frac{1}{2}(1 - \sigma_z)V_\downarrow(\mathbf{q}) + \mathbf{d} \cdot \mathbf{E}(\sigma_+ + \sigma_-), \quad \text{(A.209)}$$

where $\mathbf{K}$ is the kinetic energy of the atoms, $\mathbf{E}$ is the electric field, and the σs are the usual Pauli operators for an effective spin one-half describing the two electronic states. To this we should add the usual Hamiltonian for the electromagnetic field.

We are interested in computing the rate at which photons of energy $\hbar\Omega$ are absorbed, we will do this as a perturbation expansion in the term $\sim\mathbf{d}$. The result of such a calculation can be presented as the "Golden rule" for transition rates, but this formulation hides the underlying dynamics. So, at the risk of being pedantic, I'll go through the steps that usually lead to the Golden rule and take a detour that leads us to a formula in which the dynamics of atomic motions are more explicit.[5]

We start our system in the ground state of the electrons $(|\downarrow\rangle)$, in some initial state $(|i\rangle)$ of the atomic coordinates, and in the presence of one photon of wavevector $\mathbf{k}$ and frequency $\Omega = c|\mathbf{k}|$ (polarization is an unnecessary complication here). As the system evolves under the Hamiltonian $\mathbf{H}$, at some time t we want to measure the probability of finding the system in the excited state $|\uparrow\rangle$, in some other state of the atoms $|f\rangle$, and absent the photon. The general statement is that quantum states evolve as[6]

$$|\psi(0)\rangle \rightarrow |\psi(t)\rangle = \mathbf{T} \exp\left[-\frac{i}{\hbar}\int_0^t d\tau\mathbf{H}(\tau)\right]|\psi(0)\rangle, \quad \text{(A.210)}$$

where $\mathbf{T}$ is the time-ordering operator. Thus, for our particular problem, the probability of starting in state $|\downarrow, i, \mathbf{k}\rangle$ and ending in state $|\uparrow, f, \emptyset\rangle$ is given by

$$p_{i \rightarrow f}(t) = \left|\langle\emptyset, f, \uparrow |\mathbf{T} \exp\left[-\frac{i}{\hbar}\int_0^t d\tau\mathbf{H}(\tau)\right]| \downarrow, i, \mathbf{k}\rangle\right|^2. \quad \text{(A.211)}$$

In fact, we do not care about the final state of the atoms, and we can't select their initial state—this comes out of the Boltzmann distribution. So we really should compute

$$P(t) = \sum_{i,f}\left|\langle\emptyset, f, \uparrow |\mathbf{T} \exp\left[-\frac{i}{\hbar}\int_0^t d\tau\mathbf{H}(\tau)\right]| \downarrow, i, \mathbf{k}\rangle\right|^2 p_i, \quad \text{(A.212)}$$

where p_i is the probability of being in the initial atomic state i.

As usual, we break the Hamiltonian into two pieces, $\mathbf{H} = \mathbf{H}_0 + \mathbf{H}_1$, and do perturbation theory on $\mathbf{H}_1$. We choose $\mathbf{H}_1 = \mathbf{d}\cdot\mathbf{E}(\sigma_+ + \sigma_-)$, which is the only term that connects the states $|\downarrow\rangle$ and $|\uparrow\rangle$. The leading term in the perturbation series thus becomes

$$P(t) \approx \frac{1}{\hbar^2}\sum_{i,f}\left|\langle\emptyset, f, \uparrow |\mathbf{T}e^{-\frac{i}{\hbar}\int_0^t d\tau\mathbf{H}_0(\tau)}\int_0^t d\tau'\mathbf{H}_1(\tau')| \downarrow, i, \mathbf{k}\rangle\right|^2 p_i. \quad \text{(A.213)}$$

---

5. I am assuming that the Golden rule is well known, but that the general tools relating cross-sections and transition rates to correlation functions are less well digested.

6. Please be careful to distinguish the state $|i\rangle$ from $i = \sqrt{-1}$.

If we look more carefully at the amplitude, we have

$$\langle \emptyset, f, \uparrow | \mathbf{T} e^{-\frac{i}{\hbar} \int_0^t d\tau \mathbf{H}_0(\tau)} \int_0^t d\tau' \mathbf{H}_1(\tau') | \downarrow, i, \mathbf{k} \rangle$$

$$= \int_0^t d\tau' \langle f | \mathbf{T} \left(e^{-\frac{i}{\hbar} \int_{\tau'}^t d\tau \mathbf{H}_\uparrow(\tau)} \right) \mathbf{d} \cdot \langle \emptyset | \mathbf{E} | \mathbf{k} \rangle \mathbf{T} \left(e^{-\frac{i}{\hbar} \int_0^{\tau'} d\tau \mathbf{H}_\downarrow(\tau)} \right) | i \rangle e^{-i\Omega\tau'}, \tag{A.214}$$

where τ' is the time at which the term $\mathbf{H}_1 \sim \sigma_+$ acts to flip the state from $|\downarrow\rangle$ to $|\uparrow\rangle$, and the factor $e^{-i\Omega\tau'}$ comes from the time dependence of the electric field (or equivalently, from the energy of the absorbed photon). The term $\mathbf{H}_\downarrow$ is defined by

$$\mathbf{H}_\downarrow = \mathbf{K} + V_\downarrow(\mathbf{q}) \tag{A.215}$$

and similarly for $\mathbf{H}_\uparrow$. When we square this amplitude and sum over final states, we can identify a sum over a complete set of states,

$$\sum_f |f\rangle\langle f| = \mathbf{1}, \tag{A.216}$$

the unit operator, which allows us to collapse the result into an expectation value over the initial states of the system. In this way, the probability of a transition can be related to an average of some operators—a correlation function—in the initial state of the system.

The main points we want to explore here do not depend on the quantum mechanical character of the atomic motion. So, to keep things simple, let's assume that the motion of the atoms is classical. In practice, because the terms $\mathbf{H}_{\uparrow, \downarrow}$ depend only on the atomic coordinates and momenta, the classical approximation means that we do not have to worry about the noncommutativity of these operators at different times, and we can drop the formalities of time ordering. Putting all the terms together, we can rewrite $P(t)$ from Eq. (A.213):

$$P(t) \approx \frac{(\mathbf{d} \cdot \langle \emptyset | \mathbf{E} | \mathbf{k} \rangle)^2}{\hbar^2} \int_0^t d\tau_1 \int_0^t d\tau_2 \, e^{+i\Omega(\tau_1 - \tau_2)} \sum_i p_i \langle i | e^{+\frac{i}{\hbar} \int_0^{\tau_1} d\tau \mathbf{H}_\downarrow(\tau)}$$

$$\times e^{+\frac{i}{\hbar} \int_{\tau_1}^t d\tau \mathbf{H}_\uparrow(\tau)} e^{-\frac{i}{\hbar} \int_{\tau_2}^t d\tau \mathbf{H}_\uparrow(\tau)} e^{-\frac{i}{\hbar} \int_0^{\tau_2} d\tau \mathbf{H}_\downarrow(\tau)} | i \rangle \tag{A.217}$$

$$= \frac{(\mathbf{d} \cdot \langle \emptyset | \mathbf{E} | \mathbf{k} \rangle)^2}{\hbar^2} \int_0^t d\tau_1 \int_0^t d\tau_2 \, e^{+i\Omega(\tau_1 - \tau_2)}$$

$$\times \sum_i p_i \langle i | \exp \left(+\frac{i}{\hbar} \int_{\tau_1}^{\tau_2} d\tau [\mathbf{H}_\uparrow(\tau) - \mathbf{H}_\downarrow(\tau)] \right) | i \rangle \tag{A.218}$$

$$\propto \int_0^t d\tau_1 \int_0^t d\tau_2 \, e^{+i\Omega(\tau_1 - \tau_2)} \left\langle \exp \left[+\frac{i}{\hbar} \int_{\tau_1}^{\tau_2} d\tau \, \epsilon[\mathbf{q}(\tau)] \right] \right\rangle, \tag{A.219}$$

where $\epsilon = \mathbf{H}_\uparrow - \mathbf{H}_\downarrow = V_\uparrow - V_\downarrow$ is the instantaneous energy difference between the ground and excited states (which fluctuates as the atomic coordinates fluctuate), and

$\langle \cdot \rangle$ denotes an average over these fluctuations; $\epsilon[\mathbf{q}(\tau)]$ is the "vertical" energy difference in Figs. 2.20 and 2.22, that is, the difference between the energies of the electronic states at fixed values of the molecular coordinates $\mathbf{q}$.

Problem 181: Missing steps.   Fill in the steps leading to Eq. (A.219). If you are more ambitious, try the case where the atomic motions are fully quantum mechanical.

The integrand in Eq. (A.219) depends only on the time difference $\tau_2 - \tau_1$. Thus, we are doing an integral of the form

$$\int_0^t d\tau_1 \int_0^t d\tau_2 \, F(\tau_2 - \tau_1). \tag{A.220}$$

It seems natural to rewrite this integral over the (τ_1, τ_2) plane in terms of an integral over the time difference and the mean. In the limit that t is large, this yields

$$\int_0^t d\tau_1 \int_0^t d\tau_2 \, F(\tau_2 - \tau_1) \to t \int_{-\infty}^{\infty} d\tau \, F(\tau). \tag{A.221}$$

Thus, we have

$$P(t) \propto t \int_{-\infty}^{\infty} d\tau \, e^{+i\Omega\tau} \left\langle \exp\left[-\frac{i}{\hbar} \int_0^\tau d\tau' \, \epsilon[\mathbf{q}(\tau')]\right] \right\rangle, \tag{A.222}$$

so that the transition rate or absorption cross-section for photons of frequency Ω becomes

$$\sigma(\Omega) \propto \left\langle \int_{-\infty}^{\infty} d\tau \, \exp\left[+i\Omega\tau - \frac{i}{\hbar} \int_0^\tau d\tau' \, \epsilon[\mathbf{q}(\tau')]\right] \right\rangle. \tag{A.223}$$

Now we can recover the intuition of Fig. 2.22 as a saddle point approximation to the integral in Eq. (A.223). The saddle point approximation[7] is

$$\int dt \, \exp\left[+i\phi(t)\right] \approx \sqrt{\frac{2\pi}{|\phi''(t_*)|}} \, \exp\left[+i\phi(t_*)\right], \tag{A.224}$$

where the time t_* is defined by

$$\left.\frac{d\phi(t)}{dt}\right|_{t=t_*} = 0. \tag{A.225}$$

The condition for validity of the approximation is that the time scale

$$\delta t \sim 1/\sqrt{|\phi''(t_*)|} \tag{A.226}$$

---

7. The intuition behind the saddle point is that, as we integrate over time, the phase $\phi(t)$ "runs" very quickly, and the sine and cosine of this phase will average to zero, except near points where the phase is not changing. For more details, see Matthews and Walker (1970).

be small compared with the intrinsic time scales for variation of $\phi(t)$. As applied to Eq. (A.223), the saddle point condition is

$$0 = \frac{d}{d\tau}\left[+i\Omega\tau - \frac{i}{\hbar}\int_0^\tau d\tau'\,\epsilon[\mathbf{q}(\tau')]\right]\Bigg|_{\tau=\tau_*} \tag{A.227}$$

$$= i\Omega - \frac{i}{\hbar}\epsilon[\mathbf{q}(\tau_*)], \tag{A.228}$$

$$\hbar\Omega = \epsilon[\mathbf{q}(\tau_*)]. \tag{A.229}$$

Thus, the saddle point condition states that the integral defining the cross-section is dominated by moments when the instantaneous difference between the ground and excited state energies matches the photon energy. But this instantaneous difference $\epsilon[\mathbf{q}]$ is exactly the vertical energy difference in Fig. 2.22. Because this integral is inside an expectation value over the fluctuations in atomic coordinates, the cross-section will be proportional to the probability that this matching condition is obeyed.

If the sketch in Fig. 2.22 is equivalent to a saddle point approximation, we have to consider conditions for the validity of this approximation. The time scale defined by Eq. (A.226) becomes

$$\delta t \sim \left|\frac{1}{\hbar}\frac{d\epsilon[\mathbf{q}(\tau)]}{d\tau}\right|^{-1/2} \sim \sqrt{\frac{\hbar}{\epsilon'v}}, \tag{A.230}$$

where ϵ' is the slope of the energy difference as a function of atomic coordinates, and v is a typical velocity for motion along these coordinates. Thus, large slopes result in small values of δt, and this time scales as $\sqrt{\hbar}$.

The natural time scale of motion along the atomic coordinates is given by vibrational periods, or $\omega_{\text{vib}}^{-1} = \tau_{\text{vib}} \sim Q/v$, where Q is a typical displacement from equilibrium. Then we can write

$$\delta t \sim \sqrt{\frac{\hbar}{\epsilon'v}} \sim \sqrt{\frac{\hbar\omega_{\text{vib}}}{\epsilon'Q}\cdot\frac{Q}{v\omega_{\text{vib}}}} \sim \tau_{\text{vib}}\sqrt{\frac{\hbar\omega_{\text{vib}}}{\epsilon'Q}}. \tag{A.231}$$

Thus, $\delta t \ll \tau_{\text{vib}}$ if the energy $\epsilon'Q$ is much larger than the energy of vibrational quanta $\hbar\omega_{\text{vib}}$. But $\epsilon'Q$ is the range of energy differences between the ground and excited states that the molecule can access as it fluctuates—and this is the width of the absorption spectrum. Thus, self-consistently, if we find that the width of the spectrum is large compared to the vibrational quanta, then our saddle point approximation is accurate.

We can go a bit further if we specialize to the case where, as in Fig. 2.24, the different potential surfaces are exactly Hookean springs, so that the dynamics of atomic motions are harmonic oscillators. In the general case there are many normal modes, so we would write

$$V_\uparrow(\mathbf{q}) = \frac{1}{2}\sum_i \omega_i^2 q_i^2, \tag{A.232}$$

$$V_\downarrow(\mathbf{q}) = \epsilon_0 + \frac{1}{2}\sum_i \omega_i^2 (q_i - \Delta_i)^2. \tag{A.233}$$

In this case,

$$\epsilon[\mathbf{q}(t)] \equiv V_\uparrow[\mathbf{q}(t)] - V_\downarrow[\mathbf{q}(t)] \tag{A.234}$$

$$= \epsilon_0 + \frac{1}{2} \sum_i \omega_i^2 \Delta_i^2 - \sum_i \omega_i^2 \Delta_i q_i(t) \tag{A.235}$$

$$= \hbar\Omega_{\text{peak}} - X(t), \tag{A.236}$$

where the generalized coordinate $X(t)$ is given by a weighted combination of all the modes:

$$X(t) = \sum_i \omega_i^2 \Delta_i q_i(t). \tag{A.237}$$

Equation (A.223) for the absorption cross-section thus becomes

$$
\sigma(\Omega) \propto \left\langle \int_{-\infty}^{\infty} d\tau \, \exp\left[+i\Omega\tau - \frac{i}{\hbar} \int_0^\tau d\tau' \, \epsilon[\mathbf{q}(\tau')] \right] \right\rangle
$$

$$
= \left\langle \int_{-\infty}^{\infty} d\tau \, \exp\left[+i\Omega\tau - \frac{i}{\hbar} \int_0^\tau d\tau' \, (\hbar\Omega_{\text{peak}} - X(\tau')) \right] \right\rangle
$$

$$
= \int_{-\infty}^{\infty} d\tau \, e^{+i(\Omega - \Omega_{\text{peak}})\tau} \left\langle \exp\left[+\frac{i}{\hbar} \int_0^\tau d\tau' X(\tau') \right] \right\rangle. \tag{A.238}
$$

The key point is that, because $X(t)$ is a sum of harmonic oscillator coordinates, its fluctuations are drawn from a Gaussian distribution when we compute the average $\langle \cdot \rangle$ over the equilibrium ensemble.

Problem 182: Gaussian averages.   Derive Eq. (A.239).

If y is a Gaussian random variable, then we have

$$\langle e^y \rangle = \exp\left[\langle y \rangle + \frac{1}{2} \langle (\delta y)^2 \rangle \right]. \tag{A.239}$$

In the present case, the role of y is played by an integral over the trajectory of $X(t)$, but this should not bother us:

$$
\left\langle \exp\left[+\frac{i}{\hbar} \int_0^\tau d\tau' X(\tau') \right] \right\rangle
$$

$$
= \exp\left[\frac{1}{2} \left\langle \left(\frac{i}{\hbar} \int_0^\tau d\tau' X(\tau') \right)^2 \right\rangle \right] \tag{A.240}
$$

$$
= \exp\left[-\frac{1}{2\hbar^2} \int_0^\tau d\tau_1 \int_0^\tau d\tau_2 \langle X(\tau_1) X(\tau_2) \rangle \right], \tag{A.241}
$$

where we start by making use of the fact that $\langle X \rangle = 0$.

We see from Eq. (A.241) that the shape of the absorption spectrum is determined by the correlation function of the modes to which the electronic transition are coupled, that is, $C_X(\tau_1 - \tau_2) = \langle X(\tau_1)X(\tau_2)\rangle$. If these modes have relatively slow dynamics, then the time scales τ that enter the integral will be much shorter than the time scales over which this correlation function varies. In this limit we can approximate the integral as

$$\int_0^\tau d\tau_1 \int_0^\tau d\tau_2 \langle X(\tau_1)X(\tau_2)\rangle \approx \int_0^\tau d\tau_1 \int_0^\tau d\tau_2 \langle X(0)X(0)\rangle$$

$$= \langle X^2\rangle \tau^2. \tag{A.242}$$

Notice also that

$$\langle X^2\rangle = \left\langle \left(\sum_i \omega_i^2 \Delta_i q_i\right)^2\right\rangle = \sum_i \omega_i^4 \Delta_i^2 \langle q_i^2\rangle; \tag{A.243}$$

in the classical limit we have $\langle q_i^2\rangle = k_B T/\omega_i^2$, and hence

$$\langle X^2\rangle = k_B T \sum_i \omega_i^2 \Delta_i^2 = 2k_B T\lambda, \tag{A.244}$$

where λ generalizes the reorganization energy or the Stokes shift (Fig. 2.24) to the case of many modes. Finally, putting these pieces together, we have

$$\sigma(\Omega) \propto \int_{-\infty}^\infty d\tau \, \exp\left[+i(\Omega - \Omega_{\text{peak}})\tau\right]$$

$$\times \exp\left[-\frac{1}{2\hbar^2}\int_0^\tau d\tau_1 \int_0^\tau d\tau_2 \langle X(\tau_1)X(\tau_2)\rangle\right] \tag{A.245}$$

$$\approx \int_{-\infty}^\infty d\tau \, \exp\left[+i(\Omega - \Omega_{\text{peak}})\tau - \frac{\tau^2\lambda k_B T}{\hbar^2}\right] \tag{A.246}$$

$$= \sqrt{\frac{\pi\hbar^2}{\lambda k_B T}} \exp\left[-\frac{(\hbar\Omega - \hbar\Omega_{\text{peak}})^2}{4\lambda k_B T}\right]. \tag{A.247}$$

This result should look familiar from Eq. (2.74).

The calculation we have done here also allows us to look more precisely at the limits to our approximation. The integral in Eq. (A.246) is a Gaussian integral over τ, which means that it is done exactly by the saddle point method. The characteristic time that emerges from this calculation is

$$\delta t \sim \frac{\hbar}{\sqrt{\lambda k_B T}}. \tag{A.248}$$

If the typical vibrational time scales that enter into $C_X(\tau)$ are $\tau_{\text{vib}} \sim \hbar/k_B T$, then the condition for validity of our approximation becomes $\lambda \gg k_B T$. Tracking the factors, the approximate result should be valid if the predicted width of the absorption spectrum is (in energy units) larger than $k_B T$, or roughly 1% of $\hbar\Omega_{\text{peak}}$. This condition is rather gentle, suggesting that when the model of harmonic normal modes is correct, something like the saddle point approximation ought to work.

This calculation also gives us insight into another way that our semiclassical intuition from Fig. 2.22 can fail. If, for example, there was just a single normal mode, we would have $X = gq(t)$, where $g = \omega^2 \Delta$. But if there is just this one mode, and no other degrees of freedom to suck energy out of this mode, we must have

$$\langle q(t)q(t')\rangle = \frac{k_B T}{\omega^2} \cos[\omega(t - t')], \qquad (A.249)$$

so the integral (Eq. (A.245)) defining the cross-section becomes

$$\sigma(\Omega) \propto \int_{-\infty}^{\infty} d\tau \, \exp\left[+i(\Omega - \Omega_{\text{peak}})\tau \right.$$
$$\left. - \frac{\Delta^2\omega^2 k_B T}{2\hbar^2} \int_0^\tau d\tau_1 \int_0^\tau d\tau_2 \, \cos(\omega(\tau_1 - \tau_2)) \right] \qquad (A.250)$$

$$= \int_{-\infty}^{\infty} d\tau \, \exp\left[+i(\Omega - \Omega_{\text{peak}})\tau - \frac{\Delta^2 k_B T}{\hbar^2}(1 - \cos(\omega\tau)) \right]. \quad (A.251)$$

Notice that the term $\exp[-(\Delta^2 k_B T/\hbar^2)\cos(\omega\tau)]$ is periodic and thus has a discrete Fourier expansion; the only frequencies that appear are integer multiples of the vibrational frequency ω. As a result, we have

$$\sigma(\Omega) = \sum_n A_n \delta(\Omega - \Omega_{\text{peak}} - n\omega). \qquad (A.252)$$

Thus, in this limit of a single undamped mode, the absorption spectrum does consist of a set of sharp lines spaced by the vibrational quanta. To recover the semiclassical picture, these resonances must be washed out by a combination of multiple modes (so that the discrete absorption lines become a dense forest) and some dissipation corresponding to a lifetime or dephasing of each individual mode.

Problem 183: Washing out resonances.   Suppose that we have just a single mode, but this mode is damped, so that

$$\langle q(t)q(t')\rangle = \frac{k_B T}{\omega^2} \cos[\omega(t - t')] \exp\left[-\gamma|t - t'|\right]. \qquad (A.253)$$

If $\gamma \ll \omega$, the integral in Eq. (A.245), which defines the absorption cross-section, is almost the integral of a periodic function. Thus, there will be multiple saddle points, the first (the one we have considered in our semiclassical approximation) being close to $\tau = 0$, and all the others being close to $\tau = 2\pi n/\omega$ for integer n. Carry out this expansion, and analyze your results. Can you see how, as $\gamma \to 0$, this sum over saddle points gives back the discrete spectral lines? At large γ, what enforces the smooth dependence of the cross-section on Ω? How big does γ need to be so that we would not see much hint of the vibrational resonances in the absorption spectrum? Is it possible that the vibrations are weakly damped ($\gamma \ll \omega$), but there are no visible resonances in the absorption spectrum?

Are we missing anything by taking a classical view of the atomic motions? One qualitative point is that, if we look carefully, we will find that the absorption lines in Eq. (A.252) occur not just at $\hbar\Omega = \epsilon_0 + n\hbar\omega$, but also at $\hbar\Omega = \epsilon_0 - n\hbar\omega$, with n a positive integer, even as $T \to 0$. This can't be right: in the absence of thermal energy, the molecule can absorb higher energy photons and excite the vibrations, but there is no energy to borrow from the heat bath and allow the absorption of lower energy photons. In a correct quantum treatment, there will be a "zero-phonon line" and no absorption at lower energies. Again, for optical absorption in big floppy molecules, this behavior is largely irrelevant, although there are other cases in which it matters a lot. An example is the Mössbauer effect, in which a nucleus can absorb or emit a high-energy photon without exciting any phonons, and as a result the width of this absorption or emission line is very narrow, limited only by the lifetime of the nuclear excited state. There are many interesting physics experiments one can do with such narrow lines, and in the context of proteins, measuring the probability of a zero-phonon transition tells us something about the dynamics of the system that is hard to determine by other means.

For a completely consistent quantum mechanical treatment, we need to describe vibrational motions that are, in the classical sense, damped. Not so long ago, the description of dissipation in quantum systems was viewed as quite puzzling: quantum mechanics starts with a Hamiltonian or Lagrangian, and damping is not described in these terms, so it is not even clear how to get started. But the answer to this problem really is a matter of classical physics—we have to find a Hamiltonian that generates an effectively damped motion of the degrees of freedom that we are interested in, and then we can proceed quantum mechanically. I think many people must have had an intuition about the answer—damping arises by coupling to a large bath of other degrees of freedom—but it was not until the early 1980s that this problem got straightened out. Importantly, these considerations led to an understanding of how dissipation does not just cause systems to lose energy, they also lose quantum mechanical coherence. This again is a beautiful subject that could occupy an entire course.

Problem 184: A Hamiltonian for damping. Consider a particle of mass m with position x, moving in a potential $V(x)$. Imagine that the motion of this particle is coupled to the motions of many harmonic oscillators, with the full Hamiltonian

$$\mathbf{H} = \frac{\dot{x}^2}{2m} + V(x) + \frac{1}{2} \sum_\mu \left[\dot{q}_\mu^2 + \omega_\mu^2 (q_\mu - g_\mu x)^2 \right]. \qquad (A.254)$$

(a) Derive the classical equations of motion, both for x and for the set of coordinates $\{q_\mu\}$.

(b) Show that you can solve the equations for $\{q_\mu\}$, at least formally, and substitute back into the equation for x to find something of the form

$$m \frac{d^2 x}{dt^2} = -\frac{dV(x)}{dx} - \int_0^\infty d\tau \, K(\tau) x(t - \tau). \qquad (A.255)$$

Relate the kernel $K(\tau)$ to the parameters $\{g_\mu, \omega_\mu\}$ describing the bath of oscillators.

(c) Can you choose the parameters $\{g_\mu, \omega_\mu\}$ such that, in the limit that there are an infinite number of oscillators, you have

$$\int_0^\infty d\tau \, K(\tau) x(t - \tau) \to \gamma \frac{dx(t)}{dt}? \qquad (A.256)$$

If you can do this, you have found a Hamiltonian that describes motion of x subject to a drag force $-\gamma v$.

(d) Equation (A.255) is not quite complete, because it ignores any contribution from the initial positions and velocities of the oscillators. Show that if you allow for some nontrivial set of the conditions $\{q_\mu(t=0), \dot{q}_\mu(t=0)\}$, then the correct dynamical equation can be written as

$$m\frac{d^2x}{dt^2} = -\frac{dV(x)}{dx} - \int_0^\infty d\tau\, K(\tau)x(t-\tau) + \delta F(t), \qquad (A.257)$$

where $\delta F(t)$ is a sum of terms proportional to $\{q_\mu(t=0), \dot{q}_\mu(t=0)\}$. If these initial conditions are chosen at random from a distribution in which each oscillator has the same average energy, what is the correlation function of $\delta F(t)$? Can you keep going down this path and prove a version of the fluctuation-dissipation theorem?

A.5   The Kramers Problem

Here we take a more technical look at the problem of thermally activated motion over a barrier, as described in Section 4.1. There are several ways of doing this, because there are several equivalent ways to formulate the classical random motion of a Brownian particle. In fact, these different approaches are analogous to the different formulations of quantum mechanics.

Elementary courses on quantum mechanics usually focus on Schrödinger's equation, which describes the amplitude for a particle to be at position x at time t. But you can also look at Heisenberg's equations of motion for the position (and momentum) operator, and finally, you can use path integrals. How do these different approaches to quantum mechanics connect with the description of Brownian motion?

The Langevin equation is a bit like Heisenberg's equation for the position operator. It seems to give us something most closely related to the equations of motion in classical (noiseless) mechanics, but it requires some interpretation. In the case of the Langevin equation, because $\zeta(t)$ in Eq. (4.1) is random, when we solve for the trajectory $x(t)$, we get something different for every realization of ζ, so "solve" should be used carefully. More precisely what we get, for example, from one simulation of the Langevin equation is a sample drawn out of the distribution of trajectories.

When passing from Heisenberg's equations of motion to the Schrödinger equation, we shift from trying to follow the time dependence of coordinates to trying to see the whole distribution of coordinates at each time, as encoded in the wavefunction. Similarly, we can pass from the Langevin equation to the diffusion equation, which governs the probability $P(x,t)$ that we will find the particle at position x at time t.

The diffusion equation is an equation for the conservation of particle number, or for the conservation of probability if there is only one particle. Consider the one-dimensional case, and focus on a small slice of size Δx, as in Fig. A.6. The current flowing into this slice from the left is $J(x,t)$, and the current flowing out is $J(x+\Delta x, t)$. Any difference between these currents causes an accumulation of particles (or probability) at x, and if we normalize to probability per unit length, we have[8]

---

8. A note about units. Often when discussing diffusion, it is natural to think about the concentration of particles, which has units of particles per volume. The current of particles then has units of particles

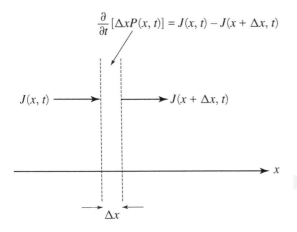

$$\frac{\partial}{\partial t}[\Delta x P(x, t)] = J(x, t) - J(x + \Delta x, t)$$

$$J(x, t) \longrightarrow \qquad \longrightarrow J(x + \Delta x, t)$$

$$x$$

$$\Delta x$$

FIGURE A.6

Conservation of probability connects spatial derivatives of flux to temporal derivatives of the local probability density, as in Eq. (A.259).

$$\frac{\partial}{\partial t}[\Delta x P(x, t)] = J(x, t) - J(x + \Delta x, t), \tag{A.258}$$

$$\rightarrow \frac{\partial P(x, t)}{\partial t} = -\frac{\partial}{\partial x}J(x, t). \tag{A.259}$$

Fick's law tells us that diffusion contributes a current that tends to reduce gradients in the concentration of particles (or equivalently, gradients in the probability of finding one particle), so that

$$J_{\text{diff}}(x, t) = -D\frac{\partial P(x, t)}{\partial x}. \tag{A.260}$$

But if there is a force $F(x) = -dV(x)/dx$ acting on the particle, it will move with an average velocity $v = F(x)/\gamma$, and hence there is also a drift current:

$$J_{\text{drift}}(x, t) = vP(x, t) = -\frac{1}{\gamma}\frac{dV(x)}{dx}P(x, t). \tag{A.261}$$

Putting these terms together, $J = J_{\text{diff}} + J_{\text{drift}}$, we have

$$\frac{\partial P(x, t)}{\partial t} = -\frac{\partial}{\partial x}\left[-D\frac{\partial P(x, t)}{\partial x} - \frac{1}{\gamma}\frac{dV(x)}{dx}P(x, t)\right]$$

$$= D\frac{\partial}{\partial x}\left[\frac{\partial P(x, t)}{\partial x} + \frac{1}{\gamma D}\frac{dV(x)}{dx}P(x, t)\right]$$

$$= D\frac{\partial}{\partial x}\left[\frac{\partial P(x, t)}{\partial x} + \frac{1}{k_B T}\frac{dV(x)}{dx}P(x, t)\right], \tag{A.262}$$

where in the last step we use the Einstein relation $D = k_B T/\gamma$. This way of writing the diffusion equation makes clear that the Boltzmann distribution $P \propto e^{-V(x)/k_B T}$ is an equilibrium ($\partial P/\partial t = 0$) solution.

---

per area per time. What we are doing here is slightly different. First, we are talking about the probability of finding *one* particle at point x. Second, we are in one dimension, and so this probability distribution has units of 1/(length), not 1/(volume). Then the current has the units of a rate, 1/(time). Check that this make the units come out right in Eq. (A.259) and Eq. (A.260).

I have said that, in looking at solutions of the Langevin equation, the signature of a "chemical reaction" with rate k is that the trajectories $x(t)$ will look like they do in Fig. 4.2. What is the corresponding signature in the solutions of the diffusion equation? More precisely, even if we solve the diffusion equation to get the full $P(x, t)$ from some initial condition, what is it about this solution that corresponds to the rate constant k? In the same way that Schrödinger's equation is a linear equation for the wavefunction, the diffusion equation is a linear equation for the probability, which we can write as

$$\frac{\partial P(x, t)}{\partial t} = \hat{L} P(x, t). \tag{A.263}$$

All the dynamics are determined by the eigenvalues of the linear operator $\hat{L}$:

$$P(x, t) = \sum_n a_n e^{\lambda_n t} u_n(x), \tag{A.264}$$

$$\hat{L} u_n(x) = \lambda_n u_n(x). \tag{A.265}$$

We know that one of the eigenvalues λ_0 has to be zero, because if $P(x, t)$ is the Boltzmann distribution, $P \propto e^{-V(x)/k_B T}$, it will not change in time. Deviations from the Boltzmann distribution should decay in time, so all the nonzero eigenvalues should be negative.

Problem 185: Positive decay rates.   We know that $P \propto \exp[-V(x)/k_B T]$ is a stationary solution of the diffusion Eq. (A.262). To study the dynamics of how this equilibrium is approached, write

$$P(x, t) = \exp\left[-\frac{V(x)}{2 k_B T}\right] Q(x, t). \tag{A.266}$$

(a) Derive the equation governing $Q(x, t)$. Show that (by introducing factors of i in the right place) this expression can be written as

$$\frac{\partial Q(x, t)}{\partial t} = -A^\dagger A Q(x, t), \tag{A.267}$$

where the combination $A^\dagger A$ is a Hermitian operator. This equation gives an explicit version of Eq. (A.263); explain why it implies that all eigenvalues $\lambda_n \leq 0$.

(b) For the case of the harmonic potential, $V(x) = \kappa x^2/2$, show that the operators $A^\dagger$ and A are the familiar creation and annihilation operators from the quantum harmonic oscillator. Use this mapping to find all eigenvalues λ_n. How do these relate to the time constant for exponential decay that you get from the noiseless dynamics, Eq. (4.1) with $\zeta(t) = 0$?

If we place the molecule in some configuration that is far from the local minima in each potential well, it will slide relatively quickly into its relaxed configuration and execute some Brownian motion around this sliding trajectory, so that it samples the Boltzmann distribution in the well. This relaxation should be described by some of the eigenvalues λ_n, and these should be large and negative, corresponding to fast relaxation. In practice, we know that molecules in solution achieve this sort of vibrational relaxation within nanoseconds if not picoseconds.

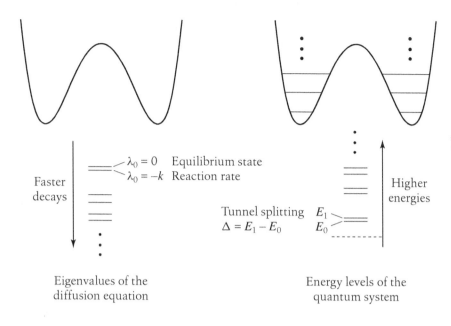

Faster
decays

$\lambda_0 = 0$ Equilibrium state
$\lambda_0 = -k$ Reaction rate

Tunnel splitting E_1
$\Delta = E_1 - E_0$ E_0

Higher
energies

Eigenvalues of the
diffusion equation

Energy levels of the
quantum system

FIGURE A.7

Decay rates in diffusion compared with energy levels in quantum mechanics. In both cases there is a small splitting between the first two eigenvalues. For the diffusion case, this splitting is the rate of thermally activated hopping over the barrier—a chemical reaction. For the quantum case the splitting is the tunneling frequency between the two wells.

The statement that there is a chemical reaction at rate k means that, as a population of molecules comes to equilibrium, all the equilibration *within* the reactant or product states is fast, corresponding to time scales much shorter than $1/k$. On the much longer time scale $1/k$, there is equilibration between the reactant and product states. Thus, if we look at the whole spectrum of eigenvalues λ_n for the diffusion equation, one eigenvalue should be zero (as noted above), almost all the others should be very large and negative, and there should be one isolated eigenvalue that is small and negative—and this will be the reaction rate k, or more precisely the sum of the rates for the forward (reactants $\rightarrow$ products) and backward (products $\rightarrow$ reactants) reactions.

We arrive, then, at a picture of the eigenvalue spectrum in which there is a small splitting (between $\lambda_0 = 0$ and $\lambda_1 = -k$) relative to the next-highest eigenvalue, as shown in Fig. A.7. This is analogous to what happens in quantum mechanical tunneling between two potential wells. The basic spacing of energy levels is set by the vibrational quanta in each well, but these states—and, in particular, the ground state—are split by a small amount corresponding to the frequency of tunneling between the two wells. It is the size of the barrier, or equivalently, the smallness of $\hbar$, which makes this splitting small. In the diffusion problem, it is the smallness of the temperature relative to the activation energy that enforces $\lambda_0 - \lambda_1 \ll \lambda_1 - \lambda_2$. We know how to solve Schrödinger's equation using the WKB approximation to extract the small tunneling amplitude, and so there should be a similar approximation to the diffusion equation that allows us to calculate the reaction rate.

The WKB approximation has a natural formulation in the path integral approach. In the limit $\hbar \rightarrow 0$, the path integral describing the amplitude for any quantum process is dominated by particular trajectories that are solutions of the classical equations of motion, although for classically forbidden processes (as with tunneling), these equations have to be continued to imaginary time. This idea of a dominant trajectory should be even clearer in the case of Brownian motion, because we do not have to deal with the continuation to imaginary time. To see how this works—and, finally, to derive the

Arrhenius law—we need to construct the probability distribution functional for the trajectories $x(t)$ that solve the Langevin equation, Eq. (4.1).

The probability that we observe a trajectory $x(t)$ can be calculated by finding the random force $\zeta(t)$ needed to generate this trajectory and then calculating the probability of this force. The random forces come from a Gaussian distribution, and we know the correlation function (Eq. (4.2)), so from the discussion in Appendix A.2, we know how to write the probability distribution functional:

$$P[\zeta(t)] \propto \exp\left[-\frac{1}{4\gamma k_B T} \int dt \, \zeta^2(t)\right]. \tag{A.268}$$

The Langevin equation (Eq. (4.1)) can be rewritten as

$$\zeta(t) = \gamma\frac{dx}{dt} + \frac{dV(x)}{dx}, \tag{A.269}$$

so it is tempting to say that the probability of observing the trajectory $x(t)$ is given by

$$P[x(t)] \sim \exp\left[-\frac{1}{4\gamma k_B T} \int dt \left(\gamma\frac{dx}{dt} + \frac{dV(x)}{dx}\right)^2\right], \tag{A.270}$$

which is almost correct.

To see what is missing in Eq. (A.270), consider the simpler case where we just have one variable x (instead of a function $x(t)$) that obeys the equation

$$f(x) = y, \tag{A.271}$$

and y is random, drawn from a distribution $P_y(y)$. It is tempting to write

$$P_x(x) = P_y(y = f(x)), \tag{A.272}$$

but this can't be right—x and y can have different units, and hence P_x and P_y must have different units. In this simple one-dimensional example, the correct statement is that the probability mass within some small region dx must be equal to the mass found in the corresponding dy:

$$P_x(x)dx = P_y(y = f(x))dy, \tag{A.273}$$

$$\Rightarrow P_x(x) = P_y(y = f(x))\left|\frac{dy}{dx}\right| \tag{A.274}$$

$$= P_y(y = f(x))\left|\frac{df(x)}{dx}\right|. \tag{A.275}$$

More generally, to equate probability distributions, we need a Jacobian for the transformation between variables. Thus, instead of Eq. (A.270), we really want to write

$$P[x(t)] \propto \exp\left[-\frac{1}{4\gamma k_B T} \int dt \left(\gamma\frac{dx}{dt} + \frac{dV(x)}{dx}\right)^2\right] \mathcal{J}, \tag{A.276}$$

where $\mathcal{J}$ is the Jacobian of the transformation between $x(t)$ and $\zeta(t)$. Importantly, the Jacobian does not depend on temperature. In contrast, the exponential term that we have written out is $\sim e^{-1/T}$, so at low temperatures, it will dominate. Thus, for this discussion, we won't worry about the Jacobian.

To make use of Eq. (A.276), it is useful to look more closely at the integral appearing in the exponential. Let's be careful to let time run from some initial time t_i up to some final time t_f:

$$\int_{t_i}^{t_f} dt \left(\gamma \frac{dx}{dt} + \frac{dV(x)}{dx} \right)^2 = \int_{t_i}^{t_f} dt \left[\left(\gamma \frac{dx}{dt} \right)^2 + 2\gamma \frac{dx}{dt} \frac{dV(x)}{dx} + \left(\frac{dV(x)}{dx} \right)^2 \right]$$

(A.277)

$$= \int_{t_i}^{t_f} dt \left[\left(\gamma \frac{dx}{dt} \right)^2 + \left(\frac{dV(x)}{dx} \right)^2 \right]$$
$$+ 2\gamma \int_{t_i}^{t_f} dt \frac{dx}{dt} \frac{dV(x)}{dx}$$

(A.278)

$$= \int_{t_i}^{t_f} dt \left[\left(\gamma \frac{dx}{dt} \right)^2 + \left(\frac{dV(x)}{dx} \right)^2 \right]$$
$$+ 2\gamma \int_{t_i}^{t_f} dt \frac{dV(x)}{dt}$$

(A.279)

$$= \int_{t_i}^{t_f} dt \left[\left(\gamma \frac{dx}{dt} \right)^2 + \left(\frac{dV(x)}{dx} \right)^2 \right]$$
$$+ 2\gamma [V(x_f) - V(x_i)],$$

(A.280)

where in the last steps we recognize one term as a total derivative; as usual, $x_i = x(t_i)$ is the initial position, and similarly $x_f = x(t_f)$ is the final position. Substituting, we can write the probability of a trajectory $x(t)$ as

$$P[x(t)] \propto \mathcal{J} e^{-S/k_B T},$$

(A.281)

where the action takes the form

$$S = \frac{V(x_f) - V(x_i)}{2} + \int_{t_i}^{t_f} dt \left[\frac{\gamma}{4} \left(\frac{dx}{dt} \right)^2 + \frac{1}{4\gamma} \left(\frac{dV(x)}{dx} \right)^2 \right].$$

(A.282)

Taking a step back, if we doing a one-dimensional problem in classical mechanics, the action would be given by

$$S_{cm} = \int_{t_i}^{t_f} dt \left[\frac{m}{2} \left(\frac{dx}{dt} \right)^2 - \mathcal{U}(x(t)) \right],$$

(A.283)

where m is the mass and $\mathcal{U}(x)$ is the potential energy. Except for a constant, then, the effective action for our problem is exactly that of a classical mechanics problem of a particle with mass m moving in a effective potential $\mathcal{U}(x)$:

$$m = \frac{\gamma}{2},$$

(A.284)

$$\mathcal{U}(x) = -\frac{1}{4\gamma} \left(\frac{dV(x)}{dx} \right)^2.$$

(A.285)

Figure A.8 shows how this effective potential relates to the original double well.

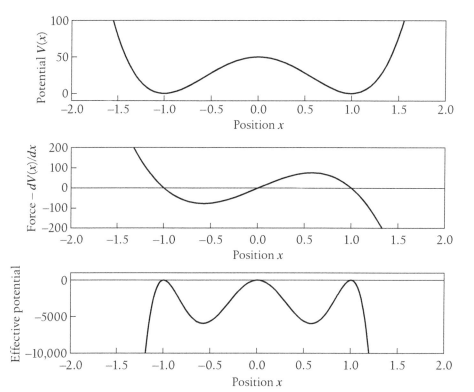

FIGURE A.8

Potentials and forces in a double well. Top panel shows the true potential, middle panel the force, and bottom panel the effective potential that enters the probability distribution of trajectories. Notice that each extremum of the potential, both maxima and minima, becomes a maximum of the effective potential, and all these maxima are degenerate at $\mathcal{U} = 0$.

At low temperatures, the distribution of trajectories will be dominated by those that extremize the action S. Clearly, one way to make the action minimal (zero, in fact) is to have the position be constant at one of the minima of the potential well. This trajectory describes a situation in which nothing happens. To have a chemical reaction, we need a trajectory that starts in the well corresponding to the reactants state, climbs up to the transition state at the top of the barrier, and then slides down the other side. Let's start with the first part of this problem: finding a trajectory that climbs the barrier. The dominant trajectory of this form will be one that minimizes the action, and from Fig. A.8 we see that this is equivalent to finding the solution to an ordinary mechanics problem in which a particle starts on top of one hill, slides down, and then gently comes to rest at the top of the next hill.

Problem 186: Zero energy? What we have just described are trajectories in the effective potential that have zero energy. There are, of course, trajectories that minimize the action but have nonzero energy. Why don't we consider them?

Taking the details of Fig. A.8 seriously, if we start at rest on top of one hill, then we start with zero energy. But energy is conserved along the trajectory, so that

$$\frac{m}{2}\left(\frac{dx}{dt}\right)^2 + \mathcal{U}(x(t)) = E = 0. \tag{A.286}$$

Thus, we have

$$\mathcal{U}(x(t)) = -\frac{m}{2} \left(\frac{dx}{dt} \right)^2, \tag{A.287}$$

$$\frac{dx}{dt} = \pm \sqrt{-\frac{2}{m} \mathcal{U}(x(t))}; \tag{A.288}$$

we are interested in trajectories that move from left to right, so we should choose the upper sign, $dx/dt > 0$. But now we can substitute into the action,

$$S_{\text{cm}} = \int_{t_i}^{t_f} dt \left[\frac{m}{2} \left(\frac{dx}{dt} \right)^2 - \mathcal{U}(x(t)) \right]$$

$$= \int_{t_i}^{t_f} dt \left[\frac{m}{2} \left(\frac{dx}{dt} \right)^2 + \frac{m}{2} \left(\frac{dx}{dt} \right)^2 \right] \tag{A.289}$$

$$= m \int_{t_i}^{t_f} dt \left(\frac{dx}{dt} \right)^2 \tag{A.290}$$

$$= m \int_{t_i}^{t_f} dt \, \frac{dx}{dt} \sqrt{-\frac{2}{m} \mathcal{U}(x(t))}, \tag{A.291}$$

so that finally we have[9]

$$S_{\text{cm}} = \int_{x_i}^{x_f} dx \, \sqrt{-2m\mathcal{U}(x)}. \tag{A.294}$$

In our case, the effective potential and mass are defined by Eq. (A.284) and Eq. (A.285), so that

$$-2m\mathcal{U}(x) = -2\frac{\gamma}{2} \left[-\frac{1}{4\gamma} \left(\frac{dV(x)}{dx} \right)^2 \right] = \frac{1}{4} \left(\frac{dV(x)}{dx} \right)^2. \tag{A.295}$$

---

9. In quantum mechanics, if we have a particle with energy E moving in a potential $V(x)$, the amplitude for tunneling through a classically forbidden region where $E < V(x)$ can be written, in the WKB approximation, as

$$A_{\text{tunnel}} \propto e^{-S/\hbar}, \tag{A.292}$$

where the action is given by

$$S = \int_{x_i}^{x_f} dx \sqrt{2m[V(x) - E]}. \tag{A.293}$$

The similarity to our problem should be clear.

It is quite nice how the factors of γ cancel. Substituting into Eq. (A.294), we find

$$S_{cm} = \frac{1}{2} \int_{x_i}^{x_f} dx \sqrt{\left(\frac{dV(x)}{dx}\right)^2} \tag{A.296}$$

$$= \pm \frac{1}{2} \int_{x_i}^{x_f} dx \, \frac{dV(x)}{dx} \tag{A.297}$$

$$= \frac{1}{2} \left| V(x_f) - V(x_i) \right|, \tag{A.298}$$

where we choose the sign in taking the root so that the action comes out positive, as it must from Eq. (A.290).

Problem 187: Extracting the dominant paths.   We have seen that, in the low-temperature limit, the reaction is dominated by trajectories that lead from one well to the other and minimize the action. Look through your simulation results from Problem 41 and collect as many examples as you can of the jumping trajectories. How do these examples compare with the theoretical prediction that comes from minimizing the action? Can you align the sample trajectories well enough to compute an average that might be more directly comparable to the theory?

To finish the calculation, we need to put some of these pieces together. The action that determines the probability of a trajectory is, from Eq. (A.282),

$$S = \frac{V(x_f) - V(x_i)}{2} + \int_{t_i}^{t_f} dt \left[\frac{\gamma}{4} \left(\frac{dx}{dt}\right)^2 + \frac{1}{4\gamma} \left(\frac{dV(x)}{dx}\right)^2 \right]$$

$$= \frac{V(x_f) - V(x_i)}{2} + S_{cm} \tag{A.299}$$

$$= \frac{V(x_f) - V(x_i)}{2} + \frac{1}{2} \left| V(x_f) - V(x_i) \right|. \tag{A.300}$$

This is a remarkably simple result. For a trajectory that climbs from the bottom of a potential well to the top of the barrier, we have $V(x_f) > V(x_i)$, and hence the action is

$$S_{climb} = V(x_f) - V(x_i) = E_{act}, \tag{A.301}$$

which is the "activation energy" for going over the barrier. In contrast, if we look at a trajectory that slides down from the barrier into the other well, we have $V(x_f) < V(x_i)$ and hence

$$S_{slide} = 0. \tag{A.302}$$

So, what we have shown is that paths that transition from reactants to products, climbing the barrier and sliding down the other side, have a minimal action $S_{react} = S_{climb} + S_{slide} = E_{act}$. Thus, the probability of seeing such a trajectory is

$$P[x_{react}(t)] \propto \mathcal{J} e^{-S_{react}/k_B T} \sim e^{-E_{act}/k_B T}, \tag{A.303}$$

which is the essence of the Arrhenius law (at last).

One could legitimately complain that we have not really solved our problem. All we have done is to show that, in some window of time, trajectories that jump from reactants to products are suppressed in probability by a factor $e^{-E_{act}/k_B T}$. This is the basic idea of the Arrhenius law, but we have not actually calculated a rate constant. In truth, this last step requires rather more technical apparatus, in the same way that finding the tunneling rate in the WKB approximation is harder than getting the exponential suppression, so we will skip it. It is worth noting, though, that deriving the exponential suppression of the rate constant is much easier than determining the absolute number or prefactor in front of the exponential.

A.6   Berg and Purcell, Revisited

In the spirit of Berg and Purcell's original discussion (see Section 4.2), the simplest example of noise in a chemical system is to consider the fluctuations in concentration as seen in a small volume. To treat this process rigorously, remember that diffusion in and out of the volume keeps the system at equilibrium. Thus, fluctuations in the concentration should be just like Brownian motion or Johnson noise. What is a little odd is that although the strength of Johnson noise is proportional to the absolute temperature, our intuition about counting molecules and the $\sqrt{N}$ rule does not seem to have a place for T. So, let's see how this works.[10]

If we measure the current flowing across a resistor in thermal equilibrium at temperature T, we will find a noise in the current that has a spectral density $S_I = 2k_B T/R$, where R is the resistance. More generally, if we measure between two points in a circuit and find a frequency-dependent complex impedance $\tilde{Z}(\omega)$, then the spectral density of current noise will be

$$S_I(\omega) = 2k_B T \operatorname{Re}\left[\frac{1}{\tilde{Z}(\omega)}\right], \tag{A.304}$$

where Re denotes the real part. In a mechanical system it is more natural to talk about positions and forces instead of currents and voltages. Now if we measure the position and apply a force, we have a mechanical response function $\tilde{\alpha}(\omega)$ analogous to the (inverse) impedance,

$$\tilde{x}(\omega) = \tilde{\alpha}(\omega)\tilde{F}(\omega), \tag{A.305}$$

where $\tilde{x}(\omega)$ is the Fourier component[11] of $x(t)$,

$$x(t) = \int_{-\infty}^{\infty} \frac{d\omega}{2\pi} e^{-i\omega t} \tilde{x}(\omega), \tag{A.306}$$

and similarly for the force $\tilde{F}(\omega)$. The analog of Eq. (A.304) for Johnson noise is that the fluctuations in position x have a spectral density

$$S_x(\omega) = \frac{2k_B T}{\omega} \operatorname{Im}\left[\tilde{\alpha}(\omega)\right]. \tag{A.307}$$

where Im denotes the imaginary part.

---

10. In what follows I make free use of the concepts of correlation functions, power spectra, and all that. See Appendix A.2 for a review of these ideas.

11. Here, more than in other sections, our conventions in defining the Fourier transform are important. Be careful about the sign of i in the exponential!

Problem 188: Some details about noise spectra. You may remember the formula for Johnson noise as $S_I = 4k_B T/R$, rather than the factor of 2 given above. Also, there are a few obvious differences between Eq. (A.304) and Eq. (A.307). Be sure you understand all these differences. The key ingredients of the explanation are that all our integrals run over positive and negative frequencies, and although voltage is analogous to force, current is analogous to velocity, not position. Check carefully that all the details work out.

In any system at thermal equilibrium, if we apply a small force we can observe a proportionally small displacement, which is described by a linear response function. In a mechanical system we have the function $\tilde{\alpha}(\omega)$, sometimes called a complex compliance. In magnetic systems the force is an applied magnetic field, and the analog of position is the magnetization; the response function is called the susceptibility. Electrical systems are a bit odd, because we usually discuss the current response to voltage, but we can also think about charge movements (see Problem 188). In all these cases, once we know the linear response function, we can predict the spectral density of fluctuations in the relevant position-like variable using Eq. (A.307). This result is called the fluctuation-dissipation theorem.

Problem 189: Recovering equipartition. If frequency is zero, then $\tilde{x} = \tilde{\alpha}(0)\tilde{F}$, which means that $\tilde{\alpha}(0) = 1/\kappa$, where κ is the stiffness of the system. We know from the equipartition theorem that the variance in position must be related to the stiffness:

$$\frac{1}{2}\kappa\langle x^2 \rangle = \frac{1}{2}k_B T. \qquad (A.308)$$

But we can also write the variance in position as an integral over the spectral density:

$$\langle x^2 \rangle = \int_{-\infty}^{\infty} \frac{d\omega}{2\pi} S_x(\omega). \qquad (A.309)$$

For these equations to be consistent, we must have

$$2\int_{-\infty}^{\infty} \frac{d\omega}{2\pi} \frac{1}{\omega} \text{Im}\left[\tilde{\alpha}(\omega)\right] = \tilde{\alpha}(0), \qquad (A.310)$$

which looks quite remarkable.

(a) Equation (A.305), in the frequency domain, is equivalent to

$$x(t) = \int_{-\infty}^{\infty} d\tau \alpha(\tau) F(t-\tau), \qquad (A.311)$$

where

$$\alpha(\tau) = \int_{-\infty}^{\infty} \frac{d\omega}{2\pi} e^{-i\omega\tau} \tilde{\alpha}(\omega). \qquad (A.312)$$

Causality means that $\alpha(\tau < 0) = 0$. What does this condition imply about the analytic properties of $\tilde{\alpha}(\omega)$ in the complex ω plane?

(b) Use your result in part (a) to verify Eq. (A.310).

Position and force, magnetization and magnetic field, charge and voltage; all of these are thermodynamically conjugate pairs of variables. More precisely, if we consider an ensemble in which the force is held fixed, then the derivative of the free energy with respect to the force is the mean position, and conversely. The fluctuation-dissipation theorem always refers to these pairs of variables. So, to describe fluctuations in chemical systems, we need to know the "force" that is conjugate to the concentration (or the number of molecules). This quantity is the chemical potential μ: the derivative of the free energy with respect to the number of molecules. To compute the response of the concentration to changes in chemical potential, we consider the diffusion equation for the concentration $c(\mathbf{x}, t)$ in the presence of a varying chemical potential $\mu(\mathbf{x}, t)$:

$$\frac{\partial c(\mathbf{x}, t)}{\partial t} = D\nabla \cdot \left[\nabla c(\mathbf{x}, t) - \frac{\nabla \mu(\mathbf{x})}{k_B T} c(\mathbf{x}, t) \right]. \tag{A.313}$$

Problem 190: Connecting back. Explain how Eq. (A.313) relates to the equation for diffusion in the presence of an external potential, Eq. (A.262). Be sure you understand the signs.

Linearizing Eq. (A.313) around a mean concentration $\bar{c}$, we have

$$\frac{\partial c(\mathbf{x}, t)}{\partial t} = D\nabla^2 c(\mathbf{x}, t) - \frac{D\bar{c}}{k_B T}\nabla^2\mu(\mathbf{x}). \tag{A.314}$$

We can solve by Fourier transforming in both space and time,

$$c(\mathbf{x}, t) = \int \frac{d^3k}{(2\pi)^3} \int \frac{d\omega}{2\pi} e^{-i\omega t} e^{+i\mathbf{k}\cdot\mathbf{x}} \tilde{c}(\mathbf{k}, \omega), \tag{A.315}$$

$$\mu(\mathbf{x}, t) = \int \frac{d^3k}{(2\pi)^3} \int \frac{d\omega}{2\pi} e^{-i\omega t} e^{+i\mathbf{k}\cdot\mathbf{x}} \tilde{\mu}(\mathbf{k}, \omega), \tag{A.316}$$

to find

$$\tilde{c}(\mathbf{k}, \omega) = \frac{D\bar{c}}{k_B T} \frac{k^2}{-i\omega + Dk^2} \tilde{\mu}(\mathbf{k}, \omega), \tag{A.317}$$

where $k = |\mathbf{k}|$. Thus, there is a $\mathbf{k}$-dependent response function,

$$\frac{\tilde{c}(\mathbf{k}, \omega)}{\tilde{\mu}(\mathbf{k}, \omega)} \equiv \tilde{\alpha}(\mathbf{k}, \omega) = \frac{D\bar{c}}{k_B T} \frac{k^2}{-i\omega + Dk^2} \tag{A.318}$$

from which we can use the fluctuation-dissipation theorem to calculate the spatiotemporal power spectrum of concentration fluctuations,

$$S_c(\mathbf{k}, \omega) = \frac{2k_B T}{\omega} \text{Im}\left[\tilde{\alpha}(\mathbf{k}, \omega)\right] = 2\bar{c}\frac{Dk^2}{\omega^2 + (Dk^2)^2}. \tag{A.319}$$

The factors of $k_B T$ cancel: the fluctuations are proportional to the temperature, but the response function—the susceptibility of the concentration to changes in chemical potential—is inversely proportional to the temperature.

How does the result in Eq. (A.319) relate to our intuition about the $\sqrt{N}$ rule? Let's think about measuring the average concentration in a small volume, which corresponds to the heuristic calculation by Berg and Purcell, as discussed in Section 4.2. To do this, we construct a variable

$$C(t) = \int d^3x \, W(\mathbf{x}) c(\mathbf{x}, t), \qquad (A.320)$$

where the weighting function $W(\mathbf{x})$ is $1/V$ inside a volume V, and zero outside. Then the correlation function of C is given by

$$\langle C(t)C(t')\rangle = \int d^3x \, W(\mathbf{x}) \int d^3x' \, W(\mathbf{x}')\langle c(\mathbf{x}, t)c(\mathbf{x}', t)\rangle \qquad (A.321)$$

$$= \int d^3x \, W(\mathbf{x}) \int d^3x' \, W(\mathbf{x}')$$

$$\times \int \frac{d^3k}{(2\pi)^3} e^{i\mathbf{k}\cdot(\mathbf{x}-\mathbf{x}')} \int \frac{d\omega}{2\pi} e^{-i\omega(t-t')} S_c(\mathbf{k}, \omega) \qquad (A.322)$$

$$= \int \frac{d\omega}{2\pi} e^{-i\omega(t-t')} \int \frac{d^3k}{(2\pi)^3} |\tilde{W}(\mathbf{k})|^2 S_c(\mathbf{k}, \omega), \qquad (A.323)$$

where as usual $\tilde{W}$ denotes the Fourier transform of W. In Fourier space the definition of $W(\mathbf{x})$ implies that $\tilde{W}(0) = 1$, and $\tilde{W}(\mathbf{k})$ will decay to zero for $k \gg 1/\ell$, where ℓ is the characteristic linear dimension of the region over which we are averaging.

Equation (A.323) allows us to identify the power spectrum of fluctuations in C:

$$S_C(\omega) = \int \frac{d^3k}{(2\pi)^3} |\tilde{W}(\mathbf{k})|^2 S_c(\mathbf{k}, \omega) \qquad (A.324)$$

$$= \bar{c} \int \frac{d^3k}{(2\pi)^3} |\tilde{W}(\mathbf{k})|^2 \frac{2Dk^2}{\omega^2 + (Dk^2)^2}. \qquad (A.325)$$

If we want the compute the variance in C, we have

$$\langle (\delta C)^2 \rangle \equiv \int \frac{d\omega}{2\pi} S_C(\omega) \qquad (A.326)$$

$$= \bar{c} \int \frac{d^3k}{(2\pi)^3} |\tilde{W}(\mathbf{k})|^2 \int \frac{d\omega}{2\pi} \frac{2Dk^2}{\omega^2 + (Dk^2)^2} \qquad (A.327)$$

$$= \bar{c} \int \frac{d^3k}{(2\pi)^3} |\tilde{W}(\mathbf{k})|^2 \qquad (A.328)$$

$$= \bar{c} \int d^3x \, W^2(\mathbf{x}), \qquad (A.329)$$

where in the last step we use Parseval's theorem. Because $W = 1/V = 1/\ell^3$ inside the averaging volume, and $W = 0$ outside, we have (with no approximations)

$$\langle (\delta C)^2 \rangle = \bar{c} V \left(\frac{1}{V} \right)^2 = \frac{\bar{c}}{V}. \qquad (A.330)$$

Because $\bar{C} = \bar{c}$, we can also write this expression as

$$\frac{\langle(\delta C)^2\rangle}{\bar{C}^2} = \frac{1}{\bar{c}\ell^3},\tag{A.331}$$

and we recognize $N = \bar{c}\ell^3$ as the mean number of molecules in the sampling volume. Thus, the rigorous calculation from the fluctuation-dissipation theorem confirms our intuition about the fractional variance in concentration being $1/N$.

To get the rest of the Berg-Purcell result (Eq. (4.63)), let's go back to Eq. (A.325) and finish computing the power spectrum of C. To simplify things, assume that the region over which we average is spherical, so we can write

$$S_C(\omega) = \bar{c} \int \frac{d^3k}{(2\pi)^3} |\tilde{W}(\mathbf{k})|^2 \frac{2Dk^2}{\omega^2 + (Dk^2)^2}$$
$$= 2\bar{c}\frac{1}{(2\pi)^3} \int_0^\infty dk\, 4\pi k^2 |\tilde{W}(\mathbf{k})|^2 \frac{Dk^2}{\omega^2 + (Dk^2)^2}.\tag{A.332}$$

If the size of the averaging region ℓ is small, then the characteristic time for diffusion across the averaging volume, $\tau \sim \ell^2/D$, is also small, and hence any frequencies that are likely to be relevant for the cell's measurements of concentration are low compared with the scales on which $S_C(\omega)$ has structure. Thus, we can confine our attention to the low-frequency limit:

$$S_C(\omega \to 0) \sim \frac{\bar{c}}{\pi^2} \int_0^\infty dk\, |\tilde{W}(\mathbf{k})|^2 \frac{Dk^4}{(Dk^2)^2} = \frac{\bar{c}}{\pi^2 D} \int_0^\infty dk\, |\tilde{W}(\mathbf{k})|^2.\tag{A.333}$$

Now we can appeal to the approximation that $\tilde{W}(\mathbf{k})$ falls to zero on the scale $|\mathbf{k}| \sim 2\pi/\ell$, so that

$$S_C(\omega \to 0) \sim \frac{\bar{c}}{\pi^2 D} \int_0^\infty dk\, |\tilde{W}(\mathbf{k})|^2 \sim \frac{2\bar{c}}{\pi D\ell}.\tag{A.334}$$

The first important result is that the concentration, averaged over a sampling volume of linear dimension ℓ, has effectively white noise in time for times $\tau \gg \ell^2/D$. If we average over a time τ_{avg}, then the result is sensitive to a bandwidth $1/\tau_{avg}$, and the variance is

$$\langle(\delta C)^2\rangle_{\tau_{avg}} \sim \frac{2\bar{c}}{\pi D\ell\tau_{avg}}.\tag{A.335}$$

Rewriting this expression as a fractional standard deviation, we have

$$\frac{\delta C_{rms}}{\bar{C}} = \frac{1}{\bar{C}}\sqrt{\langle(\delta C)^2\rangle_{\tau_{avg}}} \sim \left(\frac{2}{\pi}\right)^{1/2} \frac{1}{\sqrt{D\ell\bar{c}\tau_{avg}}},\tag{A.336}$$

which is (except for the trivial factor $\sqrt{2/\pi}$) exactly the Berg-Purcell result.

Problem 191: Concentration fluctuations in one dimension. Repeat the analysis we have just done, but in one dimension. Before going through a detailed calculation, you should try to anticipate the answer. We still expect (from the $\sqrt{N}$ intuition) that $\langle(\delta C)^2\rangle \propto \bar{c}$, but because concentration has units of molecules per length in one dimension, the other factors must be different. Try, for example,

$$\langle(\delta C)^2\rangle \sim \frac{\bar{c}}{(D\tau_{\mathrm{avg}})^n \ell^m}. \tag{A.337}$$

How are n and m constrained by dimensional analysis? Can you argue, qualitatively, for particular values of these exponents? Finally, do the real calculation and get the analog of Eq. (A.335) in one dimension. Are you surprised by the role of ℓ (i.e., by the value of m)? Or by the dependence on τ_{avg}? Can you explain why things come out this way?

Problem 192: Correlations seen by a moving observer.   Generalize the discussion above to the case where the volume in which we measure the concentration is moving at speed v_0 in some direction. Provide a formula for the correlations across time in the observed noise. Show, in particular, that there is a correlation time $\tau_c \sim D/v_0^2$. How does this relate to the qualitative argument, discussed in Section 4.2, that bacteria must integrate for a minimum time $\sim D/v_0^2$ if they are to outrun diffusion?

So the Berg-Purcell argument certainly gives the right answer for the concentration fluctuations in a small volume. But biological systems do not actually count the molecules in a volume. Instead, the molecules bind to specific sites, and it is this binding that is detected (e.g., by activating an enzymatic reaction). The Berg-Purcell formula suggests that there is a limit to the accuracy of sensing or signaling that comes from the physics of diffusion alone, independent of these details. To see how this can happen, we need to analyze fluctuations in the binding of molecules to receptor sites, coupled to their diffusion. Let's start just with the binding events.

Consider a binding site for signaling molecules, and let the fractional occupancy of the site be n. If we do not worry about the discreteness of this one site, or about the fluctuations in concentration c of the signaling molecule, we can write a kinetic equation

$$\frac{dn(t)}{dt} = k_+ c[1 - n(t)] - k_- n(t). \tag{A.338}$$

This expression describes the kinetics whereby the system comes to equilibrium. The free energy F associated with binding is determined by detailed balance:

$$\frac{k_+ c}{k_-} = \exp\left(\frac{F}{k_B T}\right). \tag{A.339}$$

Supposing that thermal fluctuations can lead to small changes in the rate constants, we can linearize Eq. (A.338) to obtain

$$\frac{d\delta n}{dt} = -(k_+ c + k_-)\delta n + c(1 - \bar{n})\delta k_+ - \bar{n}\delta k_-. \tag{A.340}$$

But from Eq. (A.339) we have

$$\frac{\delta k_+}{k_+} - \frac{\delta k_-}{k_-} = \frac{\delta F}{k_B T}. \tag{A.341}$$

Applying this constraint to Eq. (A.340), we find that the individual rate constant fluctuations cancel, and all that remains is the fluctuation in the thermodynamic binding energy δF,

$$\frac{d\delta n}{dt} = -(k_+c + k_-)\delta n + k_+c(1 - \bar{n})\frac{\delta F}{k_B T}. \tag{A.342}$$

Fourier transforming, we can solve Eq. (A.342) to find the frequency-dependent susceptibility of the coordinate n to its conjugate force F,

$$\tilde{\alpha}(\omega) \equiv \frac{\delta\tilde{n}(\omega)}{\delta\tilde{F}(\omega)} = \frac{1}{k_B T}\frac{k_+c(1 - \bar{n})}{-i\omega + (k_+c + k_-)}. \tag{A.343}$$

Now we can compute the power spectrum of fluctuations in the occupancy n using the fluctuation-dissipation theorem:

$$S_n(\omega) = \frac{2k_B T}{\omega}\text{Im}\left[\frac{\delta\tilde{n}(\omega)}{\delta\tilde{F}(\omega)}\right] \tag{A.344}$$

$$= \frac{2k_+c(1 - \bar{n})}{\omega^2 + (k_+c + k_-)^2}. \tag{A.345}$$

It is convenient to rewrite this equation as

$$S_n(\omega) = \langle(\delta n)^2\rangle\frac{2\tau_c}{1 + (\omega\tau_c)^2}, \tag{A.346}$$

where the total variance is

$$\langle(\delta n)^2\rangle = \int\frac{d\omega}{2\pi}S_n(\omega) = k_B T\left.\frac{\delta\tilde{n}(\omega)}{\delta\tilde{F}(\omega)}\right|_{\omega=0} \tag{A.347}$$

$$= \frac{k_+c(1 - \bar{n})}{k_+c + k_-} \tag{A.348}$$

$$= \bar{n}(1 - \bar{n}), \tag{A.349}$$

and the correlation time is given by

$$\tau_c = \frac{1}{k_+c + k_-}. \tag{A.350}$$

To make sense out of these results, remember what happens if we flip a coin that is biased to produce heads a fraction f of the time. On each trial we count either one or zero heads, so the mean count is f, and the mean-square count is also f; the variance is $f(1 - f)$, exactly as in Eq. (A.349). Thus, when we check the occupancy of the receptor, the outcome is determined by the equivalent of flipping a biased coin, where the bias is determined by the Boltzmann distribution.

The Lorentzian form of the power spectrum in Eq. (A.346) is equivalent to an exponential decay of correlations:

$$\langle \delta n(t) \delta n(t') \rangle = \int \frac{d\omega}{2\pi} e^{-i\omega(t-t')} S_n(\omega) \tag{A.351}$$

$$= \langle (\delta n)^2 \rangle \int \frac{d\omega}{2\pi} e^{-i\omega(t-t')} \frac{2\tau_c}{1 + (\omega\tau_c)^2} \tag{A.352}$$

$$= \langle (\delta n)^2 \rangle e^{-|t-t'|/\tau_c}. \tag{A.353}$$

The exponential decay of correlations is what we expect when the transitions between the available states have no memory. To be precise about this, imagine that a system is in one state at time $t = 0$, and there is some constant probability per unit time k of transitions out of this state (with, in the simplest case, no returns to the initial state). Then the probability $p(t)$ of still being in the initial state at time t must obey

$$\frac{dp(t)}{dt} = -kp(t), \tag{A.354}$$

and hence $p(t) = e^{-kt}$. This intuition about the connection of exponential decays to the lack of memory is very general, and should remind you of the exponential distribution of times between transitions in the calculation of chemical reaction rates (see Section 4.1) and of the exponential distribution of times between events in a Poission process (see Appendix A.1). In the present context, the exponential decay of correlations tells us that the spontaneous transitions between the occupied and unoccupied states of the receptor occur with constant probability per unit time, or as Markovian jumps. The jumping rates are just the rates k_+ and k_-, which means that when writing chemical kinetics models for a whole ensemble of molecules, we also can interpret them as Markov models for transitions among the states of individual molecules in the ensemble.

It is interesting that we recover the results for Markovian jumping between two states without making this microscopic model explicit. All we assume is the macroscopic kinetics and that the system is in thermal equilibrium, so that we can apply the fluctuation-dissipation theorem. In principle many different microscopic models can describe the molecular phenomena that are at the basis of some observed macroscopic behavior; many aspects of behavior in thermal equilibrium are independent of these details. The statistics of fluctuations in a chemical kinetic system are an example of this independence, at least near equilibrium.

The good news, then, is that fluctuations in receptor occupancy are an inevitable consequence of the *macroscopic*, or average, behavior of receptor-ligand interactions, independent of hypotheses about molecular details. The bad news is that the form of the results does not seem too related to the ideas of Berg and Purcell about the precision of concentration measurements. To make these connections clear, we need to couple the dynamics of receptor occupancy to the diffusion of the ligand.

When the concentration is allowed to fluctuate, we write

$$\frac{dn(t)}{dt} = k_+ c(\mathbf{x}_0, t)[1 - n(t)] - k_- n(t), \tag{A.355}$$

where the receptor is located at $\mathbf{x}_0$, and

$$\frac{\partial c(\mathbf{x}, t)}{\partial t} = D\nabla^2 c(\mathbf{x}, t) - \delta(\mathbf{x} - \mathbf{x}_0)\frac{dn(t)}{dt}. \tag{A.356}$$

The first equation is as before, but uses notation to remind us that the concentration c is dynamic. The second equation states that the ligand diffuses with diffusion constant D, and when the receptor located at $\mathbf{x}_0$ increases its occupancy it removes exactly one molecule from solution at that point.

Problem 193: Coupling diffusion and binding. In this problem you fill in the details needed for the analysis of Eq. (A.355) and Eq. (A.356).

(a) Begin by noticing that Eq. (A.356) is linear, so you should be able to solve it exactly. Use Fourier transforms, both in space and time, and then transform back to give a formal expression for

$$\tilde{c}(\mathbf{x}, \omega) = \int dt\, e^{+i\omega t} c(\mathbf{x}, t). \tag{A.357}$$

(b) Linearize Eq. (A.355), in the same way as in the preceding derivation, leading from Eq. (A.338) to Eq. (A.343). Along the way you will need an expression for $\tilde{c}(\mathbf{x}_0, \omega)$, which you can take from part (a). When the dust settles, you should find Eq. (A.358) and Eq. (A.359).

Following the same steps as above, we find the linear response function

$$\frac{\delta\tilde{n}(\omega)}{\delta\tilde{F}(\omega)} = \frac{k_+ c(1 - \bar{n})}{k_B T} \frac{1}{-i\omega[1 + \Sigma(\omega)] + (k_+\bar{c} + k_-)}, \tag{A.358}$$

$$\Sigma(\omega) = k_+(1 - \bar{n})\int \frac{d^3 k}{(2\pi)^3} \frac{1}{-i\omega + Dk^2}. \tag{A.359}$$

The self-energy $\Sigma(\omega)$ is ultraviolet divergent, which can be traced to the delta function in Eq. (A.356); we have assumed that the receptor is infinitely small. A more realistic treatment would give the receptor a finite size, which is equivalent to cutting off the k integrals at some (large) $\Lambda \sim \pi/a$, with a the linear dimension of the receptor. If we imagine mechanisms that read out the receptor occupancy average over a time τ long compared with the correlation time τ_c of the noise, then the relevant quantity is the low-frequency limit of the noise spectrum. Hence, we have

$$\Sigma(\omega \ll D/a^2) \approx \Sigma(0) = \frac{k_+(1 - \bar{n})}{2\pi Da}, \tag{A.360}$$

and

$$\frac{\delta\tilde{n}(\omega)}{\delta\tilde{F}(\omega)} = \frac{k_+\bar{c}(1 - \bar{n})}{k_B T}\left[-i\omega\left(1 + \frac{k_+(1 - \bar{n})}{2\pi Da}\right) + (k_+\bar{c} + k_-)\right]^{-1}, \tag{A.361}$$

where $\bar{c}$ is the mean concentration. Applying the fluctuation-dissipation theorem once again, we find the spectral density of occupancy fluctuations:

$$S_n(\omega) \approx 2k_+\bar{c}(1-\bar{n})\frac{1+\Sigma(0)}{\omega^2(1+\Sigma(0))^2+(k_+\bar{c}+k_-)^2}. \qquad (A.362)$$

If we do the integral over frequencies, we will find that the total variance in occupancy is unchanged, because this is an equilibrium property of the system; coupling to concentration fluctuations serves only to change the spectral shape, or temporal correlation of the fluctuations.

Problem 194: Reading off the results.   You should be able to verify the statements in the last sentence without detailed calculation. Explain how to "read" Eq. (A.362) and identify the total variance and correlation time.

Coupling to concentration fluctuations does serve to renormalize the correlation time of the noise:

$$\tau_c \to \tau_c[1+\Sigma(0)]. \qquad (A.363)$$

The new τ_c can be written as

$$\tau_c = \frac{1-\bar{n}}{k_-} + \frac{\bar{n}(1-\bar{n})}{2\pi\,Da\bar{c}}, \qquad (A.364)$$

so there is a lower bound on τ_c, independent of the kinetic parameters $k_\pm$:

$$\tau_c > \frac{\bar{n}(1-\bar{n})}{2\pi\,Da\bar{c}}. \qquad (A.365)$$

As discussed previously, the relevant quantity is the low-frequency limit of the noise spectrum,

$$S_n(\omega=0) = 2k_+\bar{c}(1-\bar{n}) \cdot \frac{1+\Sigma(0)}{(k_+\bar{c}+k_-)^2} \qquad (A.366)$$

$$= \frac{2\bar{n}(1-\bar{n})}{k_+\bar{c}+k_-} + \frac{[\bar{n}(1-\bar{n})]^2}{\pi\,Da\bar{c}}. \qquad (A.367)$$

If we average for a time τ, then the root-mean-square error in our estimate of n will be

$$\delta n_{\mathrm{rms}} = \sqrt{S_n(0)\cdot\frac{1}{\tau}}, \qquad (A.368)$$

and from Eq. (A.367), this noise level has a minimum value independent of the kinetic parameters $k_\pm$,

$$\delta n_{\mathrm{rms}} > \frac{\bar{n}(1-\bar{n})}{\sqrt{\pi\,Da\bar{c}\tau}}. \qquad (A.369)$$

To relate these results back to the discussion by Berg and Purcell, note that the $\omega=0$ response of the mean occupancy to changes in concentration can be written as

$$\frac{d\bar{n}}{d\ln c} = \bar{n}(1-\bar{n}). \qquad (A.370)$$

Thus, the fluctuations in n are equivalent to fluctuations in c,

$$\frac{\delta c_{\text{eff}}}{\bar{c}} = (\delta \ln c)_{\text{eff}} = \delta n_{\text{rms}} \left(\frac{d\bar{n}}{d \ln c} \right)^{-1} = \frac{1}{\sqrt{\pi D a \bar{c} \tau}}. \tag{A.371}$$

Except for the factor of $\sqrt{\pi}$, this is the Berg-Purcell result, Eq. (4.63), once again.

A startling feature of the Berg-Purcell argument is that (it seems) it can be used both when a is the size of a single receptor molecule and when it is the size of the entire bacterium. Naively, we might expect that if there are N receptors on the surface of the cell, then the signal-to-noise ratio for concentration measurements should be N times better, and correspondingly the threshold for reliable detection should be $\sqrt{N}$ times smaller:

$$\frac{\delta c_{\text{eff}}}{\bar{c}} \sim \frac{1}{\sqrt{D N a \bar{c} \tau}}. \tag{A.372}$$

In contrast, if we use the Berg-Purcell limit and take the linear dimensions of the detector to be the radius R of the bacterium, we should obtain

$$\frac{\delta c_{\text{eff}}}{\bar{c}} \sim \frac{1}{\sqrt{D R \bar{c} \tau}}. \tag{A.373}$$

What is going on? Does something special happen when $N \sim R/a$, so there is a crossover between the two results?

If we imagine a very large cell and place $N = 2$ receptors on opposite sides of the cell surface, it is hard to see that there is anything wrong with the argument leading to Eq. (A.372). More generally, if the receptors are far apart, it is plausible that they report independent measurements of the concentration, and so Eq. (A.372) should be correct. However, if we imagine bringing two receptors closer and closer together, at some point they will start to interact—a molecule released from one receptor can diffuse over and bind to the other receptor—and this interaction might lead to correlations in the noise, and a breakdown of the simple $\sqrt{N}$ improvement in the threshold for reliable detection.

To understand how diffusive interactions lead to correlations among receptors, it is useful to think about a simpler problem. Suppose that we have two balls in a fluid. If they are far apart, each one experiences a drag force and undergoes Brownian motion, and the Brownian fluctuations in the position of one ball are independent of those for the other. If we bring the two balls together, however, we know that they can influence each other through the fluid. If one ball moves at velocity v_1, it not only experiences a drag force $-\gamma v_1$, it also applies a coupling force $\gamma_c(v_1 - v_2)$ to the other ball (which may be moving at velocity v_2; clearly if $v_1 = v_2$, there should be no coupling force). If the balls are close enough that γ_c is significant, then in fact the Brownian motions of the two balls become correlated. This correlation can be derived from the fluctuation-dissipation theorem, and it also makes intuitive sense, because a random Brownian step of one object applies a force to the other. We can also see this effect experimentally, as shown in Fig. A.9.

Problem 195: Correlated Brownian motion. To make the description in the previous paragraph precise, consider the case where the particles are bound by springs (so they can't diffuse away from each other and reduce the coupling). Then, in the overdamped case, the equations of motion are

(a)

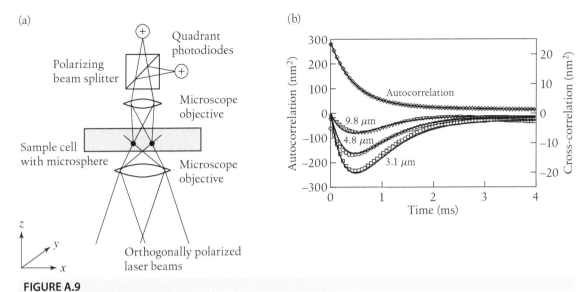

(b)

FIGURE A.9

Correlated Brownian motion. Redrawn from Meiners and Quake (1999). (a) A schematic of the experiment. The laser beams from the bottom of the figure create two optical traps, which hold the microspheres in approximately harmonic potential wells. The optics at the top allow for measurements of the spheres' positions with nanometer precision. (b) Measurements of the auto- and cross-correlations of the spheres' positions. Different curves for the cross-correlation correspond to different mean separations of the particles, which is expected to modulate the coupling between them through the fluid.

$$\gamma \frac{dx_1}{dt} = -\kappa x_1 - \gamma_c \left(\frac{dx_1}{dt} - \frac{dx_2}{dt} \right) + F_1(t), \tag{A.374}$$

$$\gamma \frac{dx_2}{dt} = -\kappa x_2 - \gamma_c \left(\frac{dx_2}{dt} - \frac{dx_1}{dt} \right) + F_2(t), \tag{A.375}$$

where κ is the stiffness of the springs (assumed identical, for simplicity), and $F_i(t)$ is an external force applied to each particle i.

(a) Derive the linear response function matrix, $\tilde{\alpha}_{ij}(\omega)$ such that

$$\tilde{x}_i(\omega) = \sum_j \tilde{\alpha}_{ij}(\omega) \tilde{F}_j(\omega). \tag{A.376}$$

(b) The generalization of the fluctuation-dissipation theorem to many degrees of freedom states that the cross-spectrum of variables x_i and x_j, defined by

$$\langle x_i(t) x_j(t') \rangle = \int \frac{d\omega}{2\pi} e^{-i\omega(t-t')} S_{ij}(\omega), \tag{A.377}$$

is given by

$$S_{ij}(\omega) = \frac{2k_B T}{\omega} \text{Im} \left[\tilde{\alpha}_{ij}(\omega) \right]. \tag{A.378}$$

Use this equation to derive the cross-spectrum of the position fluctuations for the two particles.

(c) Despite the viscous coupling, the potential energy is just the sum of contributions from the two particles. From the Boltzmann distribution, then, the positions should be independent variables. Use your results in part (b) to show that $\langle x_i x_j \rangle = \delta_{ij} k_B T / \kappa$. Notice that this corresponds to the *instantaneous* positions of the particles, as we would measure by taking a snapshot (with a fast camera).

(d) Suppose that instead of taking snapshots of the positions, we average (as in the discussion above) for a long time, so what is relevant is the low-frequency limit of the power spectra. Show that now the correlations are nonzero, and give an explicit formula for the covariance matrix of fluctuations in the temporally averaged positions.

If we imagine that positions of the Brownian particles are like receptor occupancies, and an applied force on all the particles is like a change in concentration of the relevant ligand, then diffusion of the ligand serves the same coupling effect as the viscosity of the fluid and will generate correlations among the occupancy fluctuations of nearby receptors. These correlations mean that using the positions or velocities of N Brownian particles to infer the applied force is not $\sqrt{N}$ more accurate than using one particle, and similarly using N receptors will not generate a concentration measurement that is $\sqrt{N}$ times more accurate than what is obtained with one receptor.

If we have N receptors, each of size a arrayed on a structure of linear dimension R, such as a ring or a sphere, then as N becomes large the receptors are coming closer and closer together, and we expect that correlations become stronger. If we have two detectors making measurements with noise that becomes more and more strongly correlated, at some point they start to act like one big detector. Working through the details of the calculations for the case of multiple receptors indeed shows that as N becomes large,[12] the correlations among the different receptors become limiting. In fact, the threshold for reliable detection approaches Eq. (A.373); the $N \to \infty$ receptors packed into a structure with linear dimension R acts like one receptor of size R. Considering the intuitive Berg-Purcell argument about counting molecules in a volume and getting a fresh count each time the volume clears from diffusion, what this means is that packing many receptor sites into a region of size R eventually means that all molecules in a volume $\sim R^3$ are counted. There are geometrical factors for different spatial arrangements of the receptors, but like the $\sqrt{\pi}$ in Eq. (A.371), these do not affect the basic result.

A.7   Maximum Entropy

The maximum entropy method is an approach to making models: the goal is to make a model that reproduces some particular set of observations but otherwise has as little structure as possible. In this sense, it is the opposite of what we usually do in making theories. Rather than imposing our theoretical prejudices, we try to remove everything that is not absolutely necessary for matching the data. "Little structure" means that the states of the system should be as random as possible, and as explained in Section 6.1, the only consistent way of measuring this is to use the entropy, the same quantity that appears in statistical mechanics. Importantly, this mathematical identity exists whether or not we have any reason to think that we are really looking at a system in thermal equilibrium.

Let us imagine that the system under study has states described by $\mathbf{x}$. Further, assume that the system has reached a statistically stationary point, so there is a well-defined probability distribution $P(\mathbf{x})$ out of which the states $\mathbf{x}$ are chosen. Typically, the space of states is so large that we can't "measure" this distribution in any meaningful

---

12. See the references in the Annotated Bibliography for this section for details.

sense. What experiments do is to estimate the expectation values of various functions. A familiar example is the study of a gas; we can't observe the states of all molecules, but we can observe the average force (pressure) that the gas exerts on the walls of its container.

Label the functions that we can measure reliably as $f_\mu(\mathbf{x})$. Experiments give us estimates of the expectation values, $\langle f_\mu(\mathbf{x})\rangle_{\mathrm{expt}}$, so we want to be sure that we reproduce these measurements,

$$\langle f_\mu(\mathbf{x})\rangle \equiv \sum_{\mathbf{x}} P(\mathbf{x}) f_\mu(\mathbf{x}) = \langle f_\mu(\mathbf{x})\rangle_{\mathrm{expt}}, \tag{A.379}$$

where there may be several expectation values known ($\mu = 1, 2, \cdots, K$). Actually there is one more expectation value that we always know—the average value of one is one; the distribution is normalized, so that

$$\langle f_0 \rangle = \sum_{\mathbf{x}} P(\mathbf{x}) = 1. \tag{A.380}$$

Given the set of numbers $\{\langle f_0\rangle_{\mathrm{expt}}, \langle f_1\rangle_{\mathrm{expt}}, \cdots, \langle f_K\rangle_{\mathrm{expt}}\}$ as constraints on the probability distribution $P(\mathbf{x})$, we would like to know the largest possible value for the entropy and to find explicitly the distribution that provides this maximum.

The problem of maximizing a quantity subject to constraints is formulated using Lagrange multipliers. In this case, we want to maximize $S = -\sum_{\mathbf{x}} P(\mathbf{x}) \ln P(\mathbf{x})$, so we introduce a function $\tilde{S}$, with one Lagrange multiplier λ_μ for each constraint:

$$\tilde{S}[P(\mathbf{x})] = -\sum_{\mathbf{x}} P(\mathbf{x}) \ln P(\mathbf{x}) - \sum_{\mu=0}^{K} \lambda_\mu \left[\langle f_\mu(\mathbf{x})\rangle - \langle f_\mu(\mathbf{x})\rangle_{\mathrm{expt}} \right] \tag{A.381}$$

$$= -\sum_{\mathbf{x}} P(\mathbf{x}) \ln P(\mathbf{x}) - \sum_{\mu=0}^{K} \lambda_\mu \left[\sum_{\mathbf{x}} P(\mathbf{x}) f_\mu(\mathbf{x}) - \langle f_\mu(\mathbf{x})\rangle_{\mathrm{expt}} \right]. \tag{A.382}$$

Our problem, then, is to find the maximum of the function(al) $\tilde{S}[P(\mathbf{x})]$.

As usual, we look for a maximum by differentiating and setting the result to zero:

$$0 = \frac{\partial \tilde{S}}{\partial P(\mathbf{x})} = -\left[\ln P(\mathbf{x}) + 1\right] - \sum_{\mu=0}^{K} \lambda_\mu f_\mu(\mathbf{x}). \tag{A.383}$$

Rearranging, we have

$$\ln P(\mathbf{x}) = -1 - \sum_{\mu=0}^{K} \lambda_\mu f_\mu(\mathbf{x}) \tag{A.384}$$

$$P(\mathbf{x}) = \frac{1}{Z} \exp\left[-\sum_{\mu=1}^{K} \lambda_\mu f_\mu(\mathbf{x}) \right], \tag{A.385}$$

where $Z = \exp(1 + \lambda_0)$ is a normalization constant. If we differentiate $\tilde{S}$ with respect to the Lagrange multipliers, we obtain equations that fix the expectation values to the corresponding experimental measurements, Eq. (A.379). Thus, Eq. (A.385) gives us the *form* of the maximum entropy distribution, but we still have to adjust the constants $\{\lambda_\mu\}$ to match the measured values of the expectation values.

There are several things worth saying about maximum entropy distributions. First, recall that if $\mathbf{x}$ is the state of a physical system with energy $E(\mathbf{x})$, and we know only the expectation value of the energy,

$$\langle E \rangle = \sum_{\mathbf{x}} P(\mathbf{x})E(\mathbf{x}), \tag{A.386}$$

then the maximum entropy distribution is

$$P(\mathbf{x}) = \frac{1}{Z} \exp\left[-\lambda E(\mathbf{x})\right], \tag{A.387}$$

which is the Boltzmann distribution (as promised). In this case the Lagrange multiplier λ has physical meaning—it is the inverse temperature. Further, the function $\tilde{S}$ introduced for convenience is the difference between the entropy and λ times the energy; if we divide through by λ and flip the sign, then we have the energy minus the temperature times the entropy, or the free energy. Thus, the distribution that maximizes entropy at fixed average energy is also the distribution minimizing the free energy.

In a magnetic system, for example, if we know not just the average energy but also the average magnetization, then a new term appears in the exponential of the probability distribution, and we can interpret this term as the magnetic field multiplied by the magnetization. More generally, for every order parameter that we assume is known, the probability distribution acquires a term that adds to the energy and can be thought of as a product of the order parameter with its conjugate force. These ideas should be familiar from a statistical mechanics course.

Equation (A.385) shows us that the maximum entropy distribution looks like the Boltzmann distribution, effectively assigning an energy

$$E_{\text{eff}}(\mathbf{x}) = \sum_{\mu=1}^{K} \lambda_\mu f_\mu(\mathbf{x}) \tag{A.388}$$

to every state of the system. For each expectation value that we try to match, there is a separate term in the effective energy. This is a mathematical equivalence, not an analogy.

Consider the situation in which $\mathbf{x}$ is just one real number x. Suppose that we know the mean value of x and its variance. This is equivalent to knowing the expectation values of two functions, $f_1(x) = x$ and $f_2(x) = x^2$. Thus, from Eq. (A.385), the maximum entropy distribution is of the form

$$P(x) = \frac{1}{Z} \exp(-\lambda_1 x - \lambda_2 x^2). \tag{A.389}$$

This expression is a funny way of writing a more familiar object. Identifying the parameters $\lambda_2 = 1/(2\sigma^2)$ and $\lambda_1 = -\langle x \rangle / \sigma^2$, then we can rewrite the maximum entropy distribution as the usual Gaussian,

$$P(x) = \frac{1}{\sqrt{2\pi\sigma^2}} \exp\left[-\frac{1}{2\sigma^2}(x - \langle x \rangle)^2\right]. \tag{A.390}$$

Gaussian distributions usually arise through the central limit theorem: if the random variable of interest can be thought of as the sum of many independent events, then its distribution converges to a Gaussian as "many" becomes infinite. This behavior provides a mechanistic or reductionist view of why Gaussians are so important. What we have here is very different; if all we know about a variable is the mean and the variance, then the Gaussian distribution is the maximum entropy distribution consistent with this knowledge. Because the entropy measures (returning to our physical intuition) the randomness or disorder of the system, the Gaussian distribution describes the most random (or least structured) distribution that can generate the known mean and variance.

Problem 196: Less than maximum entropy.   Many natural signals are strongly non-Gaussian. In particular, exponential (or nearly exponential) distributions are common in studies on the statistics of natural images and natural sounds. With the same mean (which you can call zero) and variance, what is the difference in entropy between the exponential $[P(x) \propto \exp(-\lambda|x|)]$ and Gaussian distributions? If we imagine that this difference is relevant to every pixel (or to every Fourier component) in an image, is this significant compared to the 8 bits/pixel of a standard digital image? What if $P(x) \propto \exp(-\lambda|x|^{\mu})$, with $\mu < 1$?

Probability distributions that have the maximum entropy form of Eq. (A.385) are special not only because of their connection to statistical mechanics, but also because they form what the statisticians call an exponential family, which seems like an obvious name. To make this notion precise, we will think of the distribution as conditional on the values of the parameters $\{\lambda_{\mu}\}$, $P(\mathbf{x}|\{\lambda_{\mu}\})$. The important point is that exponential families of distributions are (almost) unique in having sufficient statistics. To understand what this means, consider the following problem. We observe a set of samples $\mathbf{x}_1, \mathbf{x}_2, \cdots, \mathbf{x}_N$, each of which is drawn independently and at random from a distribution $P(\mathbf{x}|\{\lambda_{\mu}\})$. Assume that we know the form of this distribution (from Eq. (A.385)) but not the values of the parameters $\{\lambda_{\mu}\}$. How can we estimate these parameters directly from the set of observations $\{\mathbf{x}_n\}$?

No finite amount of data will determine the exact values of the parameters, so we need a probabilistic formulation; we want to compute the distribution of parameters given the data, $P(\{\lambda_{\mu}\}|\{\mathbf{x}_n\})$. We do this using Bayes' rule,

$$P(\{\lambda_i\}|\{\mathbf{x}_n\}) = \frac{1}{P(\{\mathbf{x}_n\})} \cdot P(\{\mathbf{x}_n\}|\{\lambda_{\mu}\}) P(\{\lambda_{\mu}\}), \qquad (A.391)$$

where $P(\{\lambda_{\mu}\})$ is the distribution from which the parameter values themselves are drawn. Then because each datum $\mathbf{x}_n$ is drawn independently, we have

$$P(\{\mathbf{x}_n\}|\{\lambda_{\mu}\}) = \prod_{n=1}^{N} P(\mathbf{x}_n|\{\lambda_{\mu}\}). \qquad (A.392)$$

For probability distributions of the maximum entropy form, we can proceed further, using Eq. (A.385):

$$P(\{\lambda_\mu\}|\{\mathbf{x}_n\}) = \frac{1}{P(\{\mathbf{x}_n\})} \cdot P(\{\mathbf{x}_n\}|\{\lambda_\mu\}) P(\{\lambda_\mu\})$$

$$= \frac{P(\{\lambda_\mu\})}{P(\{\mathbf{x}_n\})} \prod_{n=1}^{N} P(\mathbf{x}_n|\{\lambda_\mu\}) \tag{A.393}$$

$$= \frac{P(\{\lambda_\mu\})}{Z^N P(\{\mathbf{x}_n\})} \prod_{n=1}^{N} \exp\left[-\sum_{\mu=1}^{K} \lambda_\mu f_\mu(\mathbf{x}_n) \right] \tag{A.394}$$

$$= \frac{P(\{\lambda_\mu\})}{Z^N P(\{\mathbf{x}_n\})} \exp\left[-N \sum_{\mu=1}^{K} \lambda_\mu \frac{1}{N} \sum_{n=1}^{N} f_\mu(\mathbf{x}_n) \right]. \tag{A.395}$$

We see that all information that the data points $\{\mathbf{x}_n\}$ can give about the parameters $\{\lambda_\mu\}$ is contained in the average values of the functions f_μ over the data set, or the empirical means:

$$\bar{f}_\mu = \frac{1}{N} \sum_{n=1}^{N} f_\mu(\mathbf{x}_n). \tag{A.396}$$

More precisely, the distribution of possible parameter values consistent with the data depends not on all details of the data but rather only on the empirical means $\{\bar{f}_\mu\}$,

$$P(\{\lambda_\mu\}|\mathbf{x}_1, \mathbf{x}_2, \cdots, \mathbf{x}_N) = P(\{\lambda_\mu\}|\{\bar{f}_\mu\}), \tag{A.397}$$

and a consequence of this is the information-theoretic statement

$$I(\mathbf{x}_1, \mathbf{x}_2, \cdots, \mathbf{x}_N \to \{\lambda_\mu\}) = I(\{\bar{f}_\mu\} \to \{\lambda_\mu\}). \tag{A.398}$$

This situation is described by saying that the reduced set of variables $\{\bar{f}_\mu\}$ constitutes sufficient statistics for learning the distribution. Thus, for distributions of this form, the problem of compressing N data points into $K \ll N$ variables that are relevant for parameter estimation can be solved explicitly; if we keep track of the running averages $\bar{f}_\mu$, we can compress the data as we go along, and we are guaranteed that we will never need to go back and examine the data in more detail. A familiar example is that if we know data are drawn from a Gaussian distribution, running estimates of the mean and variance contain all the information available about the underlying parameter values.

The Gaussian example might leave the impression that the concept of sufficient statistics is trivial. Of course if we know that data are chosen from a Gaussian distribution, then to identify the distribution, all we need to do is to keep track of two moments. Far from trivial, this situation is quite unusual. Most distributions that we might write down do not have this property—even if they are described by a finite number of parameters, we cannot guarantee that a comparably small set of empirical expectation values captures all information about the parameter values. If we insist further that the sufficient statistics be additive and permutation symmetric, then it is a theorem that *only* exponential families have sufficient statistics.

Once we understand that the relevant probability distributions are exponential families, we could leave aside the maximum entropy perspective and take a more conventional statistical point of view. Thus, if the probability distributions considered are of the form shown in Eq. (A.385), and we collect N independent samples $\mathbf{x}_1, \mathbf{x}_2, \cdots, \mathbf{x}_N$, the probability that these data arose from the model is given by

$$P(\{\mathbf{x}_i\}|\{g_\mu\}) = \left(\frac{1}{Z(\{g_\mu\})}\right)^N \prod_{i=1}^N \exp\left[-\sum_{\mu=1}^K g_\mu f_\mu(\mathbf{x}_i)\right] \qquad (A.399)$$

$$= \exp\left[-N \ln Z(\{g_\mu\}) - \sum_{i=1}^N \sum_{\mu=1}^K g_\mu f_\mu(\mathbf{x}_i)\right]. \qquad (A.400)$$

The conditions for maximizing the likelihood of the data,

$$\frac{\partial P(\{\mathbf{x}_i\}|\{g_\mu\})}{\partial g_\nu} = 0, \qquad (A.401)$$

then become

$$0 = -N \frac{\partial \ln Z(\{g_\mu\})}{\partial g_\nu} - \sum_{i=1}^N f_\nu(\mathbf{x}_i) \qquad (A.402)$$

$$\Rightarrow -\frac{\partial \ln Z(\{g_\mu\})}{\partial g_\nu} = \frac{1}{N} \sum_{i=1}^N f_\nu(\mathbf{x}_i). \qquad (A.403)$$

But as in statistical mechanics, the logarithm of the partition function is a free energy, and derivatives of the free energy are expectation values, so that we have

$$\langle f_\nu(\mathbf{x}_i)\rangle = -\frac{\partial \ln Z(\{g_\mu\})}{\partial g_\nu}. \qquad (A.404)$$

Thus, Eq. (A.403) is equivalent to the statement that the expectation values in the model (maximum entropy) distribution must match the expectation values measured over the N samples:

$$\langle f_\nu(\mathbf{x}_i)\rangle = \frac{1}{N} \sum_{i=1}^N f_\nu(\mathbf{x}_i). \qquad (A.405)$$

The generic problem of information processing, by the brain or by a machine, is that a huge quantity of data is available and those pieces that are of interest must be extracted. The idea of sufficient statistics is intriguing in part because it provides an example where this problem of extracting interesting information can be solved completely. If the points $\mathbf{x}_1, \mathbf{x}_2, \cdots, \mathbf{x}_N$ are chosen independently and at random from some distribution, the only thing that could possibly be interesting is the structure of the distribution itself (everything else is random, by construction). This structure is described by a finite number of parameters, and there is an explicit algorithm for compressing the N data points $\{\mathbf{x}_n\}$ into K numbers that preserve all interesting information. The crucial point is that this procedure cannot exist in general, but only for certain classes of probability distributions. This discussion is an introduction to the idea that some kinds of structure in data are learnable from random examples, whereas other structures are not.

Consider the (Boltzmann) probability distribution for the states of a system in thermal equilibrium. If we expand the Hamiltonian as a sum of terms (operators), then the family of possible probability distributions is an exponential family in which the coupling constants for each operator are the parameters analogous to the λ_μ mentioned above. In principle there could be an infinite number of these operators, but for a given class of systems we usually find that only a finite set are relevant in the renormalization group sense: if we write an effective Hamiltonian for coarse-grained degrees of freedom, then only a finite number of terms will survive the coarse-graining procedure. If there are only a finite number of terms in the Hamiltonian, then the family of Boltzmann distributions has sufficient statistics, which are just the expectation values of the relevant operators. Thus, the expectation values of the relevant operators carry all information that the (coarse-grained) configuration of the system can provide about the coupling constants, which in turn is information about the identity or microscopic structure of the system. Thus, the statement that there are only a finite number of relevant operators is equivalent to the statement that a finite number of expectation values carries all information about the microscopic dynamics. The "if" part of this statement is obvious: if there are only a finite number of relevant operators, then the expectation values of these operators carry all information about the identity of the system. The statisticians, through the theorem about the uniqueness of exponential families, give us the "only if": a finite number of expectation values (or correlation functions) can provide all information about the system *only if* the effective Hamiltonian has a finite number of relevant operators. I suspect that there is more to say along these lines.

An important example of the maximum entropy idea arises when the state $\mathbf{x}$ is an integer n, as happens when we are counting. It is natural to imagine that what we know is the mean count $\langle n \rangle$. One way this problem can arise is that we are trying to communicate and are restricted to sending discrete or quantized units. An obvious case is in optical communication, where the quanta are photons. In the brain, quantization abounds. Most neurons do not generate continuous analog voltages but rather communicate with one another through stereotyped pulses or spikes. Even if the voltages vary continuously, transmission across a synapse involves the release of a chemical transmitter that is packaged into discrete vesicles. It can be relatively easy to measure the mean rate at which discrete events are counted, and we might want to know what bounds this mean rate places on the ability of the cells to convey information. Alternatively, there is an energetic cost associated with these discrete events (generating the electrical currents that underlie the spike, constructing and filling the vesicles, etc.), and we might want to characterize the mechanisms by their cost per bit rather than their cost per event.

If we know the mean count, there is (as for the Boltzmann distribution) only one function $f_1(n) = n$ that can appear in the exponential of the distribution, so that

$$P(n) = \frac{1}{Z} \exp(-\lambda n). \tag{A.406}$$

Normalization requires that

$$Z = \sum_{n=0}^{\infty} \exp(-\lambda n) = \frac{1}{1 - e^{-\lambda}}. \tag{A.407}$$

We need to choose the Lagrange multiplier to fix the mean count, which we do using the relation of averages to the derivative of the (log) partition function:

$$\langle n \rangle = -\frac{\partial \ln Z}{\partial \lambda} \tag{A.408}$$

$$= \frac{1}{e^{\lambda} - 1}, \tag{A.409}$$

which is equivalent to

$$\lambda = \ln(1 + 1/\langle n \rangle). \tag{A.410}$$

We can also compute the entropy,

$$S \equiv -\sum_{n=0}^{\infty} P(n) \log_2 P(n) \tag{A.411}$$

$$= \log_2 Z + \frac{\lambda}{\ln 2} \langle n \rangle, \tag{A.412}$$

and we can express this (maximum!) entropy in terms of the mean count,

$$S_{\max}(\text{counting}) = \log_2(1 + \langle n \rangle) + \langle n \rangle \log_2(1 + 1/\langle n \rangle). \tag{A.413}$$

The information conveyed by counting something can never exceed the entropy of the distribution of counts, and if we know the mean count, then the entropy can never exceed the bound in Eq. (A.413). Thus, if we have a system in which information is conveyed by counting discrete events, the simple fact that we count only a limited number of events (on average) sets a bound on how much information can be transmitted. We will see, near the end of Section 6.2, that real neurons and synapses approach this fundamental limit.

You might suppose that if information is coded in the counting of discrete events, then each event carries a certain amount of information, but this is not quite right. In particular, if we count a large number of events, then the maximum counting entropy becomes

$$S_{\max}(\text{counting}; \langle n \rangle \to \infty) \sim \log_2(\langle n \rangle e), \tag{A.414}$$

and so we are guaranteed that the entropy (and hence the information) per event goes to zero, although the approach is slow. In contrast, if events are rare, so that the mean count is much less than one, we find the maximum entropy per event is

$$\frac{1}{\langle n \rangle} S_{\max}(\text{counting}; \langle n \rangle \ll 1) \sim \log_2\left(\frac{e}{\langle n \rangle}\right), \tag{A.415}$$

which is arbitrarily large for small mean count. This result makes sense—rare events have an arbitrarily large capacity to surprise us and hence to convey information. It is important to note, though, that the maximum entropy per event is a monotonically decreasing function of the mean count. Thus, if we are counting spikes from a neuron, counting in larger windows (hence larger mean counts) is always less efficient in terms of bits per spike.

If it is more efficient to count in small time windows, perhaps we should think not about counting but about measuring the arrival times of the discrete events. If we look at a total (large) time interval $0 < t < T$, then we will observe arrival times $t_1, t_2, \cdots, t_N$

in this interval; note that the number of events N is also a random variable. We want to find the distribution $P(t_1, t_2, \cdots, t_N)$ that maximizes the entropy while holding fixed the average event rate. We can write the entropy of the distribution as a sum of two terms, one from the entropy of the arrival times given the count and one from the entropy of the counting distribution,

$$S = - \sum_{N=0}^{\infty} \int d^N t_n P(t_1, t_2, \cdots, t_N) \log_2 P(t_1, t_2, \cdots, t_N) \qquad (A.416)$$

$$= \sum_{N=0}^{\infty} P(N) S_{\text{time}}(N) - \sum_{N=0}^{\infty} P(N) \log_2 P(N), \qquad (A.417)$$

where we have made use of

$$P(t_1, t_2, \cdots, t_N) = P(t_1, t_2, \cdots, t_N | N) P(N), \qquad (A.418)$$

and the (conditional) entropy of the arrival times is given by

$$S_{\text{time}}(N) = - \int d^N t_n P(t_1, t_2, \cdots, t_N | N) \log_2 P(t_1, t_2, \cdots, t_N | N). \quad (A.419)$$

If all we fix is the mean count, $\langle N \rangle = \sum_N P(N) N$, then the conditional distributions for the locations of the events given the total number of events, $P(t_1, t_2, \cdots, t_N | N)$, are unconstrained. We can maximize the contribution of each of these terms to the entropy (the terms in the first sum of Eq. (A.417)) by making the distributions $P(t_1, t_2, \cdots, t_N | N)$ uniform, but it is important to be careful about normalization. When we integrate over all the times $t_1, t_2, \cdots, t_N$, we are forgetting that the events are all identical, and hence that permutations of the times describe the same events. Thus, the normalization condition is *not*

$$\int_0^T dt_1 \int_0^T dt_2 \cdots \int_0^T dt_N P(t_1, t_2, \cdots, t_N | N) = 1, \qquad (A.420)$$

but rather

$$\frac{1}{N!} \int_0^T dt_1 \int_0^T dt_2 \cdots \int_0^T dt_N P(t_1, t_2, \cdots, t_N | N) = 1. \qquad (A.421)$$

Thus, the uniform distribution must be

$$P(t_1, t_2, \cdots, t_N | N) = \frac{N!}{T^N}, \qquad (A.422)$$

and hence the entropy (substituting into Eq. (A.417)) becomes

$$S = - \sum_{N=0}^{\infty} P(N) \left[\log_2 \left(\frac{N!}{T^N} \right) + \log_2 P(N) \right]. \qquad (A.423)$$

Now to find the maximum entropy, we proceed as before. Introducing Lagrange multipliers to constrain the mean count and the normalization of the distribution $P(N)$, which leads to the function

$$\tilde{S} = - \sum_{N=0}^{\infty} P(N) \left[\log_2 \left(\frac{N!}{T^N} \right) + \log_2 P(N) + \lambda_0 + \lambda_1 N \right], \qquad (A.424)$$

we maximize this function by varying $P(N)$. As before the different Ns are not coupled, so the optimization conditions are simple:

$$0 = \frac{\partial \tilde{S}}{\partial P(N)} \tag{A.425}$$

$$= -\frac{1}{\ln 2}\left[\ln\left(\frac{N!}{T^N}\right) + \ln P(N) + 1\right] - \lambda_0 - \lambda_1 N, \tag{A.426}$$

$$\ln P(N) = -\ln\left(\frac{N!}{T^N}\right) - (\lambda_1 \ln 2)N - (1 + \lambda_0 \ln 2). \tag{A.427}$$

Combining terms and simplifying, we have

$$P(N) = e^{-(1+\lambda_0 \ln 2)} \frac{(e^{-\lambda_1 \ln 2}T)^N}{N!}, \tag{A.428}$$

$$= e^{-rT}\frac{(rT)^N}{N!}. \tag{A.429}$$

This is the Poisson distribution.

The Poisson distribution usually is derived (as in Appendix A.1) by assuming that the probability of occurrence of an event in any small time bin of size $\Delta\tau$ is independent of events in any other bin and then letting $\Delta\tau \to 0$ to obtain a distribution in the continuum. This result is not surprising: we have found that the maximum entropy distribution of events given the mean number of events (or their density $\langle N \rangle/T$) is given by the Poisson distribution, which corresponds to the events being thrown down at random with some probability per unit time (again, $\langle N \rangle/T$) and no interactions among the events. This process describes an "ideal gas" of events along a line (time); the ideal gas is the gas with maximum entropy given its density, and interactions among the gas molecules always reduce the entropy if we hold the density fixed.

The examples of maximum entropy distributions that we have discussed so far are useful but simple. In particular, we can go from the experimentally observed averages to the corresponding parameters or coupling constants in the probability distribution quite simply, with compact analytic formulas. Things get more interesting when this is not possible.

We first encountered the more sophisticated version of the maximum entropy idea in the context of amino acid sequences (Section 5.1). In this problem, we have at each site i along the sequence a variable s_i that can take on 20 possible values, corresponding to the 20 amino acids. Alternatively, we can describe the sequence by variables s_i^α, where $\alpha = 1, 2, \cdots, 20$, and $s_i^\alpha = 1$ if the amino acid at site i is of type α and zero otherwise. Then if we examine a large ensemble of sequences and measure the probability p_i^α that there is amino acid α at site i, this probability is an expectation value,

$$p_i^\alpha = \langle s_i^\alpha \rangle. \tag{A.430}$$

Correspondingly, if we measure these probabilities for all N sites and all 20 amino acids, the maximum entropy distribution that is consistent with these one-body statistics is given by

$$P_1(\{s_i^\alpha\}) = \frac{1}{Z_1(\{h_i^\alpha\})} \exp\left[\sum_{i=1}^{N}\sum_{\alpha=1}^{20} h_i^\alpha s_i^\alpha\right]. \tag{A.431}$$

As before, we have one term in the exponential for each quantity whose expectation value is measured, and each of these terms has a separate field h_i^α. If we now measure the joint distribution of amino acids at pairs of sites,

$$p_{ij}^{\alpha\beta} = \langle s_i^\alpha s_j^\beta \rangle, \qquad (A.432)$$

then the maximum entropy distribution is given by

$$P_2(\{s_i^\alpha\}) = \frac{1}{Z_2(\{h_i^\alpha, J_{ij}^{\alpha\beta}\})} \exp\left[\sum_{i=1}^{N}\sum_{\alpha=1}^{20} h_i^\alpha s_i^\alpha + \frac{1}{2}\sum_{i,j=1}^{N}\sum_{\alpha,\beta=1}^{20} J_{ij}^{\alpha\beta} s_i^\alpha s_j^\beta\right], \qquad (A.433)$$

where there are interactions of strength $J_{ij}^{\alpha\beta}$ between every pair of amino acids along the chain. With the sign convention chosen here, $J_{ij}^{\alpha\beta} > 0$ means that the combination of amino acid α at site i and β at site j is favored. The difficulty is that there is no analytic formula that connects (for example) $J_{ij}^{\alpha\beta}$ with $p_{ij}^{\alpha\beta}$ once we have more than two sites along the chain. If we imagine starting with the fields and interactions, $\{h_i^\alpha, J_{ij}^{\alpha\beta}\}$, computing the correlations $p_{ij}^{\alpha\beta}$ is a standard problem in statistical mechanics. Actually constructing the maximum entropy model involves solving the inverse of this statistical mechanics problem.

The corresponding problem for neurons is a bit simpler. If we slice time into bins of width $\Delta\tau$, then for sufficiently small $\Delta\tau$, each cell either generates an action potential or does not, so that the state of a neuron at one moment is binary, as in the Ising model: $\sigma_i = \pm 1$, with $+1$ for a spike and -1 for silence. The mean probability that cell i generates a spike is then

$$p_i = (1 + \langle \sigma_i \rangle)/2. \qquad (A.434)$$

The probability that cells i and j spike in the same small window is then

$$p_{ij} = \frac{1}{4}\langle (1 + \sigma_i)(1 + \sigma_j) \rangle. \qquad (A.435)$$

Thus, measuring the mean spike probabilities and the correlations between cells taken in pairs is equivalent to measuring $\langle \sigma_i \rangle$ and $\langle \sigma_i \sigma_j \rangle$. The maximum entropy model consistent with these measurements is then

$$P(\{\sigma_i\}) = \frac{1}{Z} \exp\left[\sum_{i=1}^{N} h_i \sigma_i + \frac{1}{2}\sum_{i\neq j=1}^{N} J_{ij}\sigma_i\sigma_j\right], \qquad (A.436)$$

where the fields h_i and the interactions J_{ij} need to be adjusted to match the measured sets of p_i and p_{ij}.

These ideas were first used to analyze responses from populations of neurons in the retina as it responds to naturalistic stimuli. Importantly, in this system as in many other areas of the brain, the correlations between neurons (that is, $p_{ij} - p_i p_j$) tend to be small but widespread. For two neurons it would be tempting to ignore the correlations all together, but already for $N = 10$ neurons this is a disastrous approximation, as seen in the upper left of Fig. A.10.

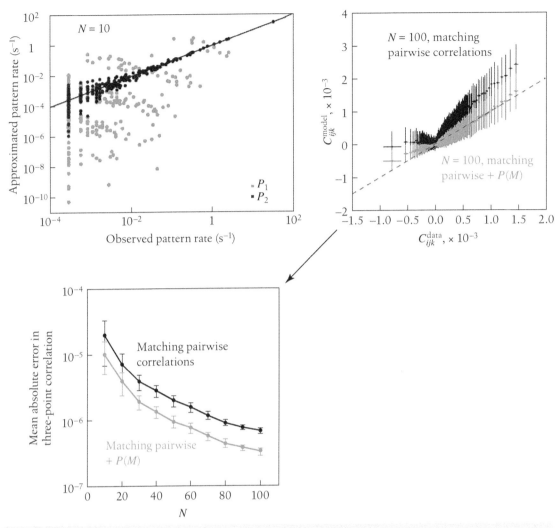

FIGURE A.10

Testing maximum entropy models for neurons. (a) In small networks (here, $N = 10$) we can check the probability of every possible state, here normalized by the size of the bin $\Delta \tau$ to give the rate at which different patterns occur. Black points show the comparison of measured and predicted probabilities for a maximum entropy model P_2 that includes pairwise correlations as well as mean spike probabilities for each neuron. Orange points show the result for a model P_1 that neglects the correlations. Notice that the failures of this model are large, even though the individual pairwise correlations are weak. Redrawn from Schneidman et al. (2006). (b) Predicted versus observed three-neuron correlations in a large network. Black points show the predictions of the maximum entropy model that matches the pairwise correlations, and orange points are for the model that also matches the overall probability of M out of the N neurons spiking together, Eq. (A.347). Note the small scale of the average errors. These errors decrease as we look at more of the neurons in this small patch of the retina. Redrawn from Tkačik et al. (2012).

Perhaps surprisingly, matching the pairwise correlations in a maximum entropy model largely repairs this problem, correctly describing the full distribution of states taken on by the system. As we look at ever larger groups of neurons, we see small but systematic deviations from the predictions of the pairwise models. The spirit of the maximum entropy approach is that we take seriously some relatively small set of measured expectation values and hope that they capture the essence of the system. When faced with discrepancies, we try to find a minimal solution, and one possibility that has been explored is to fix the probability that M out of the N neurons fire simultaneously, which is equivalent to

$$P(\{\sigma_i\}) = \frac{1}{Z} \exp\left[\sum_{i=1}^{N} h_i \sigma_i + \frac{1}{2} \sum_{i \neq j=1}^{N} J_{ij} \sigma_i \sigma_j - V\left(\sum_{i=1}^{N} \sigma_i\right)\right] \qquad (A.437)$$

for some global potential V; importantly, this form adds no more than N parameters to a model that already has $\sim N^2$ parameters, and so is a minimal increase in complexity. Although this extra constraint does not differentiate among the identities of different neurons, it goes a long way toward correcting the small deviations in, for example, the prediction of correlations among specific triplets of neurons (Fig. A.10b). I think it is fair to say that these methods give us a family of accurate statistical mechanics models for the activity in these networks, realizing an old dream, but this is just the start. To make progress, we need to understand what the models are telling us about these networks (e.g., are there multiple stable states, as in the Hopfield model, and are these used in the representation of information?). Even simple questions (e.g., where are real networks in the phase diagram of possible networks?) may have interesting answers. See the references in the Annotated Bibliography for this section for recent work.

A.8   Measuring Information Transmission

When studying classical mechanics, we can make a direct connection between the positions and momenta that appear in the equations of motion and the positions and momenta of the particles that we observe, as in the planetary orbits. This connection is a little bit subtle, because we do not actually measure particle positions; more likely, we count the photons arriving at some detector, forming an image, measure the delay in propagation of a pulse used in radar, or something similar. But in some sense classical mechanics, in contrast to quantum mechanics, is the domain of physics in which these subtleties are not important. In statistical physics the connection between equations and what is observed in the world becomes more abstract. The fundamental objects in statistical physics are probability distributions, and as a matter of definition a distribution *cannot* be measured. Instead, Nature (or even a controlled experiment) provides us with samples taken out of these distributions. This has very serious consequences for any attempt to "measure" information flow.

In thermodynamics, entropy changes are connected to heat flow, and so we can at least measure the difference in entropy between two states by tracking these heat flows. Indeed, there is a long tradition of integrating these changes from some convenient reference to measure the entropy of states at intermediate temperatures. As far as I know,

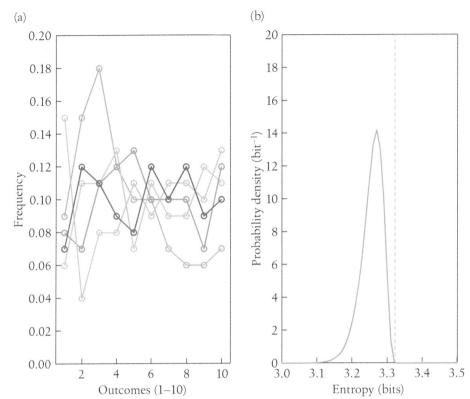

FIGURE A.11

The sampling problem in entropy estimation. (a) The frequency of occurrence found from five examples of $N = 100$ samples drawn out of $K = 10$ bins; the true probability distribution is flat, $p_i = 0.1$ for all i. (b) We estimate the entropy of the distribution by identifying the observed frequencies with probabilities. The distribution of entropies obtained in this way, from many "experiments" with $N = 100$ and $K = 10$, is shown as a solid line. It should be compared with the true entropy, shown by a dashed line at $S_{\text{true}} = \log_2(10)$.

there is no analog of this approach in the information-theoretic context. Thus, although Shannon tells us that the entropy is a fundamental property of the distribution out of which signals are drawn, there is no universal entropy meter.

To get a feeling for the problem, consider Fig. A.11. Here we have a variable that can take on ten possible values (i = 1, 2, $\cdots$, 10), all equally likely ($p_i = 0.1$ for all i), and we draw $N = 100$ samples. If we look at the frequency with which each possibility occurs, of course we do not see a perfectly flat distribution. Because with 10 bins and 100 samples we expect 10 samples per bin, it is not surprising that the fluctuations are on the scale of $1/\sqrt{10} \sim 30\%$. These fluctuations, however, are random—they average to zero if we do the same experiment many times. The problem is that if we identify the frequencies of occurrence as the best estimates of the underlying probabilities, and use these estimates to compute the entropy, we make a systematic error, as is clear from the results shown in Fig. A.11b.

Problem 197: Experiment with sampling.   Generate the analog of Fig. A.11 but with different values for the number of possible values K, where i = 1, 2, $\cdots$, K. You should also try different probability distributions (e.g., $p_i \propto 1/i$, Zipf's law). Experiment. Convince yourself that, by identifying probabilities with the observed frequencies of occurrence, you always underestimate the entropy.

The problem illustrated in Fig. A.11 might seem very specific to the conditions of that simulation (e.g., that the true distribution is flat, and hence the entropy is maximal, so perhaps all errors have to be biased downward?), but in fact it is very general. Let's consider drawing samples out of a discrete set of possibilities, $i = 1, 2, \cdots, K$, with probabilities $\mathbf{p} \equiv \{p_1, p_2, \cdots, p_K\}$. If we draw N samples in all, we will find n_i examples of the outcome i, and $\langle n_i \rangle = N p_i$. Because these are random events, we expect that the variance of the number of events of type i will be equal to the mean: $\langle (\delta n_i)^2 \rangle = \langle n_i \rangle = N p_i$. Defining the frequency of events in the usual way as $f_i = n_i / N$, we have

$$\langle f_i \rangle = p_i \quad \text{and} \quad \langle (\delta f_i)^2 \rangle = \frac{p_i}{N}. \tag{A.438}$$

But if frequencies are considered as the best estimate of probabilities (and we'll see below in what sense this familiar identification is correct), we can construct a "naive" estimate of the entropy,

$$S_{\text{naive}} = - \sum_{i=1}^{K} f_i \log_2 f_i. \tag{A.439}$$

Because the frequencies are close to the true probabilities when the number of samples is large, we can do a Taylor expansion around the point $f_i = \langle f_i \rangle = p_i$,

$$\begin{aligned}
S_{\text{naive}} &= - \sum_{i=1}^{K} f_i \log_2 f_i \\
&= - \sum_{i=1}^{K} (p_i + \delta f_i) \log_2 (p_i + \delta f_i) \tag{A.440} \\
&= - \sum_{i=1}^{K} p_i \log_2 p_i - \sum_{i=1}^{K} \left[\log_2 p_i + \frac{1}{\ln 2} \right] \delta f_i \\
&\quad - \frac{1}{2} \sum_{i=1}^{K} \left[\frac{1}{(\ln 2) p_i} \right] (\delta f_i)^2 + \cdots. \tag{A.441}
\end{aligned}$$

The first term in the series is the true entropy. The second term is a random error that averages to zero. The third term, however, has a nonzero mean, because it depends on the square of the fluctuations δf_i. Thus, when computing the average of our naive entropy estimate, we find

$$\langle S_{\text{naive}} \rangle = S_{\text{true}} - \frac{1}{2 \ln 2} \sum_{i=1}^{K} \frac{\langle (\delta f_i)^2 \rangle}{p_i} + \cdots \tag{A.442}$$

$$= S_{\text{true}} - \frac{1}{2 \ln 2} \sum_{i=1}^{K} \frac{p_i}{N p_i} + \cdots \tag{A.443}$$

$$= S_{\text{true}} - \frac{K}{(2 \ln 2) N} + \cdots. \tag{A.444}$$

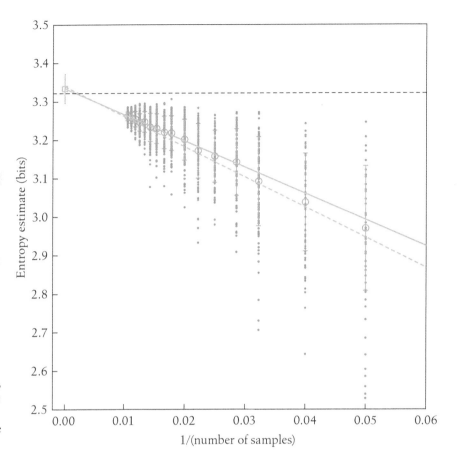

FIGURE A.12

Entropy versus number of samples.
Starting with $N = 100$ samples, as in
Fig. A.11, we draw smaller numbers
of samples at random, compute the
entropy, and search for the systematic
behavior predicted in Eq. (A.444).
Orange points are from different
subsamplings; blue circles show
means and standard deviations. The
solid green line is a linear fit for
$n > N/2$, and the dashed green line
is a quadratic fit to all data shown.
The green square is the extrapolation,
with an error bar $\sqrt{2}$ smaller than the
standard deviation found empirically
at $n = N/2$, and the dashed black line
is $S_{\text{true}} = \log_2(10)$.

Thus, no matter what the underlying true distribution, identifying frequencies with
probabilities leads to a *systematic* (not random!) underestimate of the entropy, and the
size of this systematic error is proportional to the number of accessible states (K) and
inversely proportional to the number of samples (N).

That the systematic errors have a very definite structure suggests that we should
be able to correct them. Let us see what happens to our entropy estimates in the
"experiment" shown in Fig. A.11 as we change the number of samples N. More precisely,
suppose we have only the $N = 100$ samples, but we choose $n < 100$ points out of
these samples and estimate the entropy based only on this more limited data. Equation
(A.444) suggests that if we plot the entropy estimate versus $1/n$, we should see a straight
line; a higher order version of the same calculation shows that there are quadratic
corrections. Indeed, as shown in Fig. A.12, this works. It is important to note that,
for all the accessible range of sample sizes, the entropy estimate is smaller than the true
entropy, and this error is larger than our best estimate of the error bar, which really is
troublesome. However, once we recognize the systematic dependence of the entropy
estimate on the number of samples, we can extrapolate to recover an estimate that is
correct within error bars. What we have seen here about entropy is also true about
information, which is a difference between entropies.

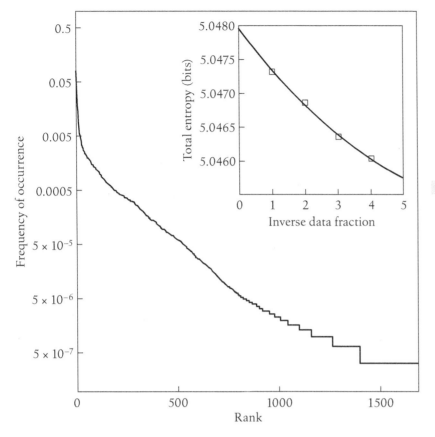

FIGURE A.13

Entropy extrapolation with real neural data. From the experiment on fly motion-sensitive neurons discussed in Figs. 6.5 and 6.6, we look at 10-letter words with time resolution $\Delta\tau = 3\,\text{ms}$. The main figure shows the Zipf plot of frequency versus rank from the full data set. Note that because there are sometimes (but rarely) two spike in one 3 ms bin, there are more than 1024 words. The inset shows the estimated entropy as a function of the (inverse) fraction of the full data set used. The line through the data is from Eq. (A.444). Redrawn from Strong et al. (1998).

It is also important to show that this extrapolation procedure also works for real data, not just for the idealized case where we choose samples independently out of a known distribution. One example is shown in Fig. A.13. These data are from the experiment on the fly motion-sensitive neuron H1 described in Section 6.2. The frequency of 10-letter words in the neural response to random motions across the visual field is shown, along with the dependence of the estimated entropy on the fraction of the data included in the calculation. This data set is large, so the systematic errors are small, but they are well described by Eq. (A.444). Importantly, we do not really know the full number of possible responses (some may have genuinely zero probability, because spikes are forbidden from coming too close together, "refractoriness"), so K is a parameter that needs to be fit to the dependence of entropy estimates on sample size.

You might worry that entropy estimates based on extrapolations are a bit heuristic. If we can really convince ourselves that we see a clean linear dependence on $1/N$, things are likely to be fine, but this approach leaves room for considerable murkiness. Also, because the expansion of the entropy estimate in powers of $1/N$ obviously is not fully convergent, there is always the problem of choosing the regime over which the asymptotic behavior is observed, a widespread problem in fitting to such asymptotic series. Although for many purposes these problems can be dismissed, it would be nice

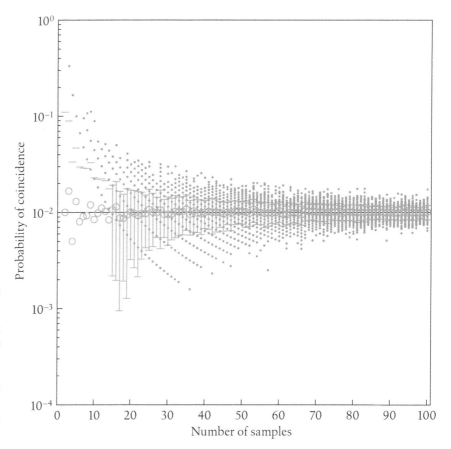

FIGURE A.14

Estimating coincidence probability. Samples are drawn from a distribution that is uniform over $K = 100$ possible states. Orange dots show examples, blue circles show the mean and standard deviation across many draws of N samples, and the black line is the exact answer. The estimate is quite good even when $N \sim \sqrt{K} \ll K$.

to do better. It also is an interesting mathematical challenge to ask whether we can estimate the entropy of a probability distribution even when the number of samples is small, perhaps even smaller than the number of possible states for the system.

When we do a Monte Carlo simulation of a physical system in thermal equilibrium, we are in the undersampled limit, where the number of samples collected must be much smaller than the number of possible states. To estimate entropy from a Monte Carlo simulation, we usually use the identity relating entropy to an integral of the heat capacity, because heat capacity is related to energy fluctuations, and these are easy to compute at each temperature. Of course, if you just have samples of the state of the system and do not actually know the Hamiltonian, you can't compute the energy, and so this method does not work. Ma suggested another approach, asking how often the system revisits the same state. In the simple case (relevant for the microcanonical ensemble) where all K possible states are equally likely, the probability that two independent samples are in the same state is $1/K$. But if we have N samples, there are $\sim N^2$ pairs that can be tested. Thus, we can get a good estimate of the probability of occupying the same state once we observe $N \sim \sqrt{K}$ independent samples, far less than the number of states. As an illustration, Fig. A.14 shows the frequency of coincidences when drawing N samples from a uniform distribution with $K = 100$ states.

Recall the classic problem of how many people need to be in the room before there is a good chance of two people having the same birthday. The answer is not 365, but more nearly $\sqrt{365}$. Put another way, if we did not know the length of the year, we could estimate this chance by polling people about their birthdays and keeping track of coincidences. Long before we have sampled all possible birthdays, Fig. A.14 shows that our estimate of this coincidence probability will stabilize—which birthdays are represented will vary from sample to sample, but the fraction of coincidences will vary much less.

In these simple examples, the probability distribution is uniform, and so the entropy is just the logarithm of the number of possible states, and this in turn is inversely proportional to the probability of a coincidence. So, being able to estimate this probability is equivalent to being able to estimate the entropy. Thus, we can generate reliable entropy estimates even in the undersampled regime, just by counting coincidences. This is a beautiful idea. The challenge is to generalize this idea to nonuniform distributions.

A better understanding of the entropy estimation problem has come through a Bayesian approach. Rather than identifying frequencies with probabilities, we imagine that the distribution itself is drawn from a distribution. To be formal, let the possible states of the system be $i = 1, 2, \cdots, K$, and let the probability distribution over these states be $p_1, p_2, \cdots, p_K \equiv \mathbf{p}$. This distribution itself is drawn from some distribution function $\mathcal{P}(\mathbf{p})$. The distribution has to be normalized, but it is tempting to think that, other than normalization, all distributions should be equally likely, so that

$$\mathcal{P}(\mathbf{p}) = \frac{1}{Z} \delta \left(\sum_{i=1}^{K} p_i - 1 \right). \tag{A.445}$$

If we observe n_1 samples in the first state, n_2 samples in the second state, and so on, then the probability of this occurring assuming some distribution $\mathbf{p}$ is

$$P(\{n_i\}|\mathbf{p}) \propto \prod_{i=1}^{K} p_i^{n_i}, \tag{A.446}$$

and so by Bayes' rule, we have

$$\mathcal{P}(\mathbf{p}|\{n_i\}) = \frac{P(\{n_i\}|\mathbf{p})\mathcal{P}(\mathbf{p})}{P(\{n_i\})} \tag{A.447}$$

$$\propto \frac{1}{Z} \left(\prod_{i=1}^{K} p_i^{n_i} \right) \delta \left(\sum_{i=1}^{K} p_i - 1 \right). \tag{A.448}$$

If we want to compute the best estimate of the distribution $\hat{p}_i$, we have to do the integral

$$\hat{p}_i = \frac{1}{Z} \int d^K \mathbf{p} \, p_i^{n_i+1} \left(\prod_{j \neq i} p_j^{n_j} \right) \delta \left(\sum_{j=1}^{K} p_j - 1 \right), \tag{A.449}$$

where the normalization $\mathcal{Z}$ is given by

$$\mathcal{Z} = \int d^K \mathbf{p} \left(\prod_{j=1}^{K} p_j^{n_j} \right) \delta \left(\sum_{j=1}^{K} p_j - 1 \right). \tag{A.450}$$

To make progress, we introduce the Fourier representation of the delta function, so that, for example,

$$\mathcal{Z} = \int d^K \mathbf{p} \left(\prod_{j=1}^{K} p_j^{n_j} \right) \int \frac{d\lambda}{2\pi} \exp \left(+i\lambda \sum_{j=1}^{K} p_j - i\lambda \right) \tag{A.451}$$

$$= \int \frac{d\lambda}{2\pi} e^{-i\lambda} \prod_{j=1}^{K} \int dp_j \, p_j^{n_j} e^{i\lambda p_j}. \tag{A.452}$$

Because we have the delta function, we are free to let the integrals over p_j run from 0 to ∞; the delta function will enforce the constraint that $p_i \leq 1$ for all i. Then the key ingredient of the calculation is the integral

$$f(n; \lambda) = \int_0^\infty dp \, p^n e^{i\lambda p}. \tag{A.453}$$

At the end of the calculation we will have to integrate over λ. Let's assume that we will be able to deform the contour of this integral into the complex λ plane in such a way that the p integral in Eq. (A.453) is well behaved. Then we can write

$$f(n; \lambda) = \int_0^\infty dp \, p^n e^{i\lambda p} \tag{A.454}$$

$$= \int_0^\infty dp \, p^n e^{-(-i\lambda)p} = \frac{n!}{(-i\lambda)^{n+1}}. \tag{A.455}$$

Putting these pieces together, we have

$$\mathcal{Z} = \int \frac{d\lambda}{2\pi} e^{-i\lambda} \prod_{j=1}^{K} \frac{n_j!}{(-i\lambda)^{n_j+1}} \tag{A.456}$$

$$= \left(\prod_{j=1}^{K} n_j! \right) \int \frac{d\lambda}{2\pi} \frac{e^{-i\lambda}}{(-i\lambda)^{\sum_{j=1}^{K}(n_j+1)}} \tag{A.457}$$

$$= \left(\prod_{j=1}^{K} n_j! \right) \int \frac{d\lambda}{2\pi} \frac{e^{-i\lambda}}{(-i\lambda)^{N+K}}, \tag{A.458}$$

where $N = \sum_j n_j$ is the total number of samples, and as before K is the number of possible states. A similar argument gives

$$\hat{p}_i = \frac{1}{\mathcal{Z}} (n_i + 1)! \left(\prod_{j \neq i} n_j! \right) \int \frac{d\lambda}{2\pi} \frac{e^{-i\lambda}}{(-i\lambda)^{N+K+1}} \tag{A.459}$$

$$= \frac{(n_i + 1)! \left(\prod_{j \neq i} n_j! \right)}{\prod_{j=1}^{K} n_j!} \times \frac{\int \frac{d\lambda}{2\pi} \frac{e^{-i\lambda}}{(-i\lambda)^{N+K+1}}}{\int \frac{d\lambda}{2\pi} \frac{e^{-i\lambda}}{(-i\lambda)^{N+K}}} \tag{A.460}$$

$$= (n_i + 1) \frac{\int \frac{d\lambda}{2\pi} \frac{e^{-i\lambda}}{(-i\lambda)^{N+K+1}}}{\int \frac{d\lambda}{2\pi} \frac{e^{-i\lambda}}{(-i\lambda)^{N+K}}} . \tag{A.461}$$

Thus, $\hat{p}_i \propto n_i + 1$, so to get the normalization right, we must have

$$\hat{p}_i = \frac{n_i + 1}{N + K} . \tag{A.462}$$

This should be contrasted with the naive estimate of probabilities based on counting frequencies, $\hat{p}_i = n_i / N$. The Bayesian estimate, with a flat prior on the space of distributions, is equivalent to the naive approach but with one extra count in every bin. This estimate never predicts probability zero, even in states never observed to occur, and is in some sense smoother than the frequencies. The trick of adding such pseudocounts to the data goes back, it seems, to Laplace, although I don't think he had the full Bayesian justification.

Problem 198: Normalization. Derive Eq. (A.462) directly by doing the integrals in Eq. (A.461).

What do these ideas about distributions of distributions have to do with entropy estimation? Somewhat heuristically, it has been suggested that by using different numbers of pseudocounts, one can improve the quality of entropy estimation. More deeply, I think, the Bayesian estimate gives us a very different view of *why* we make systematic errors when trying to compute entropies from data. Recall that when using the naive identification of frequencies with probabilities, we underestimate the entropy, as in Eq. (A.444). It is tempting to think that we are underestimating the entropy simply because, in a finite sample, we have not seen all possibilities. With the Bayesian approach and a flat prior, however, the probability distributions being estimated are smoother than the true distribution, and correspondingly we expect that the entropy will be overestimated. In fact this is true, but the problem really is more serious than this.

Suppose that we do not yet have any data. Then all we know is that the probability distribution **p** will be chosen out of the distribution $\mathcal{P}(\mathbf{p})$. This seems innocuous, because this distribution is flat and hence presumably unbiased. But we can calculate the average entropy in this distribution,

$$\langle S \rangle_{\text{prior}} \equiv \int d^K \mathbf{p} \left(- \sum_{i=1}^{K} p_i \log_2 p_i \right) \mathcal{P}(\mathbf{p}), \tag{A.463}$$

using the same tricks used above, and we find

$$\langle S \rangle_{\text{prior}} = \psi_0(K + 1) - \psi_0(1), \tag{A.464}$$

where $\psi_0(x)$ is a polygamma function,

$$\psi_m(x) = \left(\frac{d}{dx}\right)^{m+1} \Gamma(x). \tag{A.465}$$

The details of the special functions are not so important. What is important is that, when the number of states K is large, we have

$$\langle S \rangle_{\text{prior}} = \log_2 K - \mathcal{O}(1). \tag{A.466}$$

Thus, although we are choosing distributions from a flat prior, the entropies of these distributions are biased toward the maximum possible value. This bias is actually very strong. The entropy is the average of many terms, and although these terms can't be completely independent (the probabilities must sum to one), one might expect the central limit theorem to apply here, in which case the fluctuations in the entropy will be $\sigma_S \sim 1/\sqrt{K}$, which for large K is very small indeed. Thus, the distributions chosen out of $\mathcal{P}(\mathbf{p})$ are overwhelmingly biased toward having nearly maximal entropy. Although the prior on the distributions is flat, the prior on entropies is narrowly concentrated around an average entropy which, for large K, is almost $\log_2 K$.

Problem 199: Entropies in a flat prior.   Derive the mean and standard deviation of the entropy in the flat prior, $\mathcal{P}(\mathbf{p})$, from Eq. (A.445). Verify Eq. (A.466).

Just to make the problem clear, suppose that our system has only two states, as with heads and tails for a coin. Let the probability of heads be q, so that the entropy is

$$S(q) = -q \log_2(q) - (1 - q) \log_2(1 - q). \tag{A.467}$$

If we assume that q is chosen from some distribution $\mathcal{P}(q)$, then the distribution of entropies can be found from

$$P(S)dS = \mathcal{P}(q)dq. \tag{A.468}$$

Because $dS/dq = 0$ at the point where $S = 1$ bit, the distribution $P(S)$ must be singular there unless the prior on q itself has a compensating singularity. Thus, a prior that is flat in q is strongly biased in S. The situation is even worse for systems with many states because of phase-space considerations. If we want to have a low entropy distribution, then many of the p_i must be confined to very small values, which means that the volume in $\mathbf{p}$ space associated with low entropy is small. Although only one distribution has precisely the maximum entropy, there are many distributions that are close.

Problem 200: A flat prior on S.   Show that, for the problem of coin flips, having a flat prior on the entropy S is equivalent to a prior

$$\mathcal{P}(q) = \left| \log_2 \left(\frac{q}{1-q} \right) \right|. \tag{A.469}$$

If we flip a coin N times and observe n heads, then Bayes' rule states that

$$\mathcal{P}_N(q|n) \propto \mathcal{P}(q) q^n (1-q)^{N-n}, \tag{A.470}$$

and we can use this to estimate the entropy:

$$\hat{S}(n, N) = \int_0^1 dq \, \mathcal{P}_N(q|n) \left[-q \log_2(q) - (1-q) \log_2(1-q) \right]. \tag{A.471}$$

(a) For $N = 10$, plot $\hat{S}(n, N)$ versus n. Compare your results with the naive estimate:

$$S_{\text{naive}}(n, N) = -\frac{n}{N} \log_2 \left(\frac{n}{N} \right) - \left(1 - \frac{n}{N} \right) \log_2 \left(1 - \frac{n}{N} \right). \tag{A.472}$$

(b) Suppose that you are actually flipping a coin in which the probability of heads is $q_{\text{true}} \neq 1/2$. Simulate N such flips, and use your results to estimate the entropy according to both Eq. (A.471) and Eq. (A.472). How do these estimators evolve as a function of N? Hint: Remember that we have seen the results for the naive estimator already; because this is a small system (only two states), the interesting behavior is at smaller N.

This discussion suggests that we could do a much better job of entropy estimation in a Bayesian framework where the $\mathcal{P}(\mathbf{p})$ is chosen to be flat is S. I don't know of anyone who has given a complete solution to this problem. A partial solution has been proposed by noticing that there is a well-known generalization of the flat prior, the Dirichlet family of priors,

$$\mathcal{P}_\beta(\mathbf{p}) = \frac{1}{Z(\beta)} \left(\prod_{i=1}^K p_i^{\beta-1} \right) \delta \left(\sum_{i=1}^K p_i - 1 \right). \tag{A.473}$$

Evidently the flat prior corresponds to $\beta = 1$ which is biased toward large entropies, as we have seen. As β decreases, the average entropy $\bar{S}(\beta)$ of a distribution drawn out of $\mathcal{P}_\beta(\mathbf{p})$ becomes smaller, but for each value of β the distribution of entropies remains quite narrow. Thus, if we form the prior

$$\mathcal{P}(\mathbf{p}) = \int_0^1 d\beta \left| \frac{d\bar{S}(\beta)}{d\beta} \right|^{-1} \mathcal{P}_\beta(\mathbf{p}), \tag{A.474}$$

it will be approximately flat in entropy. This approach seems to work, although it is computationally intensive. As far as I know, it gives the best results of any estimation procedure so far in, for example, the analysis of neural spike trains. If we dig into the integrals that define the entropy estimate, it turns out that the key pieces of data are coincidences, in which more than one sample falls into the same bin, and in this sense we seem to have found a generalization of Ma's ideas to nonuniform distributions.

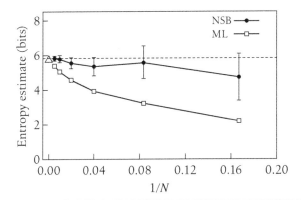

FIGURE A.15

Estimating entropies at one slice of time in the neural response to naturalistic stimuli. Neural responses are discretized with $\Delta\tau = 2$ ms resolution, and we look at 8-letter words. The stimulus is motion outdoors, and the motion is repeated many times. Here we focus on the distribution of responses at one moment relative to this repeat, for which we can collect up to 196 samples from the repetitions. The open symbols show the naive or maximum likelihood (ML) estimate, in which we identify the observed frequencies with probabilities and plug in to the computation of entropy. As expected, this estimate has a significant dependence on the number of samples but extrapolates smoothly according to Eq. (A.444). In contrast, the estimator based on the prior in Eq. (A.474), labeled the NSB estimator, remains constant within error bars, always agreeing with the extrapolation (filled symbols). Redrawn from Nemenman et al. (2004).

Again it is important to ask whether these ideas actually work with real data. In experiments on the motion–sensitive neuron H1 in the fly visual system (see Sections 4.4 and 6.2), we can in many cases collect enough data to sample the underlying distributions of neural responses, so we have ground truth. At the same time, we can look at only a small fraction of these data and ask how well our estimation procedure works. An example is shown in Fig. A.15.

Annotated Bibliography

1 Introduction

1.1 About Our Subject

If you are excited about problems at the interface of physics and biology, you must read Schrödinger's (1944) "little book" *What Is Life?*. To get a sense of the excitement and spirit of adventure that our intellectual ancestors brought to the subject, you should also look at the remarkable essays by Bohr (1933) and Delbrück (1949). Delbrück reflected on those early ideas some years later (1970), as did his colleagues and collaborators (Cairns et al. 1966). For a more professional history of the emergence of modern molecular biology from these physicists' musings, see Judson (1979). The remarkable essay about dictionaries (and so much more) is Wallace (2006). Feyerabend's (1975, 1978) books are classics. For some of the great classical physicists and their relation to the phenomena of life, see Helmholtz (1924–1925, 1954), portions of Rayleigh (1945), and Ohm (1843).

Bohr 1933. Light and life. N Bohr, *Nature* **131**, 421–423 (1933).

Cairns et al. 1966. *Phage and the Origins of Molecular Biology,* J Cairns, GS Stent, and JD Watson, eds. (Cold Spring Harbor Press, Cold Spring Harbor NY, 1966).

Delbrück 1949. A physicist looks at biology. M Delbrück, *Trans Conn Acad Arts Sci* **38**, 173–190 (1949). Reprinted in Cairns et al. (1966), pp. 9–22.

Delbrück 1970. A physicist's renewed look at biology: Twenty years later. M Delbrück, *Science* **168**, 1312–1315 (1970).

Feyerabend 1975. *Against Method. Outline of an Anarchistic Theory of Knowledge.* P Feyerabend (New Left Books, London, 1975). Revised editions in 1988 and 1993 (Verso, London).

Feyerabend 1978. *Science in a Free Society.* P Feyerabend (Verso, London, 1978).

Helmholtz 1924–1925. *Helmholtz' Treatise on Physiological Optics.* Translated from the third German edition of 1909–1911, by JPC Southall. HLF von Helmholtz (Optical Society of America, Rochester NY, 1924–1925).

Helmholtz 1954. *On the Sensations of Tone as a Physiological Basis for the Theory of Music.* Second English edition, translated from the fourth and last German edition of 1877, by AJ Ellis. HLF von Helmhotz (Dover, New York, 1984).

Judson 1979. *The Eighth Day of Creation.* HF Judson (Simon and Schuster, New York, 1979).

Ohm 1843. Über die Definition des Tones, nebst daran geknüpfter Theorie der Sirene und ähnlicher tonbildender Vorrichtungen. GS Ohm, *Ann Physik* **59**, 513–566 (1843).

Rayleigh 1945. *The Theory of Sound*. Second revised edition, with a historical introduction by RB Lindsay. JW Strutt, Baron Rayleigh (Dover, New York, 1945).

Schrödinger 1944. *What Is Life?* E Schrödinger (Cambridge University Press, Cambridge, 1944).

Wallace 2006. Authority and American usage, in *Consider the Lobster and Other Essays*. DF Wallace (Back Bay Books, New York, 2006). An earlier version appears as: Tense present: Democracy, English, and the wars over usage. *Harper's Magazine* April 2001, pp. 39–58.

1.2 About This Book

If the reference to "more is different" didn't ring a bell, by all means read Anderson (1972). I emphasized in the main text that a great deal of what we will be doing in this book is wrestling with rather raw phenomena in an attempt to extract some deeper physics questions. As different areas of physics have matured, this has become an increasingly rare activity, so it can be inspiring to look back. It is difficult to remember, for example, that cosmology as a branch of physics is quite young, and that phenomena that today constitute clear signatures of our understanding were once much more controversial; for perspective on this, see Peebles et al. (2009), and for a glimpse of the original struggles, see Peebles (1971). Anderson (1984) includes lecture notes written originally during the period when the phases of superfluid ^{3}He were being sorted out, and it again captures some of the flavor of raw theory-experiment interaction. For understanding as successive approximation, see Wilson (1993).

Anderson 1972. More is different. PW Anderson, *Science* **177,** 393–396 (1972).

Anderson 1984. *Basic Notions of Condensed Matter Physics*. PW Anderson (Benjamin Cummings, Menlo Park CA, 1984)

Peebles 1971. *Physical Cosmology*. PJE Peebles (Princeton University Press, Princeton NJ, 1971).

Peebles et al. 2009. *Finding the Big Bang*. PJE Peebles, LA Page, Jr, and RB Partridge, eds. (Cambridge University Press, Cambridge, 2009).

Wilson 1993. The renormalization group and critical phenomena. KG Wilson, in *Nobel Lectures, Physics 1981–1990*, T Grängsmyr and G Ekspang, eds., pp. 102–132 (World Scientific, Singapore, 1993). Also available at http://www.nobelprize.org.

Books in biophysics still are very personal efforts. There is an older but still interesting effort to present freshman physics with biological examples, and these wander into a wonderful array of topics (Benedek and Villars 1973–1979). Nelson (2004) provides both an introduction to biological problems for physicists and an introduction to central physics concepts for biologists, roughly at the junior/senior undergraduate level. Phillips et al. (2008) have provided a quantitative physics-style introduction to molecular and cellular biology, following the outline of classic biology books. Segel (1984), Nossal and Lecar (1991), Dayan and Abbott (2001), Boal (2002), and Dill and Bromberg (2003) provide systematic introductions to different slices of the biological world, from molecules to brains. Sneppen and Zocchi (2005), Alon (2007), and Frauenfelder (2010) provide more personal views of topics at the molecular and cellular level.

Alon 2007. *An Introduction to Systems Biology: Design Principles of Biological Circuits*. U Alon (Chapman and Hall/CRC, Boca Raton FL, 2007).

Benedek and Villars 1973–1979. *Physics, with Illustrative Examples from Medicine and Biology*, Volumes 1–3. GB Benedek and FM Villars (Addison-Wesley, Reading MA, 1973–1979).

Boal 2002. *Mechanics of the Cell*. DH Boal (Cambridge University Press, Cambridge, 2002). Second edition (2011).

Dayan and Abbott 2001. *Theoretical Neuroscience: Computational and Mathematical Modeling of Neural Systems.* P Dayan and LF Abbott (MIT Press, Cambridge MA, 2001).

Dill and Bromberg 2003. *Molecular Driving Forces: Statistical Thermodynamics in Chemistry and Biology.* KA Dill and S Bromberg (Garland Science, New York, 2003).

Frauenfelder 2010. *The Physics of Proteins.* H Frauenfelder, with SS Chan and WS Chan, eds. (Springer-Verlag, Berlin, 2010).

Nelson 2004. *Biological Physics: Energy, Information, Life.* P Nelson (WH Freeman, San Francisco, 2004).

Nossal and Lecar 1991. *Molecular and Cell Biophysics.* RJ Nossal and H Lecar (Addison-Wesley, Redwood City CA, 1991).

Phillips et al. 2008. *Physical Biology of the Cell.* R Phillips, J Kondev, and J Theriot (Garland Science, New York, 2008).

Segel 1984. *Modeling Dynamic Phenomena in Molecular and Cellular Biology.* LA Segel (Cambridge University Press, Cambridge, 1984).

Sneppen and Zocchi 2005. *Physics in Molecular Biology.* K Sneppen and G Zocchi (Cambridge University Press, Cambridge, 2005).

2 Photon Counting in Vision

2.1 A First Look

It is a pleasure to read classic papers, and surely Hecht et al. (1942) and van der Velden (1944) are classics, as is the discussion of dark noise by Barlow (1956). The prehistory of the subject, including the story about Lorentz, is covered by Bouman (1961). The general idea that our perceptual "thresholds" really are thresholds for discrimination against background noise with some criterion level of reliability made its way into quantitative psychophysical experiments in the 1950s and 1960s. It is now (happily) a standard part of experimental psychology; the canonical treatment is by Green and Swets (1966). The origins of these ideas are an interesting mix of physics and psychology, developed largely for radar during World War II; a summary of this early work is in the MIT Radar Lab series (Lawson and Uhlenbeck 1950). Another nice mix of physics and psychology is the revisiting of the original photon counting experiments using light sources with non-Poisson statistics (Teich et al. 1982). For a textbook account of coherent states, see Cohen-Tannoudji et al. (1978), and for the original, see Glauber (1963). The idea that random arrival of photons could limit our visual perception beyond the just visible was explored, early on, by de Vries (1943) and Rose (1948). Some of the early work by de Vries and co-workers on the physics of the sense organs (not just vision) is described in a lovely review (de Vries 1956). As a sociological note, de Vries was an experimental physicist with very broad interests, from biophysics to radiocarbon dating; for a short biography, see de Waard (1960).

Barlow 1956. Retinal noise and absolute threshold. HB Barlow, *J Opt Soc Am* **46,** 634–639 (1956).

Bouman 1961. History and present status of quantum theory in vision. MA Bouman, in *Sensory Communication*, W Rosenblith, ed., pp. 377–401 (MIT Press, Cambridge MA, 1961).

Cohen-Tannoudji et al. 1978. *Quantum Mechanics.* Two volumes. C Cohen-Tannoudji, B Diu and F Laloë (Wiley, New York, 1978).

Glauber 1963. Coherent and incoherent states of the radiation field. RJ Glauber, *Phys Rev* **131,** 2766–2788 (1963).

Green and Swets 1966. *Signal Detection Theory and Psychophysics.* DM Green and JA Swets (Wiley, New York, 1966).

Hecht et al. 1942. Energy, quanta and vision. S Hecht, S Shlaer, and MH Pirenne, *J Gen Physiol* **25,** 819–840 (1942).

Lawson and Uhlenbeck 1950. *Threshold Signals*. MIT Radiation Laboratory Series 24. JL Lawson and GE Uhlenbeck (McGraw-Hill, New York, 1950).

Rose 1948. The sensitivity performance of the human eye on an absolute scale. A Rose, *J Opt Soc Am* **38,** 196–208 (1948).

Teich et al. 1982. Multiplication noise in the human visual system at threshold. III. The role of non-Poisson quantum fluctuations. MC Teich, PR Prucnal, G Vannucci, ME Breton, and WJ McGill, *Biol Cybern* **44,** 157–165 (1982).

van der Velden 1944. Over het aantal lichtquanta dat nodig is voor een lichtprikkel bij het menselijk oog. HA van der Velden, *Physica* **11,** 179–189 (1944).

de Vries 1943. The quantum character of light and its bearing upon threshold of vision, the differential sensitivity and visual acuity of the eye. Hl de Vries, *Physica* **10,** 553–564 (1943).

de Vries 1956. Physical aspects of the sense organs. Hl de Vries, *Prog Biophys Biophys Chem* **6,** 207–264 (1956).

de Waard 1960. Hessel de Vries, physicist and biophysicist. H de Waard, *Science* **131,** 1720–1721 (1960).

Single-photon responses in receptor cells of the horseshoe crab were reported by Fuortes and Yeandle (1964). The papers from Baylor and co-workers (Baylor et al. 1979a,b, 1980, 1984) on single-photon responses in vertebrate rod cells, first from toads and then from monkeys, again are classics, well worth reading today, not least as examples of how to do quantitative experiments on biological systems. Aho, Donner, Reuter, and co-workers (Aho et al. 1987, 1988, 1993) have made a major effort to connect measurements on rod cells and ganglion cells with the behavior of the whole organism, using the frogs and toads as examples; among their results are the temperature dependence of dark noise (Fig. 2.8) and the latency/anticipation results in Section 2.4. The remarkable experiments showing that people really can count every photon are by Sakitt (1972). We will learn more about currents and voltages in cells in subsequent sections, starting with Section 2.3, but for background I have always liked Aidley's (1998) text, now in multiple editions; as is often the case, the earlier editions can be clearer and more compact. Even more compact is Katz (1966), which is also very beautifully written.

Aidley 1998. *The Physiology of Excitable Cells,* 4th edition. DJ Aidley (Cambridge University Press, Cambridge, 1998).

Aho et al. 1987. Retinal noise, the performance of retinal ganglion cells, and visual sensitivity in the dark-adapted frog. A-C Aho, K Donner, C Hydén, T Reuter, and OY Orlov, *J Opt Soc Am A* **4,** 2321–2329 (1987).

Aho et al. 1988. Low retinal noise in animals with low body temperature allows high visual sensitivity. A-C Aho, K Donner, C Hydén, LO Larsen, and T Reuter, *Nature* **334,** 348–350 (1988).

Aho et al. 1993. Visual performance of the toad (*Bufo bufo*) at low light levels: Retinal ganglion cell responses and prey-catching accuracy. A-C Aho, K Donner, S Helenius, LO Larsen, and T Reuter, *J Comp Physiol A* **172,** 671–682 (1993).

Baylor et al. 1979a. The membrane current of single rod outer segments. DA Baylor, TD Lamb, and K-W Yau, *J Physiol (Lond)* **288,** 589–611 (1979).

Baylor et al. 1979b. Rod responses to single photons. DA Baylor, TD Lamb, and K-W Yau, *J Physiol (Lond)* **288,** 613–634 (1979).

Baylor et al. 1980. Two components of electrical dark noise in toad retinal rod outer segments. DA Baylor, G Matthews, and K-W Yau, *J Physiol (Lond)* **309,** 591–621 (1980).

Baylor et al. 1984. The photocurrent, noise and spectral sensitivity of rods of the monkey *Macaca fascicularis*. DA Baylor, BJ Nunn, and JF Schnapf, *J Physiol (Lond)* **357,** 575–607 (1984).

Fuortes and Yeandle 1964. Probability of occurrence of discrete potential waves in the eye of *Limulus*. MGF Fuortes and S Yeandle, *J Gen Physiol* **47,** 443–463 (1964).

Katz 1966. *Nerve, Muscle and Synapse.* B Katz (McGraw-Hill, New York, 1966).

Sakitt 1972. Counting every quantum. B Sakitt, *J Physiol* **223,** 131–150 (1972).

For the discussion of compound eyes, useful background is contained in Stavenga and Hardie (1989), and in the beautiful compilation of insect brain anatomy by Strausfeld (1976), although this is hard to find; as an alternative there is an online atlas, described by Armstrong et al. (1995). There is also the more recent Land and Nilsson (2002). Although usually a reliable guide to such matters, Larson (2003) might lead you to think that multiple lenses generate multiple images, especially in the frightening example shown on his p. 584 of volume one. Everyone should have a copy of the Feynman lectures (Feynman et al. 1963); check the chapters on vision. The early work by Barlow (1952) deserves more appreciation, as noted in the main text, and the realization that diffraction must be important for insect eyes goes back to Mallock (1894). For a gentle introduction to the wider set of ideas about scaling relations between different body parts, see McMahon and Bonner (1983).

Armstrong et al. 1995. Flybrain, an online atlas and database for the *Drosophila* nervous system. JD Armstrong, K Kaiser, A Müller, K-F Fischbach, N Merchant, and NJ Strausfeld, *Neuron* **15,** 17–20 (1995). See http://flybrain.neurobio.arizona.edu/.

Barlow 1952. The size of ommatidia in apposition eyes. HB Barlow, *J Exp Biol* **29,** 667–674 (1952).

Feynman et al. 1963. *The Feynman Lectures on Physics.* RP Feynman, RB Leighton, and M Sands (Addison-Wesley, Reading MA, 1963).

Larson 2003. *The Complete Far Side,* two volumes. G Larson (Andrews McNeel, Kansas City MO, 2003).

Land and Nilsson 2002. *Animal Eyes.* MF Land and D-E Nilsson (Oxford University Press, Oxford, 2002).

Mallock 1894. Insect sight and the defining power of composite eyes. A Mallock, *Proc R Soc Lond* **55,** 85–90 (1894).

McMahon and Bonner 1983. *On Size and Life.* TA McMahon and JT Bonner (Scientific American Library, New York, 1983).

Stavenga and Hardie 1989. *Facets of Vision.* DG Stavenga and RC Hardie, eds. (Springer-Verlag, Berlin, 1989).

Strausfeld 1976. *Atlas of an Insect Brain.* NJ Strausfeld (Springer-Verlag, Berlin, 1976).

The experiments on signal-to-noise ratio in fly photoreceptors are by de Ruyter van Steveninck and Laughlin (1996a,b). For a review of relevant ideas in Fourier analysis and related matters, see Appendix A.2 and Lighthill (1958). You should come back to the ideas of Snyder and co-workers (Snyder 1977; Snyder et al. 1977) in Chapter 6, after we have covered some of the basics of information theory.

Lighthill 1958. *Introduction to Fourier Analysis and Generalized Functions.* MJ Lighthill (Cambridge University Press, Cambridge, 1958).

de Ruyter van Steveninck and Laughlin 1996a. The rate of information transfer at graded-potential synapses. R de Ruyter van Steveninck and SB Laughlin, *Nature* **379,** 642–645 (1996).

de Ruyter van Steveninck and Laughlin 1996b. Light adaptation and reliability in blowfly photoreceptors. R de Ruyter van Steveninck and SB Laughlin, *Int J Neural Syst* **7**, 437–444 (1996)

Snyder 1977. Acuity of compound eyes: Physical limitations and design. AW Snyder, *J Comp Physiol* **116**, 161–182 (1977).

Snyder et al. 1977. Information capacity of compound eyes. AW Snyder, DS Stavenga, and SB Laughlin, *J Comp Physiol* **116**, 183–207 (1977).

Finally, the following three reviews place the results on photon counting into a broader context.

Barlow 1981. Critical limiting factors in the design of the eye and visual cortex. HB Barlow, *Proc R Soc Lond B* **212**, 1–34 (1981).

Bialek 1987. Physical limits to sensation and perception. W Bialek, *Annu Rev Biophys Biophys Chem* **16**, 455–478 (1987).

Rieke and Baylor 1998. Single-photon detection by rod cells of the retina. F Rieke and DA Baylor, *Rev Mod Phys* **70**, 1027–1036 (1998).

2.2 Dynamics of Single Molecules

To get some of the early history of work on the visual pigments, one can do worse than to read Wald's Nobel lecture (Wald 1972). Wald himself (along with his wife and collaborator Ruth Hubbard) was quite an interesting fellow, much involved in politics; to connect with the previous section, his PhD adviser was Selig Hecht. For a measurement of dark noise in cones, see Sampath and Baylor (2002), and for the temperature dependence of sensitivity (Fig. 2.23), see Luo et al. (2011). The remarkable result that the quantum yield of fluorescence in rhodopsin is $\sim 10^{-5}$ is due to Doukas et al. (1984); it is worth noting that measuring this small quantum yield was possible at a time when one could not directly observe the ultrafast processes that are responsible for making the branching ratio this small. Direct measurements were finally made by Mathies et al. (1988), Schoenlein et al. (1991), and Wang et al. (1994), the last paper making clear that the initial events are quantum mechanically coherent. A detailed analysis of the Raman spectra of rhodopsin has been done by Loppnow and Mathies (1988). The question of how receptor cells choose among multiple rhodopsins has received renewed attention in the fly's eye (Sood et al. 2012).

Doukas et al. 1984. Fluorescence quantum yield of visual pigments: Evidence for subpicosecond isomerization rates. AG Doukas, MR Junnarkar, RR Alfano, RH Callender, T Kakitani, and B Honig, *Proc Natl Acad Sci (USA)* **81**, 4790–4794 (1984).

Loppnow and Mathies 1988. Excited-state structure and isomerization dynamics of the retinal chromophore in rhodopsin from resonance Raman intensities. GR Loppnow and RA Mathies, *Biophys J* **54**, 35–43 (1988).

Luo et al. 2011. Activation of visual pigments by light and heat. DG Luo, WW Yue, P Ala-Lauria, and K-W Yau, *Science* **332**, 1307–1312 (2011).

Mathies et al. 1988. Direct observation of the femtosecond excited-state cis-trans isomerization in bacteriorhodopsin. RA Mathies, CH Brito Cruz, WT Pollard, and CV Shank, *Science* **240**, 777–779 (1988).

Sampath and Baylor 2002. Molecular mechanisms of spontaneous pigment activation in retinal cones. AP Sampath and DA Baylor, *Biophys J* **83**, 184–193 (2002).

Schoenlein et al. 1991. The first step in vision: Femtosecond isomerization of rhodopsin. RW Schoenlein, LA Peteanu, RA Mathies, and CV Shank, *Science* **254**, 412–415 (1991).

Sood et al. 2012. Stochastic de-repression of rhodopsins in single photoreceptors of the fly retina. P Sood, RJ Johnston, Jr, and E Kussell, *PLoS Comput Biol* **8**, e1002357 (2012).

Wald 1972. The molecular basis of visual excitation. G Wald, in *Nobel Lectures: Physiology or Medicine 1963–1970,* J Lindsten, ed., pp. 292–315 (Elsevier, Amsterdam, 1972). Also available at http://www.nobelprize.org.

Wang et al. 1994. Vibrationally coherent photochemistry in the femtosecond primary event of vision. Q Wang, RW Schoenlein, LA Peteanu, RA Mathies, and CV Shank, *Science* **266,** 422–424 (1994).

The Born-Oppenheimer approximation is discussed in almost all quantum mechanics textbooks. For a collection of the key papers, with commentary, on the rich phenomena that can emerge in such adiabatic approximations, see Shapere and Wilczek (1989). Models for coupling of electron hopping to bond stretching (as in Problem 24) were explored by Su et al. (1980) in relation to polyacetylene. Importantly, these models predict that the excitations (e.g., upon photon absorption) are not just electrons and holes in the usual ladder of molecular orbitals, but that there are localized, mobile objects with unusual quantum numbers. These mobile objects can be generated by doping, which is the basis for conductivity in these quasi-one-dimensional materials. The original work is in Su et al. (1980); a good review is Heeger et al. (1988). Many people must have realized that the dynamical models being used by condensed matter physicists for (ideally) infinite chains might also have something to say about finite chains. For ideas in this direction, including some specifically relevant to rhodopsin, see Bialek et al. (1987), Vos et al. (1996), and Aalberts et al. (2000).

Aalberts et al. 2000. Quantum coherent dynamics of molecules: A simple scenario for ultrafast photoisomerization. DP Aalberts, MSL du Croo de Jongh, BF Gerke, and W van Saarloos, *Phys Rev A* **61,** 040701 (2000).

Heeger et al. 1988. Solitons in conducting polymers. AJ Heeger, S Kivelson, JR Schrieffer, and W–P Su, *Rev Mod Phys* **60,** 781–850 (1988).

Bialek et al. 1987. Simple models for the dynamics of biomolecules: How far can we go? W Bialek, RF Goldstein, and S Kivelson, in *Structure, Dynamics and Function of Biomolecules: The First EBSA Workshop,* A Ehrenberg, R Rigler, A Gräslund, and LJ Nilsson, eds., pp. 65–69 (Springer-Verlag, Berlin, 1987).

Shapere and Wilczek 1989. *Geometric Phases in Physics.* A Shapere and F Wilczek (World Scientific, Singapore, 1989).

Su et al. 1980. Soliton excitations in polyacetylene. W-P Su, JR Schrieffer, and AJ Heeger, *Phys Rev B* **22,** 2099–2111 (1980).

Vos et al. 1996. Su-Schrieffer-Heeger model applied to chains of finite length. FLJ Vos, DP Aalberts, and W van Saarloos, *Phys Rev B* **53,** 14922–14928 (1996).

2.3 Biochemical Amplification

General reviews of the cGMP cascade in rods are given by Burns and Baylor (2001) and Arshavsky et al. (2002). An early but very thorough account of the kinetics of the cascade was given by Pugh and Lamb (1993). Rieke and Baylor (1996) set out to understand the origins of the continuous noise in rods, but along the way they provide a beautifully quantitative dissection of the enzymatic cascade. For an explanation of how similarity to rhodopsin (and other G-protein coupled receptors) drove the discovery of the olfactory receptors, see Buck (2005). For some general background on ion channels, you can try Aidley (1998) (see the Annotated Bibliography for Section 2.1), Johnston and Wu (1995), or Hille (2001). A starting point for learning about how different choices of channels shape the dynamics of responses in insect photoreceptors is the review by Weckström and Laughlin (1995).

Arshavsky et al. 2002. G proteins and phototransduction. VY Arshavsky, TD Lamb, and EN Pugh, Jr, *Annu Rev Physiol* **64,** 153–187 (2002).

Buck 2005. Unraveling the sense of smell. LB Buck, *Angew Chem Int Ed Engl* **44**, 6128–6140 (2005). Also available at http://www.nobelprize.org.

Burns and Baylor 2001. Activation, deactivation and adaptation in vertebrate photoreceptor cells. ME Burns and DA Baylor, *Annu Rev Neurosci* **24,** 779–805 (2001).

Hille 2001. *Ion Channels of Excitable Membranes,* 3rd edition. B Hille (Sinuaer, Sunderland MA, 2001).

Johnston and Wu 1995. *Foundations of Cellular Neurophysiology.* D Johnston and SM Wu (MIT Press, Cambridge MA, 1995).

Pugh and Lamb 1993. Amplification and kinetics of the activation steps in phototransduction. EN Pugh, Jr, and TD Lamb, *Biochim Biophys Acta* **1141,** 111–149 (1993).

Rieke and Baylor 1996. Molecular origin of continuous dark noise in rod photoreceptors. F Rieke and DA Baylor, *Biophys J* **71**, 2553–2572 (1996).

Weckström and Laughlin 1995. Visual ecology and voltage gated ion channels in insect photoreceptors. M Weckström and SB Laughlin, *Trends Neurosci* **18,** 17–21 (1995).

Rieke and Baylor (1998a) provide a review of photon counting in rods with many interesting observations, including an early outline of the problem of reproducibility. An early effort to analyze the signals and noise in enzymatic cascades is by Detwiler et al. (2000). The idea that restricted, saturable domains can arise dynamically and tame the fluctuations in the output of the cascade is described by the same authors (Ramanathan et al. 2005). For invertebrate photoreceptors, it seems that reproducibility of the response to single photons can be traced to positive feedback mechanisms that generate a stereotyped pulse of concentration changes, localized to substructures analogous to the disks in vertebrate rods (Pumir et al. 2008). A textbook account of enzyme kinetics is given by Fersht (1977).

Detwiler et al. 2000. Engineering aspects of enzymatic signal transduction: Photoreceptors in the retina. PB Detwiler, S Ramanathan, A Sengupta, and BI Shraiman, *Biophys J* **79,** 2801–2817 (2000).

Fersht 1977. *Enzyme Structure and Mechanism.* AR Ferhst (WH Freeman, San Francisco, 1977).

Pumir et al. 2008. Systems analysis of the single photon response in invertebrate photoreceptors. A Pumir, J Graves, R Ranganathan, and BI Shraiman, *Proc Natl Acad Sci (USA)* **105,** 10354–10359 (2008).

Ramanathan et al. 2005. G-protein-coupled enzyme cascades have intrinsic properties that improve signal localization and fidelity. S Ramanathan, PB Detwiler, AM Sengupta, and BI Shraiman, *Biophys J* **88,** 3063–3071 (2005).

Rieke and Baylor 1998a. Single-photon detection by rod cells of the retina. F Rieke and DA Baylor, *Rev Mod Phys* **70,** 1027–1036 (1998).

The classic paper on Hill functions for cooperative binding is Hill (1910). There is some suggestion that Hill might have been the first to derive the simpler description of independent binding, often called the Langmuir isotherm; for this and more related history as seen through the lens of drug-receptor interactions, see Colquhoun (2006). The MWC model is due to Monod et al. (1965), and a contemporary competing model is due to Koshland et al. (1966). Late in his life, Perutz (1990) provided some perspective on his long adventure with hemoglobin. A key step in understanding was to show, convincingly, that there really is no direct interaction between the binding sites, and the cooperativity was mediated entirely by the shifting equilibrium between the R and T states (Shulman et al. 1975). The MWC model leaves open the question of where the energy for cooperativity is stored in the molecule; for a hypothesis very much ahead of its time, see Hopfield (1973).

Colquhoun 2006. The quantitative analysis of drug-receptor interactions: A short history. D Colquhoun, *Trends Pharm Sci* **27,** 149–157 (2006).

Hill 1910. The possible effects of the aggregation of the molecules of haemoglobin on its dissociation curves. AV Hill, *J Physiol (Lond)* **40,** Suppl iv–vii (1910).

Hopfield 1973. Relation between structure, co-operativity and spectra in a model of hemoglobin action. JJ Hopfield, *J Mol Biol* **77,** 207–222 (1973).

Koshland et al. 1966. Comparison of experimental binding data and theoretical models in proteins containing subunits. DE Koshland, Jr, G Némethy, and D Filmer, *Biochemistry* **5,** 365–385 (1966).

Monod et al. 1965. On the nature of allosteric transitions: A plausible model. J Monod, J Wyman, and JP Changeux, *J Mol Biol* **12,** 88–118 (1965).

Perutz 1990. *Mechanisms of Cooperativity and Allosteric Regulation in Proteins.* MF Perutz (Cambridge University Press, Cambridge, 1990).

Shulman et al. 1975. Allosteric interpretation of haemoglobin properties. RG Shulman, JJ Hopfield, and S Ogawa, *Q Rev Biophys* **8,** 325–420 (1975).

One of the early systematic efforts to test different models of reproducibility was by Rieke and Baylor (1998b), which pointed to the importance of multiple steps in the rhodopsin shut-off as a likely scenario. Many of the same ideas were revisited in mammalian rods by Field and Rieke (2002), setting the stage for experiments on genetic engineering of the phosphorylation sites by Doan et al. (2006); the path to identifying the multiple steps with the multiple phosphorylation events was not uncontroversial (Hamer et al. 2003). More recent work from the same group explores the competition between the kinase and the arrestin molecule, which binds to the phosphorylated rhodopsin to terminate its activity, showing that this competition influences both the mean and the variability of the single-photon response (Doan et al. 2009). For a discussion of signal-to-noise ratios in multistate systems, inspired in part by the case of rhodopsin, see Escola et al. (2009), which is related to Problems 30 and 31.

Doan et al. 2006. Multiple phosphorylation sites confer reproducibility of the rod's single-photon responses. T Doan, A Mendez, PB Detwiler, J Chen, and F Rieke, *Science* 313, 530–533 (2006).

Doan et al. 2009. Arrestin competition influences the kinetics and variability of the single-photon responses of mammalian rod photoreceptors. T Doan, AW Azevedo, JB Hurley, and F Rieke, *J Neurosci* **29,** 11867–11879 (2009).

Escola et al. 2009. Maximally reliable Markov chains under energy constraints. S Escola, M Eisele, K Miller, and L Paninski, *Neural Comp* **21,** 1863–1912 (2009).

Field and Rieke 2002. Mechanisms regulating variability of the single photon responses of mammalian rod photoreceptors. GD Field and F Rieke, *Neuron* **35,** 733–747 (2002b).

Hamer et al. 2003. Multiple steps of phosphorylation of activated rhodopsin can account for the reproducibility of vertebrate rod single-photon responses. RD Hamer, SC Nicholas, D Tranchina, PA Liebman, and TD Lamb, *J Gen Physiol* **122,** 419–444 (2003).

Rieke and Baylor 1998b. Origins of reproducibility in the responses of retinal rods to single photons. F Rieke and DA Baylor, *Biophys J* **75,** 1836–1857 (1998).

2.4 The First Synapse and Beyond

For a general overview of the retina, a good source is Dowling (1987, 2012). For the experiments on nonlinear summation at the rod-bipolar synapse, along with a discussion of the theoretical issues of noise and reliability, see Field and Rieke (2002). The analysis of optimal filtering is presented in Bialek and Owen (1990) and Rieke et al. (1991). For a discussion of how our experience of a dark night translates into photons per rod per second, see Walraven et al. (1990).

Bialek and Owen 1990. Temporal filtering in retinal bipolar cells: Elements of an optimal computation? W Bialek and WG Owen, *Biophys J* **58,** 1227–1233 (1990).

Dowling 1987, 2012. *The Retina: An Approachable Part of the Brain.* JE Dowling (Harvard University Press, Cambridge MA, 1987). Revised edition (2012).

Field and Rieke 2002. Nonlinear signal transfer from mouse rods to bipolar cells and implications for visual sensitivity. GD Field and F Rieke, *Neuron* **34,** 773–785 (2002).

Rieke et al. 1991. Optimal filtering in the salamander retina. F Rieke, WG Owen, and W Bialek, in *Advances in Neural Information Processing 3,* R Lippman, J Moody, and D Touretzky, eds., pp. 377–383 (Morgan Kaufmann, San Mateo CA, 1991).

Walraven et al. 1990. The control of visual sensitivity. J Walraven, C Enroth-Cugell, DC Hood, DIA MacLeod, and JL Schnapf, in *Visual Perception: The Neurophysiological Foundations*, L Spillmann and SJ Werner, eds., pp. 53–101 (Academic Press, San Diego, 1990).

The classic presentations of filtering, estimation, and prediction are by Kolmogorov (1939, 1941) and Wiener (1949). The long Problem 27, about optimal filtering, is based on Potters and Bialek (1994). My own thinking about these matters was very much influenced by the unpublished work of Lackner and Zweig (1988).

Kolmogoroff 1939. Sur l'interpolation et extrapolations des suites stationnaires. A Kolmogoroff, *C R Acad Sci Paris* **208,** 2043–2045 (1939).

Kolmogorov 1941. Interpolation and extrapolation of stationary random sequences (in Russian). AN Kolmogorov, *Izv Akad Nauk USSR Ser Mat* **5,** 3–14 (1941). English translation in *Selected Works of AN Kolmogorov,* Volume II, AN Shiryagev, ed., pp. 272–280 (Kluwer Academic, Dordrecht, 1992).

Lackner and Zweig 1988. Approximating functions from measured values and prior knowledge. KS Lackner and G Zweig, unpublished manuscript (1988).

Potters and Bialek 1994. Statistical mechanics and visual signal processing. M Potters and W Bialek, *J Phys I France* **4,** 1755–1775 (1994); arXiv:cond-mat/9401072 (1994).

Wiener 1949. *Extrapolation, Interpolation and Smoothing of Time Series.* N Wiener (Wiley, New York, 1949).

The idea of maximizing information transmission across the first visual synapse is something we discuss at greater length in Chapter 4. Still, you might like to look ahead, so here are three references for how these ideas developed in the context of fly vision.

van Hateren 1992. Real and optimal neural images in early vision. JH van Hateren, *Nature* **360,** 68–70 (1992).

Laughlin 1981. A simple coding procedure enhances a neuron's information capacity. SB Laughlin, *Z Naturforsch* **36c,** 910–912 (1981).

Srinivasan et al. 1982. Predictive coding: A fresh view of inhibition in the retina. MV Srinivasan, SB Laughlin, and A Dubs, *Proc R Soc Lond B* **216,** 427–459 (1982).

The classic paper about single-photon responses in retinal ganglion cells is Barlow et al. (1971); it has quite a lot of detail and still makes great reading. Modern efforts for recording simultaneously from many ganglion cells include Litke et al. (2004), Segev et al. (2004), and Amodei (2011). The idea that single molecular events can drive bursts, generating non-Poisson statistics, reappears 30 years later in the context of gene expression; see, for example, Ozbudak et al. (2002). The early papers on intensity discrimination using spikes from single neurons are Barlow (1965) and Barlow and Levick (1969); see also the even earlier works of FitzHugh (1957, 1958).

Amodei 2011. *Network-Scale Electrophysiology: Measuring and Understanding the Collective Behavior of Neural Circuits* (Doctoral dissertation, Princeton University, Princeton NJ, 2011).

Barlow 1965. Optic nerve impules and Weber's law. HB Barlow, *Cold Spring Harb Symp Quant Biol* **30**, 539–546 (1965).

Barlow and Levick 1969. Three factors limiting the reliable detection of light by retinal ganglion cells of the cat. HB Barlow and WR Levick, *J Physiol (Lond)* **200**, 1–24 (1969).

Barlow et al. 1971. Responses to single quanta of light in retinal ganglion cells of the cat. HB Barlow, WR Levick, and M Yoon, *Vision Res Suppl* **3**, 87–101 (1971).

FitzHugh 1957. The statistical detection of threshold signals in the retina. R FitzHugh, *J Gen Physiol* **40**, 925–948 (1957).

FitzHugh 1958. A statistical analyzer for optic nerve messages. R FitzHugh, *J Gen Physiol* **41**, 675–692 (1958).

Litke et al. 2004. What does the eye tell the brain?: Development of a system for the large scale recording of retinal output activity. AM Litke, N Bezayiff, EJ Chichilnisky, W Cunningham, W Dabrowski, AA Grillo, M Grivich, P Grybos, P Hottowy, S Kachiguine, RS Kalmar, K Mathieson, D Petrusca, M Rahman, and A Sher, *IEEE Trans Nucl Sci* **51**, 1434–1440 (2004).

Ozbudak et al. 2002. Regulation of noise in the expression of a single gene. E Ozbudak, M Thattai, I Kurtser, AD Grossman, and A van Oudenaarden, *Nat Genet* **31**, 69–73 (2002).

Segev et al. 2004. Recording spikes from a large fraction of the ganglion cells in a retinal patch. R Segev, J Goodhouse, JL Puchalla, and MJ Berry, II, *Nat Neurosci* **7**, 1155–1162 (2004).

The observation that neurons gradually diminish their response to constant stimuli goes back to Adrian's (1928) first experiments recording the spikes from single cells; he immediately saw the connection to the fading of our perceptions when inputs are constant, and this sort of direct mapping from neural responses to human experience is now the common language we use in thinking about the brain and mind. Adaptation in cones and its connection to the enzymatic cascade was discussed in a remarkable series of papers by Baylor, Hodgkin, and Lamb (Baylor and Hodgkin 1974; Baylor et al. 1974a,b). An early paper about adaptation to the distribution of inputs is Smirnakis et al. (1997). Since then a number of papers have explored more complex versions of this adaptation, as well as trying to tease apart the underlying mechanisms; some examples are Kim and Rieke (2001, 2003), Rieke (2001), Baccus and Meister (2002), and Olveczky et al. (2007).

Adrian 1928. *The Basis of Sensation.* ED Adrian (Christoper's, London, 1928).

Baccus and Meister 2002. Fast and slow adaptation in retinal circuitry. SA Baccus and M Meister, *Neuron* **36**, 909–919 (2002).

Baylor and Hodgkin 1974. Changes in time scale and sensitivity in turtle photoreceptors. DA Baylor and AL Hodgkin, *J Physiol (Lond)* **242**, 729–758 (1974).

Baylor et al. 1974a. The electrical response of turtle cones to flashes and steps of light. DA Baylor, AL Hodgkin, and TD Lamb, *J Physiol (Lond)* **242**, 685–727 (1974).

Baylor et al. 1974b. Reconstruction of the electrical responses of turtle cones to flashes and steps of light. DA Baylor, AL Hodgkin, and TD Lamb, *J Physiol (Lond)* **242**, 759–791 (1974).

Kim and Rieke 2001. Temporal contrast adaptation in the input and output signals of salamander retinal ganglion cells. KJ Kim and F Rieke, *J Neurosci* **21**, 287–299 (2001).

Kim and Rieke 2003. Slow Na$^+$ inactivation and variance adaptation in salamander retinal ganglion cells. KJ Kim and F Rieke, *J Neurosci* **23**, 1506–1515 (2003).

Olveczky et al. 2007. Retinal adaptation to object motion. BP Olveczky, SA Baccus, and M Meister, *Neuron* **56**, 689–700 (2007).

Rieke 2001. Temporal contrast adaptation in salamander bipolar cells. F Rieke, *J Neurosci* **21**, 9445–9454 (2001).

Smirnakis et al. 1997. Adaptation of retinal processing to image contrast and spatial scale. S Smirnakis, MJ Berry, II, DK Warland, W Bialek, and M Meister, *Nature* **386**, 69–73 (1997).

There is a decent demonstration of the Pulfrich effect available on the web (Newbold 1999). The experiments on reaction times in toads and the connection to retinal delays are from the work of Aho et al. (1993).

Aho et al. 1993. Visual performance of the toad (*Bufo bufo*) at low light levels: Retinal ganglion cell responses and prey-catching accuracy. A-C Aho, K Donner, S Helenius, LO Larsen, and T Reuter, *J Comp Physiol A* **172**, 671–682 (1993).

Newbold 1999. The Pulfrich illusion. M Newbold. Available at http://dogfeathers.com/java/pulfrich.html (1999).

3 Lessons, Problems, Principles

For recent views of evolutionary dynamics from a physicist's perspective, see Goldenfeld and Woese (2011) and Neher and Shraiman (2011). The classic reference on distributed relaxation in adaptation is Thorson and Biederman-Thorson (1974); related ideas about multiple time scales in synaptic plasticity, motivated by the problems of memory storage, were introduced by Fusi and Abbott (2007). For an early review that wrestled with generality versus particularity in nonequilibrium dynamics, see Langer (1980). For one view on the origins of life, see Dyson (1985); he may not be right, but he is eloquent and interesting, here as always, and he provides a short history of relevant ideas.

Dyson 1985. *Origins of Life.* F Dyson (Cambridge University Press, Cambridge, 1985).

Fusi and Abbott 2007. Limit on the storage capacity of bounded synapses. S Fusi and LF Abbott, *Nat Neurosci* **10**, 485–493 (2007).

Goldenfeld and Woese 2011. Life is physics: Evolution as a collective phenomenon far from equilibrium. N Goldenfeld and C Woese, *Annu Rev Cond Matt Phys* **2**, 375–399 (2011).

Langer 1980. Instabilities and pattern formation in crystal growth. JS Langer, *Rev Mod Phys* **52**, 1–28 (1980).

Neher and Shraiman 2011. Statistical genetics and evolution of quantitative traits. RA Neher and BI Shraiman, *Rev Mod Phys* **83**, 1283–1300 (2011).

Thorson and Biederman-Thorson 1974. Distributed relaxation processes in sensory adaptation. J Thorson and M Biederman-Thorson, *Science* **183**, 161–172 (1974).

4   Noise Is Not Negligible

4.1   Fluctuations and Chemical Reactions

If you need a review of the Langevin equation, I like the treatment in the little book by Kittel (1958), as well as the somewhat longer discussion by Pathria (1972). The original discussion of diffusion (even with inertia) over a barrier is due to Kramers (1940); for a modern perspective, see Hänggi et al. (1990).

Hänggi et al. 1990. Reaction-rate theory: Fifty years after Kramers. P Hänggi, P Talkner, and M Borkovec, *Rev Mod Phys* **62**, 251–341 (1990).

Kittel 1958. *Elementary Statistical Physics.* C Kittel (Wiley, New York, 1958).

Kramers 1940. Brownian motion in a field of force and the diffusion model of chemical reactions. HA Kramers *Physica* **7**, 284–304 (1940).

Pathria 1972. *Statistical Mechanics.* RK Pathria (Pergamon, Oxford, 1972).

Myoglobin was the first protein whose structure was solved by X-ray diffraction. Aspects of X-ray analysis are described in Appendix A.3. For a perspective on myoglobin, see Kendrew (1964). The experiments on myoglobin are by Austin et al. (1975), which touched off a huge follow-up literature. A clear discussion of the interplay between the reaction coordinate and a protein coordinate was given by Agmon and Hopfield (1983). The demonstration of tunneling in this system is by Alberding et al. (1976).

Agmon and Hopfield 1983. Transient kinetics of chemical reactions with bounded diffusion perpendicular to the reaction coordinate. N Agmon and JJ Hopfield, *J Chem Phys* **78**, 6947–6959 (1983).

Alberding et al. 1976. Tunneling in ligand binding to heme proteins. N Alberding, RH Austin, KW Beeson, SS Chan, L Eisenstein, H Frauenfelder, and TM Nordlund, *Science* **192**, 1002–1004 (1976).

Austin et al. 1975. Dynamics of ligand binding to myoglobin. RH Austin, KW Beeson, L Eisenstein, H Frauenfelder, and IC Gunsalus, *Biochemistry* **14**, 5355–5373 (1975).

Kendrew 1964. Myoglobin and the structure of proteins. JC Kendrew, in *Nobel Lectures in Chemistry 1942–1962*, pp. 676–698 (Elsevier, Amsterdam, 1964). Also available at http://nobelprize.org.

Classical overviews of the photosynthetic reaction center are provided by Feher and Okamura (1978) and Okamura et al. (1982). As with many biological molecules, many questions about the reaction center were sharpened once the structure was determined at atomic resolution (Deisenhoffer et al. 1984); this work was important also as a demonstration that one could use the classical methods of X-ray crystallography (see Appendix A.3) for proteins that are normally embedded in membranes. It should be emphasized, however, that the electron-transfer reactions leave an enormous variety of spectroscopic signatures— separating charges not only changes optical properties of the molecules, it also generates unpaired spins that can be seen using electron paramagnetic resonance (EPR), and the distribution of the spin across multiple atoms at the donor and acceptor sites can be mapped using electron-nuclear double resonance (ENDOR). An early view of the uses of EPR and ENDOR in biological systems is given by Feher (1970); this article appears in the proceedings of the first Les Houches physics summer school to be devoted to questions at the interface with biology. For a synthesis of structural and spectroscopic data in relation to function, see Feher et al. (1989).

Deisenhoffer et al. 1984. X-ray structure analysis of a membrane protein complex: Electron density map at 3 Å resolution and a model of the chromophores of the photosynthetic reaction center from *Rhodopseudomonas viridis*. J Deisenhoffer, O Epp, K Miki, R Huber, and H Michel, *J Mol Biol* **180,** 385–398 (1984).

Feher 1970. Electron paramagnetic resonance with applications to selected problems in biology. G Feher, in *Physical Problems in Biological Systems*, C DeWitt and J Matricon, eds., pp. 251–365 (Gordon and Breach, Paris, 1970).

Feher and Okamura 1978. Chemical composition and properties of reaction centers. G Feher and MY Okamura, in *The Photosynthetic Bacteria*, RK Clayton and WR Sistrom, eds., pp. 349–386 (Plenum, New York, 1978).

Feher et al. 1989. Structure and function of reaction centers. G Feher, JP Allen, MY Okamura, and DC Rees, *Nature* **339,** 111–116 (1989).

Okamura et al. 1982. Reaction centers. MY Okamura, G Feher, and N Nelson, in *Photosynthesis: Energy Conversion by Plants and Bacteria*, Volume 1, Govindjee, ed., pp. 195–272 (Academic Press, New York, 1982).

The original experiments that provided evidence for tunneling in photosynthetic electron transfer were done by DeVault and Chance (1966) on samples that were a bit messier than the purified reaction centers that emerged in subsequent years. The kinetics of the initial charge-separation reactions became accessible as picosecond spectroscopy matured; for a modern overview, see Blankenship (2002). The modern view of biological electron-transfer reactions, including the role of tunneling in the vibrational degrees of freedom, is due to Hopfield (1974). Exploration of the energy gap dependence of reaction rates was pioneered by Gunner et al. (1986), and the evidence for frozen distributions of electron transfer rates was provided by Kleinfeld et al. (1984). For a review of efforts to calculate electronic matrix elements in real protein structures, see Onuchic et al. (1992), and for a more recent view of the role of structural fluctuations in electron transfer, see Skourtis et al. (2010).

Blankenship 2002. *Molecular Mechanisms of Photosynthesis*. RE Blankenship (Blackwell Science, Oxford, 2002).

DeVault and Chance 1966. Studies of photosynthesis using a pulsed laser. I. Temperature dependence of cytochrome oxidation in *Chromatium*. Evidence for tunneling. D DeVault and B Chance, *Biophys J* **6,** 825–847 (1966).

Gunner et al. 1986. Kinetic studies on the reaction center protein from *Rhodopseudomonas sphaeroides*: The temperature and free energy dependence of electron transfer between various quinones in the Q_A site and the oxidized bacteriochlorophyll dimer. MR Gunner, DE Robertson, and PL Dutton, *J Phys Chem* **90,** 3783–3795 (1986).

Hopfield 1974. Electron transfer between biological molecules by thermally activated tunneling. JJ Hopfield, *Proc Natl Acad Sci (USA)* **71,** 3640–3644 (1974).

Kleinfeld et al. 1984. Electron-transfer kinetics in photosynthetic reaction centers cooled to cryogenic temperatures in charge-separated state: evidence for light-induced structural changes. D Kleinfeld, MY Okamura, and G Feher, *Biochemistry* **23,** 5780–5786 (1984).

Onuchic et al. 1992. Pathway analysis of protein electron-transfer reactions. JN Onuchic, DN Beratan, JR Winkler, and HB Gray, *Annu Rev Biophys Biomol Struct* **21,** 349–377 (1992).

Skourtis et al. 2010. Fluctuations in biological and bioinspired electron-transfer reactions. SS Skourtis, DH Waldeck, and DN Beratan, *Annu Rev Phys Chem* **61,** 461–485 (2010).

For an introduction to the interesting physics in the very fast initial events of photosynthesis, see Fleming and van Grondelle (1994). The first experiments hinting at coherence in this reaction looked at systems that had been genetically modified so that the electron, once excited, had no place to go (Vos et al. 1991, 1993); this made it possible to see the coherent vibrational motion of the molecule more clearly in spectroscopic experiments. Subsequent experiments used more-intact systems, but looked first at low temperatures (Vos et al. 1994a) and finally at room temperature (Vos et al. 1994b). Eventually it was even possible to show that photo-triggering of electron transfer in other systems could reveal coherent vibrational motions (Liebl et al. 1999). A more sophisticated version of the ideas in Fig. 4.12 is given by Skourtis et al. (1992), which came at more or less the same time as the Vos et al. experiments. Most recently, it has been discovered that when energy is trapped in the "antenna pigments" of photosynthetic systems, the migration of energy toward the reaction center (where the electron transfer occurs) is coherent (Engel et al. 2007). This discovery has generated a huge literature, both on a wider range of experiments and on the associated theory, exploring the role that coherence might have in enhancing efficiency. Although the jury still is out, I think the current state of our understanding is not so different from Fig. 4.12, namely that being *almost* coherent results in the fastest possible reactions, if not the most dramatic experimental signatures; see, for example, Lloyd et al. (2011).

Engel et al. 2007. Evidence for wavelike energy transfer through quantum coherence in photosynthetic systems. GS Engel, TR Calhoun, EL Read, T-K Ahn, T Mančal, Y-C Cheng, RE Blankenship, and GR Fleming, *Nature* **446**, 782–786 (2007).

Fleming and van Grondelle 1994. The primary steps of photosynthesis. GR Fleming and R van Grondelle, *Physics Today,* pp. 48–55, February 1994.

Liebl et al. 1999. Coherent reaction dynamics in a bacterial cytochrome c oxidase. U Liebl, G Lipowski, M Négrerie, JC Lambry, JL Martin, and MH Vos, *Nature* **401**, 181–184 (1999).

Lloyd et al. 2011. The quantum Goldilocks effect: On the convergence of timescales in quantum transport. S Lloyd, M Mohsenu, A Shabani, and H Rabitz, arXiv.org:1111.4982 [quant–ph] (2011).

Skourtis et al. 1992. A new look at the primary charge separation in bacterial photosynthesis. SS Skourtis, AJR DaSilva, W Bialek, and JN Onuchic, *J Phys Chem* **96**, 8034–8041 (1992).

Vos et al. 1991. Direct observation of vibrational coherence in bacterial reaction centers using femtosecond absorption spectroscopy. MH Vos, JC Lambry, SJ Robles, DC Youvan, J Breton, and JL Martin, *Proc Natl Acad Sci (USA)* **88,** 8885–8889 (1991).

Vos et al. 1993. Visualization of coherent nuclear motion in a membrane protein by femtosecond spectroscopy. MH Vos, F Rappaport, JC Lambry, J Breton, and JL Martin, *Nature* **363**, 320–325 (1993).

Vos et al. 1994a. Coherent dynamics during the primary electron-transfer reaction in membrane-bound reaction centers of *Rhodobacter sphaeroides*. MH Vos, MR Jones, CN Hunter, J Breton, JC Lambry, and JL Martin, *Biochemistry* **33,** 6750–6757 (1994).

Vos et al. 1994b. Coherent nuclear dynamics at room temperature in bacterial reaction centers. MH Vos, MR Jones, CN Hunter, J Breton, JC Lambry, and JL Martin, *Proc Natl Acad Sci (USA)* **91,** 12701–12705 (1994).

The papers that reignited interest in proton tunneling in enzymes were Cha et al. (1989) and Grant and Klinman (1989). The idea that these experiments should be understood in terms of coupling between quantum motion of the proton and classical motion of the protein was developed by Bruno and Bialek (1992). It took roughly a decade for these ideas to solidify, as described in reviews by Sutcliffe and Scrutton (2002) and Knapp and Klinman (2002).

Bruno and Bialek 1992. Vibrationally enhanced tunneling as a mechanism for enzymatic hydrogen transfer. WJ Bruno and W Bialek, *Biophys J* **63,** 689–699 (1992).

Cha et al. 1989. Hydrogen tunneling in enzyme reactions. Y Cha, CJ Murray, and JP Klinman, *Science* **243**, 1325–1330 (1989).

Grant and Klinman 1989. Evidence that both protium and deuterium undergo significant tunneling in the reaction catalyzed by bovine serum amine oxidase. KL Grant and JP Klinman, *Biochemistry* **28**, 6597–6695 (1989).

Knapp and Klinman 2002. Environmentally coupled hydrogen tunneling: Linking catalysis to dynamics. MJ Knapp and JP Klinman, *Eur J Biochem* **269**, 3113–3121 (2002).

Sutcliffe and Scrutton 2002. A new conceptual framework for enzyme catalysis: Hydrogen tunneling coupled to enzyme dynamics in flavoprotein and quinoprotein enzymes. MJ Sutcliffe and NS Scrutton, *Eur J Biochem* **269**, 3096–3102 (2002).

4.2 Motility and Chemotaxis in Bacteria

The study of chemotaxis has a long history. From a biologist's point of view, the modern era starts when Adler (1965, 1969) demonstrated, using mutants, that chemosensing is independent of metabolism; the modern version of these experiments in Fig. 4.18 is from Girgis et al. (2007). From a physicist's point of view, the modern era starts when Berg builds his tracking microscope and observes, quantitatively, the paths of individual bacteria (Berg 1971; Berg and Brown 1972). The experiments demonstrating the temporal character of the computations involved in chemotaxis were done by Macnab and Koshland (1972) and by Brown and Berg (1974). A nice discussion of how these temporal comparisons translate into mobility up the gradient of attractive chemicals is given by Schnitzer et al. (1990).

Adler 1965. Chemotaxis in *Escherichia coli*. J Adler, *Cold Spring Harb Symp Quant Biol* **30**, 289–292 (1965).

Adler 1969. Chemoreceptors in bacteria. J Adler, *Science* **166**, 1588–1597 (1969).

Berg 1971. How to track bacteria. HC Berg, *Rev Sci Instrum* **42**, 868–871 (1971).

Berg and Brown 1972. Chemotaxis in *Escherichia coli* analyzed by three-dimensional tracking. HC Berg and DA Brown *Nature* **239**, 500–504 (1972).

Brown and Berg 1974. Temporal stimulation of chemotaxis in *Escherichia coli*. DA Brown and HC Berg, *Proc Natl Acad Sci (USA)* **71**, 1388–1392 (1974).

Girgis et al. 2007. A comprehensive genetic characterization of bacterial motility. HS Girgis, Y Liu, WS Ryu, and S Tavazoie, *PLoS Genet* **3**, e154 (2007).

Macnab and Koshland 1972. The gradient-sensing mechanism in chemotaxis. R Macnab and DE Koshland, *Proc Natl Acad Sci (USA)* **69**, 2509-2512 (1972).

Schnitzer et al. 1990. Strategies for chemotaxis. M Schnitzer, SM Block, HC Berg, and EM Purcell, *Symp Soc Gen Microbiol* **46**, 15–34 (1990).

For fluid mechanics in general, see Landau and Lifshitz (1987). For the history of the Reynolds number, see Rott (1990). For the admonishment about Green functions, see Jackson (1975). That bacteria live at low Reynolds number, and that this affects their lifestyle, surely was known to many people for many years. But Berg's experiments on *E. coli* provided a stimulus to think about this, and it resulted in a beautiful exposition by Purcell (1977), which has been hugely influential. The appreciation that self-propulsion at low Reynolds number has a gauge theory description is due to Shapere and Wilczek (1987). The dramatic discovery that bacteria swim by rotating their flagella was made by Berg and Anderson (1973), and then Silverman and Simon (1974) succeeded in tethering cells by their flagella to see the rotation of the cell body.

Berg and Anderson 1973. Bacteria swim by rotating their flagellar filaments. *Nature* **245,** 380–382 (1973).

Jackson 1975. *Classical Electrodynamics,* 2nd edition. JD Jackson (Wiley, New York, 1975)

Landau and Lifshitz 1987. *Fluid Mechanics.* LD Landau and EM Lifshitz (Pergamon, Oxford, 1987).

Purcell 1977. Life at low Reynolds number. EM Purcell, *Am J Phys* **45,** 3–11 (1977).

Rott 1990. Note on the history of the Reynolds number. N Rott, *Annu Rev Fluid Mech* **22,** 1–11 (1990).

Shapere and Wilczek 1987. Self-propulsion at low Reynolds number. A Shapere and F Wilczek, *Phys Rev Lett* **58,** 2051–2054 (1987).

Silverman and Simon 1974. Flagellar rotation and the mechanism of bacterial motility. M Silverman and M Simon, *Nature* **249,** 73–74 (1974).

The discussion of conversion between chemical and mechanical energy as being analogous to "off-diagonal" motion on a tilted potential surface is something that emerged in the early 1990s, largely through the work of Magnasco (1993, 1994a,b). For recent work on models of the flagellar motor specifically, see Mora et al. (2009a,b).

Magnasco 1993. Forced thermal ratchets. M Magnasco, *Phys Rev Lett* **71,** 1477–1481 (1993).

Magnasco 1994a. Brownian combustion engines. M Magnasco, in *Fluctuations and Order: The New Synthesis*, M Millonas, ed. pp. 307–320 (Springer-Verlag, Berlin, 1994).

Magnasco 1994b. Molecular combustion motors. M Magnasco, *Phys Rev Lett* **72,** 2656–2659 (1994).

Mora et al. 2009a. Steps in the bacterial flagellar motor. T Mora, H Yu, Y Sowa, and NS Wingreen, *PLoS Comput Biol* **5,** e1000540 (2009).

Mora et al. 2009b. Modeling torque versus speed, shot noise, and rotational diffusion of the bacterial flagellar motor. T Mora, H Yu, and NS Wingreen, *Phys Rev Lett* **103,** 248102 (2009).

The classic, intuitive account of the physical limits to chemical sensing is by Berg and Purcell (1977). Measurements on the impulse response of the system were reported by Block et al. (1982), and these experiments, along with Segall et al. (1986), provide a more compelling demonstration that a bacterium is sensitive to single molecular events. Another interesting paper from this period is Block et al. (1983). The idea of deriving the impulse response as the solution to an optimization problem, in the spirit of the Berg-Purcell discussion but more rigorously, has been explored by several groups: Strong et al. (1998); Andrews et al. (2006); and most recently, Celani and Vergassola (2010), who introduced a novel game-theoretic approach.

Andrews et al. 2006. Optimal noise filtering in the chemotactic response of *Escherichia coli.* BW Andrews, T-M Yi, and PA Iglesias, *PLoS Comput Bio* **2,** e154 (2006).

Berg and Purcell 1977. Physics of chemoreception. HC Berg and EM Purcell, *Biophys J* **20,** 193–219 (1977).

Block et al. 1982. Impulse responses in bacterial chemotaxis. SM Block, JE Segall, and HC Berg, *Cell* **31,** 215–226 (1982).

Block et al. 1983. Adaptation kinetics in bacterial chemotaxis. SM Block, JE Segall and HC Berg, *J Bacteriol* **154,** 312–323 (1983).

Celani and Vergassola 2010. Bacterial strategies for chemotaxis. A Celani and M Vergassola, *Proc Natl Acad Sci (USA)* **107,** 1391–1396 (2010).

Segall et al. 1986. Temporal comparisons in bacterial chemotaxis. JE Segall, SM Block, and HC Berg, *Proc Natl Acad Sci (USA)* **83,** 8987–8991 (1986).

Strong et al. 1998. Adaptation and optimal chemotactic strategy in *E coli*. SP Strong, B Freedman, W Bialek, and R Koberle, *Phys Rev E* **57,** 5604–5617 (1998).

The experiments on the response of the flagellar motor to the CheY∼P concentration are by Cluzel et al. (2000). For measurements on the diffusion constant of proteins in *E. coli*, see Elowitz et al. (1999), and for observations on the structure of the motor in relation to its regulation by CheY∼P, see Thomas et al. (1999).

Cluzel et al. 2000. An ultrasensitive bacterial motor revealed by monitoring signaling proteins in single cells. P Cluzel, M Surette, and S Leibler, *Science* **287,** 1652–1655 (2000).

Elowitz et al. 1999. Protein mobility in the cytoplasm of *Escherichia coli*. MB Elowitz, MG Surette, P-E Wolf, JB Stock, and S Leibler, *J Bacteriol* **181,** 197–203 (1999).

Thomas et al. 1999. Rotational symmetry of the C ring and a mechanism for the flagellar rotary motor. DR Thomas, DG Morgan, and DJ DeRoiser, *Proc Natl Acad Sci (USA)* **96,** 10134–10139 (1999).

The idea that interesting things happen in larger and larger collections of cooperative subunits probably occurred to many people who thought in terms of statistical mechanics (Thompson 1972), and it seems to get rediscovered periodically, including in the context of bacterial chemotaxis, where it raises the possibility of criticality as a mechanism of near-infinite sensitivity (Bray et al. 1998; Duke and Bray 1999; Duke et al. 2001). In another view of receptor interactions, there are clusters that behave more nearly in MWC fashion, as discussed in Section 5.3. Statistical mechanics models for interactions among binding events also play a key role in thinking about protein-DNA interactions and the regulation of gene expression (Bintu et al. 2005a,b; Kinney et al. 2010), as we will see more clearly in subsequent sections.

Bintu et al. 2005a. Transcriptional regulation by the numbers: models. L Bintu, NE Buchler, HG Garcia, U Gerland, T Hwa, J Kondev, and R Phillips, *Curr Opin Genet Dev* **15,** 116–124 (2005).

Bintu et al. 2005b. Transcriptional regulation by the numbers: applications. L Bintu, NE Buchler, HG Garcia, U Gerland, T Hwa, J Kondev, T Kuhlman, and R Phillips, *Curr Opin Genet Dev* **15,** 125–136 (2005).

Bray et al. 1998. Receptor clustering as a cellular mechanism to control sensitivity. D Bray, MD Levin, and CJ Morton-Firth, *Nature* **393,** 85–88 (1998).

Duke and Bray 1999. Heightened sensitivity of a lattice of membrane receptors. TAJ Duke and D Bray, *Proc Natl Acad Sci (USA)* **96,** 10104–10108 (1999).

Duke et al. 2001. Conformational spread in a ring of proteins: A stochastic view of allostery. TAJ Duke, N Le Novère, and D Bray, *J Mol Biol* **308,** 541–553 (2001).

Kinney et al. 2010. Using deep sequencing to characterize the biophysical mechanism of a transcriptional regulatory sequence. JB Kinney, A Murugan, CG Callan, Jr, and EC Cox, *Proc Natl Acad Sci (USA)* **107,** 9158–9163 (2010).

Thompson 1972. *Mathematical Statistical Mechanics.* CJ Thompson (Macmillan, New York, 1972).

4.3 Molecule Counting, More Generally

In thinking about transcriptional regulation, it is useful to review some basic facts about molecular biology, for which the classic reference is the Watson et al. book *Molecular Biology of the Gene*. This volume has been through many editions, and at times flirted with being more of an encyclopedia than a textbook. I reference the current edition here, which seems a bit more compact than some of the intermediate editions, but I also encourage you to look back at earlier editions, written by Watson alone. A beautiful account of gene regulation, using the bacteriophage λ as an example, was given by Ptashne (1986), which has also evolved with time (Ptashne 1992); see also Ptashne (2001).

Ptashne 1986. *A Genetic Switch: Gene Control and Phage λ*. M Ptashne (Cell Press, Cambridge MA, 1986).

Ptashne 1992. *A Genetic Switch: Phage λ and Higher Organisms*. M Ptashne (Cell Press, Cambridge MA, 1992).

Ptashne 2001. *Genes and Signals*. M Ptashne (Cold Spring Harbor Laboratory Press, New York, 2001).

Watson et al. 2008. *Molecular Biology of the Gene*, sixth edition. JD Watson, TA Baker, SP Bell, A Gann, M Levine, and R Losick (Benjamin Cummings, San Francisco, 2008).

To make our discussion quantitative, we need to know the absolute concentration at which transcription factors act. Ptashne's books give some discussion of this, although the estimates are a bit indirect. Several groups have made measurements on the binding of transcription factors to DNA, trying to measure the concentration at which binding sites are half occupied. Sometimes these measurements are done by direct physical-chemical methods in vitro, and sometimes by less direct methods in vivo. Examples include Oehler et al. (1994), Ma et al. (1996), Pedone et al. (1996), Burz et al. (1998), and Winston et al. (1999). A modern version of the in vitro binding experiment examines the molecules one at a time, as in the work by Wang et al. (2009). An additional problem concerns the mechanism by which transcription factors find their targets. In the analysis of bacterial chemotaxis, it seemed sensible to assume that the molecules being sensed arrive at their receptors by diffusion in the surrounding three-dimensional solution. For proteins binding to DNA, there is the additional possibility of sliding (effectively in one dimension) along the DNA itself after binding. A recent paper that addresses the evidence on this point quantitatively is Li et al. (2009). The perhaps surprising result that such sliding does not really change the effective noise levels very much is given by Tkačik and Bialek (2009); see also Appendix A.6.

Burz et al. 1998. Cooperative DNA binding by Bicoid provides a mechanism for threshold dependent gene activation in the *Drosophila* embryo. DS Burz, R Pivera-Pomar, H Jackle, and SD Hanes, *EMBO J* **17,** 5998–6009 (1998).

Li et al. 2009. Effects of macromolecular crowding and DNA looping on gene regulation kinetics. G-W Li, OG Berg, and J Elf, *Nature Physics* **5,** 294–297 (2009).

Ma et al. 1996. The *Drosophila* morphogenetic protein Bicoid binds DNA cooperatively. X Ma, D Yuan, K Diepold, T Scarborough, and J Ma, *Development* **122,** 1195–1206 (1996).

Oehler et al. 1994. Quality and position of the three lac operators of *E coli* define efficiency of repression. S Oehler, M Amouyal, P Kolkhof, B von Wilcken-Bergmann, and B Müller-Hill, *EMBO J* **13,** 3348–3355 (1994).

Pedone et al. 1996. The single Cys2-His2 zinc finger domain of the GAGA protein flanked by basic residues is sufficient for high-affinity specific DNA binding. PV Pedone, R Ghirlando, GM Clore, AM Gronenborn, G Felsenfeld, and JG Omichinski, *Proc Natl Acad Sci (USA)* **93,** 2822–2826 (1996).

Tkačik and Bialek 2009. Diffusion, dimensionality, and noise in transcriptional regulation. G Tkačik and W Bialek, *Phys Rev E* **79,** 051901 (2009).

Wang et al. 2009. Quantitative transcription factor binding kinetics at the single molecule level. Y Wang, L Guo, I Golding, EC Cox, and NP Ong, *Biophys J* **96,** 609–620 (2009).

Winston et al. 1999. Characterization of the DNA binding properties of the bHLH domain of Deadpan to single and tandem sites. RL Winston, DP Millar, JM Gottesfeld, and SB Kent. *Biochemistry* **38,** 5138–5146 (1999).

An important development in the field has been the construction of fusion proteins, combining transcription factors with fluorescent proteins, and the reinsertion of these fusions into the genome; for a review, see Tsien (1998). When cells divide, their contents are partitioned, and one can observe the noise from the finite number of molecules being assigned at random to one of the two daughter cells. Rosenfeld et al. (2005) and more recently Teng et al. (2010) have shown how this randomness can be used to make very precise estimates of the number of copies of the protein in the mother cell, thus providing a calibration that converts fluorescence intensity back into copy number. Gregor et al. (2007a) discuss a case where it was possible to quantitatively test in detail that the fusion construct replaces the function of the original transcription factor. In Gregor et al. (2007b) they exploit this construct to analyze the noise in one step of transcriptional regulation, as well as making estimates of absolute concentration by comparing the fluorescence intensity to a purified standard.

Gregor et al. 2007a. Stability and nuclear dynamics of the Bicoid morphogen gradient. T Gregor, EF Wieschaus, AP McGregor, W Bialek, and DW Tank, *Cell* **130,** 141–152 (2007).

Gregor et al. 2007b. Probing the limits to positional information. T Gregor, DW Tank, EF Wieschaus, and W Bialek, *Cell* **130,** 153–164 (2007).

Rosenfeld et al. 2005. Gene regulation at the single cell level. N Rosenfeld, JW Young, U Alon, PS Swain, and MB Elowitz, *Science* **307,** 1962–1965 (2005).

Teng et al. 2010. Measurement of the copy number of the master quorum-sensing regulator of a bacterial cell. S-W Teng, Y Wang, KC Tu, T Long, P Mehta, NS Wingreen, BL Bassler, and NP Ong, *Biophys J* **98,** 2024–2031 (2010).

Tsien 1998. The green fluorescent protein. RY Tsien, *Annu Rev Biochem* **67,** 509–544 (1998).

In contrast to bacteria, many eukaryotic cells are large enough, or move slowly enough, to obtain a reliable signal by measuring gradients across the length of their bodies (Song et al. 2006; Gregor et al. 2010); for a discussion of the limits to these measurements and some of the relevant experiments, see Endres and Wingreen (2009a,b). The measurements on extreme precision of axon guidance were reported by Rosoff et al. (2004).

Endres and Wingreen 2009a. Accuracy of direct gradient sensing by single cells. RG Endres and NS Wingreen, *Proc Natl Acad Sci (USA)* **105,** 15749–15754 (2008).

Endres and Wingreen 2009b. Accuracy of direct gradient sensing by cell-surface receptors. RG Endres and NS Wingreen, *Prog Biophys Mol Biol* **100,** 33–39 (2009).

Gregor et al. 2010. The onset of collective behavior in social amoebae. T Gregor, K Fujimoto, N Masaki, and S Sawai, *Science* **328,** 1021–1025 (2010).

Rosoff et al. 2004. A new chemotaxis assay shows the extreme sensitivity of axons to molecular gradients. WJ Rosoff, JS Urbach, MA Esrick, RG McAllister, LJ Richards, and GJ Goodhill, *Nat Neurosci* **7,** 678–682 (2004).

Song et al. 2006. *Dictyostelium discoideum* chemotaxis: Threshold for directed motion. L Song, SM Nadkarni, HU Bödeker, C Beta, A Bae, C Franck, W-J Rappel, WF Loomis, and E Bodenschatz, *Eur J Cell Bio* **85,** 981–989 (2006).

It is only in the past decade that it has been possible to make direct measurements of the noise in gene expression, and even more recently that it has been possible to focus on noise in the control process itself. The initial experiment separating intrinsic from extrinsic noise sources using the two-color plasmid was by Elowitz et al. (2002), which touched off a series of experiments on both bacterial (Ozbudak et al. 2002; Pedraza and van Oudenaarden 2005) and eukaryotic (Blake et al. 2003; Raser and O'Shea 2004) systems. The experiments on noise in the Bcd-Hb system are by Gregor et al. (2007b), cited above. A review of methods for measuring Bcd concentration profiles is given by Morrison et al. (2011), and in particular, they discuss the comparison of live GFP-based imaging with antibody staining methods in fixed samples. A more detailed analysis of the data on Bcd-Hb noise is given by Tkačik et al. (2008), which also provides a broader context on the role of different noise sources in the control of gene expression. Models based on transcriptional bursting are inspired by the direct observation of these bursts in *E. coli* by Golding et al. (2005). It is worth thinking about whether the observed bursts necessarily result from the kinetics of switching between states of the transcriptional apparatus or could be traced to the binding and unbinding of transcription factors.

Blake et al. 2003. Noise in eukaryotic gene expression. WJ Blake, M Kaern, CR Cantor, and JJ Collins, *Nature* **422,** 633–637 (2003).

Elowitz et al. 2002. Stochastic gene expression in a single cell. MB Elowitz, AJ Levine, ED Siggia, and PD Swain, *Science* **297,** 1183–1186 (2002).

Golding et al. 2005. Real-time kinetics of gene activity in individual bacteria. I Golding, J Paulsson, SM Zawilski, and EC Cox, *Cell* **123,** 1025–1036 (2005).

Morrison et al. 2011. Quantifying the Bicoid morphogen gradient in living embryos. AH Morrison, M Scheeler, J Dubuis, and T Gregor, in *Imaging in Developmental Biology: A Laboratory Manual,* J Sharpe, and R Wong, eds., pp. 398–406 (Cold Spring Harbor Press, Woodbury NY, 2011); arXiv.org:1003.5572 [q-bio.QM] (2010).

Ozbudak et al. 2002. Regulation of noise in the expression of a single gene. E Ozbudak, M Thattai, I Kurtser, AD Grossman, and A van Oudenaarden, *Nat Genet* **31,** 69–73 (2002).

Pedraza and van Oudenaarden 2005. Noise propagation in gene networks. J Pedraza and A van Oudenaarden, *Science* **307,** 1965–1969 (2005).

Raser and O'Shea 2004. Control of stochasticity in eukaryotic gene expression. JM Raser and EK O'Shea, *Science* **304,** 1811–1814 (2004).

Tkačik et al. 2008. The role of input noise in transcriptional regulation. G Tkačik, T Gregor, and W Bialek, *PLoS One* **3,** e2774 (2008).

There are many cases in which it seems likely that the control of gene expression really does involve a switch between two stable states rather than a continuous adjustment that depends steeply on input signals. Classical examples are the decision of a bacteriophage to live inside its host or to hijack the cell's synthetic machinery and make many copies of itself, destroying the host—see Ptashne (1986) above—and lactose metabolism in *E. coli* (Ozbudak et al. 2004). There are also biochemical networks that involve multiple stable states, notably the calmodulin-dependent kinase involved in synaptic plasticity (Lisman 1985; Miller and Kennedy 1986; Lisman and Goldring 1988; Kennedy 1994); see also Problem 119. There are interesting questions about the role of noise in such systems, in particular the rate at which noise will trigger spontaneous transitions among the stable states. This question is in the background of the work by Lisman and Goldring on the kinase switch and is explicit in attempts by Arkin, Bhalla, and others (Arkin et al. 1998; McAdams and Arkin 1998; Bhalla and Iyengar 1999) to simulate phage λ and other circuits. I tried to emphasize the generality of this question some years ago (Bialek 2001), and around the same time several groups took much more powerful analytic approaches, focusing on phage λ again (Aurell and Sneppen 2002; Aurell et al. 2002; Roma et al. 2005). The formalism for more general analysis of such nonperturbative effects of noise in chemical kinetics is quite pretty (Elgart and Kamenev 2004), but I am still not sure we fully understand phage λ.

Arkin et al. 1998. Stochastic kinetic analysis of developmental pathway bifurcation in phage-infected *Escherichia coli* cells. A Arkin, J Ross, and HH McAdams, *Genetics* **149,** 1633–1648 (1998).

Aurell and Sneppen 2002. Epigenetics as a first exit problem. E Aurell and K Sneppen, *Phys Rev Lett* **88,** 048101 (2002).

Aurell et al. 2002. Stability puzzles in phage lambda. E Aurell, S Brown, J Johanson, and K Sneppen, *Phys Rev E* **65,** 051914 (2002).

Bhalla and Iyengar 1999. Emergent properties of biological signalling pathways. US Bhalla and R Iyengar, *Science* **283,** 381–387 (1999).

Bialek 2001. Stability and noise in biochemical switches. W Bialek, in *Advances in Neural Information Processing 13,* TK Leen, TG Dietterich, and V Tresp, eds., pp. 103–109 (MIT Press, Cambridge, 2001); arXiv:cond-matt/0005235 (2000).

Elgart and Kamenev 2004. Rare event statistics in reaction-diffusion systems. V Elgart and A Kamenev, *Phys Rev E* **70,** 041106 (2004); arXiv:cond-mat/0404241.

Kennedy 1994. The biochemistry of synaptic regulation in the central nervous system. MB Kennedy, *Annu Rev Biochem* **63,** 571–600 (1994).

Lisman 1985. A mechanism for memory storage insensitive to molecular turnover: A bistable autophosphorylating kinase. JE Lisman, *Proc Natl Acad Sci (USA)* **82,** 3055–3057 (1985).

Lisman and Goldring 1988. Feasibility of long-term storage of graded information by the Ca^{2+}/calmodulin-dependent protein kinase molecules of the postsynaptic density. JE Lisman and MA Goldring, *Proc Natl Acad Sci (USA)* **85,** 5320–5324 (1988).

McAdams and Arkin 1998. Simulation of prokaryotic genetic circuits. HH McAdams and A Arkin, *Annu Rev Biophys Biomol Struct* **27,** 199–224 (1998).

Miller and Kennedy 1986. Regulation of brain type II Ca^{2+}/calmodulin-dependent protein kinase by autophosphorylation: A Ca^{2+}-triggered molecular switch. SG Miller and MB Kennedy, *Cell* **44,** 861–870 (1986).

Ozbudak et al. 2004. Multistability in the lactose utilization network of *Escherichia coli.* EM Ozbudak, M Thattai, HN Lim, BI Shraiman, and A van Oudenaarden, *Nature* **427,** 737–740 (2004).

Roma et al. 2005. Optimal path to epigenetic switching. DM Roma, RA O'Flanagan, AE Ruckenstein, AM Sengupta, and R Mukhopadhay, *Phys Rev E* **71,** 011902 (2005); arXiv:q-bio/0406008 [q-bio.MN] (2004).

4.4 More about Noise in Perception

Although there were many precursors reaching back across centuries, the conclusive demonstration that bats navigate by echolocation, with sounds beyond the range of human hearing, was by Griffin and Galambos (1941). Griffin (1958) gives a beautiful presentation of the history and basic facts about the system. The charming demonstration with dusted mealworms is described by Simmons (1989), based on the thesis of Trappe (1982). The first suggestion of submicrosecond precision in this system was from Simmons (1979). Perhaps not surprisingly, these observations (and the provocative title of the paper in which they were presented) touched off a flurry of controversy; for different views, see Altes (1981) and Menne and Hackbarth (1986). The astonishing results on nanosecond precision and the optimality of performance in background noise were presented by Simmons et al. (1990). For context, it is interesting to look at examples of precise timing measurements in binaural hearing (Konishi 1973; Knudsen and Konishi 1978; Carr and Konishi 1990) and in weakly electric fish (Rose and Heiligenberg 1985; Carr et al. 1986; Heiligenberg 1991).

Altes 1981. Echo phase perception in bat sonar? RA Altes, *J Acoust Soc Am* **69,** 505–508 (1981).

Carr and Konishi 1990. A circuit for detection of interaural time differences in the brain stem of the barn owl. CE Carr and M Konishi, *J Neurosci* **10,** 3227–3246 (1990).

Carr et al. 1986. A time-comparison circuit in the electric fish midbrain. I. Behavior and physiology. CE Carr, W Heiligenberg, and GJ Rose, *J Neurosci* **6,** 107–119 (1986).

Griffin 1958. *Listening in the Dark.* DR Griffin (Yale University Press, New Haven CT, 1958).

Griffin and Galambos 1941. The sensory basis of obstacle avoidance by flying bats. DR Griffin and R Galambos, *J Exp Zool* **86,** 481–506 (1941).

Heiligenberg 1991. *Neural Nets in Electric Fish.* WF Heiligenberg (MIT Press, Cambridge MA, 1991).

Knudsen and Konishi 1978. A neural map of auditory space in the owl. EI Knudsen and M Konishi, *Science* **200,** 795–797 (1978).

Konishi 1973. How the owl tracks its prey: Experiments with trained barn owls reveal how their acute sense of hearing enables them to catch prey in the dark. M Konishi, *Am Scientist* **61,** 414–424 (1973).

Menne and Hackbarth 1986. Accuracy of distance measurement in the bat *Eptesicus fuscus:* Theoretical aspects and computer simulations. D Menne and H Hackbarth, *J Acoust Soc Am* **79,** 386–397 (1986).

Rose and Heiligenberg 1985. Temporal hyperacuity in the electric sense of fish. G Rose and W Heiligenberg, *Nature* **318,** 178–180 (1985).

Simmons 1979. Perception of echo phase information in bat sonar. JA Simmons, *Science* **204,** 1336–1338 (1979).

Simmons 1989. A view of the world through the bat's ear: The formation of acoustic images in echolocation. JA Simmons, *Cognition* **33,** 155–199 (1989).

Simmons et al. 1990. Discrimination of jittered sonar echoes by the echolocating bat, *Eptesicus fuscus:* The shape of target images in echolocation. JA Simmons, M Ferragamo, CF Moss, SB Stevenson, and RA Altes, *J Comp Physiol A* **167,** 589–616 (1990).

Trappe 1982. *Verhalten und Echoortung der Grossen Hufeisennase (*Rhinolophus ferrumequinum*) beim Insektenfang* (PhD thesis, Universität Marburg, Marburg, Germany, 1982).

The program of comparing human performance with statistical limits in the context of higher level perception was outlined by Barlow (1980). The experiments on symmetry in random dot patterns are by Barlow and Reeves (1979), and an analysis of optimality in motion perception using random dot stimuli was given by Barlow and Tripathy (1997). For a review of how these stimuli have been used to probe the connections between neural activity and perception, see Newsome et al. (1995). Note that, as discussed in *Spikes* (Rieke et al. 1997; see below), these experiments connecting neural activity with perception in primates have been done largely in a regime where the subject is integrating imperfectly over very long periods of time, much longer than we would expect to see constant-velocity motion in a natural setting (see also Osborne et al. 2004). This complicates efforts to compare either neural or behavioral performance with the physical limits, and indeed I do not know of any effort to measure the responses of visual cortex in a regime (e.g., photon counting in the dark) where we understand fully the sources of noise limiting our perception. There is an opportunity here for future research.

Barlow 1980. The absolute efficiency of perceptual decisions. HB Barlow, *Phil Trans R Soc B* **290,** 71–82 (1980).

Barlow and Reeves 1979. The versatility and absolute efficiency of detecting mirror symmetry in random dot displays. HB Barlow and BC Reeves, *Vision Res* **19,** 783–793 (1979).

Barlow and Tripathy 1997. Correspondence noise and signal pooling in the detection of coherent visual motion. HB Barlow and SP Tripathy, *J Neurosci* **17,** 7954–7966 (1997).

Newsome et al. 1995. Visual motion: Linking neuronal activity to psychophysical performance. WT Newsome, MN Shadlen, E Zohary, KH Britten, and JA Movshon, in *The Cognitive Neurosciences,* M Gazzaniga, ed., pp. 401–414 (MIT Press, Cambridge MA, 1995).

Osborne et al. 2004. Time course of information about motion direction in visual area MT of macaque monkeys. LC Osborne, W Bialek, and SG Lisberger, *J Neurosci* **24,** 3210–3222 (2004).

The classical work on motion estimation in insect vision was by Hassenstein and Reichardt (1956); perspectives on these early ideas are given by Reichardt (1961) and by Reichardt and Poggio (1976). A crucial piece of data in this discussion concerns the speed of a flying insect's motor response to visual motion, and a first estimate of this was given by Land and Collett (1974) in a beautiful analysis of natural flight trajectories; subsequent work was done by Wagner (1986a–c), van Hateren and Schilstra (1999), and Schilstra and van Hateren (1999).

Hassenstein and Reichardt 1956. Systemstheoretische Analyse der Zeit-, Reihenfolgen-, und Vorzeichenauswertung bei der Bewegungsperzeption des Rüsselkäfers Chlorophanus. S Hassentsein and W Reichardt, *Z Naturforsch* **11b,** 513–524 (1956).

van Hateren and Schilstra 1999. Blowfly flight and optic flow. II. Head movements during flight. JH van Hateren and C Schilstra, *J Exp Biol* **202,** 1491–1500 (1999).

Land and Collett 1974. Chasing behavior of houseflies (*Fannia canicularis*): A description and analysis. MF Land and TS Collett, *J Comp Physiol* **89,** 331–357 (1974).

Reichardt 1961. Autocorrelation, a principle for the evaluation of sensory information by the central nervous system. W Reichardt, in *Sensory Communication*, WA Rosenblith, ed., pp. 303–317 (MIT Press, Cambridge MA, 1961).

Reichardt and Poggio 1976. Visual control of orientation behavior in flies. I. A quantitative analysis. W Reichardt and T Poggio, *Q Rev Biophys* **9,** 311–375 (1976).

Schilstra and van Hateren 1999. Blowfly flight and optic flow. I: Thorax kinematics and flight dynamics. C Schilstra and JH van Hateren, *J Exp Biol* **202,** 1481–1490 (1999).

Wagner 1986a. Flight performance and visual control of flight in the free-flying house fly (*Musca domestica* L.). I: Organization of the flight motor. H Wagner, *Phil Trans R Soc Lond B* **312,** 527–551 (1986).

Wagner 1986b. Flight performance and visual control of flight in the free-flying house fly (*Musca domestica* L.). II: Pursuit of targets. H Wagner, *Phil Trans R Soc Lond B* **312,** 553–579 (1986).

Wagner 1986c. Flight performance and visual control of flight in the free-flying house fly (*Musca domestica* L.). I: Interactions between angular movement induced by wide- and small-field stimuli. H Wagner, *Phil Trans R Soc Lond B* **312,** 581–595 (1986).

Motion-sensitive neurons in the fly visual system were discovered by Bishop and Keehn (1966), around the same time that Barlow et al. (1964) discovered motion-sensitive neurons in the rabbit retina. Today we take for granted that individual neurons can be selective for very complicated things, culminating in face- and object-selective neurons in the far reaches of the visual cortex (Desimone 1991; Gross 1992), but in their earliest versions (Gross et al. 1969; Bruce et al. 1981; Perrett et al. 1982) these measurements were surprising. Indeed, in Barlow's hands, the observation of motion sensitivity played a key role in helping to shape the idea that cells respond to successively more complex conjunctions of features as we move through successive layers of processing; see the discussion in Barlow and Levick (1965). An early experiment showing that some motion-sensitive neurons are a necessary link in optomotor behavior is by Hausen and Wehrhahn (1983).

Barlow and Levick 1965. The mechanism of directionally selective units in rabbit's retina. HB Barlow and WR Levick, *J Physiol (Lond)* **178,** 477–504 (1965).

Barlow et al. 1964. Retinal ganglion cells responding selectively to direction and speed of image motion in the rabbit. HB Barlow, RM Hill, and WR Levick, *J Physiol (Lond)* **173,** 377–407 (1964).

Bishop and Keehn 1966. Two types of neurones sensitive to motion in the optic lobe of the fly. LG Bishop and DG Keehn, *Nature* **212,** 1374–1376 (1966).

Bruce et al. 1981. Visual properties of neurons in a polysensory area in superior temporal sulcus of the macaque. C Bruce, R Desimone, and CG Gross, *J Neurophysiol* **46,** 369–384 (1981).

Desimone 1991. Face-selective cells in the temporal cortex of monkeys. R Desimone, *J Cog Neurosci* **3,** 1–8 (1991).

Gross 1992. Representation of visual stimuli in inferotemporal cortex. CG Gross, *Phil Tran R Soc Lond B* **335,** 3–10 (1992).

Gross et al. 1969. Visual receptive fields of neurons in inferotemporal cortex of the monkey. CG Gross, DB Bender, and CE Rocha-Miranda, *Science* **166,** 1303–1306 (1969).

Hausen and Wehrhahn 1983. Microsurgical lesion of horizontal cells changes optomotor yaw responses in the blowfly *Calliphora erythrocephala*. K Hausen and C Wehrhahn, *Proc R Soc Lond B* **219,** 211–216 (1983).

Perrett et al. 1982. Visual neurons responsive to faces in the monkey temporal cortex. DI Perrett, ET Rolls, and W Caan, *Exp Brain Res* **47,** 329–342 (1982).

The experiments on the precision of motion discrimination using the output of H1 are from de Ruyter van Steveninck and Bialek (1995), and the reconstruction of velocity waveforms was done in Bialek et al. (1991); a review of these ideas and results is given in *Spikes* (Rieke et al. 1997). A detailed calculation of the physical limits to motion estimation in this system is in my lecture notes from the Santa Fe Summer School (Bialek 1990). For a general discussion of hyperacuity in vision, see Westheimer (1981), and for the relation of hyperacuity to physical limits, see Geisler (1984). The discussion of optimal motion estimation given here is largely based on the thesis work of Marc Potters (see Potters and Bialek 1994); related work was done by Simoncelli (1993). Application of these ideas to human visual motion perception can be found in Weiss et al. (2002). Problem 72c about third-order statistics is inspired by Fitzgerald et al. (2011). All this theoretical work, as noted, makes assumptions about noise levels and the statistical structure of sensory inputs. The experiments in Fig. 4.36 are an attempt to make direct measurements of these quantities in the natural environment. Alternatively, one can try to invert the discussion and extract from experiments apparent prior expectations and noise levels that would be consistent with optimal performance, and then see whether there is some way to test these conclusions independently. This strategy for the case of motion estimation in humans is described by Stocker and Simoncelli (2006).

Bialek 1990. Theoretical physics meets experimental neurobiology. W Bialek, in *1989 Lectures in Complex Systems, SFI Studies in the Sciences of Complexity,* Volume II, E Jen, ed., pp. 513–595 (Addison-Wesley, Menlo Park CA, 1990).

Bialek et al. 1991. Reading a neural code. W Bialek, F Rieke, RR de Ruyter van Steveninck, and D Warland, *Science* **252,** 1854–1857 (1991).

Fitzgerald et al. 2011. Symmetries in stimulus statistics shape the form of visual motion estimators. JE Fitzgerald, AY Katsov, TR Clandinin, and MJ Schnitzer, *Proc Natl Acad Sci (USA)* **108,** 12909–12914 (2011).

Geisler 1984. Physical limits of acuity and hyperacuity. WS Geisler, *J Opt Soc Am A* **1,** 775–782 (1984).

Potters and Bialek 1994. Statistical mechanics and visual signal processing. M Potters and W Bialek, *J Phys I France* **4,** 1755–1775 (1994).

Rieke et al. 1997. *Spikes: Exploring the Neural Code.* F Rieke, D Warland, RR de Ruyter van Steveninck, and W Bialek (MIT Press, Cambridge MA, 1997).

de Ruyter van Steveninck and Bialek 1995. Reliability and statistical efficiency of a blowfly movement-sensitive neuron. RR de Ruyter van Steveninck and W Bialek, *Phil Trans R Soc Lond B* **348,** 321–340 (1995).

Simoncelli 1993. *Distributed Analysis and Representation of Visual Motion.* EP Simoncelli (PhD dissertation, Massachusetts Institute of Technology, Cambridge MA, 1993).

Stocker and Simoncelli 2006. Noise characteristics and prior expectations in human visual speed perception. AA Stocker and EP Simoncelli, *Nat Neurosci* **9,** 578–585 (2006).

Weiss et al. 2002. Motion illusions as optimal percepts. Y Weiss, EP Simoncelli, and EH Adelson, *Nat Neurosci* **5,** 598–604 (2002).

Westheimer 1981. Visual hyperacuity. G Westheimer, *Prog Sens Physiol* **1,** 1–30 (1981).

The classical evidence for the systematic errors of motion estimation predicted by the correlator model are discussed by Reichardt and Poggio (1976; see above). The demonstration that quadratic behavior at low contrasts coexists with unambiguous responses to velocity at high contrast is given by de Ruyter van Steveninck et al. (1994, 1996). These experiments were done with randomly textured images, whereas classical studies of visual motion have used periodic gratings. The correlator model also predicts that velocity will be confounded with the spatial frequency of these gratings, and this error persists even under high signal-to-noise ratio conditions (Haag et al. 2004). It is not clear whether this behavior represents a genuine failure of optimal estimation, a by-product of strategies for gain control and efficient coding (Borst 2007), or simply a behavior that would never be seen under natural conditions. There are several experiments, especially in bees (Srinivasan et al. 1991, 1996; Baird et al. 2005), indicating that insects have access to signals that allow them to control their flight speed without any of the systematic errors predicted by the correlator model. Recent work confirms this conclusion in *Drosophila* using sophisticated tracking and virtual reality to allow controlled experiments under free flight conditions (Fry et al. 2009). Some experiments have shown that the responses of motion-sensitive neurons are also very different under more natural conditions (Lewen et al. 2001; de Ruyter van Steveninck et al. 2001), although most of the analysis has focused on the nature of coding in spike trains rather than the nature of the motion computation itself. An attempt to dissect the motion computation represented by the spiking output of H1 is described in Bialek and de Ruyter van Steveninck (2005).

Baird et al. 2005. Visual control of flight speed in honeybees. E Baird, MV Srinivasan, S Zhang, and A Cowling, *J Exp Biol* **208,** 3895–3905 (2005).

Bialek and de Ruyter van Steveninck 2005. Features and dimensions: Motion estimation in fly vision. W Bialek and RR de Ruyter van Steveninck, arXiv:q-bio/0505003 (2005).

Borst 2007. Correlation versus gradient type motion detectors: The pros and cons. A Borst, *Phil Trans R Soc Lond B* **362**, 369–374 (2007).

Fry et al. 2009. Visual control of flight speed in *Drosophila melanogaster*. SN Fry, N Rohrseitz, AD Straw, and MH Dickinson, *J Exp Biol* **212**, 1120–1130 (2009).

Haag et al. 2004. Fly motion vision is based on Reichardt detectors regardless of the signal-to-noise ratio. J Haag, W Denk, and A Borst, *Proc Natl Acad Sci (USA)* **101**, 16333–16338 (2004).

Lewen et al. 2001. Neural coding of naturalistic motion stimuli. GD Lewen, W Bialek, and RR de Ruyter van Steveninck, *Network* **12**, 317–329 (2001); arXiv:physics/0103088 (2001).

de Ruyter van Steveninck et al. 1994. Statistical adaptation and optimal estimation in movement computation by the blowfly visual system. RR de Ruyter van Steveninck, W Bialek, M Potters, and RH Carlson, in *Proceedings of the IEEE Conference on Systems, Man, and Cybernetics*, pp. 302–307 (1994).

de Ruyter van Steveninck et al. 1996. Adaptive movement computation by the blowfy visual system. RR de Ruyter van Steveninck, W Bialek, M Potters, RH Carlson, and GD Lewen in *Natural and Artificial Parallel Computation: Proceedings of the Fifth NEC Research Symposium*, DL Waltz, ed., pp. 21–41 (SIAM, Philadelphia, 1996).

de Ruyter van Steveninck et al. 2001. Real time encoding of motion: Answerable questions and questionable answers from the fly's visual system. RR de Ruyter van Steveninck, A Borst, and W Bialek, in *Processing Visual Motion in the Real World: A Survey of Computational, Neural and Ecological Constraints,* JM Zanker and J Zeil, eds., pp. 279–306 (Springer-Verlag, Berlin, 2001); arXiv:physics/0004060 (2000).

Srinivasan et al. 1991. Range perception through apparent image speed in freely flying honeybees. MV Srinivasan, M Lehrer, WH Kirchner, and SW Zhang, *Vis Neurosci* **6**, 519–535 (1991).

Srinivasan et al. 1996. Honeybee navigation en route to the goal: Visual flight control and odometry. MV Srinivasan, SW Zhang, M Lehrer, and TS Collett, *J Exp Biol* **199**, 237–244 (1996).

Almost all attempts to analyze the structure of computations represented by the outputs of real neurons rely on the idea of dimensionality reduction: although sensory inputs are described by many degrees of freedom, interesting computations involve a collapse onto a smaller set of relevant dimensions. In favorable cases these few dimensions might be naturally expressed as Euclidean (linear) projections in the original space of inputs. Although not always aimed at the problems of noise discussed here, methods for uncovering such simplifications are an important part of our tool kit for exploring these complex systems, so it is worth providing a short guide. Early on de Boer and Kuyper (1968) used this idea to try and separate the (hypothetically linear) mechanical filtering in the inner ear from the necessary nonlinearities of generating spikes in the auditory nerve. There is also a tradition of searching for low dimensionality in the structure of our movements and in the activity of neurons in the motor system (d'Avella and Bizzi 1998; Sanger 2000; Stephens et al. 2008). In the control of eye movements, we can even see how the reduced dimensionality of movement variations is related to the noise in sensory processing (Osborne et al. 2005). Whereas the reverse-correlation methods of de Boer and Kuyper allow us to search for a single relevant dimension, the ideas of spike-triggered covariance (de Ruyter van Steveninck and Bialek 1988; Bialek and de Ruyter van Steveninck 2005; see above) allow us to count the number of relevant dimensions, so long as the input signals are chosen from a Gaussian distribution; this approach led, both in the retina (Fairhall et al. 2006) and in primary visual cortex (Rust et al. 2005) to the discovery that cells thought to be well described by single receptive fields in fact are sensitive to multiple stimulus dimensions. It is possible to go beyond the case of Gaussian inputs by searching for Euclidean projections that capture the maximum mutual information between sensory inputs and neural outputs (Sharpee et al. 2004). Very similar ideas have emerged in the analysis of sequence specificity in DNA-binding proteins (Kinney et al. 2007, 2010), as will be discussed more thoroughly in Section 5.1. Finally, there is a substantial literature on the search for nonlinear low-dimensional structures in data—manifolds; examples include Bishop et al. (1998), Roweis and Saul (2000), and Tenenbaum et al. (2000). There are connections to the classical problem of measuring the dimensionality of attractors in dynamical systems (Grassberger and Proccacia 1983), and there are information-theoretic formulations as well (Chigirev and Bialek 2004), but as far as I know these ideas have not yet been converted into methods that would use neural responses to measure, for example, the dimensionality of the relevant manifold in the space of sensory inputs.

d'Avella and Bizzi 1998. Low dimensionality of supraspinally induced force fields. A d'Avella and E Bizzi, *Proc Natl Acad Sci (USA)* **95,** 7711–7714 (1998).

Bishop et al. 1998. GTM: The generative topographic mapping. C Bishop, M Svensen, and C Williams, *Neural Comp* **10,** 215–234 (1998).

de Boer and Kuyper 1968. Triggered correlation. E de Boer and P Kuyper, *IEEE Trans Biomed Eng* **15,** 169–179 (1968).

Chigirev and Bialek 2004. Optimal manifold representation of data: An information theoretic perspective. DV Chigirev and W Bialek, in *Advances in Neural Information Processing 16,* S Thrun, L Saul, and B Schölkopf, eds., pp. 161–168 (MIT Press, Cambridge MA, 2004).

Fairhall et al. 2006. Selectivity for multiple stimulus features in retinal ganglion cells. AL Fairhall, CA Burlingame, R Narasimhan, RA Harris, JL Puchalla, and MJ Berry, II, *J Neurophysiol* **96,** 2724–2738 (2006).

Grassberger and Proccacia 1983. Characterization of strange attractors. P Grassberger and I Proccacia, *Phys Rev Lett* **50,** 346–349 (1983).

Kinney et al. 2007. Precise physical models of protein-DNA interaction from high-throughput data. JB Kinney, G Tkačik, and CG Callan, Jr, *Proc Natl Acad Sci (USA)* **104,** 501–506 (2007).

Kinney et al. 2010. Using deep sequencing to characterize the biophysical mechanism of a transcriptional regulatory sequence. JB Kinney, A Murugan, CG Callan, Jr, and EC Cox, *Proc Natl Acad Sci (USA)* **107,** 9158–9163 (2010).

Osborne et al. 2005. A sensory source for motor variation. LC Osborne, SG Lisberger, and W Bialek, *Nature* **437,** 412–416 (2005).

Roweis and Saul 2000. Nonlinear dimensionality reduction by locally linear embedding. S Roweis and L Saul, *Science* **290,** 2323–2326 (2000).

Rust et al. 2005. Spatiotemporal elements of macaque V1 receptive fields. NC Rust, O Schwartz, JA Movshon, and EP Simoncelli, *Neuron* **46,** 945–956 (2005).

de Ruyter van Steveninck and Bialek 1988. Real-time performance of a movement sensitive neuron in the blowfly visual system: Coding and information transfer in short spike sequences. RR de Ruyter van Steveninck and W Bialek, *Proc R Soc Lond B* **234,** 379–414 (1988).

Sanger 2000. Human arm movements described by a low-dimensional superposition of principal components. TD Sanger, *J Neurosci* **20,** 1066–1072 (2000).

Sharpee et al. 2004. Analyzing neural responses to natural signals: Maximally informative dimensions. T Sharpee, NC Rust, and W Bialek, *Neural Comp* **16,** 223–250 (2004); arXiv:physics/0212110 (2002).

Stephens et al. 2008. Dimensionality and dynamics in the behavior of *C. elegans*. GJ Stephens, B Johnson-Kerner, W Bialek, and WS Ryu, *PLoS Comp Bio* **4,** e1000028 (2008); arXiv:0705.1548 [q-bio.OT] (2007).

Tenenbaum et al. 2000. A global geometric framework for nonlinear dimensionality reduction. J Tenenbaum, V de Silva, and J Langford, *Science* **290,** 2319–2323 (2000).

Let me also add an aside about the different organisms that have provided examples for our discussion thus far. Since that formative year of having the office next door to Rob de Ruyter van Steveninck when I was a postdoc in Groningen, the fly visual system has seemed to me an ideal testing ground for physicists' ideas. However, if you think that brains are interesting because you want to understand your own brain, you might believe that insects are a bit of a side show relative to animals that share more of our brain structures— monkeys, cats, or even mice. There are obvious questions of strategy here, including the fact that (perhaps paradoxically) it can be easier to control the behavior of a primate than that of an insect, creating opportunities for certain kinds of quantitative experiments. There also are questions about how much universality we should expect. Are there things to be learned about brains in general, or is everything about our brain different from that of "lower" animals? Can careful quantitative analyses of "simpler" systems sharpen the questions that we ask about bigger brains (even if the answers are different), or does each case present such unique challenges? I think it is fair to say that for several decades there has been a strong consensus of the mainstream neuroscience community that the answers to these questions point away from the study of insect brains. Recently, however, there has been substantial growth in a community of scientists interested in exploiting the tools of modern molecular biology to study the brain, and this group is attracted to model organisms with well-developed methods of genetic manipulation, such as the fruit fly *Drosophila melanogaster* and its close relatives. Thus, the coming years are likely to see a resurgence of interest in insect brains, which should create more opportunities for physicists.

The rather astonishing results in Fig. 4.40 are from Griffiths and Tenenbaum (2006). The original work on optimal cue combination was by Ernst and Banks (2002), and recent work demonstrates near optimal performance in cases where the optimal combination rules are strongly nonlinear (Ma et al. 2011). Serious discussion of ambiguous percepts goes back to the early 1800s (Necker 1832; Wheatstone 1838); for brief historical reviews, see Fisher (1968) and Long and Toppino (2004). There was a resurgence of interest in these phenomena when it was realized that they offered an opportunity to distinguish between neurons responding to visual inputs and neurons that encode our subjective experiences (Logothetis and Schall 1989; Leopold and Logothetis 1996; Logothetis 1998; Blake and Logothetis 2002). In the analysis of these experiments, and more broadly in the search for models that might explain the phenomena, a key role is played by very quantitative measurements on human behavior; for an early example, see Borsellino et al. (1972) and for recent work, see Moreno-Bote et al. (2010). Models seeking to explain the alternation of our perceptions under nominally static inputs have appealed to adaptation dynamics that result in oscillations (Lehky 1988; Laing and Chow 2002), to noise that drives spontaneous transitions among attractor states in the relevant neural networks (Ditzinger and Haken 1990; Riani and Simonotto 1994; Moreno-Bote et al. 2007), and to subtle features in the optimal processing of ambiguous data (Bialek and DeWeese 1995).

Bialek and DeWeese 1995. Random switching and optimal processing in the perception of ambiguous signals. W Bialek and M DeWeese, *Phys Rev Lett* **74**, 3077–3080 (1995).

Blake and Logothetis 2002. Visual competition. R Blake and NK Logothetis, *Nat Rev Neurosci* **3**, 13–21 (2002).

Borsellino et al. 1972. Reversal time distribution in the perception of visual ambiguous stimuli. A. Borsellino, A Marco, A Allazetta, S Rinesi, and B Bartolini, *Biol Cybern* **10**, 139–144 (1972).

Ditzinger and Haken 1990. The impact of fluctuations on the recognition of ambiguous patterns. T Ditzinger and H Haken, *Biol Cybern* **63**, 453–456 (1990).

Ernst and Banks 2002. Humans integrate visual and haptic information in a statistically optimal fashion. MO Ernst and MS Banks, *Nature* **415**, 429–433 (2002).

Fisher 1968. Ambiguity of form: Old and new. GH Fisher, *Attention Percept and Psychophys* **4**, 189–192 (1968).

Griffiths and Tenenbaum 2006. Optimal predictions in everyday cognition. TL Griffiths and JB Tenenbaum, *Psych Sci* **17**, 767–773 (2006).

Laing and Chow 2002. A spiking neuron model for binocular rivalry. CR Laing and CC Chow, *J Comp Neurosci* **12**, 39–53 (2002).

Lehky 1988. An astable multivibrator model of binocular rivalry. SR Lehky, *Perception* **17**, 215–228 (1988).

Leopold and Logothetis 1996. Activity changes in early visual cortex reflect monkeys' percepts during binocular rivalry. DA Leopold and NK Logothetis, *Nature* **379**, 549–553 (1996).

Logothetis 1998. Single units and conscious vision. NK Logothetis, *Phil Trans R Soc Lond B* **353**, 1801–1818 (1998).

Logothetis and Schall 1989. Neuronal correlates of subjective visual perception. NK Logothetis and JD Schall, *Science* **245**, 761–763 (1989).

Long and Toppino 2004. Enduring interest in perceptual ambiguity: Alternating views of reversible figures. GM Long and TC Toppino, *Psych Bull* **130**, 748–768 (2004).

Ma et al. 2011. Behavior and neural basis of near optimal visual search. WJ Ma, V Navalpakkam, JM Beck, R van den Berg, and A Pouget, *Nat Neurosci* **14**, 783–790 (2011).

Moreno-Bote et al. 2007. Noise-induced alternations in an attractor network model of perceptual bistability. R Moreno-Bote, J Rinzel, and N Rubin, *J Neurophys* **98**, 1125–1139 (2007).

Moreno-Bote et al. 2010. Alternation rate in perceptual bistability is maximal at and symmetric around equi-dominance. R Moreno-Bote, A Shpiro, J Rinzel, and N Rubin, *J Vision* **10**, 1–18 (2010).

Necker 1832. Observations on some remarkable phaenomena seen in Switzerland: and on an optical phaenomenon which occurs on viewing a crystal or geometrical solid. LA Necker, *Phil Mag 3* **1**, 329–337 (1832).

Riani and Simonotto 1994. Stochastic resonance in the perceptual interpretation of ambiguous figures: A neural networks model. M Riani and E Simonotto, *Phys Rev Lett* **72**, 3120–3123 (1994).

Wheatstone 1838. Contributions to the physiology of vision. Part the first. On some remarkable, and hitherto unobserved, phenomena of binocular vision. C Wheatstone, *Phil Trans R Soc Lond* **128**, 371–394 (1838).

4.5 Proofreading and Active Noise Reduction

Now is a good time to look back at Schrödinger's (1944) remarkable little book. The results that motivated him were presented by Timoféef-Ressovsky et al. (1935). For some later perspectives, see Delbrück's 1970 Nobel lecture; the title refers to an earlier lecture, also very much worth reading for its eloquence and prescience (Delbrück 1949). A review of DNA structure is given in Appendix A.3, and some general references on molecular biology are listed in the Annotated Bibliography for Section 4.3. Our understanding of the genetic code emerged through a remarkable, decade-long interplay between theory and experiment. The key ideas can be traced through several papers by Crick (1958, 1963, 1967). The 1958 paper presents explicit descriptions of the "sequence hypothesis" and the "central dogma," perhaps for the first time in print. The 1963 paper includes a detailed overview of many theoretical ideas, most discarded along the way but nonetheless fascinating. Finally, by the 1967 paper, the presentation is more or less that of complete understanding. An extraordinary saga. Although many people played key roles, I think it is fair to say that Crick provided the most steady theoretical guidance throughout the long march to a complete codebook. Many of his papers, including personal correspondence during this period, are available at http://profiles.nlm.nih.gov/ps/retrieve/Collection/CID/SC. The ideas of kinetic proofreading—and, as emphasized in the text, the idea that there is a general physics problem cutting across a wide range of biological phenomena—were presented by Hopfield (1974) and Ninio (1975). Hopfield (1980) constructed a scenario in which the basic idea of paying (energetically) for increased accuracy still operates, but with none of the experimental signatures of the original proofreading scheme.

Crick 1958. On protein synthesis. FHC Crick, *Symp Soc Exp Biol* **12,** 138–163 (1958).

Crick 1963. Recent excitement in the coding problem. FHC Crick, *Prog Nucleic Acid Res Mol Biol* **1,** 163–217 (1963).

Crick 1967. The Croonian Lecture, 1966: The genetic code. FHC Crick, *Proc R Soc Lond* B **167,** 331–347 (1967).

Delbrück 1949. A physicist looks at biology. M Delbrück, *Trans Conn Acad Arts Sci* **38,** 173–190 (1949). Reprinted in *Phage and the Origins of Molecular Biology*, J Cairns, GS Stent, and JD Watson, eds., pp. 9–22 (Cold Spring Harbor Press, Cold Spring Harbor NY, 1966).

Delbrück 1970. A physicist's renewed look at biology: Twenty years later. M Delbrück, *Science* **168,** 1312–1315 (1970). Also available at http://nobelprize.org.

Hopfield 1974. Kinetic proofreading: A new mechanism for reducing errors in biosynthetic processes requiring high specificity. JJ Hopfield, *Proc Natl Acad Sci (USA)* **71,** 4135–4139 (1974).

Hopfield 1980. The energy relay: A proofreading scheme based on dynamic cooperativity and lacking all characteristic symptoms of kinetic proofreading in DNA replication and protein synthesis. JJ Hopfield, *Proc Natl Acad Sci (USA)* **77,** 5248–5252 (1980).

Ninio 1975. Kinetic amplification of enzyme discrimination. J Ninio *Biochimie* **57,** 587–595 (1975).

Schrödinger 1944. *What Is Life?* E Schrödinger (Cambridge University Press, Cambridge, 1944).

Timoféef-Ressovsky et al. 1935. Über die Natur der Genmutation und der Genstruktur. NW Timoféef-Ressovsky, KG Zimmer, and M Delbrück, *Nachrichten Gesellschaft Wissenschaften Göttingen* **1,** 190–245 (1935). Translated with historical commentary in *Creating a Physical Biology: The Three-Man Paper and Early Molecular Biology*, PR Sloan and Fogel, eds., pp. 221–271 (University of Chicago Press, Chicago, 2011).

Experimental evidence that proofreading actually occurs, more or less according to the kinetic scheme outlined by Hopfield, probably is clearest in the case of the tRNA charging enzymes (Fersht and Kaethner 1976; Hopfield et al. 1976; Yamane and Hopfield 1977). Perhaps the most dramatic qualitative evidence for proofreading is the fact that large classes of antibiotics act by disrupting the proofreading steps on the ribosome, a topic that has received renewed attention in the context of ribosome structure (Yonath 2005). As noted in the main text, the existence of proofreading implies that organisms face a trade-off between making errors and consuming energy. Attempts at a quantitative formulation of this problem begin with Ehrenberg and Kurland (1984), and echoes of this problem can be found even in very recent work (Scott et al. 2010). A further corollary of proofreading is that the mutation rate itself becomes (more easily) mutable, and this could have dramatic consequences for the pace of evolutionary change (Magnasco and Thaler 1996).

Ehrenberg and Kurland 1984. Costs of accuracy determined by a maximal growth rate constraint. M Ehrenberg and CG Kurland, Q *Rev Biophys* **17,** 45–82 (1984).

Fersht and Kaethner 1976. Enzyme hyperspecificity. Rejection of threonine by the valyl-tRNA synthetase by misacylation and hydrolytic editing. AR Fersht and MM Kaethner, *Biochemistry* **15,** 3342–3346 (1976).

Hopfield et al. 1976. Direct experimental evidence for kinetic proofreading in amino-acylation of tRNAIle. JJ Hopfield, T Yamane, V Yue, and SM Coutts, *Proc Natl Acad Sci (USA)* **71,** 1164–1168 (1976).

Magnasco and Thaler 1996. Changing the pace of evolution. MO Magnasco and DS Thaler, *Phys Lett A* **221,** 287–292 (1996).

Scott et al. 2010. Interdependence of cell growth and gene expression: Origins and consequences. M Scott, CW Gunderson, E Mateescu, Z Zhang, and T Hwa, *Science* **330,** 1009–1102 (2010).

Yamane and Hopfield 1977. Experimenal evidence for kinetic proofreading in the aminoacylation of tRNA by synthetase. T Yamane and JJ Hopfield, *Proc Natl Acad Sci (USA)* **74,** 2246–2250 (1977).

Yonath 2005. Antibiotics targeting ribosomes: Resistance, selectivity, synergism, and cellular regulation. A Yonath, *Annu Rev Biochem* **74,** 649–679 (2005).

The basic idea of kinetic proofreading—an enzymatic mechanism dissipating energy to stabilize a better-than-Boltzmann distribution of molecular states—has by now been applied in several different contexts. For the disentangling of DNA strands, see Yan et al. (1999, 2001), who were inspired in part by the experiments of Rybenkov et al. (1997). This is a good entry point for a beautiful subject, otherwise neglected here, namely, the topology of DNA and the enzymes that catalyze changes in this topology. For an introduction to the subject see Bauer et al. (1980), Wasserman and Cozzarelli (1986), and Vologodskii (1992). It should be emphasized that, absent topological enzymes, it would be difficult to believe that DNA really is a double helix, because of the problems of entanglement that were appreciated already by Watson and Crick in their original work on DNA structure (see Appendix A.3 and associated references). Thus, it is sobering to look at the title of Crick et al. (1979) and realize that this came more than 25 years after the double helix was first proposed. For the sensitivity and specificity of initial events in the immune response, see McKeithan (1995) and Altan-Bonnet and Germain (2005), and for a more general view of signal transduction specificity, see Swain and Siggia (2002).

Altan-Bonnet and Germain 2005. Modeling T cell antigen discrimination based on feedback control of digital ERK responses. G Altan-Bonnet and RN Germain, *PLoS Biology* **3,** 1925–1938 (2005).

Bauer et al. 1980. Supercoiled DNA. WR Bauer, FHC Crick, and JH White, *Scientific American* **243,** 118–133 (1980).

Crick et al. 1979. Is DNA really a double helix? FHC Crick, JC Wang, and WR Bauer, *J Mol Biol* **129**, 449–457 (1979).

McKeithan 1995. Kinetic proofreading in T-cell receptor signal transduction. TW McKeithan, *Proc Natl Acad Sci (USA)* **92**, 5042–5046 (1995).

Rybenkov et al. 1997. Simplification of DNA topology below equilibrium values by type II topoisomerases. VV Rybenkov, C Ullsperger, AV Vologodskii, and NR Cozzarelli, *Science* **277**, 690–693 (1997).

Swain and Siggia 2002. The role of proofreading in signal transduction specificity. PS Swain and ED Siggia, *Biophys J* **82**, 2928–2933 (2002).

Vologodskii 1992. *Topology and Physics of Circular DNA.* A Vologodskii (CRC Press, Boca Raton, FL, 1992).

Wasserman and Cozzarelli 1986. Biochemical topology. SA Wasserman and NR Cozzarelli, *Science* **232**, 951–960 (1986).

Yan et al. 1999. A kinetic proofreading mechanism for disentanglement of DNA by topoisomerases. J Yan, MO Magnasco, and JF Marko, *Nature* **401**, 932–935 (1999).

Yan et al. 2001. Kinetic proofreading can explain the suppression of supercoiling of circular DNAs by type-II topoisomerases. J Yan, MO Magnasco, and JF Marko, *Phys Rev E* **63**, 031909 (2001).

The use of optical forces to manipulate biological systems goes back to work by Ashkin and Dziedzic (1987), which in turn grew out of Ashkin's (1978, 1980) earlier work. It is worth noting that the same ideas of optical trapping for neutral dielectric particles were a key step in the development of atomic cooling, as described, for example, by Chu (1998). For the state of the art in single-molecule experiments, see Greenleaf et al. (2007). The experiments on single RNA polymerase molecules were by Shaevitz et al. (2003) and Abbondanzieri et al. (2005).

Abbondanzieri et al. 2005. Direct observation of base-pair stepping by RNA polymerase. EA Abbondanzieri, WJ Greenleaf, JW Shaevitz, R Landick, and SM Block, *Nature* **438**, 460–465 (2005).

Ashkin 1978. Trapping of atoms by resonance radiation pressure. A Ashkin, *Phys Rev Lett* **40**, 729–732 (1978).

Ashkin 1980. Applications of laser radiation pressure. A Ashkin, *Science* **210**, 1081–1088 (1980).

Ashkin and Dziedzic 1987. Optical trapping and manipulation of viruses and bacteria. A Ashkin and JM Dziedzic, *Science* **235**, 1517–1520 (1987).

Chu 1998. The manipulation of neutral particles. S Chu, *Revs Mod Phys* **70**, 685–706 (1998). Also available at http://nobelprize.org.

Greenleaf et al. 2007. High-resolution, single-molecule measurements of biomolecular motion. WJ Greenleaf, MT Woodside, and SM Block. *Annu Rev Biophys Biomol Struct* **36**, 171–190 (2007).

Shaevitz et al. 2003. Backtracking by single RNA polymerase molecules observed at near-base-pair resolution. JW Shaevitz, EA Abbondanzieri, R Landick, and SM Block, *Nature* **426**, 684–687 (2003).

For a review of structure and mechanics of the mammalian ear, see Dallos (1984). Early experiments on the mechanics of the stereocilia are by Flock and Strelioff (1984) and Strelioff and Flock (1984). The remarkable vibration sensitivity of the frog inner ear is described in Narins and Lewis (1984) and Lewis and Narins (1985). Estimates of the power flowing into the inner ear at threshold, as well as the minimum detectable power in other sensory systems, were given some time ago by Khanna and Sherrick (1981). The image of the cochlear hair cells in Fig. 4.51 appears in Ryan (2000).

Dallos 1984. Peripheral mechanisms of hearing. P Dallos, in *Comprehensive Physiology 2011, Supplement 3: Handbook of Physiology, The Nervous System, Sensory Processes,* pp. 595–637 (Wiley-Blackwell, 1984). Also available at: http://www.comprehensivephysiology.com/WileyCDA/Section/id420607.html.

Flock and Strelioff 1984. Studies on hair cells in isolated coils from the guinea pig cochlea. Å Flock and D Strelioff, *Hearing Res* **15,** 11–18 (1984).

Khanna and Sherrick 1981. The comparative sensitivity of selected receptor systems. SM Khanna and CE Sherrick, in *The Vestibular System: Function and Morphology,* T Gualtierotti, ed., pp. 337–348 (Springer-Verlag, New York, 1981).

Lewis and Narins 1985. Do frogs communicate with seismic signals? ER Lewis and PM Narins, *Science* **227,** 187–189 (1985).

Narins and Lewis 1984. The vertebrate ear as an exquisite seismic sensor. PM Narins and ER Lewis, *J Acoust Soc Am* **76,** 1384–1387 (1984).

Ryan 2000. Protection of auditory receptors and neurons: Evidence for interactive damage. AF Ryan, *Proc Natl Acad Sci (USA)* **97,** 6339–6340 (2000).

Strelioff and Flock 1984. Stiffness of sensory-cell hair bundles in the isolated guinea pig cochlea. D Strelioff and Å Flock, *Hearing Res* **15,** 19–28 (1984).

Recent examples of active cooling are by Corbitt et al. (2007) and Abbott et al. (2009), who are aiming at improving the sensitivity of gravitational wave detection. For discussion of the quantum limits to mechanical measurements, see Caves et al. (1980), Caves (1985), and Braginsky and Khalili (1992). For the quantum limits to amplifier noise (which has a long history), see Caves (1982). For discussions of thermal noise, I have always liked the treatment in Kittel's little book (cited in the Annotated Bibliography for Section 4.1), which includes a discussion of thermal noise power.

Abbott et al. 2009. Observation of a kilogram-scale oscillator near its quantum ground state. B Abbott, R Abbott, R Adhikari, P Ajith, B Allen, et al. (LIGO collaboration), *New J Phys* **11,** 073032 (2009).

Braginksy and Khalili 1992. *Quantum Measurement.* VB Bragnisky and FYa Khalili (Cambridge University Press, Cambridge, 1992).

Caves 1982. Quantum limits on noise in linear amplifiers. CM Caves, *Phys Rev D* **26,** 1817–1839 (1982).

Caves 1985. Defense of the standard quantum limit for free-mass position. CM Caves, *Phys Rev Lett* **54,** 2465–2468 (1985).

Caves et al. 1980. On the measurement of a weak classical force coupled to a quantum-mechanical oscillator. I: Issues of principle. CM Caves, KS Thorne, RWP Drever, VD Sandberg, and M Zimmermann, *Revs Mod Phys* **52,** 341–392 (1980).

Corbitt et al. 2007. Optical dilution and feedback cooling of a gram-scale oscillator to 6.9 mK. T Corbitt, C Wipf, T Bodiya, D Ottaway, D Sigg, N Smith, S Whitcomb, and N Mavalvala, *Phys Rev Lett* **99,** 160801 (2007).

The physics of hearing is a subject that goes back to Helmholtz (1863) and Rayleigh (1877). An early measurement using tone-on-tone masking to probe the sharpness of frequency selectivity in the cochlea was done by Wegel and Lane (1924); a modern account of psychophysical measurements on effective detection bandwidths is given by Moore (2003). By the mid-1960s, it became possible to compare these perceptual bandwidths with the tuning curves of individual sensory neurons emerging from the cochlea, as in Fig. 4.53 (Kiang et al. 1965), strengthening Helmholtz's idea that our sensations of tone have their roots in the dynamics of the ear itself. The classical measurements on cochlear mechanics by von Békésy were collected in 1960. The modern era of such measurements begins with the use of the Mössbauer effect (Frauenfelder 1962) to measure much smaller displacements of the basilar membrane (Johnstone and Boyle 1967) and to demonstrate that these motions have decidedly sharper frequency tuning in response to quieter sounds (Rhode 1971). These experiments triggered renewed interest in theories of cochlear mechanics, and some beautiful papers from this period are by Zweig (1976), Zweig et al. (1976), and Lighthill (1981), which connect the dynamics of the inner ear to more general physical considerations. There then followed a second wave of mechanical measurements, using interferometry (Khanna and Leonard 1982), Doppler velocimetry (Ruggero and Rich 1991) and improved Mössbauer methods (Sellick et al. 1982). Some perspective on these measurements and models as they emerged can be found in Lewis et al. (1985), which also makes connections to the mechanics of other inner ear organs. Basic quantities in models for basilar membrane motion are the pressure across the membrane and the mechanical impedance of the membrane itself. It is only recently that it has become possible to make direct measurements of these quantities (Olson 1999; Dong and Olson 2009).

von Békésy 1960. *Experiments in Hearing.* G von Békésy; EG Wever, ed. (McGraw-Hill, New York, 1960).

Dong and Olson 2009. In vivo impedance of the gerbil cochlear partition at auditory frequencies. W Dong and ES Olson, *Biophys J* **97,** 1233–1243 (2009).

Frauenfelder 1962. *The Mössbauer Effect.* H Frauenfelder (WA Benjamin, New York, 1962).

Helmholtz 1863. *Die Lehre von den Tonempfindungen als physiologische Grundlage für die Theorie der Musik.* H von Helmholtz (Vieweg und Sohn, Braunschweig, Germany, 1863). The most widely used translation is the second English edition, based on the fourth (and last) German edition of 1977; translated by AJ Ellis with an introduction by H Margenau, *On the Sensations of Tone as a Physiological Basis for the Theory of Music* (Dover, New York, 1954).

Johnstone and Boyle 1967. Basilar membrane vibration examined with the Mössbauer technique. BM Johnstone and AJF Boyle, *Science* **158,** 390–391 (1967).

Khanna and Leonard 1982. Basilar membrane tuning in the cat cochlea. SM Khanna and DGB Leonard, *Science* **215,** 305–306 (1982).

Kiang et al. 1965. *Discharge Patterns of Single Fibers in the Cat's Auditory Nerve.* Research Monograph 35. NYS Kiang, with T Watanabe, EC Thomas, and LF Clark (MIT Press, Cambridge MA, 1965).

Lewis et al. 1985. *The Vertebrate Inner Ear.* ER Lewis, EL Leverenz, and WS Bialek (CRC Press, Boca Raton FL, 1985).

Moore 2003. *An Introduction to the Psychology of Hearing,* fifth edition. BCJ Moore (Academic Press, San Diego, 2003).

Lighthill 1981. Energy flow in the cochlea. J Lighthill, *J Fluid Mech* **106,** 149–213 (1981).

Olson 1999. Direct measurement of intra-cochlear pressure waves. ES Olson, *Nature* **402,** 526–529 (1999).

Rayleigh 1877. *Theory of Sound.* JW Strutt, Baron Rayleigh (Macmillan, London, 1877). More commonly available is the revised and enlarged second edition, with a historical introduction by RB Lindsay (Dover, New York, 1945).

Rhode 1971. Observations of the vibration of the basilar membrane in squirrel monkeys using the Mössbauer technique. WS Rhode, *J Acoust Soc Am* **49,** 1218–1231 (1971).

Ruggero and Rich 1991. Application of a commercially-manufactured Doppler-shift laser velocimeter to the measurement of basilar membrane vibration. MA Ruggero and NC Rich, *Hearing Res* **51**, 215–230 (1991).

Sellick et al. 1982. Measurement of basilar membrane motion in the guinea pig using the Mössbauer technique. PM Sellick, R Patuzzi, and BM Johnstone, *J Acoust Soc Am* **72**, 131–141 (1982).

Wegel and Lane 1924. The auditory masking of one pure tone by another and its probable relation to the dynamics of the inner ear. RL Wegel and CE Lane, *Phys Rev* **23**, 266–285 (1924).[1]

Zweig 1976. Basilar membrane motion. G Zweig, *Cold Spring Harb Symp Quant Biol* **40**, 619–633 (1976).

Zweig et al. 1976. The cochlear compromise. G Zweig, R Lipes, and JR Pierce, *J Acoust Soc Am* **59**, 975–982 (1976).

The idea of active filtering in the inner ear goes back to a remarkably prescient paper by Gold (1948), who is better known, perhaps, for his contributions to astronomy and astrophysics; see Burbidge and Burbidge (2006). The idea that active elements are at work in the mechanics of the mammalian cochlea gained currency as experiments showed the "vulnerability" of frequency selectivity (Evans 1972), and with the dramatic observation of acoustic emissions from the ear (Kemp 1978; Zurek 1981; van Dijk et al. 2011). Importantly, these emissions are observed not only from the rather complex mammalian cochlea but also from simpler ears of amphibians. The idea that active filtering is essential for noise reduction is discussed in Bialek (1987). The view of active filtering as an approach to the Hopf bifurcation begins with Eguíluz et al. (2000) and Camalet et al. (2000); further developments include Ospeck et al. (2001), Duke and Jülicher (2003), Magnasco (2003), and Nadrowski et al. (2004). For a general discussion of dynamical systems and bifurcation theory, see Guckenheimer and Holmes (1983).

Bialek 1987. Physical limits to sensation and perception. W Bialek, *Annu Rev Biophys Biophys Chem* **16**, 455–478 (1987).

Burbidge and Burbidge 2006. Thomas Gold, 1920–2004. G Burbidge and M Burbidge, *Bio Mem Natl Acad Sci (USA)* **88**, 1–15 (2006).

Camalet et al. 2000. Auditory sensitivity provided by self-tuned critical oscillations of hair cells. S Calamet, T Duke, F Jülicher, and J Prost, *Proc Natl Acad Sci (USA)* **97**, 3183–3187 (2000).

van Dijk et al. 2011. The effect of static ear canal pressure on human spontaneous otoacoustic emissions: Spectral width as a measure of the intra-cochlear oscillation amplitude. P van Dijk, B Maat, and E de Kleine, *J Assoc Res Otolaryngol* **12**, 13–28 (2011).

Duke and Jülicher 2003. Active travelling wave in the cochlea. T Duke and F Jülicher, *Phys Rev Lett* **90**, 158101 (2003).

Eguíluz et al. 2000. Essential nonlinearities in hearing. VM Eguíluz, M Ospeck, Y Choe, AJ Hudspeth, and MO Magnasco, *Phys Rev Lett* **84**, 5232–5235 (2000).

Evans 1972. The frequency response and other properties of single fibres in the guinea-pig cochlear nerve. EF Evans, *J Physiol (Lond)* **226**, 263–287 (1972).

Gold 1948. Hearing. II: The physical basis of the action of the cochlea. T Gold, *Proc R Soc Lond B* **135**, 492–498 (1948).

---

1. It is amusing to note that this paper sometimes is cited in the biological literature as having been published in the journal *Physiological Reviews*. Presumably this reflects authors or editors copying the reference to "*Phys Rev*" and "correcting" it to "*Physiol Rev.*"

Guckenheimer and Holmes 1983. *Nonlinear Oscillations, Dynamical Systems and Bifurcation of Vector Fields.* J Guckenheimer and P Holmes (Springer-Verlag, Berlin, 1983).

Kemp 1978. Stimulated acoustic emissions from within the human auditory system. DT Kemp, *J Acoust Soc Am* **64,** 1386–1391 (1978).

Magnasco 2003. A traveling wave over a Hopf bifurcation shapes the cochlear tuning curve. MO Magnasco, *Phys Rev Lett* **90,** 058101 (2003).

Nadrowski et al. 2004. Active hair-bundle motility harnesses noise to operate near an optimum of mechanosensitivity. B Nadrowski, P Martin, and F Jülicher, *Proc Natl Acad Sci (USA)* **101,** 12195–12199 (2004).

Ospeck et al. 2001. Evidence of a Hopf bifurcation in frog hair cells. M Ospeck, VM Equíluz, and MO Magnasco, *Biophys J* **80,** 2597–2607 (2001).

Zurek 1981. Spontaneous narrowband acoustic signals emitted by human ears. PM Zurek, *J Acoust Soc Am* **69,** 514–523 (1981).

5 No Fine Tuning

5.1 Sequence Ensembles

A good general reference about proteins is Fersht (1998). The proposal of helical structures for polypeptides goes back to the classic paper of Pauling and Corey (1951). For a modern introduction to polymer physics, see de Gennes (1979). The small simulation in the problems is not a substitute for exploring the theory of spin glasses; the classic papers are collected, with an introduction, by Mézard et al. (1986), and a textbook account is given by De Dominicis and Giardina (2006). Early efforts to apply these methods to the random heteropolymer were made by Shakhnovich and Gutin (1989).

De Dominicis and Giardina 2006. *Random Fields and Spin Glasses.* C De Dominicis and I Giardina (Cambridge University Press, Cambridge, 2006).

Fersht 1998. *Structure and Mechanism in Protein Science: A Guide to Enzyme Catalysis and Protein Folding.* AR Ferhst (WH Freeman, San Francisco, 1998).

de Gennes 1979. *Scaling Concepts in Polymer Physics.* PG de Gennes (Cornell University Press, Ithaca NY, 1979).

Mézard et al. 1986. *Spin Glass Theory and Beyond.* M Mézard, G Parisi, and MA Virasoro (World Scientific, Singapore, 1986).

Pauling and Corey 1951. Atomic coordinate and structure factors for two helical configurations of polypeptide chains. L Pauling and RB Corey, *Proc Natl Acad Sci (USA)* **37,** 235–240 (1951).[2]

Shakhnovich and Gutin 1989. Formation of unique structure in polypeptide chains: Theoretical investigation with the aid of a replica approach. EI Shakhnovich and AM Gutin, *Biophys Chem* **34,** 187–199 (1989).

Models that incorporate only native interactions, with no frustration, have their origin in work by Gō, reviewed in Gō (1983). A more explicit discussion of minimizing frustration as a principle was given by Bryngelson and Wolynes (1987), and the funnel landscape of Fig. 5.5 is described in Onuchic et al. (1995). Detailed simulations based on the Gō model are described by Clementi et al. (2000a,b).

---

2. This paper was the first in a remarkable series on protein structure. For some perspective, see http://www.pnas.org/site/misc/classics1.shtml.

Bryngleson and Wolynes 1987. Spin glasses and the statistical mechanics of protein folding. JD Bryngelson and PG Wolynes, *Proc Natl Acad Sci (USA)* **84,** 7524–7528 (1987).

Clementi et al. 2000a. How native-state topology affects the folding of dihydrofolate reductase and interleukin-1β. C Clementi, PA Jennings, and JN Onuchic, *Proc Natl Acad Sci (USA)* **97,** 5871–5876 (2000).

Clementi et al. 2000b. Topological and energetic factors: What determines the structural details of the transition state ensemble and "en-route" intermediates for protein folding? An investigation for small globular proteins. C Clementi, H Nymeyer, and JN Onuchic, *J Mol Biol* **298,** 937–953 (2000).

Gō 1983. Theoretical studies of protein folding. N Gō, *Annu Rev Biophys Bioeng* **12,** 183–210 (1983).

Onuchic et al. 1995. Toward an outline of the topography of a realistic protein-folding funnel. JN Onuchic, PG Wolynes, Z Luthey-Schultern, and ND Socci, *Proc Natl Acad Sci (USA)* **92,** 3626–3630 (1995).

The lattice simulations exploring protein designability were by Li et al. (1996). The analytic argument connecting designability to the eigenvalues of the contact matrix was given by England and Shakhnovich (2003), and Li et al. (1998) gave the argument relating folding to error correction in the HP model.

England and Shakhnovich 2003. Structural determinant of protein designability. JL England and EI Shakhnovich, *Phys Rev Lett* **90,** 218101 (2003).

Li et al. 1996. Emergence of preferred structures in a simple model of protein folding. H Li, R Helling, C Tang, and N Wingreen, *Science* **273,** 666–669 (1996).

Li et al. 1998. Are protein folds atypical? H Li, C Tang, and NS Wingreen, *Proc Natl Acad Sci (USA)* **95,** 4987–4990 (1998).

Recognizing from sequence data alone that different proteins belong in the same family is one of the central problems in bioinformatics. It is intimately connected to the problem of sequence alignment and became more urgent as sequence data began to emerge more rapidly. Classic papers include Smith and Waterman (1981), Lipman and Pearson (1985), and Altschul et al. (1990). Using common assumptions, Hwa and Lässig (1996) showed that the successful alignment of long sequences involves a phase transition at a critical value of the degree of similarity. Importantly, all these approaches involve some hypotheses about how to measure similarity, and (from a Bayesian point of view) this is equivalent to a claim about the distribution of sequences that can arise in a given family; making this claim consistent with what we are learning about these distributions is still challenging.

Altschul et al. 1990. Basic local alignment search tool. S Altschul, W Gish, W Miller, E Myers, and D Lipman, *J Mol Biol* **215,** 403–410 (1990).

Hwa and Lässig 1996. Similarity detection and localization. T Hwa and M Lässig, *Phys Rev Lett* **76,** 2591–2594 (1996).

Lipman and Pearson 1985. Rapid and sensitive protein similarity searches. DJ Lipman and WR Pearson, *Science* **227,** 1431–1435 (1985).

Smith and Waterman 1981. Identification of common molecular subsequences. TF Smith and MS Waterman, *J Mol Biol* **147,** 195-197 (1981).

The idea of protein families was essential in the experiments that searched for, and found, the receptors in the olfactory system (Buck and Axel 1991); see Axel (2005) and also Buck

(2005) in Section 2.3. The structural correspondence between bacterial serine proteases and their mammalian counterparts is from Brayer et al. (1978, 1979) and Fujinaga et al. (1985). Experiments on the sampling of sequence space while preserving one-point and two-point correlations were done by Socolich et al. (2005) and Russ et al. (2005). The equivalence of these ideas to the maximum entropy method was shown in Bialek and Ranganthan (2007). For more on maximum entropy approaches to sequence ensembles, see Halabi et al. (2009), Weigt et al. (2009), Mora et al. (2010), and Marks et al. (2011). For a broader view of maximum entropy models applied to biological systems, see Appendix A.7 and Mora and Bialek (2011).

Axel 2005. Scents and sensibility: A molecular logic of olfactory perception. R Axel, in *Les Prix Nobel 2004*, T Frängsmyr, ed., pp. 234–256 (Nobel Foundation, Stockholm, 2004). Also available at http://www.nobelprize.org.

Bialek and Ranganthan 2007. Rediscovering the power of pairwise interactions. W Bialek and R Ranganathan, arXiv:0712.4397 [q-bio.QM] (2007).

Brayer et al. 1978. Molecular structure of crystalline *Streptomyces gresius* protease A at 2.8 Å resolution: II. Molecular conformation, comparison with α-chymotrypsin, and active-site geometry. GD Brayer, LTJ Delbaere, and MNG James, *J Mol Biol* **124,** 261–283 (1978).

Brayer et al. 1979. Molecular structure of the α-lytic protease from *Myxobacter 495* at 2.8 Å resolution. GD Brayer, LTJ Delbaere, and MNG James, *J Mol Biol* **131,** 743–775 (1979).

Buck and Axel 1991. A novel multigene family may encode odorant receptors: A molecular basis for odor recognition. L Buck and R Axel, *Cell* **65,** 175–187 (1991).

Fujinaga et al. 1985. Refined structure of α-lytic protease at 1.7 Å resolution: Analysis of hydrogen bonding and solvent structure. M Fujinaga, LTJ Delbaere, GD Brayer, and MNG James, *J Mol Biol* **183,** 479–502 (1985).

Halabi et al. 2009. Protein sectors: Evolutionary units of three-dimensional structure. N Halabi, O Rivoire, S Leibler, and R Ranganathan, *Cell* **138,** 774–786 (2009).

Marks et al. 2011. Protein 3D structure computed from evolutionary sequence variation. DS Marks, LJ Colwell, R Sheridan, TA Hopf, A Pagnani, R Zecchina, and C Sander, *PLoS One* **6,** e28766 (2011).

Mora and Bialek 2011. Are biological systems poised at criticality? T Mora and W Bialek. *J Stat Phys* **144,** 268–302 (2011); arXiv:1012.2242 [q-bio.QM] (2010).

Mora et al. 2010. Maximum entropy models for antibody diversity. T Mora, AM Walczak, W Bialek, and CG Callan, *Proc Natl Acad Sci (USA)* **107,** 5405–5410 (2010).

Russ et al. 2005. Natural-like function in artificial WW domains. WP Russ, DM Lowery, P Mishra, MB Yaffe, and R Ranganathan, *Nature* **437,** 579–583 (2005).

Socolich et al. 2005. Evolutionary information for specifying a protein fold. M Socolich, SW Lockless, WP Russ, H Lee, KH Gardner, and R Ranganathan, *Nature* **437,** 512–518 (2005).

Weigt et al. 2009. Identification of direct residue contacts in protein-protein interaction by message passing. M Weigt, RA White, H Szurmant, JA Hoch, and T Hwa, *Proc Natl Acad Sci (USA)* **106,** 67–72 (2009).

The modern picture of transcriptional regulation traces its origins to Jacob and Monod (1961), another of the great and classic papers that still are rewarding to read decades after they were published. Their views were motivated primarily by studies of the *lac* operon, and the origins of these reach back to Monod's thesis (1942), which was concerned with the phenomenology of bacterial growth. As recounted in Judson (1979), for example, the idea that genes turn on because of the release from repression was due to Szilard; the written record of these ideas is not as clear as it could be, but one can try Szilard (1960). For a modern view, faithful to the history, see Müller-Hill (1996). The other "simple" paradigmatic example of protein-DNA interactions in the regulation of gene expression is the case of bacteriophage λ, which is reviewed by Ptashne (1986), which has also evolved with time (Ptashne 1992); see also Ptashne (2001). These systems provided the background for the pioneering discussion of sequence specificity in protein-DNA interactions (von Hippel and Berg 1986; Berg and von Hippel 1987, 1988). In parallel to this statistical approach, there were direct biochemical measurements of binding energies; an early attempt to bring these different literatures into correspondence was by Stormo and Fields (1998).

Berg and von Hippel 1987. Selection of DNA binding sites by regulatory proteins. I: Statistical-mechanical theory and application to operators and promoters. OG Berg and PH von Hippel, *J Mol Biol* **193,** 723–743 (1987).

Berg and von Hippel 1988. Selection of DNA binding sites by regulatory proteins. II: The binding specificity of cyclic AMP receptor protein to recognition sites. OG Berg and PH von Hippel, *J Mol Biol* **200,** 709–723 (1988).

von Hippel and Berg 1986. On the specificity of DNA-protein interactions. PH von Hippel and OG Berg, *Proc Natl Acad Sci (USA)* **83,** 1608–1612 (1986).

Jacob and Monod 1961. Genetic regulatory mechanisms in the synthesis of proteins. F Jacob and J Monod, *J Mol Biol* **3,** 318–356 (1961).

Judson 1979. *The Eighth Day of Creation* HF Judson (Simon and Schuster, New York, 1979).

Monod 1942. *Recherche sur la Croissance des Cultures Bactériennes.* J Monod (Hermann, Paris, 1942).

Müller-Hill 1996. *The lac Operon: A Short History of a Genetic Paradigm.* B Müller-Hill (W de Gruyter and Co, Berlin, 1996).

Ptashne 1986. *A Genetic Switch: Gene Control and Phage λ.* M Ptashne (Cell Press, Cambridge MA, 1986).

Ptashne 1992. *A Genetic Switch: Phage λ and Higher Organisms.* M Ptashne (Cell Press, Cambridge MA, 1992).

Ptashne 2001. *Genes and Signals.* M Ptashne (Cold Spring Harbor Laboratory Press, New York, 2001).

Stormo and Fields 1998. Specificity, free energy and information content in protein-DNA interactions. GD Stormo and DS Fields, *Trends Biochem Sci* **23,** 109–113 (1998).

Szilard 1960. The control of the formation of specific proteins in bacteria and in animal cells. L Szilard, *Proc Natl Acad Sci (USA)* **46,** 277–292 (1960).

The emergence of whole genome sequences opened several new approaches to the problem of specificity. One important idea is that sequences that are targets for protein binding should have a nonrandom structure, and we should be able to find this structure in a relatively unsupervised fashion (Busemaker et al. 2000a,b). Several experimental methods have been developed that make it possible to survey, with varying degrees of precision, the binding of particular proteins to all possible sites along the genome. Among these methods are protein-binding microarrays (Mukherjee et al. 2004), chromatin immunoprecipitation (Lee et al. 2002), and microfluidics (Maerkl and Quake 2007). For an approach to the analysis of such measurements making explicit use of dimensionality reduction methods (as discussed in relation to neural computation in Section 4.4), see Kinney et al. (2007). This approach inspired experiments aimed at wider exploration of sequence space (Kinney et al. 2010). For other such explorations, see Ligr et al. (2006) and Gertz et al. (2009).

Bussemaker et al. 2000a. Building a dictionary for genomes: Identification of presumptive regulatory sites by statistical analysis. H Bussemaker, H Li, and ED Siggia, *Proc Natl Acad Sci (USA)* **97,** 10096–10100 (2000).

Bussemaker et al. 2000b. Regulatory element detection using a probabilistic segmentation algorithm. H Bussemaker, H Li, and ED Siggia, *Proc Int Conf Intell Sys Mol Biol* **8,** 67–74 (2000).

Gertz et al. 2009. Analysis of combinatorial cis-regulation in synthetic and genomic promoters. J Gertz, ED Siggia, and BA Cohen, *Nature* **457,** 215–218 (2009).

Kinney et al. 2007. Precise physical models of protein-DNA interaction from high-throughput data. JB Kinney, G Tkačik, and CG Callan, Jr, *Proc Natl Acad Sci (USA)* **104,** 501–506 (2007).

Kinney et al. 2010. Using deep sequencing to characterize the biophysical mechanism of a transcriptional regulatory sequence. JB Kinney, A Murugan, CG Callan, Jr, and EC Cox, *Proc Natl Acad Sci (USA)* **107,** 9158–9163 (2010).

Lee et al. 2002. Transcriptional regulatory networks in *Saccharomyces cerevisae.* TI Lee, NJ Rinaldi, F Robert, DT Odom, Z Bar-Joseph, et al., *Science* **298,** 799–804 (2002).

Ligr et al. 2006. Gene expression from random libraries of yeast promoters. M Ligr, R Siddharthan, FR Cross, and ED Siggia, *Genetics* **172,** 2113–2122 (2006).

Maerkl and Quake 2007. A systems approach to measuring the binding energy landscape of transcription factors. SJ Maerkl and SR Quake, *Science* **315,** 233–237 (2007).

Mukherjee et al. 2004. Rapid analysis of the DNA binding specificities of transcription factors with DNA microarrays. S Mukherjee, MF Berger, G Jona, XS Wang, D Muzzey, M Snyder, RA Young, and ML Bulyk, *Nat Genet* **36,** 1331–1339 (2004).

If we can take seriously the linear model of binding energies, then we can start to ask how the many transcription factors tile the space of possible specificities (Sengupta et al. 2002), and there are more general questions to be asked about the way in which regulatory signals come to specify particular genes (Wunderlich and Mirny 2009). We can look at the evolution of binding sites by comparing closely related organisms, testing the hypothesis that all that matters is the predicted binding energy (Mustonen et al. 2008), and as an alternative we can look at the evolution of the transcription factors themselves (Maerkl and Quake 2009).

Maerkl and Quake 2009. Experimental determination of the evolvability of a transcription factor. SJ Maerkl and SR Quake, *Proc Natl Acad Sci (USA)* **106,** 18650–18655 (2009).

Mustonen et al. 2008. Energy-dependent fitness: A quantitative model for the evolution of yeast transcription factor binding sites. V Mustonen, J Kinney, CG Callan, Jr, and M Lässig, *Proc Natl Acad Sci (USA)* **105,** 12376–12381 (2008).

Sengupta et al. 2002. Specificity and robustness in transcription control networks. A Sengupta, M Djordjevic, and BI Shraiman, *Proc Natl Acad Sci (USA)* **99,** 2072–2077, (2002).

Wunderlich and Mirny 2009. Different strategies for transcriptional regulation are revealed by information-theoretical analysis of binding motifs. Z Wunderlich and LA Mirny, *Trends Genet* **25,** 434–440 (2009); arXiv.org:0812.3910 [q-bio.GN] (2008).

5.2 Ion Channels and Neuronal Dynamics

Our understanding of ion channels goes back to the classic papers of Hodgkin and Huxley (1952a–d), still very much worth reading. The series of papers (of which the first really is Hodgkin et al. 1952) describes many ingenious experiments, culminating in a mathematical model that predicts the form and speed of the action potential. For some perspective on this work, see Hodgkin (1958, 1979). For a modern textbook account, see Dayan and Abbott (2001). The Hodgkin-Huxley model is complicated, so over the years there have been various attempts at simplifying to the point where one can gain analytic insight. The original approach to this problem was by FitzHugh (1961) and Nagumo et al. (1962); for modern approaches, see Abbott and Kepler (1990) and in Nelson (2004).

Abbott and Kepler 1990. Model neurons: From Hodgkin-Huxley to Hopfield. LF Abbott and T Kepler, in *Statistical Mechanics of Neural Networks*, L Garrido, ed., pp. 5–18 (Springer-Verlag, Berlin, 1990).

Dayan and Abbott 2001. *Theoretical Neuroscience.* P Dayan and LF Abbott (MIT Press, Cambridge MA, 2001).

FitzHugh 1961. Impulses and physiological states in theoretical models of nerve membrane. R FitzHugh, *Biophys J* **1,** 445–466 (1961).

Hodgkin 1958. The Croonian Lecture: Ionic movement and electrical activity in giant nerve fibers. AL Hodgkin, *Proc R Soc Lond B* **148,** 1–37 (1958).

Hodgkin 1979. Chance and design in electophysiology: an informal account of certain experiments on nerve carried out between 1934 and 1952. AL Hodgkin, in *Pursuit of Nature: Information Essays on the History of Physiology*, pp. 1–21 (Cambridge University Press, Cambridge, 1979).

Hodgkin and Huxley 1952a. Currents carried by sodium and potassium ions through the membrane of the giant axon of *Loligo*. AL Hodgkin and AF Huxley, *J Physiol (Lond)* **116,** 449–472 (1952).

Hodgkin and Huxley 1952b. The components of membrane conductance in the giant axon of *Loligo*. AL Hodgkin and AF Huxley, *J Physiol (Lond)* **116,** 473–496 (1952).

Hodgkin and Huxley 1952c. The dual effect of membrane potential on sodium conductance in the giant axon of *Loligo*. AL Hodgkin and AF Huxley, *J Physiol (Lond)* **116,** 497–506 (1952).

Hodgkin and Huxley 1952d. A quantitative description of membrane current and its application to conduction and excitation in nerve. AL Hodgkin and AF Huxley, *J Physiol (Lond)* **117,** 500–544 (1952).

Hodgkin et al. 1952. Measurement of the current-voltage relations in the membrane of the giant axon of *Loligo*. AL Hodgkin, AF Huxley, and B Katz, *J Physiol (Lond)* **117,** 442–448 (1952).

Nagumo et al. 1962. An active pulse transmission line simulating nerve axon. J Nagumo, S Arimoto, and S Yoshizawa, *Proc IRE* **50,** 2061–2070 (1962).

Nelson 2004. *Biological Physics: Energy, Information, Life.* P Nelson (WH Freeman, San Francisco, 2004).

For a modern view of ion channels, see Hille (2001). For a detailed discussion of a system in which the effective resonance generated by channel kinetics has functional importance, see Wu and Fettiplace (2001). For one example of the added complexities of splicing and phosphorylation, see Tian et al. (2001).

Hille 2001. *Ion Channels of Excitable Membranes,* third edition. B Hille (Sinauer, Sunderland MA, 2001).

Tian et al. 2001. Alternative splicing switches potassium channel sensitivity to protein phosphorylation. L Tian, RR Duncan, MS Hammon, LS Coghill, H Wen, R Rusinova, AG Clark, IB Levitan, and MJ Shipston, *J Biol Chem* **276,** 7717–7720 (2001).

Wu and Fettiplace 2001. A developmental model for generating frequency maps in the reptilian and avian cochleas. YC Wu and R Fettiplace, *Biophys J* **70,** 2557–2570 (2001).

The problem of setting the numbers of each kind of ion channel emerged in attempts to make quantitative models of individual neurons in the stomatogastric ganglion. For a recent overview of the stomatogastric ganglion, emphasizing its role as a model system for studying network dynamics, see Marder and Bucher (2007). These models reached a very high degree of sophistication, as described in the series of papers by Buchholtz et al. (1992), Golowasch and Marder (1992), and Golowasch et al. (1992). The basic idea of regulating the number of ion channels by feedback from the electrical activity of the cell was described by LeMasson et al. (1993); see Abbott and LeMasson (1993) for a more complete account. Dramatic experimental evidence for self-tuning of channel numbers came (quickly) from Turrigiano et al. (1994). For feedback mechanisms sensitive to multiple time scales, see Liu et al. (1998). A more complete, global view of single neuron dynamics in the space of channel copy numbers, including the nonconvexity of regions in parameter space that implement similar functions, was presented by Goldman et al. (2001). More recently it has become possible to connect the problem of robustness in neuronal dynamics more directly to ideas about the control of gene expression by measuring, directly, the mRNA levels for individual channel types (Schultz et al. 2006, 2007). Finally, everything we have said about single neurons can be revisited at the level of networks (Prinz et al. 2004).

Abbott and LeMasson 1993. Analysis of neuron models with dynamically regulated conductances. LF Abbott and G LeMasson, *Neural Comp* **5,** 823–842 (1993).

Buchholtz et al. 1992. Mathematical model of an identified stomatogastric ganglion neuron. F Buchholtz, J Golowasch, IR Epstein, and E Marder, *J Neurophysiol* **67,** 332–340 (1992).

Goldman et al. 2001. Global structure, robustness, and modulation of neuronal models. MS Goldman, J Golowasch, E Marder, and LF Abbott, *J Neurosci* **21,** 5229–5238 (2001).

Golowasch and Marder 1992. Ionic currents of the lateral pyloric neuron of the stomatogastric ganglion of the crab. J Golowasch and E Marder, *J Neurophysiol* **67,** 318–331 (1992).

Golowasch et al. 1992. The contribution of individual ionic currents to the activity of a model stomatogastric ganglion neuron. J Golowasch, F Buchholtz, IR Epstein, and E Marder, *J Neurophysiol* **67,** 341–349 (1992).

LeMasson et al. 1993. Activity-dependent regulation of conductances in model neurons. G LeMasson, E Marder, and LF Abbott, *Science* **259,** 1915–1917 (1993).

Liu et al. 1998. A model neuron with activity-dependent conductances regulated by multiple calcium sensors. Z Liu, J Golowasch, E Marder, and LF Abbott, *J Neurosci* **18,** 2309–2320 (1998).

Marder and Bucher 2007. Understanding circuit dynamics using the stomatogastric nervous system of lobsters and carbs. E Marder and D Bucher, *Annu Rev Physiol* **69,** 291–316 (2007).

Prinz et al. 2004. Similar network activity from disparate circuit parameters. AA Prinz, D Bucher, and E Marder, *Nat Neurosci* **7,** 1345–1352 (2004).

Schultz et al. 2006. Variable channel expression in identified single and electrically coupled neurons in different animals. DJ Schultz, JM Goaillard, and E Marder, *Nat Neurosci* **9,** 356–362 (2006).

Schultz et al. 2007. Quantitative expression profiling of identified neurons reveals cell-specific constraints on highly variable levels of gene expression. DJ Schultz, JM Goaillard, and E Marder, *Proc Natl Acad Sci (USA)* **105,** 690–702 (2007).

Turrigiano et al. 1994. Activity-dependent changes in the intrinsic properties of cultured neurons. G Turrigiano, LF Abbott, and E Marder, *Science* **264,** 974–977 (1994).

5.3 The States of Cells

Some basic ideas about adaptation in sensory neurons were established early on by Adrian (1926) and Adrian and Zotterman (1926a,b); for a review, see Rieke et al. (1997).

Adrian 1926. The impulses produced by sensory nerve endings: Part I. ED Adrian, *J Physiol (Lond)* **61,** 49–72 (1926).

Adrian and Zotterman 1926a. The impulses produced by sensory nerve endings: Part II. The response of a single end organ. ED Adrian and Y Zotterman, *J Physiol (Lond)* **61,** 151–171 (1926).

Adrian and Zotterman 1926b. The impulses produced by sensory nerve endings: Part III. Impulses set up by touch and pressure. ED Adrian and Y Zotterman, *J Physiol (Lond)* **61,** 465–483 (1926).

Rieke et al. 1997. *Spikes: Exploring the Neural Code.* F Rieke, D Warland, R de Ruyter van Steveninck, and W Bialek (MIT Press, Cambridge MA, 1997).

Quantitative measurements on adaptation in bacterial chemotaxis go back at least to Block et al. (1983). Renewed interest in this system was triggered by the work of Barkai and Leibler (1997), who used adaptation in chemotaxis as an example for the more general problem of robustness. The idea that perfect adaptation could be achieved even in the presence of variations in protein copy numbers was then tested more directly by Alon et al. (1999). Recent work suggests that, although the biochemical network responsible for chemotaxis may allow for robustness against variations in protein copy numbers, under natural conditions there is relatively precise control over (at least) relative copy numbers, even for proteins on different operons (Kollmann et al. 2005). In competition experiments, one can even show that tight correlations between protein concentrations improve chemotactic performance (Løvdok et al. 2009). For an example of more detailed calculations on the precision of adaptation by methylation, see Meir et al. (2010), and for the experiments of Fig. 5.37, see MacLean et al. (2003).

Alon et al. 1999. Robustness in bacterial chemotaxis. U Alon, MG Surette, N Barkai, and S Leibler, *Nature* **397,** 168–171 (1999).

Barkai and Leibler 1997. Robustness in simple biochemical networks. N Barkai and S Leibler, *Nature* **387,** 913–917 (1997).

Block et al. 1983. Adaptation kinetics in bacterial chemotaxis. SM Block, JE Segall, and HC Berg, *J Bacteriol* **154,** 312–323 (1983).

Kollmann et al. 2005. Design principles of a bacterial signalling network. M Kollmann, L Løvdok, K Bartholome, J Timmer, and V Sourjik *Nature* **438,** 504–507 (2005).

Løvdok et al. 2009. Role of translational coupling in robustness of bacterial chemotaxis pathway. L Løvdok, K Bentele, N Vladimirov, A Müller, FS Pop, D Lebiedz, M Kollmann, and V Sourjik, *PLoS Biology* **7,** e1000171 (2009).

MacLean et al. 2003. Activity-independent homeostasis in rhythmically active neurons. JN MacLean, Y Zhang, BR Johnson, and RM Harris-Warrick, *Neuron* **37,** 109–120 (2003).

Meir et al. 2000. Precision and kinetics of adaptation in bacterial chemotaxis. Y Meir, V Jakovljevic, O Oleksiuk, V Sovnik, and NS Wingreen, *Biophys J* **99,** 2766–2774 (2010).

The effort to schematize the cell cycle as a Boolean network is described by Li et al. (2004), Zhang et al. (2006), and Lau et al. (2007). A contrasting approach involves exploiting modern single-cell imaging and manipulation to get at the dynamics of particular molecular components in the system, showing how small pieces of the network ensure that transitions between successive states are deterministic despite noise (Bean et al. 2006; DiTalia et al. 2007; Skotheim et al. 2008). The most recent work along these lines shows that following the concentration of a single component in the network is sufficient to predict the fate of the cell as it "decides" to make a transition from one state to another (Doncic et al. 2011).

Bean et al. 2006. Coherence and timing of cell cycle start examined at single-cell resolution. JM Bean, ED Siggia, and FR Cross, *Mol Cell* **21,** 3–14 (2006).

DiTalia et al. 2007. The effects of molecular noise and size control on variability in the budding yeast cell cycle. S DiTalia, JM Skotheim, JM Bean, ED Siggia, and FR Cross, *Nature* **448,** 947–952 (2007).

Doncic et al. 2011. Distinct interactions select and maintain a specific cell fate. A Doncic, M Falleur-Fettig, and JM Skotheim, *Mol Cell* **43** 528–539 (2011).

Lau et al. 2007. Function constrains network architecture and dynamics: A case study on the yeast cell cycle Boolean network. K Lau, S Ganguli, and C Tang, *Phys Rev E* **75,** 051907 (2007).

Li et al. 2004. The yeast cell cycle network is robustly designed. F Li, Y Lu, T Long, Q Ouyang, and C Tang, *Proc Natl Acad Sci (USA)* **101,** 4781–4786 (2004).

Skotheim et al. 2008. Positive feedback of G1 cyclins ensures coherent cell cycle entry. JM Skotheim, S DiTalia, ED Siggia, and FR Cross, *Nature* **454,** 291–297 (2008).

Zhang et al. 2006. A stochastic model of the yeast cell cycle network. Y Zhang, M Qian, Q Ouyang, M Deng, F Li, and C Tang, *Physica D* **219,** 35–39 (2006).

A modern textbook account of development in the fly embryo is provided by Lawrence (1992). We know which genes are relevant to the earliest events in patterning because of pioneering experiments first by E. B. Lewis and then by E. F. Wieschaus and C. Nüsslein-Vollhard. Lewis identified a series of puzzling mutant flies where a mutation in a single gene could generate flies that were missing segments, or had extra segments. It is as if the "program" of embryonic development has subroutines. Wieschaus and Nüsslein-Vollhard decided to search for all the genes such that mutations in those genes would perturb the formation of spatial structure in the embryo, and they found that there are surprisingly few such genes, on the order of 100. To get a feeling for all this, one can certainly do worse than to read the Nobel lectures from 1995 (Lewis 1997; Nüsslein-Volhard 1997; Wieschaus 1997).

Lawrence 1992. *The Making of a Fly: The Genetics of Animal Design.* PA Lawrence (Blackwell, Oxford, 1992).

Lewis 1997. The bithorax complex: The first fifty years. EB Lewis, in *Nobel Lectures, Medicine or Physiology 1991–1995,* N Ringertz, ed., pp. 247–272 (World Scientific, Singapore, 1997). Also available at http://www.nobelprize.org.

Nüsslein-Volhard 1997. The identification of genes controlling development in flies and fishes. C Nüsslein-Volhard, in *Nobel Lectures, Medicine or Physiology 1991–1995,* N Ringertz, ed., pp. 285–306 (World Scientific, Singapore, 1997). Also available at http://www.nobelprize.org.

Wieschaus 1997. Molecular patterns to morphogenesis: The lessons from *Drosophila.* EF Wieschaus, in *Nobel Lectures, Medicine or Physiology 1991–1995,* N Ringertz, ed., pp. 314–326 (World Scientific, Singapore, 1997). Also available at http://www.nobelprize.org.

The classical ideas about pattern formation in nonequilibrium systems were presented by Turing (1952), who was aiming specifically at an understanding of embryonic development. Modern views are given by Cross and Hohenberg (1993) and by Cross and Greenside (2009).

Cross and Greenside 2009. *Pattern Formation and Dynamics in Nonequilibrium Systems.* M Cross and H Greenside (Cambridge University Press, Cambridge 2009).

Cross and Hohenberg 1993. Pattern formation outside of equilibrium. MC Cross and PC Hohenberg, *Rev Mod Phys* **65,** 851–1112 (1993).

Turing 1952. The chemical basis of morphogenesis. AM Turing, *Phil Trans R Soc Lond B* **237,** 33–72 (1952).

The general idea that cells determine their position, and hence their fate, in an embryo by responding to the concentration of some special morphogen molecule is very old, and it did not take too long before people started to think about the role of diffusion in establishing morphogen gradients. Some milestones are Wolpert's (1969) discussion of positional information and Crick's (1970) surprisingly influential discussion of diffusion. The transcription factor Bcd, in the *Drosophila* embryo, provides a clear example of these ideas (Driever and Nüsslein-Vollhard 1988a,b; Ephrussi and St Johnston 2004).

Crick 1970. Diffusion in embryogenesis. F Crick, *Nature* **225,** 420–422 (1970).

Driever and Nüsslein-Vollhard 1988a. A gradient of Bicoid protein in *Drosophila* embryos. W Driever and C Nüsslein-Vollhard, *Cell* **54,** 83–93 (1988).

Driever and Nüsslein-Vollhard 1988b. The Bicoid protein determines position in the *Drosophila* embryo in a concentration-dependent manner. W Driever and C Nüsslein-Vollhard, *Cell* **54,** 95–104 (1988).

Ephrussi and St Johnston 2004. Seeing is believing: The Bicoid morphogen gradient matures. A Ephrussi and D St Johnston, *Cell* **116,** 143–152 (2004).

Wolpert 1969. Positional information and the spatial pattern of cellular differentiation. L Wolpert, *J Theor Biol* **25,** 1–47 (1969).

Houchmandzadeh et al. (2002) drew attention to the problem of variability in morphogen gradients, and their suggestion that the emergence of reproducible patterns was an example of robustness in biochemical networks attracted considerable attention. Among the models that emerged in an attempt to flesh out the idea of robustness, some make specific use of gradients from the two ends of the embryo to compensate for global parameter variations and allow for scaling with the size of the egg (Houchmandzadeh et al. 2005; McHale et al. 2006). Others use nonlinearities in degradation reactions (Eldar et al. 2002) or in the transport process (Bollenbach et al. 2005) to generate spatial profiles that are robust against variations in source strength. Although much of this discussion focuses on early events in embryonic development, there is also the idea that the final patterns of gene expression, which are more closely tied to cell fate, should be robust steady states of the relevant biochemical networks (von Dassow et al. 2000). Even earlier work emphasized the similarity of these networks to neural nets, with stable patterns being analogous to stored memories (Mjolsness et al. 1991), and one can see this as a modern formulation of the ideas of "canalization" (Waddington 1942). The topological approach shown in Fig. 5.49 is from Corson and Siggia (2012).

Bollenbach et al. 2005. Robust formation of morphogen gradients. T Bollenbach, K Kruse, P Pantazis, M Gonzalés-Gaitán, and F Jülicher, *Phys Rev Lett* **94,** 018103 (2005).

Corson and Siggia 2012. Geometry, epistasis, and developmental patterning. F Corson and ED Siggia, *Proc Natl Acad Sci (USA)* **109,** 5568–5575 (2012).

von Dassow et al. 2000. The segment polarity network is a robust developmental module. G von Dassow, E Meir, EM Munro, and GM Odell, *Nature* **406,** 188–192 (2000).

Eldar et al. 2002. Robustness of the BMP morphogen gradient in *Drosophila* embryonic development. A Eldar, R Dorfman, D Weiss, H Ashe, B-Z Shilo, and N Barkai, *Nature* **419,** 304–308 (2002).

Houchmandzadeh et al. 2002. Establishment of developmental precision and proportions in the early *Drosophila* embryo. B Houchmandzadeh, E Wieschaus, and S Leibler, *Nature* **415,** 798–802 (2002).

Houchmandzadeh et al. 2005. Precise domain specification in the developing *Drosophila* embryo. B Houchmandzadeh, E Wieschaus, and S Leibler, *Phys Rev E* **72,** 061920 (2005).

McHale et al. 2006. Embryonic pattern scaling achieved by oppositely directed morphogen gradients. P McHale, W-J Rappel, and H Levine, *Phys Biol* **3,** 107–120 (2006).

Mjolsness et al. 1991. A connectionist model of development. E Mjolsness, DH Sharp, and J Reinitz, *J Theor Biol* **152,** 429–453 (1991).

Waddington 1942. Canalization of development and the inheritance of acquired characters. CH Waddington, *Nature* **150,** 563–565 (1942).

For recent measurements on the reproducibility of the early events in the fly embryo, see Dubuis et al. (2012). As noted above, and in Section 2.3, the overall precision and reproducibility of pattern formation in the fruit fly embryo is equivalent to $\sim 10\%$ accuracy in the concentration of Bcd. Whether or not the actual Bcd profiles are reproducible at this level, the overall reproducibility of the system does suggest a standard for making measurements. For a recent discussion of the state of the art in these experiments, see Dubuis et al. (2010). The measurements on reproducibility of the Bcd profiles shown in Fig. 5.47 are from Gregor et al. (2007b), using the Bcd-GFP fusion developed in Gregor et al. (2007a). Experiments on the scaling of Bcd profiles across species include Gregor et al. (2005, 2008). Complementary to the problem of scaling is the problem of size control; see Shraiman (2005).

Dubuis et al. 2010. Quantifying the Bicoid morphogen gradient in living fly embryos. JO Dubuis, AH Morrison, M Scheeler, and T Gregor, arXiv:1003.5572 [q-bio.QM] (2010).

Dubuis et al. 2012. Positional information, in bits. JO Dubuis, G Tkačik, EF Wieschaus, T Gregor, and W Bialek, arXiv.org:1201.0198 [q-bio.MN] (2012).

Gregor et al. 2005. Diffusion and scaling during early embryonic pattern formation. T Gregor, W Bialek, DW Tank, RR de Ruyter van Steveninck, DW Tank, and EF Wieschaus, *Proc Natl Acad Sci (USA)* **102,** 18403–18407 (2005).

Gregor et al. 2007a. Stability and nuclear dynamics of the Bicoid morphogen gradient. T Gregor, EF Wieschaus, AP McGregor, W Bialek, and DW Tank, *Cell* **130,** 141–152 (2007).

Gregor et al. 2007b. Probing the limits to positional information. T Gregor, DW Tank, EF Wieschaus, W Bialek, *Cell* **130,** 153–164 (2007).

Gregor et al. 2008. Shape and function of the Bicoid morphogen gradient in dipteran species with different sized embryos. T Gregor, AP McGregor, and EF Wieschaus, *Dev Biol* **316,** 350–358 (2008).

Shraiman 2005. Mechanical feedback as a possible regulator of tissue growth. BI Shraiman, *Proc Natl Acad Sci (USA)* **102,** 3318–3323 (2005).

5.4 Long Time Scales in Neural Networks

It is hard to remember that our modern picture of the brain as composed of discrete cells, connected by synapses, is only ~ 100 years old (Cajal 1906).[3] By the mid-twentieth century researchers knew a bit about electrical signaling (though it would take another decade for Hodgkin and Huxley to sort things out), and people started to think about how these signals might implement computations; the foundational paper from this period is McCulloch and Pitts (1943). Many people must have had the idea that such networks could be described

---

3. The remarkable observations by Ramón y Cajal remain a foundation for our thinking about the brain; his Nobel lecture cited here is just a sampling. His methods included extensions of the staining techniques first developed by Golgi. Golgi, however, did not see neurons in his experiments, but rather a continuous structure. It seems likely that 1906 was the only year in which two people who disagreed so deeply shared the Nobel Prize for their contributions to the same subject.

using concepts from statistical physics, but the development took some time. A major milestone is Cooper (1973),[4] and then the dam broke with Hopfield's papers (Hopfield 1982, 1984). Textbook accounts are given by Amit (1989) and by Hertz et al. (1991).

Amit 1989. *Modeling Brain Function: The World of Attractor Neural Networks*. DJ Amit (Cambridge University Press, Cambridge, 1989).

Cajal 1906. The structure and connexions of neurons. S Ramon y Cajal. Nobel lecture 1906, in *Nobel Lectures, Physiology or Medicine 1901–1921* (Elsevier, Amsterdam, 1967). Also available at http://www.nobelprize.org.

Cooper 1973. A possible organization of animal memory and learning. LN Cooper, in *Collective Properties of Physical Systems: Proceedings of Nobel Symposium 24,* B Lundqvist and S Lundqvist, eds., pp. 252–264 (Academic Press, New York, 1973).

Hertz et al. 1991. *Introduction to the Theory of Neural Computation* J Hertz, A Krogh, and RG Palmer (Addison-Wesley, Redwood City, CA, 1991).

Hopfield 1982. Neural networks and physical systems with emergent collective computational abilities. JJ Hopfield, *Proc Natl Acad Sci (USA)* **79,** 2554–2558 (1982).

Hopfield 1984. Neurons with graded response have collective properties like those of two-state neurons. JJ Hopfield, *Proc Natl Acad Sci (USA)* **81,** 3088–3092 (1984).

McCulloch and Pitts 1943. A logical calculus of ideas immanent in nervous activity. WS McCulloch and W Pitts, *Bull Math Biophys* **5,** 115–133 (1943).

A central feature of the Hopfield model is that the stable states are maintained by loops of neurons exciting one another throughout the network. This idea goes back at least to Lorente de Nó (1938). The idea of long-term storage of information being based on changing the strengths of synapses, and that these changes would be based on correlations, has its origins in James (1892) and Hebb (1949). For an overview of synapses, including plasticity and its relation to learning and memory, see Cowan et al. (2003). The experiments shown in Fig. 5.52 were done by Markram et al. (1997); see also Markram et al. (2011) for a review of how ideas about synaptic plasticity have developed.

Cowan et al. 2003. *Synapses*. WM Cowan, TC Südhof, and CF Stevens, eds. (Johns Hopkins University Press, Baltimore, 2003).

Hebb 1949. *The Organization of Behavior: A Neuropsychological Theory*. DO Hebb (Wiley, New York, 1949).

James 1892. *Psychology: The Briefer Course*. W James (Henry Holt and Company, 1892). There is a modern edition from Dover Publications (New York, 2001), based on the 1961 abridged version from Harper and Row (New York, 1961).

Lorente de Nó 1938. Analysis of the activity of the chains of internuncial neurons. R Lorente de Nó, *J Neurophysiol* **1,** 207–244 (1938).

Markram et al. 1997. Regulation of synaptic efficacy by coincidence of postsynaptic APs and EPSPs. H Markram, J Lübke, M Frotscher, and B Sakmann, *Science* **275,** 213–215 (1997).

Markram et al. 2011. A history of spike-timing-dependent plasticity. H Markram, W Gerstner, and PJ Sjöstrom, *Frontiers Syn Neuro* **3,**4 (2011).

The Hopfield model is mathematically equivalent to a spin glass and was introduced just as powerful new analytic methods were being developed to deal with spin glasses and the

---

4. The volume in which Cooper's paper appears is quite remarkable, and very much worth exploring.

statistical mechanics of other disordered systems. These threads came together in a beautiful series of papers (Amit et al. 1985, 1987; Crisanti et al. 1986). The persistent activity predicted by such models had in fact been observed some years previously (Fuster and Alexander 1971; Funahashi et al. 1989). It continues to be a central topic in explorations of brain function (Prut and Fetz 1999; Romo et al. 1999; Major and Tank 2004; Huk and Shadlen 2005).

Amit et al. 1985. Spin-glass models of neural networks. DJ Amit, H Gutfreund, and H Sompolinsky, *Phys Rev A* **32,** 1007–1018 (1985).

Amit et al. 1987. Statistical mechanics of neural networks near saturation. DJ Amit, H Gutfreund, and H Sompolinsky, *Ann Phys* **173,** 30–67 (1987).

Crisanti et al. 1986. Saturation level of the Hopfield model for neural networks. A Crisanti, DJ Amit, and H Gutfreund, *Europhys Lett* **2,** 337–341 (1986).

Funahashi et al. 1989. Mnemonic coding of visual space in the monkey's dorsolateral prefrontal cortex. S Funahashi, CJ Bruce, and PS Goldman-Rakic, *J Neurophysiol* **61,** 331–349 (1989).

Fuster and Alexander 1971. Neuron activity related to short-term memory. JM Fuster and GE Alexander, *Science* **173,** 652–654 (1971).

Huk and Shadlen 2005. Neural activity in macaque parietal cortex reflects temporal integration of visual motion signals during perceptual decision making. AC Huk and MN Shadlen, *J Neurosci* **25,** 10420–10436 (2005).

Major and Tank 2004. Persistent neural activity: prevalence and mechanisms. G Major and D Tank, *Curr Opin Neurobiol* **14,** 675–684 (2004).

Prut and Fetz 1999. Primate spinal interneurons show pre-movement instructed delay activity. Y Prut and EE Fetz, *Nature* **401,** 590–594 (1999).

Romo et al. 1999. Neuronal correlates of parametric working memory in the prefrontal cortex. R Romo, CD Brody, A Hernández, and L Lemus, *Nature* **399,** 470–473 (1999).

For the maximum entropy method, see Appendix A.7 and its references in the Annotated Bibliography. There is a long history of work on the oculomotor system, with the view gradually emerging of this system as an integrator in the mathematical sense. For background on the mechanics of the vestibular system, see Lewis et al. (1985), and for some perceptive thoughts on the problem of integration, see Robinson (1989). Seung (1996) took this idea seriously and showed how models of the relevant neural network were constrained, which inspired a new generation of experimental work, notably on the self-tuning of the system (Major et al. 2004a,b).

Lewis et al. 1985. *The Vertebrate Inner Ear.* ER Lewis, EL Leverenz, and WS Bialek (CRC Press, Boca Raton FL, 1985).

Major et al. 2004a. Plasticity and tuning by visual feedback of the stability of a neural integrator. G Major, R Baker, E Aksay, B Mensh, HS Seung, and DW Tank, *Proc Natl Acad Sci (USA)* **101,** 7739–7744 (2004).

Major et al. 2004b. Plasticity and tuning of the time course of analog persistent firing in a neural integrator. G Major, R Baker, E Aksay, HS Seung, and DW Tank, *Proc Natl Acad Sci (USA)* **101,** 7745–7750 (2004).

Robinson 1989. Integrating with neurons. DA Robinson, *Annu Rev Neurosci* **12,** 33–45 (1989).

Seung 1996. How the brain keeps the eyes still. HS Seung, *Proc Natl Acad Sci (USA)* **93,** 13339–13344 (1996).

5.5 Perspectives

The idea of choosing parameters at random in biochemical networks was explored by Barkai and Leibler (1997) and by von Dassow et al. (2000), among others, using simulations. Much earlier, Sompolinsky et al. (1988) had analyzed the dynamics of random neural networks, identifying a transition between a stationary phase and a chaotic phase at a critical value of the typical synaptic strength. For attempts to connect these random networks to the behavior of the cortex, see van Vreeswijk and Sompolinksy (1996, 1998). More recently, Rajan et al. (2010) have emphasized that input signals can drive random networks across the transition between chaos and order, providing a possible new view of the nature of variability in cortical responses (Abbott et al. 2011).

Abbott et al. 2011. Interactions between intrinsic and stimulus-dependent activity in recurrent neural networks. LF Abbott, K Rajan, and H Sompolinsky, in *The Dynamic Brain: An Exploration of Neuronal Variability and Its Functional Significance.* M Ding and D Glanzman, eds., pp. 65–82 (Oxford University Press, New York, 2011).

Barkai and Leibler 1997. Robustness in simple biochemical networks. N Barkai and S Leibler, *Nature* **387,** 913–917 (1997).

von Dassow et al. 2000. The segment polarity network is a robust developmental module. G von Dassow, E Meir, EM Munro, and GM Odell, *Nature* **406,** 188–192 (2000).

Rajan et al. 2010. Input-dependent suppression of chaos in recurrent neural networks. K Rajan, LF Abbott, and Sompolinsky, *Phys Rev E* **82,** 011903 (2010).

Sompolinsky et al. 1988. Chaos in random neural networks. H Sompolinsky, A Crisanti, and HJ Sommers, *Phys Rev Lett* **61,** 259–262 (1988).

van Vreeswijk and Sompolinsky 1996. Chaos in neuronal networks with balanced excitatory and inhibitory activity. *Science* **274,** 1724–1726 (1996).

van Vreeswijk and Sompolinsky 1998. Chaotic balanced state in a model of cortical circuits. *Neural Comp* **10,** 1321–1371 (1998).

6   Efficient Representation

6.1 Entropy and Information

To a remarkable extent, Shannon's original work provides a complete and accessible guide to the foundations of the subject (Shannon 1948). Seldom has something genuinely new emerged so fully in one (admittedly long, two-part) paper. For a modern textbook account, the standard is set by Cover and Thomas (1991). An fascinating if idiosyncratic treatment of Shannon's ideas is given by Brillouin (1962). A recent textbook that emphasizes connections between information theory and statistical physics is Mézard and Montanari (2009). The brief discussion of four-letter words is based on Stephens and Bialek (2010). For more about the accumulation of evidence and its connections both to diffusion processes and human behavior, see Bogacz et al. (2006, 2010) and references therein.

Bogacz et al. 2006. The physics of optimal decision making: A formal analysis of models of performance in two-alternative forced-choice tasks. R Bogacz, E Brown, J Moehlis, P Holmes, and JD Cohen, *Psych Rev* **113,** 700–765 (2006).

Bogacz et al. 2010. Do humans produce the speed-accuracy trade-off that maximizes reward rate? R Bogacz, PT Hu, PJ Holmes, and JD Cohen, *Q J Exp Psych* **63,** 863–891 (2010).

Brillouin 1962. *Science and Information Theory.* L Brillouin (Academic, New York, 1962).

Cover and Thomas 1991. *Elements of Information Theory.* TM Cover and JA Thomas (Wiley, New York, 1991); there is also a second edition (2006).

Mézard and Montanari 2009. *Information, Physics and Computation.* M Mézard and A Montanari (Oxford University Press, Oxford, 2009).

Shannon 1948. A mathematical theory of communication, CE Shannon, *Bell Sys Tech J* **27,** 379–423, and 623–656 (1948). Reprinted in *The Mathematical Theory of Communication,* CE Shannon and W Weaver, pp. 31–125 (University of Illinois Press, Urbana, 1949).

Stephens and Bialek 2010. Statistical mechanics of letters in words. GJ Stephens and W Bialek, *Phys Rev E* **81,** 066119 (2010); arXiv:0801.0253 [q-bio.NC] (2008).

6.2 Noise and Information Flow

The exploration of neural coding using ideas from information theory rests on a large literature, starting with Adrian's first experiments recording the spikes from individual sensory neurons in the 1920s. For a guide to the field up to the mid-1990s, see Rieke et al. (1997). The idea of using ergodicity to make direct estimates of the entropy and information in spike trains as they encode dynamic signals is presented by Strong et al. (1998a) and de Ruyter van Steveninck et al. (1997). For the original discussion of the limits to information transmission by spike trains, see MacKay and McCulloch (1952). For the analysis of information carried by single events, see DeWeese and Meister (1999) and Brenner et al. (2000).

Brenner et al. 2000. Synergy in a neural code. N Brenner, SP Strong, R Koberle, W Bialek, and RR de Ruyter van Steveninck, *Neural Comp* **12,** 1531–1552 (2000).

DeWeese and Meister 1999. How to measure the information gained from one symbol. MR DeWeese and M Meister, *Network* **10,** 325–340 (1999).

MacKay and McCulloch 1952. The limiting information capacity of a neuronal link. D MacKay and WS McCulloch, *Bull Math Biophys* **14,** 127–135 (1952).

Rieke et al. 1997. *Spikes: Exploring the Neural Code.* F Rieke, D Warland, RR de Ruyter van Steveninck, and W Bialek (MIT Press, Cambridge MA, 1997).

de Ruyter van Steveninck et al. 1997. Reproducibility and variability in neural spike trains. RR de Ruyter van Steveninck, GD Lewen, SP Strong, R Koberle, and W Bialek, *Science* **275,** 1805–1808 (1997).

Strong et al. 1998a. Entropy and information in neural spike trains. SP Strong, R Koberle, RR de Ruyter van Steveninck, and W Bialek, *Phys Rev Lett* **80,** 197–200 (1998); arXiv:cond-mat/9603127 (1996).

The experiments connecting motion perception to the activity of individual neurons in the visual cortex were reported in a series of beautiful papers by Newsome, Movshon, and their collaborators starting with Newsome et al. (1989) and reviewed by Newsome et al. (1995); of particular relevance here is Britten et al. (1993). That the neurons in these experiments generated reproducible responses to stimuli that repeated their temporal details was emphasized by Bair and Koch (1996). The analysis in Fig. 6.8 is from Strong et al. (1998b), unpacking a footnote in Strong et al. (1998a) above. Experiments designed to look more specifically at these problems of information transmission for dynamic signals in MT were done by Buračas et al. (1998).

Bair and Koch 1996. Temporal precision of spike trains in extrastriate cortex of the behaving macaque monkey. W Bair and C Koch, *Neural Comp* **8,** 1185–1192 (1996).

Britten et al. 1993. Response of neurons in macaque MT to stochastic motion signals. KH Britten, MN Shadlen, WT Newsome, and JA Movshon, *Vis Neuroci* **10,** 1157–1169 (1993).

Buračas et al. 1998. Efficient discrimination of temporal patterns by motion-sensitive neurons in primate visual cortex. GT Buračas, AM Zador, MR DeWeese, and TD Albright, *Neuron* **20,** 959–969 (1998).

Newsome et al. 1989. Neuronal correlates of a perceptual decision. WT Newsome, KH Britten, and JA Movshon, *Nature* **341,** 52–54 (1989).

Newsome et al. 1995. Visual motion: Linking neuronal activity to psychophysical performance. WT Newsome, MN Shadlen, E Zohary, KH Britten, and JA Movshon, in *The Cognitive Neurosciences*, M Gazzaniga, ed., pp. 401–444 (MIT Press, Cambridge MA, 1995).

Strong et al. 1998b. On the application of information theory to neural spike trains. SP Strong, RR de Ruyter van Steveninck, W Bialek, and R Koberle, in *Pacific Symposium on Biocomputing '98*, RB Altman, AK Dunker, L Hunter, and TE Klein, eds., pp. 621–632 (World Scientific, Singapore, 1998).

The classic discussion of information transmission in the presence of Gaussian noise is by Shannon (1949), and the standard modern textbook account is in Cover and Thomas (1991). The discussion of positional information is based on Dubuis et al. (2012). The idea of decoding neural spike trains to recover the underlying continuous time-varying signal (stimulus reconstruction) was introduced in Section 4.4 and is reviewed by Rieke et al. (1997) above; the first use of this approach to compare information and entropy was by Rieke et al. (1993). This led to experiments showing that coding efficiencies are higher for more naturalistic stimuli (Rieke et al. 1995), and this problem was eventually revisited in the context of much more natural stimuli using the direct methods of information estimation coupled with more sophisticated strategies for dealing with finite sampling (Nemenman et al. 2008), as explained in Appendix A.8. The measurements on information capacity in the fly retina are by de Ruyter van Steveninck and Laughlin (1996). The comparison of this information rate with the limits set by counting vesicles is discussed by Rieke et al. (1997) above.

Cover and Thomas 1991. *Elements of Information Theory*. TM Cover and JA Thomas (Wiley, New York, 1991); there is also a second edition (2006).

Dubuis et al. 2012. Positional information, in bits. JO Dubuis, G Tkačik, EF Wieschaus, T Gregor, and W Bialek, arXiv.org:1201.0198 [q-bio.MN] (2012).

Nemenman et al. 2008. Neural coding of a natural stimulus ensemble: Information at sub-millisecond resolution. I Nemenman, GD Lewen, W Bialek, and RR de Ruyter van Steveninck, *PLoS Comp Bio* **4,** e1000025 (2008); arXiv:q-bio.NC/0612050 (2006).

Rieke et al. 1993. Coding efficiency and information rates in sensory neurons. F Rieke, D Warland, and W Bialek, *Europhys Lett* **22,** 151–156 (1993).

Rieke et al. 1995. Naturalistic stimuli increase the rate and efficiency of information transmission by primary auditory neurons. F Rieke, DA Bodnar, and W Bialek, *Proc R Soc Lond B* **262,** 259–265 (1995).

de Ruyter van Steveninck and Laughlin 1996. The rate of information transfer at graded-potential synapses. RR de Ruyter van Steveninck and SB Laughlin, *Nature* **379,** 642–645 (1996).

Shannon 1949. Communication in the presence of noise. CE Shannon, *Proc IRE* **37,** 10–21 (1949).

The measurement of information and coding efficiency in neural codes has become a substantial effort, as has the study of coding under more natural conditions. Here is a sampling from early stages of mammalian visual processing (Kara et al. 2000; Reinagel and Reid 2000; Yu et al. 2005; Koch et al. 2006), mammalian auditory processing (Attias and Schreiner 1998; Liu et al. 2001; Escabi et al. 2003), and invertebrate sensors of various types (Rokem et al. 2006; Simmons and de Ruyter van Steveninck 2010).

Attias and Schreiner 1998. Coding of naturalistic stimuli by auditory midbrain neurons. H Attias and CE Schreiner, in *Advances in Neural Information Processing Systems 10*, MI Jordan, MJ Kearns, and SA Solla, eds., pp. 103–109 (MIT Press, Cambridge MA, 1998).

Escabi et al. 2003. Naturalistic auditory contrast improves spectrotemporal coding in the cat inferior colliculus. MA Escabi, LM Miller, HL Read, and CE Schreiner, *J Neurosci* **23,** 11489–11504 (2003).

Kara et al. 2000. Low response variability in simultaneously recorded retinal, thalamic, and cortical neurons. P Kara, P Reinagel, and RC Reid, *Neuron* **27,** 635–646 (2000).

Koch et al. 2006. How much the eye tells the brain. K Koch, J McLean, R Segev, MA Freed, MJ Berry II, V Balasubramanian, and P Sterling, *Curr Biol* **16,** 1428–1434 (2006).

Liu et al. 2001. Variability and information in a neural code of the cat lateral geniculate nucleus. RC Liu, S Tzonev, S Rebrik, and KD Miller, *J Neurophysiol* **86,** 2789–2806 (2001).

Reinagel and Reid 2000. Temporal coding of visual information in the thalamus. P Reinagel and RC Reid, *J Neurosci* **20,** 5392–5400 (2000).

Rokem et al. 2006. Spike-timing precision underlies the coding efficiency of auditory receptor neurons. A Rokem, S Watzl, T Gollisch, M Stemmler, AVM Herz, and I Samengo, *J Neurophysiol* **95,** 2541–2552 (2006).

Simmons and de Ruyter van Steveninck 2010. Sparse but specific temporal coding by spikes in an insect sensory-motor ocellar pathway. PJ Simmons and RR de Ruyter van Steveninck, *J Exp Biol* **213,** 2629–2639 (2010).

Yu et al. 2005. Preference of sensory neural coding for $1/f$ signals. Y Yu, R Romero, and TS Lee, *Phys Rev Lett* **94,** 108103 (2005).

6.3 Does Biology Care about Bits?

The connection between information and gambling goes back to Kelly (1956). Connections of these ideas to fitness in fluctuating environments are discussed by Bergstrom and Lachmann (2005), Kussell and Leibler (2005), and more generally by Rivoire and Leibler (2011). The specific case of persistence in bacteria has been explored by Balaban et al. (2004) and Kussell et al. (2005); for a review, see Gefen and Balaban (2009). The analogy to rate-distortion theory, demonstrating a minimum number of bits required to achieve a criterion mean growth rate, is from Taylor et al. (2007). For a treatment of rate-distortion theory itself, see Cover and Thomas (1991), above. Although they did not explicitly use the language of rate-distortion theory, Park and Levitt (1995) explored the compression of protein structures into a small set of local discrete states, asking how the complexity of this representation related to its accuracy. The idea of infotaxis is presented by Vergassola et al. (2007). For background on the way in which insects and other organisms find their way through odor plumes, see Murlis et al. (1992), Mafra-Neto and Cardé (1994), and Balkovsky and Shraiman (2002). The beautiful convection patterns in Fig. 6.18 are discussed in Bodenschatz et al. (1991).

Balaban et al. 2004. Bacterial persistence as a phenotypic switch. NQ Balaban, J Merrin, R Chait, L Kowalik, and S Leibler, *Science* **305,** 1622–1625 (2004).

Balkovsky and Shraiman 2002. Olfactory search at high Reynolds number. E Balkovsky and BI Shraiman, *Proc Natl Acad Sci (USA)* **99,** 12589–12593 (2002).

Bergstrom and Lachmann 2005. The fitness value of information. CT Bergstrom and M Lachmann, axXiv:q-bio.PE/0510007 (2005).

Bodenschatz et al. 1991. Transitions between patterns in thermal convection. E Bodenschatz, JR de Bruyn, G Ahlers, and DS Cannell, *Phys Rev Lett* **67**, 3078–3081 (1991).

Gefen and Balaban 2009. The importance of being persistent: Heterogeneity of bacterial populations under antibiotic stress. O Gefen and NQ Balaban, *FEMS Microbiol Rev* **33**, 704–717 (2009).

Kelly 1956. A new interpretation of information rate. JL Kelly, Jr, *Bell Sys Tech J* **35**, 917–926 (1956).

Kussell and Leibler 2005. Phenotypic diversity, population growth, and information in fluctuating environments. EL Kussell and S Leibler, *Science* **309**, 2075–2078 (2005).

Kussell et al. 2005. Bacterial persistence: A model of survival in changing environments. EL Kussell, R Kishony, NQ Balaban, and S Leibler, *Genetics* **169**, 1807–1814 (2005).

Mafra-Neto and Cardé 1994. Fine-scale structure of pheromone plumes modulates upwind orientation of flying moths. A Mafra-Neto and RT Cardé, *Nature* **369**, 142–144 (1994).

Murlis et al. 1992. Odor plumes and how insects use them. J Murlis, JS Elkinton, and RT Cardé, *Annu Rev Entomol* **37**, 505–532 (1992).

Park and Levitt 1995. The complexity and accuracy of discrete state models of protein structure. BH Park and M Levitt, *J Mol Biol* **249**, 493–507 (1995).

Rivoire and Leibler 2011. The value of information for populations in varying environments. O Rivoire, and S Leibler, *J Stat Phys* **142**, 1124–1166 (2011); arXiv.org:1010.5092 [q-bio.PE] (2010).

Taylor et al. 2007. Information and fitness. SF Taylor, N Tishby, and W Bialek, arXiv:0712.4382 [q-bio.PE] (2007).

Vergassola et al. 2007. "Infotaxis" as a strategy for searching without gradients. M Vergassola, E Villermaux, and BI Shraiman, *Nature* **445**, 406–409 (2007).

The information bottleneck was introduced by Tishby et al. (1999). It has connections with statistical mechanics approaches to clustering (Rose et al. 1990) and has more immediate antecedents in the idea of clustering distributions of words (Pereira et al. 1993). For more about predictive information, see Bialek et al. (2001, 2007). The possibility that even familiar programs of coordinated changes in bacterial gene expression may reflect (implicit) predictions is discussed by Tagkopoulos et al. (2008). Energy efficiency in neural coding is discussed by Laughlin et al. (1998) and Balasubramanian et al. (2001).

Balasubramanian et al. 2001. Metabolically efficient information processing. V Balasubramanian, D Kimber, and MJ Berry II, *Neural Comp* **13**, 799–815 (2001).

Bialek et al. 2001. Predictability, complexity and learning. W Bialek, I Nemenman, and N Tishby, *Neural Comp* **13**, 2409–2463 (2001); arXiv:physics/0007070 (2000).

Bialek et al. 2007. Efficient representation as a design principles for neural coding and computation. W Bialek, RR de Ruyter van Steveninck, and N Tishby, in *2006 IEEE International Symposium on Information Theory*, pp. 659–663; arXiv.org:0712.4381 [q-bio.NC] (2007).

Laughlin et al. 1998. The metabolic cost of neural information. SB Laughlin, RR de Ruyter van Steveninck, and JC Anderson, *Nat Neurosci* **1**, 36–41 (1998).

Pereira et al. 1993. Distributional clustering of English words. FC Pereira, N Tishby, and L Lee, in *31st Annual Meeting of the Association for Computational Linguistics,* LK Schubert, ed., pp. 183–190 (1993).

Rose et al. 1990. Statistical mechanics and phase transitions in clustering. K Rose, E Gurewitz, and GC Fox, *Phys Rev Lett* **65**, 945–948 (1990).

Tagkopoulos et al. 2008. Predictive behavior within microbial genetic networks. I Tagkopoulos, Y Liu, and S Tavazoie, *Science* **320,** 1313–1317 (2008).

Tishby et al. 1999. The information bottleneck method. N Tishby, FC Pereira, and W Bialek, in *Proceedings of the 37th Annual Allerton Conference on Communication, Control and Computing*, B Hajek and RS Sreenivas, eds., pp. 368–377 (1999); arXiv:physics/0004057 (2000).

The relationship between information theory and language is old and fraught. Shannon used written texts as an example, often approximating them by simple models, though I think it is clear that he did not take the models too seriously as descriptions of real languages. Chomsky, however, took the limitations of these early probabilistic models very seriously, and one of the foundational papers of modern linguistics begins essentially as a critique of Shannon (Chomsky 1956). There is a persistent belief that there must be "more than just statistics" to our understanding of language, and hence that information theory misses something essential; Abney (1996:19) summarizes this well: "in one's introductory linguistics course, one learns that Chomsky disabused the field once and for all of the notion that there was anything of interest to statistical models of language. But one usually comes away a little fuzzy on the question of what, precisely, he proved." One of Chomsky's most famous arguments (Chomsky 1957) is the claim that statistical approaches could not distinguish between sentences that have never occurred but are instantly recognizable as grammatical (e.g., Colorless green ideas sleep furiously) and sentences that never occurred because they are forbidden by grammatical rules (Furiously sleep ideas green colorless). Among other interesting things, Pereira (2000) constructs a simple probabilistic model, based on a modest body of newspaper texts, to show that these sentences have probabilities that differ by a factor of more than 10^5.

Abney 1996. Statistical methods and linguistics. S Abney, in *The Balancing Act: Combining Statistical and Symbolic Approaches to Language*, JL Klavans and P Resnik, eds., pp. 1–26 (MIT Press, Cambridge, MA, 1996).

Chomsky 1956. Three models for the description of language. N Chomsky, *IRE Trans Inf Theory* **IT-2,** 113–124 (1956).

Chomsky 1957. *Syntactic Structures*. N Chomsky (Mouton, The Hague, 1957).

Pereira 2000. Formal grammar and information theory: Together again? F Pereira, *Phil Trans R Soc Lond A* **358,** 1239–1253 (2000).

6.4 Optimizing Information Flow

Laughlin's classic paper on matching input-output relations to the distribution of inputs still is very much worth reading 30 years later (Laughlin 1981). The corresponding analysis for a genetic regulatory element is by Tkačik et al. (2008a), with more theoretical exploration in Tkačik et al. (2008b). The literature on information transmission in biochemical and genetic networks is growing rapidly; for examples see Ziv et al. (2007), Mugler et al. (2008), Yu et al. (2008), and Tostvein and ten Wolde (2009). For a detailed model of calcium signaling by means of the protein calmodulin (of relevance to Problem 159), see Pepke et al. (2010).

Laughlin 1981. A simple coding procedure enhances a neuron's information capacity. SB Laughlin, *Z Naturforsch* **36c,** 910–912 (1981).

Mugler et al. 2008. Quantifying evolvability in small biological networks. A Mugler, E Ziv, I Nemenman, and CH Wiggins, *IET Sys Biol* **3,** 379–387 (2009); arXiv:0811.2834 (2008).

Pepke et al. 2010. A dynamic model of interactions of Ca^{2+}, calmodulin, and catalytic subunits of Ca^{2+}/calmodulin-dependent protein kinase II. S Pepke, T Kinzer-Ursem, S Mihalas, and MB Kennedy, *PLoS Comp Biol* **6,** e1000675 (2010).

Tkačik et al. 2008a. Information flow and optimization in transcriptional regulation. G Tkačik, CG Callan, Jr, and W Bialek, *Proc Natl Acad Sci (USA)* **105,** 12265–12270 (2008).

Tkačik et al. 2008b. Information capacity of genetic regulatory elements. G Tkačik, CG Callan, Jr, and W Bialek, *Phys Rev E* **78,** 011910 (2008).

Tostvein and ten Wolde 2009. Mutual information between input and output trajectories of biochemical networks. F Tostvein and PR ten Wolde, *Phys Rev Lett* **102,** 21801 (2009).

Yu et al. 2008. Negative feedback that improves information transmission in yeast signalling. RC Yu, CG Pesce, A Colman-Lerner, L Lok, D Pincus, E Serra, M Holl, K Benjamin, A Gordon, and R Brent, *Nature* **456,** 755–761 (2008).

Ziv et al. 2007. Optimal signal processing in small stochastic biochemical networks. E Ziv, I Nemenman, and CH Wiggins, *PLoS* **2,** e1007 (2007).

The idea of matching input-output relations to the statistics of inputs had a big impact on neuroscience, in particular driving the exploration of input statistics under natural conditions. An early paper in this direction was by Field (1987), who noted that the power spectra of natural images were approximately scale invariant. Natural images are strongly non-Gaussian, so we expect that scaling should mean much more than an appropriately shaped power spectrum, which is true (Ruderman 1994; Ruderman and Bialek 1994). Exploration of these statistical structures beyond the Gaussian approximation led to the ideas of variance normalization and the prediction of adaptation to the local variance; even the power spectrum constrains the information collected by the retina, as in Problem 163. Well before these analyses there was a substantial body of work on contrast gain control at various levels of the visual system; see, for example, Shapley and Victor (1981). The work on image statistics prompted a more explicit search for adaptation to the distribution of visual inputs beyond the mean light intensity (Smirnakis et al. 1997). Brenner et al. (2000) describe experiments mapping the input-output relations of the fly's motion-sensitive visual neurons when inputs are drawn from different distributions, demonstrating that this adaptation serves to optimize information transmission, and Fairhall et al. (2001) explored the dynamics of this process, showing that one could catch the system using the wrong code and transmitting less information. It seems likely that we have not exhausted the subtleties of adaptation mechanisms in the retina; see, for example, the recent work of Kastner and Baccus (2011).

Brenner et al. 2000. Adaptive rescaling optimizes information transmission. N Brenner, W Bialek, and RR de Ruyter van Steveninck, *Neuron* **26,** 695–702 (2000).

Fairhall et al. 2001. Efficiency and ambiguity in an adaptive neural code. AL Fairhall, GD Lewen, W Bialek, and RR de Ruyter van Steveninck, *Nature* **412,** 787–792 (2001).

Field 1987. Relations between the statistics of natural images and the response properties of cortical cells. DJ Field, *J Opt Soc Am A* **4,** 2379–2394 (1987).

Kastner and Baccus 2011. Correlated dynamic encoding in the retina using opposing forms of plasticity. DB Kastner and SA Baccus, *Nat Neurosci* **14,** 1317–1322 (2011).

Ruderman 1994. The statistics of natural images. DL Ruderman, *Network* **5,** 517–548 (1994).

Ruderman and Bialek 1994. Statistics of natural images: Scaling in the woods. DL Ruderman and W Bialek, *Phys Rev Lett* **73,** 814–817 (1994).

Shapley and Victor 1981. How the contrast gain control modifies the frequency responses of cat retinal ganglion cells. RM Shapley and JD Victor, *J Physiol (Lond)* **318,** 161–179 (1981).

Smirnakis et al. 1997. Adaptation of retinal processing to image contrast and spatial scale. S Smirnakis, MJ Berry, II, DK Warland, W Bialek, and M Meister, *Nature* **386,** 69–73 (1997).

Adaptation to the distribution of inputs has now been reported in many neural systems: in the songbird (Nagel and Doupe 2006) and mammalian (Dean et al. 2005, 2006) auditory systems, in the visual cortex (Sharpee et al. 2006), and in the somatosensory system (Maravall et al. 2007). For a review, including the connections to information transmission, see Wark et al. (2007). The retina offers an accessible model system in which to explore the mechanisms of such statistical adaptation, and these mechanisms seem quite rich and diverse (Kim and Rieke 2001, 2003; Rieke 2001; Baccus and Meister 2002). For more on the idea that adaptation happens as fast as possible given the need to accumulate evidence regarding a change in the input distribution, see Wark et al. (2009). Indeed, many of these changes in the effective input-output relation happen so quickly that some are loathe to call them "adaptation," and it might be that these phenomena are better viewed as intrinsic to the nonlinear behavior of neurons on short time scales, rather than as additional mechanisms on longer times scales. But such models may have difficulty accounting for apparent adaptation to the variance in signals with long correlation time, for which see Brenner et al. (2000), above. This still is an active area of research. For ideas about the adaptation and/or evolution of biochemical and genetic networks, see Francois et al. (2007) and Francois and Siggia (2010).

Baccus and Meister 2002. Fast and slow contrast adaptation in retinal circuitry. SA Baccus and M Meister, *Neuron* **36,** 900–919 (2002).

Dean et al. 2005. Neural population coding of sound level adapts to stimulus statistics. I Dean, NS Harper, and D McAlpine, *Nat Neurosci* **8,** 1684–1689 (2005).

Dean et al. 2006. Rapid neural adaptation to sound level statistics. I Dean, BL Robinson, NS Harper, and D McAlpine, *J Neurosci* **28,** 6430–6438 (2006).

Francois and Siggia 2010. Predicting embryonic patterning using mutual entropy fitness and in silico evolution. P Francois and ED Siggia, *Development* **137,** 2385–2395. (2010).

Francois et al. 2007. A case study of evolutionary computation of biochemical adaptation. P Francois, V Hakim, and ED Siggia, *Phys Biol* **5,** 026009 (2007).

Kim and Rieke 2001. Temporal contrast adaptation in the input and output signals of salamander retinal ganglion cells. KJ Kim and F Rieke, *J Neurosci* **21,** 287–299 (2001).

Kim and Rieke 2003. Slow Na^+ inactivation and variance adaptation in salamander retinal ganglion cells. KJ Kim and F Rieke, *J Neurosci* **23,** 1506–1516 (2003).

Maravall et al. 2007. Shifts in coding properties and maintenance of information transmission during adaptation in barrel cortex. M Maravall, RS Petersen, AL Fairhall, E Arabzadeh, and ME Diamond, *PLoS Biology* **5,** e19 (2007).

Nagel and Doupe 2006. Temporal processing and adaptation in the songbird auditory forebrain. KI Nagel and AJ Doupe, *Neuron* **51,** 845–859 (2006).

Rieke 2001. Temporal contrast adaptation in salamander bipolar cells. F Rieke, *J Neurosci* **21,** 9445–9454 (2001).

Sharpee et al. 2006. Adaptive filtering enhances information transmission in visual cortex. TO Sharpee, H Sugihara, AV Kurgansky, SP Rebik, MP Stryker, and KD Miller, *Nature* **439,** 936–942 (2006).

Wark et al. 2007. Sensory adaptation. B Wark, BN Lundstrom, and A Fairhall, *Curr Opin Neurobiol* **17,** 423–429 (2007).

Wark et al. 2009. Time scales of inference in visual adaptation. B Wark, A Fairhall, and F Rieke, *Neuron* **12,** 750–761 (2009).

The ideas about efficient coding in the population of retinal ganglion cells go back to Barlow (1959, 1961), in delightfully original papers that I still find very inspiring. The precise mathematical formulation of these ideas took until Atick and Redlich (1990); van Hateren (1992) worked out essentially the same principles with invertebrate rather than vertebrate retinas in mind, and there are important precursors in the work of Srinivasan et al. (1982), Snyder (1977), and Snyder et al. (1977) on compound eye design. The possibility that opponent color processing is an example of efficient coding was suggested by Buchsbaum and Gottschalk (1983), and the full analysis based on hyperspectral images is due to Ruderman et al. (1998). The beginnings of a program to derive gene regulatory networks that optimize information transmission are described in Tkačik et al. (2009, 2011), and Walczak et al. (2010). The results on the transfer function of the photoreceptor–large monopolar cell synapse (shown in Fig. 6.35) are from Laughlin and de Ruyter van Steveninck (1996). The map of retinal receptive fields (shown in Fig. 6.32) is from Shlens et al. (2006).

Atick and Redlich 1990. Toward a theory of early visual processing. JJ Atick and AN Redlich, *Neural Comp* **2,** 308–320 (1990).

Barlow 1959. Sensory mechanisms, the reduction of redundancy, and intelligence. HB Barlow, in *Proceedings of the Symposium on the Mechanization of Thought Processes,* volume 2, DV Blake and AM Utlley, eds., pp. 537–574 (HM Stationery Office, London, 1959).

Barlow 1961. Possible principles underlying the transformation of sensory messages. HB Barlow, in *Sensory Communication*, W Rosenblith, ed., pp. 217–234 (MIT Press, Cambridge MA, 1961).

Buchsbaum and Gottschalk 1983. Trichromacy, opponent colour coding and optimum colour information transmission in the retina. G Buchsbaum and A Gottschalk, *Proc R Soc Lond B* **220,** 89–113 (1983).

van Hateren 1992. Real and optimal neural images in early vision. JH van Hateren, *Nature* **360,** 68–70 (1992).

Laughlin and de Ruyter van Steveninck 1996. Measurements of signal transfer and noise suggest a new model for graded transmission at an adapting retinal synapse. SB Laughlin and RR de Ruyter van Steveninck, *J Physiol (Lond)* **494,** P19 (1996).

Ruderman et al. 1998. Statistics of cone responses to natural images: Implications for visual coding. DL Ruderman, TW Cronin, and C-C Chiao, *J Opt Soc Am A* **15,** 2036–2045 (1998).

Shlens et al. 2006. The structure of multi-neuron firing patterns in primate retina. J Shlens, GD Field, JL Gaulthier, MI Grivich, D Petrusca, A Sher, AM Litke, and EJ Chichilnisky, *J Neurosci* **26,** 8254–8266 (2006).

Snyder 1977. Acuity of compound eyes: Physical limitations and design. AW Snyder, *J Comp Physiol* **116,** 161–182 (1977).

Snyder et al. 1977. Information capacity of compound eyes. AW Snyder, DS Stavenga, and SB Laughlin, *J Comp Physiol* **116,** 183–207 (1977).

Srinivasan et al. 1982. Predictive coding: A fresh view of inhibition in the retina. MV Srinivasan, SB Laughlin, and A Dubs, *Proc R Soc Lond B* **216,** 427–459 (1982).

Tkačik et al. 2009. Optimizing information flow in small genetic networks. I. G Tkačik, AM Walczak, and W Bialek, *Phys Rev E* **80,** 031920 (2009).

Tkačik et al. 2011. Optimizing information flow in small genetic networks. III. A self-interacting gene. G Tkačik, AM Walczak, and W Bialek, arXiv.org:1112.5026 [q-bio.MN] (2011).

Walczak et al. 2010. Optimizing information flow in small genetic networks. II: Feed-forward interaction. AM Walczak, G Tkačik, and W Bialek, *Phys Rev E* **81,** 041905 (2010).

Perhaps the most obvious evidence that the energy costs of spiking are significant is that functional magnetic resonance imaging (fMRI) of the brain actually works: what one sees in these experiments are the changes in blood oxygenation that reflect the metabolic load associated with neural activity (Ogawa et al. 1990, 1992). The importance of fMRI has been one of the stimuli for a more detailed accounting of the energy budget of the brain (Attwell and Laughlin 2001; Raichle and Gusnard 2002). The description of sparse or efficient coding in the auditory system is based on the work of Lewicki (2002) and colleagues (Smith and Lewicki 2005, 2006). This work grows out of earlier ideas of Lewicki and Sejnowksi (2000). For an overview of sparse coding and spikes, see Olshausen (2002). The results on efficient coding of predictive information are from Palmer et al. (2012).

Atwell and Laughlin 2001. An energy budget for signaling in the grey matter of the brain. D Atwell and SB Laughlin, *J Cereb Blood Flow Metab* **21,** 1133–1145 (2001).

Lewicki 2002. Efficient coding of natural sounds. MS Lewicki, *Nat Neurosci* **5,** 356–363 (2002).

Lewicki and Sejnowski 2000. Learning overcomplete representations. MS Lewicki and TJ Sejnowski, *Neural Comp* **12,** 337–365 (2000).

Ogawa et al. 1990. Brain magnetic resonance imaging with contrast dependent on blood oxygenation. S Ogawa, TM Lee, AR Kay, and DW Tank, *Proc Natl Acad Sci (USA)* **87,** 9868–9872 (1990).

Ogawa et al. 1992. Intrinsic signal changes accompanying sensory stimulation: Functional brain mapping with magnetic resonance imaging. S Ogawa, DW Tank, R Menon, JM Ellerman, SG Kim, H Merkle, and K Ugurbil, *Proc Natl Acad Sci (USA)* **89,** 5951–5955 (1992).

Olshausen 2002. Sparse codes and spikes. BA Olshauen, in *Probabilistic Models of the Brain: Perception and Neural Function*, RPN Rao, BA Olshausen, and MS Lewicki, eds., pp. 257–272 (MIT Press, Cambridge MA, 2002).

Palmer et al. 2012. Predictive information in a neural population. SE Palmer, O Marre, MJ Berry, II, and W Bialek, in preparation (2012).

Raichle and Gusnard 2002. Appraising the brain's energy budget. *Proc Natl Acad Sci (USA)* **99,** 10237–10239 (2002).

Smith and Lewicki 2005. Efficient coding of time-relative structure using spikes. EC Smith and MS Lewicki, *Neural Comp* **17,** 19–45 (2005).

Smith and Lewicki 2006. Efficient auditory coding. EC Smith and MS Lewicki, *Nature* **439,** 978–982 (2006).

6.5 Gathering Information and Making Models

Our modern understanding of the preference for simple models, as explained here, is quite important and well known in certain circles, but less widely appreciated than it should be. Part of the difficulty is the presence of many independent threads in the literature. Rissanen had a very clear point of view, which is essentially that presented here, although in different language; sources go back at least to Rissanen (1978), with a summary in Rissanen (1989). Schwarz (1978) did essentially the calculations shown here, demonstrating the asymptotic behavior of the total probability of a model class generating the observed data, and in particular finding the crucial $(K/2) \log N$ term that provides an N-dependent penalty for complexity. All these problems became more urgent with the emergence of neural networks, which could be viewed as models with very large numbers of parameters. In this context, MacKay (1992) understood the critical role of Occam factors, the integrals over parameter values that favor simpler models; see also his marvelous textbook (MacKay 2003). Balasubramanian (1997) generalized these ideas and translated them into the language of physicists, showing how the Occam factors can be thought of as entropy in the space of models. I learned a lot from talking to Balasubramanian and from working out these ideas in the context of a field-theoretic approach to learning distributions (Bialek et al. 1996). For the case of the spherical horse, see Devine and Cohen (1992).

Balasubramanian 1997. Statistical inference, Occam's razor, and statistical mechanics on the space of probability distributions. V Balasubramanian, *Neural Comp* **9,** 349–368 (1997).

Bialek et al. 1996. Field theories for learning probability distributions. W Bialek, CG Callan, and SP Strong, *Phys Rev Lett* **77,** 4693–4697 (1996).

Devine and Cohen 1992. *Absolute Zero Gravity: Science Jokes, Quotes and Anecdotes.* B Devine and JE Cohen (Fireside Press, Philadelphia, 1992).

MacKay 1992. A practical Bayesian framework for backpropagation networks. DJC MacKay, *Neural Comp* **4,** 448–472 (1992).

MacKay 2003. *Information Theory, Inference, and Learning Algorithms.* DJC MacKay (Cambridge University Press, Cambridge, 2003).

Rissanen 1978. Modeling by shortest data description. J Rissanen, *Automatica* **14,** 465–471 (1978).

Rissanen 1989. *Stochastic Complexity and Statistical Inquiry* J Rissanen (World Scientific, Singapore, 1989).

Schwarz 1978. Estimating the dimensionality of a model. G Schwarz, *Ann Stat* **6,** 461–464 (1978).

The study of neural networks also led to a very explicit formulation of learning as a statistical mechanics problem (Levin et al. 1990). In this framework it was discovered that there could be phase transitions in the learning of large models (Seung et al. 1992), and these can be understood as a competition between energy (goodness of the fit) and entropy in the space of models; the effective temperature is the inverse of the number of examples observed, so the system cools as more data is collected. Meanwhile the computer scientists have developed approaches to learning rules and distributions that focus on rigorous bounds—given that we have seen N examples, can we guarantee that our inferences are within ϵ of the correct model with probability $1 - \delta$? These ideas have their origins in Vapnik and Chervonenkis (1971) and Valiant (1984). The rapprochement between the different formulations is described by Haussler et al. (1996). For an early discussion about sudden versus gradual learning, see Spence (1938). For a modern example, emphasizing the need for a unified approach to sudden and gradual learning, see Rubin et al. (1997).

Haussler et al. 1996. Rigorous learning curve bounds from statistical mechanics. D Haussler, M Kearns, HS Seung, and N Tishby, *Machine Learning* **25,** 195–236 (1996).

Levin et al. 1990. A statistical approach to learning and generalization in layered neural networks. E Levin, N Tishby, and SA Solla, *Proc IEEE* **78,** 1568–1574 (1990).

Rubin et al. 1997. Abrupt learning and retinal size specificity in illusory-contour perception. N Rubin, K Nakayama, and R Shapley, *Curr Biol* **7,** 461–467 (1997).

Seung et al. 1992. Statistical mechanics of learning from examples. HS Seung, H Sompolinsky, and N Tishby, *Phys Rev A* **45,** 6056–6091 (1992).

Spence 1938. Gradual versus sudden solution of discrimination problems by chimpanzees. KW Spence, *J Comp Psych* **25,** 213–224 (1938).

Valiant 1984. A theory of the learnable. LG Valiant, *Commun ACM* **27,** 1134–1142 (1984).

Vapnik and Chervonenkis 1971. On the uniform convergence of relative frequencies of events to their probabilities. VN Vapnik and AY Chervonenkis, *Theory Probability Applications* **16,** 264–280 (1971).

The analysis of the response to sudden changes in probability is by Gallistel et al. (2001), and the difference between individual and ensemble learning curves was emphasized by Gallistel et al. (2004). The experiments on fluctuating probabilities with primate subjects are by Sugrue et al. (2004), and the analysis described in Fig. 6.42 is by Corrado et al. (2005). For views on the emerging ideas of neuroeconomics, see Glimcher (2003) and Camerer et al. (2005). Measurements on learning through trial-by-trial analysis of continuous-movement trajectories were pioneered by Thoroughman and Shadmehr (2000), who considered human arm movements. They had a particular view of the class of models that subjects use in these experiments, which simplified their analysis, but the idea is much more general. For measurements on the precision of tracking eye movements, see Osborne et al. (2005, 2007). For the beautiful relationship between latency and target probability in saccadic eye movements, see Carpenter and Williams (1995), who analyze not just the mean latency but also the whole distribution. Regarding the connection between learning and predictive information, see Bialek et al. (2001).

Bialek et al. 2001. Predictability, complexity and learning. W Bialek, I Nemenman, and N Tishby, *Neural Comp* **13**, 2409–2463 (2001); arXiv:physics/0007070 (2000).

Camerer et al. 2005. Neuroeconomics: How neuroscience can inform economics. C Camerer, G Lowenstein, and D Prelec, *J Econ Lit* **43**, 9–64 (2005).

Carpenter and Williams 1995. Neural computation of log likelihood in control of saccadic eye movements. RHS Carpenter and MLL Williams, *Nature* **377**, 59–62 (1995).

Corrado et al. 2005. Linear-nonlinear-Poisson models of primate choice dynamics. GS Corrado, LP Sugure, HS Seung, and WT Newsome, *J Exp Psych: Animal Behav* **84**, 581–617 (2005).

Gallistel et al. 2001. The rat approximates an ideal detector of changes in rates of reward: Implications for the law of effect. CR Gallistel, TA Mark, AP King, and P Latham, *J Exp Psych: Animal Behav Proc* **27**, 354–372 (2001).

Gallistel et al. 2004. The learning curve: Implications of a quantitative analysis. CR Gallistel, S Fairhurst, and P Balsam, *Proc Natl Acad Sci (USA)* **101**, 13124–13131 (2004).

Glimcher 2003. *Decisions Uncertainty and the Brain: The Science of Neuroeconomics.* PW Glimcher (MIT Press, Cambridge MA, 2003).

Osborne et al. 2005. A sensory source for motor variation. LC Osborne, SG Lisberger, and W Bialek, *Nature* **437**, 412–416 (2005).

Osborne et al. 2007. Time course of precision in smooth pursuit eye movements of monkeys. LC Osborne, SS Hohl, W Bialek, and SG Lisberger, *J Neurosci* **27**, 2987–2998 (2007).

Sugrue et al. 2004. Matching behavior and the representation of value in the parietal cortex. LP Sugrue, GS Corrado, and WT Newsome, *Science* **304**, 1782–1787 (2004).

Thoroughman and Shadmehr 2000. Learning of action through adaptive combination of motor primitives. KA Thoroughman and R Shadmehr, *Nature* **407**, 742–747 (2000).

Appendix: Some Further Topics

A.1 Poisson Processes

Portions of this section were adapted from Rieke et al. (1997). The connection between power spectra and the shape of single-photon (or more general Poisson) events is sometimes called Campell's theorem, and there is a classic discussion by Rice (1944–1945), reprinted in the marvelous book edited by Wax (1954); the other articles in this book (by Chandrasekar and others) also are very much worth reading! Feynman and Hibbs (1965) give a beautiful discussion of how a Poisson stream of pulses comes to approximate continuous Gaussian noise; of course there is much more in this book as well. For a more complete discussion of photon statistics and the role of coherent states, you can look to yet another classic paper: Glauber (1963).

Feynman and Hibbs 1965. *Quantum Mechanics and Path Integrals.* RP Feynman and AR Hibbs (McGraw-Hill, New York, 1965).

Glauber 1963. Coherent and incoherent states of the radiation field. RJ Glauber, *Phys Rev* **131,** 2766–2788 (1963).

Rice 1944–1945. Mathematical analysis of random noise. SO Rice, *Bell Sys Tech J* **23,** 282–332 (1944) and **24,** 46–156 (1945).

Rieke et al. 1997. *Spikes: Exploring the Neural Code.* F Rieke, D Warland, RR de Ruyter van Steveninck, and W Bialek (MIT Press, Cambridge MA, 1997).

Wax 1954. *Selected Papers on Noise and Stochastic Processes.* N Wax, ed. (Dover Publications, New York, 1954).

A.3 Diffraction and Biomolecular Structure

You should read the classic trio of papers on DNA structure, which appeared one after the other in the April 25, 1953, issue of *Nature*: Watson and Crick (1953a), Wilkins et al. (1953), and Franklin and Gosling (1953). The foundations of helical diffraction theory had been established just a year before by Cochran et al. (1952); a brief account is given in Holmes and Blow (1965). The astonishing realization that the structure of DNA implies a mechanism for the transmission of information from generation to generation was presented by Watson and Crick (1953b). It is especially interesting to read their account of the questions raised by their proposal, and to see how their brief list became the agenda for the emerging field of molecular biology over the next two decades. The rest is history, as the saying goes, so you should read at least one history book (Judson 1979). More a memoir than a history, Watson's account of the events is also fascinating, especially in the edition that includes responses from many of the other characters in the story (Watson 1980).

Cochran et al. 1952. The structure of synthetic polypeptides. I. The transform of atoms on a helix. W Cochran, FHC Crick, and V Vand, *Acta Cryst* **5,** 581–586 (1952).

Franklin and Gosling 1953. Molecular configuration in sodium thymonucleate. RE Franklin and RG Gosling. *Nature* **171,** 740–741 (1953).

Holmes and Blow 1965. The use of X-ray diffraction in the study of protein and nucleic acid structure. KC Holmes and DM Blow, *Meth Biochem Anal* **13,** 116–239 (1965).

Judson 1979. *The Eighth Day of Creation.* HF Judson (Simon and Schuster, New York, 1979).

Watson 1980. *The Double Helix: A Personal Account of the Discovery of the Structure of DNA.* JD Watson, Norton Critical Edition, G Stent, ed. (Norton, New York, 1980).

Watson and Crick 1953a. A structure for deoxyribose nucleic acid. JD Watson and FHC Crick, *Nature* **171,** 737–738 (1953).

Watson and Crick 1953b. Genetical implications of the structure of deoxyribonucleic acid. JD Watson and FHC Crick, *Nature* **171,** 964–967 (1953).

Wilkins et al. 1953. Molecular structure of deoxypentose nucleic acids. MHF Wilkins, AR Stokes, and HR Wilson, *Nature* **171,** 738–740 (1953).

The foundational work on the determination of protein structures was by Kendrew (1964) and Perutz (1964). For a modern textbook account, see Drenth (1999), as well as the classic by Holmes and Blow (1965), above.

Drenth 1999. *Principles of Protein X-Ray Crystallography.* J Drenth (Springer-Verlag, Berlin, 1999).

Kendrew 1964. Myoglobin and the structure of proteins. JC Kendrew, in *Nobel Lectures: Chemistry 1942–1962,* pp. 676–698 (Elsevier, Amsterdam, 1964). Also available at http://www.nobelprize.org.

Perutz 1964. X-ray analysis of haemoglobin. MF Perutz, in *Nobel Lectures: Chemistry 1942–1962,* pp. 653–673 (Elsevier, Amsterdam, 1964). Also available at http://www.nobelprize.org.

A.4 Electronic Transitions in Large Molecules

For a good account of the saddle point approximation, see Matthews and Walker (1970). The idea that coupling to a bath of oscillators could describe dissipation in quantum mechanics goes back, at least, to Feynman and Vernon (1963). These ideas were revitalized by Caldeira and Leggett (1981, 1983), who were especially interested in the impact of dissipation on quantum tunneling.

Caldeira and Leggett 1981. Influence of dissipation on quantum tunnelling in macroscopic systems. AO Caldeira and AJ Leggett, *Phys Rev Lett* **46,** 211–214 (1981).

Caldeira and Leggett 1983. Quantum tunnelling in a dissipative system. AO Caldeira and AJ Leggett, *Ann Phys (NY)* **149,** 374–456 (1983).

Feynman and Vernon 1963. The theory of a general quantum system interacting with a linear dissipative system. RP Feynman and FL Vernon, Jr, *Ann Phys (NY)* **24,** 118–173 (1963).

Matthews and Walker 1970. *Mathematical Methods of Physics,* second edition. J Matthews and RL Walker (WA Benjamin, New York, 1970).

A.5 The Kramers Problem

The original discussion of diffusion (even with inertia) over a barrier is due to Kramers (1940); for a modern perspective, see Hänggi et al. (1990). Every physics student should understand the basic instanton calculation of tunneling, as an illustration of the power of path integrals. There is no better treatment than that given by Coleman (1988) in his justly famous Erice lectures. If you read Coleman, you will not only get a deeper view of what we have covered here, you will also find all the missing pieces about the prefactor of the rate constant, and much more. For more general background on path integrals, including some discussion of how to use them for classical stochastic processes, the standard reference is Feynman and Hibbs (1965). For more rigorous accounts of many of these issues, see Zinn-Justin (1989).

Coleman 1988. *Aspects of Symmetry.* S Coleman (Cambridge University Press, Cambridge, 1988).

Feynman and Hibbs 1965. *Quantum Mechanics and Path Integrals.* RP Feynman and AR Hibbs (McGraw-Hill, New York, 1965).

Hänggi et al. 1990. Reaction-rate theory: Fifty years after Kramers. P Hänggi, P Talkner, and M Borkovec, *Rev Mod Phys* **62,** 251–341 (1990).

Kramers 1940. Brownian motion in a field of force and the diffusion model of chemical reactions. HA Kramers, *Physica* **7,** 284–304 (1940).

Zinn-Justin 1989. *Quantum Field Theory and Critical Phenomena.* J Zinn-Justin (Oxford University Press, Oxford, 1989).

A.6 Berg and Purcell, Revisited

Almost all PhD students in physics have seen some cases of the fluctuation-dissipation theorem somewhere in their statistical mechanics courses. Whether you have seen the general formulation depends a bit on who taught the course, and how far you went. As usual, an excellent discussion can be found in Landau and Lifshitz (1977). A later volume in the Landau and Lifshitz series (Lifshitz and Pitaevskii 1980) provides a clear discussion of concentration fluctuations. Many people find the idea of correlations between Brownian particles to be surprising, so it is worth looking at real experiments that measure these correlations (Meiners and Quake 1999).

Landau and Lifshitz 1977. *Statistical Physics*. LD Landau and EM Lifshitz (Pergamon, Oxford, 1977).

Lifshitz and Pitaevskii 1980. *Statistical Physics, Part 2*. EM Lifshitz and LP Pitaevskii (Pergamon, Oxford, 1980).

Meiners and Quake 1999. Direct measurement of hydrodynamic cross correlations between two particles in an external potential. JC Meiners and SR Quake, *Phys Rev Lett* **82,** 2211–2214 (1999).

The idea that fluctuations in certain chemical systems could be described using the fluctuation-dissipation theorem must have occurred to many people, and I remember discussing it long ago (Bialek 1987). The emergence of experiments on noise in the control of gene expression made it more interesting to get everything straight, so my colleagues and I tried to do this (Bialek and Setayeshgar 2005, 2008; Tkačik and Bialek 2009). Interestingly, before this work, many simulations based on the master equation or its equivalent were advertised as exact, although they implicitly assumed that all molecules were well stirred. Around the same time as our work, van Zon et al. (2006) carried out simulations that included diffusion, using novel methods to make the computations more efficient, and they saw a substantial contribution from the noise that we identified with the Berg-Purcell limit. There is more to be done here, I suspect.

Bialek 1987. Physical limits to sensation and perception. W Bialek, *Annu Rev Biophys Chem* **16,** 455–478 (1987).

Bialek and Setayeshgar 2005. Physical limits to biochemical signaling. W Bialek and S Setayeshgar, *Proc Natl Acad Sci (USA)* **102,** 10040–10045 (2005).

Bialek and Setayeshgar 2008. Cooperativity, sensitivity and noise in biochemical signaling. W Bialek and S Setayeshgar, *Phys Rev Lett* **100,** 258101 (2008).

Tkačik and Bialek 2009. Diffusion, dimensionality and noise in transcriptional regulation. G Tkačik and W Bialek, *Phys Rev E* **79,** 051901 (2009).

van Zon et al. 2006. Diffusion of transcription factors can drastically enhance the noise in gene expression. JS van Zon, M Morelli, S Tănase-Nicola, and PR ten Wolde, *Biophys J* **91,** 4350–4367 (2006).

A.7 Maximum Entropy

The idea of using maximum entropy methods (Jaynes 1957) to think about correlations in networks of neurons arose from a very practical problem—if we observe correlations among pairs of neurons, should we be surprised if we observe, for example, three or four neurons generating action potentials simultaneously? For continuous variables, we can separate different orders of correlations quite simply; this is the idea of cumulants in statistics, or connected diagrams in field theory. For discrete variables, pairwise correlations imply higher order correlations, even without any further assumptions. One touchstone for this idea is in statistical mechanics—the usual Ising model has only interactions between two spins at a time, but when we coarse grain this model to give the Landau-Ginzburg Hamiltonian, we generate ϕ^4 interaction terms, so that the magnetization ϕ (which is a spatially smoothed version of the original spins) must have nontrivial fourth-order correlations. Schneidman et al. (2003) showed how the maximum entropy construction could be used to generalize the idea of connected correlations to discrete variables.

Jaynes 1957. Information theory and statistical mechanics. ET Jaynes, *Phys Rev* **106,** 620–630 (1957).

Schneidman et al. 2003. Network information and connected correlations. E Schneidman, S Still, MJ Berry, II, and W Bialek, *Phys Rev Lett* **91,** 238701 (2003).

When setting out to use the maximum entropy method to analyze the responses of real neurons in the vertebrate retina, we expected we would clean out the pairwise correlations and uncover the higher order effects responsible for the known tendency of many neurons to fire simultaneously (Schnitzer and Meister 2003). The surprising result was that the pairwise Ising model provides an accurate description of the combinatorial patterns of spiking and silence in ganglion cells of the salamander retina as they respond to natural and artificial movies, and in cortical cell cultures (Schneidman et al. 2006). After the initial success in the salamander retina, similarly encouraging results were obtained in the primate retina, under very different stimulus conditions (Shlens et al. 2006, 2009); in visual cortex (Ohiorhenuan et al. 2010; Yu et al. 2008); and in networks grown in vitro (Tang et al. 2008). Most of these detailed comparisons of theory and experiment were done for groups of $N \sim 10$ neurons, small enough that the full distribution $P_{expt}(\{\sigma_i\})$ could be sampled experimentally and used to assess the quality of the pairwise maximum entropy model. The push to larger networks is described in a series of papers by Tkačik et al. (2006, 2009, 2012). The use of these ideas in the context of flocking is described by Bialek et al. (2012), and this is (I think) an interesting example, because it is a relatively simple and direct implementation of the maximum entropy principle, in which matching a single expectation value (the directional correlations among neighboring birds) provides a nearly complete quantitative theory for propagation of directional ordering throughout the flock.

Bialek et al. 2012. Statistical mechanics for natural flocks of birds. W Bialek, A Cavagna, I Giardina, T Mora, E Silvestri, M Viale, and A Walczak, *Proc Natl Acad Sci (USA)* **109,** 4786–4791 (2012); arXiv.org:1107.0604 [physics.bio-ph] (2011).

Ohiorhenuan et al. 2010. Sparse coding and higher-order correlations in fine-scale cortical networks. IE Ohiorhenuan, F Mechler, KP Pupura, AM Schmid, Q Hu, and JD Victor, *Nature* **466,** 617–621 (2010).

Schneidman et al. 2006. Weak pairwise correlations imply strongly correlated network states in a neural population. E Schneidman, MJ Berry, II, R Segev, and W Bialek, *Nature* **440,** 1007–1012 (2006).

Schnitzer and Meister 2003. Multineuronal firing patterns in the signal from eye to brain. MJ Schnitzer and M Meister, *Neuron* **37,** 499–511 (2003).

Shlens et al. 2006. The structure of multi-neuron firing patterns in primate retina. J Shlens, GD Field, JL Gaulthier, MI Grivich, D Petrusca, A Sher, AM Litke, and EJ Chichilnisky, *J Neurosci* **26,** 8254–8266 (2006).

Shlens et al. 2009. The structure of large-scale synchronized firing in primate retina. J Shlens, GD Field, JL Gaulthier, M Greschner, A Sher, AM Litke, and EJ Chichilnisky, *J Neurosci* **29,** 5022–5031 (2009).

Tang et al. 2008. A maximum entropy model applied to spatial and temporal correlations from cortical networks *in vitro.* A Tang, D Jackson, J Hobbs, W Chen, JL Smith, H Patel, A Prieto, D Petruscam, MI Grivich, A Sher, P Hottowy, W Dabrowski, AM Litke, and JM Beggs, *J Neurosci* **28,** 505–518 (2008).

Tkačik et al. 2006. Ising models for networks of real neurons. G Tkačik, E Schneidman, MJ Berry, II, and W Bialek, arXiv:q-bio.NC/0611072 (2006).

Tkačik et al. 2009. Spin glass models for networks of real neurons. G Tkačik, E Schneidman, MJ Berry, II, and W Bialek, arXiv:0912.5409 [q-bio.NC] (2009).

Tkačik et al. 2012. Searching for collective behavior in a large network of real neurons. G Tkačik, O Marre, D Amodei, E Schneidman, MJ Berry, II, and W Bialek, in preparation (2012).

Yu et al. 2008. A small world of neuronal synchrony. S Yu, D Huang, W Singer, and D Nikolic, *Cereb Cortex* **18**, 2891–2901 (2008).

The success of maximum entropy methods in describing real biological networks has led to considerable interest in better algorithms for solving the inverse statistical mechanics problem that is at the heart of the maximum entropy construction (Broderick et al. 2007), which is part of the broader connection between statistical physics and problems usually in the domain of computer science; for a perspective, see Mézard and Mora (2009). One possibility is to use perturbation theory (Sessak and Monasson 2009) or other standard approximations from the forward statistical mechanics problem (Nguyen and Berg 2012). If we know, or suspect, that the nonzero J_{ij} form a sparse set, or correspond to local connectivity, then it is possible to exploit this feature (Cocco and Monasson 2011); an extreme case is if the interactions have a one-dimensional geometry, although in this limit one is not restricted by locality (Gori and Trombettoni 2011). In a very different limit the couplings may be widespread but of low rank, as in a Hopfield model that stores relatively few memories. This simplification should also be detectable and exploitable as a generalization of principal components analysis (Cocco et al. 2011). The inverse problem may also be more easily solvable in the context of different models, in particular dynamical models where the trajectories of states provide more information than just samples out of a distribution (Cocco et al. 2009). Much of the difficulty of the problem comes from maintaining consistency of the probability distribution, adjusting the partition function Z as we adjust the fields and couplings $\{h_i, J_{ij}\}$. One can try to relax this in various ways and hope that the model is accurate enough that it will be restored automatically (or almost so) in fitting the data; for ideas along this line see Bethge and Berens (2008), Aurell and Ekeberg (2011), and Ganmor et al. (2011).

Aurell and Ekeberg 2011. Inverse Ising inference using all the data. E Aurell and M Ekeberg, arXiv:1107.3536 [cond-mat.dis-nn] (2011).

Bethge and Berens 2008. Near-maximum entropy models for binary neural representations of natural images. M Bethge and P Berens, in *Advances in Neural Information Processing Systesm 20*, J Platt, D Koller, Y Singer, and S Roweis, eds., pp. 97–104 (MIT Press, Cambridge MA, 2008).

Broderick et al. 2007. Faster solutions of the inverse pairwise Ising problem. T Broderick, M Dudík, G Tkačik, RE Schapire, and W Bialek, arXiv:0712.2437 [q-bio.QM] (2007).

Cocco and Monasson 2011. Adaptive cluster expansion for inferring Boltzmann machines with noisy data. S Cocco and R Monasson, *Phys Rev Lett* **106**, 090601 (2011).

Cocco et al. 2009. Neuronal couplings between retinal ganglion cells inferred by efficient inverse statistical physics methods. S Cocco, S Leibler, and R Monasson, *Proc Natl Acad Sci (USA)* **106**, 14058–14062 (2009).

Cocco et al. 2011. High-dimensional inference with generalized Hopfield model: Principal component analysis and corrections. S Cocco, R Monasson, and V Sessak, *Phys Rev E* **83**, 051123 (2011).

Ganmor et al. 2011. Sparse low-order interaction network underlies a highly correlated and learnable neural population code. E Ganmor, R Segev, and E Schneidman, *Proc Natl Acad Sci (USA)* **108**, 9679–9684 (2011).

Gori and Trombettoni 2011. Inverse Ising problem for one-dimensional chains with arbitrary finite-range couplings. G Gori and A Trombettoni, arXiv:1109.5529 [cond-mat.stat-mech] (2011).

Mézard and Mora 2009. Constraint satisfaction problems and neural networks: A statistical physics perspective. M Mézard and T Mora, *J Physiol Paris* **103,** 107–113 (2009)

Nguyen and Berg 2012. Bethe-Peierls approximation and the inverse Ising problem. HC Nguyen and J Berg, *J Stat Mech: Theory Expt* P03004 (2012).

Sessak and Monasson 2009. Small-correlation expansions for the inverse Ising problem. V Sessak and R Monasson, *J Phys A* **42,** 055001 (2009).

A.8 Measuring Information Transmission

The idea that naive counting leads to systematic errors in entropy estimation goes back, at least, to Miller (1955). The importance of this problem for the analysis of information transmission in neurons was emphasized by Treves and Panzeri (1995), who also brought more sophistication to the calculation of the series expansion that we have started here. Shortly after, Strong et al. (1998) showed how these extrapolation methods could be used to estimate entropy and information in neural responses to complex, dynamic sensory inputs. An important technical point is that Strong et al. took seriously the $1/N$ behavior of the entropy estimate, but did not use an analytic calculation of the slope of S_{est} versus $1/N$; the reason is that some seemingly possible neural responses are expected to have probability zero, because there is a hard core repulsion ("refractoriness") between spikes, but we do not know in advance exactly how big this effect will be. As a result, the actual number of possible states K is uncertain, and in addition it is not true that all samples collected in the experiment will be independent. Both these effects leave the $1/N$ behavior intact but change the slope.

Miller 1955. Note on the bias of information estimates. GA Miller, in *Information Theory in Psychology: Problems and Methods II-B*, H Quastler, ed., pp. 95–100 (Free Press, Glencoe IL, 1955).

Strong et al. 1998. Entropy and information in neural spike trains. SP Strong, R Koberle, RR de Ruyter van Steveninck, and W Bialek, *Phys Rev Lett* **80,** 197–200 (1998).

Treves and Panzeri 1995. The upward bias in measures of information derived from limited data samples. A Treves and S Panzeri, *Neural Comp* **7,** 399–407 (1995).

The idea of estimating entropy by counting coincidences is presented in Ma (1981), although it has precursors in Serber (1973). The attempts to build a flat prior on the entropy are described in papers by Nemenman (2002) and Nemenman et al. (2002). Moments of the entropy distribution in Dirichlet priors were calculated by Wolpert and Wolf (1995), so you can see where the polygamma functions come from. Within the family of Dirichlet priors, we have noted that the flat prior on distributions ($\beta = 1$) was discussed by Laplace (1814) as the idea of starting with one pseudocount in every bin; Jeffreys (1946), and later Krichevskii and Trofimov (1981) proposed using half a pseudocount ($\beta = 1/2$); Schurmann and Grassberger (1996) suggested that entropy estimates could be improved by adjusting the number of pseudocounts to the number of bins, $\beta = 1/K$. The procedure developed by Nemenman et al. integrates over all β, allowing a dominant β to emerge that is matched to the structure of the data and to the number of samples, as well as to K. The example shown in Fig. A.15 is from Nemenman et al. (2004).

Jeffreys 1946. An invariant form for the prior probability in estimation problems. H Jeffreys, *Proc R Soc Lond A* **186,** 453–461 (1946).

Krichevskii and Trofimov 1981. The performance of universal encoding. R Krichevskii and V Trofimov, *IEEE Trans Inf Thy* **27,** 199–207 (1981).

Laplace 1814. *Essai philosophique sur les probabilités* (Courcier, Paris, 1814). Available in translation as *A Philosophical Essay on Probabilities*, translated by F Truscott and F Emory (Dover, New York, 1951).

Ma 1981. Calculation of entropy from data of motion. S-K Ma, *J Stat Phys* **26,** 221–240 (1981).

Nemenman 2002. Inference of entropies of discrete random variables with unknown cardinalities. I Nemenman, arXiv:physics/0207009 (2002).

Nemenman et al. 2002. Entropy and inference, revisited. I Nemenman, F Shafee, and W Bialek, in *Advances in Neural Information Processing 14*, TG Dietterich, S Becker, and Z Ghahramani, eds., pp. 471–478 (MIT Press, Cambridge MA, 2002); arXiv:physics/0108025 (2001).

Nemenman et al. 2004. Entropy and information in neural spike trains: Progress on the sampling problem. I Nemenman, W Bialek, and RR de Ruyter van Steveninck, *Phys Rev E* **69,** 056111 (2004); arXiv:physics/0306063 (2003).

Schurmann and Grassberger 1996. Entropy estimation of symbol sequences. T Schurmann and P Grassberger, *Chaos* **6,** 414–427 (1996).

Serber 1973. *Estimation of Animal Abundance and Related Parameters.* GAF Serber (Griffin, London, 1973).

Wolpert and Wolf 1995. Estimating functions of probability distributions from a finite set of samples. DH Wolpert and DR Wolf, *Phys Rev E* **52,** 6841–6854 (1995).

The specific problem of estimating entropy in neural responses has the added feature that spikes are discrete, but they can occur at any time, so there is a question of whether we should view the whole problem as discrete (with bins along the time axis) or continuous (with a metric along the time axis); the discussion here has focused on the discrete formulation. For metric space approaches to spike trains, see Victor (2002). For a general overview of the entropy estimation problem, with particular attention to the challenges posed by neural data, see Paninski (2003), who tries to make many of the heuristic arguments in the field more rigorous. One can also view entropy estimation as a problem in computational complexity—how many samples, and hence how many computational steps, are needed to approximate the entropy to some level of accuracy? For an approach in this spirit, see Batu et al. (2002).

Batu et al. 2002. The complexity of approximating the entropy. T Batu, S Dasgupta, R Kumar, and R Rubinfeld, in *Proceedings of the 34th Symposium on the Theory of Computing (STOC)*, J Reig, ed., pp. 678–687 (ACM, New York, 2002).

Paninski 2003. Estimation of entropy and mutual information. L Paninski, *Neural Comp* **15,** 1191–1253 (2003).

Victor 2002. Binless strategies for estimation of information from neural data. J Victor, *Phys Rev E* **66,** 051903 (2002).

Index

Page numbers for entries in figures are followed by an *f* and those for entries in problems are followed by a *p*.